Atlas of Invertebrate Viruses

Editors

Jean R. Adams, Ph.D.

Insect Biocontrol Laboratory
Plant Science Institute
Agricultural Research Service
United States Department of Agriculture
Beltsville Agricultural Research Center-West
Beltsville, Maryland

Jean R. Bonami, Dr. ès Sc.

Laboratoire de Pathologie Comparée
Université des Sciences et Techniques du Languedoc
Montpellier, France

CRC Press
Taylor & Francis Group
Boca Raton London New York

CRC Press is an imprint of the
Taylor & Francis Group, an **informa** business

DEDICATION

This atlas is dedicated to pioneers in invertebrate pathology, Drs. Edward A. Steinhaus, Arthur M. Heimpel, Constantin Vago, and Thomas W. Tinsley, whose leadership has inspired and guided those of us who have followed in the quest of unraveling the mysteries of invertebrate viruses for the benefit of mankind.

PREFACE

The purpose of our efforts in compiling this atlas has been to provide a helpful reference for invertebrate pathologists, virologists, and electron microscopists on invertebrate viruses. Investigators from around the world have generously shared of their expertise in order to introduce scientists to the exciting advances in invertebrate virology.

The electron microscope has proven to be a tool that has afforded exciting experiences as we have seen new viruses for the first time and have studied the ultrastructural cytopathology of these new diseases. The senior editor has been continually impressed with the precision, the detail, the beauty that the Creator of life has endowed within living matter. (Ps. 104: 24-25). B. J. Conner* has referred to the Creation as a treasure hunt suggesting that each of us can "be alert and perservere in seeking the truth and beauty the Creator has waiting for us to uncover for His glory and our benefit."

As we have delved into studies with these fascinating microorganisms and their replication processes, more questions have arisen than can be resolved in any one investigator's career and so the challenge is passed on to those who follow us. We have come to the realization that each one has a choice, i.e., to take action to protect our environment or to pollute it. Those in invertebrate virology have made the choice to direct their efforts toward studies that will provide data which will contribute to the making of a safer environment today — and for future generations.

We wish to express our sincere gratitude to the contributors for their provision of excellent electron micrographs enabling the reader to see clearly the beautiful ultrastructure that they have discovered.

Jean Ruth Adams
Jean Robert Bonami

---

* Daily Guideposts, *Guideposts*, Carmel, New York, 69, 1989.

THE EDITORS

Jean R. Adams, Ph.D., is a Research Entomologist with the U.S. Department of Agriculture, Agricultural Research Service, Plant Sciences Institute, Insect Biocontrol Laboratory at Beltsville, Maryland.

Dr. Adams graduated from Rutgers — the State University of New Jersey in 1950 with a B.S. in Agricultural Research with a minor in horticulture, then worked as a laboratory technician in the development of insecticide products at Rohm & Haas Co. in Bristol, Pennsylvania from 1951—1957. In 1957 she received a Trubeck Fellowship to pursue graduate studies in Entomology at Rutgers University and received a Ph.D. in 1962 after spending a year of additional training at the University of Pennsylvania in the Department of Biology. She then came to the U.S. Dept. of Agriculture, ARS, Insect Pathology Laboratory and has been serving there since 1962.

Dr. Adams is a founding member of the Society for Invertebrate Pathology, Entomological Society of America, American Registry of Professional Entomologists (A.R.P.E.), Electron Microscopy Society of America (E.M.S.A.), AAAS, American Society for Microbiology (ASM), American Society for Cell Biology, Entomological Society of Washington, Maryland Entomology Society, Chesapeake chapters of the Electron Microscopy Society (C.S.E.M.) and A.R.P.E., Sigma Xi, Sigma Delta Epsilon, U.S.D.A. Organization of Professional Employees and ASM Washington Branch.

Dr. Adams has served as secretary of the Society for Invertebrate Pathology and on the Editorial Board of the *Journal of Invertebrate Pathology,* and has served in various positions on the Council of the C.S.E.M. and the Governing Board of the Chesapeake Chapter of A.R.P.E. Awards received include one from the Electron Microscopy Society of America in 1982 for a Scientific Exhibit entitled "A Close Up Look at Microbials for Insect Control" and selection for its Traveling Exhibit 1982—1983.

Dr. Adams has published more than 80 papers. Her current research involves isolation and identification of new insect pathogens; ultrastructural histopathological and immunocytochemical studies on insect pathogens in insect tissues; and the identification and study of diseases encountered in insect mass rearing facilities and the development of techniques to ensure insect rearings that are disease-free.

Jean-Robert Bonami, Dr. ès Sc., is Head of the Department of Marine Invertebrate Pathology in the Laboratory of Comparative Pathology at the University of Montpellier II, Sciences et Techniques du Languedoc, France.

Dr. Bonami received his Licence and Maitrise ès Sciences from the University of Montpellier II in 1968 and his Doctorat d'Etat ès Sciences in 1980.

In 1970 he was appointed the title of scientist by the Centre National de la Recherche Scientifique while serving at the Laboratory of Comparative Pathology. Dr. Bonami is a member of several professional organizations, among them: the French Society of Microbiology, Asian Fisheries Society (Fish Health Section), and European Association of Fish Pathologists. He is also a member of the Working Group on Pathology of Marine Organisms of the ICES (International Council for Exploration of the Sea) and co-promoter of the international PAMAQ (Pathology in Marine Aquaculture) meetings.

Dr. Bonami has been the recipient of numerous research grants (CNEXO, IFREMER, MIR) and has published approximately 70 papers on infectious diseases of marine animals, i.e., molluscan parasitic and viral diseases, fish diseases, and crustacean virosis.

Dr. Bonami's current research interests relate to viral diseases of marine crustacea, particularly those of economic importance, with a particular interest in cell specificity as it relates to reovirus and baculovirus diseases.

CONTRIBUTORS

Jean R. Adams, Ph.D.
Insect Biocontrol Laboratory
Plant Science Institute
Agricultural Research Service
United States Department of Agriculture
Beltsville Agricultural Research
  Center-West
Beltsville, Maryland

Darrell W. Anthony, M.S., (Retired)
Insects Affecting Man and Animal
  Research Laboratory
Agricultural Research Service
United States Department of Agriculture
Gainesville, Florida

Max Arella, Ph.D.
Institut Armand-Frappier
Medicine Comparée Research Center
Laval-des-Rapides
Quebec, Canada

Leslie Bailey, Sc.D.
Department of Entomology and
  Nematology
Rothamsted Experimental Station
Harpenden, England

Brenda V. Ball
Department of Entomology and
  Nematology
Rothamsted Experimental Station
Harpenden, England

Colin D. Beaton
Division of Entomology
CSIRO
Canberra, Australia

Jean R. Bonami, Dr. ès Sc.
Laboratoire de Pathologie Comparée
Université des Sciences et Techniques du
  Languedoc
Montpellier, France

Michèle Bouloy, Dr. ès Sc.
Unité de Virologie Moléculaire
Institut Pasteur
Paris, France

Gilbert Brun, Dr. ès Sc.
Laboratoire de Génétique des Virus
CNRS
Gif-Sur-Yvette, France

Guy Charpentier, Ph.D.
Departement de Chimie-Biologie
Université du Quebec à Trois-Rivières
Trois-Rivières, Canada

Michel Comps, Dr. ès Sc.
GIE-RA, IFREMER
Palavas-les-Flots, France

John A. Couch, Ph.D.
Environmental Research Laboratory
United States Environmental Protection
  Agency
Gulf Breeze, Florida

S. M. Eley
Chemical Defense Establishment
Porton Down, Salisbury Wilts, England

Brian A. Federici, Ph.D.
Department of Entomology
University of California — Riverside
Riverside, California

R. A. Gardner, Ph.D.
Department of Biological Science
University of Salford
Salford, United Kingdom

Simon Garzon, Ph.D.
Departement de Microbiologie et
  d'Immunologie
Université de Montréal
Montréal, Canada

R. H. Goodwin, Ph.D.
Rangeland Insect Laboratory
Agricultural Research Service
United States Department of Agriculture
Montana State University
Bozeman, Montana

Philip M. Grimley, M.D.
Department of Pathology
Uniformed Services University of the
  Health Sciences
Bethesda, Maryland

David Guzo, Ph.D.
Insect Biocontrol Laboratory
Agricultural Research Service
United States Department of Agriculture
Beltsville, Maryland

John J. Hamm, Ph.D.
Insect Biological Population Management
Research Laboratory
United States Department of Agriculture
Agriculture Research Service
Tifton, Georgia

Roberta T. Hess
Department of Entomological Sciences
University of California
Berkeley, California

Dr. Alois M. Huger, Ph.D.
Federal Biological Research Center for
  Agriculture and Forestry
Institute of Biological Research Pest
  Control
Darmstadt, Germany

T. Hukuhara, Ph.D.
Faculty of Agriculture
Tokyo University of Agriculture and
  Technology
Tokyo, Japan

Peter J. Krell, Ph.D.
Department of Microbiology
University of Guelph
Guelph, Canada

Dr. Aloysius Krieg, Ph.D.
Federal Biological Research Center for
  Agriculture and Forestry
Institute for Biological Control
Darmstadt, Germany

Donald V. Lightner, Ph.D.
Department of Veterinary Science
University of Arizona
Tucson, Arizona

J. Thomas McClintock, Ph.D.
Office of Pesticide Programs
Health Effects Division
United States Environmental Protection
  Agency
Washington, D.C.

Richard J. Milner, Ph.D.
Entomology Division
CSIRO
Canberra, Australia

David H. Molyneux , Ph.D.
Department of Biology
University of Salford
Salford, United Kingdom

Norman F. Moore, Ph.D.
Ministry of Defence
White Hall
London, England

Gilles Morel, Ph.D.
Laboratoire de Pathologie des Invertébrés
Faculté des Sciences
Université de la Réunion
Ste Clotilde, France

Carl Reinganum, M.Sc.
(Deceased)
Department of Agriculture and Rural
  Affairs
Plant Research Institute
Burnley, Australia

Eloise L. Styer, Ph.D.
Veterinary Diagnostic and Investigational
  Laboratory
University of Georgia
Tifton, Georgia

Yoshinori Tanada, Ph.D.
Department of Entomological Science
University of California
Berkeley, California

Peter Tijssen, Ph.D.
Institut Armand-Frappier
Medicine Comparée Research Center
Laval-des-Rapides
Quebec, Canada

George J. Tompkins, Ph.D.
Office of Pesticide Programs
Environmental Fate and Effects Division
United States Environmental Protection
  Agency
Washington, D.C.

Hitoshi Watanabe, Ph.D.
Nodai Research Institute
Tokyo University of Agriculture
Tokyo, Japan

TABLE OF CONTENTS

Chapter 1

INTRODUCTION AND CLASSIFICATION OF VIRUSES OF INVERTEBRATES

Jean R. Adams

The beginnings of invertebrate pathology have been recorded by Steinhaus who describes the great contributions of such pioneers as White who studied bee diseases; Glaser and Chapman who described the wilt disease of gypsy moth larvae and other insects; and Paillot and Metalnikhov who contributed greatly to all phases of insect pathology.[1-4] One of the pioneers, Vago[5] has recently prepared a very interesting account of Pasteur's life and general work including contributions in the area of invertebrate pathology. Pasteur's studies on bacterial diseases of *Bombyx* initiated invertebrate bacterial pathology, and his studies on microsporidia revealed many facts concerning their pathogenesis in silkworms. Steinhaus, whom many regard as the founding father of invertebrate pathology, prepared the first text on insect pathology and served as editor of a new journal entitled *Journal of Insect Pathology* in 1959 which later became the *Journal of Invertebrate Pathology*. He edited the two-volume series on *Insect Pathology, An Advanced Treatise*,[6,7] initiated the formation of the Society for Invertebrate Pathology in 1969, and served as its first president. Other texts and chapters have followed which have traced the development of this field with its explosion of knowledge.[8-84] It is not possible in an atlas to present exhaustive lists of references on invertebrate viruses; rather key references should serve to guide the reader to a particular area of interest.

Many earlier studies reported on new isolations of viruses naming them for the host from which they were isolated. As new techniques in nucleic acid and protein chemistry were developed, biochemical differences between groups of viruses were identified which greatly contributed to laying a foundation for a system of classification. Matthews[85] describes the historical background and development of the organization of a classification system in 1966 when the International Committee on Nomenclature of Viruses met. In 1974 the committee became the International Committee on Taxonomy of Viruses (ICTV). Four reports have been presented, the last appearing in 1982.[86-89] A fifth report is due to appear in 1990 or 1991. Since these reports, Longworth[90] has presented historical background on insect virus taxonomy as well as the current problems, and Tinsley and Kelly[91] have reviewed the taxonomy and nomenclature of insect pathogenic viruses. The genetic basis for describing viruses as species has been discussed by Bishop.[92]

Several very helpful atlases of insect viruses and insect diseases have been published;[93-96] the latest are in Russian[95] and Chinese.[96]

Particle morphology is an important character in assigning a virus to a taxonomic group and is very helpful in diagnostic work since partially purified unknown viruses may be negatively stained for morphological examinations in the electron microscope. Then further studies can be planned in order to characterize the virus. Figures 1 and 2 illustrate the morphology and relative sizes of the viruses of invertebrates with the polyhedra, granulosis virus, and spheroid bodies at a different scale due to limitation of space. Table 1 lists the general dimensions according to the ICTV report[89] or in the chapters that follow.

Invertebrate viruses are unique in that in some groups the virions or nucleocapsids are occluded in a crystalline protein matrix called a viral occlusion body (VO), i.e., a polyhedron in nuclear polyhedrosis virus, a capsule in granulosis virus, and a spheroid in entomopoxvirus. The protein protects the virions when they are released from the host cadaver into a harsh environment. At a later time an insect host will consume the viral occlusion, or occlusion body, perhaps washed up from the soil onto a leaf. In the insect's gut (approximately pH

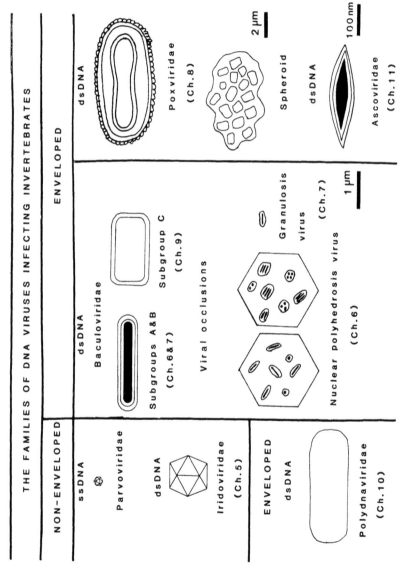

FIGURE 1. The families of DNA viruses infecting invertebrates. The viral structures are drawn to the following scale: (1) virions, bar = 100 nm (shown by the Ascoviridae); (2) the viral occlusions - nuclear polyhedrosis virus (NPV) and granulosis virus (GV), and the spheroid are shown at the scales indicated. Spheroids may vary greatly in size.[47] The spheroid shown is of comparable dimensions to the entomopoxvirus reported from *Choristoneura conflicta*.[100] The nonoccluded baculovirus shown is of comparable dimensions to the virus reported from *Oryctes rhinoceros*. The tail-like appendage (not shown) measures 10 × 270 nm.[101] The polydnavirus shown is of comparable dimensions to those found in the Ichneumonidae[34] (see Chapter 10). (Adapted from Matthews, R. E. F., *Intervirology*, 17, 7, 1982.)

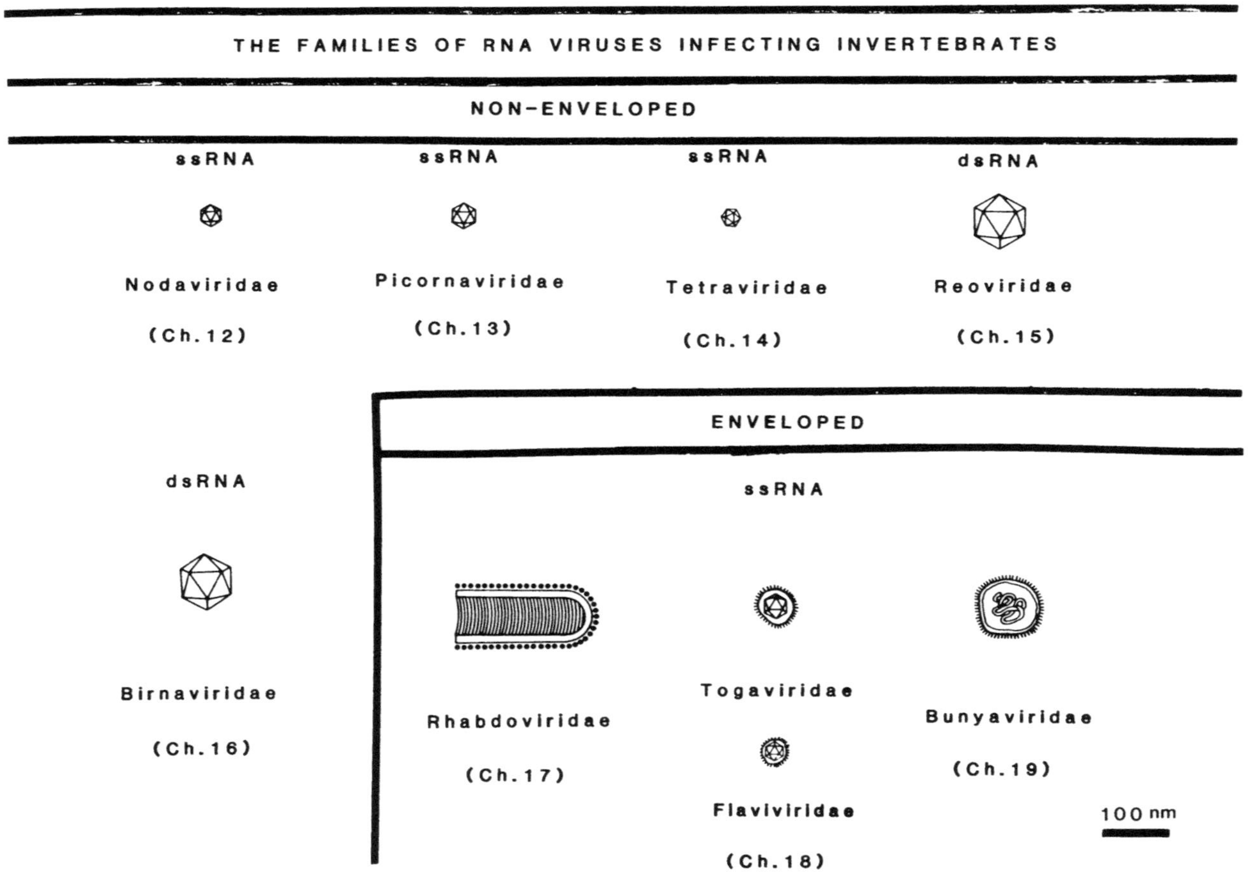

FIGURE 2. The families of RNA viruses infecting invertebrates. The Reoviridae contains a virus occlusion, cytoplasmic polyhedrosis (CPV), which is not shown due to space limitations. CPVs may range from 1 to 15 μm in diameter. The virions are drawn to scale, bar = 100 nm. (Adapted from Matthews, R. E. F., *Intervirology*, 17, 7, 1982.)

TABLE 1
Taxonomic Groups of Invertebrate Viruses

Family	Nucleic acid	Particle morphology Shape	Dimension	Enveloped virion	Viral occlusion
Parvoviridae	ssDNA	Isometric	18—26 nm	—	—
Iridoviridae	dsDNA	Icosahedral	125—300 nm	—	—
Baculoviridae					
Nuclear polyhedrosis virus	dsDNA	Bacilliform	40—60 × 200—400 nm	+	+
Granulosis virus	dsDNA	Bacilliform	30—60 × 260—360 nm	+	+
Nonoccluded virus	dsDNA	Bacilliform	100—200 × 200—225 nm	+	—
Poxviridae	dsDNA	Brickshaped or ovoid	165—300 × 150—470 nm	+	+
Polydnaviridae	dsDNA	Ovoid	150 × 350 nm	+	—
		Nucleocapsid surrounded by 2 envelopes	85 × 330 nm	+	—
Ascoviridae	dsDNA	Allantoid to bacilliform nucleocapsid surrounded by 2 or 3 envelopes	130 × 400 nm	+	—
Nodaviridae	ssRNA	Icosahedral	29 nm, diam	—	—
Picornaviridae	ssRNA	Spherical	22—30 nm, diam	—	—
Tetraviridae	ssRNA	Icosahedral	35—39 nm, diam	—	—
Reoviridae	dsRNA	Icosahedral with 12 surface projections	55—69 nm, diam	—	+
Birnaviridae	dsRNA	Icosahedral	60 nm, diam	—	—
Rhabdoviridae	ssRNA	Bullet shaped or bacilliform	50—95 × 130—380 nm	+	—
Togaviridae	ssRNA	Spherical with peplomers	60—65 nm, diam	+	—
Flaviviridae	ssRNA	Spherical with peplomers	35—45 nm, diam	+	—
Bunyaviridae	ssRNA	Spherical, oval with peplomers	90—100 nm, diam	+	—

10) the polyhedron protein matrix dissolves, releasing the virions which attach to the membranes of the microvilli. The virions are taken in by a process of fusion, move to host cell nuclei, and begin to replicate viral DNA, eventually producing more occlusion bodies. The details of these events will be described in detail in Chapters 6 through 8.

Another interesting observation about invertebrate viruses is that they are relatively specific for their target hosts; thus they may be used safely in microbial control as part of integrated pest management since no other known animals in the environment will be affected adversely.[10,12,21,31,37,40,43,45,55,58,61,64,73,77,82,97]

To date, more than 1100 viruses of invertebrates have been reported.[21,98] This number of viruses may be the tip of the iceberg since more than 80% of the animals are invertebrates and have not been thoroughly studied.[32] Future investigations will surely reveal new members of the virus families listed as well as uncover new viruses which will require the establishment of new families or groups. The virus families listed are those accepted by the ICTV as of August 1990.[99]

REFERENCES

1. **Steinhaus, E. A.**, *Principles of Insect Pathology*, McGraw-Hill, New York, 1949.
2. **Steinhaus, E. A.**, *Insect Microbiology*, Cornell University Press, Hafner Publishing, New York, 1947.
3. **Steinhaus, E. A.**, Microbial control-the emergence of an idea: a brief history of insect pathology through the nineteenth century, *Hilgardia*, 26, 107, 1956.
4. **Steinhaus, E. A.**, *Disease in a Minor Chord*, Ohio State University Press, Columbus, 1975.
5. **Vago, C.**, A tribute to Louis Pasteur, *J. Invertebr. Pathol.*, 51, 179, 1988.
6. **Steinhaus, E. A., Ed.**, *Insect Pathology, An Advanced Treatise*, Vol. 1, Academic Press, New York, 1963.
7. **Steinhaus, E. A., Ed.**, *Insect Pathology, An Advanced Treatise*, Vol. 2, Academic Press, New York, 1963.
8. **Bergold, G. H.**, Insect viruses, *Adv. Virus Res.*, 1, 91, 1953.
9. **Franz, J. M.**, Biological control of pest insects in Europe, *Annu. Rev. Entomol.*, 6, 183, 1961.
10. **Ignoffo, C. M.**, Specificity of insect viruses, *Bull. Entomol. Soc. Am.*, 14, 265, 1968.
11. **Vago, C. and Bergoin, M.**, Viruses of invertebrates, *Adv. Virus Res.*, 13, 247, 1968.
12. **Weiser, J.**, Recent advances in insect pathology, *Annu. Rev. Entomol.*, 15, 245, 1970.
13. **Burges, D. and Hussey, N. W., Eds.**, *Microbial Control of Insects and Mites*, Academic Press, London, 1971.
14. **Aruga, H. and Tanada, Y., Eds.**, *The Cytoplasmic-Polyhedrosis Virus of the Silkworm*, University of Tokyo Press, Tokyo, 1971.
15. **Tinsley, T. W. and Harrap, K. A., Eds.**, *Moving Frontiers in Invertebrate Virology, Monographs in Virology*, Vol. 6, S. Karger, Basel, 1972.
16. **Gibbs, A. J., Ed.**, *Viruses and Invertebrates, North Holland Research Monographs-Frontiers of Biology*, Vol. 31, North-Holland, Amsterdam, 1973.
17. **Ignoffo, C. M.**, Development of a viral insecticide: concept to commercialization, *Exp. Parasitol.*, 33, 380, 1973.
18. **Kelly, D. C. and Robertson, J. S.**, Icosahedral cytoplasmic deoxyriboviruses, *J. Gen. Virol.*, 20, 17, 1973.
19. **Cantwell, G. E., Ed.**, *Insect Diseases*, Vol. 1, Marcel Dekker, New York, 1974.
20. **Cantwell, G. E.**, Honey bee diseases, parasites, and pests, in *Insect Diseases*, Vol. 2, Cantwell, G. E., Ed., Marcel Dekker, New York, 1974, 501.
21. **David, W. A. L.**, The status of viruses pathogenic for insects and mites, *Annu. Rev. Entomol.*, 20, 97, 1975.
22. **Paschke, J. D. and Summers, M. D.**, Early events in the infection of the arthropod gut by pathogenic insect viruses, in *Invertebrate Immunity*, Maramorosch, K. and Shope, R. E., Eds., Academic Press, New York, 1975, 75.
23. **Summers, M., Engler, R., Falcon, L. A., and Vail, P. V., Eds.**, *Baculoviruses for Insect Pest Control: Safety Considerations*, American Society for Microbiology, Washington, D.C., 1975.

24. **Payne, C. C. and Rivers, C. F.,** A provisional classification of cytoplasmic polyhedrosis viruses based on the size of the RNA genome segments, *J. Gen. Virol.,* 33, 71, 1976.
25. **Smith, K. M.,** *Virus-Insect Relationships,* Longman, London, 1976.
26. **Granados, R. R.,** Infection and replication of insect pathogenic viruses in tissue culture, *Adv. Virus Res.,* 20, 189, 1979.
27. **Hostetter, D. L. and Ignoffo, C. M., Eds.,** Environmental stability of microbial insecticides, *Misc. Publ. Entomol. Soc. Am.,* 10 (3), 1977.
28. **David, W. A. L.,** The granulosis virus of *Pieris brassicae* (L.) and its relationship with its host, *Adv. Virus Res.,* 22, 112, 1978.
29. **Tinsley, T. W.,** Properties and replication of insect baculoviruses, *Beltsville Symp. Agric. Res., I. Virol. Agric.,* 1977, 117.
30. **Longworth, J. W.,** Small isometric viruses of invertebrates, *Adv. Virus Res.,* 23, 103, 1978.
31. **Kurstak, E. T. and Maramorosch, K.,** Safety considerations and development problems make an ecological approach of biocontrol by viral insecticides imperative, in *Viruses and Environment,* Kurstak, E. and Maramorosch, Eds., Academic Press, New York, 1978, 571.
32. **Tinsley, T. W. and Harrap, K. A.,** Viruses of invertebrates, in *Comprehensive Virology,* Vol. 12, Fraenkel-Conrat, H. and Wagner, R. P., Eds., Plenum Press, New York, 1978, 1.
33. **Harrap, K. A. and Payne, C. C.,** The structural properties and identification of insect viruses, *Adv. Virus Res.,* 25, 273, 1979.
34. **Stoltz, D. B. and Vinson, S. B.,** Viruses and parasitism in insects, *Adv. Virus Res.,* 24, 125, 1979.
35. **Brun, G. and Plus, N.,** The viruses of *Drosophila,* in *The Biology and Genetics of Drosophila,* Ashburner, M. and Wright, T. R. F., Eds., Academic Press, New York, 1980, 625.
36. **Kalmakoff, K. J. and Longworth, J. F.,** Microbial Control of Insect Pests, N. Z. Dep. Sci. Ind. Res. Bull., 228, 1980.
37. **Miltenburger, H. G., Ed.,** Safety aspects of baculoviruses as biological insecticides, *Symp. Proc., Bundesministerium fur Forschung und Technologie,* 1980, 11.
38. **Shieh, T. R. and Bohmfalk, G. T.,** Production and efficacy of baculoviruses, *Biotechnol. Bioeng.,* 22, 1357, 1980.
39. **Bailey, L.,** *Honey Bee Pathology,* Academic Press, London, 1981.
40. **Burgess, H. D., Ed.,** *Microbial Control of Pests and Plant Diseases, 1970—1980,* Academic Press, London, 1981.
41. **Davidson, E. W., Ed.,** *Pathogenesis of Invertebrate Microbial Diseases,* Allanheld, Osmun Publishers, Towata, NJ, 1981.
42. **Miller, L. K.,** A virus vector for genetic engineering in invertebrates, in *Genetic Engineering in the Plant Sciences,* Panopoulos, N. J., Ed., Praeger, New York, 203.
43. **Kurstak, E., Ed.,** *Microbial and Viral Pesticides,* Marcel Dekker, New York, 1982.
44. **Reavy, B. and Moore, N. F.,** The replication of small RNA containing viruses of insects, *Microbiologica,* 5, 63, 1982.
45. **Deacon, J. W.,** Microbial control of plant pests and diseases, in *Aspects of Microbiology,* Vol. 7, Van Nostrand Reinhold, Berkshire, 1983.
46. **Sturrock, J. W. and Ulbricht, T. L. V., Eds.,** Special Issue: Biological Control, *Agric. Ecosystems Environ.,* 10, 99, 1983.
47. **Arif, B. M.,** The entomopoxviruses, *Adv. Virus Res.,* 29, 195, 1984.
48. **Hunter, F. R., Crook, N. E., and Entwistle, P. F.,** Viruses as pathogens for the control of insects, in *Microbial Methods for Environmental Biotechnology,* Grainger, J. M. and Lynch, J. M., Eds., Academic Press, New York, 1984, 323.
49. **Johnson, P. T.,** Viral diseases of marine invertebrates, *Helgol. Meeresunters,* 37, 65, 1984.
50. **Poinar, G. O., Jr. and Thomas, G. M.,** *Laboratory Guide to Insect Pathogens and Parasites,* Plenum Press, New York, 1984.
51. **Summers, M. D.,** The biology and molecular biology of baculoviruses, *Proc. Chin. Acad. Sci. — U.S. Natl. Acad. Sci., Joint Symp. Biol. Control Insects,* Adkisson, P. L. and Shijun, M. A., Eds., Science Press, Beijing, 1984, 142.
52. **Tanada, Y. and Hess, R. T.,** The cytopathology of baculovirus infections in insects, in *Insect Ultrastructure,* Vol. 2, King, R. D. and Akai, H., Eds., Plenum Press, New York, 1984, 517.
53. **Faulkner, P. and Boucias, D. G.,** Genetic improvement of insect pathogens: emphasis on the use of baculoviruses, in *Biological Control in Agriculture IPM Systems,* Hoy, M. A. and Herzog, D. C., Eds., Academic Press, Orlando, FL, 1985, 263.
54. **Hoy, M. A. and Herzog, D. C., Eds.,** *Biological Control in Agricultural IPM Systems,* Academic Press, Orlando, FL, 1985.
55. **Maramorosch, K. and Sherman, K. E., Eds.,** *Viral Insecticides for Biological Control,* Academic Press, Orlando, FL, 1985.

56. **Moore, N. F., Reavy, B., and King, L. A.,** General characteristics, gene organization and expression of small RNA viruses of insects, *J. Gen. Virol.,* 66, 647, 1985.

57. **Doerfler, W. and Bohm, P., Ed.,** The molecular biology of baculoviruses, in *Current Topics in Microbiology and Immunology,* Springer-Verlag, Heidelberg, 1985, 131.

58. **Granados, R. R. and Federici, B. A., Eds.,** *The Biology of Baculoviruses,* Vol. 1, CRC Press, Boca Raton, FL, 1986.

59. **Granados, R. R. and Federici, B. A., Eds.,** *The Biology of Baculoviruses,* Vol. 2, CRC Press, Boca Raton, FL, 1986.

60. **Harrap, K. A. and Payne, C. C.,** The structural properties and identification of insect viruses, *Adv. Virus Res.,* 25, 273, 1986.

61. **Morris, O. N., Cunningham, J. C., Finney-Crawley, J. R., Jaques, R. P., and Kinoshita, G.,** Microbial insecticides in Canada: their registration and use in agriculture, forestry and public and animal health, *Bull. Entomol. Soc. Can.,* 18, 1986.

62. **Miller, D. W., Safer, P., and Miller, L. K.,** An insect baculovirus host-vector system for high level expression of foreign genes, in *Genetic Engineering, Vol. 8, Principles and Methods,* Setlow, J. K. and Hollander, A., Eds., Plenun Press, New York, 1986, 277.

63. **Bitton, G., Maruniak, J. E., and Zettler, W. F.,** Virus survival in natural ecosystems, in *Survival and Dormancy of Microorganisms,* Hemis, V., Ed., John Wiley & Sons, New York, 1987, 301.

64. **Burges, H. D. and Pillai, J. S.,** Microbial bioinsecticides, in *Microbial Technology in the Developing World,* DaSilva, E. J., Dommergues, Y. R., Nyns, E. J., and Ratledge, C., Eds., Oxford University Press, Oxford, 1987, 121.

65. **Evans, H. F. and Entwistle, P. F.,** Viral diseases, in *Epizootiology of Insect Diseases,* Fuxa, J. R. and Tanada, Y., Eds., John Wiley & Sons, New York, 1987, 257.

66. **Krell, P. J. and Beveridge, T. J.,** The structure of bacteria and the molecular biology of viruses, in *Int. Rev. Cytol.,* Suppl. 17, 15, 1987.

67. **Maramorosch, K., Ed.,** *Biotechnology in Invertebrate Pathology and Cell Culture,* Academic Press, San Diego, CA, 1987.

68. **Rowlands, D. J., Mayo, M. A., and Mahy, B. W. J., Eds.,** *The Molecular Biology of the Positive Strand RNA Viruses,* FEMS Symposium No. 32, Academic Press, London, 1987.

69. **Summers, M. D. and Smith, G. E.,** *A Manual of Methods for Baculovirus Vectors and Insect Cell Culture Procedures,* Texas Agric. Exp. Stn. Bull. No. 1555, 1987.

70. **Fuxa, J. R. and Tanada, Y., Eds.,** *Epizootiology of Insect Diseases,* John Wiley & Sons, New York, 1987.

71. **Arif, B. M. and Jamieson, P.,** The baculoviruses, in *Biocontrol of Plant Diseases,* Vol. 1, Nukerji, K. G. and Garg, K. L., Eds., CRC Press, Boca Raton, FL, 1988, 9370.

72. **Carbonell, L. F. and Miller, L. K.** Genetic engineering of viral pesticides — expression of foreign genes in nonpermissive cells, in *Molecular Strategies for Crop Protection UCLA Symposia on Molecular and Cellular Biology,* New Series Vol. 48, Arntzen, C. J. and Ryan, C., Eds., Alan R. Liss, New York, 1987, 235.

73. **Huber, J.,** Safety of baculoviruses used as biological insecticides, in *Risk Assessment for Deliberate Releases: The Possible Impact of Genetically Engineered Microorganisms on the Environment,* Springer-Verlag, Berlin, 1988, 65.

74. **Kang, C. Y.,** Baculovirus vectors for expression of foreign genes, *Adv. Virus Res.,* 35, 177, 1988.

75. **Miller, D. W.,** Genetically engineered viral insecticides, *Biotechnology for Crop Protection,* ACS Symposium Series American Chemical Society 379, Washington, D.C., 1988, 405.

76. **Miller, L. K.,** Baculoviruses as gene expression vectors, *Annu. Rev. Microbiol.,* 42, 177, 1988.

77. **Sussman, M., Collins, C. H., Skinner, F. A., and Stewart-Tull, D. E., Eds.,** *The Release of Genetically-Engineered Micro-organisms,* Academic Press, San Diego, CA, 1988.

78. **Maeda, S.,** Expression of foreign genes in insects using baculovirus vectors, *Annu. Rev. Entomol.,* 34, 351, 1989.

79. **Shieh, T. R.,** Industrial production of viral pesticides, *Adv. Virus Res.,* 36, 315, 1989.

80. **Blissard, G. W. and Rohrmann, G. F.,** Baculovirus diversity and molecular biology, *Annu. Rev. Entomol.,* 35, 127, 1990.

81. **Casida, J. E., Ed.,** *Pesticides and Alternatives: Innovative Chemical and Biological Approaches to Pest Control.* Elsevier, Amsterdam, 1990.

82. **Laird, M., Lacey, L. A., and Davidson, E. W.,** *Safety of Microbial Insecticides,* CRC Press, Boca Raton, FL, 1990.

83. **Kawanishi, C. Y. and Held, G. A.,** Viruses and bacteria as sources of insecticides, in *Safer Insecticides Development and Use,* Hodgson, E. and Kuhn, R. J., Eds., Marcel Dekker, New York, 1990, 351.

84. **Volkman, L. and Keddie, B. A.,** Nuclear polyhedrosis virus pathogenesis, *Semin. Virol.,* 1, 249, 1990.

85. **Matthews, R. E. F.,** The history of viral taxonomy, in *A Critical Appraisal of Viral Taxonomy,* Matthews, R. E. F., Ed., CRC Press, Boca Raton, FL, 1983.

86. **Wildy, P.,** *Classification and Nomenclature of Viruses, First Report of the International Committee on Nomenclature of Viruses, Monographs in Virology,* Vol. 5, S. Karger, Basel, 1971.
87. **Fenner, F.,** Classification and nomenclature of viruses, second report of the International Committee on Taxonomy of Viruses, *Intervirology,* 7, 5, 1976.
88. **Matthews, R. E. F.,** Classification and nomenclature of viruses, the third report of the International Committee on Taxonomy of Viruses, *Intervirology,* 12, 133, 1979.
89. **Matthews, R. E. F.,** Classification and nomenclature of viruses, fourth report of the International Committee on Taxonomy of Viruses, *Intervirology,* 17, 7, 1982.
90. **Longworth, J. F.,** Current problems in insect virus taxonomy, in *A Critical Appraisal of Viral Taxonomy,* Matthews, R. E. F., Ed., CRC Press, Boca Raton, FL, 1983, 123.
91. **Tinsley, T. W. and Kelly, D. C.,** Taxonomy and nomenclature of insect pathogenic viruses, in *Viral Insecticides for Biological Control,* Maramorosch, K. and Sherman, K. E., Eds., Academic Press, Orlando, FL, 1985, 3.
92. **Bishop, D. H. L.,** The genetic basis for describing viruses as species, *Intervirology,* 24, 79, 1985.
93. **Weiser, J.,** *An Atlas of Insect Diseases,* Dr. W. Junk, B. V. Publishers, Prague, 1977.
94. **Maramorosch, K.,** *The Atlas of Insect and Plant Viruses,* Academic Press, New York, 1977.
95. **Tchuchriy, M. G.,** *The Ul'trastruktura Virusov Cheshuckrylykh. Vrediteli Rastenii* (Atlas), Shtiintsa Publ., Kishinev, 1982.
96. **Liang, D. G., Cai, Y. N., Lin, D. Y., Zhang, Q. L., Hu, Y. Y., He, H. J., and Zhao, K. B.,** *The Atlas of Insect Viruses in China,* Hunan Sci. and Techn. Press, Hunan, China, 1986.
97. **Groner, A.,** Specificity and safety of baculoviruses, in *The Biology of Baculoviruses,* Vol. 1, Granados, R. R. and Federici, B. A., Eds., CRC Press, Boca Raton, FL, 1986, 275.
98. **Martignoni, M. E. and Iwai, P. J.,** A catalog of viral diseases of insects, mites, and ticks, *U.S. Dep. Agric. For. Serv. Gen. Tech. Rep., PNW-195,* 1, 1986.
99. **Francki, R. I. B.,** personal communication.
100. **Cunningham, J. C., Burke, J. M., and Arif, B. M.,** An entomopoxvirus found in populations of the large aspen tortrix, *Choristoneura conflictana, Can. Entomol.,* 105, 767, 1973.
101. **Payne, C. C., Compson, D., and DeLooze, S. M.,** Properties of the nucleocapsids of a virus isolated from *Oryctes rhinoceros, Virology,* 77, 269, 1977.

Chapter 2

PREPARATION OF INVERTEBRATE VIRUSES AND TISSUES FOR EXAMINATION

J. R. Adams and J. R. Bonami

TABLE OF CONTENTS

I. INTRODUCTION

The objective of this chapter is to present briefly several general techniques that have proven useful in our own and other laboratories. We include key references for special techniques developed for invertebrate virus studies, and the rationale for choosing them. When choosing, adapting, revising, or even developing a new technique, one must carefully consider the possible artifacts that may be produced and their effect(s) on the results. Techniques useful for general light and electron microscopic examinations of the invertebrate viruses are presented in many of the key references listed in Appendix, Section 1. An excellent review edited by Crang and Klomparens[1] has focused on the parameters that should be considered, the pitfalls encountered in certain electron microscopic techniques, and the possible solutions to these difficulties.

II. DIAGNOSIS

Helpful texts to consult concerning the diagnosis of invertebrate diseases include those by Steinhaus,[2] Aruga and Tanada,[3] Thomas,[4] Cantwell,[5] Smith,[6] Maramorosch,[7] Weiser,[8] Kalmakoff and Longworth,[9] Bailey,[10] Tchuchrity,[11] Poinar and Thomas;[12] and chapters by Bergold,[13] Steinhaus,[14] Wittig,[15] Weiser and Briggs,[16] Cantwell,[17] Vaughn,[18] Wigley,[19] Brun and Plus,[20] Johnson,[21] MacKay,[22] Doane and Anderson,[23] and Palmer and Martin.[24]

Steinhaus[14,25] described specifically the development of methods to diagnose diseases of invertebrates using the term "symptom" to refer to "any objective aberration in behavior or function indicating disease" while "sign" was referred to as "any objective physical aberration or manifestation of disease indicated by a change in structure." A larva infected with nuclear polyhedrosis virus (NPV) may appear whitish, swollen, larger or smaller than a healthy larva (signs) and show symptoms of constipation and loss of appetite. Steinhaus[14] proposed steps in diagnosis as follows: (1) physical examination by general inspection; (2) macroscopic followed by microscopic examination; (3) isolation if possible of the microbial pathogen and histopathologic examination of tissues of each system, e.g., integument, alimentary, circulatory, etc. Often these observations will guide one to properly determine the group of pathogens involved, and then additional laboratory tests may be performed to establish the genus, etc. Steinhaus[25] noted that identification of the pathogen may not be sufficient to diagnose the disease. Further laboratory tests are required including close observation of the full course of the disease, the use of Koch's postulates, and biochemical studies.

A. LIGHT MACROSCOPIC AND MICROSCOPIC EXAMINATIONS
1. Living Specimens
A sample of hemolymph or specific tissue(s) collected from living specimens may reveal the presence of invertebrate pathogens under bright field, phase, dark field, or differential interference contrast (DIC) microscopy. Larvae should be surface sterilized in 2% Clorox followed by 70% ethanol solution with a final rinse in sterile distilled water and dried on sterile filter paper.[26] A sample of hemolymph is taken with a sterile pipette following puncture with a sterile needle, placed on a slide, and covered with a coverslip. Tissues may be dissected and macerated on a slide in sterile distilled water or sterile buffer and covered with a coverslip.

2. Dead Specimens
A dead specimen may be surface sterilized if desired in 5.25% sodium hypochlorite for 3 to 5 min.[4,12] This step is not as critical for virus infected specimens since any bacteria present will probably be eliminated in the initial steps of a centrifugation procedure or early

in the isolation and purification schedule. The dead larva may be placed in a vial containing distilled water and refrigerated several days until putrefaction occurs. The virus particles, occlusion bodies (OBs) or viral occlusions (VOs), which are released may be isolated according to a schedule similar to that described next.

To prepare a specimen for examination the same day, grind it in distilled water in a tissue grinder. Small specimens are easily ground in a Kontes Duall® tissue grinder which has a ground glass pestle and tube. Larger volume samples may be filtered through several layers of cheesecloth in a funnel with a fine mesh screen which serves to remove most of the insect cuticle fragments, etc. Centrifuge the filtrate at 830 to 1480 g for 10 min. The P_1 pellet is then examined by light and electron microscopy. A fast convenient stain is nigrosine (see Appendix, Section 2) which shows microorganisms as white bodies against a black background. If available, phase contrast, dark field, or DIC microscopy may be used. A second centrifugation at 7470 g for 20 min will pellet smaller microorganisms. A third centrifugation at 13,200 g for 30 min to 1 h is sufficient to pellet viruses or microorganisms as small as granulosis virus. Higher speed centrifugations for longer times are required to sediment many of the small nonoccluded or nonocclusion body viruses. The reader should consult Chapter 3 or the following chapters for details on the isolation of specific invertebrate viruses.

Granulosis viruses are easily seen by dark field microscopy[27] or with stained smears;[28] enhanced magnification (12.5 × eye piece) in the light microscope as used by Hunter[29] is also useful.

3. Stains

Stains useful for preparation of smears of infected hemolymph or tissues include Giemsa[12,16] and Buffalo black for CPV.[30]

4. Virus Counting Techniques

Quantitation of larger insect viruses, e.g., NPV, CPV, and entomopox viruses may be done by techniques such as counting chamber methods, proportional counts, dry film counts, and electronic counting; these are described by Burges and Thomson,[31] Cantwell,[32] and Wigley.[33] A recent review has been prepared by Kaupp and Sohi.[34]

B. ELECTRON MICROSCOPIC EXAMINATION OF PATHOGENS
1. Preparation of Samples

To examine pellets of microorganisms collected by centrifugation, resuspend the pellet in a small volume (about 1 ml) of distilled water and place a small sample (up to 4 μl) on a specimen grid prepared with a formvar film or a parlodion film stabilized with a thin layer of carbon.[35] Some investigators also coat the specimen grids[35] with 0.1% polylysine[36] and may omit the carbon coat. The polylysine leaves cationic sites that combine with anionic sites on microorganisms. Glow discharging support films on specimen grids in the high vacuum evaporator will also reduce the number of negative charges on the films, thus ensuring better attachment of the microorganisms.

When the sample has dried down on the grid, it may be placed directly in the electron microscope. Small nonoccluded viruses can be revealed with a light positive stain of 1% (w/v) aqueous ammonium molybdate (AM) or a 2% (w/v) aqueous phosphotungstic acid (PTA) adjusted to pH 7 with NaOH. A heavier stain of either creates a darker background around the virus particles (negative stain). The negative stain method was first described by Hall.[37] Helpful references on the negative staining method include those by Brenner and Horne,[38] Horne,[39-41] Horne and Ronchetti,[42] Haschemeyer and Myers,[43] Doane and Anderson,[23] and Palmer and Martin.[24] The pseudoreplica technique described by Palmer and Martin[24] gives an excellent detail on the surface morphology of negatively stained viruses.

For routine diagnosis of viral diseases in marine invertebrates (crustacea especially), the following simple technique is often used in our laboratory. Take 1 mm³ of target organ (hepatopancreas for example) and put it in 2% PTA (6 to 10 drops). After crushing the tissue with a Pasteur pipette, take one drop of the fluid, put it in contact with a carbon-collodion coated grid (3 min), dry it with a filter paper, and directly observe in TEM. Hemolymph can be used, too. In this case, the hemolymph-PTA mixture is sonicated to disrupt cells before it is placed on the grid. Since marine invertebrate viruses have been found to occur in lower concentrations and are more difficult to isolate than insect viruses, routine diagnosis has been based on TEM observations that include the notation of shape, size, and presence or absence of envelope, and for some baculoviruses, the superficial structure of the nucleocapsid.

Many viruses are very fragile and require the use of buffers throughout the isolation and purification procedures. The artificial physiological medium of Pantin[44] is useful for marine invertebrates. Pathogens may be frozen in this medium. Bonami[45] also used TN buffer (0.02 M Tris-HCl; 0.5 M NaCl) at pH 7.2 especially for enveloped viral particles infecting marine invertebrates. Insect viruses may be purified in buffers such as Tris-HCl, citrate, etc. For further details the reader is referred to Chapter 3 and those that follow.

Small samples of larger insect viruses may be pelleted in microcentrifuge tubes, e.g., 1.5 or 2.0-ml tubes in a high speed microcentrifuge with a swinging bucket rotor which spins samples up to 15,000 rpm (19,500 g) such as the Tomy® MTX-150. Very small samples (0.2 ml) after partial purification may be pelleted in a Beckman Airfuge® at 100,000 g for about 30 min, resuspended, and negatively stained. Ernest Fullam, Inc. supples EFFA® ultracentrifuge tubes for deposition of very small samples on specimen grids as well as centrifuge tubes for pelleting samples in BEEM® capsules.

When virus samples in buffers are placed on sucrose density gradients, often sucrose remains even after several washings of the bands of virus, making it difficult to examine the contents of the bands in the electron microscope. This problem was solved by Webb,[46] who developed a washing technique to remove sucrose, buffers, salts, etc. from samples by placing the grids on filter paper strips that hang over the edge of a beaker of distilled water which is continually being refilled to the top. Adams[47] further modified Webb's[46] technique by using a crystallizing dish (100 mm diameter × 50 mm deep) which provided for the washing of 8 to 10 grids at one time, each on a separate thin strip of filter paper (about 20 × 90 mm). A 0.2 μm filter (such as Gelman Versacap®) on the distilled water line eliminates the possibility of introducing contamination in the washing procedure which usually takes about 10 min. Samples may then be contrast enhanced by use of AM or PTA immediately upon removal from the filter paper strips. Fragile viruses may require a brief period of fixation (about 15 to 30 min in 2 to 2.5% glutaraldehyde prior to positive or negative staining).[23,24,46]

Other effective methods for application of viruses to specimen grids include the water drop method, agar diffusion method,[23] and spray techniques.[39,41] Virus morphology may serve as an aid in the rapid diagnosis of a virus especially when combined with immuno epifluorescence microscopy (EM) which allows one to visualize samples with concentrations as low as 10² to 10³ virus particles per ml.[24] A morphological guide to the families of viruses based on virus particle diameters is listed in Appendix, Section 3. If there is a small circle on the center of the viewing screen of the electron microscope, it is possible to calibrate the diameters for each magnification, thus enabling the investigator to roughly determine the diameters of the unknown virus particles that have been isolated.

2. Quantitation of Small Virus Particles

Small virus particles may require counting in the electron microscope. Counting techniques include the latex particle reference method, spray droplet method, agar filtration, lowered drop and loop drop methods, and sedimentation techniques as described by Miller[48,49] and Doane and Anderson.[23] Microsphere size standards (10 to 100 μm) and nanosphere size standards) (100 to 900 nm) have been recently introduced by Duke Scientific Corp. to replace the polystyrene latex spheres. These particle size standards are certified for mean diameter and traceability by the National Bureau of Standards and are available from companies that sell electron microscopy supplies. DeBlois et al.[50,51] using polystyrene latex spheres as standards made measurements on three viruses (*Tipula* iridescent virus (TIV) and NPVs of gypsy moth and European pine sawfly) by three techniques, i.e., laser light scattering spectroscopy, resistive pulse analysis, and electron microscopy, to determine the distribution of the number of virions per bundle of the NPVs and the particle volumes.

Buoyant density centrifugation of viral DNA was used by Rohrmann et al.[52] to quantify the approximate percentage of the SNPV and MNPV isolated from *Orgyia pseudotsugata* in mixed samples of these viruses.

3. Electron Microscopy of Nucleic Acids

The procedure developed by Kleinschmidt and Zahn[53] and Kleinschmidt[54] has enabled investigators to visualize DNA as a DNA-cytochrome c complex forming a filament about 10 nm in diameter which can be observed easily in the electron microscope. Helpful reviews and texts which give specific directions on the developments of this technique have been prepared by Davis et al.,[55] Inman and Schnös,[56] Younghusband and Inman,[57] Chow et al.,[58] Thomas,[59] Ferguson and Davis,[60] Garon,[61,62] Nermut,[63] Allison,[64] Coggins,[65,66] and Robinson et al.[67]

A microversion of the Kleinschmidt method, as described by Delain and Brack,[68] has been used in the Montpellier Laboratory to visualize nucleic acid molecules. This method requires only 10 μl of a 2 μg/ml nucleic acid suspension. φX 174 (ss) and φX 174 RF11 (ds) DNA are used as length markers.

Electron microscopy of the DNA of nuclear polyhedrosis viruses (NPVs) have been investigated by Summers,[69] Bud and Kelly,[70,71] Brown et al.,[72] Tjia et al.,[73] Scharnhorst et al.,[74] Jurkovicova,[75] Jurkovicova et al.,[76] Loh et al.,[77] and Skuratovskaya et al.[78] The electron microscopy of the *Oryctes rhinoceros* baculovirus has been reported by Revet and Monsarrat;[79] that of iridescent viruses, by Delius et al.[80] and Schnitzler et al.;[81] and polydna viruses, by Krell and Stoltz.[82,83] Several reports have appeared on the electron microscopy of RNA from cytoplasmic polyhedrosis virus,[84,85] flacherie virus,[86] and HPS-1[87] isolated from a *Drosophila melanogaster* cell line.

The molecular weights of seven NPVs and three granulosis viruses (GVs) were determined by Burgess[88] using electron microscopic measurements of contour lengths which were compared with measured lengths of DNA marker molecules.

4. Special Techniques

Hosur et al.[89] used single crystal X-ray diffraction to determine the structure of black beetle virus (a nodavirus) at 3.0 Å resolution, and Mazzone et al.[90,91] observed *Lymantria dispar* MNPV in the high voltage electron microscope (HVEM). They compared the transmission image densities of virions and OBs with that through polystyrene latex particles of known mass taken under the same conditions in the TEM. The masses of three viruses (*Tipula* iridescent virus, *L. dispar* MNPV, and *Neodiprion sertifer* MNPV) and their molecular weights were calculated from the quantitative EM procedures of Zeitler and Bahr[92] and Bahr et al.[93]

III. EXAMINATION OF TISSUES

A. LIGHT MICROSCOPY

Techniques for the preparation of animal and invertebrate tissues are given in several texts.[3,16,94-96] Fixatives recommended for insect tissues include Bouin-Duboscq and Carnoy. The fixative most generally used for marine invertebrates is Davidson's fixative[97] although

Bouin's, Carnoy's, and Zenker's fixatives have also been used for bivalve mollusks by Comps.[98] Several fixatives were compared (including neutral buffered formalin; seawater formalin; Davidson's, Bouin's, and Helly's [Zenker-formol] fixatives) by Johnson,[99] who found Helly's fluid was the best for general use in crustacea.

Comps[98] used the Feulgen reaction and acridine orange at pH 3.8, according to the method of Dart and Turner,[100] and epifluorescence microscopy to detect the presence of cytoplasmic DNA of *Iridovirus* infecting oyster cells. This method was also used successfully to recognize S virus (ssRNA) and P virus (dsRNA) in crab tissues during multiple infections.[100a] The fluorescence of S virus (ssRNA) exhibits the same fluorescence color as ribosome rich areas while that of P virus (dsRNA) is more intense. Although the use of acridine orange is not often reported in journals because photographs are usually published in black and white, it is used extensively in laboratories and provides good results in the recognition of ss and ds DNA or RNA. However, the buffer pH needs careful monitoring in order to obtain satisfactory results.

B. PARAFFIN EMBEDDING AND SECTIONING

A general schedule follows: (1) fixation overnight, (2) washing in 70% ethanol, (3) dehydration in a graded series of ethyl alcohols to 100% ethanol, and (4) infiltration in paraffin or paraplast. Ribbons of sections, preferably 2 μm or less, are prepared (see Reference 94); floated on clean glass slides pretreated with Mayer's albumin,[94,96] Haupt's fixative,[94] or Pappas' subbing solution;[101] and dried on a slide warmer.

A very useful general Azan stain developed by Hamm[102] for OBs or VOs of NPV, CPV, GV, and entomopox spheroid bodies is a modification of Mallory's triple stain in which azocarmine is substituted for basic fuschin. VOs are stained bright red (see Appendix, Section 2). Other stains for VOs include iron hemotoxylin,[8,12,16] Huger's[103] modification of iron hemotoxylin for GV, and Vago and Amagier's[104] technique for staining NPV.

C. AUTORADIOGRAPHY

Reviews on autoradiography include those by Salpeter and Bachmann,[105] Salpeter and McHenry,[106] Maraldi,[107] Williams,[108,109] Rogers,[110] and Salpeter.[111] Procedures for autoradiographic techniques may be found in the technical publications from Ilford Ltd.[112] and *Kodak Tech. Bits,*[113] and a reference by Budd.[114] An extensive list of references covering every phase of the autoradiography technique has been published by Kodak.[115] Metabolic changes in virus infected insects which have been detected by autoradiography were reported by Watanabe,[116-122] Morris,[123-128] Watanabe and Kobayashi,[129,130] and Benz and Wäger.[131] An autoradiographic method was developed by Mix and Tomasoric[132] for observation of molluscan cells by using high specific activity of tritiated thymidine.

D. HISTOCHEMISTRY AND IMMUNOCYTOCHEMISTRY

Benz[133] reviewed early studies in insect pathology in which histochemical techniques established sites of nucleic acid metabolism in NPV-infected tissues of *Diprion hercyniae* larvae[134] and the GV-infected tissues of *Carpocapsa pomonella*.[135] Kurstak[136] and Sternberger[137,138] reviewed the development of the immunoperoxidase technique and presented procedures. This technique involves a peroxidase (usually horseradish peroxidase) which has a low molecular weight of 40,000 that is attached chemically to a viral antibody and reacted with a homologous intracellular viral antigen. The presence of the antigen is revealed by staining reactions, such as diaminobenzidine that forms a blue-brown reaction product and Nadi that forms a purple-staining reaction product.

Summers et al.[139] and Volkman et al.[140] used immunoperoxidase detection to locate baculovirus antigens in infected insect cell cultures while Keddie et al.[141] used the indirect peroxidase antiperoxidase (PAP) method with diaminobenzidine as the electron donor to determine the temporal infection sequence of AcMNPV in *Trichoplusia ni* larvae. Hilwig[142] used immunofluorescence and scanning electron microscopy (SEM) to study the changes in the nucleus and surface of AcMNPV-infected lepidopteran cells while Ohba and Aizawa[143] studied *Chilo* iridescent virus by the immunoperoxidase technique and Hukuhara and Akami[144] used immunofluorescence to study OBs in soil samples. Uchima et al.[145] used the immunoperoxidase technique to follow the uptake and enhancement of virus infection by the synergistic factor of a granulosis virus of the armyworm, *Pseudaletia unipuncta.* Allen and Ball[146] used immunodiffusion and immunosorbent electron microscopy (ISEM) to distinguish the virus particles (SAV) isolated from the aphid, *Sitobion avenae,* from other previously identified small nonoccluded viruses.

E. *IN SITU* HYBRIDIZATION

Techniques for detection of viral nucleic acids by *in situ* hybridization have been presented by Haase et al.,[147-149] Haase,[150] and Hofler et al.[151] These procedures have been very useful in the detection of low concentrations of viral genes in cell populations or tissues. Hatfill et al.[152] identified and located aphid lethal paralysis virus (ALPV) particles in 8 μm frozen sections of infected *Rhopalosiphum padi* aphids by *in situ* hybridization techniques. (For other references see Appendix, Section 1).

F. ELECTRON MICROSCOPY
1. Fixatives

Many texts are available to guide an investigator in the preparation of tissues for electron microscopy including those by Hayat,[153-156] Glauert,[157] and Todd[158] (see also Appendix, Section 1). Artifacts that occur in the fixation are discussed in a recent text,[1] and corrective measures are presented. Bowers and Maser[159] discussed such parameters as choice of buffer vehicle and concentration, fixative and concentration, osmolality, additives, and temperature and presented a general protocol for the fixation of tissues (see Appendix, Section 4). Insect and invertebrate pathologists must adapt, fine tune, or develop the protocols for their particular tissues and/or viruses in order to obtain the "perfect fixation", i.e., tissues or viruses with no artifacts. Maser et al.[160] studied relationships between pH, osmolality, and concentration of fixative solutions and presented graphs to aid in the preparations of fixatives for electron microscopy. Another helpful reference by Coetzee and Van der Merwe[161] deals with the effect of glutaraldehyde on the osmolality of the buffer vehicle. The increase, beyond the numerical addition of the glutaraldehyde and buffer vehicle, is due to a higher buffer dissociation constant.

Adams and Wilcox[162] determined the osmolality of the hemolymph of several insect species and then adapted the fixative of Smith and Smith[163] that contains 2.5% glutaraldehyde in 0.05 M sodium cacodylate buffer + 0.17 M sucrose with an osmolality of 490 milliosmols (mOsm). This is hypertonic to the tissues of most insects and insect cell cultures, but has given satisfactory results. Even the fixative containing 4% formaldehyde + 1% glutaraldehyde (4CF-1G) of McDowell and Trump[164] gave excellent preservation of insect tissues in spite of its high osmolality (>1800 mOsm). Gipson and Scott[165] studied the effects of various fixatives and embedding media on pellets of virions and sections of *Pseudoplusia includens* SNPV. The best preservation was obtained with the 4% glutaraldehyde, 1% osmium tetroxide schedule including embedding in Epon-Araldite or Araldite. Other fixation procedures are given in Appendix, Section 4.

The adjustment of osmotic pressure, which is particularly high in some marine species such as crustaceans and mollusks, is one of the most important problems in the preparation of pathogens and tissues of invertebrates other than insects. Osmotic pressure may be adjusted using NaCl or sucrose or simply by using ultrafiltered seawater. Environmental conditions (such as euryhaline or marine conditions) appear as one of the most important parameters influencing the osmotic pressure of the internal fluids of marine invertebrates despite the fact that some marine invertebrates show an osmotic and ionic regulation (osmoregulation).

This is especially true in the case of species which have a life cycle in waters of different salinities. However, in most of the cases the internal osmotic pressure of marine invertebrates is close to that of the seawater in which they are living. "Regular" seawater has a freezing point depression of $-1.872°C$ which is equivalent osmotically to 0.56 *m* NaCl (32.15 g/1).[166] In most cases, the osmotic pressure of fixatives must be adapted to the species considered.

Fixatives to prepare tissues from marine invertebrates for electron microscopy include 2.5% glutaraldehyde in 0.2 *M* cacodylate buffer followed by postfixation in 1.5% osmium tetroxide in 0.2 *M* cacodylate buffer (pH 7.2) containing 0.4 *M* NaCl (according to Bonami[45]), or 2.5 to 3% glutaraldehyde in 0.12 *M* Sorenson's phosphate buffer (pH 8) followed by postfixation in 1.5% osmium tetroxide in 0.12 *M* Sorenson's phosphate buffer with the addition of 0.3 *M* sucrose (according to Perkins[167]). Baur and Stacey[168] recommended PIPES as the buffer vehicle for fixation of mammalian and marine tissues. Kingsley and Cole[169] recently compared freezing point vs. vapor pressure osmometers and noted differences in readings obtained on glutaraldehyde and formaldehyde solutions. General fixatives for transmission electron microscopy have been discussed recently by several authors[153-159,170] and presented in many of the texts listed in Appendix, Section 1.

2. Dehydration

Most protocols include changes at 10 to 15 min intervals in increasing concentrations of alcohol or acetone to remove the water from the tissues. Many investigators also include an en bloc stain of uranyl acetate in aqueous or alcoholic solution. Although most invertebrate tissues may be changed at the intervals noted earlier if a problem should develop, a recent report by Johnson[171] might be consulted because he found that extending the length of time in the dehydration schedule, especially in the 50% ethanol, greatly improved the embedding and sectioning quality of perfused brain tissues. Artifacts in dehydration and embedding have been discussed by Mollenhauer.[172]

3. Embedding

There are many references to guide the investigator in arriving at a formula of the embedding medium for their particular tissues or pathogens. Components must be weighed and thoroughly mixed in order to ensure consistent results. The Epon epoxy resin developed by Luft[173,174] has been used universally with the aid of two references that list exact anhydride: epoxy percentages for Epon[175] and Araldite and Araldite:Epon[176,177] embedding. The chapters by Aldrich,[178] and Aldrich and Mollenhauer[179] as well as reports by Mollenhauer and Droleskey,[180] Ellis,[181] and Acetarin et al.[182] discuss embedding with epoxy resins. Many investigators prefer Spurr's[183] low viscosity resin for easier infiltration of the tissues or pellets of cell cultures or purified viruses. The senior author used Dow Epoxy Resin®332 in a formula developed by Winborn[184] until it was removed from the market due to possible carcinogenicity. It sectioned easily with glass knives. Its replacement (D.E.R. 324) proved unsatisfactory. Currently an Araldite 506® formula has been substituted (see Appendix, Section 4). The use of silicone[185] and surfactant[186] additives has improved the cutting quality of epoxy resins and extended the usefulness of glass knives.

To ensure adequate infiltration of the embedding medium, we allow the specimens to infiltrate under vacuum overnight in the final embedding medium before the oven is turned on (see Appendix, Section 4). If one finds holes in the sections of pellets of invertebrate pathogens or tissues, the times of infiltration in the embedding medium can be extended, e.g., Endo and Wergin[187] successfully embedded plant parasitic nematodes in Spurr's resin by extending the times in all steps of the embedding process. The concentration of resin was increased by 10% at each change (10% resin:90% acetone) from 10 to 100%, allowing at least 3 h per change and overnight (instead of making changes in the night), until in the third change in 100% resin when the specimens were poured into aluminum weighing pans and polymerized in an oven at 70°C for 48 h.

Acrylic resins which are water soluble such as L. R. White and Gold® and Lowicryl® have been developed for immunocytochemistry, which can be heat or low temperature cured. The advantage of low temperature curing is that viral antigens are better preserved (see Appendix, Section 1). Procedures for use are reported in texts by Robinson et al.[67] and Hayat.[188] The advantage of low temperature curing is that viral antigens are better preserved. Cryoultramicrotomes are now available for the preparation of frozen sections for immuno-cytochemistry or X-ray microanalysis.

Artifacts encountered in sectioning and staining have been discussed by Klomparens[189] and Bell,[190] respectively. References on sectioning and staining are included in Appendix, Section 1.

4. Immunocytochemistry

The recent developments in immunocytochemistry with applications in virology warrant the presentation of some of the background that led to the development of immunolabeling techniques with colloidal gold. Following the significant report of Coons et al.[191] in which fluorescent antibodies were used to identify sites of antigen-antibody reaction in the light microscope, investigators began to develop electron dense markers that would label anti-bodies at the electron microscopic level. Singer[192] introduced the ferritin-labeled antibody technique while Nakane and Pierce[193] used horseradish peroxidase (HRP) and diaminob-enzidine (DAB) to localize antigens by conjugating antibody to HRP which is a small molecule (40,000 MW) and approximately one tenth the size of ferritin. The enzyme bridge[194] and peroxidase antiperoxidase (PAP)[195] methods were then introduced. The next great ad-vance was in the adsorption of whole sera to colloidal gold by Faulk and Taylor[196] which ushered in the development of immunogold labeling — a technique which has allowed investigators to make significant advances in electron immunocytochemistry.

Texts and chapters on electron immunocytochemistry include those by Sternberger,[137,138] Singer et al.,[197] Polak and Van Noorden,[198,199] Bullock and Petrusz,[200-202] Polak and Varn-dell,[203] Kuhlmann,[204] Verkleij and Leunissen,[205] Garzon et al.,[206] Bendayan,[207-210] DeMay,[211] Hacket et al.,[212] Smit and Todd,[213] Roth,[214] Varndell and Polak,[215] and Robinson et al.,[67] and Hayat.[216-218]

Varndell and Polak[215] discussed three types of immunolabeling techniques, i.e., no embedding, preembedding, and postembedding, and gave excellent directions for each tech-nique as well as the procedures for immunolabeling. It is possible to make various sizes of colloidal gold[219,220] which may be attached to immunoglobulin (IgG) directly or indirectly by the gold labeled antigen detection (GLAD) technique. In the GLAD technique a single IgG molecule recognizes the same antigenic determinants in the tissue and those of pure antigen adsorbed to colloidal gold (see Reference 215, Figures 3 and 4). In the protein A technique, protein A with gold particles attached has a high affinity for IgG, binding to the Fc region of many immunoglobulin species (see Reference 215, Figure 5). Other gold labeling techniques include the biotinylated ligand-streptavidin-gold marker procedure and the di-nitrophenol-hapten staining.[215] The Appendix, Section 1 lists some other pertinent references on immunocytochemical techniques.

Recent experiments have revealed that microwave fixation gives excellent preservation of tissues, cells, and antigens for light and electron microscopy[221-225] (see also Appendix, Section 1).

Applications of some of immunocytochemical techniques in invertebrate virology include those of Gipson and Scott,[165] Bladon et al.,[226] Tanada et al.[227,228] Nagata and Tanada,[229] Zhu et al.,[230] Hukuhara and Zhu,[231] Van der Wilk et al.,[232] Vlak et al.,[233] Russell and Rohrmann,[234] and Van Lent et al.[235] Enzyme treatments of virions of *Pseudoplusia includens*

SNPV by Gipson and Scott[165] resulted in degradation of nucleocapsids with all enzymes tested except trypsin which was active only in combination with DNase. No evidence was found of cytoskeleton involvement in the formation and maintenance of *Tipula* iridescent virus viroplasmic centers in immunofluorescent studies on intact and fractionated *Estigmena acrea* cells.[226] Tanada et al.[227] used a ferritin conjugated antibody to locate the synergistic factor (SF) and to establish that the site of action of the SF was the cell membrane of the microvillus of the midgut. Further studies by Tanada et al.[228] and Nagata and Tanada[229] confirmed that the surfaces of the inoculated midgut microvilli contained the SF from the observations of the reaction product (aurous *p*-nitrothiolphenolate) which occurred when the esterase of the SF was reacted with *p*-nitrophenylthiol acetate and *p*-nitrophenylthiol butyrate. Zhu et al.[230] and Hukuhara and Zhu[231] used immunoelectron microscopy studies with 5-nm gold labeling to show that the SF is located on the viral nucleocapsid envelope. Van der Wilk et al.[232] used the protein A gold technique to detect the location of polyhedrin, p10, and virion antigens of AcMNPV in *Spodoptera frugiperda* cell cultures while Vlak et al.[233] used immunogold labeling to compare wild type and recombinant AcMNPV/p10Z-2 in *S. frugiperda* cells (IPLB-SF-21) and established that the p10 of AcMNPV was a structural element of the fibrous structures. Russell and Rohrmann[234] localized viral occlusion (VO) or polyhedron envelope with gold immunolabeling techniques using a polyclonal antiserum produced against OpMNPV PE-trpE while Van Lent et al.[235] demonstrated with gold immunolabeling that a 34-kDa protein (pp34) is involved in the morphogenesis of the polyhedron envelope and that the electron dense spacers are essential for polyhedron morphogenesis while the fibrillar structures are not. An artifact revealed by this technique was that anti pp34 serum localized in fibrillar structures despite the absence of pp34 in the AcMNPV mutant with inactivated pp34 gene.

5. Quantitation of Virus Particles in Thin Sections

Quantitation mapping[236] and quantitation of virus particles in sectioned pellets has been reported and reviewed by Miller et al.[237] and Miller.[48,49] Weiss et al.[238] used the formula derived by Miller[48,49] which assumes the virus particles as spherical bodies to arrive at an estimate of 23 virus particles per OB of AcMNPV of 1 μm diameter. Allaway[239] used light and electron microscopy with different reasoning and mathematical calculations than that reported by Miller et al.[237] and Miller.[48,49] Estimates of the number of virus particles per VO and number of nucleocapsids (NCs) per VO, respectively, were as follows: *Agrotis segetum* MNPV (English isolate) — 168 and 674; *A. segetum* MNPV (Polish isolate) — 55 and 215; *Mamestra brassicae* MNPV — 47 and 100; *Plusia gamma* SNPV — 39 and 39. Allaway[239] proposed that the great differences in the isolates of *A. segetum* MNPV was due to differences in mean volume and the distribution of the number of NCs per virus particles or virus bundle. Smith et al.[240] found that cell culture media supplemented with AlCl$_3$ and ZnSO$_4$ 7H$_2$O yielded the largest VOs of AcMNPV.

Tompkins et al.[241] examined thin sections of pellets of VOs to quantitate differences in isolates of MNPVs produced in many passages through homologous and heterologous host species. The number of virus particles or bundles per VO were recorded as well as the number of nucleocapsids per virus particle or bundle from sections of over 200 randomly oriented VOs for each sample. Using the same procedure, Adams[242] also quantitated differences in numbers of virions per VO and numbers of NCs per virion in samples of *Lymantria dispar* MNPV from the U.S. (Gypchek) and U.S.S.R. (VIRIN-ENSh).

6. X-Ray Microanalysis

X-ray microanalytical techniques have been developed to locate elements and quantitate differences at the ultrastructural level. Texts and reviews on energy dispersive X-ray microanalysis include those by Russ,[243,244] Hayat,[245] Vaughn,[246] Morgan,[247] Aldrich,[248] and

Johnson and Cantino.[249] Adams[250] examined sections of MNPVs to determine elemental differences. The background levels were much higher around the polyhedra or VOs than in areas of embedding medium only, indicating a probable extraction of some elements from the VOs. Minor differences between MNPVs were found in Na, Mg, Si, P, S, Cl, K, Ni, and Zn. This was in general agreement with the elemental analysis reported by Shapiro and Ignoffo[251] and Faust et al.[252] Sections of pellets of *Spodoptera frugiperda* MNPV also contained a few fragments of insect cuticle trapped between the polyhedra. X-ray microanalysis of those fragments and the VOs indicated that the Ca and much of the P detected were actually in the insect fragments between VOs rather than in the VOs. The VOs contained S and concentrations of P considerably lower than that found in the insect fragments, illustrating its value as a technique which will give the precise location of the elements in a sample.[250] The problem with extraction is likely with any sample prepared by standard fixation methods. Fast freezing techniques and the use of cryoultramicrotomy[253] with the improved efficiency of elemental detection in X-ray microanalytical equipment, should provide more accurate information on the histopathologic changes involved in viral infection process in invertebrate tissues.

A new technique[254] has been introduced recently called the Life Cell Process® in which amorphous phase tissue water is removed with a molecular distillation drying apparatus without devitrification or rehydration.

7. Cryotechniques

Exciting advances in the development of cryotechniques have been reviewed by Steinbrecht and Zierold.[255] Some of these include the use of cryofixation by immersion,[256] the Tokuyasu method,[257] metal-mirror cryofixation, and cryosubstitution; embedding at low temperatures[258,260] has been given as well as a comparison of these methods by Sitte et al.[261] Freeze substitution and freeze-drying techniques have also been developed.[262] Comparison of the results of these methods, which preserve tissues and viruses more closely to the natural state, with those results obtained by the use of conventional methods should greatly expand our understanding of the ultrastructural morphology of viruses and the location of antigens in labeling studies in immunocytochemistry as well as more precise location of elements of interest in tissues using X-ray microanalysis.

G. SCANNING ELECTRON MICROSCOPY

Texts and reviews that are very helpful in choosing preparative techniques for SEM include those by Hayes,[263] Hayat,[154,264-269] Echlin,[270] Becker and Johari,[271] Gabriel,[272] Jones,[273] Bell,[274] Murphy,[275] Beckett and Read,[276] Albrecht and Hodges,[277] and Tanaka et al.[278] Protocols such as that of Porter et al.[279] are very similar to the fixation and dehydration steps in the preparation of samples for TEM. Kelley et al.[280] and Laczko and Varga[281] have used the osmium bridging properties of thiocarbohydrazide as a ligand to increase the osmium bound in the tissues (OTO method) thus protecting the tissues under the electron beam. A thinner carbon and/or metal conductive coating is then required allowing better visualization of the specimen structure. Gunning and Crang[282] have recently introduced the GACH (glutaraldehyde-carbonhydrazide) technique in which specimens are stabilized by means of the copolymerization of glutaraldehyde with carbohydrazide prior to air drying. Additional references may be found in Appendix, Section 1.

Following the dehydration steps specimens may be critical point dried[283-286] or dried by use of hexamethyldisilazane[287,288] or a fluorocarbon[289] method. Specimens may be mounted on specimen stubs with double sticky tape and coated with carbon and/or metal in a vacuum evaporator or sputter coater.

Examinations of VOs of NPVs, CPVs, GVs, and entomopox virus spheroids by Adams and Wilcox[290] using a scanning attachment on a transmission electron microscope enabled an evaluation of preparative techniques. Double fixation including the OTO method, dehydration, and critical point drying gave definition of surface ultrastructure which was superior to that obtained by less extensive methods of preparation.

Tanaka et al.[291] have recently described the progress in obtaining ultrahigh resolution SEM on biological specimens. Recently, an environmental scanning electron microscope has been developed by ElectroScan Corp. which is reported to bridge the gap between the light and scanning electron microscope, allowing the viewing of fresh, wet tissues at magnifications of up to about 50,000×.

This may prove helpful in studies on invasion with VOs, NPVs, CPVs, and GVs and entomopox virus spheroids. It remains to be determined if the resolution will be adequate for virions of VOs and small nonoccluded viruses.

IV. CONCLUSIONS

The prospects are very bright as invertebrate pathologists continue to benefit from the technological advances in TEM, SEM, STEM, and X-ray microanalytical equipment with concomitant developments in techniques for labeling of viral antigens and location of elements of interest involved in the ultrastructural cytopathological investigations on invertebrate viral replication.

ADDENDUM

Recent reports pertinent to techniques included those of Arakawa and Schimizu[141] who developed a more sensitive agglutination test for detection of NPV by using protein A coated latex particles. Russell et al.[142] used immunoelectron microscopy with immunogold labeling and monospecific antibodies to follow the temporal expression and localization of the polyhedrin, p10 p39, and PE (polyhedron envelope) protein of OpMNPV in *L. dispar* cell cultures. Nagamine et al.[143] used rhodamine-conjugated goat anti-mouse IgG with monoclonal antibodies to detect structural polypeptides of BmNPV occluded virions. Charlton and Volkman[144] studied the sequential rearrangement and nuclear polymerization of actin in AcMNPV-infected IPBL-SF-21 cells. TRITC (tetramethylrhodamine isothiocyanate)-conjugated phalloidin was used in fluorescence microscopy to follow involvement of actin microfilaments and a monoclonal antibody Mab 39 (39P10) to localize p39, the major capsid protein.

REFERENCES

1. **Crang, R. F. E. and Klomparens, K. L., Eds.,** *Artifacts in Biological Electron Microscopy,* Plenum Press, New York, 1988.
2. **Steinhaus, E. A.,** *Principles of Insect Pathology,* McGraw-Hill, New York, 1949.
3. **Aruga, H. and Tanada, Y., Eds.,** *The Cytoplasmic Polyhedrosis Virus of the Silkworm,* University of Tokyo Press, Tokyo, 1971.
4. **Thomas, G. M.,** *Diagnostic Techniques in Insect Diseases,* Cantwell, G. E., Ed., Vol. 1, Marcel Dekker, New York, 1974.
5. **Cantwell, G. E., Ed.,** *Insect Diseases,* Vol. 1, Marcel Dekker, New York, 1974.
6. **Smith, K. M.,** *Virus-Insect Relationships,* Longman, London, 1976.
7. **Maramorosch, K., Ed.,** *The Atlas of Insect and Plant Viruses,* Academic Press, New York, 1977.
8. **Weiser, J.,** *An Atlas of Insect Diseases,* Dr. W. Junk, B. V. Publishers, The Hague, Netherlands, 1977.
9. **Kalmakoff, J. and Longworth, J. F., Eds.,** *Microbial Control of Insect Pests,* N. Z. Dep. Sci. Ind. Res. Bull., 228, 1980.
10. **Bailey, L.,** *Honey Bee Pathology,* Academic Press, London, 1981.

11. **Tchuchrity, M. G.**, *The Ultrastructure of the Viruses from the Lepidoptera-The Pests of Plants*, (in Russian) Shtiintsa Kishinev, 1982.
12. **Poinar, Jr., G. O. and Thomas, G. M.**, *Laboratory Guide to Insect Pathogens and Parasites*, Plenum Press, New York, 1984.
13. **Bergold, G. H.**, The nature of nuclear-polyhedrosis Viruses, in *Insect Pathology, An Advanced Treatise*, Vol. 1, Steinhaus, E. A., Ed., Academic Press, New York, 1963.
14. **Steinhaus, E. A.**, Background for the diagnosis of insect diseases, in *Insect Pathology-an Advanced Treatise*, Vol. 2, Steinhaus, E. A., Ed., Academic Press, New York, 1963.
15. **Wittig, G.**, Techniques in insect pathology, in *Insect Pathology—An Advanced Treatise*, Vol. 2, Steinhaus, E. A., Ed., Academic Press, New York, 1963, 591.
16. **Weiser, J. and Briggs, J. D.**, Identification of Pathogens, *Microbial Control of Insects and Mites*, Burges, H. D. and Hussey, N. W., Eds., Academic Press, New York, 1971, 13.
17. **Cantwell, G. E., Ed.**, *Insect Diseases*, Vol. 1, Marcel Dekker, New York, 1974.
18. **Vaughn, J. L.**, Virus and rickettsial diseases, in *Insect Diseases*, Vol. 1, Marcel Dekker, New York, 1974, 49.
19. **Wigley, P. J.**, Diagnosis of virus infections-staining of insect inclusion body viruses, in *Microbial Control of Insect Pests*, Kalmakoff, J. and Longworth, J. F., Eds., N. Z. Dep. Sci. Ind. Res. Bull., 228, 1980, 35.
20. **Brun, G. and Plus, N.**, The Viruses of Drosophila, in *The Biology and Genetics of Drosophila*, Ashburner, M. and Wright, T. R. F., Eds., Academic Press, New York, 1980, 625.
21. **Johnson, P. T.**, Viral diseases of marine invertebrates, *Helgol. Meeresunters.*, 37, 65, 1984.
22. **MacKay, B.**, *Introduction to Diagnostic Electron Microscopy*, Appleton-Century-Crofts, New York, 1981.
23. **Doane, F. and Anderson, W.**, *Electron Microscopy in Diagnostic Virology — a Practical Guide and Atlas*, Cambridge University Press, New York, 1987.
24. **Palmer, E. I. and Martin, M. L.**, *Electron Microscopy in Viral Diagnosis*, CRC Press, Boca Raton, FL, 1988.
25. **Steinhaus, E. A.**, Diagnosis: a central pillar of insect pathology, *Entomophaga*, 2, 7, 1962.
26. **Granados, R. R. and Lawler, K. A.**, *In vivo* pathway of *Autographa californica* baculovirus invasion and infection, *Virology*, 108, 297, 1981.
27. **Smirnoff, W. A.**, A guide for the application of dark field microscopy for routine diagnosis in insect pathology, *Nat. Can.*, 96, 261, 1969.
28. **Kaupp, W. J. and Burke, F. R.**, A staining technique for the inclusion bodies of a granulosis virus, *Can. For. Serv. For. Pest Manage. Inst., Tech. Note*, No. 1, 1984.
29. **Hunter, K.**, personal communication.
30. **Sikorowski, P. P., Goddon, P. P., and Broome, J. R.**, Presence of cytoplasmic polyhedrosis virus in the hemolymph of *Heliothis virescens* larvae and adults, *J. Invertebr. Pathol.*, 18, 167, 1971.
31. **Burges, H. D. and Thomson, E. M.**, Standardization and assay of microbial insecticides, in *Microbial Control of Insects and Mites*, Burges, H. D. and Hussey, N. E., Eds., Academic Press, New York, 1971, 591.
32. **Cantwell, G. E.**, Honey bee diseases, parasites, and pests, in *Insect Diseases*, Vol. 2, Cantwell, G. E., Ed., Marcel Dekker, New York, 1974, 501.
33. **Wigley, P. J.**, Counting microorganisms, in *Microbial Control of Insect Pests*, Kalmakoff, J. and Longworth, J. F., Eds., N. Z. Dep. Sci. Ind. Res. Bull., 228, 1980, 29.
34. **Kaupp, W. J. and Sohi, S. S.**, Quantitation of insect viruses in *Viral Insecticides for Biological Control*, Maramorosch, K. and Sherman, K. E., Eds., Academic Press, Orlando, FL, 1985, 675.
35. **Bradley, D. E.**, The preparation of specimen support films, in *Techniques for Electron Microscopy*, 2nd ed., Kay, D. H., Ed., F. A. Davis, Philadelphia, 1965, 58.
36. **Mazia, D., Schatten, G., and Sale, W.**, Adhesion of cells to surfaces coated with polylysine, *J. Cell Biol.*, 66, 198, 1975.
37. **Hall, C. E.**, Electron densitometry of stained virus particles, *J. Biophys. Biochem. Cytol.*, 1, 1, 1955.
38. **Brenner, S. and Horne, R. W.**, A negative staining method for high resolution electron microscopy of viruses, *Biochem. Biophys. Acta*, 34, 103, 1959.
39. **Horne, R. W.**, The examination of small particles, in *Techniques for Electron Microscopy*, Kay, D. H., Ed., Blackwell Scientific, F. A. Davis, Philadelphia, 1965, 311.
40. **Horne, R. W.**, Negative staining methods, in *Techniques for Electron Microscopy*, Kay, D. H., Ed., F. A. Davis, Philadelphia, 1965, 328.
41. **Horne, R. W.**, Electron microscopy of isolated virus particles and their components, in *Methods in Virology*, Maramorosch, K. and Koprowski, H., Eds., Academic Press, New York, 1967, 521.
42. **Horne, R. W. and Ronchetti, I. P.**, A negative staining-carbon film technique for studying viruses in the electron microscope, *J. Ultrastruct. Res.*, 47, 361, 1974.

43. **Haschemeyer, R. H. and Myers, R. J.,** Negative staining, in *Principles and Techniques of Electron Microscopy,* Vol. 2, Hayat, M. A., Ed., Van Nostrand Reinhold, New York, 1972, 99.

44. **Pantin, C. F. A.,** On the excitation of crustacean muscle, *J. Exp. Biol.,* 11, 1934.

45. **Bonami, J. R.,** Recherches sur les infections virales des crustacés marins: étude des maladies à étiologie simple et complexe chez les décapodes des cotes françaises, Thèse Doct. Etat., Univ. Sci. Tech. Languedoc, Montpellier, France, 1980.

46. **Webb, M. J. W.,** A method for the rapid removal of sugars and salts from virus preparations on electron microscope grids, *J. Microsc. (Oxford),* 98, 109, 1973.

47. **Adams, J. R.,** unreported data, 1975.

48. **Miller, II, M. F.,** Particle counting of viruses, in *Principles & Techniques of Electron Microscopy,* Vol. 4, Hayat, M. A., Ed., Van Nostrand Reinhold, New York, 1974, 89.

49. **Miller, II, M. F.,** Virus particle counting by electron microscopy, in *Electron Microscopy in Biology,* Vol. 2, Griffith, J. D., Ed., John Wiley & Sons, New York, 1982, 305.

50. **DeBlois, R. W., Uzgiris, E. E., Cluxton, D. H., and Mazzone, H. M.,** Comparative measurements of size and polydispersity of several insect viruses, *G. E. Corp. Res. & Dev. Tech. Info. Serv.,* Rep. No. 77CRD250, 1, 1977.

51. **DeBlois, R. W., Uzgiris, E. E., Cluxton, D. H., and Mazzone, H. M.,** Comparative measurements of size and polydispersity of several insect viruses, *Anal. Biochem.,* 90, 273, 1978.

52. **Rohrmann, G. F., Martignoni, M. E., and Beaudreau, G. S.,** Quantification of two viruses in technical preparations of *Orgyia pseudotsugata* baculovirus by means of buoyant density centrifugation of viral deoxyribonucleic acid, *Appl. Environ. Microbiol.,* 35, 690, 1978.

53. **Kleinschmidt, A. K. and Zahn, R. K.,** Uber desoxyribonucleicsaurenmolekeln in protein-mischfilmen, in *Chem. Org. Chem. Biochem. Biophys. Biol.,* 14b, 770, 1959.

54. **Kleinschmidt, A. K.,** Monolayer techniques in electron microscopy of nucleic acid molecules, *Methods Enzymol.,* 22, 361, 1968.

55. **Davis, R. W., Davidson, N., and Simon, M.,** Electron microscope heteroduplex methods for mapping regions of base sequence homology in nucleic acids, *Methods Enzymol.,* 21, 413, 1971.

56. **Inman, R. B. and Schnös, M.,** Denaturation mapping of DNA, in *Principles and Techniques of Electron Microscopy,* Vol. 4, Hayat, M. A., Ed., Van Nostrand Reinhold, New York, 1974, 64.

57. **Younghusband, H. B. and Inman, R. B.,** The electronmicroscopy of DNA, *Annu. Rev. Biochem.,* 43, 605, 1974.

58. **Chow, L. T., Scott, J. M., and Broker, T. R.,** The electron microscopy of nucleic acids, *Laboratory Manual, Cold Spring Harbor Laboratory,* New York, 1975.

59. **Thomas, J. O.,** Electron microscopy of DNA, in *Principles and Techniques of Electron Microscopy,* Vol. 9, Hayat, M. A., Ed., Van Nostrand Reinhold, New York, 1978, 64.

60. **Ferguson, J. and Davis, R. W.,** Quantitative electron microscopy of nucleic acids, in *Advanced Techniques in Biological Electron Microscopy II,* Koehler, J. K., Ed., Springer-Verlag, Wien, 1978, 23.

61. **Garon, C. F.,** Electron microscopy of nucleic acids, in *Gene Amplification and Analysis,* Vol. 2, Chirikjian, J. G. and Papas, T. S., Eds., Elsevier/North Holland, Amsterdam, 1981, 573.

62. **Garon, C. F.,** Electron microscopy of nucleic acids, in *Ultrastructure Techniques for Microorganisms,* Aldrich, H. C. and Todd, W. J., Eds., Plenum Press, New York, 1986, 161.

63. **Nermut, M. V.,** Advanced methods in electron microscopy of virses, in *New Developments in Practical Virology,* Vol. 5, Howard, C. R., Ed., Alan R. Liss, New York, 1982, 2.

64. **Allison, D. P.,** Preparation of deoxyribonucleic acid (DNA) for electron microscopy, in *Electron Microscopy,* Jones, B. R., Ed., Library Research Assoc., New York, 1983, 389.

65. **Coggins, L. W.,** Preparation of nucleic acids for electron microscopy, in *Electron Microscopy in Molecular Biology, a Practical Approach,* Sommerville, J. and Scheer, U., Eds., IRL Press, Washington, D.C., 1987, 1.

66. **Coggins, L. W.,** Denaturation and hybridization of nucleic acids in *Electron Microscopy in Molecular Biology, a Practical Approach,* Somerville, J. and Scheer, U., Eds., IRL Press, Oxford, 1987, 31.

67. **Robinson, D. G., Ehlers, U., Herken, R., Herrmann, B., Mayer, F., and Schurmann, F. W.,** *Methods of Preparation for Electron Microscopy,* Springer-Verlag, Wien, 1987, 190.

68. **Delain, E. and Brack, C.,** Visualization des molecules d'acide nucléiques. II. Microversion de la technique de diffusion, *J. Microscopie,* 21, 217, 1974.

69. **Summers, M. D.,** Biophysical and biochemical properties of baculoviruses, in *Baculoviruses for Insect Pest Control: Safety Considerations,* Summers, M. D., Engler, R., Falcon, L. A., and Vail, P. V., Eds., American Society for Microbiology, Washington, D.C., 1975, 17.

70. **Bud, H. M. and Kelly, D. C.,** The DNA contained by nuclear polyhedrosis viruses isolated from four *Spodoptera* spp. (Lepidoptera, Noctuidae); genome size and configuration assessed by electron microscopy, *J. Gen. Virol.,* 37, 135, 1977.

71. **Bud, H. M. and Kelly, D. C.,** An electron microscope study of partially lysed *Baculovirus* nucleocapsids. The intra-nucleocapsid packaging of viral DNA, *J. Ultrastruct. Res.,* 73, 361, 1980.

72. **Brown, D. A., Bud, H. M., and Kelly, D. C.,** Biophysical properties of the structural components of a granulosis virus isolated from the cabbage white butterfly *(Pieris brassicae), Virology,* 81, 317, 1977.
73. **Tjia, S. T., Carstens, E. B., and Doerfler, W.,** Infection of *Spodoptera frugiperda* cells with *Autographa californica* nuclear polyhedrosis virus. II. The viral DNA and the kinetics of its replication, *Virology,* 99, 391, 1979.
74. **Scharnhorst, D. W., Saving, K. L., Vuturo, S. B., Cooke, P. H., and Weaver, R. F.,** Structural studies on the polyhedral inclusion bodies, virions, and DNA of the nuclear polyhedrosis virus of the cotton bollworm *Heliothis zea, J. Virol.,* 21, 292, 1977.
75. **Jurkovicova, M.,** Characterization of Nuclear Polyhedrosis Viruses Obtained from *Adoxophyes orana* and from *Barathra brassicae,* Ph.D. thesis, State Agricultural University, Wageningen, the Netherlands, 1979, 1.
76. **Jurkovicova, M., Van Touw, J. H., Sussenbach, J. S., and Schegget, T. J.,** Characterization of the nuclear polyhedrosis virus DNA of *Adoxophyes orana* and of *Barathra brassicae, Virology,* 93, 8, 1979.
77. **Loh, L. C., Hamm, J. J., Kawanishi, C. C., and Huang, E.,** Analysis of the *Spodoptera frugiperda* nuclear polyhedrosis virus genome by restriction endonucleases and electron microscopy, *J. Virol.,* 44, 747, 1982.
78. **Skuratovskaya, I. N., Fodor, I., and Strokovskaya, L. I.,** Properties of the nuclear polyhedrosis virus of the great wax moth; oligomeric circular DNA and the characteristic of the genome, *Virology,* 120, 465, 1982.
79. **Revet, B. and Monsarrat, P.,** L'acide nucléique du virus du Coleoptère *Oryctes rhinoceros* L.: un ADN superhélicoidal de haut poids moléculaire, *C. R. Acad. Sci., Paris, Series D,* 278, 331, 1974.
80. **Delius, H., Darai, G., and Flugel, R. M.,** DNA analysis of insect iridescent virus 6: evidence for circular permutation and terminal redundancy, *J. Virol.,* 49, 609, 1984.
81. **Schnitzler, P., Soltau, J., Fischer, M., Reisner, H., Scholz, J., Delius, H., and Darai, G.,** Molecular cloning and physical mapping of the genome of insect iridescent virus type 6: further evidence for circular permutation of the viral genome, *Virology,* 160, 66, 1987.
82. **Krell, P. J. and Stoltz, D. B.,** Unusual baculovirus of the parasitoid wasp *Apanteles melanoscelus:* isolation and preliminary characterization, *J. Virol.,* 29, 1118, 1979.
83. **Krell, P. J. and Stoltz, D. B.,** Virus-like particles in the ovary of an ichneumonid wasp: purification and preliminary characterization, *Virology,* 101, 408, 1980.
84. **Yazaki, K. and Miura, K-I.,** Relation of the structure of cytoplasmic polyhedrosis virus and the synthesis of its messenger RNA, *Virology,* 105, 467, 1980.
85. **Yazaki, K., Mizumo, A., Sano, T., Fujii, H., and Miura, K.,** A new method for extracting circular and supercoiled genome segments from cytoplasmic polyhedrosis virus, *J. Virological Methods,* 14, 275, 1986.
86. **Hu, Y., Hashimoto, Y., and Kawase, S.,** An electron microscopic observation of RNA from infectious flacherie virus of the silkworm *Bombyx mori* L. (Lepidoptera:Bombycidae), *Appl. Entomol. Zool.,* 23, 103, 1988.
87. **Scott, M. P., Fostel, J., and Pardue, M.,** A new type of virus from cultured *Drosophila* cells: characterization and use in studies of the heat-shock response, *Cell,* 22, 929, 1980.
88. **Burgess, S.,** Molecular weights of lepidopteran baculovirus DNAs: derivation by electron microscopy, *J. Gen. Virol.,* 37, 501, 1977.
89. **Hosur, M., Schmidt, T., Tucker, R., Johnson, J., Gallagher, T. M., Selling, B. H., and Ruckert, R.,** Structure of an insect virus at 3.0 Å resolution, *Proteins: Struct., Funct. Genet.,* 2, 167, 1987.
90. **Mazzone, H. M., Engler, W. F., Wray, G., Zirmae, A., Conroy, J., Zerillo, R., and Bahr, G. F.,** High voltage electron microscopy of viral inclusion bodies, *Proc. 38th Annu. Meet. Electron Microscopy Soc. Am.,* Reno, NV, Bailey, G. W., Ed., Claitor's Publ. Div., Baton Rouge, LA, 1980, 468.
91. **Mazzone, H. M., Wray, G., Engler, W., and Bahr, G.,** High voltage electron microscopy of cells in culture and viruses in *Invertebrate Systems In Vitro,* Kurstak, E., Maramorosch, K., and Dubendorfer, A., Eds., Elsevier/North Holland, Amsterdam, 1980, 511.
92. **Zeitler, E. and Bahr, G.,** A photometric procedure for weight determination of submicroscopic particles, quantitative electron microscopy, *J. Appl. Phys.,* 33, 847, 1962.
93. **Bahr, G. F., Engler, W. F., and Mazzone, H. M.,** Determination of the mass of viruses by quantitative electron microscopy, *Q. Rev. Biophys.,* 9, 459, 1976.
94. **Humason, G. L.,** *Animal Tissue Techniques,* W. H. Freeman, San Francisco, 1962.
95. **Barbosa, P.,** *Manual of Basic Techniques in Insect Histology,* Autumn Publishers, Amherst, 1974.
96. **Luna, L. G.,** *Manual of Histologic Staining Methods of the Armed Forces Institute of Pathology,* 3rd ed., McGraw-Hill, New York, 1960.
97. **Shaw, B. L. and Battle, H. I.,** The gross and microscopic anatomy of the digestive tract of the oyster *Crassostrea virginica* (Gmelin), *Can. J. Zool.,* 35, 325, 1957.
98. **Comps, M.,** Recherches Histologiques et Cytologiques sur les Infections Intracellulaires des Mollusques Marins, Thèse Doct. Etat. Univ. Sci. Tech. Languedoc, Montpellier, France, 1983, 128.

99. **Johnson, P. T.**, *Histology of the Blue Crab, Callinectes sapidus. A model for the Decapoda*, Praeger Publ Div. of Greenwood Press, Westport, CT, 1980.

100. **Dart, L. and Turner, T. R.**, Fluorescence microscopy in exfoliative cytology, *Lab. Invest.*, 8, 1513, 1959.

100a. **Bonami, J. R. and Comps, M.**, unpublished data.

101. **Pappas, P. W.**, The use of a chrome alum-gelatin (subbing) solution as a general adhesive for paraffin sections, *Stain Technol.*, 46, 1971, 121.

102. **Hamm, J. J.**, A modified azan staining technique for inclusion body viruses, *J. Invertebr. Pathol.*, 8, 125, 1966.

103. **Huger, A.**, Methods for staining capsular virus inclusion bodies typical of granuloses of insects, *J. Insect Pathol.*, 3, 338, 1961.

104. **Vago, C. and Amagier, A.**, Coloration histologique pour la différenciation des corps d'inclusion polyédriques de virus d'insectes, *Ann. Epiphyt.*, 14, 269, 1963.

105. **Salpeter, M. M. and Bachmann, L.**, Autoradiography, in *Principles & Techniques of Electron Microscopy*, Vol. 2, Hayat, M. A., Ed., Van Nostrand Reinhold, New York, 1972, 221.

106. **Salpeter, M. M. and McHenry, F. A.**, Electron microscope autoradiography in *Advanced Techniques in Biological Electron Microscopy*, Koehler, J. K., Ed., Springer-Verlag, Wien, 1973, 113.

107. **Maraldi, N. M.**, Electron autoradiography of free specimens, in *Principles & Techniques of Electron Microscopy*, Vol. 6, Hayat, M. A., Ed., Van Nostrand Reinhold, New York, 1976, 217.

108. **Williams, M. A.**, Autoradiography & Immunocytochemistry, Part 1, *Practical Methods in Electron Microscopy*, Vol. 6, Glauert, A. M., Ed., North Holland Publishing, Amsterdam, 1977, 217.

109. **Williams, M. A.**, Autoradiography: its methodology at the present time, *J. Microsc. (Oxford)*, 128, 79, 1982.

110. **Rogers, A. W.**, *Techniques of Autoradiography*, 3rd ed., Elsevier/North Holland, Amsterdam, 1979.

111. **Salpeter, M. M.**, Electron microscopy in medical research and diagnosis present and future directions, Section V: EM Autoradiography, *J. Electron Microsc. Tech.*, 4, 125, 1986.

112. **Anon.**, Technical Information, Nuclear Research Material-a range of material for autoradiography, *Ilford Limited*, Scientific Products Business, Mobberley, Knutsford, Cheshire, England.

113. **Anon.**, Autoradiography at the light microscope level, *Kodak Tech. Bits, Publ.*, P-3-88-1, Koday Tech. Bits, Eastman Kodak, Rochester, 1988.

114. **Budd, G. C.**, Autoradiography, in *Electron Microscopy*, Jones, B. R., Ed., Library Research Assoc., New York, 1983, 295.

115. **Anon.**, *Photographic Procedures for Light and Electron Microscope Autoradiography*, Eastman Kodak Publ., No. M6-110, 1986, 20.

116. **Watanabe, H.**, Localization of ribonucleic acid synthesis in the midgut cells infected with cytoplasmic-polyhedrosis virus in the silkworm, *Bombyx mori* L. (Lepidoptera:Bombycidae), *Appl. Entomol. Zool.*, 1, 154, 1966.

117. **Watanabe, H.**, Site of viral RNA synthesis within the midgut cells of the silkworm, *Bombyx mori*, infected with cytoplasmic-polyhedrosis virus., *J. Invertebr. Pathol.*, 9, 480, 1967.

118. **Watanabe, H.**, Autoradiographic studies on the nucleic-acid synthesis in the midgut of the silkworm, *Bombyx mori* L., under normal and virus-infected conditions, *J. Sericult. Sci. Jpn.*, 36, 371, 1967.

119. **Watanabe, H.**, Autoradiographic studies on the nucleic acid synthesis in the midgut and some tissues of the silkworm, *Bombyx mori* L. (Lepidoptera: Bombycidae), infected with nuclear-polyhedrosis., *Appl. Entomol. Zool.*, 2, 371, 1967.

120. **Watanabe, H.**, Protein synthesis in the tissues of the silkworm, *Bombyx mori* infected with nuclear polyhedrosis virus, *J. Invertebr. Pathol.*, 9, 428, 1967.

121. **Watanabe, H.**, Light radioautographic study of protein synthesis in the midgut epithelium of the silkworm, *Bombyx mori*, infected with a cytoplasmic-polyhedrosis virus, *J. Invertebr. Pathol.*, 11, 310, 1968.

122. **Watanabe, H.**, An electron microscope radioautography of DNA synthesis in the fat cell of silkworm *Bombyx mori* infected with a nuclear-polyhedrosis virus, *J. Invertebr. Pathol.*, 20, 223, 1972.

123. **Morris, O. N.**, RNA changes in insect tissue infected with a nuclear polyhedrosis virus, *J. Invertebr. Pathol.*, 8, 35, 1966.

124. **Morris, O. N.**, Incorporation of radioactive uridine into RNA of the lepidopteran *Barathra brassicae*, *J. Invertebr. Pathol.*, 8, 259, 1966.

125. **Morris, O. N.**, Metabolic changes in diseased insects. I. Autoradiographic studies on DNA synthesis in normal and polyhedrosis infected Lepidoptera, *J. Invertebr. Pathol.*, 10, 28, 1968.

126. **Morris, O. N.**, Metabolic changes in diseased insects. II. Autoradiographic studies on DNA and RNA synthesis in nuclear polyhedrosis and cytoplasmic polyhedrosis virus infections, *J. Invertebr. Pathol.*, 11, 476, 1968.

127. **Morris, O. N.**, Metabolic changes in diseased insects. III. Nucleic acid metabolism in Lepidoptera infected by densonucleosis and *Tipula* iridescent viruses, *J. Invertebr. Pathol.*, 16, 180, 1970.

128. **Morris, O. N.**, Metabolic changes in diseased insects. IV. Radioautographic studies on protein changes in nuclear polyhedrosis, densonucleosis, and *Tipula* iridescent virus infections, *J. Invertebr. Pathol.*, 18, 191, 1971.

129. **Watanabe, H. and Kobayashi, M.**, Effects of a virus infection on the protein synthesis in the silk gland of *Bombyx mori*, *J. Invertebr. Pathol.*, 14, 102, 1969.

130. **Watanabe, H. and Kobayashi, M.**, Histopathology of a granulosis in the larva of the fall webworm, *Hyphantria cunea*, *J. Invertebr. Pathol.*, 16, 71, 1970.

131. **Benz, G. and Wäger, R.**, Autoradiographic studies on nucleic acid metabolism in granulosis-infected fat body of larvae of *Carpocapsa*, *J. Invertebr. Pathol.*, 18, 70, 1971.

132. **Mix, M. C. and Tomasoric, S.**, The use of high specific activity tritiated thymidine and autoradiography for studying molluscan cells., *J. Invertebr. Pathol.*, 21, 318, 1973.

133. **Benz, G.**, Physiopathology and histochemistry, in *Insect Pathology, An Advanced Treatise*, Vol. 1, Steinhaus, E. A., Ed., Academic Press, New York, 1963, 299.

134. **Benz, G.**, Histopathological changes and histochemical studies on the nucleic acid metabolism in the polyhedrosis-infected gut of *Diprion hencyniae* (Hartig), *J. Insect Pathol.*, 2, 259, 1960.

135. **Wäger, R. and Benz, G.**, Histochemical studies on nucleic acid metabolism in granulosis-infected *Carpocapsa pomonella*, *J. Invertebr. Pathol.*, 17, 107, 1971.

136. **Kurstak, E.**, The immunoperoxidase technique: localization of viral antigens in cells, in *Methods in Virology*, Vol. 5, Maramorosch, K. and Kaprowski, H., Eds., Academic Press, New York, 1971, 423.

137. **Sternberger, L. A.**, Enzyme immunohistochemistry, in *Electron Microscopy of Enzymes: Principles and Methods*, Vol. 1, Hayat, M. A., Ed., Van Nostrand Reinhold, New York, 1973.

138. **Sternberger, L. A.**, *Immunocytochemistry*, Prentice-Hall, Englewood Cliffs, NJ, 1974.

139. **Summers, M. D., Volkman, L. E., and Hsieh, C. H.**, Immunoperoxidase detection of baculovirus antigens in insect cells, *J. Gen. Virol.*, 40, 545, 1978.

140. **Volkman, L. E., Goldsmith, P. A., and Hess, R. T.**, Evidence for microfilament involvement in budded *Autographa californica* nuclear polyhedrosis virus production, *Virology*, 156, 32, 1987.

141. **Arakawa, A. and Shimizu, S.**, Agglutination test using protein-A coated latex particles for detecting the nuclear polyhedrosis virus of the silkworm, *Bombyx mori* (Lepidoptera, Bombycidae), *Appl. Entomol. Zool.*, 25, 519, 1990.

142. **Russell, R. L. Q., Pearson, M. N., and Rohrmann, G. F.**, Immunoelectron microscopic examination of *Orgyia pseudotsugata* multicapsid nuclear polyhedrosis virus-infected *Lymantria dispar* cells-time course and localization of major polyhedron-associated proteins, *J. Gen. Virol.*, 72, 275, 1991.

143. **Nagamine, T., Kobayashi, M., Saga, S., and Hoshino, M.**, Preparation and characterization of monoclonal antibodies against occluded virions of *Bombyx mori* nuclear polyhedrosis virus, *J. Invertebr. Pathol.*, 57, 311, 1991.

144. **Charlton, C. A. and Volkman, L. E.**, Sequential rearrangement and nuclear polymerization of action in baculovirus-infected *Spodoptera frugiperda* cells, *J. Virol.*, 65, 1219, 1991.

145. **Uchima, K., Egerter, D. E., and Tanada, Y.**, Synergistic factor of a granulosis virus of the armyworm, *Pseudaletia unipuncta*: its uptake and enhancement of virus infected *in vitro*, *J. Invertebr. Pathol.*, 54, 156, 1989.

146. **Allen, M. F. and Ball, B. V.**, Purification, characterization, and some properties of a virus from the aphid *Sitobion avenae*, *J. Invertebr. Pathol.*, 55, 162, 1990.

147. **Haase, A. T., Brahic, M., Stowring, I., and Blum, K.**, Detection of viral nucleic acids by *in situ* hybridization, in *Methods in Virology*, Vol. 7, Maramorosch, K. and Koprowski, H., Eds., Academic Press, New York, 1984, 189.

148. **Haase, A. T., Walker, D., Stowring, L., Ventura, P., Geballe, A., Blum, H., Brahic, M., Goldberg, R., and O'Brien, K.**, Detection of two viral genomes in single cells by double-label hybridization *in situ* and color microradioautography, *Science*, 227, 189, 1985.

149. **Haase, A. T., Ganta, D., Blum, H., Stowring, L., Ventura, P., and Geballe, A.**, Combined macroscopic and microscopic detection of viral genes in tissues, *Virology*, 140, 210, 1985.

150. **Haase, A. T.**, Analysis of viral infections by *in situ* hybridization, *J. Histochem. Cytochem.*, 34, 27, 1986.

151. **Hofler, H., DeLellis, R. A., and Wolfe, H. J.**, *In situ* hybridization and immunohistology, in *Advances in Immunohistochemistry*, DeLellis, R. A., Ed., Raven Press, New York, 1988, 47.

152. **Hatfill, S. J., Williamson, C., Kirby, R., and Von Wechmar, M. B.**, Identification and localization of aphid lethal paralysis virus particles in the tissue sections of the *Rhopalosiphum padi* aphid by *in situ* nucleic acid hybridization, *J. Invertebr. Pathol.*, 55, 265, 1990.

153. **Hayat, M. A.**, *Principles and Techniques of Electron Microscopy, Biological Applications*, Vol. I, Van Nostrand Reinhold, New York, 1970.

154. **Hayat, M. A.**, *Fixation for Electron Microscopy*, Academic Press, New York, 1981.

155. **Hayat, M. A.**, *Basic Techniques for Transmission Electron Microscopy*, Academic Press, Orlando, FL, 1986.

156. **Hayat, M. A.,** *Principles and Techniques of Electron Microscopy, Biological Applications,* 3rd ed., CRC Press, Boca Raton, FL, 1989.

157. **Glauert, A. M.,** *Practical Methods in Electron Microscopy, Fixation Dehydration and Embedding of Biological Specimens,* Vol. 3, (Part 1), American Elsevier, New York, 1974.

158. **Todd, W. J.,** Effects of specimen preparation on the apparent ultrastructure of microorganisms in *Ultrastructure Techniques for Microorganisms,* Aldrich, H. C. and Todd, W. J., Eds., Plenum Press, New York, 1986, 87.

159. **Bowers, B. and Maser, M.,** Artifacts in fixation for transmission electron microscopy, in *Artifacts in Biological Electron Microscopy,* Crang, R. F. E. and Klomparens, K. L., Eds., Plenum Press, New York, 1988, 13.

160. **Maser, M. D., Powell, III, T. E., and Philpott, C. W.,** Relationships among pH, osmolality, and concentration of fixative solutions, *Stain Technol.,* 42, 175, 1967.

161. **Coetzee, J. and Van der Merwe, C. F.,** Effect of glutaraldehyde on the osmolarity of the buffer vehicle, *J. Microsc. (Oxford),* 138, 99, 1985.

162. **Adams, J. R. and Wilcox, T. A.,** Determination of osmolalities of insect hemolymph from several species, *Ann. Entomol. Soc. Am.,* 66, 575, 1973.

163. **Smith, V. and Smith, D. S.,** Observations on the secretory processes in the corpus cardiacum of the stick insect, *Carausius morosus, J. Cell Sci.,* 1, 59, 1966.

164. **McDowell, E. M. and Trump, B. F.,** Histologic fixatives suitable for diagnostic light and electron microscopy, *Arch. Pathol. Lab. Med.,* 100, 405, 1976.

165. **Gipson, I. and Scott, H. A.,** An electron microscope study of effects of various fixatives and thin-section enzyme treatments on a nuclear polyhedrosis virus, *J. Invertebr. Pathol.,* 26, 171, 1975.

166. **Nicol, J. A. C.,** The Biology of Marine Animals, 2nd ed., Sir Isaac Pitman & Sons, London, 1967.

167. **Perkins, F. O.,** Ultrastructure of vegetative stages in *Labyrinthomyxa marina (Dermocystidium marinum),* a commercially significant oyster pathogen, *J. Invertebr. Pathol.,* 13, 199, 1969.

168. **Baur, P. S. and Stacey, T. R.,** The use of PIPES buffer in the fixation of mammalian and marine tissues for electron microscopy, *J. Microsc. (Oxford),* 109, 315, 1977.

169. **Kingsley, R. E. and Cole, R. E.,** Osmometry: a comparison of methods, *EMSA Bull.,* 17 (2), 74, 1987.

170. **Visscher, G. E. and Argentiere, G. J.,** Fundamentals of fixation: an introductory overview, *EMSA Bull.,* 17 (2), 83, 1987.

171. **Johnson, Jr., J. E.,** Problems associated with dehydration and embedding of biological samples for electron microscopy, *Chesapeake Microsc. J.,* 1, 5, 1988.

172. **Mollenhauer, H. H.,** Artifacts caused by dehydration and epoxy embedding in transmission electron microscopy, in *Artifacts in Biological Electron Microscopy,* Crang, R. F. E. and Klomparens, K. L., Eds., Plenum Press, New York, 1988, 43.

173. **Luft, L. H.,** Improvements in epoxy resin embedding methods, *J. Biophys. Biochem. Cytol.,* 9, 409, 1961.

174. **Luft, J. H.,** Embedding media-old and new in *Advanced Techniques in Biological Electron Microscopy,* Koehler, J. K., Ed., Springer-Verlag, Wien, 1973, 1.

175. **Burke, C. N. and Geiselman, C. W.,** Exact anhydride epoxy percentages for electron microscopy embedding (Epon), *J. Ultrastruct. Res.,* 36, 121, 1971.

176. **Chang, S. C.,** Compounding of Luft's epon embedding medium for use in electron microscopy with reference to anhydride: epoxide ratio adjustment, *Mikroskopie,* 29, 337, 1973.

177. **Geiselman, C. W. and Burke, C. N.,** Exact anhydride: epoxy percentages for araldite and araldite-epon embedding, *J. Ultrastruct. Res.,* 43, 220, 1973.

178. **Aldrich, H. C.,** Electron microscopy techniques, in *Cell Biology of Physarum and Didymium,* Vol. II, Aldrich, H. C. and Daniel, J. W., Eds., Academic Press, New York, 1982, 255.

179. **Aldrich, H. C. and Mollenhauer, H. H.,** Secrets of successful embedding, sectioning and imaging, in *Ultrastructure Techniques for Microorganisms,* Aldrich, H. C. and Todd, W. J., Eds., Plenum Press, New York, 1986, 101.

180. **Mollenhauer, H. H. and Droleskey, R. E.,** Some characteristics of epoxy embedding resins and how they affect contrast, cell organelle size, and block shrinkage, *J. Electron Microsc. Tech.,* 2, 557, 1985.

181. **Ellis, E. A.,** Araldites, low viscosity epoxy resins, and mixed resin embedding: formulations and uses, *EMSA Bull.,* 16 (2), 53, 1986.

182. **Acetarin, J. D., Carlemalm, E., Kellenberger, E., and Villiger, W.,** Correlation of some mechanical properties of embedding resins with their behavior in microtomy, *J. Electron Microsc. Tech.,* 7, 63, 1987.

183. **Spurr, A. R.,** A low-viscosity epoxy resin embedding medium for electron microscopy, *J. Ultrastruct. Res.,* 26, 31, 1969.

184. **Winborn, W. B.,** Dow epoxy resin with triallyl cyanurate and similarly modified Araldite and Maraglas mixtures, as embedding media for electron microscopy, *Stain Technol.,* 40, 227, 1965.

185. **Langenberg, W. G.,** Silicone additive facilities epoxy plastic sectioning, *Stain Technol.,* 57, 79, 1982.

186. **Mollenhauer, H. H.,** Surfactants as resin modifiers and their effect on sectioning., *J. Electron Microsc. Tech.,* 3, 217, 1986.

187. **Endo, B. Y. and Wergin, W. P.**, personal communication.
188. **Hayat, M. A.**, *Principles and Techniques of Electron Microscopy. Biological Applications.* 3rd ed., CRC Press, Boca Raton, FL, 1989.
189. **Klomparens, K. L.**, Artifacts in ultrathin sectioning, in *Artifacts in Biological Electron Microscopy*, Crang, R. F. E. and Klomparens, K. L., Eds., Plenum Press, New York, 1988, 65.
190. **Bell, M.**, Artifacts in staining procedures, in *Artifacts in Biological Electron Microscopy*, Crang, R. F. E. and Klomparens, K. L., Eds., Plenum Press, New York, 1988, 81.
191. **Coons, A. H., Creech, H. J., and Jones, R. N.**, Immunological properties of an antibody containing a fluorescent group, *Proc. Soc. Exp. Biol.*, 47, 200, 1941.
192. **Singer, S. J.**, Preparation of an electron dense antibody conjugate, *Nature (London)*, 183, 1523, 1959.
193. **Nakane, P. K. and Pierce, G. B.**, Enzyme labeled antibodies: preparation and application for the localization of antigens, *J. Histochem. Cytochem.*, 14, 929, 1966.
194. **Mason, T. E., Spicer, S. S., Swallow, R. A., Dreskin, R. B., and Phifer, R. F.**, An immunoglobulin-enzyme method for localizing tissue antigens, *J. Histochem. Cytochem.*, 17, 563, 1969.
195. **Sternberger, L. A., Cuculis, J., Meyer, H. G., and Hardy, Jr., P. H.**, The unlabeled antibody enzyme method of immunochemistry. Preparation and properties of soluble antigen-antibody complex (horseradish peroxidase antihorseradish peroxidase) and its use in the identification of spirochetes, *J. Histochem. Cytochem.*, 18, 315, 1970.
196. **Faulk, W. P. and Taylor, C. M.**, An immunocolloid method for the electron microscope, *Immunocytochemistry*, 8, 1081, 1971.
197. **Singer, S. J., Tokuyasu, K. T., Dutton, A. H., and Chen, W. T.**, High resolution immunoelectron microscopy of cell and tissue ultrastructure in *Electron Microscopy in Biology*, Vol. 2, Griffith, J. D., Ed., John Wiley & Sons, New York, 1982, 55.
198. **Polak, J. and Van Noorden, S.**, *Immunocytochemistry: Practical Applications*, John Wright PSG, Massachusetts, 1983.
199. **Polak, J. and Van Noorden, S.**, *An Introduction to Immunocytochemistry: Current Techniques and Problems, R M S Microscopy Handbook II*, Oxford University Press, New York, 1984.
200. **Bullock, G. R. and Petrusz, P., Eds.**, *Techniques in Immunocytochemistry*, Vol. 1, Academic Press, Orlando, FL, 1982.
201. **Bullock, G. R. and Petrusz, P., Eds.**, *Techniques in Immunocytochemistry*, Vol. 2, Academic Press, Orlando, FL, 1983.
202. **Bullock, G. R. and Petrusz, P., Eds.**, *Techniques in Immunocytochemistry*, Vol. 3, Academic Press, Orlando, FL, 1985.
203. **Polak, J. M. and Varndell, I. M., Eds.**, *Immunolabeling for Electron Microscopy*, Elsevier, Amsterdam, 1984.
204. **Kuhlmann, W. D.**, *Immuno-Enzyme Techniques in Cytochemistry*, Verlag Chemie GmbH, Weinheim, 1984.
205. **Verkleij, A. J. and Leunissen, J. L. M.**, *Immuno-Gold Labeling in Cell Biology*, Verkleij, A. J. and Leunissen, J. L. M., Eds., CRC Press, Boca Raton, FL, 1989.
206. **Garzon, S., Kurstak, E., and Bendayan, M.**, Ultrastructural localization of viral antigens using the protein A-gold technique, *J. Virol. Methods*, 5, 67, 1982.
207. **Bendayan, M.**, Protein A-gold electron microscopic immunochemistry: methods, applications, and limitations, *J. Electron Microsc. Tech.*, 1, 243, 1984.
208. **Bendayan, M.**, Enzyme-gold electron microscope cytochemistry: a new approach for the ultra-structural localization of macromolecules, *J. Electron Microsc. Tech.*, 1, 349, 1984.
209. **Bendayan, M.**, The enzyme-gold technique: a new cytochemical approach for the ultrastructural localization of macromolecules, in *Techniques in Immunocytochemistry*, Vol. 3, Bullock, G. R. and Petrusz, P., Eds., Academic Press, Orlando, FL, 1985, 179.
210. **Bendayan, M.**, Introduction of the protein G-gold complex for high resolution immunocytochemistry, *J. Electron Microsc. Tech.*, 6, 7, 1987.
211. **DeMay, J.**, Colloidal gold as marker and tracer in light and electron microscopy, *EMSA Bull.*, 14 (1), 54, 1984.
212. **Hacker, G. W., Springall, D. R., Tang, S. K., Van Noorden, S., Lackie, P., Grimelius, L., Adam, H., and Polak, J. M.**, Immunogold-silver staining (IGSS)-a review, *Mikroskopie*, 42, 318, 1985.
213. **Smit, J. and Todd, W. J.**, Colloidal gold labels for immunocytochemical analysis of microbes in *Ultrastructure Techniques for Microorganisms*, Aldrich, H. C. and Todd, W. J., Eds., Plenum Press, New York, 1986, 469.
214. **Roth, J.**, Post-embedding cytochemistry with gold-labeled reagents: a review, *J. Microsc.*, 143, 125, 1986.
215. **Varndell, I. M. and Polak, J. M.**, *EM Immunolabeling in Electron Microscopy in Molecular Biology, a Practical Approach*, Somerville, J. and Scheer, U., Eds., IRL Press, Oxford, 1987, 179.
216. **Hayat, M. A., Ed.**, *Colloidal Gold. Principles, Methods, and Applications*, Vol. 1, Academic Press, San Diego, 1989.

217. **Hayat, M. A.,** *Colloidal Gold. Principles, Methods, and Applications*, Vol. 2, Academic Press, San Diego, 1989.

218. **Hayat, M. A.,** *Colloidal Gold. Principles, Methods, and Applications*, Vol. 3, Academic Press, San Diego, 1990.

219. **Goodman, S. L., Trejdosiewicz, L. K., Livingston, D. C., and Hodges, G. M.,** Colloidal gold markers and probes for routine application in microscopy, *J. Microsc. (Oxford)*, 123, 201, 1981.

220. **Bowers, B.,** Preparation of colloidal gold probes for electron microscopy, *EMSA Bull.*, 13 (2), 101, 1983.

221. **Boon, M. E. and Kok, L. P.,** *Microwave Cookbook of Pathology: the Art of Microscopic Visualization*, Boon, M. E. and Kok, L. P., Eds., Coulomb Press, Leyden, Netherlands, 1988.

222. **Van Dort, J. B., Schneijdenberg, T. W., Boon, M. E., Kok, L. P., and De Bruijn, W. C.,** Preservation of structure and cytochemical reactivity at the ultrastructural level, using microwave irradiation, *Histochem. J.*, 20, 365, 1988.

223. **Login, G. R. and Dvorak, A. M.,** Microwave fixation provides excellent preservation of tissue, cells and antigens for light and electron microscopy, *Histochem.*, 20, 373, 1988.

224. **Login, G. R., Dwyer, B. K., and Dvorak, A. M.,** Rapid primary microwave-osmium fixation. I. Preservation of structure for electron microscopy in seconds, *J. Histochem. Cytochem.*, 38, 755, 1990.

225. **Wild, P., Krahenbuhl, M., and Schraner, E. M.,** Potency of microwave irradiation during fixation for electron microscopy, *Histochemistry*, 91, 213, 1989.

226. **Bladon, T., Frosch, M., Sabour, P. M., and Lee, P. E.,** Association of nuclear matrix proteins with cytoplasmic assembly sites of *Tipula* iridescent virus, *Virology*, 155, 524, 1986.

227. **Tanada, Y., Inoue, H., Hess, R. T., and Omi, E. M.,** Site of action of a synergistic factor of a granulosis virus of the armyworm, *Pseudaletia unipuncta*, *J. Invertebr. Pathol.*, 34, 249, 1980.

228. **Tanada, Y., Hess, R. T., Omi, E. M., and Yamamoto, T.,** Localization of a new synergistic factor of a granulosis virus by its esterase activity in the larval midgut of the armyworm, *Pseudaletia unipuncta*, *Microbios*, 37, 87, 1983.

229. **Nagata, M. and Tanada, Y.,** Origin of an alkaline protease associated with the capsule of a granulosis virus of the armyworm, *Pseudaletia unipuncta* (Haworth), *Arch. Virol.*, 76, 245, 1983.

230. **Zhu, Y., Hukuhara, T., and Tamura, K.,** Location of a synergistic factor in the capsule of a granulosis virus of the armyworm, *Pseudaletia unipuncta*, *J. Invertebr. Pathol.*, 54, 49, 1989.

231. **Hukuhara, T. and Zhu, Y.,** Enhancement of the *in vitro* infectivity of a nuclear polyhedrosis virus by a factor in the capsule of a granulosis virus, *J. Invertebr. Pathol.*, 54, 71, 1989.

232. **Van der Wilk, F., Van Lent, J. W. M., and Vlak, J. M.,** Immunogold detection of polyhedrin, p10, and virion antigens in *Autographa californica* nuclear polyhedrosis virus-infected *Spodoptera frugiperda* cells, *J. Gen. Virol.*, 68, 2615, 1987.

233. **Vlak, J. M., Klinkenberg, F. A., Zaal, K. J. M., Usmany, M., Klinge-Roode, E. C., Geervliet, J. B. F., Roosien, J., and VanLent, J. W. M.,** Functional studies on the p10 gene of *Autographa californica* nuclear polyhedrosis virus using a recombinant expressing a p-10-β-galactosidase fusion gene, *J. Gen. Virol.*, 69, 765, 1988.

234. **Russell, R. L. Q. and Rohrmann, G. F.,** A baculovirus polyhedron envelope protein Immunogold localization in infected cells and mature polyhedra, *Virology*, 174, 177, 1990.

235. **Van Lent, J. W. M., Groenen, J. T. M., Klinge-Roode, E. C., Rohrmann, G. F., Zuidema, D., and Vlak, J. M.,** Localization of the 34kDa polyhedron envelope protein in *Spodoptera frugiperda* cells infected with *Autographa californica* nuclear polyhedrosis virus, *Arch. Virol.*, 111, 103, 1990.

236. **Sterling, P.,** Quantitative mapping with the electron microscope, in *Principles and Techniques of Electron Microscopy, Biological Applications*, Vol. 5, Hayat, M. A., Ed., Van Nostrand Reinhold, New York, 1975, 1.

237. **Miller, M. F., Allen, P. T., and Dmochowski, L.,** Quantitative studies on oncornaviruses in thin sections, *J. Gen. Virol.*, 21, 57, 1973.

238. **Weiss, S. A., Smith, G. C., Kalter, S. S., Vaughn, J. L., and Dougherty, E. M.,** Improved replication of *Autographa californica* nuclear polyhedrosis in roller bottles; characterization of the progeny virus, *Intervirology*, 15, 213, 1981.

239. **Allaway, G. P.,** Virus particles packaging in baculovirus and cytoplasmic polyhedrosis virus inclusion bodies, *J. Invertebr. Pathol.*, 42, 357, 1983.

240. **Smith, C. C., Weiss, S. A., Kalter, S. S., and Vaughn, J. L.,** Quantitative studies on nuclear polyhedrosis virus produced *in vitro*, in *Proc. 39th Ann. Meet. EMSA, Electron Microscopy Soc. Am.*, Bailey, G. W., Ed., Claitor's Publ. Div., Baton Rouge, LA, 1981, 630.

241. **Tompkins, G. J., Vaughn, J. L., Adams, J. R., and Reichelderfer, C. F.,** Effects of propagating *Autographa californica* nuclear polyhedrosis virus and its *Trichoplusia ni* variant in different hosts, *Environ. Entomol.*, 10, 801, 1981.

242. **Adams, J. R.,** Electron microscopic examinations of Gypchek and VIRIN-ENsh, in *A Comparison of the US (Gypshek) and USSR (VIRIN-ENSH) Preparations of the Nuclear Polyhedrosis Virus of the Gypsy Moth,*

Lymantria dispar, Ignoffo, C. M., Martignoni, M. E., and Vaughn, J. L., Eds., American Society for Microbiology, Washington, D.C., 1983, 11.

243. **Russ, J. C.,** *Elemental X-Ray Analysis of Materials,* EXAM Methods, Walsh, C. J., Ed., Edax Laboratories International, Inc., Prairie View, IL, 1972.

244. **Russ, J. C.,** *Fundamentals of Energy Dispersive X-Ray Analysis,* Butterworths, MA, 1984.

245. **Hayat, M. A., Ed.,** *X-Ray Microanalysis in Biology,* University Park Press, Baltimore, 1980.

246. **Vaughn, D., Ed.,** *Energy Dispersive X-Ray Microanalysis: An Introduction,* Kevex Corp., California, 1983.

247. **Morgan, A. J.,** *X-Ray Microanalysis in EM for Biologists,* RMS Microscopy Handbooks, Oxford University Press, New York, 1985.

248. **Aldrich, H. C.,** X-ray microanalysis, in *Ultrastructure Techniques for Microorganisms,* Aldrich, H. C. and Todd, W. J., Eds., Plenum Press, New York, 1986, 517.

249. **Johnson, D. and Cantino, M.,** Artifacts of analysis in biological electron microscopy, in *Artifacts in Biological Electron Microscopy,* Crang, R. F. E. and Klomparens, K. L., Eds., Plenum Press, New York, 1988, 219.

250. **Adams, J. R.,** The use of electron microscopy in insect pathology, *Norelco Rep.,* 32, 11, 1985.

251. **Shapiro, M. and Ignoffo, C. M.,** Elemental analysis of polyhedral inclusion bodies of the cotton bollworm, *Heliothis zea,* nucleopolyhedrosis virus, *J. Invertebr. Pathol.,* 17, 449, 1971.

252. **Faust, R. M., Hallam, G. M., and Travers, R. S.,** Spectrographic elemental analysis of the parasporal crystals produced by *Bacillus thuringiensis* var. *dendrolimus* and the polyhedral inclusion bodies of the nucleopolyhedrosis virus of the fall armyworm, *Spodoptera frugiperda, J. Invertebr. Pathol.,* 22, 478, 1973.

253. **Tokuyasu, K. T.,** Application of cryoultramicrotomy to immunocytochemistry, *J. Microsc. (Oxford),* 143, 139, 1986.

254. **Linner, J. G., Livesay, S. A., Harrison, D. S., and Steiner, A. L.,** A new technique for removal of amorphous phase tissue water without ice crystal damage: a preparative method for ultrastructural analysis and immunoelectron microscopy, *J. Histochem. Cytochem.,* 34, 1123, 1986.

255. **Steinbrecht, R. A. and Zierold, K., Eds.,** *Cryotechniques in Biological Electron Microscopy,* Springer-Verlag, Berlin, 1987.

256. **Dubochet, J., Adrian, M., Chang, J. J., Homo, J. C., Lepault, J., McDowall, A. W., and Schultz, P.,** Cryo-electron microscopy of vitrified specimens, *Q. Rev. Biophysics.,* 21, 129, 1988.

257. **Tokuyasu, K. T.,** Application of cryoultramicrotomy to immunocyto-chemistry, *J. Microsc. (Oxford),* 143, 139, 1986.

258. **Sitte, H., Edelmann, L., and Newmann, K.,** Cryofixation without pretreatment at ambient pressure, in *Cryotechniques in Biological Electron Microscopy,* Steinbrecht, R. A. and Zierold, K., Eds., Springer-Verlag, Berlin, 1987, 87.

259. **Sitte, H., Newmann, K., and Edelmann, L.,** Cryofixation and cryosubstitution for routine work in transmission electron microscopy, in *The Science of Biological Specimen Preparation,* Muller, M., Becker, R. D., Boyde, A., and Wolosewick, J. J., Eds., SEM, AMF O'Hare, Chicago, 1986, 103.

260. **Humbel, B. and Muller, M.,** Freeze substitution and low temperature embedding, in *The Science of Biological Specimen Preparation,* Muller, M., Becker, R. P., Boyde, A., and Wolosewick, J. J., Eds., SEM, AMF O'Hare, Chicago, 1986, 175.

261. **Sitte, H., Newmann, K., and Edelmann, L.,** Cryosectioning according to Tokuyasu vs. rapid-freezing, freeze-substitution and resin embedding, in *Immuno-Gold Labeling in Cell Biology,* Verkleij, A. J. and Leunissen, J. L. M., Eds., CRC Press, Boca Raton, FL, 1989, 63.

262. **Steinbrecht, R. A. and Muller, M.,** Freeze-substitution and freeze-drying, in *Cryotechniques in Biological Electron Microscopy,* Steinbrecht, R. A. and Zierold, K., Eds., Springer-Verlag, Berlin, 1987, 149.

263. **Hayes, T. L.,** Scanning electron microscope techniques in biology, in *Advanced Techniques in Biological Electron Microscopy,* Koehler, J. K., Ed., Springer-Verlag, New York, 1973, 153.

264. **Hayat, M. A., Ed.,** *Principles and Techniques of Scanning Electron Microscopy,* Vol. 1, Van Nostrand Reinhold, New York, 1974.

265. **Hayat, M. A., Ed.,** *Principles and Techniques of Scanning Electron Microscopy,* Vol. 2, Van Nostrand Reinhold, New York, 1974.

266. **Hayat, M. A., Ed.,** *Principles and Techniques of Scanning Electron Microscopy,* Vol. 3, Van Nostrand Reinhold, New York, 1975.

267. **Hayat, M. A., Ed.,** *Principles and Techniques of Scanning Electron Microscopy,* Vol. 4, Van Nostrand Reinhold, New York, 1975.

268. **Hayat, M. A., Ed.,** *Principles and Techniques of Scanning Electron Microscopy,* Vol. 5, Van Nostrand Reinhold, New York, 1976.

269. **Hayat, M. A., Ed.,** *Principles and Techniques of Scanning Electron Microscopy,* Vol. 6, Van Nostrand Reinhold, New York, 1978.

270. **Echlin, P.,** Low temperature biological scanning electron microscopy, in *Advanced Techniques in Biological Electron Microscopy II,* Koehler, J. K., Ed., Springer-Verlag, Berlin, 1978, 89.

271. **Becker, R. P. and Johari, O., Eds.,** *Cell Surface Labeling,* SEM, AMF O'Hare, Chicago, 1979.

272. **Gabriel, B. L.,** *Biological Scanning Electron Microscopy,* Van Nostrand Reinhold, New York, 1982, 186.

273. **Jones, B. R., Ed.,** *Electron Microscopy,* Library Research Assoc., New York, 547, 1983.

274. **Bell, P.,** *SEM of Cells in Culture,* SEM, AMF O'Hare, Chicago, 1984.

275. **Murphy, J. A. and Roomans, G. M., Eds.,** *Preparation of Biological Specimens for SEM,* SEM Inc., AMF O'Hare, Chicago, 1984.

276. **Beckett, A. and Read, N. D.,** Low-temperature scanning electron microscopy in *Ultrastructure Techniques for Microorganisms,* Aldrich, H. C. and Todd, W. J., Eds., Plenum Press, New York, 1986, 45.

277. **Albrecht, R. M. and Hodges, G. M., Eds.,** *Biotechnology and Bioapplications of Colloidal Gold,* Scanning Microscopy Int., AMF O'Hare, Chicago, 1988.

278. **Tanaka, K., Mitsushima, A., Kashima, Y., Nakadora, T., and Fukudome, H.,** Ultra-high resolution scanning electron microscopy of biological materials, *Progress in Clinical and Biological Res., Vol. 295, Cells and Tissues: A Three-Dimensional Approach by Modern Techniques in Microscopy,* Molta, P. M., Ed., Alan R. Liss, New York, 1989, 21.

279. **Porter, K. R., Kelly, D., and Andrews, P. M.,** The preparation of cultured cells and soft tissue for scanning electron microscopy, *Proc. 5th Annu. Stereoscan Colloquium,* Kent Cambridge Scientific, Morton Grove, IL, 1972, 1.

280. **Kelley, R. O., Dekker, R. A. F., and Bluemink, J. G.,** Ligand-mediated osmium binding its application in coating biological specimens for scanning electron microscopy, *J. Ultrastruct. Res.,* 45, 254, 1973.

281. **Laczko, J. and Varga, S.,** Experiences with the thiocarbohydrazide mediated osmium binding in coating biological specimens for scanning electron microscopy, *Mikroskopie,* 32, 69, 1976.

282. **Gunning, W. T. and Crang, R. E.,** The usefulness of glutaraldehyde-carbohydrazide copolymerization in biological specimen stabilization for scanning electron microscopy, *J. Electron Microsc. Tech.,* 1, 131, 1984.

283. **Hayat, M. A. and Zirkin, B. R.,** Critical point-drying method, in *Principles and Techniques of Electron Microscopy,* Hayat, M. A., Ed., Van Nostrand Reinhold, New York, 3, 297, 1973.

284. **Anon.,** EMSA Education Committee, Critical Point Drying, *EMSA Bull.,* 9 (2), 67, 1979.

285. **Boyde, A.,** Pros and cons of critical point drying and freeze drying for SEM, *SEM Proc. II,* 1978, 303.

286. **Barrett, L. A. and Pendergrass, R. E.,** A method for handling free cells through critical point drying, *J. Microsc. (Oxford),* 109, 1977.

287. **Nation, J. L.,** A new method using hexamethyldisilazane for preparation of soft insect tissues for scanning electron microscopy, *Stain Technol.,* 58, 347, 1983.

288. **Lamoureaux, W.,** Prevention of outgassing when coating tissues dried with hexamethyldisilazane, *EMSA Bull.,* 18 (1), 91, 1988.

289. **Kennedy, J. R., Williams, R. W., and Gray, J. P.,** Use of Peldri II (a fluorocarbon solid at room temperature) as an alternative to point drying for biological tissues, *J. Electron Microsc. Tech.,* 11, 125, 1989.

290. **Adams, J. R. and Wilcox, T. A.,** Scanning electron microscopical comparisons of insect virus occlusion bodies prepared by several techniques, *J. Invertebr. Pathol.,* 40, 12, 1982.

291. **Tanaka, K., Mitsushima, A., Kashima, Y., Nakadena, T., and Osatake, H.,** Application of an ultra-high-resolution scanning electron microscope (UHS-TI) to biological specimens, *J. Electron Microsc. Tech.,* 12, 146, 1989.

Chapter 3

PURIFICATION OF INVERTEBRATE VIRUSES

George J. Tompkins

TABLE OF CONTENTS

I. INTRODUCTION

The majority of invertebrate viruses isolated have been from insects. This chapter describes the techniques available for the purification of viruses and reviews the most useful and successful approaches that have been utilized.

Each step in a purification process is designed to remove nonviral substances and leave progressively purer viruses. The viruses should be assayed during the purification process to verify that the virus has not been lost in the process. Viruses contain nucleic acid enclosed by protein, making viruses behave chemically and physically like proteins, and the methods of the protein chemist are most successful in purifying viruses. Standardized procedures should be the criteria for reliable purification within any virus group.

The stability of a virus is an important factor to consider in the purification of a virus. Proteins and nucleic acids are sensitive and may be denatured or destroyed by heating or extremes of pH. In some cases the viral components are stabilized by being incorporated into particles. In these instances the increased stability should be taken advantage of in the purification process. Before a purification process is started one should determine if it will harm the virus and plan means of purification which do not harm the virus.

II. PROCEDURES FOR VIRUS PURIFICATION

A. HOMOGENIZATION AND FILTRATION

The first step in virus purification is the homogenization of the cellular material from either diseased hosts or infected tissue culture cells. The material containing the virus may be suspended in distilled water, 0.01 *M* Tris (pH 7.3 to 8.0) or other suitable buffer, and homogenized in a ground glass tissue homogenizer, a blender, or other tissue grinding device. Sometimes the material is frozen before being homogenized to help disrupt the cells. After grinding, the cellular materials are removed by filtering the homogenate through several layers of cheesecloth, muslin, or glass wool. The homogenate containing the virus is ready for further purification. Sometimes excess lipids can be removed by centrifuging the virus suspension through a 40% (w/w) sucrose solution.[1,2] The treatment of the virus suspension with 0.1 to 2% sodium dodecyl sulfate (SDS) may be effective in disrupting cells and in reducing contaminating material.[1-4]

B. EXTRACTION WITH SOLVENTS

The extraction of virus solutions with a solvent immiscible in water, such as chloroform, may be used for some viruses. The virus remains in the water layer, organic materials go into the solvent, and denatured proteins and other materials may collect at the boundary between the two. Differential solubility techniques are generally applied at the start of a separation procedure when the virus constitutes a small percentage of the total material. These techniques have a high capacity but low resolving power and are used to eliminate gross impurities from the sample. The use of polyethylene glycol (PEG) precipitation constitutes a differential solubility technique that is sometimes used because PEG has little tendency to denature or interact with protein. The use of PEG does not significantly raise ionic strength, allowing fractions to be run directly on ion exchange columns.

C. CENTRIFUGATION

Most virus purification procedures are based on one or another form of centrifugation. Differential centrifugation consists of alternate periods of fast and slow speeds and is frequently used to separate viruses from crude mixtures. A high speed centrifugation will pellet virus and other large particulate matter and leave soluble, small material in the liquid. The solid pellet from the bottom of the tube is resuspended in water or buffers and recentrifuged again at a slower speed which may allow large solid matter from the cells to settle out and the virus to remain in suspension. The pellet is discarded, and the supernatant liquid containing the virus is saved and recentrifuged at high speed to pellet the virus. This cyclic procedure is repeated until no further separation is achieved.

Density gradient centrifugation is useful for small-scale purifications of viruses. This procedure involves the centrifugation in centrifuge tubes containing layers of sucrose or glycerol solution of increasing density. The virus is placed on top, the tube is centrifuged for a given time, the virus is found in a thin band at some particular density, and the virus-containing band can be removed by using a pipette or a syringe.

D. CHROMATOGRAPHY

Chromatography is a high resolution technique sometimes used in virus purification. This separation process depends on relative degrees of adsorption to a solid support material, an adsorbent, which may be an inorganic material such as alumina or silica or can be composed of synthetic plastic resins formulated as small beads. A solution containing the virus is allowed to flow through the column, and the virus is washed off with aqueous solutions or solvents that do not harm the virus. Chromatography separates on the basis of surface qualities, not shape or density, as does centrifugation.

E. ELECTROPHORESIS

Electrophoresis is another means of purifying viruses for small-scale separations. This technique involves the separation in an electrical field. The protein molecules on the outside of the virus carry electrical charges, and the balance and distribution of these charges determine how the virus behaves in an electrical field. Electrophoretic techniques utilizing liquid media, paper strips, or gels are available. This process is suitable for small-scale purification and is primarily used to achieve final purification of small samples or to establish their identity or degree of purity.

III. INSECT VIRUSES

A. NUCLEAR POLYHEDROSIS AND GRANULOSIS VIRUSES

After homogenization and filtration through cheesecloth, muslin, or glass wool, the polyhedra are pelleted at 10,000 g for 10 min and resuspended in 0.4 to 2% (w/v) SDS for 30 min to several h.[1,2,5,6] This sample is then pelleted as before and resuspended in 0.5 M NaCl for 30 min at room temperature. The sample is pelleted at 10,000 g for 10 min, rinsed twice in deionized water, and dispersed by sonication. Carbon tetrachloride or Tween 80 has been used to remove nonviral protein by mixing with the viral suspension and then centrifuging through a 30% sucrose suspension at 80,000 g for 1 h. It was found that the carbon tetrachloride decreased the viral infectivity but that a 1% concentration of Tween 80 removed most ballast proteins without loss of infectivity.[7]

The polyhedra and granules are further purified by isopycnic centrifugation on 25 to 60% (w/w) linear sucrose gradients at 65 to 96,000 g for 1 to 3 h.[1,2,6,8] The use of glycerol gradients has also been used successfully.[9,10] After centrifugation the virus band is removed with a Pasteur pipette or a syringe, diluted with several volumes of deionized water, and centrifuged at 10,000 g for 10 to 15 min. The pelleted virus is resuspended in a small volume of deionized water and stored at 4°C if it is to be used immediately or stored at -20°C for long-term storage to prevent contamination.

1. Purification of Alkali Dissociated Viruses

Purified polyhedra or granules may be solubilized under a variety of alkaline conditions. Virions can be released from polyhedra or granules by resuspending them at a concentration of 5 mg/ml in dilute alkaline saline, pH 10.8 (0.03 to 0.1 M Na_2CO_3 to 0.05 M NaCl) for 30 to 90 min at 26 to 30°C.[1,2,10-12] A small amount of a chelating agent (0.001 M EDTA) may be added to reduce oxidation that may cause denaturation or aggregation.[13] The use of 0.022 M glycine-KOH buffer with 0.1 M KCl at pH 11.5 may also be used to remove the virions from the occlusion bodies.[14] After dissolution the pH of the solution may be adjusted from pH 10.8 to 8 or lower, and the virions may be pelleted by centrifugation at 23 to 30,000 g for 1 h. The resultant pellet is resuspended in distilled water and layered onto a 30 to 60% (w/w) sucrose solution and centrifuged at 45 to 75,000 g for 1 h at 4°C. The zones of virions correspond to the number of nucleocapsids within the virion envelope.[14] The virions can be collected by a Pasteur pipette or syringe, diluted with distilled water, and centrifuged at 30,000 g for 1 h; the pellet can be washed with distilled water, and the centrifugation repeated. The pellet containing the virions is resuspended in distilled water.

2. Nucleocapsid Purification

Nucleocapsids may be isolated from the virions by removing the envelopes with 1 to 1.5% Nonidet P40 (NP40) in 0.005 to .01 M Tris (hydroxymethyl) aminomethane (Tris) buffer, pH 7.5 to 8.0 at 30°C for 1 to 18 h.[2,3,15] The time required for the envelope removal, the required pH of the buffer, and the concentration of NP40 used may vary for different viruses, and the treatment may have to be modified. The nucleocapsids can be purified with anywhere from 20 to 60% (w/w) sucrose gradients from 45 to 70,000 g for 45 min to 2 h.[2,16] The use of 30 to 70% (v/v) glycerol gradients at 70,000 g for 1 h has also been successfull.[11] Nucleocapsids have also been purified on 1.18 to 1.5 g/ml CsCl gradients at 208,000 g for 2 h.[15] The band containing nucleocapsids is removed, and the nucleocapsids are washed in distilled water. The electron microscope is used for determining the integrity of the virions and the removal by detergents of the envelope from nucleocapsids.

B. CYTOPLASMIC POLYHEDROSIS VIRUSES

Cytoplasmic polyhedrosis viruses replicate in the gut epithelium of the larval lepidopteran host. To purify these polyhedra the guts are removed and homogenized in a blender or homogenizer in 0.01 M Tris or 0.1% SDS. The suspension is filtered through gauze, cheesecloth, or organdy to remove large debris.[17,18] The virus is pelleted at 10,000 g for 15 min, and the polyhedra are resuspended in distilled water or 0.1% SDS. This solution is placed on a 20 to 60% sucrose gradient (w/w) and centrifuged at 10 to 30,000 g for 1 h. The gradient centrifugation cycle may have to be repeated to obtain suspensions of pure polyhedra.

Virions may be released from polyhedra by treating them with 0.1 M Na_2CO_3 or a solution consisting of 0.1 M NaCl and 0.05 M Na_2CO_3 at pH 10.6 for a period of 1 h at room temperature. After 1 h the solution is brought to pH 7.2 to 7.4 with 0.2 M HCl and centrifuged at 12,000 g for 10 min. The supernatant fluid is removed and centrifuged at 75,000 g for 90 min, and the pellet is then resuspended in buffer or in distilled water.

C. ENTOMOPOXVIRUSES

Virus-infected larvae may be homogenized in distilled water or in 0.2 M Tris-HCl, pH 7.5 in a glass tissue homogenizer or in a blender. The homogenate is filtered through a mesh cloth, and the filtrate is centrifuged at 12,000 g for 10 min at 4°C.[19,20] The pellet is resuspended in distilled water or in 0.2 M Tris-HCl, pH 7.5, containing 3% SDS, and sonicated for approximately 30 s at 20 W and recentrifuged at 12,000 g for 10 min. The occlusion bodies are resuspended in 0.01 M Tris-HCl, pH 7.8; layered onto a 40 to 65% sucrose step gradient (w/v); and centrifuged at 64,000 g for 1.5 h at 4°C. The occlusion bodies form a visible band at the 58 to 61% sucrose interface and can be removed with a Pasteur pipette.[20] The occlusion bodies are washed in distilled water, pelleted at 12,000 g for 20 min, and resuspended in distilled water.

To release the virus from the occlusion bodies an equal volume of occlusion bodies in distilled water is added to 1.6 M Na_2CO_3 solution containing 0.1 M mercaptoacetic acid, allowing 3 to 10 min for dissolution of the occlusion bodies at room temperature. This preparation is centrifuged at 12,000 g for 20 min to pellet the virus, which is then resuspended in 0.01 M Tris-HCl, pH 8.0, containing 0.001 M EDTA.[20]

D. IRIDESCENT VIRUSES

Dead or diseased insects are ground in distilled water, 0.05 M-phosphate buffer (pH 7.0 to 7.4) or 0.01 M sodium borate buffer (pH 7.5) in a tissue grinder.[21-24] The suspension may be filtered through cheesecloth and then three cycles of differential centrifugation at 1,000 and 17,500 g to remove debris. Pellets from the high speed centrifugation are resuspended in distilled water or buffer, placed on a sucrose gradient (5 to 50%, w/w) and centrifuged for 20 to 30 min at 15 to 29,000 g. The virus fraction is removed with a Pasteur pipette, diluted with distilled water or buffer, and concentrated by high speed centrifugation at 25 to 30,000 g for 30 min. The pellet of virus is resuspended in distilled water or the appropriate buffer and can be stored at 4°C.

E. NONOCCLUDED VIRUSES

Nonoccluded spherical DNA viruses have been isolated from several lepidopteran species. One method used to extract a nonoccluded virus of the navel orangeworm is by disrupting diseased larvae in phosphate buffered saline (pH 7.5) in a glass tissue grinder and then further homogenization in a Polytron tissue disintegrator.[25] Viral particles are segregated from tissue fragments by use of three fluorocarbon treatments, and the crude virus homogenate is combined with Gene solv-D (Allied Chemical Corp., Morristown, NJ) and shaken for 30 s. This suspension is centrifuged at 8,000 g for 15 min, and the aqueous upper layer containing the virus is collected and stored at −20°C.

Other nonoccluded viruses may be purified by taking aqueous extracts of larvae, centrifuging at 80,000 g for 20 min, and then treating with a chloroform/butanol mixture to remove insect protein, lipid, and fat.[26] Several cycles of this treatment remove most of the nonviral material. The addition of ammonium sulfate (one third saturation) to the partially purified aqueous extract for 24 h at 4°C followed by centrifugation at 72,000 g for 30 min is an effective method of purification. The pellet of virus is resuspended in distilled water.

F. BEE VIRUSES AND CRICKET PARALYSIS VIRUS

Chronic bee paralysis virus can be isolated from infected bees by grinding them in 0.01 M potassium phosphate buffer (pH 6.7) containing 0.02% sodium diethyldithiocarbamate (DIECA) and 0.1 volume of diethyl ether.[27,28] This mixture is then emulsified with carbon tetrachloride, coarsely filtered, and cleared by centrifugation at 8,000 g for 10 min; next it is centrifuged at 100,000 g for 2 h. The pellet is resuspended in 0.01 M phosphate buffer and after a second cycle of low and high speed centrifugation, the extract is layered onto sucrose density gradients (10 to 40% w/v) and centrifuged at 45,000 g for 4.5 h at 15°C. The virus is collected and dialyzed against phosphate buffer, then centrifuged at 100,000 g for 2 h, and resuspended in phosphate buffer. These techniques can also be used to purify sacbrood virus, black queen cell virus, and Kashmir bee virus.[29]

Arkansas bee virus can be purified by grinding a single infected pupa in a tissue grinder or mortar and pestle in 1 ml of 50 mM potassium phosphate buffer (pH 6.8), 0.05% DIECA, 0.5 ml ether, and 0.5 ml CCl_4. This slurry is cleared by centrifugation at 12,000 g for 10 min, and the virus is then precipitated by addition of 0.25 volume of 40% polyethylene glycol (6,000 MW) in 1 M ammonium acetate.[30] After 30 min of incubation on ice the solution is centrifuged at 12,000 g for 20 min, and the pellet is resuspended in 1 ml of phosphate buffer. This preparation is centrifuged through 30% (w/v) sucrose at 180,000 g

for 2.5 h, and the virus pellet is resuspended in 200 μl phosphate buffer. Further purification is obtained by centrifugation at 180,000 *g* for 60 min in linear-log sucrose gradients in potassium phosphate buffer.

Cricket paralysis virus is extracted by homogenizing infected larvae in 10 m*M* ammonium acetate, pH 7.0, with an equal volume of CCl_4 and separating the phases by centrifugation. The supernatant is then homogenized with 0.5 volume of ether and finally with CCl_4 to remove the ether. The aqueous phase is centrifuged at 12,000 *g* for 45 min, and the pellet is resuspended in a small volume of 10 m*M* ammonium acetate buffer.[31] This suspension is centrifuged in a 10 to 40% (w/v) sucrose gradient in 10 m*M* ammonium acetate at 80,000 *g* for 2.5 h, and the light-scattering band is removed with a Pasteur pipette. This is dialyzed against ammonium acetate buffer and centrifuged in a cesium chloride density gradient for 16 h at 95,000 *g*. The light-scattering band is removed with a Pasteur pipette and dialyzed against 20 m*M* phosphate buffer, pH 7.2.

G. FLOCK HOUSE VIRUS

Infected larvae are ground in 0.2% Triton X-100 and 0.1% 2-mercaptoethanol and centrifuged at 2400 *g* for 10 min to remove debris. 10% (v/v) PEG (6000 MW) is added, held for 1 h at 2°C, and then centrifuged at 4000 *g* for 20 min; next the resultant pellet is resuspended in $^1/_{10}$ the initial volume of 0.02 *M* sodium phosphate buffer, pH 7.2, and centrifuged at 2700 *g* for 10 min to remove insoluble material.[32] The supernatant is layered onto 12 to 30% (w/w) sucrose gradients and centrifuged at 110,000 *g* for 2.75 h at 20°C, and the gradients are fractionated. The virus is concentrated by centrifugation at 110,000 *g* for 2 h, and the virus pellet is resuspended in a small volume of sodium phosphate buffer.

H. HOUSEFLY VIRUS

Housefly virus can be purified by grinding or blending dead flies in 0.01 *M* phosphate buffer ($Na_2H PO_4$ and NaH_2PO_4 1:1 volume, pH adjusted to 7.2 with 0.1 *M* NaOH) and then filtering through muslin.[33] This suspension is then mixed with CCl_4 (4:1 v/v) which precipitates large host protein molecules but also results in the disruption of a few virions. After 10 min of gentle agitation the mixture is centrifuged at 12,000 *g* for 10 min. The supernatant fluid is then centrifuged at 142,000 *g* for 2 h. The pellet is resuspended in phosphate buffer and again mixed with CCl_4 and centrifuged. The final virus pellet is resuspended in phosphate buffer.

I. DNA VIRUS FROM TSETSE FLY

Hypertrophied salivary glands from tsetse flies are mixed with buffer to a final concentration of 0.03 *M* Tris-HCl, 0.025 *M* KCl, pH 7.5 (TK buffer) and ground in a glass tissue homogenizer. The homogenate is centrifuged at 1,000 *g* for 20 min, and the supernatant is decanted and centrifuged at 42,000 *g* for 1 h at 10°C.[34] The pellet is resuspended in the TK buffer and centrifuged at under 1,000 *g* for 5 min; the supernatant fraction then is layered onto a continuous 30 to 65% (w/w) sucrose gradient prepared in TK buffer and centrifuged at 94,000 *g* for 2 h at 10°C. The virus band, with some membranous material, is harvested, diluted with an equal volume of TK buffer, and centrifuged at 94,000 *g* for 2 h to pellet the virus. The pellet is resuspended in TK buffer and mixed with an equal volume of 0.25% sodium deoxycholate in TK buffer. After 1 h at 20°C this mixture is layered onto a 30 to 65% (w/w) sucrose gradient and centrifuged at 94,000 *g* for 2 h. The virus bands at a position corresponding to 45% (w/w) sucrose and is recovered, diluted, centrifuged as just stated, and resuspended in distilled water.

J. SMALL RNA VIRUSES

The purification of small RNA viruses isolated from *Dasychira pudibunda*, *Trichoplusia ni*, *Nudaurelia cytherea capensis*, and *Darna trima* is as follows.[35] Frozen or fresh larvae are finely ground in 0.05 *M* Tris buffer, pH 7.4; the crude extracts are gently shaken with chloroform and then centrifuged at 2000 *g* for 10 min. The aqueous phase is removed, pooled, and centrifuged at 80,000 *g* for 1 h. The pellet is resuspended in 0.05 *M* Tris and banded on a 10 to 50% (w/v) sucrose gradient in the same buffer at 65,000 *g* for 90 min. The bands are removed and diluted severalfold with cold 0.05 *M* Tris buffer prior to pelleting at 80,000 *g* for 1 h. The sucrose gradient procedure is repeated, and the virions are then further purified on 32% (w/w initial concentration) cesium chloride gradients.

An icosahedral RNA virus of *Pseudoplusia includens* is purified by blending 30 to 40 g larvae in 200 ml CCl$_4$ and 400 ml 0.01 *M* phosphate buffer pH 7.2, containing 0.2% DIECA.[36] Following a low speed centrifugation of the homogenate, the virus is precipitated by making the aqueous phase 8% PEG and 0.3 *M* NaCl, resuspended in phosphate buffer, and subjected to two alternate high (78,000 *g* for 30 min) and low (7,700 *g* for 10 min) speed centrifugations. The final step consists of density gradient centrifugation (0.2 to 0.7 *M* sucrose in 0.01 *M* phosphate buffer, pH 7.2) at 68,000 *g* for 2 h. The virus fraction is recovered and pelleted by centrifuging at 105,000 *g* for 1 h. The virus may also be centrifuged to equilibrium in CsCl, fractionated, and pelleted by high speed centrifugation.

A small RNA virus associated with Baculovirus infection in *Trichoplusia ni* is purified by homogenizing 3 g of frozen larvae in 10 ml of 0.5 *M* Tris, 0.01 *M* EDTA at pH 7.5 with 0.2% mercaptoethanol and 3 ml of CCl$_4$ in a Polytron homogenizer.[37] The homogenate is clarified by centrifugation at 8000 *g* for 15 min, and the virus is precipitated by addition of PEG 6000 to a concentration of 8% and NaCl to 0.1 *M*. The virus is then pelleted through 1 ml of 30% sucrose at 135,000 *g* for 1 h and resuspended in 0.1 *M* Tris-EDTA, pH 7.5. Further purification may be accomplished on 10 to 40% linear sucrose gradients in 0.05 *M* Tris-EDTA, pH 7.5, at 135,000 *g* for 1 h at 4°C, or by chromatography on ECTEOLA Cellex E ion-exchange cellulose columns (Bio-Rad Laboratories, Richmond, CA). The virus is eluted from ECTEOLA columns in 0.1 *M* Tris-EDTA, pH 7.5, and 0.5 *M* NaCl.

K. POLYDNAVIRUSES

Calyxes of 20 to 50 female wasps are dissected into 200 μl of standard Dulbecco phosphate-buffered saline (PBS, containing 0.7 m*M* Ca^{2+} and 0.5 m*M* Mg^{2+}) at pH 7.3; the calyx fluid is squeezed out[38] or the ovaries may be collected in cold phosphate-buffered saline.[39] The ovaries are minced with microscissors, and the cell debris is centrifuged at 1000 *g* for 15 min to sediment large tissue fragments. The supernatant fluid is then layered onto either a 5 to 20% Ficol (w/v) or a 20 to 40% (w/w) sucrose continuous gradient in PBS and centrifuged at 35,000 *g* for 35 min[38] or layered onto a 25 to 50% (w/w) sucrose gradient and centrifuged at 125,000 *g* for 60 min.[39] The virus band is removed, resuspended in PBS, pelleted at 30,000 *g* for 30 min at 5°C, and resuspended in 100 to 200 μl PBS and frozen for storage at −80°C.

L. APHID VIRUS

Aphids are frozen in liquid nitrogen and then from one to ten aphids are ground to a fine powder with a mortar and pestle. The powder is homogenized in a blender with 50 to 70 ml of 0.1 *M* sodium phosphate buffer, pH 7.0, for 1 min. The homogenate is strained through four layers of cheesecloth, adjusted to 1% Triton X-100, and stirred 10 min at room temperature.[40] One third volume of cold chloroform is added and the emulsion stirred. After 10 min on ice, the phases are separated by centrifugation at 2600 *g* for 15 min. The aqueous phase is centrifuged at 80,000 *g* for 4 h, and the pellet is resuspended in 1 to 2 ml of phosphate buffer and left overnight. The preparation is homogenized in a ground glass tissue grinder and clarified by centrifugation for 20 min at 11,000 *g* and then layered onto a linear density gradient of 10 to 40% sucrose in 0.1 *M* phosphate buffer, pH 7.0. After centrifugation at 68,000 *g* for 3 h, the virus is collected and pelleted at 120,000 *g* for 2.5 h.

IV. MARINE INVERTEBRATE VIRUSES

These viruses are generally purified in a buffer with a high ionic strength to limit damage caused by osmotic shock, especially to enveloped viruses. The W2 virus of the crab, *Carcinus mediterraneus*, is a nonenveloped virus. The hepatopancreas, gills, and gut of infected crabs are homogenized in TN buffer (0.02 *M* Tris-HCl, 0.5 *M* NaCl, pH 7.2) in a Potter tissue blender.[41] This homogenate is clarified by centrifuging at 2,500 *g* for 15 min, and the suspension is then centrifuged at 50,000 *g* for 1 h. The pellet is resuspended in TN buffer, mixed with an equal volume of Freon (1,1,2 trichloro-2,2,1 trifluoroethane), shaken vigorously, and centrifuged several min at 2500 *g*. The upper aqueous phase is removed and reextracted two or three times with Freon to remove cell debris. The final extract is layered onto a linear sucrose gradient (30 to 50% w/w) in TN buffer and centrifuged at 130,000 *g* for 2.5 h. The fraction containing the virus is diluted in TN buffer and centrifuged at 130,000 *g* for 1 h. The virus can be further purified in a preformed linear cesium chloride gradient (20 to 48% w/w) and centrifuged at 130,000 *g* for 15 h. The virions are diluted in either 0.15 *M* NaCl, 0.015 *M* sodium citrate buffer, pH 7, or 0.016 *M* Tris-HCl; centrifuged at 130,000 *g* for 1 h, and resuspended in one of these buffers.

A virus, *Baculovirus penaei*, has been isolated from infected hepatopancreatic tissue of pink shrimp, *Penaeus duorarum*. The infected hepatopancreatic tissue from the shrimp is macerated in a tissue grinder and 30% suspensions in Hanks' balanced salt solution (BSS) are prepared.[42] The suspensions are centrifuged at 1000 *g* for 30 min to remove cellular debris. The supernatant is placed onto a 40 to 80% sucrose gradient prepared in BSS over a 0.5 ml cushion of saturated sucrose and centrifuged at 60,000 *g* for 4 h. The polyhedra containing bands are removed and resuspended in BSS.

REFERENCES

1. **McCarthy, W. J. and Liu, S. Y.,** Electrophoretic and serological characterization of *Porthetria dispar*, polyhedron protein, *J. Invertebr. Pathol.*, 28, 57, 1976.
2. **Tompkins, G. J., Vaughn, J. L., Adams, J. R., and Reichelderfer, C. F.,** Effects of propagating *Autographa californica* nuclear polyhedrosis virus and its *Trichoplusia ni* variant in different hosts, *Environ. Entomol.*, 10, 801, 1981.
3. **Harrap, K. A., Payne, C. C., and Robertson, J. S.,** The properties of three baculoviruses from closely related hosts, *Virology*, 79, 14, 1977.
4. **Summers, M. D. and Egawa, K.,** Physical and chemical properties of *Trichoplusia ni* granulosis virus granulin, *J. Virol.*, 12(5), 1092, 1973.
5. **Cibulsky, R. J., Harper, J. D., and Gudauskas, R. T.,** Biochemical comparison of virion proteins from five nuclear polyhedrosis viruses infecting plusiine larvae (Lepidoptera:Noctuidae), *J. Invertebr. Pathol.*, 30, 303, 1977.
6. **Lee, H. H. and Miller, L. K.,** Isolation of genotypic variants of *Autographa californica* nuclear polyhedrosis virus, *J. Virol.*, 27(3), 754, 1978.
7. **Strokovskaya, L. I., Skuratovskaya, I. N., Zherebtsova, E. N., and Sutugina, L. P.,** Purification of free polyhedrosis virus infecting *Galleria mellonella*, *Acta Virol.*, 21(2), 157, 1977.
8. **Tweeten, K. A., Bulla, Jr., L. A., and Consigli, R. A.,** Isolation and purification of a granulosis virus from infected larvae of the Indian meal moth, *Plodia interpunctella*, *Appl. Environ. Microbiol.*, 34(3), 320, 1977.
9. **Harrap, K. A.,** The structure of nuclear polyhedrosis viruses, I. The inclusion body, *Virology*, 50, 114, 1972.

10. **Longworth, J. F., Robertson, J. S., and Payne, C. C.,** The purification and properties of inclusion body protein of the granulosis virus of *Pieris brassicae, J. Invertebr. Pathol.,* 19, 42, 1972.
11. **Tweeten, K. A., Bulla, L. A., Jr., and Consigli, R. A.,** Characterization of an extremely basic protein derived from granulosis virus nucleocapsids, *J. Virol.,* 33(2), 866, 1980.
12. **Harrap, K. A. and Longworth, J. F.,** An evaluation of purification methods for baculoviruses, *J. Invertebr. Pathol.,* 24, 55, 1974.
13. **McIntosh, A. H. and Ignoffo, C.,** Restriction endonuclease patterns of three baculoviruses isolated from species of *Heliothis, J. Invertebr. Pathol.,* 41, 27, 1983.
14. **Kawanishi, L. Y. and Paschke, J. D.,** Density gradient centrifugation of the virions liberated from *Rachiplusia ou* nuclear polyhedra, *J. Invertebr. Pathol.,* 16, 89, 1970.
15. **Summers, M. D. and Smith, G. E.,** Baculovirus structural polypeptides, *Virology,* 84, 390, 1978.
16. **Payne, C. C., Compson, D., and deLooze, S. M.,** Properties of the nucleocapsids of a virus isolated from *Oryctes rhinoceros, Virology,* 77, 269, 1977.
17. **Croizier, G., Jacquemard, P., Amargier, A., Croizier, L., and Couilloud, R.,** A cytoplasmic poly-hedrosis of *Earias insulana* Boisduval, a new reovirus disease in the Lepidoptera Noctuidae of the genus *Earias, Coton Fibres Trop.,* 38(3), 280, 1983.
18. **Lewandowski, L. J., Kalmakoff, J., and Tanada, Y.,** Characterization of a ribonucleic acid polymerase activity associated with purified cytoplasmic polyhedrosis virus of the silkworm *Bombyx mori, J. Virol.,* 4(6), 857, 1969.
19. **Jaeger, B. and Langridge, W. H. R.,** Infection of *Locusta migratoria* with Entomopoxviruses from *Arphia couspersa* and *Melanoplus sanguinipes* grasshoppers, *J. Invertebr. Pathol.,* 43, 374, 1984.
20. **Langridge, W. H. R.,** Partial characterization of DNA from five Entomopoxviruses, *J. Invertebr. Pathol.,* 42, 369, 1983.
21. **Stadelbacher, E. A., Adams, J. R., Faust, R. M., and Tompkins, G. J.,** An iridescent virus of the bollworm *Heliothis zea* (Lepidoptera:Noctuidae), *J. Invertebr. Pathol.,* 32, 71, 1978.
22. **Glitz, D. G., Hills, G. J., and Rivers, C. F.,** A comparison of the *Tipula* and *Sericesthis* iridescent viruses, *J. Gen. Virol.,* 3, 209, 1968.
23. **Tajbakhsh, S., Dove, M. J., Lee, P. E., and Seligy, V. L.,** DNA components of the *Tipula iridescent virus, Biochem. Cell Biol.,* 64, 495, 1986.
24. **Kalmakoff, J. and Tremaine, J. H.,** Physicochemical properties of *Tipula* iridescent virus, *J. Virol.,* 2(7), 738, 1968.
25. **Kellen, W. R. and Hoffman, D. F.,** A pathogenic nonoccluded virus in hemocytes of the navel orangeworm, *Amyelois transitella* (Pyralidae:Lepidoptera), *J. Invertebr. Pathol.,* 38, 52, 1981.
26. **Longworth, J. F. and Harrap, K. A.,** A nonoccluded virus isolated from four Saturniid species, *J. Invertebr. Pathol.,* 10, 139, 1968.
27. **Ball, B. V.,** Acute paralysis virus isolates from honeybee colonies infested with *Varro jacobsoni, J. Apicultural Res.,* 24(2), 115, 1985.
28. **Ball, B. V., Overton, H. A., Buck, K. W., Bailey, L., and Perry, J. N.,** Relationships between the multiplication of chronic bee-paralysis virus and its associate particle, *J. Gen. Virol.,* 66, 1423, 1985.
29. **Anderson, D. L.,** A comparison of serological techniques for detecting and identifying honeybee viruses, *J. Invertebr. Pathol.,* 44, 233, 1984.
30. **Lommel, S. A., Morris, T. J., and Pinnock, D. E.,** Characterization of nucleic acids associated with Arkansas bee virus, *Intervirology,* 23, 199, 1985.
31. **Scotti, P. D.,** Cricket paralysis virus replicates in cultured *Drosophila* cells, *Intervirology,* 6, 333, 1976.
32. **Scotti, P. D., Dearing, S., and Mossop, D. W.,** Flock House Virus: a nodavirus isolated from *Costelytra zealandica* (White) (Coleoptera: Scarabaeidae), *Arch. Virol.,* 75, 181, 1983.
33. **Moussa, A. Y.,** Studies of the Housefly virus structure and disruption during purification procedures, *Micron,* 12, 131, 1981.
34. **Odino, M. O., Payne, C. C., Crook, N. E., and Jarrett, P.,** Properties of a novel DNA virus from the Tsetse fly, *Glossina pallidipes, J. Gen. Virol.,* 67, 527, 1986.
35. **Moore, N. F., Greenwood, L. K., and Rixon, K. R.,** Studies on the capsid protein of several members of the Nudaurelia B Group of small RNA viruses of insects, *Microbiologica,* 4, 59, 1981.
36. **Chao, Y. C., Scott, H. A., and Young, S. Y. III,** An icosahedral RNA virus of the soybean looper *(Pseudoplusia includens), J. Gen Virol.,* 64, 1835, 1983.
37. **Morris, T. J., Hess, R. T., and Pinnock, D. E.,** Physicochemical characterization of a small RNA virus associated with baculovirus infection in *Trichoplusia ni, Intervirology,* 11, 238, 1979.
38. **Krell, P. J. and Stoltz, D. B.,** Virus-like particles in the ovary of an ichneumonid wasp: purification and preliminary characterization, *Virology,* 101, 408, 1980.
39. **Krell, P. J., Summers, M. D., and Vinson, S. B.,** Virus with a multipartite superhelical DNA genome from the ichneumonid parasitoid *Campoletis sonorensis, J. Virol.,* 43(3), 859, 1982.

40. **D'Arcy, C. J., Burnett, P. A., Hewings, A. D., and Goodman, R. M.,** Purification and characterization of a virus from the aphid *Rhopalosiphum padi, Virology,* 112, 346, 1981.
41. **Mari, J. and Bonami, J. R.,** W2 virus of *Carcinus mediterraneus* (Reovirus), *J. Gen. Virol.,* 69, 569, 1988.
42. **Lewis, D. H.,** An enzyme-linked immunosorbent assay (ELISA) for detecting penaeid baculovirus, *J. Fish Dis.,* 9, 519, 1986.

Chapter 4

PARVOVIRIDAE. STRUCTURE AND REPRODUCTION OF DENSONUCLEOSIS VIRUSES

P. Tijssen and M. Arella

TABLE OF CONTENTS

I. INTRODUCTION

To date, all viruses isolated containing a single-stranded, linear DNA genome belong to the Parvoviridae family. The subset of parvoviruses infecting insects are called denso-nucleosis viruses (DNVs), or densoviruses. The structural characteristics of the DNVs are very similar to those of other parvoviruses,[1-6] and for a long period it had been suspected that the replication and expression strategies of these DNVs would closely resemble those of the vertebrate parvoviruses. Recent results, however, have shown that at least an important subgroup of DNVs has a very different molecular biology.

The common name of densonucleosis virus was first given to describe the characteristic histopathologic symptoms, i.e., hypertrophied and densely stained nuclei (eosinophilic) of sensitive cells in infected larvae upon histological inspection. This name was subsequently shortened to densoviruses, although this seems to change the meaning of the name, and all insect parvoviruses were grouped into one genus, the *Densovirus.* It is not clear whether the penaeid parvovirus-like virus (infecting the shrimp)[7] also belongs to the *Densovirus.*

Only a few DNVs have been investigated in some detail, particularly those from *Galleria mellonella* and *Bombyx mori,* and most of this review will be concentrated on these viruses. Although a discussion of the molecular biology of this virus group is not the goal of this review, we would like to emphasize some striking differences with the vertebrate parvoviruses because this will open further the possibility of their use in pest management.

II. ISOLATION AND CLASSIFICATION OF DENSONUCLEOSIS VIRUSES

DNVs are believed to be widespread pathogens of insects. No systematic searches have been reported but insects which are well studied, such as the silkworm, *Bombyx mori,* were found to host several different DNVs. The first DNV was isolated about 25 years ago from *Galleria mellonella,* and since then several other DNVs have been isolated (see Table 1).[8-27] All of the DNVs isolated thus far are fatal for their host but it remains possible that less pathogenic DNVs exist.

The number of isolates from *Bombyx mori* may indicate that DNVs are very ubiquitous. The "Ina flacherie virus" was shown by Shimizu[13] to be a DNV and was classified as *Bombyx* DNV-1. Matsui[28] and Furuta[29] also isolated DNVs from the silkworm but both viruses were found to be closely related to the Ina isolate. DNV isolates from Satu and Yamanashi[14,30] resembled each other, but were found to be quite different from *Bombyx* DNV-1. They were, therefore, classified as *Bombyx* DNV-2. Finally, a third type of DNV, *Bombyx* DNV-3, was isolated from diseased silkworms in the People's Republic of China.[17] More types may still exist in *Bombyx mori,* one of which resembles the *Galleria mellonella* isolate with respect to restriction maps.[31]

In the latest accepted classification of the Parvoviridae,[32] parvoviruses are grouped into three genera: (1) autonomous vertebrate parvoviruses — *Parvovirus;* (2) vertebrate parvo-viruses requiring a helper virus — *Dependovirus;* and (3) parvoviruses of invertebrates — *Densovirus.* This classification is no longer tenable for a long list of reasons, and in the last meeting of the International Committee for Taxonomy of Viruses (ICTV, Parvoviruses) we decided to reclassify parvoviruses according to their molecular biology (expression strategies) although the definite classification is still the subject of ongoing discussions.[33]

It has been assumed that DNVs form a homogeneous group of parvoviruses, but largely unpublished new data indicate that there are at least two groups of DNVs. The characteristics of these groups are listed in Table 2. For the few DNVs investigated in this respect it is particularly striking that those with the 6-kb genomes are polytropic whereas those with the 4.9-kb genomes infect only the midgut cells of the host. As will be underscored later, DNVs

TABLE 1
Isolation of Densonucleosis Viruses

Order	Host	Other hosts	Country of isolation	Ref.
Lepidoptera	Galleria mellonella	?	France	8
	Euxoa auxiliaris Grote	+	U.S.	9
	Junonia coenia Hb	+	U.K.	10
	Aglais urticae	+	U.K.	10
	Agraulis vanillae	+	U.K.	11
	Sibine fusca Stoll		France	12
	Bombyx mori L.	+	Japan	13 (DNV-1)
	Bombyx mori L.		Japan	14 (DNV-2)
	Bombyx mori L.		P.R.C.	15 (DNV-3)
	Diatraea sacchara		Guadeloupe	16
	Pieris rapae		P.R.C.	17
	Lymantria dispar		France	18
	Casphalia extranea		Ivory Coast	19
	Pseudoplusia includens		U.S.	20
Diptera	Simulium vittatum		U.S.	21[a]
	Aedes aegypti L.	+	U.S.S.R.	22
	Aedes pseudoscutellaris		Venezuela	23
Orthoptera	Periplaneta fuligi-nosa Serville	+	Japan	24
	Acheta domestica		France	25
Odonata	Leucorrhinia dubia		Sweden	26

[a] This virus-like particle resembles strongly the particles observed by Charpentier et al.[27] Its morphology and mode of replication is in our opinion not characteristic of a DNV.

with the 6-kb genomes (at least for those from *Galleria* and *Junonia* but probably also for others, which are closely related as suggested by hybridization data) have the genes for nonstructural and structural proteins on the 5' ends of the opposite DNA strands. This property contrasts with that of the vertebrate parvoviruses and the DNVs with 4.9-kb genomes where these genes reside on the same strand. Thus, parvoviruses are homogeneous with respect to replication (hairpin transfer) but differ in expression strategies, leading to the classification into Parvoviridae and its genera, respectively.

Different procedures have been described for the isolation of DNVs. A method which was found to give high yields of pure virus follows.[34] Infected larvae are, after death, putrefied for several days in Hanks' solution. This suspension is then diluted (1:1) with a 0.025 M phosphate buffer (pH 7.5) and an equal volume of carbontetrachloride, homogenized, and centrifuged at 20,000 g. The pellet is extracted once more, and the combined supernatants are subjected to differential centrifugation (30 min, 30,000 g and then the supernatant 70 min at 100,000 g). The pellet is usually difficult to resuspend (best by using glass wool or adding 0.5 ml buffer and incubating overnight). This virus preparation is then centrifuged to equilibrium (24 h at 100,000 g) in CsCl (40%). Complete virus particles will band at 1.40 g/ml and empty capsids at 1.30 g/ml. Moreover, a third band is usually observed which contains virions of different physicochemical properties and different protein composition.[35,36]

III. MORPHOLOGICAL AND PHYSICOCHEMICAL PROPERTIES

Parvoviruses are quite homogeneous with respect to their physicochemical and morphological properties (see Table 3). A striking difference between the vertebrate and invertebrate parvoviruses is the capacity of the former to agglutinate diverse red blood cells. Thus far none of the DNVs have been shown to be able to cause hemagglutination.

TABLE 2
Subgroups of Densonucleosis Viruses

Viruses and properties	Group I	Group II
	Viruses	
Prototype parvovirus	*Galleria* DNV	*Bombyx* DNV-1
Other members and possible members[a]	*Junonia*	*Casphalia*[a]
	Agraulis	*Periplaneta*[a]
	Bombyx DNV-2[a]	*Bombyx* DNV-3
	Pseudoplusia	
	Aedes[a]	
	Genome	
Genome length	6.0 kb	4.9 kb
Size of inverted repeats	0.5 kb	<0.25 kb
Location of genes		
NS gene(s)	On " + " strand	On " + " strand
Number of NS-ORFs	3	Probably 1
S gene	On " − " strand	On " + " strand
	Structural proteins	
Profile obtained by SDS-PAGE	4 Well-separated polypeptides	2 Doublets
Peptide maps	All structural proteins closely related	Proteins in doublets closely related
	Biological properties	
Tropism	Polytropic (except midgut)	Only midgut cells

[a] Some DNVs have been characterized for some but not all of these aspects; however, a DNV which combines properties of these two groups has not yet been described.

A. VIRION

The major form of infectious virus bands at 1.40 g/ml in aqueous CsCl. It was observed for vertebrate parvoviruses that immature precursor virus (i.e., particles containing the genome but their capsid is not yet rearranged by proteolysis) bands at 1.45 to 1.47 g/ml.[39] In contrast, the dense particles (~1.44 g/ml) observed for DNV have a different physicochemical composition but are, nevertheless, infectious.[35] However, it was observed more recently that these dense particles have light-scattering properties reminiscent of elongated particles and that they are sensitive to DNase.[40] The partial encapsidation of DNA allows a higher degree of hydration of the latter and may be responsible for the lower sedimentation coefficient of this type of virions (about 85 S). Interestingly, negative staining of these two types of DNV particles yields, in certain conditions, different disruption patterns. DNV-1 (banding at 1.40 g/ml) preparations, besides the full particles, also contain empty shells and full capsomers (diameter of about 10 nm), whereas DNV-2 preparations (banding at 1.44 g/ml) lack empty shells and full capsomers but have ringlike capsomers instead (see Figure 1). It should be noted that DNV-2 lack about half of the intermediate-sized structural proteins.[35] The sizes of the capsomers and virion fit a model in which more or less spherical capsid proteins have a diameter of about 5 nm, capsomers (consisting of a ring of five capsid proteins) about 10 nm, and full particles (dodecahedron of 12 capsomers) of about 24 nm.[41]

The hydrodynamic properties of empty and full (DNV-1) particles are very similar: (1) the partial specific volumes are 0.770 and 0.705 ml/g, respectively; and (2) identical diffusion

TABLE 3
Survey of Physicochemical Properties of Parvoviruses

Virion

High density in CsCl solution (about 1.40 g/ml)
Sedimentation coefficient of about 100—110 S
Empty capsides about 60—70 S
Molecular mass about 6×10^6 Da
Extinction coefficient at 260 nm about 9 cm²/mg
Absorbance ratio 260/280 nm and 260/240 nm slightly over 1.5
High resistance to thermal inactivation
Readily inactivated by UV-irradiation, formalin, lyophilization

DNA

About 25—35% of particle mass
Single-stranded, linear molecule
4300—6000 Nucleotides long
Terminal palindromes
Some, but not all, have inverted terminal repeats (e.g., DNVs, AAVs, B19)
DNVs have polyamines (spermine, spermidine, putrescine)[37,38]

Structural proteins

About 65—75% of virion
Depending on parvovirus, 1—4 structural proteins
Structural proteins have molecular mass of 50—100 kDa
Structural proteins are coded by a nested set of mRNAs and have, therefore, extensive sequence homologies

coefficients ($D_{20,w}$ of 1.68×10^{-7} cm^{-2} s^{-1}) have been obtained by dynamic light-scattering studies.[42] These studies suggest that the DNA does not influence diffusion, i.e., no free diffusion of solvent through the protein coat, and that the stability of the structure is governed by protein-protein interactions so that the virus behaves like a hard sphere.[36,43] Disulfide bonds are not involved in the creation of the remarkably stable coat of these viruses.[35] Using the Stokes relation, the hydrodynamic radius can be calculated to be 12.7 nm. The low water content (0.21 ml/g as compared to, e.g., 1.5 ml/g for reovirus[44]) also suggests a very compact, little-hydrated particle. On the other hand, for vertebrate parvoviruses nucleic acid is at least partially neutralized by interaction with basic amino acids, for DNVs polyamines are present (although they could at most neutralize $^1/_7$ of the nucleic acid) and other cations may be involved. Continuous electrometric titrations and circular dichroic spectra have shown[43] that electrostatic interactions between protein and DNA in the virus are minimal. Finally, it has been shown that ionic conditions have an impact on the hydrodynamic properties of the virion[35] and circular dichroism studies[43] demonstrated a conformational difference between the protein coats of full and empty particles.

B. NUCLEIC ACID

In contrast to vertebrate parvoviruses, all DNVs seem to encapsidate both the " + " and " − " strands in separate capsids in about equal amounts, which may be a result of the presence of inverted terminal repeats in all DNVs (no exceptions have been found yet). Therefore, extraction of DNA always yields double-stranded molecules unless the ionic strength is very low. Circular dichroism studies[43] indicated a limited secondary structure of the DNA in the virion. Moreover, a significant hyperchromicity (15 to 25%) is observed upon addition of formaldehyde, also indicating that DNA in the capsid is largely single-stranded. Sequencing results indicate that without the terminal repeats hardly any base pairing is possible.[45]

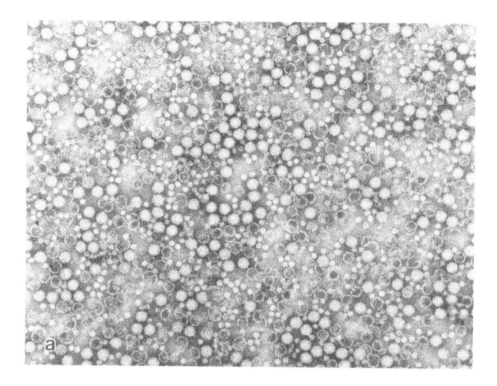

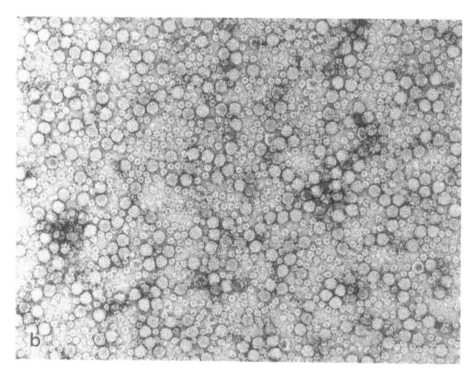

FIGURE 1. (a) Densonucleosis virus *(Galleria)* banding at 1.40 g/ml in CsCl gradients; note the presence of free capsomers and open shells in addition to the complete virus. (Negative staining in 1% uranyl acetate; magnification × 180,000.) (b) *Galleria* DNV banding at 1.44 g/ml in CsCl gradients; in this case, empty shells and full capsomers are virtually absent whereas ringlike capsomers are present. (Negative staining in 1% uranyl acetate; magnification × 180,000.)

DNAs from group I DNVs are about 6.0 kb, whereas those from group II DNVs are about 4.9 kb.[45,46] Whereas the terminal palindromes from group I DNV-DNAs are 0.47 to 0.50 kb inverted repeats, those from group II are at most 0.25 kb although little is known about the latter. The inverted repeats of group I DNV-DNAs have a high degree of secondary structure and are difficult to clone without subsequent deletions or to sequence. In our hands,[45] only JC8111 cells could be used although they require special growth requirements and reproduce slowly. These terminal hairpins occur in two alternate sequences (flip and flop) which are reverse complements of each other, indicating a hairpin transfer during DNA reproduction.

An essential difference between group I DNV-DNAs and group II DNV-DNAs is the organization and location of open reading frames (ORFs). The DNAs of group I (studied so far) have the nonstructural (NS) and structural (S) protein ORFs on the 5' halves of the complementary strands. The three NS ORFs on the left half of the "+" strand consist of two ORFs in the same frame (separated by one stop codon) and a third ORF overlapping the second ORF on another frame directly after the stop codon separating the two other NS ORFs (the start of this NS ORF only differs in a few nucleotides for the *Galleria* and the *Junonia* DNVs).[45,47] In contrast, the unique S ORF on the 5' half of the complementary strand ("−") codes for all four structural proteins.

The genome organization of the group II DNV-DNAs seems similar to that of vertebrate parvoviruses. The left ORF specifies one or more NS proteins whereas the right ORF on the same strand specifies the four structural proteins.[46] However, *Bombyx* DNV-1 DNA also has a 501 nucleotide ORF (167 codons) in the left half of the complementary strand. Although an initiation codon is present, striking TATA and CAAT boxes are missing, and no protein has been assigned yet to this ORF.[46] Moreover, ATPase conserved sequences present in the NS ORF of vertebrate parvoviruses and group I DNV-DNAs for *Bombyx* DNV-1 are reported to be present in the S ORF suggesting an alternative splicing mechanism. This point remains to be clarified.

C. STRUCTURAL PROTEINS

Proteins comprise about 65 to 80% of the virus particle. The compact virions can be disrupted with strong denaturants such as SDS.[36] Vertebrate parvoviruses may have one to four structural proteins but DNVs analyzed thus far all possess four structural proteins. The molecular mass of the polypeptides of these viruses, given in Table 4, have been established by SDS-PAGE.[24,28,36,43,48,49] It was noted[36] that the Weber and Osborn method of determining the molecular weight of proteins by continuous SDS-PAGE is not reliable for DNV structural proteins. For instance, the molecular mass of VP1 of *Galleria* DNV was estimated by SDS-PAGE to be about 110 kDa.[43,48] We observed by retardation analysis that it would be 98 kDa,[36] which was subsequently corrected to take into consideration the corrected value of β-galactosidase (sequence analysis showed its molecular weight to be 116,250 instead of 135,000) and estimated to be 92 kDa.[3] Recent completed sequence analysis predicts this protein to be 89 kDa.[45] For the DNVs of group I, VP4 is the major protein (50 to 75% of total mass), but for the DNVs of group II either VP3 or VP4 may be the major protein.

Early studies demonstrated that all structural proteins have common sequences,[34] and recent sequencing results show that they all originate from the same gene and co-terminate at the C end.[45] These proteins are, therefore, specified by a nested set of mRNAs.

Posttranslational modifications of the proteins generated have not been noted. Unpublished experiments demonstrated that no significant carbohydrate moieties are attached.[40] We have not, as yet, investigated the presence of phosphoryl groups in the proteins.

TABLE 4
Molecular Mass of DNV Structural Proteins as
Estimated by SDS-PAGE

Virus	Structural proteins (kDa)				Ref.
	VP1	VP2	VP3	VP4	
Virus Group I					
Galleria DNV	92	67	58.5	49	3, 36
	110	71	61	42	43, 48
Junonia DNV	108	71	61	43	43, 48
Aglais DNV	109	71	59	42	43, 48
Agraulis DNV	110	72	62	43	43, 48
Pseudoplusia	87	64	54	47	28
Virus Group II					
Bombyx DNV-1	77	70	57	50	49
Periplaneta DNV	76	61	52	48	24

IV. MORPHOGENESIS AND MOLECULAR BIOLOGY OF DNV REPLICATION

The morphogenesis of some parvoviruses has been characterized at the microscopic and ultrastructural levels; on the other hand, the molecular biology of these viruses is increasingly understood.

It has been shown unequivocally for vertebrate parvoviruses that they infect mitotic cells (''mitolytic viruses''). Thus, rapidly proliferating cell populations (cerebellar granular cortex, fetal tissues, transformed cells, etc.) are particular targets.[39] Although such an absolute requirement for S-phase transition is also expected for densonucleosis viruses, this has never been firmly proven. In fact, polytropic DNVs seem to infect indiscriminately all tissues, even fat and nervous cells, except for the midgut. Since most enzymes and precursors required for replication are expected to be present in high concentrations only during the S phase, it remains to be established how the polytropic invertebrate parvoviruses have adapted to this situation. Interestingly, the nonstructural gene organization of this subgroup of parvoviruses, as discussed before, is very different from group II DNVs. Particularly interesting is a 30-kDa nonstructural protein (strongly expressed by *in vitro* translation of isolated DNV transcripts),[50] which could be coded by the small overlapping ORF.

In early studies,[51-53] the time course of reproduction of parvoviruses was investigated. In this respect the complete cycle takes about 20 to 24 h for all parvoviruses studied thus far. This time cycle can be divided into three stages: a latent period of 6 h, the appearance of viral protein in the cytoplasm in the following 3 h, and then the appearance of proteins in the nucleus. About 10 to 15 h later, enough particles have been produced to cause cytolysis and the virus to be detectable in the supernatant.

Most viral particles seem to be adsorbed to the cells within 2 h of the addition of the virus.[54] The majority of these particles penetrate into the cell[53] but only become sensitive to DNase when they enter the nucleus. The uncoating of the virus thus occurs concomitantly with nucleus penetration. Interestingly, tropic variants of the same virus, e.g., MVM[55] and PPV[56] all penetrate the cells susceptible for one of the variants. Restriction in tropism then occurs after virus uptake by the cells. This is surprising, since both investigations indicate that three amino acid replacements in the capsid protein are essential. Although these aspects have not been studied for densonucleosis viruses, they probably are very similar.

Ultrastructural changes usually can be observed from about 6 h after infection. In the cytoplasm, a regression of the rough endoplasmic reticulum and the appearance of a large number of dispersed ribosomes in the cytosol are observed.[57,58] This indicates that the synthesis of proteins transported into the endoplasmic reticulum (ER), Golgi, etc. is greatly diminished in favor of synthesis of cytosol or nuclear proteins. The signal to direct viral proteins to the nucleus is conserved among all vertebrate parvoviruses. We have observed[45] that this signal is also conserved for *Galleria* DNV. However, from the sequence of *Bombyx* DNV-1[46] it appears that this virus does not use the same strategy. It should be noted that only the largest structural proteins carry this signal. One way the structural proteins lacking this signal can be transported to the nucleus is their association with proteins having this signal (a minority), e.g., by the formation of procapsids. It is interesting that Lederman et al.[59] observed empty capsid formation in the cytoplasm of bovine parvovirus-infected cells, close to the nuclear pore, suggesting a co-transport. The nucleoli undergo hypertrophy which is increasingly accompanied by a segregation of its fibrillar and granular components (see Figure 2).

This ultrastructural change coincides with the formation of a virogenic stroma and a condensation of the heterochromatin and its displacement toward the periphery of the nucleus. Little is known about the molecular processes going on at this stage. It has been observed that DNA is extruded from the capside although one end remains attached to it.[60] Probably the liberated terminal palindrome of the genome folds back and serves as a primer for DNA synthesis. This is followed by a hairpin transfer and fill-in polymerization to complete the parental strand. The replicative form of DNA thus obtained can then serve for reproduction of both strands (hairpin transfer equally effective at both ends)[61] and transcription.

In group I DNV-DNAs, in which ORFs occupy the 5' halves of the opposite strands, NS and S ORFs are separated by about 30 nucleotides. Both the NS and S ORFs are immediately followed by AATAAA polyadenylation signals. By Northern blotting, several viral poly (A)$^+$ RNA species could be detected and hybrid-selected viral RNA could be translated *in vitro*. Viral specific RNA species had sizes of 1.8, 2.4, 3.5, 4.0, and 5.0 kb, which yielded all structural proteins and at least a very abundant 30-kDa polypeptide.[50] The three longest transcripts exceeded the ORF length and are present in low concentrations. Their role remains unclear; perhaps they may play a role in the operation of DNV gene expression, e.g., by antisense regulation. Since both promoters are at the 5' end in the terminal repeats, no distinction is possible at this level; however, it remains fully possible that enhancers in the unique sequence play a role.

This replication and transcription is accompanied by a rapid hypertrophy of the nucleus and segregation of the nucleoli. These nucleoli are pushed toward the periphery by the invading virogenic stroma and ultimately disappear completely. Nuclear bodies can be observed in the virogenic stroma but their role remains unclear. New viral particles emerge from the virogenic stroma (see Figure 2) or are grouped together in islands; they cause a gradual replacement of the virogenic stroma and hypertrophy by the virus mass until the nuclear membrane ruptures. This allows a passage of the virions into the cytoplasm and, ultimately, cell lysis. During the latter stages of virus assembly, a swelling and degeneration of various organelles, such as the mitochondria and lysosomes, and a gradual development of new structures due to a progressive accumulation of small round particles inside these vesicles can be observed.

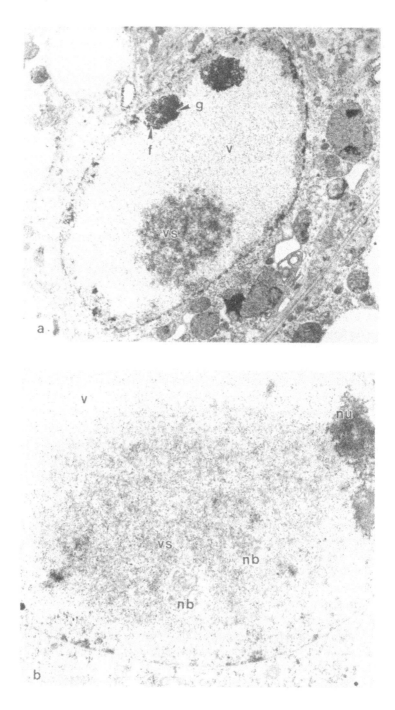

FIGURE 2.   (a) Intermediate stage of infection; the two nucleoli are pushed toward the
periphery and segregate into a fibrillar (f) and granular (g) part; the virogenic stroma (vs)
sheds virus (v) and pushes the heterochromatin also toward the periphery of the nucleus;
cytoplasmic organelles swell and disintegrate. (b) Close-up of virogenic stroma (vs); two
nuclear bodies (nb) can be easily distinguished from the stroma; the heterochromatin is
disintegrating while fibrillar (RF molecules) components are encapsidated.

V. USE OF DENSONUCLEOSIS VIRUSES IN PEST MANAGEMENT

It is known that occluded insect viruses belonging to the nuclear polyhedrosis, cytoplasmic polyhedrosis, and granulosis viruses are efficiently used as biological control agents of insect pests. Nevertheless, the high virulence and infectivity of some nonoccluded viruses such as *Oryctes* baculovirus and *Sibine* DNV may also represent a great asset in biological control.

The efficiency of control of pests is directly related to the mechanisms of transmission of DNVs in insect populations. In the case of polytropic DNVs, such as the one from *Galleria,* the horizontal viral transmission in a population is due to cannibalism of dying larvae; whereas for *Sibine* DNV, replication in the midgut epithelium cells leads to tumorlike lesions and proliferated cells are expelled into the intestine. In this case, the excretion of virus-contaminated cells is the means of virus transmission. The role of parasites of insects in transmission of *Galleria* DNV and *Sibine* DNV has also been demonstrated, thus enhancing the efficiency of insect control.

Despite their high virulence for insects, the use of densonucleosis viruses as pesticides has not yet been investigated in detail, both because of safety considerations and because of the trend of using occluded viruses in biological control experiments. The successful control of at least two insect pests by DNVs were reported.[19] In Colombia, where the leaf-eating larvae of *Sibine fusca* causes extensive damage to the oil palm trees, it was possible to control the insect by aerial spraying suspensions of the *Sibine* DNV-infected larvae at concentrations equivalent to one to five larvae per hectare.[62] Samplings at 15 d postinfection showed a mortality rate varying from 73% at the lowest dose to 97% at the highest dose. The mortality rate reached almost 100% within a month of the application, while parasites of the insects were not affected and contributed to further dissemination of *Sibine* DNV.

In Ivory Coast, an outbreak of the moth, *Casphalia extranea,* causing damage in oil palm and coconut plantations was efficiently controlled resulting in 92% mortality by aerial treatment with an infected larve concentration of 50 to 100 dead caterpillars per hectare.

VI. CONCLUSION

In many respects, DNVs resemble vertebrate parvoviruses. This has hampered early investigations toward their use in pest control. Recent results on the molecular biology of DNVs indicate that they use a different expression strategy and may be less harmful than suspected.

ACKNOWLEDGMENTS

The authors wish to acknowledge the financial support from FCAR (M. A. and P. T.) and NSERC (P. T.) for the research reviewed here.

REFERENCES

1. **Kurstak, E. and Tijssen, P.,** Animal parvoviruses: comparative aspects and diagnosis, in *Comparative Diagnosis of Viral Diseases,* Vol. 3, Kurstak, E. and Kurstak, C., Eds., Academic Press, New York, 1981, 3.
2. **Kurstak, E., Tijssen, P., and Garzon, S.,** Densonucleosis viruses (Parvoviridae), in *The Atlas of Insect and Plant Viruses,* Maramorosch, K., Ed., Academic Press, New York, 1977, 67.

3. **Tijssen, P., Kurstak, E., Su, T. M., and Garzon, S.,** Densonucleosis viruses: unique pathogens of insects, *Proc. 3rd Int. Coll. Invert. Pathol.,* University of Sussex, Brighton, 1982, 148.

4. **Kawase, S., Garzon, S., Su, D.-M., and Tijssen, P.,** Insect parvovirus diseases, in *Handbook of Parvoviruses,* Vol. 2, Tijssen, P., Ed., CRC Press, Boca Raton, FL, 1990, 213.

5. **Arella, M., Garzon, S., Bergeron, J., and Tijssen, P.,** Physicochemical properties, production, and purification of parvoviruses, in *Handbook of Parvoviruses,* Vol. 1, Tijssen, P., Ed., CRC Press, Boca Raton, FL, 1990, 11.

6. **Tijssen, P., Arella, M., and Kawase, S.,** Molecular biology of densonucleosis viruses, in *Handbook of Parvoviruses,* Vol. 1, Tijssen, P., Ed., CRC Press, Boca Raton, FL, 1990, 283.

7. **Lightner, D. V. and Redman, R. M.,** A parvo-like virus disease of penaeid shrimp, *J. Invertebr. Pathol.,* 45, 471, 1985.

8. **Meynardier, G., Vago, C., Plantevine, G., and Atger, P.,** Virose d'un type inhabituelchez le L'épidoptère, *Galleria mellonella, Rev. Zool. Agric. Appl.,* 63, 207, 1964.

9. **Sutter, G. R.,** A nonoccluded virus of the army cutworm, *J. Invertebr. Pathol.,* 21, 62, 1973.

10. **Tinsley, T. W. and Longworth, J. F.,** Parvoviruses, *J. Gen. Virol.,* 20, 7, 1973.

11. **Kelly, D. C., Ayres, M. D., Spencer, L. K., and Rivers, C. F.,** Densonucleosis virus 3: a recent parvovirus isolated from *Agraulis vanillae* (Lepidoptera:Nymphalidae), *Microbiologica,* 3, 455, 1980.

12. **Meynardier, G., Amargier, A., and Genty, Ph.,** Une virose d'une type densonucléose chez le lépidoptère *Sibine fusca* Stoll., *Oleagineux,* 32, 357, 1977.

13. **Shimizu, T.,** Pathogenicity of an infectious flacherie virus of the silkworm, *Bombyx mori,* obtained from sericultural farms in the suburbs of Ina city, *J. Sericult. Sci. Jpn.,* 44, 45, 1975.

14. **Seki, H. and Iwashita, Y.,** Histopathological features and pathogenicity of a densonucleosis virus of the silkworm, *Bombyx mori,* isolated from sericultural farms in Yamanashi prefecture, *J. Sericult. Sci. Jpn.,* 52, 400, 1983.

15. **Iwashita, Y. and Chun, C. Y.,** The development of a densonucleosis virus isolated from silkworm larvae, *Bombyx mori,* of China, in *Ultrastructure and Functioning of Insect Cells,* Akai, H., King, R. C., and Morohoshi, S., Eds., Society of Insect Cells, Tokyo, 1982, 161.

16. **Meynardier, G., Galichet, P. F., Veyrunes, J. C., and Amargier, A.,** Densonucleosis virus in *Diatraea-saccharalis* lepidoptera (Pyralidae), *Entomophaga,* 22, 115, 1977.

17. **Sun, F. L. and Chen, M. S.,** A new insect virus of *Pieris rapae* L. I. Isolation and characterization of the virus, *Acta Microbiol. Sinica,* 21, 41, 1981.

18. **Grignon, N.,** Recherches sur une infection chronique dans une lignée cellulaire de lépidoptère, Thèse, Univ. Sci. Tech. Languedoc, Montpellier, France, cited in Fediere, 1983 (Reference 19), 1982.

19. **Fediere, G.,** Recherche sur des viroses epizootiques de lépidoptères Limacodidae ravageurs de palmacées, Thèse, Univ. Sci. Tech. Languedoc, Montpellier, France, 1983.

20. **Chao, Y.-C., Young, S. Y., III, Kim, K. S., and Scott, H. A.,** A newly isolated densonucleosis virus from *Pseudoplusia-includens* lepidoptera (Noctuidae), *J. Invertebr. Pathol.,* 46, 70, 1985.

21. **Federici, B. A. and Lacey, L. A.,** Intranuclear disease of uncertain etiology in larvae of the blackfly *Simulium-vittatum, J. Invertebr. Pathol.,* 50, 184, 1987.

22. **Lebedeva, O. P., Kuznetsova, M. A., Zelenko, A. P., and Gudz-Gorban, A. P.,** Investigation of a virus disease of the densonucleosis type in a laboratory culture of *Aedes aegypti, Acta Virol.,* 17, 253, 1973.

23. **Gorziglia, M., Botero, L., Gil, F., and Esparza, J.,** Preliminary characterization of virus-like particles in a mosquito *Aedes-pseudoscutellaris* cell line mos-61, *Intervirology,* 13, 232, 1980.

24. **Suto, C.,** Characterization of a virus newly isolated from the smokey-brown cockroach *Periplaneta-fuliginosa, Nagoya J. Med. Sci.,* 42, 13, 1979.

25. **Meyardier, G., Matz, G., Veyrunes, J.-C., and Bres, N.,** Densonucleosis type virus in Orthoptera, *Ann. Soc. Entomol. Fr.,* 13, 487, 1977.

26. **Charpentier, R.,** A nonoccluded virus in nymphs of the dragonfly *Leu corrhinia-dubia* (Odonata, Anisoptera, *J. Invertebr. Pathol.,* 34, 95, 1979.

27. **Charpentier, G., Back, C., Garzon, S., and Strykowski, H.,** Observations on a new intranuclear virus-like particle infecting larvae of a blackfly *Simulium vittatum* (Diptera, Simuliidae), *Dis. Aquat. Org.,* 1, 147, 1986.

28. **Matsui, M.,** Smaller virus particles than the flacherie virus found in the diseased silkworm larvae, *Bombyx mori* L., *Jpn. J. Appl. Entomol. Zool.,* 17, 113, 1973.

29. **Furuta, Y.,** Studies on the previously unreported virus infecting the silkworm, *Bombyx mori* L., *J. Sericult. Sci. Jpn.,* 42, 443, 1973.

30. **Shimizu, T. and Watanabe, H.,** Failure to harvest cocoons attributed to an epizootic of densonucleosis caused by a new strain of the virus in sericultural farms, *J. Sericult. Sci. Jpn.,* 53, 436, 1984.

31. **Maeda, S.,** personal communication.

32. **Siegl, G., Bates, R., Berns, K. I., Carter, B. J., Kelly, D. C., Kurstak, E., and Tattersall, P.,** Characteristics and taxonomy of Parvoviridae, *Intervirology,* 23, 61, 1985.

33. **Siegl, G., Bates, R., Berns, K. I., Carter, B. J., Cotmore, S., Lederman, M., Tal, J., Tattersall, P., and Tijssen, P.**, Characteristics and taxonomy of Parvoviridae, *Intervirology*, 23, 61, 1985.
34. **Tijssen, P. and Kurstak, E.**, Biochemistry, biophysical and biological properties of densonucleosis virus (Parvovirus). III. Common sequences of structural proteins, *J. Virol.*, 37, 17, 1981.
35. **Tijssen, P., Tijssen-van der Slikke, T., and Kurstak, E.**, Biochemical, biophysical and biological properties of densonucleosis virus (Parvovirus). II. Two types of infectious virions, *J. Virol.*, 21, 225, 1981.
36. **Tijssen, P., van den Hurk, J., and Kurstak, E.**, Biochemical, biophysical and biological properties of densonucleosis virus (Parvovirus). I. Structural proteins, *J. Virol.*, 17, 686, 1981.
37. **Kelly, D. C. and Elliott, R. M.**, Polyamines contained by two densonucleosis viruses, *J. Virol.*, 21, 408, 1977.
38. **Bando, H., Nakagaki, M., and Kawase, S.**, Polyamines in densonucleosis virus from the silkworm, *Bombyx mori*, *J. Invertebr. Pathol.*, 42, 264, 1983.
39. **Cotmore, S. F. and Tattersall, P.**, The autonomously replicating parvoviruses of vertebrates, *Adv. Virus Res.*, 33, 91, 1987.
40. **Tijssen, P.**, unpublished results, 1990.
41. **Tijssen, P. and Kurstak, E.**, Studies on the structure of the two infectious types of densonucleosis virus, *Intervirology*, 11, 261, 1979.
42. **Caloin, M.**, Determination de la composition et du comportement hydrodynamique du densonucleoside virus, *J. Chim. Phys.*, 75, 978, 1978.
43. **Kelly, D. C., Moore, N. F., Spilling, C. R., Barwise, A. H., and Walker, I. O.**, Densonucleosis virus structural proteins, *J. Virol.*, 36, 224, 1980.
44. **Farrell, J. A., Harvey, J. D., and Bellamy, A. R.**, Biophysical studies of reovirus type 3. I. The molecular weights of reovirus and reovirus cores, *Virology*, 62, 145, 1974.
45. **Tijssen, P.**, Nucleotide sequence and organization of genome of *Galleria mellonella* densonucleosis virus (GmDNV), *EMBO Workshop, Molecular Biology of Parvoviruses*, Ma'ale Hachmisha, Israel, 1989, 28.
46. **Bando, H., Kusuda, J., Gojobori, T., Maruyama, T., and Kawase, S.**, Organization and nucleotide sequence of a densovirus genome imply a host-dependent evolution of parvoviruses, *J. Virol.*, 61, 553, 1987.
47. **Bergoin, M., Jourdan, M., Gervais, M., Jousset, F. X., Skory, S., and Dumas, B.**, Molecular cloning, nucleotide sequence and organization of an infectious genome of the *Junonia coenia* densovirus (JcDNV), *EMBO Workshop, Molecular Biology of Parvoviruses*, Ma'ale Hachmisha, Israel, 1989, 58.
48. **Kelly, D. C., Moore, N. F., Spilling, C. R., Barwise, A. H., and Walker, I. O.**, Densonucleosis virus structural proteins, ERRATA, *J. Virol.*, 38, 1104, 1981.
49. **Nakagaki, M. and Kawase, S.**, DNA of a new parvo-like virus isolated from the silkworm, *Bombyx mori*, *J. Invertebr. Pathol.*, 35, 124, 1980.
50. **Gros, O., Tijssen, P., and Tal, J.**, Expression of densonucleosis virus GmDNV in *Galleria mellonella* larvae: size analysis and *in vitro* translation of viral transcription products, *J. Invertebr. Pathol.*, in press.
51. **Rose, J. A.**, Parvovirus reproduction, *Comprehensive Virol.*, 3, 1, 1974.
52. **Jay, F. T., Laughlin, C. A., de la Maza, L. M., and Carter, B. J.**, Adeno-associated virus replication, in *Replication of Mammalian Parvoviruses*, Ward, D. C. and Tattersall, P., Eds., Cold Spring Harbor Press, Cold Spring Harbor, NY, 1978, 133.
53. **Berns, K. I. and Hauswirth, W. W.**, *Adv. Virus Res.*, 25, 407, 1979.
54. **Rose, J. A. and Koczot, F. J.**, Adenovirus-associated virus multiplication. VII. Helper requirement for viral deoxyribonucleic acid and ribonucleic acid synthesis, *J. Virol.*, 10, 1, 1972.
55. **Ball-Goodrich, L., Moir, R., Cotmore, S. F., and Tattersall, P.**, The fibrotropic determinant of MVMp is encoded within the capsid gene, and is located on the surface of the virion, *EMBO Workshop, Molecular Biology of Parvoviruses*, Ma'ale Hachmisha, Israel, 1989, 24.
56. **Bergeron, J. and Tijssen, P.**, Nucleotide sequence and organization of genome of the NADL-2 strain of PPV, *EMBO Workshop, Molecular Biology of Parvoviruses*, Ma'ale Hachmisha, Israel, 1989, 30.
57. **Singer, I. I. and Toolan, H. W.**, Ultrastructural studies of H-1 parvovirus replication. I. Cytopathology produced in human NB epithelial cells and hamster embryo fibroblasts, *Virology*, 65, 40, 1975.
58. **Garzon, S. and Kurstak, E.**, Ultrastructural studies on the morphogenesis of densonucleosis virus (Parvovirus), *Virology*, 70, 517, 1976.
59. **Lederman, M., Chen, K. C., Stout, E. R., and Bates, R. C.**, Possible sequences for nuclear accumulation of parvoviral proteins, *Cell. Biol. Int. Rep.*, 10, 383, 1986.
60. **Bourguignon, G. J., Tattersall, P., and Ward, D. C.**, DNA of minute virus of mice: self-priming, non-permuted, single-stranded genome with 5' terminal hairpin duplex, *J. Virol.*, 20, 290, 1976.
61. **Chen, K. C., Tyson, J. J., Lederman, M., Stout, E. R., and Bates, R. C.**, A kinetic hairpin transfer model for parvoviral DNA replication, *J. Mol. Biol.*, 208, 283, 1989.
62. **Genty, P. and Marieu, D.**, Utilisation d'une germe entomopathogène dans la lutte contre *Sibine fusca* Stoll (Limacodidae), *Oleagineux*, 30, 349, 1975.

Chapter 5

IRIDOVIRIDAE

Darrell W. Anthony (Retired) and Michel Comps

TABLE OF CONTENTS

I. IRIDOVIRUSES OF INSECTS

A. VIRUSES PRODUCING IRIDESCENCE IN INSECTS
1. Introduction

The iridescent viruses were so named became of the optical properties they exhibit when patently infected specimens are viewed by reflected light or from the appearance of the virus pellet after purification. Williams and Smith[1] were the first to apply the name *Tipula* iridescent virus (TIV) to a disease of the crane fly, *Tipula palidosa,* reported earlier by Xeros.[2] Since that time, similar iridescent viruses have been recorded from a wide variety of insects and a few crustacea. The virus particles are icosahedral (20 sided), contain DNA, and they replicate in the cytoplasm of the host cell. Large numbers of the virus particles accumulate in paracrystalline arrays, and the entire cytoplasm of infected cells may be filled. The intensity and color of iridescence is a function of size, shape, and spacing of the virus particles in these crystalline arrays.[3] It is clear that similarities exist between the iridescent insect viruses and morphologically similar viruses in amphibians, reptiles, mammals, fungi, and worms. Stoltz[4] points out that they are similar in gross morphology, nucleic acid (DNA), and site of replication (cytoplasm). He therefore suggested that icosahedral cytoplasmic deoxyriboviruses (ICDV) might be used as common terminology for the entire group, especially since not all ICDV infections produce discernable iridescence in their hosts. This presentation is concerned primarily with those viruses producing iridescence in insects.

2. Classification

By 1970 the number of iridescent virus isolates from a wide variety of insect hosts was becoming so large and confusing that Tinsley and Kelly[5] suggested an interim system of nomenclature modeled on that used for the adenoviruses. The first iridescent virus from *Tipula palidosa* (TIV) became Type 1, *Sericesthis* iridescent virus (SIV) became Type 2, mosquito iridescent virus (MIV) from *Aedes taeniorhynchus* became Type 3, and so on.

In a recent review of the invertebrate icosahedral cytoplasmic deoxyriboviruses, Hall[6] has provided an extended list of the types of iridescent viruses 1 through 32. Hall[6] also lists 23 additional iridescent virus isolates that have not been assigned type numbers.

In the first report of the International Committee on the Taxonomy of Viruses, Wildy[7] included in the genus *Iridovirus* the iridescent viruses from insects and the ICDVs from vertebrates. Later, Fenner[8] proposed family status for these viruses (Iridoviridae) with a single genus *(Iridovirus)* which was reserved for the iridoviruses (IVs) from insects. Matthews,[9] in the Committee's fourth report, separated the insect group further. The genus *Iridovirus* was reserved for the smaller viruses (particles in the 120 nm range that produced a blue iridescence in infected hosts and purified pellets, e.g., TIV), while the larger viruses (particles of 180 nm that produced a yellowish green or an orange iridescence) were assigned to the genus *Chloridovirus.* An example of the latter is the MIV regular strain from *Aedes taeniorhynchus.* The frog virus (FV) group was assigned to the genus *Ranovirus,* which includes the most thoroughly studied member of the group, frog virus 3 (FV3), and others of the FV group. Genera were not assigned to the African swine fever (ASF) group or the lymphocystis disease group.

B. VIRUS PARTICLE STRUCTURE

The structure of several IVs has been studied extensively.[1,3,4,10-13] In thin sections and in air-dried preparations, the virus particles appear hexagonal or pentagonal in shape. Electron microscope studies have shown that the particle consists of a membrane modified by the apposition of an icosahedral lattice of morphological subunits and an inner electron dense nucleoprotein core. The icosahedral symmetry of the virion was demonstrated by a double shadowing technique.[1] Stoltz[4] reported the presence of two unit membranes in regular MIV (RMIV) particles, but their presence could only be inferred in TIV and *Chironomus* ICDV.

In later studies comparing turquoise strain of MIV (TMIV) with *Corethrella* iridescent virus (CoIV), and frog virus 3 (FV3) Stoltz[10] concluded there is probably only a single unit membrane.

Membranes are best demonstrated by negative staining of partially disrupted virus particles.[4,10] However, negative stained preparations of purified virus particles also give an indication as to the presence and location of the membrane. Figure 1 is a negative stained preparation of the iridescent virus from *Heliothis zea*. Wrigley[12,13] demonstrated the presence of a surface lattice containing morphological subunits by chemical dissociation of the particle into triangular, pentagonal, and linear fragments. The pentagonal-shaped fragments were believed to have been derived from an icosahedral surface. He suggested that SIV probably had 1562 subunits, although 1292 and 1472 could not be ruled out, and TIV probably had 1472 subunits, but 1292 and 1562 could not be ruled out. Stoltz[10] found that dissociation of RMIV, TMIV, CoIV, *Chironomus* ICDV, and FV3 also produced similar fragments and proposed at least 1560 subunits for the *Chironomus* ICDV. Freeze etch preparations of TMIV (see Figure 2) indicate the presence of the subunits on the surface of virus particles, but an estimate of the number on each surface could not be made by this method.

Tubular or cylindrical structures have been associated with several viruses.[10,14-16] Darlington et al.[14] considered rod-shaped structures associated with a frog virus to be aberrant and bizarre viral forms; however, Hitchborn and Hills[15] showed that infection of the turnip yellow mosaic virus in plants led to the production of tubes as well as virus particles. Stoltz[10] reported that cylindrical or tubular forms were found in both R and T strains of MIV, in CoIV, and in FV3. Anthony and Hall[16] compared the diameters of the cylindrical forms of RMIV and TMIV and found a definite size relationship between these structures and their respective virus particles. Cylindrical forms are shown in longitudinal and cross sections (see Figure 3). Diameters of the cylinders were slightly smaller than those of the particles, and variable in length.[16] Some appeared to be capped on one or both ends by typical virus particles, while other particles in the vicinity of the tubes appeared to be incomplete and without cores. The significance of these structures is not known; however, there is little doubt that they are associated with the virus.

C. BIOCHEMICAL COMPOSITION OF GENOME

The chemical and physical properties of *Iridovirus* members TIV, SIV, and CIV have been studied extensively,[17-24] but only the MIVs of the genus *Chloridovirus* have received significant attention.[25-29] In both genera, the virus contains double-stranded DNA which appears to be linear. Estimates of the DNA content vary between 10.5 and 19% depending on the individual reports and the virus studied.[20,22,25,26,28] The difference in size of *Iridovirus* members (120 to 130 nm) and those of *Chloridovirus* (165 to 200 nm) probably accounts for some of the significant differences in particle and DNA weight. Particle weights for TIV and SIV (*Iridovirus*) have been calculated at 11.9 to 13.0 $\times$ 10^8 and 5.51 $\times$ 10^8 Da,[19,21,22] while particle weights of RMIV and TMIV (*Chloridovirus*) have been reported as 2.5 to 2.75 $\times$ 10^9 and 2.10 $\times$ 10^9 Da.[26,28] DNA molecular weights of the *Iridovirus* group ranges from 114 to 160 $\times$ 10^6 while DNA molecular weights of RMIV and TMIV have been calculated at 4.64 $\times$ 10^8 and 2.1 $\times$ 10^8, respectively. Considering the DNA molecular weights of RMIV and TMIV, these genomes are greater by two- to threefold than that reported for other iridescent viruses, and Wagner et al.[28] question whether there might be more than a single molecule of DNA per virion. Protein analysis of both *Iridovirus* and *Chloridovirus* show a range of 13 to 30 structural polypeptides with molecular weights ranging from 10 to 250 $\times$ 10^3. Both genera contain small amounts of lipid (5 to 10%); however, both genera are ether resistant.

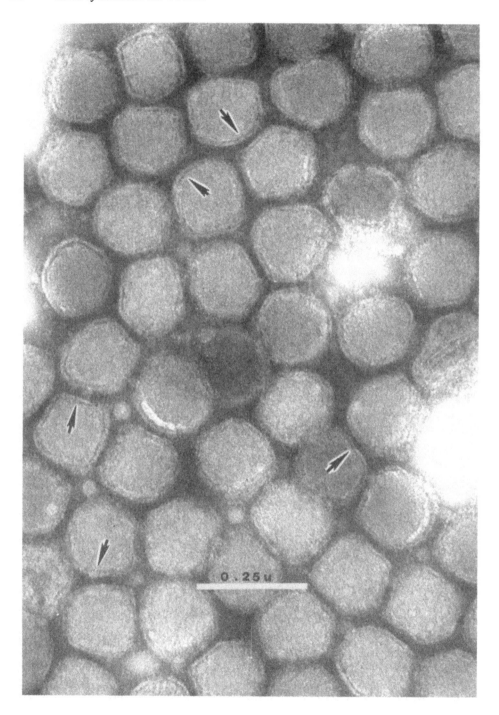

FIGURE 1.   Iridescent virus from *Heliothis zea*. Negative stained (1% ammonium molybdate) preparation from purified virus pellet. Arrows designate particles with discernible membranes. (Photograph courtesy of Dr. Jean Adams, U.S. Department of Agriculture, ARS, Insect Pathology Laboratory, Beltsville Agricultural Research Center, Beltsville, MD.)

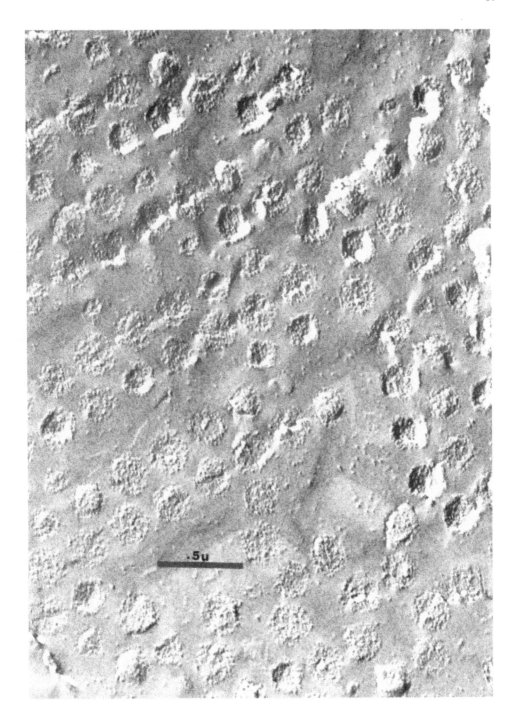

FIGURE 2. Freeze etch preparation of TMIV in *Aedes taeniorhynchus*. Subunits visible on surface of virus particles. (Preparation made in cooperation with the Central Electron Microscope Facility, IFAS, University of Florida, Gainesville, FL, Dr. Henry Aldrich, Director.)

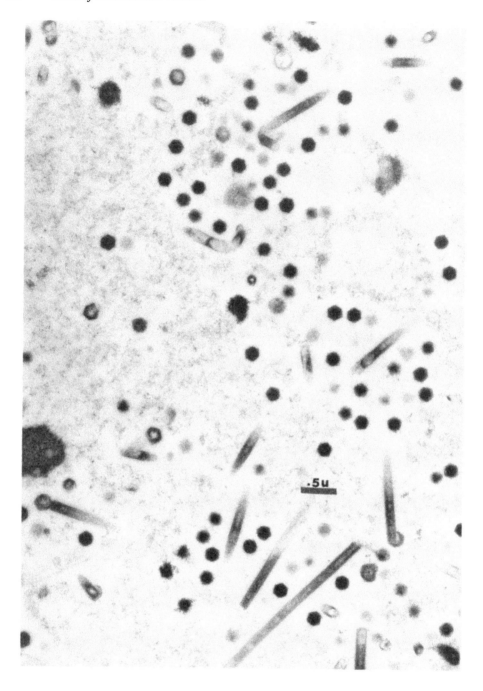

FIGURE 3. Cylindrical and rod-shaped structures associated with TMIV in remnant of fat body cell in *Aedes taeniorhynchus*.

D. HOST RANGE, SPECIFICITY, AND TRANSMISSION

In nature, the transmission of most insect viruses is usually thought to be by mouth *(per os)*. However, in the case of the iridescent viruses, field observations have shown very low infection rates in natural populations, and feeding experiments in the laboratory have also resulted in very low rates of transmission or in some cases failure to transmit the virus.[30-34] TIV, SIV, CIV, and others can easily be transmitted over a wide host range by intrahemocoelic injection of virus suspensions. Smith et al.[35] listed 7 species of Diptera, 11 species of Lepidoptera, and 3 species of Coleoptera experimentally infected with TIV. Fukuda[36] obtained p.o. transmission of CIV (maintained in larvae of *Galleria mellonella*) to 13 species of mosquitos; however, all infection rates were very low. Day and Mercer[19] also reported the transmission of SIV to the mosquito, *Aedes aegypti*.

Iridescent viruses from mosquitos appear to be much more specific. *Per os* transmissions of MIV from *A. taeniorhynchus* to other *Aedes* species have been successful, although rates of infection were very low. Similarly, MIV from *Psorophora ferox* was successfully transmitted to other *Psorophora* species, and cross transmitted to *A. taeniorhynchus*.[37] Attempts to transmit RMIV from *A. taeniorhynchus* to the corn earworm *(Heliothis zea)*, the cabbage looper *(Trichoplusia ni)*, and greater wax moth *(Galleria mellonella)* by feeding and intrahemocoelic injection were unsuccessful.[38] The iridescent virus from the mosquito predator *Corethrella brakeleyi* was transmissible to *C. appendiculata*; however, early instar larvae of *A. taeniorhynchus*, *A. sollicitans*, and *P. ferox* were not susceptible.[39]

An iridescent virus from *Heliothis zea* reported by Stadelbacher et al.[40] is interesting in that it was found in conjuction with an unusually high incidence of parasitization by a mermithid nematode *(Hexamermis* sp.). All attempts at peroral transmission of this virus resulted in failure, although injections of virus suspensions into the body cavity successfully induced disease. Since parasitic nematodes emerged from some of the larvae that developed the iridescent virus and there was a high rate of parasitism in the population, the authors suggest that infection could be carried by nematodes when they invade the body cavity.

Vertical, or transovarial, transmission is another method of dissemination whereby iridescent viruses may be maintained in natural populations. This applies especially to the MIVs. Transovarial transmission of MIV has been demonstrated by several workers.[31,41-44] Early instar mosquito larvae exposed to aqueous virus suspensions usually results in infection rates of 5 to 20%, and patent infections appear in late third or fourth instar larvae. Virtually all patently infected larvae die prior to pupation, although a few may survive until the pupal stage. Only female larvae exposed to the virus during late instars transmit it to their progeny. The percentage of females that transmit the virus transovarially has been found to be 15% or less;[31,42] however, this does provide a mechanism for the perpetuation of the virus in nature. Linley and Nielson[32] proposed a tentative cycle of natural transmission where transovarial transmission produces infected larvae which die in the fourth instar. These larvae provide a source of new infection when healthy late instar larvae feed on the diseased cadavers. This in turn leads to the presence of infected adults, which complete the cycle by depositing infected eggs. Larvae from MIV-infected eggs of *A. taeniorhynchus* and *P. ferox* stored for 26 weeks became patently infected.[41] This time interval is believed to be sufficient for survival of the disease during long droughts and periods when eggs are in diapause.

Thirteen species of mosquitoes that are hosts to MIV belong to the floodwater genera *Aedes* and *Psorophora*. Two species that breed in permanent water, *Culiseta annulata* and *Culex territans*, were reported as hosts of MIV by Butchatsky.[45] The infected larvae were collected from a small forest pool near Kiev, U.S.S.R., in July 1973. During the previous April, the same site yielded an iridescent virus from larvae of *Aedes cantans*. It is of interest that Fukuda[36] obtained very low rates of infection by CIV to *Culiseta inornata*, *C. melanura*, *Culex salinarius*, and *C. territans* in peroral transmission trials. Many attempts in the laboratory to infect these species with MIV resulted in failure.

It is evident that vertical transmission plays an important part in the perpetuation and dissemination of MIVs in nature. Hall[6] suggests that vertical (transovarial) transmission may be the major epizootiological cycle while horizontal *(per os)* transmission which occurs when early instar larvae feed on decomposing infected larvae may be only a minor cycle. The importance of vertical transmission regarding other IVs is not known; however, Hall[6] reported that vertical transmission of the TIV by field-collected females of *Tipula paludosa* has been demonstrated. It seems probable that further research may show that many of the IVs may be vertically transmitted, thus explaining, at least in part, the very low infection rates reported by many workers in p.o. transmission experiments.

E. GROSS PATHOLOGY

Under natural conditions, IVs produce a fatal disease in the immature stages of their hosts. In the case of holometabolous insects, the disease usually becomes apparent during the late larval instars, and death ensues prior to pupation. Virtually all patently infected specimens die. There are no visible signs of infection during the early stages, and larvae appear and act normal until the time that the disease is recognized by the distinctive iridescence that is characteristic for that species. TIV has been described as an "opalescent blue indigo," and shades of blue to violet and lavender to greenish blue have been used to describe the appearance of many IVs. Regular MIV (RMIV) from *Aedes taeniorhynchus* produces an iridescence that ranges from orange to brown, orange being the predominant shade. Turquoise MIV (TMIV) also from *A. taeniorhynchus* was so named because of its distinctive color. Most other MIVs have been described as producing "bluish" iridescence.

As infection by most IVs progresses, the iridescence becomes brighter and more extensive; however, as death nears infected specimens may appear somewhat swollen, become moribund, and the bright iridescence may noticeably fade. It is not unusual to find MIV-infected larvae that show little or no iridescence and appear nearly white at the time of death.[30]

There may be many ICDVs that are structurally and biochemically related to the IVs and that cause disease in their hosts, but do not produce iridescence. For example, the *Chironomus* ICDV (from *Chironomus plumosus*) shows structural features similar to TIV, but it does not form the crystals necessary for iridescence.[4] Similar ICDVs have been found in other members of the Chironomidae including *Chironomus attenuatus* and *Goeldichironomus holoprasinus*.[38] Characteristic iridescence was absent in these species, and the only indication of infection was markedly reduced activity.

F. CYTOPATHOLOGY
1. Fat Body

During the late stages of the disease, most tissues of the infected host contain virus particles. The fat body is a primary site of replication for virtually all types of IVs, and the paracrystalline arrays of virus formed therein impart the characteristic iridescence. Figure 4 shows parts of two infected fat body cells of *Heliothis zea*. A few lipid droplets identify the cells, but they are almost devoid of normal cytoplasmic organelles. The fat body is an important storage depot for protein and glycogen as well as fat, and these reserves are especially important during pupation and subsequent development to adulthood. There can be little doubt that such intensive viral infections disrupt the metabolic processes of the fat body, thereby preventing most infected larvae from completing development.

2. Epidermis

The epidermis, imaginal bud tissue, and tracheal epithelium are also common sites of viral replication. Figure 5 shows MIV infection in the epidermis of *Psorophora varipes*. Hall and Anthony[42] report that the extent of infection of epidermis was variable within each individual mosquito; however, as shown in Figure 5, many cells may be heavily infected.

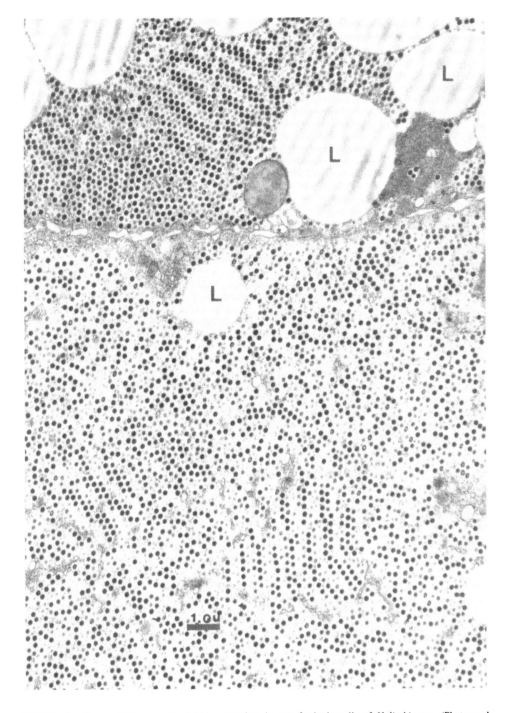

FIGURE 4. Paracrystalline arrays of iridescent virus in two fat body cells of *Heliothis zea*. (Photograph courtesy of Dr. Jean Adams, U.S. Department of Agriculture, ARS, Insect Pathology Laboratory, Beltsville Agricultural Research Center, Beltsville, MD.)

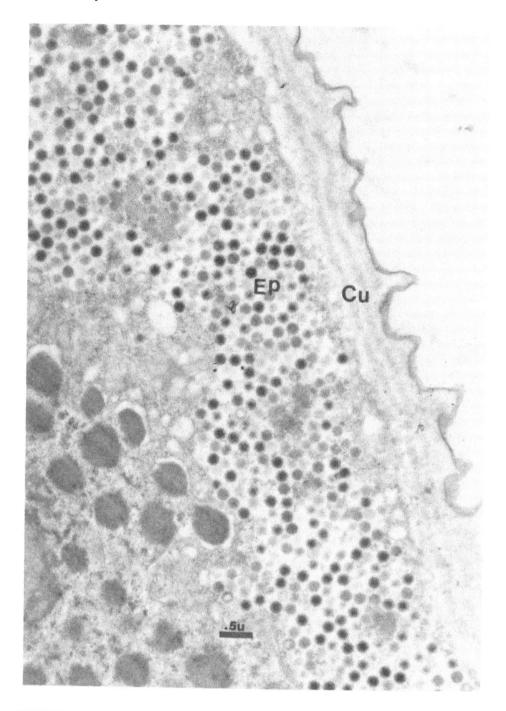

FIGURE 5.   MIV infection in epidermis of *Psorophora varipes*. Cu = cuticle, Ep = epidermis. (Infected material supplied by Dr. H. C. Chapman [retired], Gulf Coast Mosquito Research Laboratory, U.S. Department of Agriculture, ARS, Lake Charles, LA.)

Imaginal bud tissue, which is also of epidermal origin, was infected with both RMIV and TMIV in virtually all specimens examined. Figure 6 shows two thoracic imaginal bud cells with virus particles packed close to the nuclei. Many of these cells become greatly infected as the disease advances, and it is doubtful appendages could develop from these cells even if the larvae could have completed development.[42] Tracheal epithelium, which is also composed of modified epidermal tissue, was usually heavily infected. Figure 7 shows an infection of RMIV in *Aedes taeniorhynchus*.

3. Nerves

Visceral nerves were lightly infected with both strains of MIV[16,42] and in CoIV.[39] As shown in Figure 8, it is significant that these infections are usually very light and show little or no evidence of cytopathology. Infections of the abdominal or thoracic ganglia have not been observed, and this may account for the ability of most infected larvae to swim normally until shortly before death.

4. Hemocytes

Virus particles were frequently observed in the hemocytes of mosquitoes infected with MIV. This is not surprising because phagocytosis of pathogens by hemocytes is a primary defensive response in insects. Figure 9 shows a hemocyte containing MIV from *Aedes vexans*. The virus particles appear to be enclosed in vacuoles, and single particles appear to have a halo. It is possible that these were phagocytic vacuoles, but there is no evidence of limiting membranes that frequently occur when foreign substances are phagocytized. Since hemocytes move freely in the hemolymph, they may serve to spread the infection throughout the body of the host.

5. Alimentary Tract

Very low rates of infection by p.o. transmission trials is characteristic of MIV infections, and other iridescent viruses may also be difficult to transmit by feeding.[40] Extensive examinations by both light and electron microscopy were made in an attempt to identify tissues in larvae of *Aedes taeniorhynchus* that were susceptible to infection by RMIV.[16,42,46] These examinations failed to show the presence of infections in the foregut, midgut, Malpighian tubules, or hindgut. The only established infection observed was in tissue at the esophageal invagination where the esophagus enters the foregut.[42] In a later study,[47] a few virus particles were found in midgut microvilli inside the peritrophic membrane and within one midgut epithelial cell. The cuticular linings of the foregut and hindgut and the peritrophic membrane of the midgut appear to act as mechanical barriers to infection. Abrasions or discontinuities of these structures could provide a possible mode of entry for the virus, especially during the molting process when new cuticular linings are soft, and a new peritrophic membrane is being formed.

Penetration of the migut by naked DNA is another possible route of entry for the virus. Uncoating of the DNA could occur in the presence of the proper proteolytic enzymes or high pH, and the naked DNA would be of a size that could readily pass through the peritrophic membrane. Specific cellular receptor sites on the cells of the midgut would not be required for the penetration of the viral DNA. The mechanics of p.o. infection of MIV is still unknown; however, these studies indicate that low rates of transmission might be expected.

6. Reproductive Organs

The significance of MIV infection in gonadal tissue is dependent on when the larvae became infected. Ovarian tissue of fourth instar larvae of *A. taeniorhynchus* (infected during the first instar) were highly susceptible and showed varying degrees of infection.[16,42,47] No evidence of infection was found in larval testes by Hall and Anthony,[42] but Anderson[47]

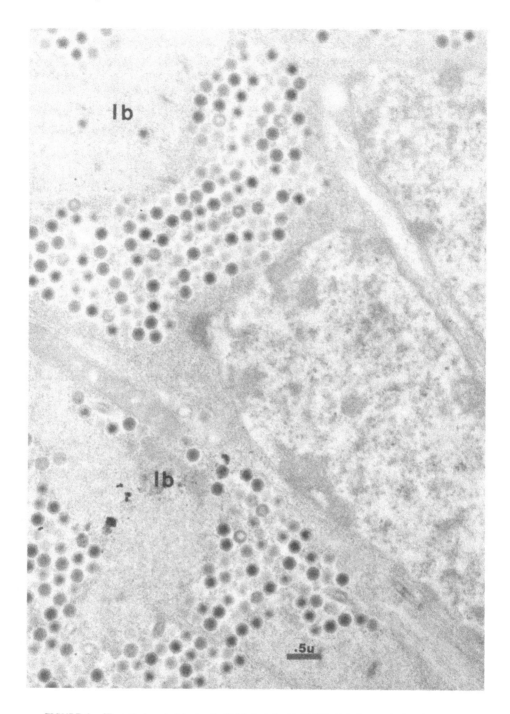

FIGURE 6.   Thoracic imaginal bud cells (Ib) infected with TMIV of *Aedes taeniorhynchus*.

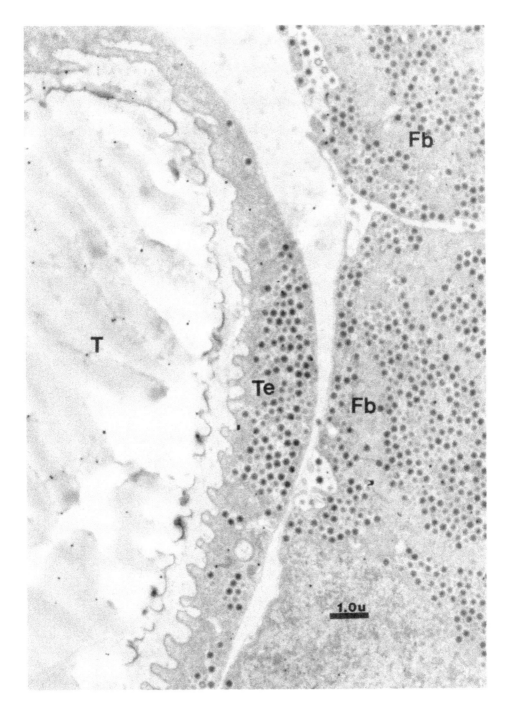

FIGURE 7. Tracheal epithelium of *Aedes taeniorhynchus* infected with RMIV. Trachea (T), tracheal epithelium (Te), fat body (Fb).

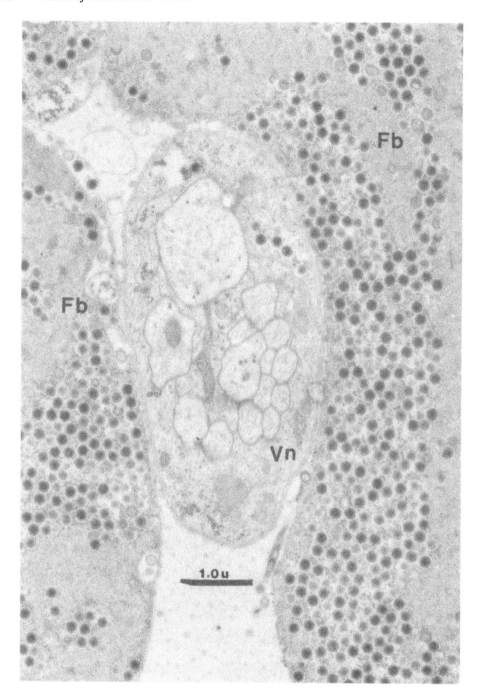

FIGURE 8.   Visceral nerve with a few MIV particles. Nerve located between two heavily infected fat body cells of *Aedes taeniorhynchus*. Visceral nerve (Vn), fat body (Fb). (Infected material supplied by Dr. H. C. Chapman [retired], Gulf Coast Mosquito Research Laboratory, U.S. Department of Agriculture, ARS, Lake Charles, LA.)

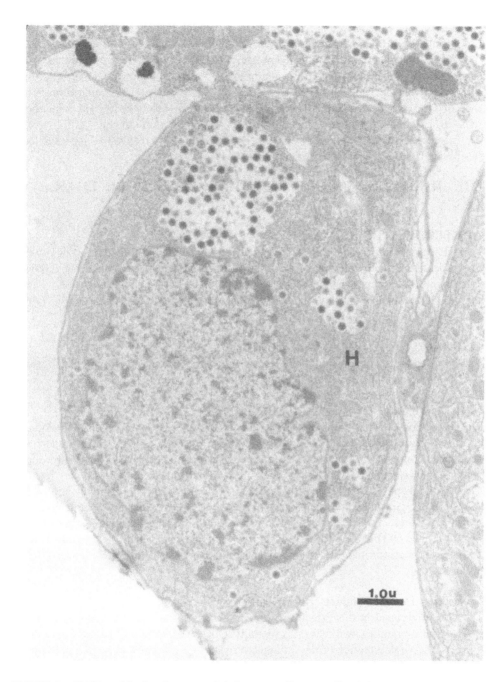

FIGURE 9.   RMIV particles in a hemocyte of *Aedes vexans*. Hemocyte (H). (Infected material supplied by Dr. H. C. Chapman [retired], Gulf Coast Mosquito Research Laboratory, U.S. Department of Agriculture, ARS, Lake Charles, LA.)

reported that a few cells of the larval testes of *A. stimulans* became infected. Electron microscope studies of ovaries from an adult *Aedes taeniorhynchus* exposed to virus as a 3-day-old larva, which had produced infected progeny, revealed numerous virus particles. The follicular epithelium and nurse cells were the primary sites of infection. In some cells, the virus particles were packed in groups suggesting an active site of viral replication. The presence of virus particles in follicular cells and their nurse cells supports the contention that MIV is transmitted by infected females to their progeny within the egg.[31,37,42,44]

For more detailed information on specific IVs, the reader is referred to the excellent reviews by Hall[6] and Tinsley and Harrap.[48]

II. IRIDOVIRUSES OF INVERTEBRATES OTHER THAN INSECTS

A. MEMBERS OF THE GROUP

Several icosahedral cytoplasmic deoxyriboviruses generally proposed as iridoviruses have been described in invertebrates other than insects. The main cases currently known concern the mollusks, crustaceans, and worms, the infected hosts coming from continental and marine environments. These virus diseases and their abbreviations are listed in Table 1.

B. SYMPTOMS

According to the gross signs observed during the infections caused by these pathogenic agents, two groups of viruses can be distinguished.

1. Iridescent Viruses

The characteristic iridescence observed in insects is also observed in other invertebrates because of the paracrystalline array of the virions. Devauchelle and Durchon[68] report an iridescent appearance of the Polychete *Nereis diversicolor* infected by NIV. The infections of *Armadillidium vulgare* by AIV (Iridovirus, Type 31) and of *Porcellio dilatatus* by the PIV (Iridovirus, Type 32) are associated with a high mortality rate.[71] Moreover, a blue to purple color of the cuticle and the appearance of blue-green flecks beneath some membranes are observed in these Isopods.[71] The phenomenon of opalescence, a unique sign of the infection described by Federici and Hazard[70] of the Daphnid, *Simocephalus expinosus*, could be due to a peculiar form of iridescence induced by the presence of SIV (Iridovirus, Type 20) in the tissues.

2. Viruses that Do Not Produce Iridescence

The viruses grouped in this category cause various characteristic symptoms associated sometimes with mortalities or direct mass mortalities of the affected animals without other gross signs. The virus infection of the Portuguese oyster, *Crassostrea angulata*, called gill disease is characterized by an evolutive ulceration of the gills (see Figure 10A) and the labial palps which can spread and affect the major part of the gill.[61,62] At this stage, about 30% of the oysters died in some production areas.[64]

The clinical signs associated with the oyster velar virus disease affecting the larval *Crassostrea gigas* reared in hatcheries constitute a more complex syndrome. The infected larvae exhibit a retraction of the visceral mass and a disturbed mobility, according to Elston and Wilkinson,[66] by the loss of velar cells and by the deciliation of the epithelium.

A third pattern of symptoms has been observed in specimens of *Octopus vulgaris* from the bay of Naples. In this case, the infected animals exhibit an edematous nodular tumor in the muscle tissue of the tentacles; these lesions can also spread to the siphon and the ventral surfaces of the mantle.[69] Signs of apathy and self-multilation are observed in the most affected animals which finally die.

TABLE 1
Viral Infections of Mollusks, Crustaceans, and Worms

Virus name	Particle size (nm)	Host	Ref.
Hemocytic infection virus (HIV)	380	*Crassostrea angula, Crassostrea gigas*	58, 59, 60
Gill necrosis virus (GNV)	380	*Crassostrea angulata*	58, 61—64
Oyster velar virus (OVV)	228	*Crassostrea gigas*	65, 66
Nereis iridescent virus (NIV)	160—180	*Nereis diversicolor*	67, 68
Octopus infection virus (OIV)	120—140	*Octopus vulgaris*	69
Simocephalus iridescent virus (SIV)	136	*Simocephalus expinosus*	70
Armadillidium iridescent virus (AIV)	125	*Armadillidium vulgare*	71
Porcellio iridescent virus (PIV)	125	*Porcellio dilatatus*	71
Lymnaea infection virus (LIV)	180—200	*Lymnaea truncatula*	72

The gills disease, or gill necrosis virus (GNV), caused an epizootic affecting up to 40% of the stocks of Portuguese oysters in France between 1967 and 1969;[63] the oyster velar virus disease appeared to be a yearly recurrent phenomenon (from 1976 to 1984), causing mortalities up to 50%, mainly during the months of April and May.[66] On the other hand, the presence of tumors associated with an iridovirus seems to be an enzootic infection since there were only 8.4% of the diseased animals in a relatively small population of *O. vulgaris*. Finally, some iridovirus infections can induce mortalities without the appearance of characteristic clinical signs in the animals. The mass mortality of the Portuguese oyster (from 1970 to 1973), and the 1977 summer mortality of a small stock of *C. gigas* was caused by Hemocytic Infection Virus (HIV*).[59,60] Similar conditions characterizing the LIV infection were reported by Barthe et al.[72] who observed abnormal mortalities in *L. truncatula* of up to 90%.

C. VIRUS STRUCTURE AND COMPOSITION
The viruses mentioned earlier belong to the icosahedral cytoplasmic deoxyriboviruses (ICDV) group. Using the bidirectional shadowing method, Devauchelle[67] demonstrated the icosahedral structure of NIV. Furthermore, the observation on ultrathin sections of particle profiles exhibiting five-, three- and twofold axes of rotational symmetry established that HIV and GNV also had this structure.[58] As for the other viruses, there is virtually no information on this point, most likely because the authors considered the icosahedral structure as evident without demonstration.

The fine structure of the virions corresponds to that demonstrated by Stoltz.[10] There is a dense core, generally spheroid (see Figure 10B) or an elongated profile (160 nm in length, 103 nm in width; point to point, 228 nm) (see Figure 11) in the mature OVV particles.[65] In the electron micrographs of the OIV presented by Rungger et al.[69] some sections of the central core also seem to have a pararectangular profile. The dense core is separated from the shell by a less dense layer, surrounded by a distinct trilaminar element (see Figure 12A)

---

* The acronym "HIV" as used in this text is *not* to be confused with the Human Immunodeficiency Virus.

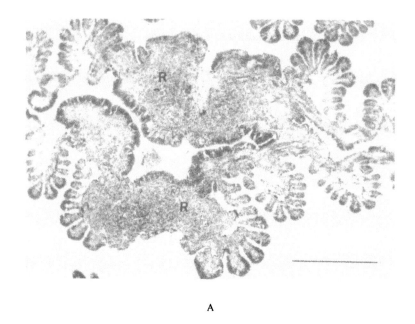

A

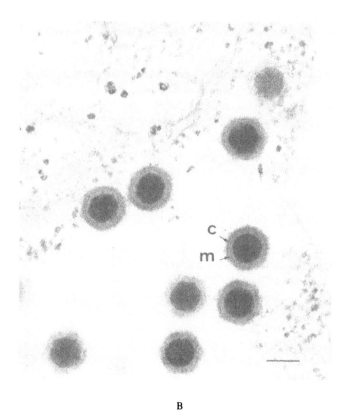

B

FIGURE 10.   (A) Histological section through typical gill lesions of diseased *Crassostrea angulata;* the necrosis of the filaments is associated with an important inflammatory reaction (R). (Bar = 0.5 mm.) (B) SIV infection; mature particles exhibiting dense core are limited by a unit membrane (m) and the external capsid (c). (Bar = 100 nm.) (Courtesy of Dr. B. A. Federici.)

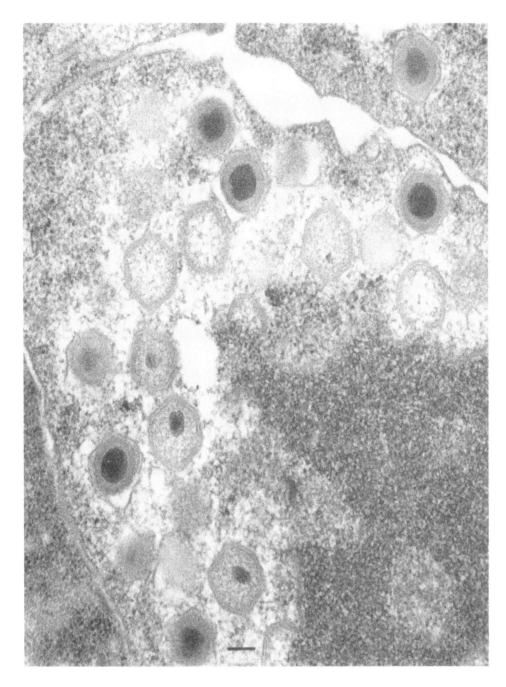

FIGURE 11.   OVV particles exhibiting a pararectangular core. (Bar = 100 nm.) (Courtesy of Dr. R. Elston.)

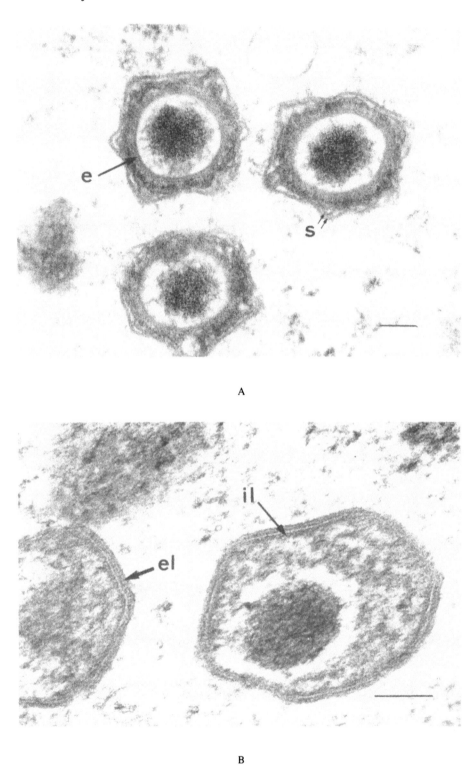

A

B

FIGURE 12.   (A) HIV mature virions scattered in the cytoplasm; note trilaminar element surrounding the central core (e) and shell appears to be constituted of two unit membranes (s). (Bar = 100 nm.) (B) GNV immature particles are enveloped by two major layers; the external layer (el) consists of subunits while the inner one (il) corresponds to a unit membrane. (Bar = 100 nm.)

for HIV and GNV as shown by Comps.[58] A similar structure was also noted in LIV by Barthe et al.[72]

The structural details of the shell of these viruses have not been clearly indicated in all cases. A complex capsid and associated membrane were reported for AIV and PIV by Federici[71] and a unit-membrane surrounded by the capsid (see Figure 10B) for SIV[70] while LIV is delimited by a "dense material".[72] No information has been given about the OIV shell. On the other hand, the examination of NIV by shadowing and negative staining of virus suspension and by applying cytochemical methods to thin sections has shown that the shell is constituted of a unit-membrane and an external layer of subunits arranged in a precise pattern. Fine fibrils can also be observed at the periphery of the layer of subunits.[67] A similar structure occurs in the iridoviruses of oysters GNV and HIV (see Figure 12B). Generally, on thin sections, the layer of subunits appear as a trilaminar element (like a unit membrane) (see Figure 13); the shell appears to be formed of two bilayered membranous structures.[58,66,68]

To date, knowledge of the physical properties and chemical composition of these viruses is very incomplete. Thus, the nature of the nucleic acid has generally been determined only by indirect methods showing the presence of DNA in the cytoplasm of infected cells, e.g., strongly Feulgen positive reaction (PIV, AIV), characteristic staining by acridine orange (OVV), with DAPI (HIV, GNV), or with malachite green (OIV). Concerning NIV, more precise information has been obtained with electron microscopic cytochemical techniques. The detection of DNA by the HATAG method of Thiery[73] revealed that the genome of NIV was not located in the central area but at the periphery of the dense core.[67]

D. CYTOPATHOLOGY

The infections are characterized by the appearance of an inclusion of electron dense material in which DNA is present in the cytoplasm of infected cells. Virus particles are formed at the periphery of the virogenic stroma. The first morphological event of the HIV morphogenesis is the development of a unit membrane enclosing a part of the virogenic material and, in juxtaposition to these membranes, a layer of subunits appears which induces an angular profile in the incomplete particles (see Figures 13, 14).

The immature particles with a polygonal outline become detached from the viroplasm and undergo a process of maturation in the cytoplasm consisting of the condensation of the internal component (see Figures 15, 16). Elston and Wilkinson[66] mention the presence of empty capsids in the cytoplasm of the host cell during the replication of OVV.

At an advanced stage of infection, the biggest viruses (HIV, GNV, OVV, and LIV) have few virions in the cytoplasm (see Figure 17A), and they are irregularly spread while the smallest have very numerous virions (NIV, AIV) and form paracrystalline arrays (see Figure 20). This configuration is less evident for SIV and OIV.

Among the most typical cytopathological effects, cell hypertrophy is observed in the infections caused by GNV, AIV, and PIV (see Figures 17, 18). In some cases (GNV, AIV, and LIV), an increase of the size of the nucleolus can be noted while viruses such as HIV and OIV induce a nuclear pycnosis.

Under all circumstances, there is a degradation of the cytopasmic organelles. Myelinic structures develop in LIV infections, while spheroidal inclusions consisting of numerous concentric layers of dense material or membranous structures associated with the disaggregation of the muscle fibrils of OIV appear in infected *O. vulgaria*. During the HIV infection of *C. gigas* and the OMV infection of *O. vulgaris,* structures develop showing some similarities to the paracrystalline inclusions described in numerous reports on iridescent viruses.

Finally, the NIV infection of the spermatocytes shows peculiar cytopathology as the intercellular walls disappear, nucleoli aggregate within a membrane, and the cell organelles move to the periphery[68] (see Figure 19).

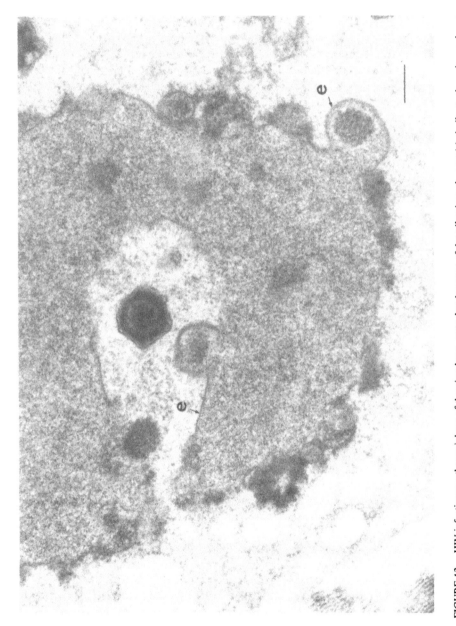

FIGURE 13.   HIV infection; on the periphery of the viroplasm, note development of the trilaminar element (e) similar to the unit membrane which constitutes the inner component of the shell. (Bar = 300 nm.)

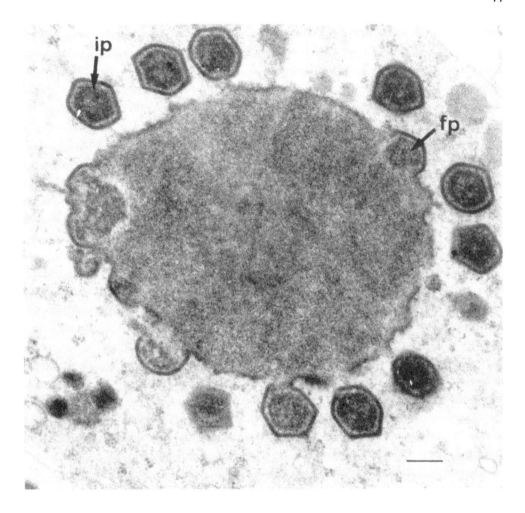

FIGURE 14. Section of HIV lesion in part of cytoplasm shows a single viroplasm, which is finely granular and different stages of the virus morphogenesis. In the complete immature particles (ip) as in forming particles (fp), the granular material is obviously more condensed. (Bar = 300 nm.)

E. TRANSMISSION

To date, the experimental approach of studies on these various infections remains very limited. The only data obtained about the transmission of the infectious agents concerns AIV and PIV. Injection with purified virions reproduced the disease with typical signs. Only one specimen of *P. dilatatus* developed the infection with blue discoloration by p.o. inoculation.

ADDENDUM

Since the original preparation of this manuscript, a number of papers that pertain to the Iridoviridae have been published. These include several new studies on the biochemistry and characterization of the genome of the *Chilo* iridescent virus (CIV, Type 6)[49-51] and recent work on the replication and maintenance of *Tipula* iridescent virus (TIV, Type 1) in mosquito cells.[52,53] Orange et al.[54] have identified viral antigenic polypeptides with specific monoclonal antibodies obtained against CIV, and in a very recent study, Tajbakhsh[55] reported on molecular cloning, characterization, and expression of the TIV capsid gene. An apparently new iridescent virus (IV) has been described from the mole cricket, *Scapteriscus borellii*

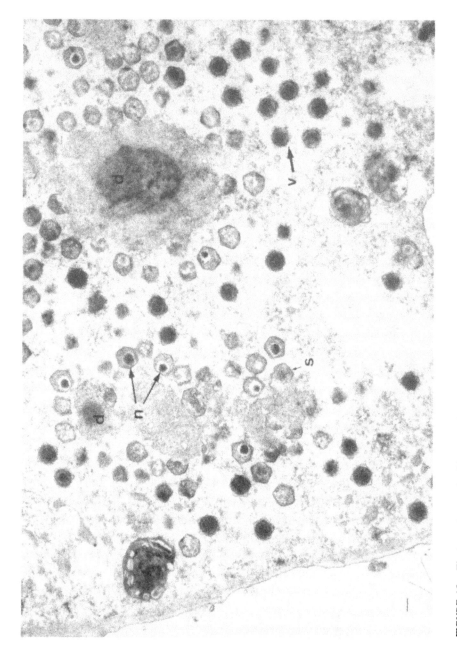

FIGURE 15.   Electron micrograph of intracytoplasmic GNV lesion. The virogenic stroma includes several areas of denser material (d) and numerous shells (s) are forming around the viroplasm. The process of maturation of the particles is characterized by the development of dense nucleoids (n) which pull away from the inner surface of the shells and further condense to become mature virions (v). (Bar = 300 nm.)

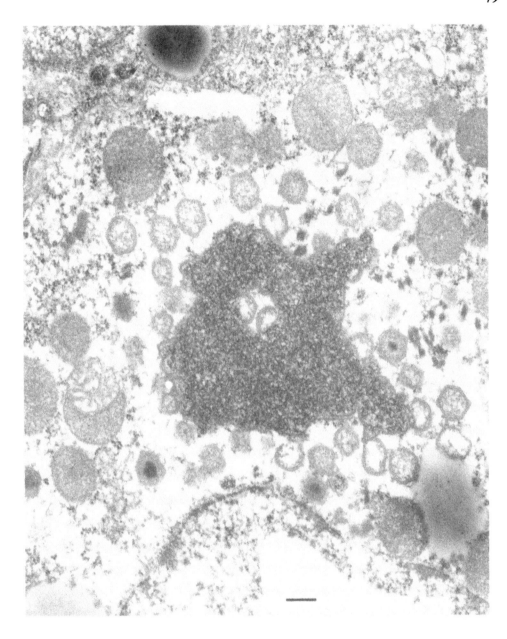

FIGURE 16. OVV infection in the cytoplasm of larval oyster velar epithelium showing the process of maturation of the viral particles in which virus particles detach from the viroplasm and undergo condensation of the internal component. (Bar = 200 nm.) (Courtesy of Dr. R. Elston.)

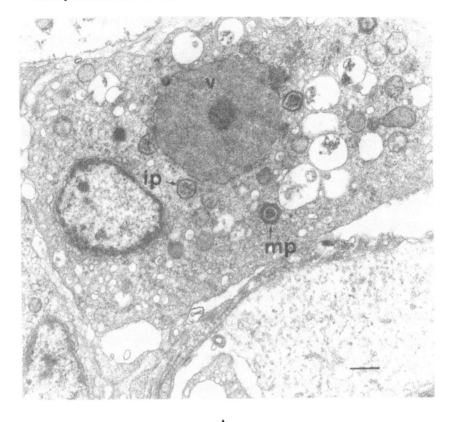

A

B

FIGURE 17.   (A) Hemocyte of *Crassostrea gigas* with a typical HIV lesion. Note viroplasm (v), immature particles (ip), mature particles (mp), the vacuolated cytoplasm, and the pycnotic nucleus. (Bar = 500 nm.) (B) GNV infection; histological section showing polymorphic hypertrophic cells (ch); some of them contain a basophilic inclusion (i) corresponding to the viral lesion; note the hypertrophied nuclei (n) exhibiting a large nucleolus. (Bar = 10 pm.)

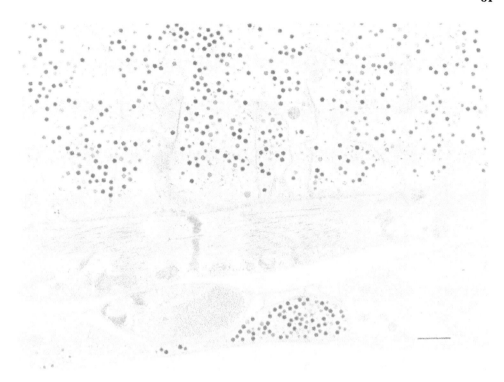

FIGURE 18. Hypertrophic AIV infected cells. (Bar = 1 μm.) (Courtesy of Dr. B. A. Federici.)

(Orthoptera:Gryllotalpidae).[56] The disease is highly pathogenic and was found in epizootic proportions in local populations. Laboratory tests showed that dry wood termites of the genus *Cryptotermes* were also highly susceptible. A new natural occurrence host record has also been reported for MIV in *Aedes sollicitans*.[57]

ACKNOWLEDGMENTS

The support and assistance of Dr. Gary Mount, Director, and Dr. David Dame, Research Leader for Insect Pathology at the Insects Affecting Man and Animals Research Laboratory, U.S. Department of Agriculture, ARS, Gainsville, FL 32604, are gratefully acknowledged. The author also wishes to thank Dr. Donald W. Hall, Professor of Entomology, University of Florida, Gainesville, FL 32611, and Dr. Donald Weidhaas (retired), IAMARL, for their review and helpful suggestions regarding the manuscript. Thanks are also due to Dr. Albert Undeen and Ms. Susan Avery, Insect Pathology Section, IAMARL, for their assistance; and finally the author is most grateful to Dr. Jean Adams, Insect Pathology Laboratory, U.S. Department of Agriculture, ARS, Beltsville, MD 20705, for furnishing photographs for this manuscript.

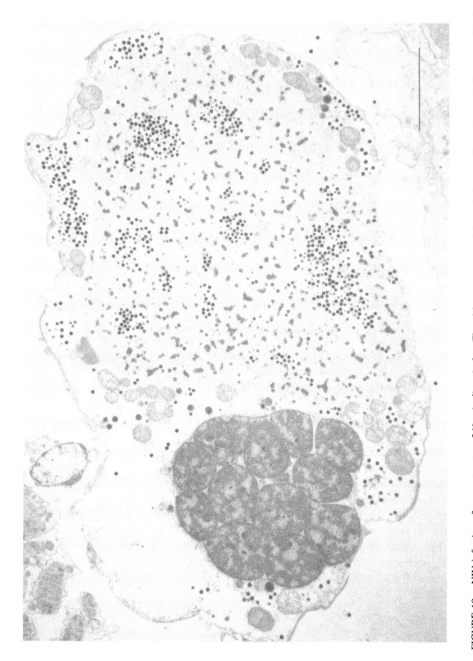

FIGURE 19.   NIV infection of spermatocytes of *Nereis diversicolor*. The virions are scattered through the cytoplasm; the nucleoi are assembled within a common membrane. Note the mitochondria are moved to the peripheral areas. (Bar = 5 μm.) (Courtesy of Dr. G. Devauchelle.)

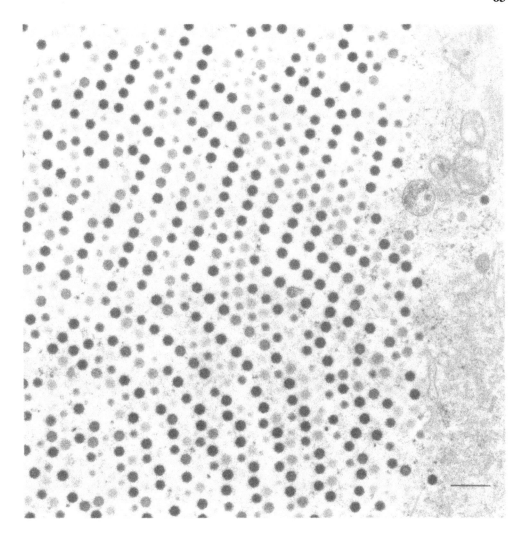

FIGURE 20. Section of an AIV infected *Armadillidium vulgare* cell showing a paracrystalline array of virions in the cytoplasm. (Bar = 500 nm.) (Courtesy of Dr. B. A. Federici.)

REFERENCES

1. **Williams, K. R. and Smith, K. M.,** The polyhedral form of the *Tipula* iridescent virus, *Biochim. Biophys. Acta,* 28, 464, 1958.
2. **Xeros, N.,** A second virus disease of the leather jacket *Tipula paludosa, Nature (London),* 174, 562, 1954.
3. **Williams, R. C. and Smith, K. M.,** A crystalline insect virus, *Nature (London),* 179, 119, 1957.
4. **Stoltz, Donald B.,** The structure of icosahedral cytoplasmic deoxyriboviruses, *J. Ultrastruct. Res.,* 37, 219, 1971.
5. **Tinsley, T. W. and Kelly, D. C.,** An interim nomenclature system for the iridescent group of insect viruses, *J. Invertebr. Pathol.,* 16, 470, 1970.
6. **Hall, D. W.,** Pathobiology of invertebrate icosahedral cytoplasmic deoxyriboviruses (Iridoviridae), in *Viral Insecticides for Biological Control,* Maramorosch, K. and Sherman, K. E., Eds., Academic Press, New York, 1985, 163.
7. **Wildy, P.,** Classification and nomenclature of viruses, in *Monogr. Virol.,* 5, 1, 1971.
8. **Fenner, F.,** Second report of the international committee on the taxonomy of viruses, *Intervirology,* 6, 1, 1976.

9. **Matthews, R. E. F.,** Fourth report of the international committee on taxonomy of viruses, *Intervirology*, 17, 1, 1982.

10. **Stoltz, D. B.,** The structure of icosahedral cytoplasmic deoxyriboviruses. II. An alternative model, *J. Ultrastruct. Res.*, 43, 58, 1973.

11. **Smith, K. M. and Hills, G. T.,** Replication and ultrastructure of insect viruses, *Proc. IX Int. Congr. Entomol.*, Vienna, 2, 1962, 823.

12. **Wrigley, N. G.,** An electron microscope study of the structure of *Sericesthis* iridescent virus, *J. Gen. Virol.*, 5, 123, 1969.

13. **Wrigley, N. G.,** An electron microscope study of the structure of *Tipula* iridescent virus, *J. Gen. Virol.*, 6, 169, 1970.

14. **Darlington, R. W., Granoff, A., and Breeze, D. C.,** Viruses and renal carcinoma of *Rana pipiens*. II. Ultrastructural studies and sequential development of virus isolated from normal and tumor tissue, *Virology*, 29, 149, 1966.

15. **Hitchborn, J. H. and Hills, G. J.,** A study of tubes produced in plants infected with a strain of turnip yellow mosaic virus, *Virology*, 35, 50, 1968.

16. **Anthony, D. W. and Hall, D. W.,** Electron microscope studies of the "R" and "T" strains of mosquito iridescent virus in *Aedes taeniorhynchus* (Wied.) larvae, *Proc. IV Int. Coll. Insect Pathol.*, Society for Invertebrate Pathology, College Park, MD, 1970, 386.

17. **Thomas, R. S.,** The chemical composition and particle weight of *Tipula* iridescent virus, *Virology*, 14, 240, 1961.

18. **Thomas, R. S. and Williams, R. C.,** Localization of DNA and protein in *Tipula* iridescent virus (TIV) by enzymatic digestion and electron microscopy, *J. Biophys. Biochem. Cytol.*, 11, 15, 1961.

19. **Day, M. F. and Mercer, E. H.,** Properties of an iridescent virus from the beetle, *Sericesthis pruinosa*, *Aust. J. Biol. Sci.*, 17, 892, 1964.

20. **Bellet, A. J. D. and Inman, R. B.,** Some properties of deoxyribonucleic acid preparations from *Chilo*, *Sericesthis*, and *Tipula* iridescent viruses, *J. Mol. Biol.*, 25, 425, 1967.

21. **GLitz, D. G., Hills, G. J., and Rivers, C. F.,** A comparison of *Tipula* and *Sericesthis* iridescent viruses, *J. Gen. Virol.*, 3, 209, 1968.

22. **Kalmakoff, J. and Tremaine, J. H.,** Physico-chemical properties of *Tipula* iridescent virus, *J. Virol.*, 2, 738, 1968.

23. **Cerutti, M. and Devauchelle, G.,** Characterization and localization of CIV polypeptides, *Virology*, 145, 123, 1985.

24. **Tajbakhsh, S., Dove, M. J., Lee, P. E., and Seligy, V. L.,** DNA components of *Tipula* iridescent virus, *Biochem. Cell Res.*, 64, 495, 1986.

25. **Faust, R. M., Dougherty, E. M., and Adams, J. R.,** Nucleic acid in the blue-green and orange mosquito iridescent viruses (MIV) isolated from larvae of *Aedes taeniorhynchus*, *J. Invertebr. Pathol.*, 10, 160, 1967.

26. **Matta, J. F.,** The characterization of a mosquito iridescent virus. II. Physico-chemical characterization, *J. Invertebr. Pathol.*, 16, 157, 1970.

27. **Hall, D. W. and Lowe, R. E.,** Physical and serological comparison of "R" and "T" strains of a mosquito iridescent virus affecting *Aedes taeniorhynchus*, *J. Invertebr. Pathol.*, 19, 317, 1972.

28. **Wagner, G. W., Paschke, J. D., Campbell, W. R., and Webb, S. R.,** Biochemical and biophysical properties of two strains of mosquito iridescent virus, *Virology*, 52, 72, 1973.

29. **Lowe, R. E., Hall, D. W., and Matta, J. F.,** Comparison of the mosquito iridescent viruses with other iridescent viruses, *Proc. IV Int. Coll. Insect Pathol.*, College Park, MD, 1970, 163.

30. **Clark, T. B., Kellen, W. R., and Lum, P. T. M.,** A mosquito iridescent virus (MIV) from *Aedes taeniorhynchus* (Wied.), *J. Invertebr. Pathol.*, 7, 519, 1965.

31. **Linley, J. R. and Nielson, H. T.,** Transmission of a mosquito iridescent virus in *Aedes taeniorhynchus*. I. Laboratory experiments, *J. Invertebr. Pathol.*, 12, 7, 1968.

32. **Linley, J. R. and Nielson, H. T.,** Transmission of a mosquito iridescent virus in *Aedes taeniorhynchus*, II. Experiments related to transmission in nature, *J. Invertebr. Pathol.*, 12, 17, 1968.

33. **Matta, J. F. and Lowe, R. E.,** The characterization of a mosquito iridescent virus (MIV). I. Biological characteristics, infectivity and pathology, *J. Invertebr. Pathol.*, 16, 38, 1970.

34. **Smith, K. M.,** The iridescent viruses, in *Virus-Insect Relationships*, Longman, New York, 1976, chap. 5.

35. **Smith, K. M., Hills, G. J., and Rivers, C. F.,** Studies on the cross-inoculation of the *Tipula* iridescent virus, *Virology*, 13, 233, 1961.

36. **Fukuda, T.,** *Per os* transmission of *Chilo* iridescent virus to mosquitoes, *J. Invertebr. Pathol.*, 18, 152, 1971.

37. **Woodard, D. B. and Chapman, H. C.,** Laboratory studies with the mosquito iridescent virus (MIV), *J. Invertebr. Pathol.*, 11, 296, 1968.

38. **Anthony, D. W.**, Use of viruses for control of aquatic insect pests, in *Impact of the Use of Microorganisms on the Aquatic Environment*, EPA 660-3-75-001 Ecological Research Series, January 1975, p. 23.

39. **Chapman, H. C., Clark, T. B., Anthony, D. W., and Glenn, Jr., F. E.**, An iridescent virus from larvae of *Corethrella brakeleyi* (Diptera: Chaoboridae) in Louisiana, *J. Invertebr. Pathol.*, 18, 284, 1971.

40. **Stadelbacher, E. A., Adams, J. R., Faust, R. M., and Tompkins, G. J.**, An iridescent virus from the bollworm, *Heliothis zea* (Lepidoptera: Noctuidae), *J. Invertebr. Pathol.*, 32, 71, 1978.

41. **Woodard, D. B. and Chapman, H. C.**, Laboratory studies with the mosquito iridescent virus (MIV), *J. Invertebr. Pathol.*, 11, 296, 1968.

42. **Hall, D. W. and Anthony, D. W.**, Pathology of a mosquito iridescent virus (MIV) infecting *Aedes taeniorhynchus*, *J. Invertebr. Pathol.*, 18, 61, 1971.

43. **Fukuda, T. and Clark, T. B.**, Transmission of the iridescent virus (RMIV) by adult mosquitoes of *Aedes taeniorhynchus* to their progeny, *J. Invertebr. Pathol.*, 25, 275, 1975.

44. **Hembree, S. C.**, Non-participation of males of *Aedes taeniorhynchus* (Weidemann) in vertical transmission of regular mosquito iridescent virus, *Mosq. News*, 39, 672, 1979.

45. **Butchatsky, L. P.**, An iridovirus from larvae of *Culiseta annulata* and *Culex territans*, *Acta Virol.*, 21, 85, 1977.

46. **Hembree, S. C. and Anthony, D. W.**, Possible site of entry of the regular mosquito iridescent virus (RMIV) in *Aedes taeniorhynchus* larvae, *Mosq. News*, 40, 449, 1980.

47. **Anderson, J. F.**, An iridescent virus infecting the mosquito *Aedes stimulans*, *J. Invertebr. Pathol.*, 5, 219, 1970.

48. **Tinsley, T. W. and Harrap, K. A.**, Newly characterized protist and invertebrate viruses, in *Comprehensive Virology*, Vol. 12, Fraenkel-Conrat, H. and Wagner, R. E., Eds., Plenum Press, New York, 1978.

49. **Soltau, J. B., Fischer, M., Schnitzler, P., Scholz, J., and Darai, G.**, Characterization of the genome of insect iridescent virus type 6 by physical mapping, *J. Gen. Virol.*, 68, 2717, 1987.

50. **Fischer, M., Schnitzler, P., Scholz, J., Rose-Wolf, A., Delius, H., and Darai, G.**, DNA nucleotide sequence analysis of the PvuII DNA fragment L of the genome of insect iridescent virus type 6 reveals a complex cluster of multiple tandem, overlapping, and interdigitated repetitive DNA elements, *Virology*, 167, 497, 1988.

51. **Fischer, M., Schnitzler, P., Delius, H., and Darai, G.**, Identification and characterization of the repetitive DNA element in the genome of insect iridescent virus type 6, *Virology*, 167, 485, 1988.

52. **Bertin, J., Frosch, M., and Lee, P. E.**, Formation and maintenance of viroplasmic centers in *Tipula* iridescent virus-infected mosquito cells with deranged cytoskeletons, *Eur. J. Cell Biol.*, 43, 215, 1987.

53. **Tajbakhsh, S., Kiss, G., Lee, P. E., and Seligy, V. L.**, Semipermissive replication of *Tipula* iridescent virus in *Aedes albopictus* C 6/36 cells, *Virology*, 174, 264, 1990.

54. **Orange, N., Guerillon, J., and Devauchelle, G.**, Identification of viral antigenic determinants by non-oclonal antibodies directed against *Chilo* iridescent virus (*Iridovirus* type 6), *Arch. Virol.*, 99, 243, 1988.

55. **Tajbakhsh, S., Lee, P. E., Watson, D. C., and Seligy, V. L.**, Molecular cloning characterization and expression of the *Tipula* iridescent virus capsid gene, *J. Virol.*, 64, 125, 1990.

56. **Fowler, H. G.**, An epizootic iridovirus of Orthoptera Gryllotalpidae *(Scapteriscus borellii)* and its pathogenicity to termites Isoptera, *Cryptotermes*, *Rev. Microbiol.*, 20, 115, 1989.

57. **Becnel, J. J. and Fukuda, T.**, Natural occurrence of a mosquito iridescent virus in *Aedes sollicitans*, *J. Am. Mosq. Control Assoc.*, 5, 610, 1989.

58. **Comps, M.**, Recherches histologiques et cytologiques sur les infections intracellulaires des molloscques bivalves marins, Doctoral dissertation, Univ. Sci. Tech. Languedoc, Montpellier, France, 1983.

59. **Comps, M., Bonami, J. R., Vag, C., and Campillo, A.**, Une virose de l'huître portugaise *(Crassostrea angulata)*, *C. R. Acad. Sci. Paris, Ser. D*, 282, 1991, 1976.

60. **Comps, M. and Bonami, J. R.**, Infection virale associée à des mortalités chez l'huître *Crassostrea gigas* Th., *C. R. Acad. Sci. Paris, Ser. D*, 285, 1139, 1977.

61. **Comps, M.**, La maladie des branchies chez les huîtres du genre *Crassostrea*, caractéristiques et évolution des altérations, processus de cicarisation, *Rev. Trav. Inst. Peches Marit.*, 34(1), 23, 1970.

62. **Marteil, L.**, La maladie des branchies, *Cons. Inter. Explor. Mer.*, C. M. K5, 1968.

63. **Marteil, L.**, Données générales sur la maladie des branchies, *Rev. Trav. Inst. Peches Marit.*, 33, 145, 1969.

64. **Deltreil, J. P.**, Remarques sur la croissance en élevage suspendu de *Crassostrea angulata* Lmk affectée par la maladie des branchies dans le bassin d'Arcachon, *Rev. Trav. Inst. Peches Marit.*, 33, 176, 1969.

65. **Elston, R.**, Viruslike particles associated with lesions in larval Pacific oysters, *(Crassostrea gigas)*, *J. Invertebr. Pathol.*, 33, 71, 1979.

66. **Elston, R. and Wilkinson, M. T.**, Pathology, management and diagnosis of oyster velar virus disease (OVVD), *Aquaculture*, 48, 189, 1985.

67. **Devauchelle, G.**, Ultrastructural characterization of an iridovirus from the marine worm *Nereis diversicolor* (O. F. Muller), *Virology*, 81, 237, 1977.

68. **Devauchelle, G. and Durchon, M.,** Sur la presence d'un virus de type iridovirus dans les cellules males de *Nereis diversicolor* (O. F. Muller), *C. R. Acad. Sci. Paris, Ser. D,* 277, 463, 1973.

69. **Rungger, D., Rastelli, M., Braendle, E., and Malsberger, R. G.,** A viruslike particle associated with lesions in the muscles of *Octopus vulgaris, J. Invertebr. Pathol.,* 17, 72, 1971.

70. **Federici, B. A. and Hazard, E. I.,** Iridovirus and cytoplasmic polyhedrosis virus in the fresh water daphnid *Simocephalus expinosus, Nature (London),* 254, 327, 1975.

71. **Federici, B. A.,** Isolation of an iridovirus from two terrestrial Isopods, the pill bug, *Armadillidium vulgare,* and the sow bug, *Porcellio dilatatus, J. Invertebr. Pathol.,* 36, 373, 1980.

72. **Barthe, D., Faucher, Y., and Vago, C.,** Infection virale chez le Mollusque pulmoné *Lymnaea truncatula* Muller, *C. R. Acad. Sci. Paris, Ser. III,* 298(17), 513, 1984.

73. **Thiery, J. P.,** Mise en évidence de L'ADN par la methode "HATAG", *J. Microsc. (Paris),* 14, 95a, 1972.

Chapter 6

BACULOVIRIDAE. NUCLEAR POLYHEDROSIS VIRUSES

PART 1. NUCLEAR POLYHEDROSIS VIRUSES OF INSECTS

Jean R. Adams and J. Thomas McClintock

PART 2. NUCLEAR POLYHEDROSIS VIRUSES OF INVERTEBRATES OTHER THAN INSECTS

John A. Couch

TABLE OF CONTENTS

Part 1

Part 2

PART 1.
NUCLEAR POLYHEDROSIS VIRUSES OF INSECTS

I. INTRODUCTION

A. MEMBERS OF THE GROUP

The baculoviruses are classified in the family Baculoviridae which is divided into three subgroups: (1) nuclear polyhedrosis viruses (NPVs), (2) granulosis virus (GV), and (3) nonoccluded baculoviruses.[1] To date only the Baculoviridae can be considered as being unique to arthropods.[2] More than 520 NPVs have been identified in insects.[3] The families of insects in which NPVs have been found include many pests of economic importance. Table 1 lists the families from which NPVs have been isolated. Many NPVs appear to be specific for the families or genera from which they are isolated. Because of this specificity, NPVs are ideal microbial candidates for use in an integrated pest management program. In addition to their specificity, NPVs are not harmful to animals or humans, will not pollute the environment, and since these viruses occur in nature, are safe for insect pest suppression.[4-7] In fact, viruses which overwinter and still remain viable may serve as a source of inoculum for the host insect the next year. In nature, these viruses occur in an insect population and are amplified by biotic and abiotic persistence until an epizootic is produced in which large numbers of larvae die as a result of viral infections.[8-10]

Many excellent reviews have been prepared on baculoviruses which will serve to guide those interested to probe specific areas of interest. The baculoviruses have been studied more thoroughly than any other groups of invertebrate viruses, and it will be possible to include only key references in discussing them although the general concepts we now have represent the contributions of many investigators over more than 40 years. Benz has recently presented an interesting historical perspective.[10] Helpful reviews have been prepared on baculovirus structure (NPV), biochemistry and cytopathology,[11-63,67-72,74-76,78-82,86] serology,[64-66,73,83-85] microbial control,[77,87-151,160] and safety.[4-6,152-159]

B. HOST RANGE

Early studies indicated that nuclear polyhedrosis viruses (NPV) replicate in the nuclei of susceptible cells causing mortality in only the species or genus from which they were isolated. There have been discussions of "latent" viruses which were activated in times of stress or upon viral inoculation in the test insects. The question about whether the test insects were virus free was not certain; thus early cross-transmission studies were suspect.[601] Unfortunately, the biochemical techniques were not yet developed to test the viruses produced from passages through alternate hosts.

The specificity of insect viruses was reviewed by Ignoffo[95] and recently by Groner[4] and McIntosh et al.[148] In studies on *Trichoplusia ni* NPVs collected from five areas in the U.S., electron microscopic examinations revealed different morphologies and differences in the number of virus rods occluded per bundle.[161] Polyhedra containing single enveloped virus rods were called single-embedded viruses (SEV or SNPV), and those with more than one enveloped virus rod were called multiple-embedded viruses (MEV or MNPV). Polyhedra or viral occlusions (VOs) containing single nucleocapsids per envelope are now designated SNPVs while VOs containing multiple nucleocapsids per envelope are designated MNPVs. Tompkins et al.[162] found that corn earworm larvae, *Heliothis zea* (Boddie), were susceptible to *T. ni* MNPV (TnMNPV) as well as their homologous SNPV. However, this SNPV only infects other *Heliothis* sp. Since the MNPVs were less specific in activity, Heimpel searched for an NPV with an extended host range. Although host specificity was clearly an advantage, commercial companies were not interested in producing nuclear polyhedrosis viruses for small markets. Vail et al.[163] found an MEV-NPV which was isolated from the alfalfa looper,

TABLE 1
Nuclear Polyhedrosis Viruses Isolated
from Insect Species[3]

Order and family	Species from which NPVs have been isolated
Coleoptera	
Cerambycidae	2
Curculionidae	1
Dermestidae	2
Diptera	
Calliphoridae	1
Chironomidae	1
Culicidae	20
Sciaridae	3
Tachinidae	1
Tipulidae	1
Hymenoptera	
Argidae	1
Diprionidae	19
Pamphylidae	3
Tenthridinidae	8
Lepidoptera	
Anthelidae	2
Arctiidae	22
Argyresthiidae	1
Bombyciidae	4
Brassolidae	1
Carposinidae	1
Coleophoridae	1
Cossidae	1
Cryptophasidae	1
Dioptidae	1
Gelechiidae	3
Geometridae	63
Hepialidae	3
Hesperidae	5
Lasiocampidae	34
Limacodidae	11
Lymantriidae	49
Lyonetiidae	1
Noctuidae	107
Notodontidae	12
Nymphalidae	15
Papillionidae	6
Pieridae	9
Plutellidae	1
Psychidae	5
Pyralidae	23
Saturniidae	22
Sphingidae	14
Thaumetopoeidae	3
Thyatiridae	1
Tineidae	2
Tortricidae	26
Yponomeutidae	4
Zygaenidae	1
Neuroptera	
Chrysopidae	1

TABLE 1 (continued)
Nuclear Polyhedrosis Viruses Isolated
from Insect Species[3]

Order and family	Species from which NPVs have been isolated
Hemerobiidae	1
Siphonoptera	
Pulicidae	1[a]
Thysanura	
Phaemachilidae	1
Trichoptera	
Limnephilidae	1

[a] Beard, C. B., Butler, J. F., and Maruniak, J. E., *J. Invertebr. Pathol.*, 54, 128, 1989.

Autographa californica (AcMNPV), which multiplied in six other lepidopteran species. Heimpel[164] mass produced large quantities of AcMNPV, distributed samples around the world for testing, and compiled data required for registration by the Food and Drug Administration (FDA) and later the Environmental Protection Agency (EPA). AcMNPV was shown to successfully infect over 25 insect species. An MNPV of the celery looper, *Syngrapha falcifera*, isolated by Hostetter[165] (U.S. Department of Agriculture, ARS, Entomology Laboratory, Kimberly, ID) was reported to be pathogenic to 30 of 39 lepidopteran species in eight families tested. Studies of AcMNPV by scientists around the world have contributed much to our knowledge about baculoviruses today. Table 2 presents the host range reported for AcMNPV while Table 3 lists some reports of cross transmissions with other NPVs. A review which includes cross-transmission data has been prepared recently by Groner.[4]

In the light of our present knowledge the problem with many NPVs collected years ago is that their history is unknown. When it was established that MNPVs could cross infect other closely related insect species, some investigators proposed on the basis of equal virulence in other host species that NPVs could be produced by passage in the insect easiest to rear and also yielding high concentration of polyhedra.[168] Tompkins et al.[199] demonstrated that the insect of choice for virus propagation was very important since the virulence of the NPVs tested varied according to the host in which they were produced. Differences were noted in the sizes of the polyhedra produced, the number of virions in cross sections of the polyhedra section, and the number of nucleocapsids per virion. Tompkins et al.[199] found that AcMNPV increased in size after several passages in *Spodoptera exigua* and was more virulent to *T. ni* larvae. Electron microscope examinations of sections of the polyhedra revealed that there were more virions per cross section and a greater mean and mode of nucleocapsids per bundle of virions than contained in the original AcMNPV sample. Recent studies have shown changes in virulence of NPVs that occur with passage in alternate noctuid cell lines and hosts.[201]

Serial passages of most MNPVs through other host species cause the virulence or the activity to decrease when compared with the activity of the passage in the native host. One noted exception was reported in *Plusia gamma* inoculated with *Mamestra brassicae* MNPV.[189] Investigations with NPVs from several different *Heliothis* species revealed interesting results which are not yet fully understood. The SNPVs from *H. armigera* and *H. zea* were more virulent to *H. armigera* larvae than the *H. armigera* MNPV.[191] A 5992-fold difference was observed between larvae of *H. virescens* and *H. zea* when challenged with AcMNPV.[174] Ignoffo et al.[190] reported a 1000-fold difference in susceptibility to *H. zea* SNPV between larvae of *H. virescens* and *H. subflexa* and suggested that resistance may be controlled by a single gene.

TABLE 2
Insect Species Susceptible to *Autographa californica* MNPV

Family	Species	Ref.
Arctiidae	*Estigmene acrea* (Drury)	164, 168
	Halisodota caryae Harris	169
	Hyphantria cunea (Drury)	172
Gelechiidae	*Pectinophora gossypiella* (Saunders)	167
Geometridae	*Alsophila pometaria* (Harris)	172
	Ennomos subsignarius (Hubner)	172
Lasiocampidae	*Malacosoma americanum* (F)	172
Lyonetiidae	*Bucculatrix thurberiella* Busck	164
Noctuidae	*Agrotis ipsilon* (Hufnagel)	176, 178, 180
	Agryogramma basigera (Walker)	170
	Autographa biloba Stephens	170
	Autographa californica (Speyer)	164, 170
	Eupsilia sp.	172
	Heliothis virescens	177, 179, 181
	Heliothis zea (Boddie)	164, 174, 177, 179
	Melanchra picta (Harris)	175
	Plusia balluca Geyer	170
	Pseudoplusia includens (Walker)	170
	Rachiplusia ou (Guenée)	170
	Spodoptera exigua (Hubner)	164, 168
	Trichoplusia ni (Hubner)	164, 166, 168, 170, 181
	Xylena curvimacula Morrison	166
Phycitidae	*Cactoblastis cactorum* (Berg)	180
Pyralidae	*Galleria mellonella* (L.)	180
	Ostrinia nubilalis (Hubner)	173
Saturniidae	*Anisota senatoria* J. E. Smith	172
Yponomeutidae	*Plutella xylostella* (L.)	164

Further tests on three *Heliothis* spp. with *H. zea* SNPV, *H. armigera* MNPV, and *H. armigera* granulosis virus (GV) revealed that *H. zea* larvae were 4 to 6 times more susceptible to *H. armigera* MNPV than were *H. virescens* or *H. armigera* larvae. Larvae of *H. virescens* were about 20 and 35 times more susceptible to *H. armigera* GV than were the larvae of *H. zea* and *H. armigera*, respectively. Larvae of *H. armigera* were more susceptible to *H. zea* SNPV than to *H. armigera* MNPV.[191] Other studies using 6-d-old *H. armigera* larvae indicated that *H. armigera* MNPV was less virulent than *H. armigera* SNPV and *H. zea* SNPV.[192] This test and the tests of Allaway et al.[189] were supported by restriction enzyme comparisons of the original viruses with the passaged viruses which demonstrated host-induced modification following serial passages in alternate hosts. McKinley et al.[197] found that in *Spodoptera* spp. genuine cross infections occurred only in *S. littoralis* and *S. fru-giperda* which were infected by the heterologous *Spodoptera* MNPVs. The cross-infection tests conducted on the other *Spodoptera* sp. *(S. exempta)* with its MNPV and *H. armigera* SNPV activated the homologous NPVs. The isolated viruses were analyzed by structural polypeptide profiles on SDS-polyacrylamide gels, gel immunodiffusion tests using specific antisera, and G:C content of the DNA.

Changes in virulence following exposure to chemical mutagens and serial passage have been reported.[202] In a recent study, chemical treatment and serial passage of AcMNPV in alternate lepidopteran hosts revealed a 100-fold increase in virulence if the virus was treated initially in *T. ni* and assayed in *S. frugiperda*. Significant changes were also observed in the lethal time (LT_{50}) of several treated virus samples, but none of these changes persisted after several serial passages. Even though the effect of *in vivo* mutagenesis was not stable with passage, the method demonstrated the possible use of low-passage virus isolates, which

TABLE 3
Cross-Infection Studies with M Nuclear Polyhedrosis
Viruses

MNPV	Susceptible host	Ref.
Trichoplusia ni	Heliothis zea	162
	Spodoptera exigua	164
	Orgyia pseudotsugata	183
Cadra cautella	Plodia interpunctella	184
Anticarsia gemmatalis	Heliothis zea[a]	185
	Trichoplusia ni[a]	
	Pseudoplusia includens[a]	
	Spodoptera ornithogalli[a]	
Melanchra picta	Mamestra configurata	186
	Amanthes c-nigrum	
	Peridromia saucia	
	Spodoptera praefica	
	Trichoplusia ni	
Rachiplusia ou	Agrotis ipsilon	180, 187
Heliothis punctigera	Heliothis punctigera	188
	Heliothis armigera	
Mamestra brassicae	Plusia gamma	189
	Noctua pronuba	
	Mamestra brassicae	
	Lacanobia oleracea	
Agrotis segetum	Agrotis segetum	189
	Noctua pronuba	
	Agrotis exclamationis	
Heliothis armigera	Heliothis zea	191, 192
	Heliothis virescens	
	Heliothis armigera	
Galleria mellonella	Galleria mellonella	200
	Trichoplusia ni	
	Manduca sexta	
Choristoneura fumiferana	C. fumiferana	193
	Trichoplusia ni	
	Galleria mellonella	
Mamestra brassicae	Euxoa scandans	194
Euxoa messoria	Euxoa scandans	
Agrotis segetum	Euxoa scandans	
Spodoptera littoralis	Locusta migratoria	195
	Schistocerca gregaria	
Lymantria monacha	Lymantria monacha	196
	Panolis flammea	
Spodoptera littoralis	Spodoptera frugiperda	197
Spodoptera frugiperda	Spodoptera littoralis	
Panolis flammea	Mamestra brassicae	198

[a]   High dosages required for mortality.

demonstrated significant increase in virulence, on crops infested with a particular insect pest.

In vivo infectivity tests of plaque-purified MNPV isolates, variants, and natural recombinants of *Rachiplusia ou, T. ni,* and *Galleria mellonella* in *T. ni* and *H. virescens* larvae have shown minor differences in virulence. This was not unexpected since Smith and Summers[203] and Summers et al.[204] have demonstrated by construction of physical maps that these MNPVs are variants of AcMNPV.

C. GROSS PATHOLOGY (SYMPTOMS)

Insect larvae are most susceptible to viral infection during the early instars. Typical symptoms of viral infection include lethargy, loss of appetite, negative geotrophism, increasing whitish appearance as fat and hypodermal nuclei become filled with NPVs, and cadavers of infected larvae which often hang upside down attached by the posterior prolegs in an inverted V position (wilt). A puncture of the cuticle reveals a milky hemolymph which is filled with polyhedra. An infected insect larva may produce as many as 1×10^9 viral occlusions (VOs).

II. VIRUS STRUCTURE AND COMPOSITION

A. VIRAL OCCLUSIONS
1. Morphotypes of NPVs

The SNPV contain many single bacilliform virions while the MNPVs contain many bundles of bacilliform virions. These virions are embedded in a crystalline protein lattice called polydedrin which forms the viral occlusion (VO) that was formerly called the occlusion body. An example of the SNPVs is shown in Figures 1 and 2 while examples of MNPVs are shown in Figures 3 to 9 and 11. Polyhedra have been reported to range in size from 0.5 to 15 μm with the predominant size range between 0.6 to 2.5 μm in diameter. Since so many NPVs have been discovered, it is no longer possible to distinguish between them based on morphology of the shapes which may be cuboidal, tetrahedral, dodecahedral, or irregular.[161,205,206] In the past most NPVs were named according to the original host from which they were isolated since early studies indicated that many NPVs were host specific. However, the recent studies noted above have indicated that passage through alternate hosts is possible with some MNPVs. Therefore, techniques such as restriction enzyme analyses of the viral DNA are now required to properly identify NPVs.

For general identification of NPVs by light microscopy, a rapid technique is the use of nigrosin smears of hemolymph or a homogenate of the diseased insect in which the polyhedra appear as angular white bodies against the black background. Other microscopic techniques that are useful to identify NPVs include the use of dark field, Vago's technique,[207] Wigley's technique,[208] phase or differential interference contrast, or electron microscopy. These techniques are discussed in more detail in Chapter 2.

2. Viral Occlusion Composition

The occlusion of viruses in a crystalline protein lattice occurs only in invertebrates infected with viruses in the following virus families: Baculoviridae, Poxviridae (Entomopoxvirinae), and Reoviridae (cytoplasmic polyhedrosis virus group). The viral occlusion serves to protect the virus particles from harsh environmental conditions. The viral occlusion protein called polyhedrin for NPV and granulin for GV, is present in a multimeric conformation forming a paracrystalline lattice around the virions. It consists of a single polypeptide ranging in relative molecular weight (M_r) from 29,000 to 31,000.[35] Excellent reviews on polyhedrin have been prepared recently.[42,45]

The protein molecules are probably oriented in a face-centered cubic lattice. Based on X-ray diffraction studies Bergold[209,210] proposed that the lattice is crystalline in nature. Harrap[211] later modified Bergold's model by proposing that the protein molecules of the lattice may be the shape of six-armed nodal units rather than spheres, which would better fit the amount of space and matter evident in electron micrographs of sections. It would also aid in the understanding of how symmetry would be preserved if there is only one way that the six-armed nodal unit would link with its neighbors. Hughes[212] noted differences in measurements of center-to-center distances between molecules of protein in the crystalline lattices of VOs from nine species. From studies using electron microscopy at high magnifications, Hughes suggested that the 300 to 400 Å sections revealed stacks of possibly four

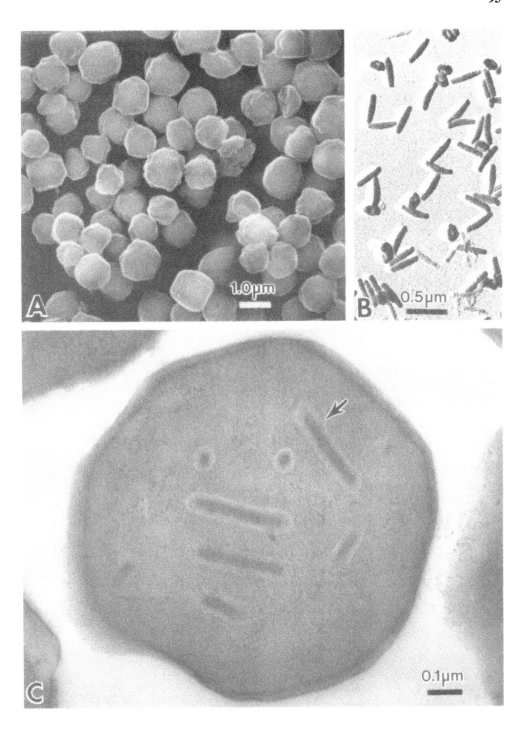

FIGURE 1. (A) Scanning electron micrograph (SEM) of the corn earworm, *Heliothis zea*, nuclear polyhedrosis virus (NPV) polyhedra or viral occlusions (VOs). (B) Virions of *H. zea* NPV may be released from the virus occlusions by treatment with alkali (Na_2CO_3). (C) Section of *H. zea* SNPV; the virions containing single nucleocapsids per envelope (arrow) are designated as S and are embedded in the crystalline protein matrix of the viral occlusion.

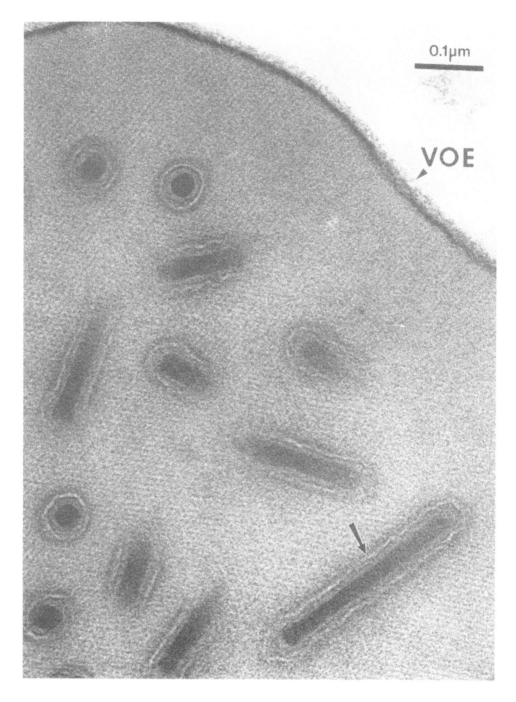

FIGURE 2.   A portion of a section of SNPV isolated from the cabbage looper, *Trichoplusia ni*, showing single nucleocapsids per envelope (arrow). Note VO envelope or "membrane" (VOE).

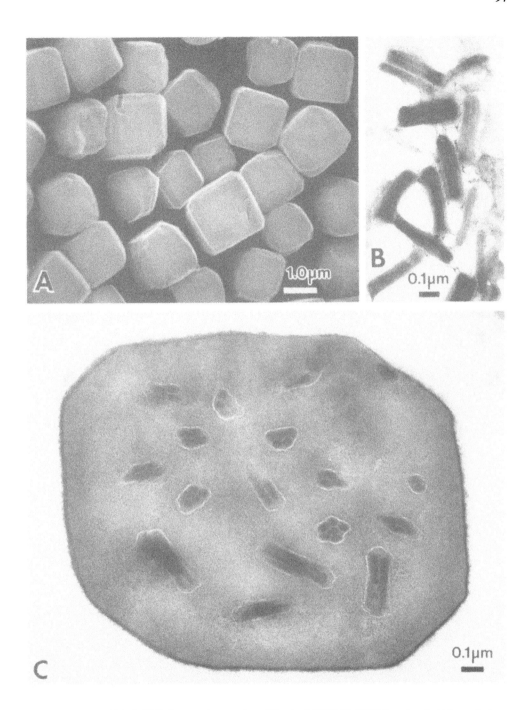

FIGURE 3. (A) SEM of alfalfa looper, *Autographa californica* MNPV (AcMNPV); note cuboidal morphology. (B) Bundles of virions or polyhedra-derived virus (PDV) released by alkali treatment of the polyhedra or VOs of *Rachiplusia ou*. (C) Section of a VO of AcMNPV showing bundles of virions embedded in the polyhedron protein matrix; these virions which have been referred to as ENCs or MEV (multiple-embedded virions) are now referred to as M or multiple nucleocapsids per envelope.

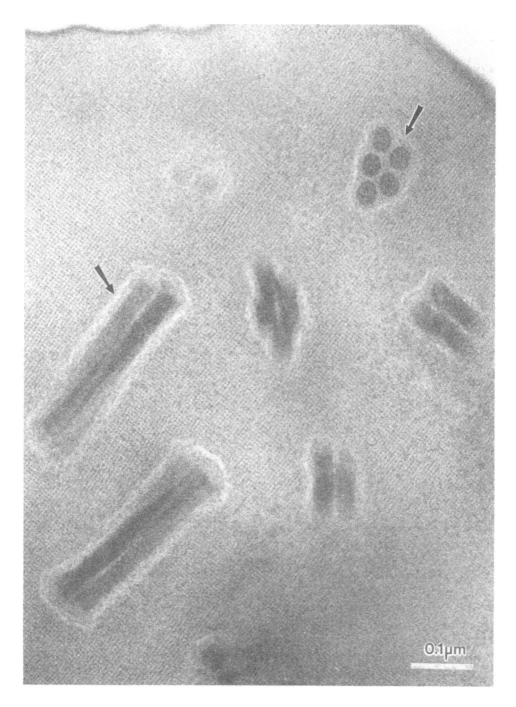

FIGURE 4.   Part of a section of a MNPV of *Heliothis armigera* showing the crystalline protein lattice surrounding the bundles of virions (arrows).

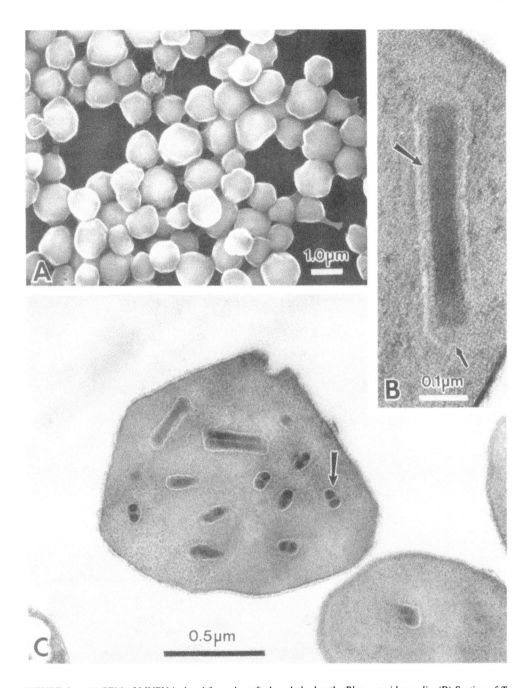

FIGURE 5. (A) SEM of MNPV isolated from the tufted apple budmoth, *Platynota ideausalis*. (B) Section of *T. ni* SNPV-VO showing a virion with protein fibers extending from nucleocapsid (NC) in the intermediate layer (arrow) between the capsid and the envelope; a long fiber (arrow) extends from NC to the viral envelope. (C) Section of *P. ideausalis* VO showing two NCs per virion bundle (arrow).

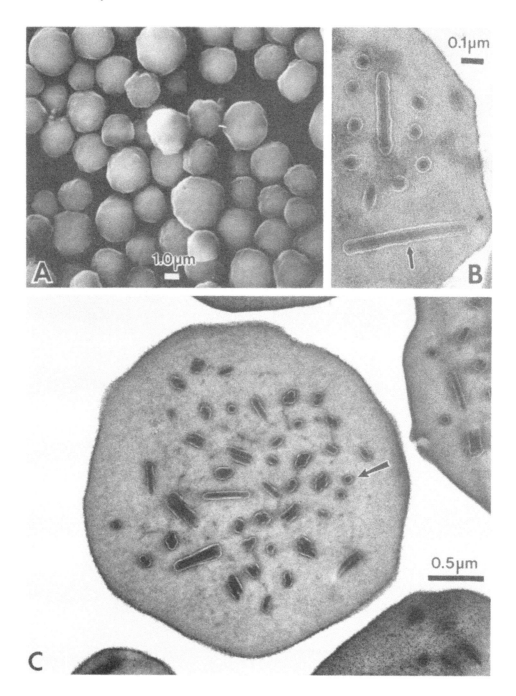

FIGURE 6.   (A) SEM of the MNPV isolated from the gypsy moth, *Lymantria dispar;* (B) section of a VO of *H. zea* SNPV showing a virion of normal length and a virion approximately twice the normal length (arrow); (C) section of *L. dispar* MNPV showing bundles of virions as well as single virions (arrow).

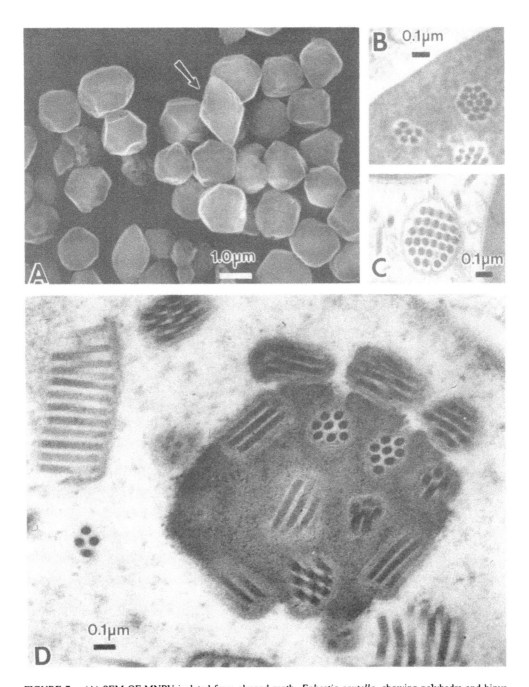

FIGURE 7. (A) SEM OF MNPV isolated from almond moth, *Ephestia cautella*, showing polyhedra and bipyramidal protein bodies (arrow). (From Adams, J. R. and Wilcox, T. A., *J. Invertebr. Pathol.*, 40, 12, 1982. With permission.) (B) and (C) Sections of bundles of virions of an MNPV infecting the brown tail moth, *Euproctis similis*. ([B] courtesy of F. Kawamoto. [C] From Kawamoto, F. and Asayama, T., *J. Invertebr. Pathol.*, 26, 47, 1975. With permission.) (D) Section of MNPV infecting *E. cautella* showing bundles of virions occluded in the developing VO and a row of NCs attached at one end to a *de novo* viral envelope.

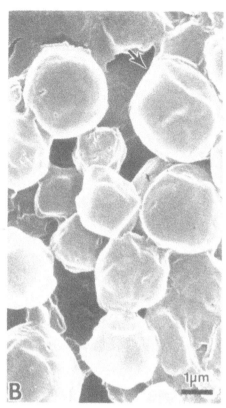

FIGURE 8.   Gypsy moth MNPV. A method for detecting defective VOs. (A) STEM-DF image (negative image); the dark areas indicate intact parts of the VOs while the light areas indicate defective parts. (B) The SEM of the same image fails to detect the defective polyhedra or VOs; for orientation the arrows point to the same VO in each field; part of the SEM image was clipped due to the use of Polaroid 665 film which has a smaller size print and negative than Polaroid 55 film. (From Adams, J. R., *Norelco Rep.*, 32, 11, 1985. With permission.)

to six layers of protein molecules, thus interpreting the variation in the opacity and shapes of protein molecules of the crystalline lattice (see Reference 212, Figure 3). Unfortunately his untimely death prevented further studies.

Surrounding the capsid of the virion in well-preserved specimens, a fibrous network is continuous with the crystalline polyhedrin in longitudinal sections which are cut near the edge of the bundle or in cross sections. It appears that the envelope is penetrated by polyhedrin fibers (see Figures 2 and 4). The separation of the enveloped nucleocapsid from the crystalline protein matrix which is noted in many sections of polyhedra may be due to an artifact in fixation and embedding, e.g., extraction of lipids, in which the membrane is poorly preserved or the VO was fixed prior to maturation (see Figures 2, 3C, 4, 5B and C, 11).

3. Virus Occlusion Envelope

Surrounding the polyhedrin protein is a coat or envelope which serves to protect the VOs and fragile virions within from the harsh elements of the environment. Minion et al.[213] showed that it contained polysaccharides. Although referred to as the polyhedron "membrane" by some, it is not a true membrane. When ruptured, cracks will often develop within resulting in virion degradation (see Figures 8 and 9). Ultrastructural studies by Harrap[211] revealed that holes 60 Å in diameter with a central core were arranged in a hexagonal pattern on the surface of the polysaccharide coat. Interspersed randomly were holes of 150 Å in

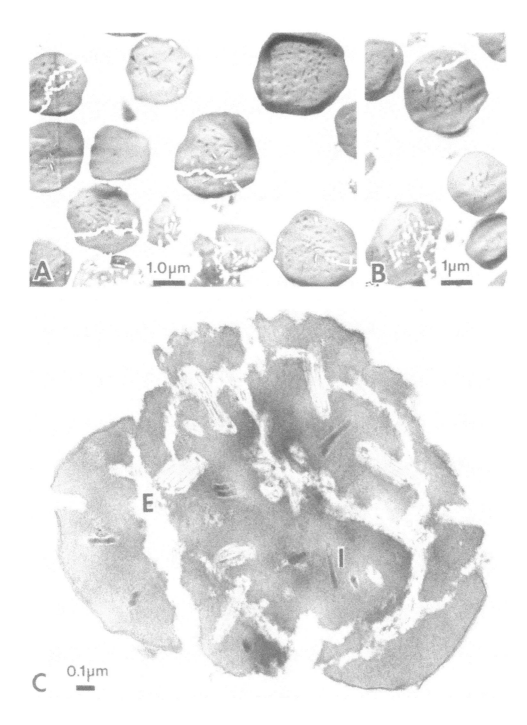

FIGURE 9. (A) and (B) Sections of intact and cracked VOs of *L. dispar* MNPV showing several stages of degradation; (C) section of MNPV, note empty capsids (E) and intact virions (I).

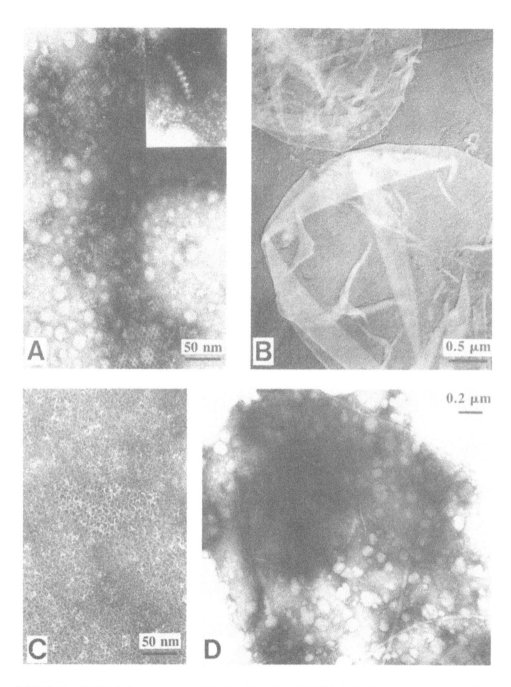

FIGURE 10.   (A) Polyhedron protein or polyhedrin of *L. dispar* MNPV in lattice-like aggregation, inset shows end-on appearance of the protein units. (B) VO membranes of *L. dispar* MNPV shadowed with gold and palladium. (C) VO membrane of *L. dispar* MNPV. (Negatively stained with uranyl acetate.) (D) VO membrane of *L. dispar* MNPV. (Negatively stained with uranyl acetate.) Note holes. (From Harrap, K. A., *Virology*, 50, 114, 1972. With permission.)

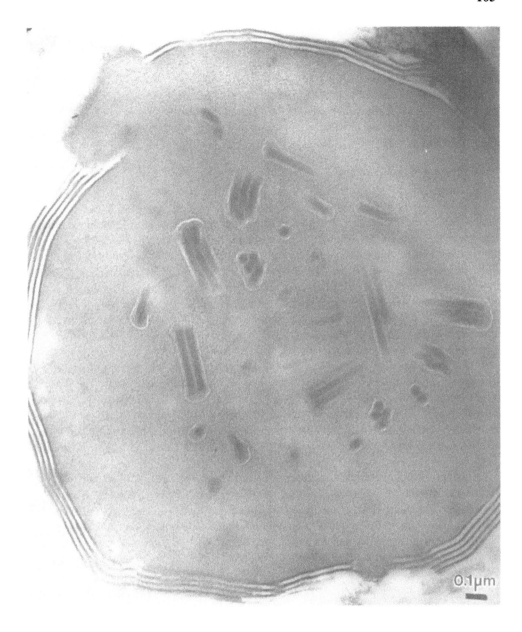

FIGURE 11.   Section of MNPV of *Agrotis segetum* showing multiple polyhedron membranes.

diameter (see Figure 10). The polyhedron envelope seems to be penetrated by protein fibers of the crystalline lattice (see Figures 2 and 4). An excellent review on the polysaccharide coat has been prepared by Federici.[50] Some polyhedra have multiple envelopes or coats (see Figure 11). Studies by Whitt and Manning[214] on the carbohydrate, which they called calyx, revealed a phosphorylated 34 K protein component (pp34) of the AcMNPV-VO envelope which is associated with the calyx via a thiol linkage. Gombart et al.[215] recently found a BV polyhedral envelope-associated protein of 32.4 K in *Orgyia pseudotsugata* (OpMNPV) and determined its genetic location and nucleotide sequence. Western blot analysis and immunofluorescence studies using an anti-TrpE-p32 antiserum detected the PDV associated protein of 32 K at 24 h postinoculation (p.i.) in the nucleus and cytoplasm which increased in concentration in the cytoplasm at later times p.i. By these techniques, Gombart et al.[215]

also established with polyhedra solubilized according to certain conditions that p32 was associated with the polyhedron envelope. (For further discussion see Section III.C.5.)

4. Production of Protein Bodies

In addition to the production of viral occlusions some insect viruses produce additional protein bodies and crystals which contain no virions. These often accumulate in the endoplasmic reticulum near the nuclear envelope and/or may occur in lower numbers in the nucleus (see Figure 12).[216,217]

B. PARTICLE STRUCTURE
1. Nucleocapsids

Over the years, several reviews on baculovirus particle structure have been presented.[2,13,15,17,19,24-28,30,32,37,43,44,46,49,50] Federici has listed the terms used by many authors in reference to the parts of the virion.[50] In this discussion, the terminology used in Matthews' classification will be employed[1] unless modified by Francki.[218] Two phenotypic forms of baculovirus virions exist; the invasive form is enveloped in the nucleus and occluded in polyhedra and may be referred to as polyhedra-derived virus (PDV) (see Figures 13 and 14), while the hemocoelic form acquires an envelope by budding from the plasma or nuclear membrane, is involved in the secondary infection cycle, and is now referred to as the extracellular virus (ECV) (see Figure 14).[219-225] The envelope of the invasive form or PDV is produced *de novo* within the nucleus.[226,227]

In vivo the virions are released from polyhedra in the highly alkaline gut of susceptible insect larvae. *In vitro* the virions may be released by mild alkali treatment.[15] The effect of NaOH treatment on a polyhedron which was terminated before dissolution of the virions is shown in Figure 13A. The chemically released virions have been called larval occluded virus alkali-liberated (LOVAL) or polyhedral-derived virus (PDV), while the budded virions which have been called nonoccluded virus (NOV) or budded virus (BV) are now called ECV.

The bacilliform nucleocapsid produced measures approximately 40 to 60 nm × 250 to 300 nm. The number of nucleocapsids may range from 1 to 29 per virion (see Figures 1 to 9, 11, 15).[228] The dimensions of the virion are shown in Figure 16. The length of the virion bundles are generally uniform but occasionally a bundle will contain a nucleocapsid nearly twice the length of the other nucleocapsids (see Figure 6B). Double and triple lengths of normal virions have been studied in *Galleria mellonella* NPVs.[229] Curved virions occur in the NPV isolated from larvae of *Wiseana cervinata*.[230]

Optical diffraction studies of NPV and GV nucleocapsids have revealed similar capsid structures which are composed of rings of protein subunits stacked 4.5 nm apart.[231] The cylindrical portion of the nucleocapsid (NC) is capped at both ends by regular arrangements of protein subunits which stain differently from capsid sheath. The DNA is contained within a uniformly packed cylinder about 32 nm in diameter which is surrounded by 12 elliptical protein subunits (see Reference 232, Figure 5). This model was determined from low angle X-ray scattering studies. Electron microscope studies of partially degraded NCs show extrusions of nonsupercoiled DNA. In extensively degraded NCs, rosettes are formed at one end or in fewer instances, at both ends of the NC[233] (see Figure 17A). Bud and Kelly[233] have also demonstrated that when poly-L-lysine was added to purified *H. zea* NPV DNA in a positively charged environment, particles of DNA were formed of similar length but only half as wide.[233] The different steps of unwinding of the viral DNA of *Tipula paludosa* NCs have been observed by Revet and Guelpa[234] who found that this viral DNA has a low superhelical density. From these data Kelly[31] proposed that it is possible that the low superhelical density, approaching zero, allows the viral DNA to be packaged as a rod rather than an icosahedron late in infection and results as a consequence of the interaction with the basic proteins uniformly distributed throughout the NCs.[31,233-235]

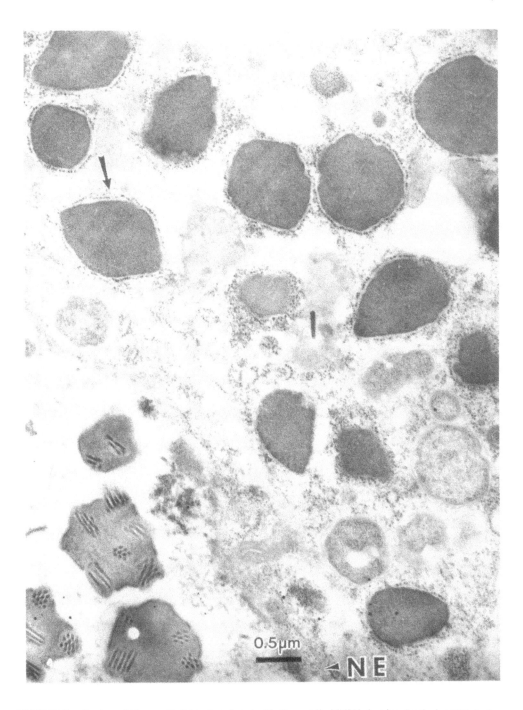

FIGURE 12.   Section of *E. cautella* fat body infected with *E. cautella* MNPV showing developing VOs in the nucleus with protein bodies in the cytoplasm (arrow). Nuclear envelope (NE).

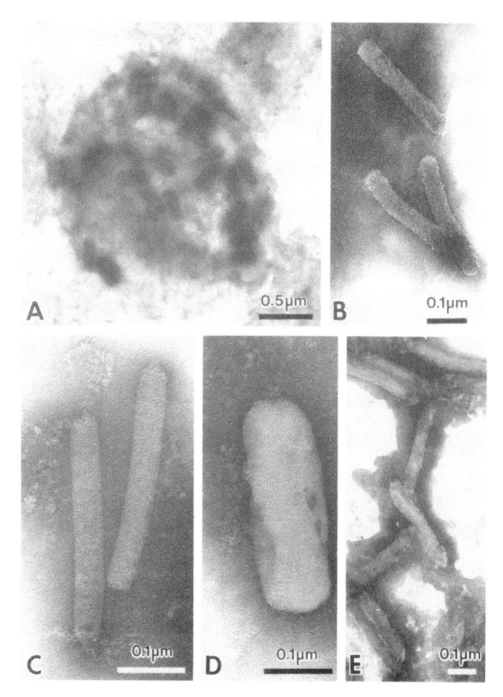

FIGURE 13.   Invasive form of virions of *Autographa californica* MNPV released from polyhedra or viral occlusions (VOs). These are referred to as polyhedra derived virus (PDV). (A) Virions that were trapped within the polyhedron envelope following alkali treatment; (B) to (D) NCs removed from the crystalline protein matrix or polyhedron after dissolution; (E) virion envelopes or capsids after release of nucleocapsids. ([B] to [E] were stained with 1% aqueous ammonium molybdate.)

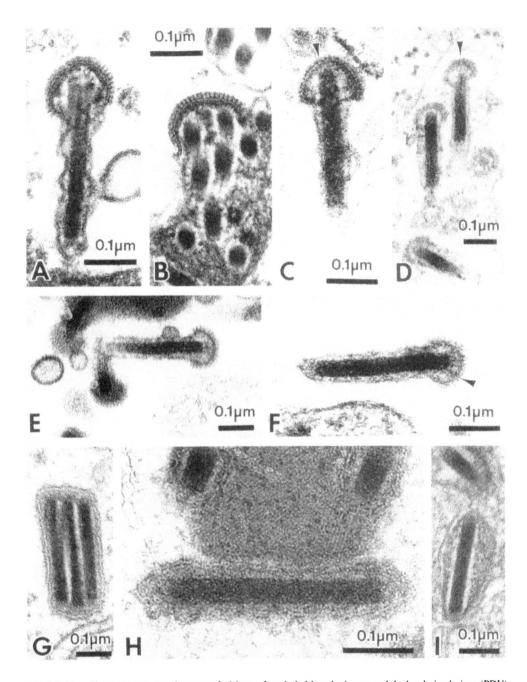

FIGURE 14. There are two morphotypes of virions of occluded baculoviruses: polyhedra derived virus (PDV) and extracellular virus (ECV). (A) to (F) Sections of ECV; as the nucleocapsids bud from the plasma membrane the anterior end of the viral envelope is modified with glycoprotein spikes or peplomers about 10 to 14 nm long; presumably fiber or filaments from the NCs attach to the plasma membrane prior to the budding event (note fiber attachments to viral envelope directly below points in [C], [D], and [F]). (G) to (I) Sections of polyhedra derived virus (PDV); the external surface of the viral envelopes are "spiny" but do not have peplomers; (A), (B), (G), (I) = *in vitro* studies; (C), (F), (H) = *in vivo* studies; (A) and (B), *L. dispar* MNPV virions; (C), (E), (F), *H. zea* SNPV virions; (D), (H), *T. ni* SNPV virions; (G), (I), *A. californica* MNPV virions. ([A] to [D] and [G] to [H] from Adams, J. R., Goodwin, R. H., and Wilcox, T. A., *Biol. Cell.*, 28, 261, 1977. With permission.)

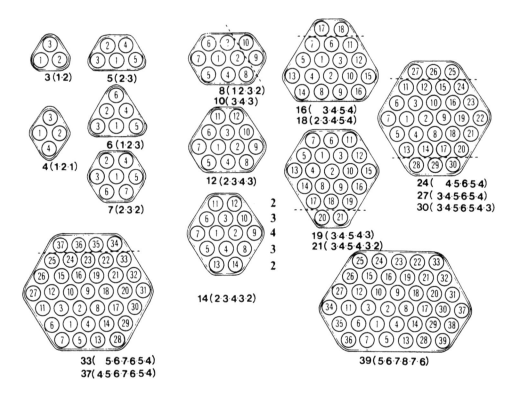

FIGURE 15.   Two-dimensional diagram of various arrangement patterns of NCs with the viral envelopes. (From Kawamoto, F. and Asayama, T., *J. Invertebr. Pathol.*, 26, 47, 1975. With permission.)

The theories proposed for the packaging of viral DNA have been summarized by Bud and Kelly[233] who presented a folded DNA rather than the supercoiled model postulated by earlier investigators. They suggested that the detection of polyamines in the NCs may, together with specific base sites, play a role *in vivo* similar to polylysine *in vitro* in folding the DNA.[233-236] Recently a model of nucleocapsid morphogenesis has been proposed by Fraser[237] in which the cap and nucleoprotein core complex formed in the virogenic stroma enter the capsid sheath in such a way that the viral DNA passes through the cap structure and winds the circular nucleoprotein strand as it enters (see Figure 17B). This is the reverse of the uncoating mechanism observed by Granados and Lawler.[225] Electron microscopic observations of NCs in various stages of uncoating and assembly tend to support this model.[238-242]

The number of viral encoded structural proteins is yet unknown but three have been identified. Pearson et al.[243] described p39 which they demonstrated by the use of a monoclonal antibody directed against a p39 K virion structural protein revealing that it is a major component of both PDV and EV phenotypes. Blissard et al.[244] then determined the nucleotide sequence, transcriptional mapping, and temporal expression of the gene encoding p39. The nucleotide sequence of AcMNPV gene encoding p39 has been sequenced and characterized by Guarino and Smith.[245] Thiem and Miller[246] then identified, sequenced, and mapped p39, the major capsid protein. Recently Müller et al.[247] have also sequenced, mapped, and immunochemically characterized a protein of 87 K which is also a component of the virions of ECV, OV, and capsids of OpMNPV, thus establishing that the OpMNPV capsid is composed of two proteins. The OpMNPV p87 polyclonal antiserum did not cross react with any AcMNPV specific proteins in AcMNPV-infected *Spodoptera frugiperda* cells indicating that a homolog of p87 was not present in AcMNPV. The p6.9 protein is an arginine-serine-rich protein which has been located on the viral genome by Wilson et al.[248] and Wilson[249]

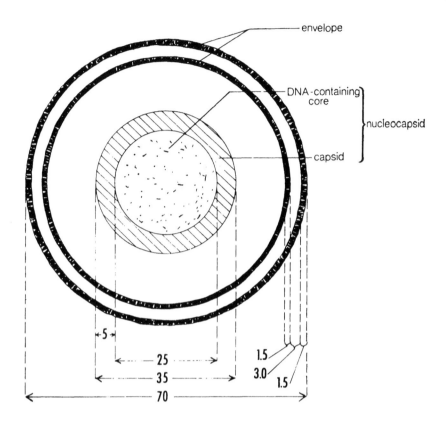

FIGURE 16.   Diagram of a virion. (From Hughes, K. M., *J. Invertebr. Pathol.*, 19, 198, 1972.
With permission.)

who found it associated with NCs and postulated that it may be involved in the condensation
of viral DNA prior to or during packaging (see also References 250, 251).

2. Viral Envelope

The nucleocapsid is enclosed within an envelope of typical trilaminar structure composed
of a lipid layer between two layers of protein. Although the envelopes of ECV and PDV
are derived from different sources (i.e., nuclear or plasma and *de novo* synthesis in the
nucleus, respectively) similar structural protein compositions are observed with only minor
quantitative and qualitative differences in the envelope. However, the ECV contains considerably more of the 64-K glycoprotein, an envelope antigen.[64] This difference in protein
composition is assessed by SDS-polyacrylamide gel electrophoresis (PAGE).[41,64,252,253] Immunoelectron microscopy of virions of ECV and PDV indicate that the antigens reacting
with neutralizing antibody are dispersed throughout the viral envelope but appear concentrated at the end containing the peplomers.[254] (Further details are presented in Section III.)

Several models of the NPV-virion structure have been proposed.[226,255,256] Recently,
Federici[50] has attempted to clarify the appropriate terminology when referring to the morphological structures of the virion by presenting a model (see Reference 50, Table 3, Figures
4 and 5). Observations of newly formed virions reveal the loosely fitting envelope, the
nucleocapsid sheath, and the intermediate layer of granular material surrounding the NC
sheath (see Figure 14).

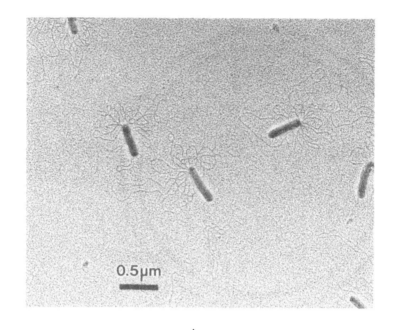

A

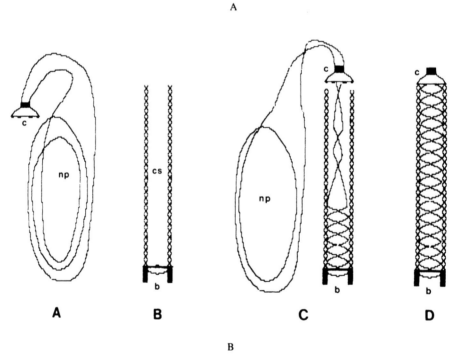

A   B   C   D

B

FIGURE 17.   (A) Deoxyribonucleoprotein exuding from nucleocapsids of *H. zea* SNPV-NCs; note that loops of DNP emerge from the capsids and that there is no beaded structure typical of nucleosomes associated with chromatin. (Courtesy of Timothy F. Booth, N.E.R.C., Institute of Virology and Environmental Microbiology, Oxford, U.K.) (B) Proposed model of nucleocapsid morphogenesis. Cap and nucleoprotein core complex associating the virogenic stroma (A); base and capsid sheath are assembled separately in pockets of the virogenic stroma (B); nucleoprotein core entering preassembled capsid sheaths through the cap structure and winding the circular nucleoprotein strand as it enters (C); fully assembled nucleocapsid (D). (From Fraser, M. J., *J. Ultrastruct. Mol. Struct. Res.*, 95, 189, 1986. With permission.)

C. GENOME STRUCTURE

1. Viral Genome

The account of early biochemical and electron microscopic studies presented by Benz[10] describes the confusion that arose among early investigators who did not expect to find bacilliform virus rods within polyhedra and granulosis capsules. The contributions of several "pioneer" investigators finally established that the "infections units" were bacilliform rods containing DNA which accumulated in the nuclei of infected cells and were occluded in polyhedrin for NPVs or granulin in the case of GV (capsules).[257-266] Reviews discussing the biochemical investigations on NPVs and GVs include those of Bergold,[13-15] Vago and Bergoin,[17] Summers,[267] Smith,[24] Harrap and Payne,[28] Miller,[34] Kelly,[31,37] Vaughn and Dougherty,[38] Arif,[43] Cochran et al.,[49] and Blissard and Rohrmann.[80]

AcMNPV-DNA has been characterized by Vlak and Odink.[268] Summers[269] reported that highly purified preparations of baculovirus virions contained 7 to 15% DNA. Summers and Anderson[269-271] also showed that DNA was liberated from NPV and GV virions as linear, relaxed, and covalently closed double-stranded molecules. Following extraction, the molecular weight of the viral genomes was calculated from measurements of the DNA molecules recovered from sucrose gradient and analytical centrifugation techniques, and by the Kleinschmidt technique (see Reference 267, Table 2). Differences in the molecular weights among the NPVs or GVs have been discussed in several reviews.[37,267]

For identification and understanding of some of the molecular differences which occur in baculoviruses, restriction endonuclease (RENS) analyses have been used. Following REN digestion, specific fragments of DNA are produced which are separated by agarose gel electrophoresis. The resulting fragmentation profiles provide characteristic fingerprints of the DNA studied[272-274] (see Figure 18). Multiple RENS are usually employed for identification and obtaining individual DNA fragments from which the size and order within the molecule are determined. The efforts of many investigators have established that NPVs contain double-stranded circular, supercoiled DNA ranging in size from about 80 to 200 kilobase pairs (kbp). The first physical maps for AcMNPV were prepared by several techniques; however, the results were in general agreement.[275-279] Such techniques which enable the preparation of a physical map include double digestions of isolated DNA fragments with RENs (see review by Cochran et al.[49]), cloning in plasmid or cosmid vectors, and Southern cross-blot hybridizations.[272] Since then physical maps have been prepared for 20 MNPVs and SNPVs (see Reference 49, Table 1).

NPV-DNAs have also been analyzed using methylation-sensitive RENs. Methylation of DNA at specific sequences serves to inactivate gene expression. Knebel et al.[280] demonstrated that the *in vitro* methylation of the AcMNPV 10K promoter sequence led to inactivation of gene expression. However, extensive methylation in AcMNPV following replication in permissive[281] or semipermissive[282] cell-virus systems has not been detected. Using HpaII and MspI, McClintock and Dougherty[272] demonstrated that the DNA of *Lymantria dispar* MNPV did not appear to be extensively methylated, at least not at the nucleotide sequences recognized by these methylation sensitive isoschizomers (see Figure 19).

Cochran et al.[49] also compared the DNAs from a variant of AcMNPV with *S. exigua* MNPV-25 noting REN site order and map coordinates as well as intragenic homologous regions. To date many of the MNPVs have been called variants of AcMNPV since only small differences occur in the restriction profiles between variants isolated from the same host. Minor differences in the REN patterns have been observed in the MNPVs of *G. mellonella, S. exigua,* and *Rachiplusia ou;* where the latter NPV appeared to be the most divergent.[49,276] Small genotypic variations have also been found among NPVs isolated from *Heliothis spp.,*[283,284] *Mamestra spp.,*[285,286] *T. ni,*[287] *S. littoralis,*[288-290] *Agrotis segetum,*[291] and from among seven plaque-purified isolates of *S. frugiperda.*[292] Smith and Crook[293] found eight distinct genotypes from a field isolate of *Artogeia (Pieris) rapae* GV by low mortality

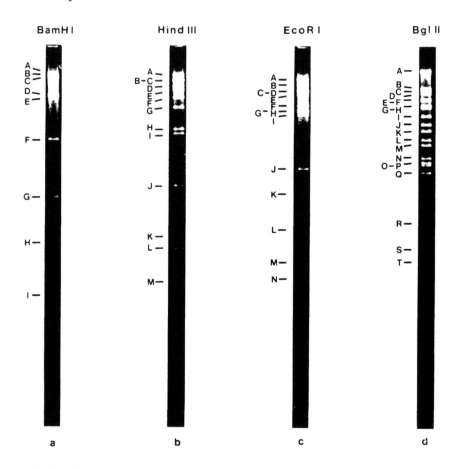

FIGURE 18.   Restriction endonuclease profiles of LdMNPV DNA. Samples of viral DNA were digested with *Bam*HI (a), *Hind*III (b), *Eco*RI (c), or *Bgl*II (d), and analyzed in a 0.7% agarose gel. Following electrophoresis, the gel was stained with ethidium bromide and photographed under UV illumination. Individual fragments for each enzymatic digest were assigned an alphabetical designation on the basis of size such that the largest fragment was designated A. The size of each fragment was determined by comparing its mobility with *Eco*RI- or *Hind*III-digested λ-phage DNA fragments of known size. (From McClintock, J. T. and Dougherty, E. M., *J. Gen. Virol.*, 69, 2303, 1988. With permission.)

dose infections of late instar larvae from which REN profiles were prepared from the individual infected larvae.

Jewell and Miller[294] have examined the DNA sequence homology among six MNPVs. Using REN analysis and Southern blot hybridization techniques, Smith and Summers[295] found that DNAs from subgroups A, B, and C had sequences homologous to the *Eco*RI-I fragment of AcMNPV, the polyhedrin gene-containing fragment. Using the cloned *Hind*III-V fragment (which contains the polyhedrin gene) of AcMNPV, fragments with sequence homology to the polyhedrin gene of AcMNPV were identified for the MNPVs isolated from *Spodoptera* and *Heliothis sp.*[296] Rohrmann et al.[297] compared the DNA sequence homology between AcMNPV and *Orgyia pseudotsugata* NPVs. Arif et al.[298,299] found that cloned *Eco*RI fragments of *Choristoneura fumiferana* MNPV (CfMNPV) hybridized to the Ac-MNPV *Eco*RI map in approximately the same order as they occurred on the CfMNPV *Eco*RI physical map. These data suggest that the relative gene order is similar between these two viruses.

During experiments to map the genome of CfMNPV, Arif and Doerfler[300] observed that the cloned *Eco*RI-A fragment of CfMNPV hybridized not only to itself, but also to three

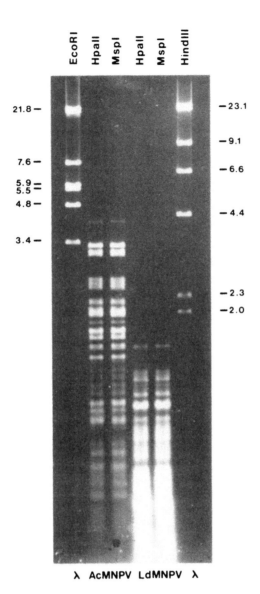

FIGURE 19. Digestion of AcMNPV and LdMNPV DNA with methylation-sensitive restriction enzymes. Viral DNA was digested with *Hpa*II and *Msp*I, electrophoresed in a 1.0% agarose gel, stained, and photographed under UV illumination. The numbers on the left and right represent the relative size (kb) of λ-phage DNA fragments digested with *Eco*RI or *Hin*dIII. (From McClintock, J. T. and Dougherty, E. M., *J. Gen. Virol.*, 69, 2303, 1988. With permission.)

other fragments. These data suggested that the additional sites of hybridization represented certain DNA sequences which were repeated in the genome in at least four regions. Four repeated sequences in the CfMNPV genome[301] were found to correspond to four of the five homologous regions (hrs) of the AcMNPV studied by Cochran and Faulkner[302] (see Reference 43, Figure 2). More recently, using cosmid cloning and Southern cross-blot hybridization to map the genome of *L. dispar* MNPV (LdMNPV), McClintock and Dougherty[272] identified at least four hrs interspersed along the genome in a fashion similar to that reported for other baculoviruses. In some respects the four hrs of LdMNPV appeared to be similar to those described for AcMNPV and CfMNPV. The origin of hrs has been of interest, and their conserved nature suggests that they are important in the replication and expression of the NPV genome; however, the significance of this is not yet understood.[272,300-305]

To determine whether the hrs enhanced delayed-early gene expression, the various hrs were cloned into plasmids containing chloramphenicol acetyltransferase (CAT, 39CAT plasmid) upstream and under the control of the AcMNPV 39-K promoter. Interestingly, all regions enhanced CAT activity to varying degrees. Based on these data, Guarino et al.[306] proposed that hrs function as enhancers by *cis* activating the transcription of regulatory genes, which, in turn, *trans* activated the transcription of delayed-early genes. Recently, the complete nucleotide sequence of the five hrs of AcMNPV was determined by Guarino et al.[306] and recently reviewed by Guarino[307] who presented evidence that the hrs function as enhancers of delayed-early gene transcription. Differences between OpMNPV and AcMNPV were presented by Gombart et al.[308] while the recent review by Blissard and Rohrmann[80] contains an excellent discussion of the genetic organization and diversity of baculovirus genomes.

2. Acquisition of Host Cellular DNA

Baculovirus genomes may acquire host cellular DNA during the infection cycle. Blissard and Rohrmann[80] have suggested that transposable elements may play a major role in the variation that occurs in baculoviruses. Miller and Miller[309] demonstrated such host cell insertion when AcMNPV was passaged 25 times in a *Trichoplusia ni* cell line. Following passage, several mutants were isolated, one of which produced few polyhedra plaques (FP-D). REN analysis of this mutant revealed new *Hin*dIII-a, -b, and -c fragments with a corresponding loss of the *Hin*dIII-K restriction fragment. The mutant FP-D was later found to be a mixture of FP-DL (identical to FP-D) and FP-DS which had lost the *Hin*dIII-a, -b, and -K fragments but appeared to produce higher levels of the *Hin*dIII-c fragment. The *Hin*dIII-a and -b fragments which were contiguous, were actually *Hin*dIII-K. The *Hin*dIII-c fragment was actually the *Hin*dIII-K fragment with a 0.27 kbp insertion. FP-DL contained a 7.3 kbp insertion plus terminal direct repeats of 0.27 kbp which was called a *copia*-like transposable element (TED). The site of integration of the TED in the viral genome was 86.7 map units.[309] Studies by Fraser et al.[310] on five and three FP mutants of AcMNPV and GMNPV, respectively, revealed that 0.8 to 2.8 kbp insertions of cell DNA occurred in the region between 35.0 and 37.7 map units of the AcMNPV genome. Marker rescue studies and analysis of infected cell-specific proteins indicated that the insertion of the cell DNA into the viral genome produced the FP mutant phenotype, with a corresponding loss of expression of a 25 kDa viral protein.[310]

Further studies revealed that some of these mutants contained transposable elements from the *T. ni* 368 cell genome.[311] Sequencing studies of the viral/host DNA junctions of Gm FP1 and Gm FP3 revealed a 750 bp host sequence with 7 bp perfect inverted terminal repeats inserted within the DNA of Gm FP3. Friesen et al.[312] found that the retrotransposon TED transposed DNA from the *T. ni* host cell to the AcMNPV genome during the infection process. This transposition promoted the transcription of many new RNAs from the element into the flanking *Hin*dIII-K sequences which subsequently altered the expression of adjacent early genes.[312] In further studies the organization of the viral genes within the 3.7 kbp *Hin*dIII-K/*Eco*RI-S region of the AcMNPV genome was found to contain two nonoverlapping genes which extended in opposite directions and encoded the 35 K and 94 K proteins, genes which appeared to be in the immediate-early class.[313] The integration of the retrotransposon TED into the viral genome disrupted the coding region of the 94-kDa gene.[313] The nucleotide sequence of the regions of the AcMNPV genome carrying insertion elements from *S. fru-giperda* has been identified.[314] Friesen and Nissen[315] recently reported on the complete nucleotide sequence of TED as integrated within the AcMNPV mutant FP-D. Their data demonstrated that the long terminal repeats of TED possess a U3-R-U5 structure analogous to the retroproviruses and together with a gene similar to *pol*, that TED transposition occurs via reverse transcription of an RNA intermediate. Fraser[59,316] has recently reviewed the potential significance of transposon-mediated mutagenesis of baculoviruses vectoring genetic elements.

Beames and Summers[317] isolated a host cell DNA sequence from *T. ni* and one from *S. frugiperda* which were transcribed in infected cells. Northern and S1 nuclease analysis revealed altered transcripts which were produced at the site of mutation as the viral DNA sequences were deleted and host cell sequences were inserted. This produced two newly mapped overlapping poly $(A)^+$ transcripts which were transcribed from a region having an open reading frame for a 25-K protein. Recently, Beames and Summers[318] have made sequence comparisons of cellular and viral copies of host cell DNA insertions found in AcMNPV. Neither of the DNA insertions, IFP1.6 or IFP2.2, encoded a large open reading frame, suggesting that new genes were not inserted in the baculovirus genome. The IFP1.6 insertion is structurally similar to other insertion elements reported in baculoviruses which have short terminal inverted repeats flanked by a duplicated target site sequence TTAA (see also References 314, 319, 320). The IFP2.2 insertion, unlike any other host DNA insertion element reported in baculovirus genomes, lacked terminal direct or inverted repeats and had a flanking 8 bp duplicated target site.[318] Transposable elements identified in baculovirus genomes may have originated from highly or moderately repeated host all DNA.[80,310,317]

In summary, the biochemical studies described before have shown how the viral genome may be altered by passage in insect hosts. The most susceptible insect hosts appear to be those challenged with virus originally propagated or isolated in the same or closely related host species. Viruses passaged in alternate or distantly related host(s) may show evidence of decreased virulence due to some host-induced modification or absence of viral replication due to the inability of the virus to attach, penetrate, or replicate.

3. Polyhedrin Gene

Van der Beek et al.[321] first established that the polyhedral protein is virus encoded. Using specific mRNAs the polyhedrin gene was located on the *Eco*RI-I fragment of AcMNPV.[279] Recently, sequencing data for the polyhedrin gene of AcMNPV revealed a 5′ leader sequence of 58 bp and a coding sequence of 732 bp in length.[322] When the DNA sequence of a morphology mutant (M5)[323] of AcMNPV was compared to another AcMNPV morphology mutant (MS), Carstens et al.[324] showed that a single-point mutation caused a substitution of leucine for proline at amino acid 58 in the M5 polyhedrin and that this mutation was shown to be responsible for the formation of cubic polyhedra with virtually no virions occluded into the VOs (see Reference 324, Figure 4.) A second mutant (Ac-M5poly1) studied carried a single-point mutation at position 172 of the polyhedrin gene changing thymine to cytosine. The polyhedrin was expressed as cubic and pyramidal poly-hedra which had few or no virions.[325]

From recombinant studies, using mutants AcM5poly1 and M29, Carstens et al.[325] further determined by direct DNA sequencing of the *Bam*HI-F fragment of M29 that a single-point mutation occurred at position 253 of the polyhedrin gene which caused a substitution of phenylalanine for leucine at amino acid 84. This substitution eliminated viral occlusion formation. Carstens et al.[324] noted that the predicted secondary structure analysis of wild type (wt) polyhedrin indicated that the region from amino acids 80 to 86 possibly forms a β-sheet[324] so that the mutation at position 84 of the M29 polyhedrin would not alter the underlying secondary structure but might have exerted an effect on the tertiary protein structure.[325]

To determine transcriptional/initiation and processing, Howard et al.[326] synthesized a 102 base cDNA fragment of the 5′ leader sequence of the AcMNPV polyhedrin mRNA. They were able to demonstrate that posttranscriptional processing and subsequent splicing of the 3′ end of the mRNA did not occur. S1 nuclease analysis and high resolution polyacrylamide gel electrophoresis of the 3′ end of the mRNA revealed that transcription terminated at 372 to 376 bp upstream from the end of the polyhedrin reading frame.[326] These results suggested that transcription of the polyhedrin mRNA was initiated at bp 501 rather than the position reported earlier (8 bases upstream).[322] Recent studies by several laboratories have identified and characterized specific sequences which are essential for the efficient expression of the polyhedrin gene.[327-332] This information is important in the design of virus expression vectors.[326]

Luckow and Summers[333] recently reviewed their findings on the nucleotide sequence spanning the polyhedrin start codon in the wild type AcMNPV and in several transfer vectors. The most recent report on the nucleotide sequence of AcMNPV polyhedrin is given in the manual of methods for baculovirus expression vectors prepared by Summers and Smith.[58] The polyhedrin genes of *Bombyx mori* NPV (BmNPV)[334] and *Orgyia pseudotsugata* MNPV (OpMNPV)[335,336] have also been sequenced. Recent reviews by Rohrmann[45] and Vlak and Rohrmann[42] on polyhedrin give further details on this protein which is produced in abundance late in the infection cycle by NPVs.

III. CYTOPATHOLOGY

A. EARLY STUDIES

Early studies on cytopathology consisted of light microscopic examinations of NPV-infected larvae. These served mostly to illustrate the extensive infections produced by NPVs which resulted in high mortalities.[11,12,26,337] The use of the electron microscope greatly enabled investigators to visualize the virus particles or virions and begin to unravel the mysteries of the replicative cycle of NPVs by observations of the cytopathic effects (CPE) on the host tissue.[13,14,16,338-359] Early reviews on NPVs which included cytopathology were prepared by some of the "pioneers" in insect pathology.[13-15,17,24] It was possible to observe the effect of the virus on infected nuclei, to see the virogenic stroma, the virus rods or virions, and the formation of polyhedra.[347-349] Bird[350] and Day et al.[351] reported that polyhedra dissolved in the gut and proposed cycles of viral development in susceptible insect hosts. Some of the challenging unknowns at that time were the definition of an infectious unit, the events that occur during each stage of viral replication in infected cells, and the fate of virions which are not occluded in polyhedra. Through the efforts of many investigators some of these questions are being resolved. The infection pathways of NPVs *in vivo* and *in vitro* have been illustrated by Fraser[316] (see Figures 20 to 21).

Table 4 shows some of the observations in early histological studies. Drake and McEwen[353] reported that the nuclei of *T. ni* fat body, hypodermis, tracheal matrix, and blood cells were filled with polyhedra. Heimpel and Adams[161] reported the finding of polyhedra in midgut nuclei of *T. ni* larvae, while Mathad et al.[359] did not. One difference was that in the former test, insects were reared on foliage while in the latter test insects were reared on a semi-synthetic diet. Another difference was the viral inoculum. Perhaps the inoculum used by Mathad et al.[359] was very high in SNPVs since Heimpel and Adams,[161] using a mixture of SNPV and MNPV, found only the MNPV polyhedra in the infected midgut cells. Baugher and Yendol[361] found AcMNPV was more virulent to fourth instar *T. ni* when fed with diet plugs than when fed with cabbage leaf discs. In contrast to the NPVs that infect many tissues, two NPVs have been reported that infect only specific tissues causing infection only in hypodermal cells.[360,362] The *Pseudaletia unipuncta* HNPV strain which causes hypertrophy of susceptible tissues also infects the tracheal matrix but not the fat body.[356] The infections

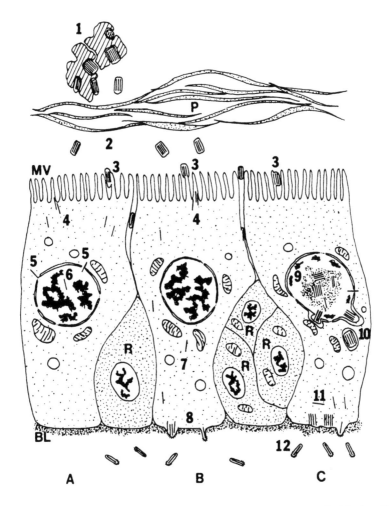

FIGURE 20. Infectious pathways of NPV in insect larvae. VOs are ingested and dissolved in the midgut of the larvae, releasing the embedded virions (1). Released virions pass through the peritrophic membrane (P) and associate with columnar epithelial cells or the midgut wall (2). Virions contact microvilli or lateral surfaces of the columnar epithelial cells of the midgut and gain entry by membrane fusion (3). Once inside the cell, two routes may be followed: initiation of replication (cell A) or translocation to the basal membrane (cell B). Epithelial regenerative cells (R). Cell A: nucleocapsids are transported through the cytoplasm to the nucleus (4), probably through association with microtubules; nucleocapsids enter the nucleus, probably through nuclear pores (5); uncoating of the viral genome and initiation of viral replication occurs in the nucleus (6). Cell B: in some cases nucleocapsids bypass the nucleus and migrate directly to the basal membrane of the columnar epithelial cells (7); nucleocapsids bypassing the nucleus can bud through the basal membrane and basal lamina (BL) into the hemolymph, acquiring a cell membrane-derived envelope in the process (8). Cell C: virus replication in the nucleus of columnar epithelial cells produces progeny nucleocapsids but no VOs; a diffuse virogenic stroma is formed in which nucleocapsid assembly takes place (9); nucleocapsids exit the nucleus by budding through the nuclear envelope, forming transport vacuoles (10); transport vacuoles are lost during transport of nucleocapsids to the basal membrane (11) where budding of progeny virus takes place; extracellular virus in the hemolymph (12) transmits the disease throughout the larval body. (From Fraser, M. J., *Ann. Entomol. Soc. Am.*, 79, 773, 1986. With permission.)

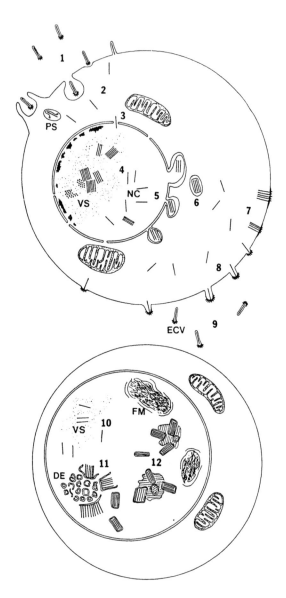

FIGURE 21. Biphasic replication process in cell cultures. (A) First phase: DNA replication, nucleocapsid assembly, and extracellular virus production; virus enters the cell either by phagocytosis (viropexis) or fusion of virion envelope with cell membrane (1); envelopes are lost in phagosomes (PS) or during membrane fusion, and the naked nucleocapsids migrate to the nucleus (2); nucleocapsids enter nucleus through nuclear pores (3) and begin replication which results in breakdown of cellular chromatin and elaboration of a diffuse virogenic stroma (VS) (4); nucleocapsids (NC) are made in the virogenic stroma and migrate to the nuclear periphery (5); NC escape the nucleus by budding through the nuclear envelope, forming transport vacuoles in the process (6); transport vacuoles are lost and naked nucleocapsids align at the inner surface of the modified cytoplasmic membrane in preparation for envelope acquisition (7); cytoplasmic membrane derived envelopes are gained through budding (8); extracellular virions (ECV) (9) have peplomer protrusions at one end which are thought to facilitate adsorption to cell membranes; ECV production continues until 20 to 24 h postinfection. (B) Second phase: *de novo* envelope synthesis; VO maturation; virogenic stroma (VS) shrinks as nucleocapsid synthesis and assembly are completed (10); *de novo* envelope (DE) synthesis and nucleocapsid envelopment (11) precede encapsulation in enlarging VO matrix (12); envelope-like material produced by fibrous material (FM) surrounds mature VO. (From Fraser, M. J., *Ann. Entomol. Soc. Am.*, 79, 773, 1986. With permission.)

TABLE 4
Early Studies on Nuclear Polyhedrosis Virus Infection in Several Insect Species

Type	Polyhedra (μm)	Virions (nm)	Susceptible tissues	Host	Ref.
M	2	160 × 350	FB, TR, HY, H	*Choristoneura fumiferana*	352
			FB, HY, TR, H	*Trichoplusia ni*	353
M	0.6—4.5	35—41 × 270—370	FB, TR, MS, NS, MU, G, PC	*Malacosoma alpicola*	340
M + S			FB, TR, HY, (Gut—M NPV only)	*Trichoplusia ni*	161
M	1.5—2.2	40 × 312	FB, TR, HY, NS, BR, OM	*Melanchra picta*	186
M		21—32 × 235—300	HY, FB, TR, MT, TFS, MS, AM	*Ephestia cautella*	217
M			FB, HY, TR, FG, HG, TS, ID, H, SG, MT, BR, G	*Spodoptera frugiperda*	354
S			FB, TR, HY, H, MU, MT, NT, TS, IWD	*Trichoplusia ni*	357
M + S			HY, FB, TR, H, IWD, TS, TL, MGC, SG, MT, NL, TS	*Trichoplusia ni*	359
M			HY, FB, SG, MT, NS, FG, MG, IB	*Tineola bisselliella*	346
M			FB, HY, TR, NS, MT, CV, FG, MG, PV, HG	*Tineola bisselliella*	358

Note: AM = anterior midgut (region of peritrophic membrane formation); BR = brain; CV = cardiac value; F = fat body; FG = foregut; G = ganglia; H = hemocytes; HG = hindgut; HY = hypodermis; IB = imaginal buds; ID = imaginal discs; IWD = imaginal wing discs; MG = midgut; MGC = midgut connective tissue; MS = muscle sheath; MT = Malpighiar tubules; MU = muscle; NL = neural lamella; NS = neural sheath; NT = nerve tissue; OM = ommatidia; PC = pericardial cells; PV = pyloric valve; SG = silk gland; TFS = testicular and follicular epithelium; TL = thoracic legs; TR = trachea; TS = testicular sheath.

of mosquitoes[363,364] and sawflies[338,342] by NPVs are also restricted to midgut nuclei, gastric caecae[363] and cardia.[364]

A synergistic factor (SF) has been found in the capsule matrix of the Hawaiian strain of the GV of the armyworm, *P. unipuncta,* which enhances the infectivities of certain baculoviruses *in vivo* and *in vitro.* It is a 126-K lipoprotein whose activity depends mainly on its phospholipid component. It increases virion attachment to midgut cells as it attaches to virion envelopes and microvilli membranes.[365-379] Using immunoelectron microscopy with colloidal gold markers in *in vitro* studies, Hukuhara and Zhu[379] found that the SF is adsorbed on the virion membranes and on the plasma membranes of BV-infected cell cultures. Two interesting reviews of the studies conducted on the SF has been presented by Tanada[380,381] and are discussed in greater depth in Chapter 7.

B. INVASION

Excellent reviews have appeared on the cytopathology of baculovirus invasion and replication.[18,22,24,25,29-31,36,40,51,52,82] Light microscopic examinations of cell cultures reveal the cytopathic effect (CPE), e.g., the hypertrophy of the nucleus, virogenic stroma, and formation of viral occlusions (VOs). Through the efforts of many investigators, the infection pathways are being revealed. Early studies showed that polyhedra were dissolved in the high alkaline conditions of the midgut liberating the virions. This may occur within 12 min.[382]

The virions then pass through the loose fibrous peritrophic membrane.[22,29,225,240] Since the foregut and hindgut contain cuticular linings, the insect is protected from viral invasion in those tissues. Recent studies by Derkson and Granados[383] on the peritrophic membrane (PM) in *T. ni* larvae inoculated with AcMNPV indicated that a specific reaction took place between the virus and the PM. The PM became fragile; a 68 K-glycoprotein of the PM disappeared within 15 min p.i. The protein composition of the PM was affected in both *in vivo* and *in vitro* assays. The viral factor causing the striking structural change was associated with the polyhedrin fraction but was not the VO protein itself. *T. ni* larval bioassays showed that a factor present in the AcMNPV polyhedrin enhanced the infectivity of AcMNPV. Granados and Corsaro[384] recently cloned a *T. ni* GV gene product known as viral enhancing factor (VEF) which is a protein with a molecular weight or M_r of 101 K. VEF has the ability to disrupt the structural integrity of the insect peritrophic membrane and aid in the attachment of virions to midgut columnar cells. ECVs were first observed in midgut columnar cell nuclei[382,383,385,386] while goblet cells were uninfected. Studies by Summers[238,387] and Tanada and Leutenegger[388] revealed infection pathways in GV infections which appeared to be similar to NPVs. Virions fuse with the microvilli in the apical region; (see Figure 22A, B) the naked nucleocapsid (NC) is taken into the cytoplasm and moves toward the nucleus,[238,387] perhaps along a gradient associated with the microtubules (see Figure 22C) although there is no direct experimental evidence to prove this.[225,240] The NCs probably enter the nucleus through a nuclear pore and are uncoated[225] (see Figure 22D). Tanada et al.[377] found NCs at the nuclear envelope and within the nucleus but no evidence of uncoating at the nuclear pore in studies on HNPV. This is in contrast to studies on GV infections in which the NCs uncoat leaving the capsid at nuclear pores while the viral DNA enters through that site.[22,29,238]

The first evidence of viral infection in the midgut cells is the hypertrophy of nuclei as the process of cell division is altered[347] and viral DNA synthesis is initiated. The chromatin marginates and the nucleus becomes filled with virogenic stroma as progeny virions are produced.[348-352] Virions are enveloped *de novo*[227] and occluded in polyhedra or viral occlusions in the midgut columnar cell nuclei of hymenopteran larvae.[342] A similar cycle of viral replication occurs in the midgut cells of lepidopteran larvae except that the virions are not occluded in polyhedra although polyhedrin is synthesized and a few VOs form which contain few or no virions (see Figure 23).

The replication of SNPVs as reviewed by Granados and Williams[51] show the time sequence of six events in the replication cycle of *H. zea* SNPV (HzSNPV) and AcMNPV (see Reference 51, Tables 2 and 3, Figure 2). Briefly they found the following times postinoculation (p.i.) for these events over a 48-h period.[225,240]

	HzSNPV	AcMNPV
1. Virus fusion and entry	1—4 h p.i.	$^1/_4$—4 h p.i.
2. NCs in cell cytoplasm	2—4 h p.i.	$^1/_2$—6 h p.i.
3. NCs uncoating in cell nucleus	2—4 h p.i.	1—6 h p.i.
4. Virogenic stroma and progeny	8—48 h p.i.	8—48 h p.i.
5. OB formation	16—48 h p.i.	24—48 h p.i.
6. Virions budding at basement membrane	16—48 h p.i.	$^1/_2$—6, 12—48 h p.i.

In vivo the time to mortality varies from 2 d up to 9 d according to dosage and insect host. Third instar *Heliothis armigera* inoculated with *H. armigera* SNPV and analyzed by enzyme-linked immunosorbent assay, showed virus particle antigens in extracts of whole larvae at 12 h p.i.[389] Viral entry is by fusion to the microvillar membrane.[240,369,390] By measuring the infectivity of the hemolymph and calculating the virus plaque-forming unit per milliliter of hemolymph sample, Granados and Lawler[225] determined that many NCs pass directly through the columnar cells to the basement membrane within 30 min. These findings, supported by additional electron microscopic examinations, confirmed that there

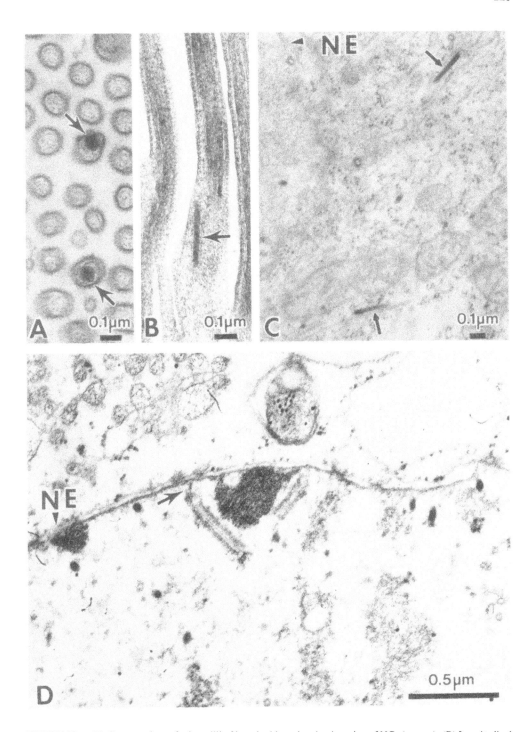

FIGURE 22. (A) Cross sections of microvilli of larval midgut showing invasion of NCs (arrows). (B) Longitudinal section through microvilli showing a NC (arrow) which has invaded the gut and is passing down the microvillus toward the midgut cytoplasm. (C) NCs free in the cytoplasm appear to be associated with microtubules (arrows); *T. ni* SNPV infection in *T. ni* midgut cells; nuclear envelope (NE). (D) Section of *T. ni* midgut showing early stage of *T. ni* SNPV viral replication; a viral envelope lacking a nucleocapsid is attached to the nuclear envelope by a filament (arrow); a filament within the lumen of the viral envelope terminates outside the envelope.

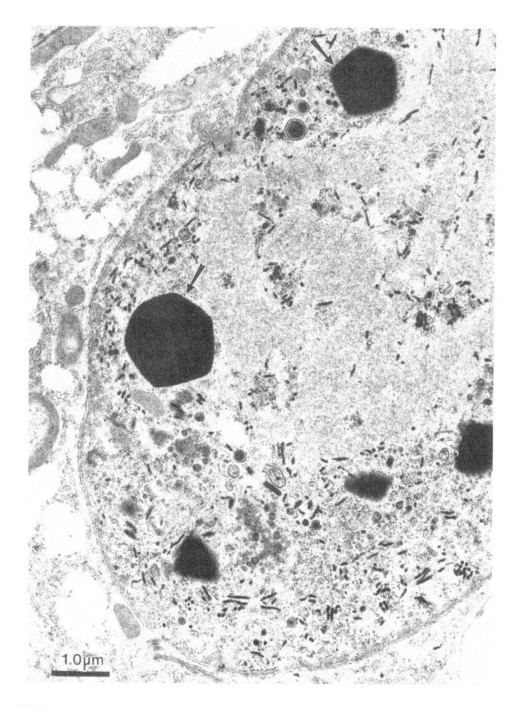

FIGURE 23.   Section of midgut of *T. ni* infected with *T. ni* NPV. Viral replication of NCs has occurred. A few virions are formed but virions were not occluded in the polyhedron protein or VOs (arrows).

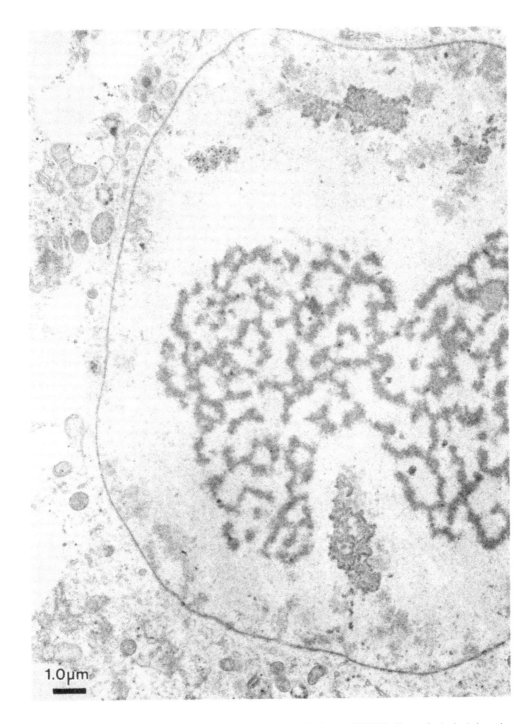

FIGURES 24—29.   Sections showing stages of infection of *Heliothis zea* SNPV in *H. zea* fat body 4 d postin-oculation (p.i.). Part of nucleus is shown in order to view the NCs at higher magnification. FIGURE 24. Note hypertrophied nucleus and margination of chromatin.

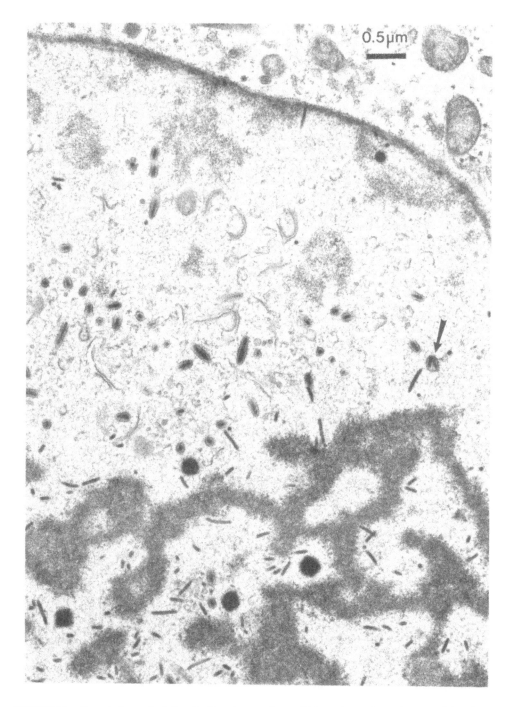

FIGURE 25.   Several stages of viral replication show nucleocapsid formation and envelopment of the virions scattered throughout the virogenic stroma. Note enveloped V-shaped NC (arrow).

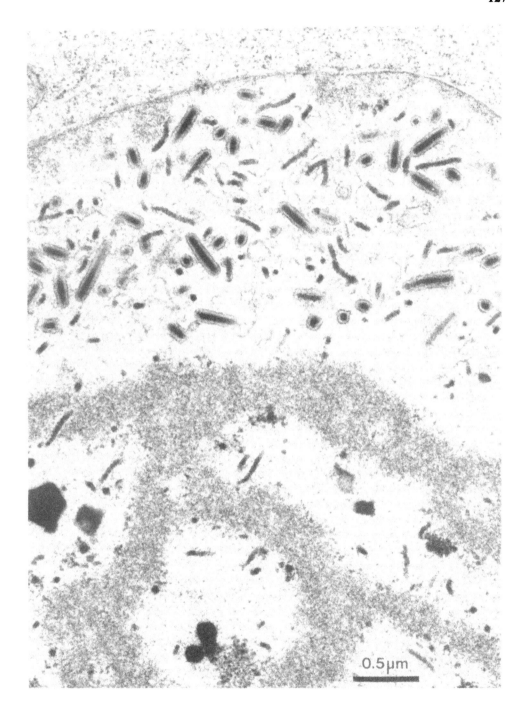

FIGURE 26. High magnification of NCs and virions.

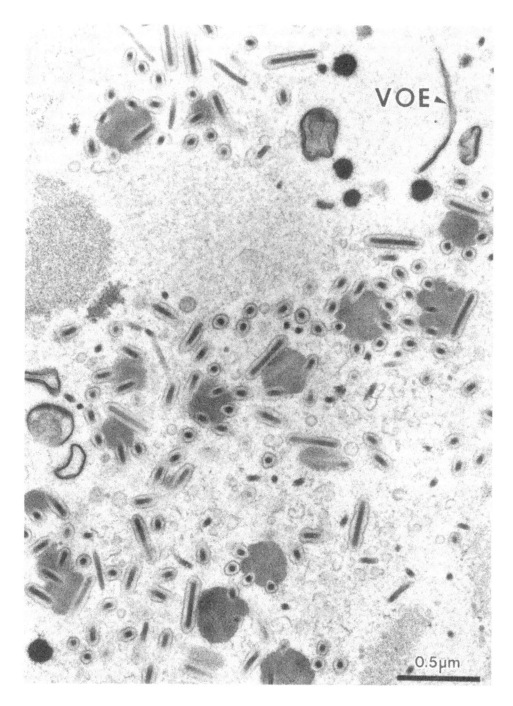

FIGURE 27. Early stage of polyhedron formation. Also note polyhedron or VO envelope (PE).

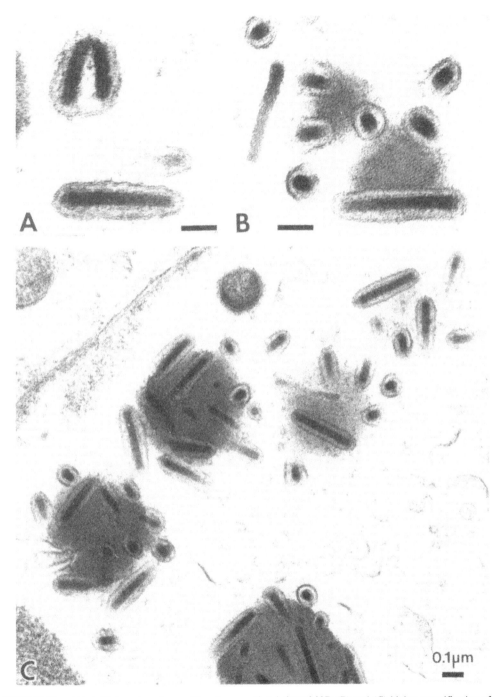

FIGURE 28. (A) A V-shaped nucleocapsid and a normal rod-shaped NC; (B) and (C) higher magnification of polyhedra or VO formation. ([A] and [B] bars = 0.1 μm.)

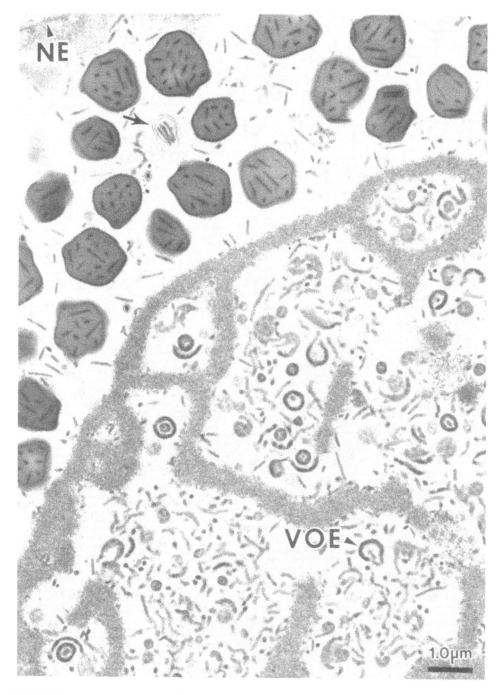

FIGURE 29.   Later stage of viral replication showing mature polyhedra between the virogenic stroma and the nuclear envelope (NE). The virogenic stroma contains polyhedron or VO envelope (PE) fragments, NCs, and virions. Bundle of virions (arrow) is surrounded by VOs.

are possibly two routes of virion invasion following virus fusion and entry[240,387] once inside the microvilli: (1) passage directly through columnar cell cytoplasm[225] and (2) initial replication in columnar cell nuclei followed by budding through the plasma membrane into the canalicular spaces.[219,225,240,386,391] Passage between cell membranes has been shown in some GV infection studies[382] but not in NPVs.[373] Some of these events are illustrated in Figures 22, 23. The progeny NCs that are produced in midgut nuclei may be released through a rupture in the nuclear envelope or may bud through the nuclear envelope, pinching off a vesicle of nuclear membrane which may be shed later releasing the NC. At the basement membrane, progeny NCs acquire an envelope by budding through the plasma membrane. The plasma membrane is modified with peplomers (glycoprotein spikes) at the area of attachment (see Figures 14, 41).[51,52,219-225,240,391] Electron micrographs usually show NCs surrounded by loosely fitting envelopes (see Figure 14). These are enveloped nucleocapsids with peplomers or ECVs.[222] ECVs and NCs have been observed in the hemolymph of NPV-infected larvae.[36]

A recent immunohistochemical study on AcMNPV infection pathways was conducted in *T. ni* larvae by Keddie et al.[392] to detect viral antigen expression before polyhedra production in individual cells. An indirect peroxidase-antiperoxidase (PAP) method was used with diaminobenzidine as an electron donor. Two monoclonal antibodies were used: MabB12D5, which is specific for the 64-K peplomer antigen (gp64) found in ECVs and Mab39P10, which is specific for a 39-K major capsid protein present in both phenotypes.

Initial infection occurred in the midgut of both columnar epithelial and regenerative cells. The regenerative cells were infected by NCs or virions that passed through or between the cells of the columnar epithelium. No evidence indicated direct passage into the hemocoel. While virus-infected columnar epithelial cells were sloughed off, the infected nidi of the regenerative cells remained to maintain the systemic infection. The infection spread from the midgut to the migut connective tissue sheath which contained muscle cells, tracheoblasts, and fibroblast-like cells.

C. SECONDARY AND LATER INFECTIONS

The secondary infection cycle in insect tissues, *in vivo*, is similar to that which occurs in insect cell cultures, *in vitro* (see Table 5). Excellent reviews have been prepared by Paschke and Summers,[22] Granados,[23,29] Faulkner,[30] Tanada and Hess,[36] Volkman,[52,64] Volkman and Keddie,[82] and Kelly.[37] Insect cell cultures are easily infected with NPVs when inoculated with infectious hemolymph or infectious cell culture supernatants containing virions. Several methods have been developed to prepare OB-derived virions for cell culture studies.[394-396] Two pathways have been proposed for the uncoating of an enveloped virus, i.e., fusion and viropexis or adsorptive endocytosis. Leland and Miller[397] suggest that both pathways are incomplete in themselves since only paramyxoviruses have the required ability to enter by fusion and viropexis while most other enveloped animal viruses appear to infect cells by adsorptive endocytosis; however, the latter pathway does not explain how uncoating occurs. They proposed that most enveloped viruses enter host cells by adsorptive endocytosis in coated vesicles which later release NCs as fusion occurs with lysosomal membranes in such a way that the NC is protected from the lysosomal enzymes and released into the cytoplasm. White et al.[398] proposed such a pathway (see Reference 398, Figure 3) which has been modified by Volkman[64] for BV entry and uncoating (see Reference 64, Figure 6). According to Volkman,[64] the virions that enter host cells via the endocytic pathway are gathered into invaginations of the plasma membrane (called clathrin coated pits) which pinch off to become coated vesicles. The clathrin disassociates from the vesicles, and fusion of two vesicles may occur to form an endosome. The pH of the endosome decreases, presumably by an ATP-driven proton pump and induces the fusion of the viral envelope to the endosomal membrane, thus allowing the nucleocapsid to escape to the cytoplasm.

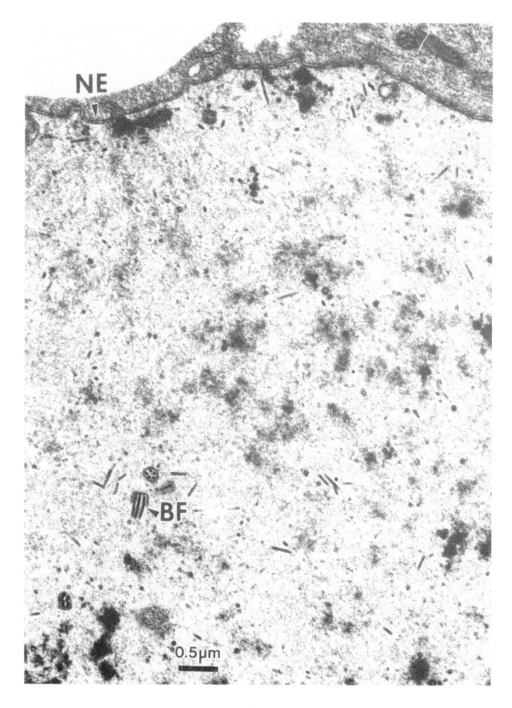

FIGURE 30.

FIGURES 30—36.   Sections showing stages of infection of *Autographa californica* MNPV in fat body of beet armyworm, *Spodoptera exigua,* at 6 d p.i. FIGURES 30—32. NC replication occurs and multiple nucleocapsid per envelope or bundle formation (BF) begins as the virion envelope surrounds the NCs. Some virions are attached at one end to viral envelope material. Nuclear envelope (NE).

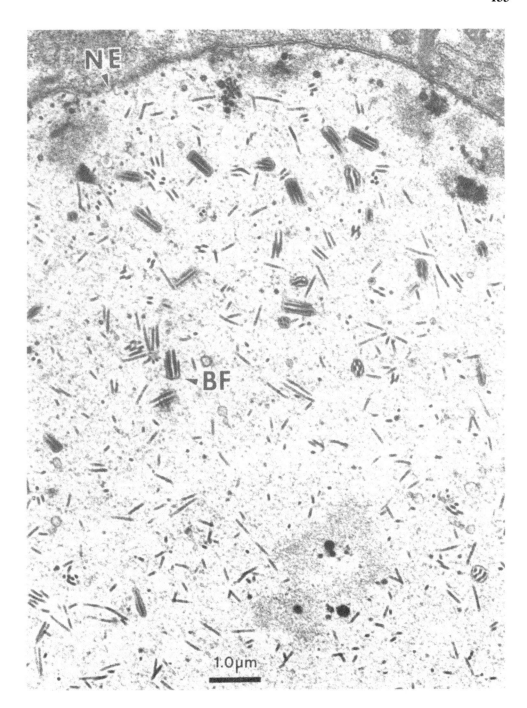

FIGURE 31.

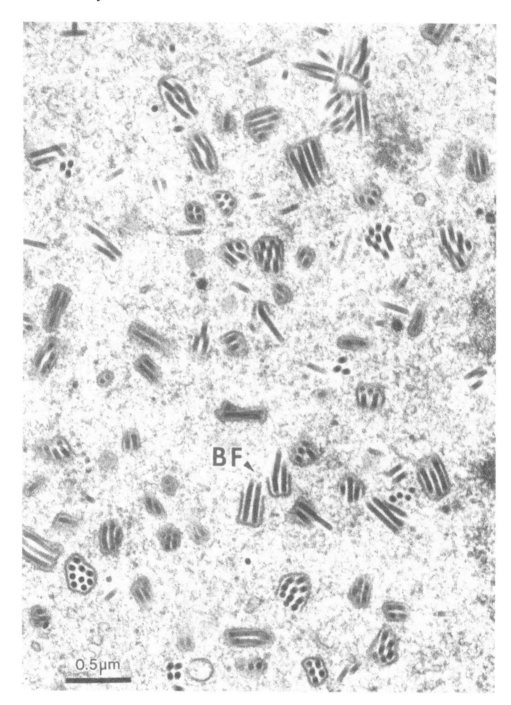

FIGURE 32.

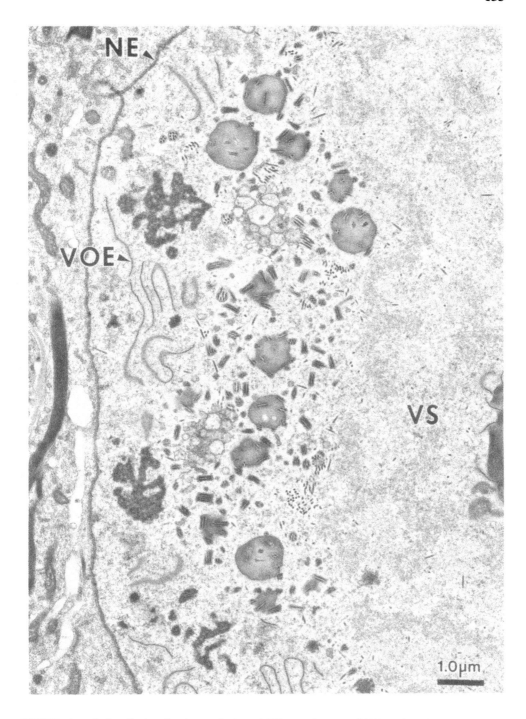

FIGURE 33.   Viral replication showing newly formed NCs, envelopment of NCs, and occlusion of virions in VOs. Note also polyhedron or VO envelope (PE), nuclear envelope (NE), and virogenic stroma (VS).

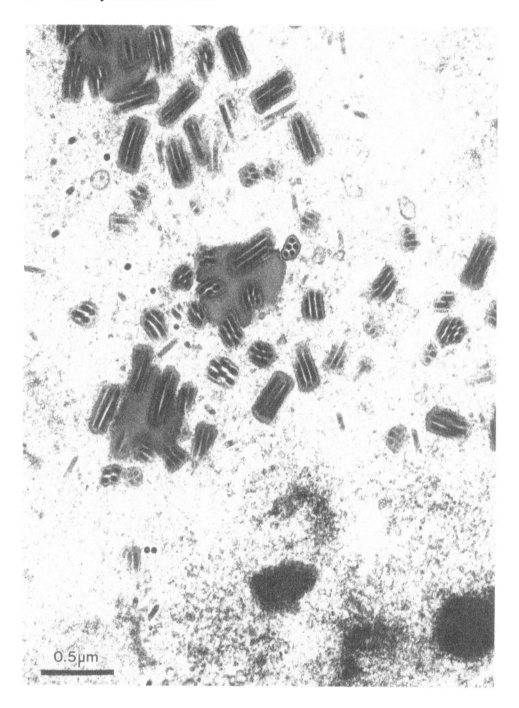

FIGURE 34.   Early stage of polyhedron formation.

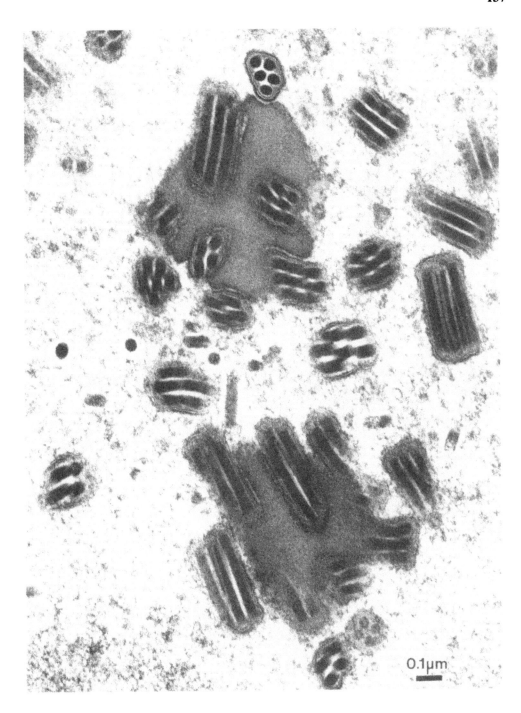

FIGURE 35. Higher magnification of a part of Figure 34.

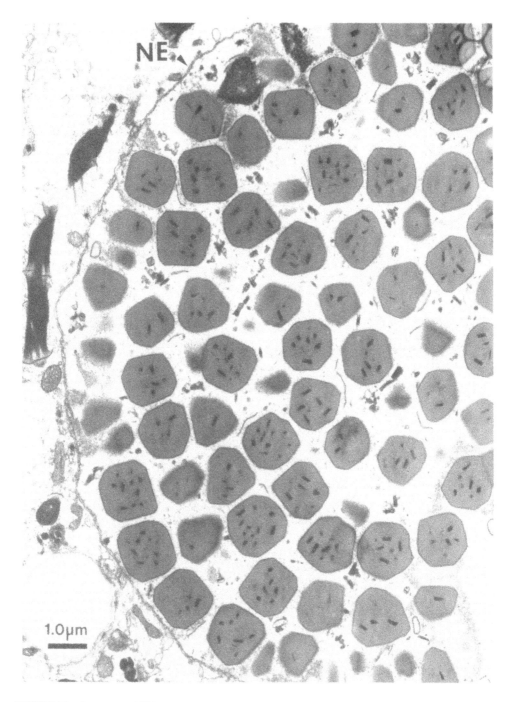

FIGURE 36.   Late stage of VO maturation. Nucleus is filled with mature polyhedra or VOs. Nuclear envelope (NE).

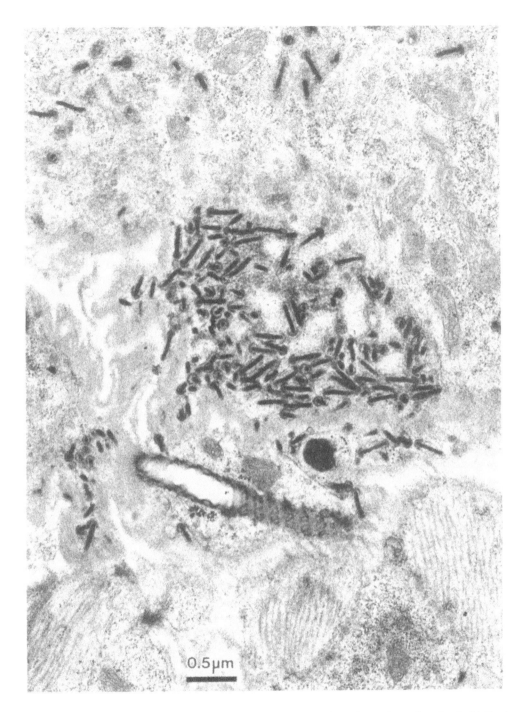

FIGURE 37.   ECVs with peplomers in vesicles in tracheal cell cytoplasm of *T. ni* larvae inoculated with *T. ni* NPV. Examinations of several areas revealed that the virions occurred singly in vesicles in large numbers in tracheal tissue surrounding the midgut.

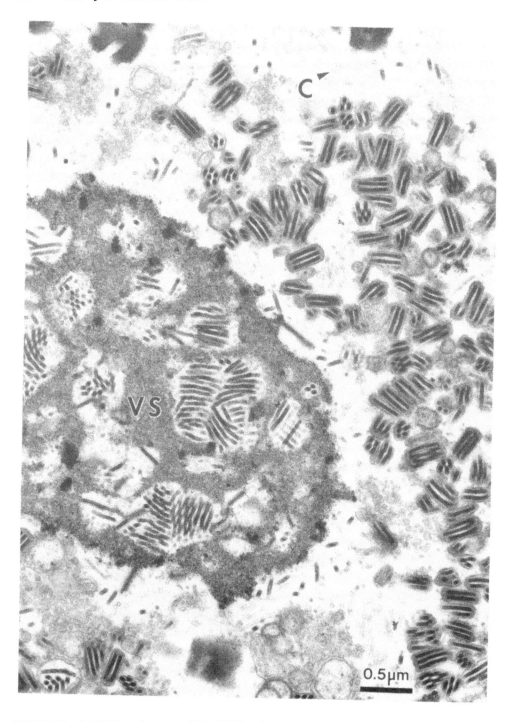

FIGURE 38.   AcMNPV replication in IPLB-TN-368 cells. NCs replicate in virogenic stroma (VS) and are enveloped nearby. Note capsid sheath or membrane (C). This is comparable to the structure labeled "capsid" in Figure 16.

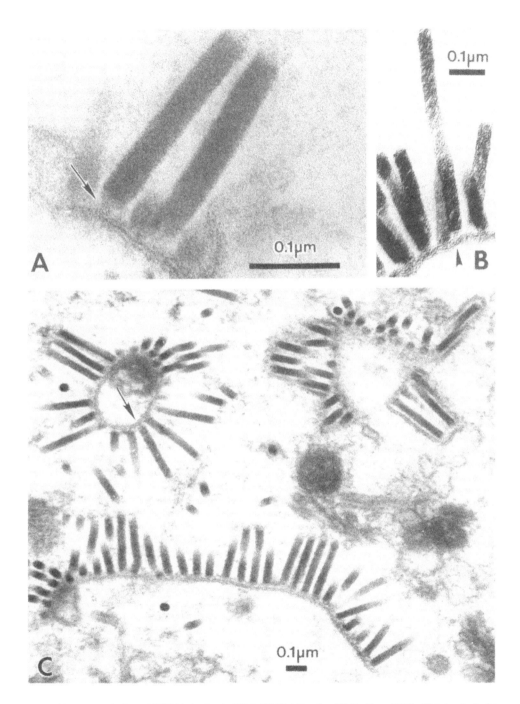

FIGURE 39.   (A) and (B) AcMNPV infection in IPLB-SF 21 AE cells. (A) Section of NC which is attached to *de novo* viral envelope apparently through the cap structure (arrow). (From Fraser, M. J., *J. Ultrastruct. Mol. Struct. Res.*, 95, 189, 1986. With permission.) (B) Note abnormally long NC attached to *de novo* viral envelope (point). (C) Replication of AcMNPV in IPLB-TN 368 cells; some sections show the attachment of NCs to *de novo* formed envelope (↑) prior to virion bundle formation.

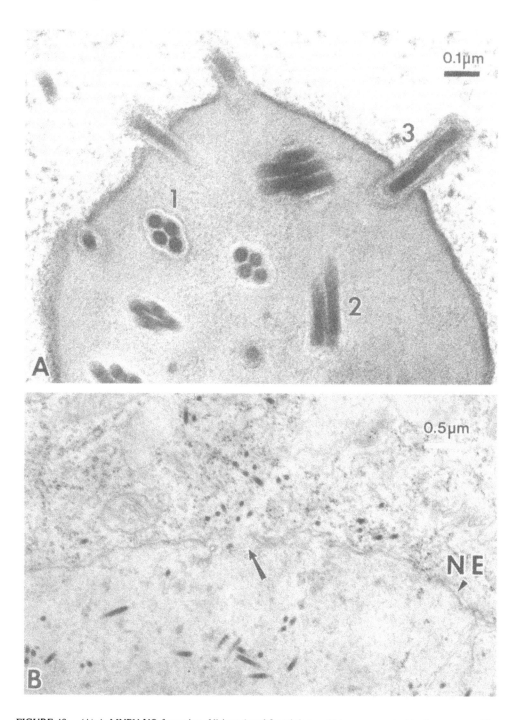

FIGURE 40.   (A) AcMNPV-VO formation. Virions 1 and 2 and those within are enmeshed in the paracrystalline lattice of the polyhedrin while virion-3 is just becoming occluded. (B) Following virion replication some NCs may exit from the nucleus through a rupture (arrow) in the nuclear envelope (NE); *T. ni* BV in *T. ni* larvae. (From Adams, J. R., Goodwin, R. H., and Wilcox, T. A., *Biol. Cell.*, 28, 261, 1977. With permission.)

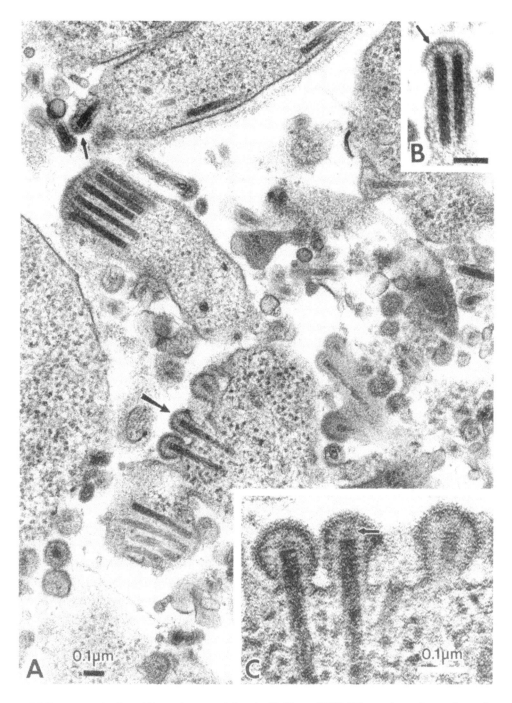

FIGURE 41. (A) Section of *H. zea* larva gut infected with *H. zea* SNPV NCs acquire peplomers (arrows) as they bud through the plasma membrane of the midgut columnar cells into the canalicular spaces. (B) Peplomer formation of *T. ni* MNPV in *T. ni*; fibers (arrow) connect the NCs to the peplomer modified plasma membrane as it becomes transformed to a viral envelope. (Bar = 0.1 μm.) (C) Enlargement of a portion of Figure 41A showing fibers (arrow) which connect the NCs to the peplomer modified plasma membrane in the process of acquisition of viral envelopes.

TABLE 5
Observations of Events of NPV Replication *(in vitro)* by
Microscopy

Events	Hrs	Ref.
Virus attachment and entry	<1	400
	0—8	239
	1	401
	1—7	241
	1—2	417
Virions or NCs in vesicles in cytoplasm	1—7	241
	1	401
	1—2	417
NCs at nuclear envelope	1—7	241
	3	400
NCs in nucleus	3	400
	1	401
Hypertrophy of nucleus	6—12	400
	12	401
	10—18	417
Virogenic stroma	7	241
	9	418
	10—18	417
	12	401
	24	239
Progeny NCs	9	401
	10	241
	12	418
Progeny NCs	16	417
	24—48	400
Envelopment of NCs	12	241, 401
	24	417
	36	239
Budding from nuclear envelope	13	241
	15	401
	18	418
	48	400
VO formation	18 or later	401
	24 or later	400, 418
	36 or later	417
	40 or later	239
Budding at plasma membrane	12	241
	15	401
	18	418
	24	417

Electron micrographs of BV-infected tissues show NCs in vesicles and NCs free in the cytoplasm in early stages of secondary infection *in vivo* and *in vitro* (Figures 14, 22C, 37, 43C). The primary route of infection is by the adsorptive endocytosis pathway.[29,36,222,225,239,240,377,379,399-401] Using neutralizing monoclonal antibody AcV[1], Volkman and Goldsmith[402] found that ECV entered cells predominantly by adsorptive endocytosis as AcV[1] acted to prevent AcMNPV-BV from using this pathway. However, not all the infectivity was neutralized. Further studies[403] showed that some AcMNPV-BV entered by fusion, not by phagocytosis. Fusion was detected as a less efficient pathway by immunoelectron microscope studies. Cytochalasin B inhibited phagocytosis but did not reduce virion infectivity.

Two electron microscopic studies showed fusion of virions to plasma membranes.[222,241] Wang and Kelly[404] concluded from their studies that virions enter predominantly by binding

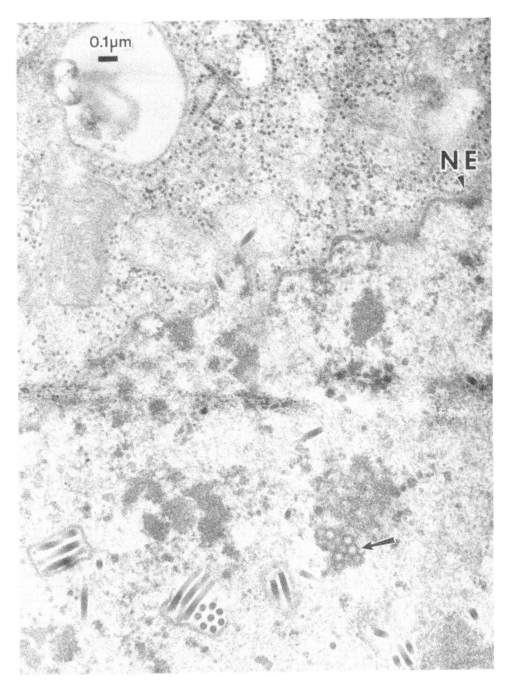

FIGURE 42. NCs may exit from infected nuclei in transport vesicles that pinch off from hypertrophied nuclear envelope by exocytosis. Normal virions are formed as well as rings of membrane or *de novo* envelope (arrow) without NCs. The section of *T. ni* 10B3C5 cell culture shows infection with *T. ni* MNPV, but cells were actually inoculated with *T. ni* GV. Nuclear envelope (NE).

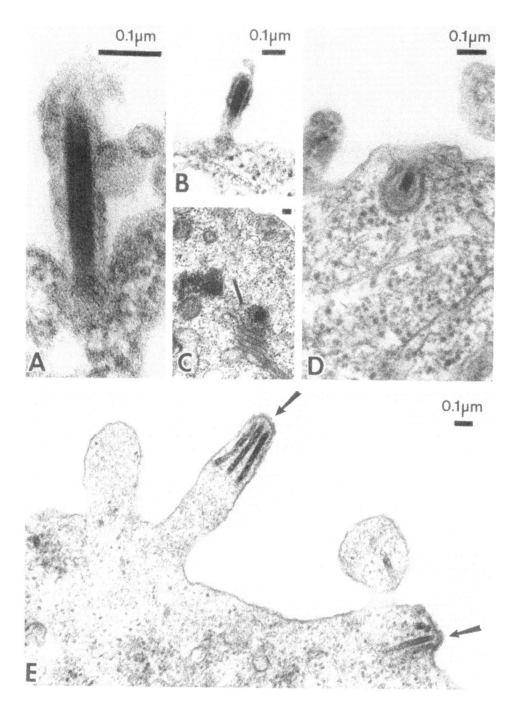

FIGURE 43.   (A) to (D) Secondary infection of AcMNPV in *L. dispar* cell line. (A) and (B) Virions with peplomers or ECV (extracellular virus) appear to fuse to plasma membrane and enter by that alternate route of infection; (C) NC in cell cytoplasm. (Bar = 0.1 μm.) (D) Many ECVs enter susceptible cells by the adsorptive endocytosis route of secondary infection; virions are taken in an invagination of plasma membrane called clathrin coated pits which pinch off to become coated vesicles. (E) At the budding site, the plasma membrane is modified with peplomers (arrow) just prior to the acquisition of the viral envelope; IPLB-HZ 1075 cell line inoculated with *H. zea* SNPV. ([A] to [E] from Adams, J. R., Goodwin, R. H., and Wilcox, T. A., *Biol. Cell.*, 28, 261, 1977. With permission.)

and fusion with the plasma membrane to release the NCs into the cytoplasm where uncoating occurs. Since virions and NCs were resistant to DNase activity they measured the sensitivity of [^{32}P]labeled cytoplasmic and nuclear fractions to DNase. Electron microscopic examinations of AcMNPV-infected cells at 10 min p.i. revealed virus particles near the plasma membrane and in the cytoplasm and capsids in the cytoplasm. They proposed that virions may also be taken up by micropinocytosis. Dougherty et al.[405] have shown that virions are absorbed rapidly, e.g., in plaque-assay studies virus was adsorbed within the first 40 s of exposure to AcMNPV-virions by 45% of *T. ni* cells.

1. Entry into Host Cell Nuclei

The pathway of NC entry has been determined for GV infections. Entrance occurs at the nuclear pores with release of the viral genome into the nucleus leaving the capsid at the nuclear membrane.[238] The *in vitro* pathway in NPV infections is not as evident, although Granados[29,225] did propose *in vivo* and *in vitro* pathways and an uncoating mechanism for nucleocapsids.

Once the viral genome is within the nucleus, viral replication is initiated. The nucleus hypertrophies (Figure 24), the virogenic stroma fills the center of the nucleus, and the metabolic activity of the cell is diverted to the production of progeny NCs.[29,225,391] The NCs appear to be produced directly from the substance of the virogenic stroma.[406] SNPV NCs appear to be replicated individually (see Figures 25 to 29) while MNPV NCs appear to develop attached to a *de novo* protein tubule or envelope (see Figures 31 to 33, 39).[184,217,237,400] Recently Fraser[237] showed the attachment of NCs by a cap structure to the *de novo* envelope. (See Reference 237, Figure 2A to C.) He also proposed a model (see Figure 17B) for NC morphogenesis which is the reverse of the uncoating mechanism observed earlier by Granados and Lawler.[225]

EM observations of replicating NCs show open ended envelopes[223,231] as well as rows of thin tubular membranes the diameter of capsid sheaths (see Figures 30, 32, 38). Bassemir et al.[241] have determined in studies on AcMNPV in *Mamestra brassicae* cell lines that short empty capsids are found at 10 h p.i. while partial and complete capsids were produced 11 to 12 h p.i. At 12 h p.i. most capsids were filled followed by budding virus progeny at 13 h p.i. Electron micrographs by Kawamoto et al.[223,407] of the virions of the Oriental tussock moth, *Euproctis subflava,* revealed electron dense particle-like structures in the matrix between the envelope and the capsid of the NC. They concluded from their measurements of the stripe pattern, with regular intervals of about 22 nm and angles of about 60° to the axis, that the intermediate layer had surrounded the NC in a spiral fashion during envelopment "interlaminating between envelope and nucleocapsid". The intermediate layer which they observed (see Reference 402, Figures 1b; 2e, f, and j) is shown on the virion structure model in Federici's[50] model (see Figure 4) as a space with dots. The 22 nm discs were presumed to degenerate once the virion matured. Federici[50] compared these structures to the circular subunits observed by Harrap[408] who found them only in negative stained preparations of virions and surmised that they were lost in the fixation and dehydration procedures used for sections of infected tissues. In most virions the intermediate layer is of equal thickness (see Figure 5B); however, virions isolated from NPVs that infect mosquitoes appear to be bulbous on one end.[257] Kawamoto et al.[407] also noted a "claw" at one end of the NC, an observation similar to that reported by Teakle.[409]

Studies to investigate microfilament involvement during NPV replication showed that when AcMNPV-infected cell cultures were treated with cytochalasin B and D, inhibitors of microfilament elongation, the production of infectious ECV was prevented by inhibiting the synthesis of complete virions (see Figure 44).[410] In further studies Volkman[411] observed in the same virus-cell culture system that microfilaments are involved in AcMNPV nucleocapsid assembly as capsid antigen is synthesized and transported to the nucleus. In the presence of cytochalasin D both capsid antigen and viral DNA were synthesized but nucleocapsid as-

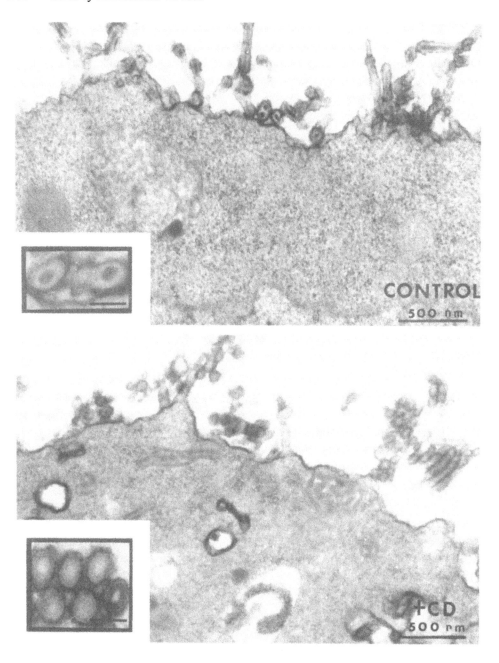

FIGURE 44. Budding particles from AcMNPV infected cells, 28 h p.i. grown with and without cytochalasin D (CD) at 0.5 mg/ml. Virus budding from the surface of the control cell shows typical morphological structure. Particles budding from the cytochalasin-treated cells lack electron dense nucleic acid cores and associated capsids. (Insets: bar = 100 nm.) (From Volkman, L. E., Goldsmith, P. A., and Hess, R. T., *Virology*, 156, 32, 1987. With permission.)

sembly was blocked. Immunoelectron microscopic studies showed that monoclonal antibody (anticapsid Mab39P10) reacted positively with capsid antigen in aberrant tubular structures in the nucleus associated with the inner nuclear membrane and in the cytoplasm (see Figure 45). These studies did not determine whether the nuclear actin was synthesized in the nucleus, transported to the nucleus from the cytoplasm, or both.[411]

Recent studies by Hess et al.[412] established that AcMNPV NC assembly is an actin-dependent process. When AcMNPV infected and control IPLB-SF-21 cells were incubated in 5 μg/ml and 0.5 μg/ml cytochalasin D (CD), many abnormalities in cell morphology were noted as well as the inhibition of proper NC assembly. In a low percentage of CD-treated cells, rodlike particles with hollow cores occurred near or were budding from cell plasma membranes. These contained viral structural proteins but lacked nucleocapsids and did not react with antibody specific for the 64-kDa surface antigen of AcMNPV. Instead many cells contained membranous profiles lacking viral DNA and spider-like electron-dense reticulate fibers in the virogenic stroma (see Reference 412, Figures 18 to 21). These structures have frequently been observed by the senior author in examinations of numerous insect cell cultures not incubated with CD due to factors yet unknown. (See Figure 50.) The findings of these structures clearly indicate interference in the normal process of viral replication.

Recently Volkman and Zaal[393] showed that microtubules maintain the normal fibroblastic shape of IPLB-Sf-21 cells. In the process of viral replication of AcMNPV, they found progressive reorganization and depolymerization of microtubules with the rounding of infected cells. In the process of viral replication, microtubules reorganized in response to early virus gene product synthesis and depolymerized in response to late gene product synthesis. These events were followed using antibody to α-tubulin in indirect immunofluorescence microscopy. A late virus protein (p10) was not a mediator of changes in the microtubules. They found that disruption of microtubules with colchicine did not interfere with AcMNPV infection and replication indicating that NC transport and replication were independent of microtubules.[393] The dramatic alterations in actin microfilament organization during viral infection were visualized by indirect immunofluorescence microscopy using rhodamine-conjugated phalloidin which showed that AcMNPV mobilizes the actin microfilaments to the nucleus where they apparently are involved in nucleocapsid assembly as they fill the ring zone and colocalize with newly assembled and forming viral NCs. Cytochalasin D, which interferes with microfilament function, also interferes with nucleocapsid morphogenesis. It prevents the polymerization of actin in the nuclei of infected cells. Thus, AcMNPV appears to utilize actin microfilaments in the nucleus in the process of nucleocapsid assembly.[393]

The number of NCs per virion (bundle) or multiple nucleocapsids per envelope varies from one in SNPVs to up to 39 for the MNPV of *Euproctis similis*.[228] Kawamoto and Asayama[228] studied the arrangement patterns of NCs in *E. similis* and presented a diagram which is shown in Figure 15. A method to calculate the number of virions per OB from data obtained from light microscopy and thin sections was devised by Allaway[413] and is discussed further in Chapter 2. The details of the process of polyhedron crystallization are still not fully understood. The polyhedrin crystallizes around the virions without penetrating through the virions or distorting the face-centered 4-nm cubic lattice. There is a characteristic size range for each NPV. Some of the larger polyhedra are joined during the developmental process. This has been observed *in vivo* and *in vitro*.[32,46,414-416] Only enveloped NCs virions are occluded in polyhedra in most native host infections. The mature polyhedra or VOs, enclosed in envelopes which provide protection, are finally released when the nuclear envelope ruptures.

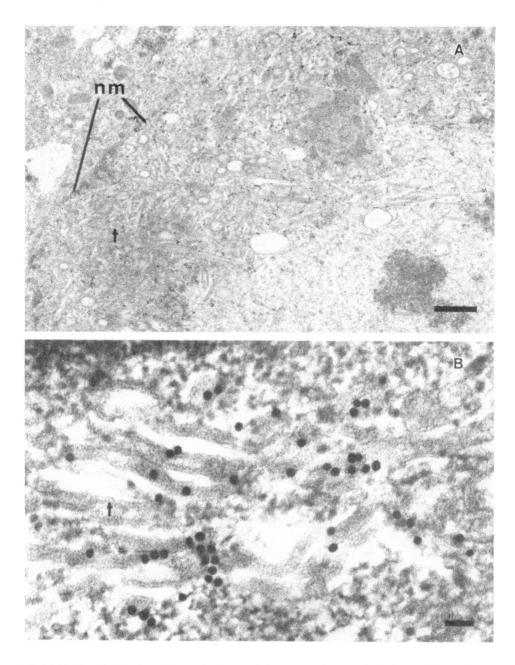

FIGURE 45.   Localization of capsid antigen in aberrant tubular structures juxtaposed to the inner nuclear membrane in cells infected with *Autographa californica* M nuclear polyhedrosis virus (AcMNPV) and grown in the presence of 5 mg/ml cytochalasin D (CD) for 51 h. The primary antibody was anticapsid Mab39P10, and the secondary antibody was anti-mouse IgG conjugated to 15-nm gold particles. The tubular structures (t) and the nuclear membrane (nm) are indicated ([A] bar = 500 nm; [B] bar = 50 nm.) (From Volkman, L. E., *Virology*, 163, 547, 1988. With permission.)

2. Exit from the Nucleus, Budding from Plasma Membrane, and Spread of Infection

The NCs that are not occluded in VOs may exit from the nucleus through a rupture of the nuclear membrane or by budding.[29,36,219-225,377,399-401,417-420] Possible routes include: (1) budding into a vesicle which later fuses to a lysosome with eventual destruction; (2) passage to the plasma membrane within a vesicle of nuclear envelope or transport vesicle[237] and exit by exocytosis; or (3) approach to the plasma membrane in an outpouching of nuclear envelope which touches the plasma membrane and releases the vesicle and/or NC by exocytosis[222] (see Figures 40B, 41, 42, 43E, 46).

Immunohistochemical staining patterns of some AcMNPV infected *T. ni* midgut columnar epithelial cells using MabB12D5 showed a polar movement of this viral protein component toward the basolateral areas which Keddie et al.[392] speculated could aid in directing the virus to bud from those membranes. After ECVs budded into the hemolymph, staining patterns indicated that some cell types were more susceptible to infection than others, e.g., one or two foci of infection were observed in epidermal or hypodermal cells 6 to 12 h p.i. while staining of hemocytes and tracheal epithelium was observed 31 to 40 h p.i. and staining of fat body occurred 45 to 50 h p.i. The initial infection of fat body occurred adjacent to infected tracheal epithelial cells. These findings support earlier studies that showed the initial infection of the tracheal matrix cells and their involvement in the spread of viral infection.[222,385] In addition infected hemocytes which adhered to basal lamina of various organs apparently augmented penetration of ECVs for the systemic spread of infection.[387]

A recent review has been prepared on the infectivity of BVs to cultured cells by Granados and Hashimoto.[421] Many new cell lines of different origins are being developed which will greatly aid in adding to our knowledge of viral replication.[422-426] Reviews on insect cell culture include those by Granados,[23] Vaughn,[427] Kuroda et al.,[428] and Hink and Hall,[429] and Mitsuhashi.[86] Goodwin et al.[430] reported on newly developed lepidopteran insect cell lines and the enriched media used for baculovirus VO production. Normal (Figure 47) and some BV-infected insect cell cultures are shown in Figures 38, 39, 41 to 43, 44 to 46, and 48.

3. Morphology Mutants

Prolonged serial passages of NPVs in cell culture results in the production of morphology mutants, e.g., in AcMNPV studies FP phenotypes (few polyhedra or <10 polyhedra per infected nucleus) are produced rather than the MP phenotype or wild type (approximately 30 or more polyhedra per nucleus).[419,431-436] Excellent reviews have been prepared recently which include details on morphology mutants.[52,53,80]

Prolonged passaged NPVs in homologous or heterologous cell lines have shown abnormal cellular responses which included: (1) aberrant virions, (2) rings of membrane, (3) lower numbers of NCs occluded/polyhedra, (4) NCs not occluded in polyhedra, (5) absence or low percentage of enveloped NCs, (6) elaboration of large quantities of viral envelope in the absence of NCs, (7) elaboration of large quantities of polyhedron envelope in the absence of polyhedron crystallization, and (8) high production of fibrous strands and crystals of protein (see Figures 42, 49 to 51, 55, 56).[434] Several of these morphology mutants have been characterized[314,436-439] with these abnormalities apparently resulting from the insertion of host cell DNA or repeated viral sequences into the viral genome.[59,74,297,300,301,304,309,310,440] (See Section II.C.2.)

Ultrastructural studies by MacKinnon et al.[419] revealed that after 20 passages abnormalities in NPV morphogenesis occurred such as the presence of virogenic stroma and capsids but no NCs. Virions were produced but VOs were absent and the number of virions was greatly reduced. In addition to the abnormalities described before, aberrant virions were formed approximately one third the normal length after 40+ passages.[419] In S. *frugiperda*

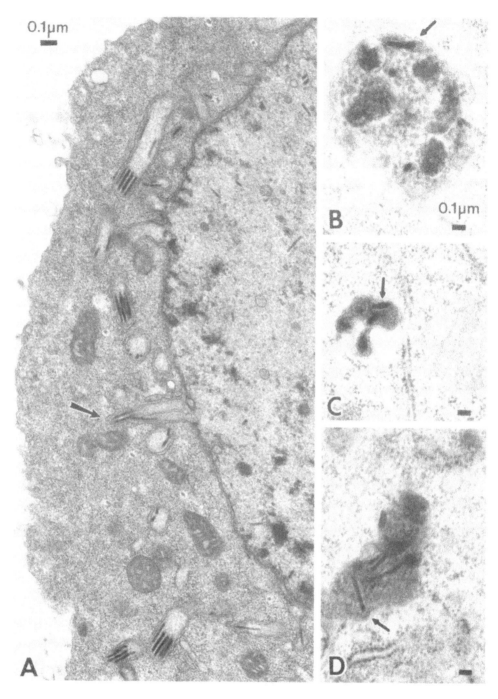

FIGURE 46.   (A) Release of NCs from infected nucleus; NCs may exit in an outpouching (arrow) of the nuclear envelope which pinches off into a vesicle or may touch the plasma membrane releasing the NC by budding or NC may exit in a vesicle by exocytosis, *L. dispar* MNPV infection in IPLB-LD 652 cells. (B) to (D) NCs and virions (arrows) may be trapped in phagosomes or lysosomes and be destroyed or may be released into the cytoplasm as NCs, AcMNPV infection in IPLB-TN-368 cells. (Bar = 0.1 μm.)

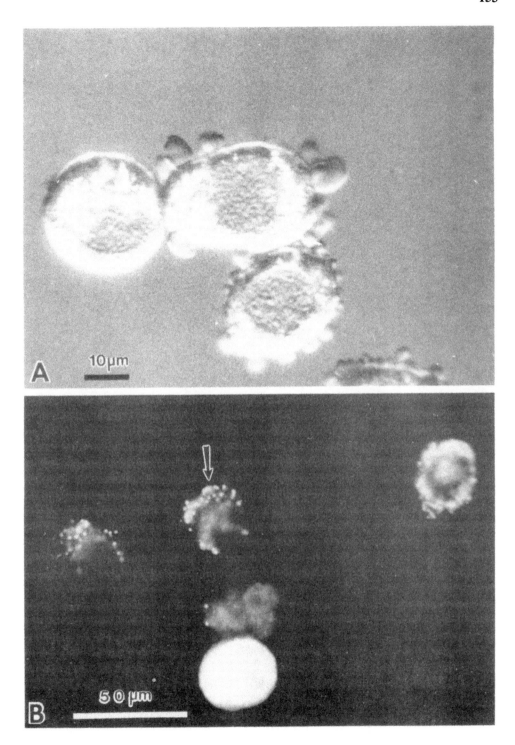

FIGURE 47. Gypsy moth *(Lymantria dispar)* fat body cell line, IPLB-LdFB. (A) Differential interference contrast (Nomarski) photomicrograph of cells 7 d postsubculture. (B) Fluorescence photomicrograph of IPLB-LdFB cells (stained with lipophilic stain Nile red). Arrow shows lipid-containing vacuoles. (Courtesy of D. E. Lynn.)

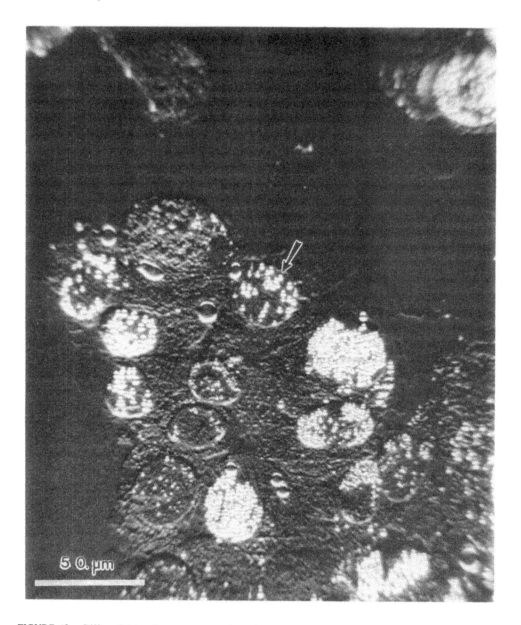

FIGURE 48.   Differential interference contrast photomicrograph of gypsy moth IPLB-LdFB cells 4 d p.i. with LdMNPV. Arrow shows polyhedra. (Courtesy of D. E. Lynn.)

and *T. ni* cells, a mutant of AcMNPV M-29 was unable to form VOs but produced only small particles (95 to 180 nm diameter). REN analysis revealed that the *Hind*III restriction site at the F/V junction of the viral DNA was absent.[439] Another mutant of AcMNPV (M5), which resulted from chemical treatment contained no virions.[436] Recent studies on the effects of multiple independent passages of AcMNPV in *T. ni* (TN-368) and *S. frugiperda* (IPLB-SF-21) cell lines revealed that two regions of the genome (the *Pst*I-G [7.6 to 13.1%] and *Pst*I-I [14.4 to 17.9%]) were deleted repeatedly. The deletions in the *Pst*I-G region and two insertions were mapped to a 1 kb *Pvu*II-*Bgl*II fragment. Insertional mutations were also observed within the *Pst*I-E/*Hind*III-I region of the viral genome. The suggestion was made that the deletions might confer some growth advantages in cell culture.[441]

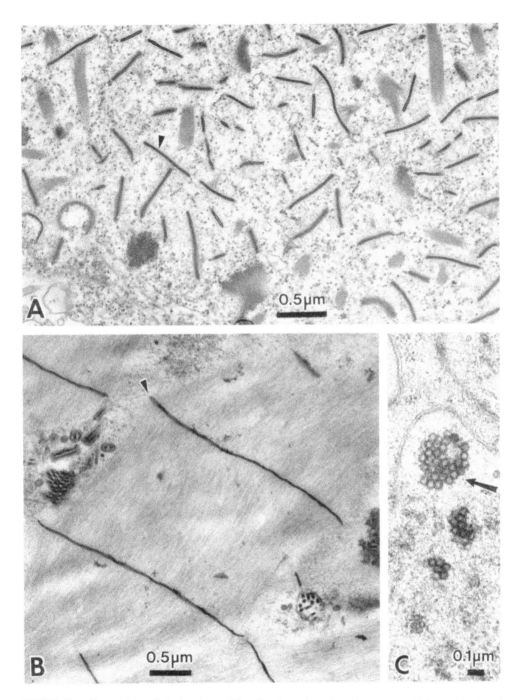

FIGURE 49.   Abnormal morphologies observed in cell cultures in prolonged passages. (A) Overproduction of electron dense "spacer" or polyhedra "membranes" (point); *Spodoptera frugiperda* MNPV in IPLB-SF-1254 cells. (B) Overproduction of fibrous sheets of protein (p10) containing electron dense spacers (point); IPLB-TN 112AB cells infected with *T. ni* MNPV. (C) Rings of membrane or viral envelope (arrow) without NCs; IPLB-HZ 1079 cells infected with AcMNPV.

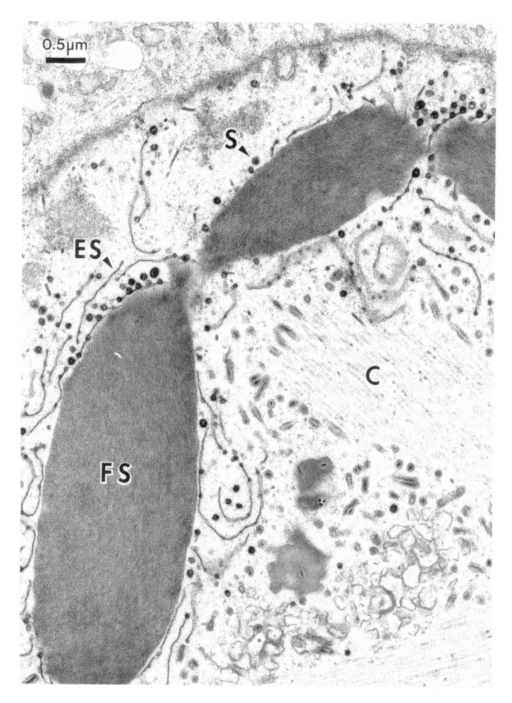

FIGURE 50.   Abnormal morphology observed in a fat body nucleus of beet armyworm, *Spodoptera exigua*, inoculated with AcMNPV. A few polyhedra or VOs were formed (area not shown) but nucleus contains high production of condensed fibrous sheets (FS), capsids (C), electron dense spacers (ES), and electron dense spheres (S) on the electron spacers which were about 0.5 to 1.2 nm diameter. Progeny NCs and virions were produced but few were occluded in VOs.

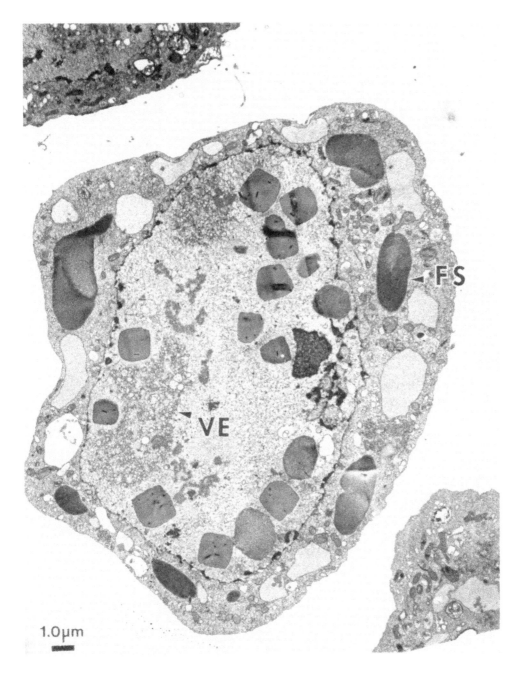

FIGURE 51.   Abnormal morphology observed. VOs contain only a few virions. Few progeny NCs were produced. Fibrous sheets (FS) of protein (p10) were produced in nucleus and six large FS bodies formed in the cytoplasm. The cytoplasm also contains many bodies of lower electron density than the FS bodies which are presumably protein. Viral envelope (VE) with an abnormal morphology has accumulated in two areas of the nucleus. IPLB-SF 21-AE cells infected with AcMNPV.

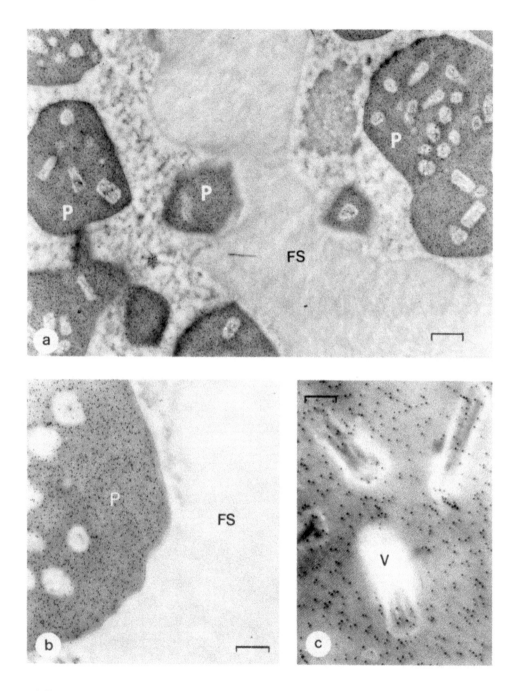

FIGURE 52.   Section of AcMNPV-infected *Spodoptera frugiperda* cells (50 h p.i.) treated with antiserum against polyhedrin and complexed with protein A gold. Polyhedra (P), virion (V), and fibrous structures (FS) are indicated. ([a] to [c] Bars = 500 nm, 130 nm, and 100 nm, respectively.) (From Van der Wilk, F., Van Lent, J. W. M., and Vlak, J. M., *Virology*, 68, 2615, 1987. With permission.)

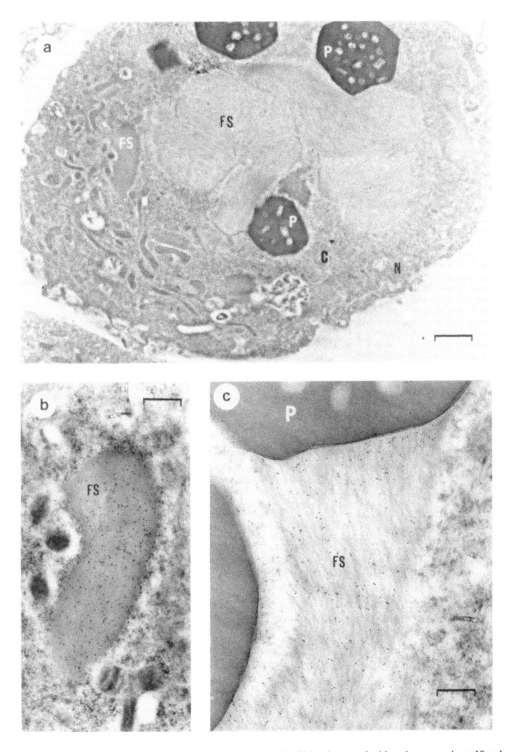

FIGURE 53. Section of AcMNPV-infected *S. frugiperda* cells (50 h p.i.) treated with antiserum against p10 and complexed with protein A gold. Enlargements show: cytoplasmic structures (a), nuclear structures (b), and fibrous structures (c). The nucleus (N), polyhedra (P), cytoplasm (C), and fibrous structures (FS) are indicated. ([a] Bar = 2.5 μm; [b], [c] bar = 250 nm.) (From Van der Wilk, F., Van Lent, J. W. M., and Vlak, J. M., *Virology*, 68, 2615, 1987. With permission.)

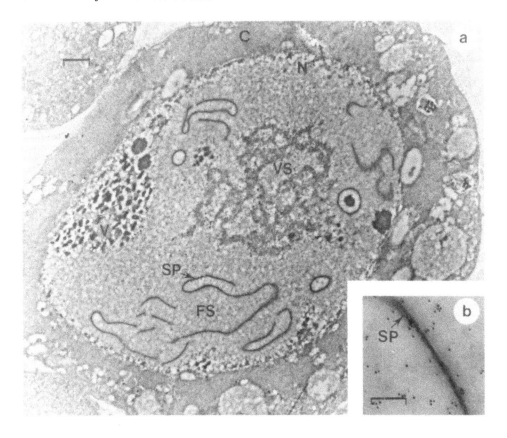

FIGURE 54.   Section of *S. frugiperda* cells infected with AcMNPV deletion mutant (d 16-G) and treated with
antiserum against p10 and complexed with protein A gold. The nucleus (N), cytoplasm (C), virogenic stroma (VS)
fibrous structures (FS), electron-dense spacers (SP), and virus particles (V) are indicated. ([a] Bar = 1 μm; [b
bar = 100 nm.) (From Van der Wilk, F., Van Lent, J. W. M., and Vlak, J. M., *Virology*, 68, 2615, 1987. With
permission.)

O'Reilly et al.[442] recently characterized the DA26 gene in a hypervariable region of the
AcMNPV genome which is an early gene that is not essential for viral replication and its
role is not yet understood.

Viral isolations from insects contain a mixture of genotypic variants.[148,253,276,441-443]
Electron microscopic studies of two plaque varients of AcMNPV revealed that when *T. ni*
larvae were injected or fed *per os* with either of the variants (MP or FP), the infected tissues
always contained both MP and FP polyhedra although infected nuclei contained only one
phenotype. In cell culture one plaque type always developed plaques of that type.[444]

Beames and Summers[445] have recently determined the location and nucleotide sequence
of the 25-K protein missing from AcMNPV-FP mutants. The *StuI*-*Hind*III fragment (36.6
to 37.7 m.u.) encodes the 25 K protein. Primer extension data indicated that in AcMNPV-
infected cells, transcripts within the *StuI*-*Hind*III fragment were initiated within a 12-base
consensus sequence often found near the transcription start sites of late and hyperexpressed
genes in BVs. The insertion of the host cell DNA sequence moved the first 10 bases of the
12-base sequence approximately 1.6 kbp upstream. Transposons also occur in other regions
of the BV genome[80] as Oellig et al.[78] found in comparisons of OpMNPV with AcMNPV
which would not be detected by the FP detection system noted before. If such mutants
produce no morphological or phenotypic effect, detection systems will have to be devised
for identifying them.[78,80] The changes in genetic composition of *Mamestra brassicae* MNPV

were followed through 25 successive passages in fifth instar *M. brassicae* larvae.[446] Of 82 restriction sites detected, 81 were maintained during passages while 1 site within a DNA fragment of 2.7 kbp (separating fragments *Hind*III-G and *Hind*III-J) was progressively deleted.[446]

Cusack and McCarthy[447] found three classes of variants in their studies of LdMNPV in IPLB-LD-6524 cells (i.e., Ld-S, Ld-F, and Ld-V, based on the number of polyhedra produced per cell). After 20 passages of the Ld-S variant, CPE indicated that genomic alterations had occurred. Cells with no polyhedra and cells with few polyhedra occurred as well as a low number of cells of the Ld-S variant. This had a restriction enzyme profile similar to the pre-high multiplicity passage Ld-S variant, while the remainder of the cells could not be plaque purified of submolar amounts of the Ld-S population. Probes of blots of variants with labeled total cellular DNA showed that specific hybridization did not occur, indicating that the passage of 20 variants arose from rearrangements of viral DNA. They proposed that apparently a small amount of the standard Ld-S or "helper" virus is necessary for the mutant to complement its missing function although it is able to replicate more efficiently than the helper virus in a mixed population.

4. p10 Gene

The second most abundant protein produced late in the infection cycle is the 10-K protein[440,448-451] which has a relative molecular weight of 10,000 and is encoded by the *Eco*RI-P fragment (p10 gene). The 5' end of the p10 gene lies in the *Hind*III-Q fragment while the 3' end spans the *Hind*III-P fragment.[440,452] The p10 genes of AcMNPV and OpMNPV have been sequenced by Kuzio et al.[453] and Leisy et al.,[454] respectively. Comparisons of OpMNPV with AcMNPV by Leisy et al.[454] showed that while the p10 genes were similar in size (282:288; OpMNPV:AcMNPV), only 54% nucleotide sequence homology and 41% amino acid sequence homology were shared. Comparisons of hydropathy plots noted differences in sections which were hydrophilic or hydrophobic in nature.[454] In a later study Quant-Russell et al.[455] demonstrated a 14 K-band by SDS-PAGE rather than the 10 K in AcMNPV which supported the report by McClintock et al.[456] that an 8 to 10-K band did not occur in the LdMNPV-BV gene, a baculovirus closely related to OpMNPV.

The p10 genes of AcMNPV[457,458] and OpMNPV[454] have been shown to be transcribed by overlapping RNAs. Friesen and Miller[457] proposed that overlapping mRNAs may regulate transcription initiated at one promoter which extends through another promoter preventing initiation of transcription at the second promoter. RNA transcription in several regions of the AcMNPV genome was under "strict temporal control". One type of transcriptional unit was identified and shown to share a common 5' end while another type shared a common 3' end. Lubbert and Doerfler[458] speculated that overlapping transcripts might substitute for alternate splicing to provide a "variety of transcription products" since there are no introns in the AcMNPV genome.

Electron microscopic observations of NPV-infected tissues and cell cultures show moderate quantities of fibrous protein also synthesized in the nucleus and cytoplasm[25,36,419,459-462] (see Figures 12 and 49B). Recent studies using virus expression vectors and protein A gold immunocytochemistry have revealed that the nonstructural fibrous protein is p10. Croizier et al.[463] demonstrated that the insertion of a foreign gene downstream of the p10 promoter, previously described by Knebel et al.,[280] prevented the formation of the fibrous proteins.

Using protein A gold labeling with polyhedra and p10 antiserum, Van der Wilk et al.[464] detected and localized these proteins in sections of AcMNPV-infected *S. frugiperda* cells (see Figures 52 and 53). The p10 antigen reacted with p10 antiserum and labeled only the fibrous protein structures in the nucleus and cytoplasm while the polyhedrin antiserum reacted only with the polyhedrin protein of the VOs. When *S. frugiperda* cells were infected with a polyhedrin defective mutant of AcMNPV (d16-G) and subsequently treated with p10

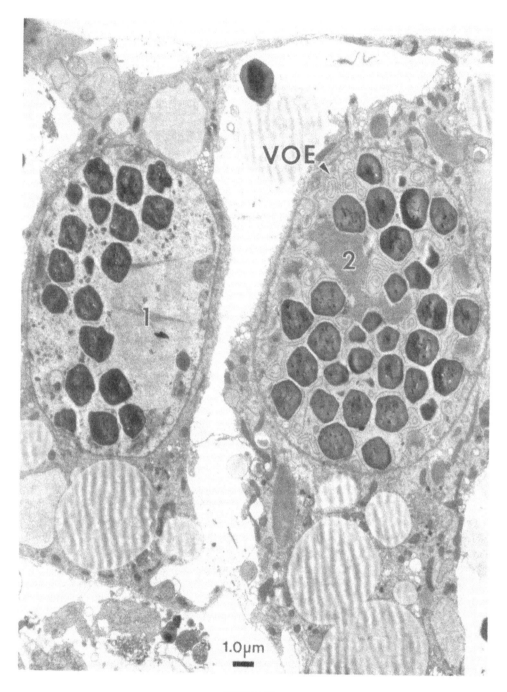

FIGURE 55.

FIGURES 55, 56.   Section of fat body of black cutworm, *Agrotis ipsilon* infected with *Rachiplusia ou* MNPV. The 3 cells lie adjacent to each other. Cell 1 shows normal polyhedra or VOs while cells 2 and 3 contain polyhedra or VOs with lower numbers of virions occluded and overproduction of VO or polyhedron envelope (VOE) and fibrous protein (FP) in nucleus and cytoplasm. Observations of many NPV infected cell cultures have shown infections with varying degree of abnormal viral morphology. This also has been observed in some infected insect tissues with host NPV and more often with nonhost NPVs. It is significant that viral expression varies from cell to cell within the same tissues of the host insect with some NPVs indicating that the host insect is not readily succumbing to the infection.

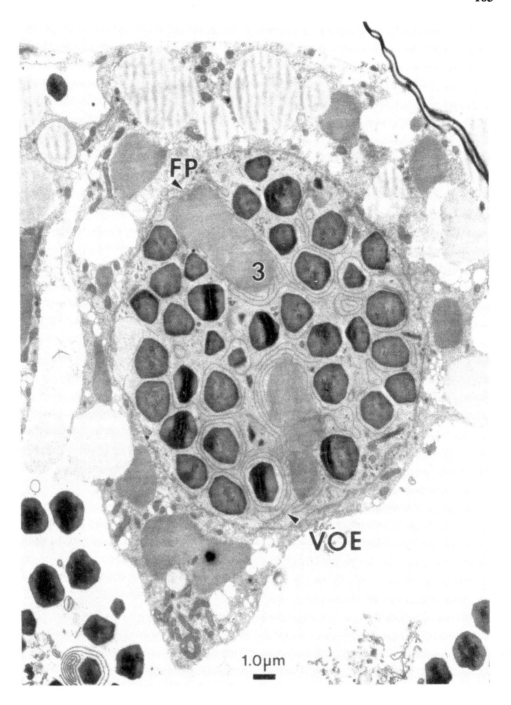

FIGURE 56.

antiserum and gold label, the fibrous structures and electron-dense spacers were labeled intensely in a manner similar to AcMNPV infected cells (see Figure 54). Such data established that these structures were synthesized in the absence of polyhedrin.[464]

Vlak et al.[465] then constructed an AcMNPV recombinant which expressed the p10/LacZ fusion protein and demonstrated that the p10 gene was not necessary for viral replication. In recombinant virus infected cells, the fibrous structures and the electron dense "spacers" were absent. The polyhedra also lacked an envelope. Granular structures were found in the nucleus in the same regions where fibrous structures normally occurred. Sections treated with antiserum against LacZ protein showed high numbers of gold label over the granular structure. The effect of the recombinant on the p10 gene was to prevent the formation of fibrous structures and adversely affected the assembly of the polyhedron or VO envelope. These studies demonstrated the potential use of the p10 promoter to regulate the expression of foreign genes.

Weyer and Possee[466,467] revealed the sequence requirements for high level activity of the AcMNPV p10 promoter and determined that a sequence of about 101 nucleotides upstream from the p10 ATG codon was sufficient for high level promoter activity. In a more recent study Qin et al.[468] demonstrated specific sequence requirements for optimal chloramphenicol acetyltransferase (CAT) expression under the control of the p10 promoter. Using 5' deletion mutants of the AcMNPV p10 gene promoter, activity was unaffected by 5' deletion to position −77, which lies approximately 11 bp upstream from the p10 cap site. A 5' deletion of 12 or more bp in the p10 promoter completely eliminated activity. These studies suggest that the essential portion of the p10 promoter lies between positions −77 and −65 and that the region appears to be highly conserved in a variety of late baculovirus genes.[45]

Another cytopathological investigation on AcMNPV p10 gene function was recently conducted by Williams et al.[469] They constructed two p10 deletion mutants with the β-gal gene (LacZ) (Ac 228 and Ac 229 [low and high producers of β-galactosidase, respectively]) and one without it (Ac 231). The LacZ sequence was out of frame in Ac228 and in frame in Ac229. The SDS-PAGE profiles of Ac229 and Ac231 showed no production of p10 but Ac229 produced large quantities of β-gal with reduced synthesis of polyhedrin. Light microscopic studies using fluorescence and peroxidase labeling with either polyclonal or monoclonal antibodies failed to indicate recognizable p10 in infected *S. frugiperda* or *T. ni* cells. Cells infected with wild type (wt) AcMNPV began to lyse at 2 d p.i. At 4 d p.i. cells infected with Ac231 and Ac228 displayed large hypertrophied nuclei filled with VOs while Ac229 infected cells were hypertrophied but contained few VOs.

Comparison of wt AcMNPV infection to Ac229 and 231 infections over a time course of 18 to 48 h p.i. showed that the mutants lacked the large cytoplasmic and nuclear fibrillar masses seen in the wt infection "suggesting that p10 is a major constituent of these bodies". VO membrane sheets were present in Ac229 and Ac231 following the same time course as wt infection indicating that p10 gene expression was not required for their formation. The cells infected with Ac229 also contained large aggregates of spherical particles (37 to 55 nm diameter) with an electron dense core and an outer halo of lower electron density (100 to 120 nm diameter). Cells infected with Ac231 resembled the wt AcMNPV infection except that the virogenic stroma stained poorly compared to wt virus and aberrant nucleocapsid formation was observed. Differences were also noted in the VOs produced by the mutant viruses such as defects in the attachment of viral occlusion envelope causing shedding of occlusion matrix. Williams et al.[469] concluded that the electron dense bilaminar sheets or FS and VO membranes are composed of the same structures which are synthesized at different stages of development since the deletion of the p10 gene caused a loss of the fibrillar structures but did not affect the production of VOs and VO envelope. However, the transport and attachment of FS to VOs were adversely affected. These studies further support the role of the VO membrane in protecting the VOs in the environment. Differences in the viral morphogenesis using p10/LacZ deletion mutants from that reported by Vlak et al.[465] included the presence of FS sheets while Williams et al.[469] found none. Williams et al.[469] concluded that the *de novo* granules that occurred near VOs probably interfered with the attachments of the fibrous sheets to the VOs and may result from the accumulation of the p10/LacZ fusion product. Vlak et al.[465] reported that the granules described before reacted with antisera raised against B-gal and p10 in immunogold labeling experiments.

A recombinant containing a DNA sequence coding for the enzyme phosphotransferase, which inactivates neomycin, was inserted into the *Bgl*II site of a pVC9-*Eco*RI P vector. This construct PGP-46-Neo and *Galleria mellonella* viral DNA were cotransfected in *L. dispar* cells, and recombinant viruses were selected for resistance to Geneticin (G418). These experiments demonstrated that the p10 gene of baculoviruses may be utilized as an insertion site for foreign DNA as has been shown previously for the polyhedrin gene.[470] Recently Weyer et al.[471] obtained normal appearing polyhedra (in thin sections) with a recombinant virus in which the polyhedrin coding sequence was placed under the control of the p10 promoter. Thus, it appears that the polyhedrin gene does not have to be expressed under the control of the polyhedrin promoter and at its normal location for the production of VOs.

5. Virus Occlusion Formation

The process of VO formation has been studied by electron microscopy.[462] It is possible to observe stages of this process as crystallization occurs around a few virions until mature VOs are produced (see Figures 28B to C, 33 to 35, 40A). The polyhedrin lattice appears to crystallize around the virions which are scattered at random without loss of its integrity.[15,25] The numbers of virions contained within polyhedra have been estimated for three NPVs; *Agrotis segetum* and *M. brassicae* MNPVs contained 17.2 to 19.6 virions per μm^3 while *Plusia gamma* SNPVs contained 59.6 virions per μm^3 of VO.[413]

The polyhedron envelope or membrane is not a true membrane since it is composed primarily of carbohydrates[213] and protein.[214] Minion et al.[213] found that "ghost" of polyhedron envelopes of *H. zea* SNPV contained solely carbohydrate moieties (see Figures 2, 10B and D, 13A, 33, and 40A). It is derived from condensed fibrous sheets which are synthesized in the process of secondary BV infections.[462] The fibrous sheets are comparable to structures called fibrous material or elements,[25,459,472] sheets with electron dense spacers[419] and cisternae[472] (see Figures 27, 49A and B, 50, 53, 54). Two proposals were made concerning the formation of the polyhedron envelope or membrane. Hess and Falcon[472] proposed that the envelope consists of two closely appressed membranes while Chung et al.[462] showed electron micrographs from which they proposed that it is one half of a condensed double fibrous sheet, i.e., three lamellae separated by electron lucid layers. Recent studies by Vlak et al.[465] using recombinant viruses demonstrated that the production of normal and condensed fibrous sheets were under the control of the p10 gene. Fibrous sheets and polyhedron envelope were not produced, and polyhedra formed without membranes or envelopes in AcMNPV/p10Z-2 infected *S. frugiperda* cells. Williams et al.[469] concluded that electron dense bilaminar sheets or FS spacers and VO envelopes are developmental forms of the same structure based on their studies with p10 deletion mutants.

Russell and Rohrmann,[473] using a polyclonal antiserum produced against an OpMNPV PE-trpE fusion protein to localize the polyhedron envelope (PE) protein in intact and dissolved polyhedra, and in *L. dispar* cells infected with wt OpMNPV, found by immunocytochemical techniques that immunogold particles were associated with the polyhedron envelope. Untreated polyhedra and polyhedra treated with proteinase K, followed by PE-trpE antiserum with gold label, revealed numerous labels on the polyhedron envelope or envelope fragments in the case of the enzyme treated polyhedra, establishing that the protein detected was an integral part of the polyhedron envelope.[473]

A mutant of AcMNPV lacking the polyhedron envelope was constructed and analyzed by Zuidema et al.[474] who showed that a 34-K protein (pp34) is involved in the morphogenesis of the polyhedron envelope. The polyhedra lacking envelopes were infectious *per os* and were less resistant to dissolution in 0.05 M Na_2CO_3 than wild type (wt) AcMNPV. Comparisons by electron microscopy of the AcMNPV wt and polyhedron envelope lacking mutant revealed that electron-dense "spacers" normally present in wt infections were absent in the pp34 lacking mutants. Further studies by van Lent et al.[475] have employed immunoelectron microscopy to localize the 34-K polyhedron envelope phosphoprotein (pp34). This protein was found associated with fibrillar structures, electron dense spacers, and polyhedron envelopes in wt AcMNPV infected *S. frugiperda* cells while in cells infected with the AcMNPV mutant containing the β-galactosidase (LacZ) gene in place of the p10 gene, the pp34 was localized in the electron-dense spacers and the polyhedron envelope.[474] However, in cells infected with the AcMNPV mutant (with inactivated pp34 gene), anti-pp34 serum localized in fibrillar structures despite the absence of pp34 in these cells (thus producing an artifact of the technique). The use of OpMNPV polyhedron envelope (pp32) antiserum on the wt AcMNPV, AcMNPV mutant (+ pp34), and AcMNPV mutant (+ p10) localized pp34 in the same structures as described before. This excellent study showed that the electron-dense spacers are essential for polyhedron morphogenesis, while the fibrillar structures are not.[474]

6. Other Viral Proteins

In addition to polyhedrin and p10, the p6.5, p26, p34.8, and p39 genes of AcMNPV and OpMNPV have been mapped, sequenced, and compared.[243,244,476-478] In a recent study, Müller et al.[247] described a p87 capsid-associated protein of OpMNPV, the location of its gene, and the characterization of its expression. When compared to AcMNPV, none of the AcMNPV-specific proteins cross reacted with the OpMNPV p87 polyclonal antiserum. In addition, no DNA sequences homologous to the p87 gene were detected in the genome of AcMNPV. Other essential genes and their products, e.g., p42, p67, p74, DNA polymerase, and ecdysteroid UDP-glucosyl transferase have been identified and their role in the replicative and infectious process characterized by various laboratories.[246,327,442,479-484] Together, approximately 40 other virus-encoded proteins are produced by NPVs which are regulated at the transcription and translational level.[440,485-487]

7. Persistant Viral Infections

Persistent baculovirus infections have recently been reviewed by Burand et al.[47] Granados et al.[488] were the first to report defective interfering viral particles (HZ-1) which caused persistent infections in *H. zea* IMC-HZ-1 cells. Infected nuclei showed hypertrophy; viral replication of virions occurred but no VOs were formed. Some persistently infected cell lines showed abnormal morphology and CPE while others appeared similar to healthy cell lines. The host range of HZ-1 *in vitro* now includes *T. ni* (TN-368), *S. frugiperda* (IPLB-SF-21), *H. zea* (IPLB-1075), and *M. brassicae* and *L. dispar* (IPLB-652).[488-492] Persistent viral infections have also been obtained in *T. ni* and *S. frugiperda* cell lines.[490,493-495] Of the persistently infected *S. frugiperda* cells, 70% contained antigens for SfMNPV but only about 1% of the cells showed nuclei with viral infection.[493] Crawford and Sheehan[494] found that cured cells (virus free) were partially resistant to infection by SfMNPV or AcMNPV. Cured cells from the AcMNPV persistent infection were highly resistant to both SfMNPV and AcMNPV.

The HZ-1 particles have been characterized[489,495,496] and are larger than standard *H. zea* virions, i.e., approximately 85 × 487 nm (negatively stained) or 363 × 96 nm in sections.[488] Differences in the cytopathology from normal BV replication included the uncoating at the nuclear pore similar to granulosis virus infection. The virus particles are enveloped within the nucleus rather than budding through the nuclear or plasma membrane[47] and are released following rupture of the nuclear membrane. When compared to standard HzNPV, the DNA of HZ-1 particles contained deletions of up to 91 kbp.[496] Standard virions appeared to be required for replication of HZ-1 particles yet interfered with infection or replication of standard NPVs. Burand and Wood[496] also found that variations occurred in the synthesis of 14 proteins (reduced production) while that of 5 proteins increased when compared to protein synthesis in standard HzNPV infections. Recently Chao et al.[497] prepared a physical map of the Hz-1 baculovirus genome which contains 228 kb and does not contain a *Not*I restriction site. The areas of the genome of five virus isolates that contained deletions were in the 22 to 45 map unit region, and the deletions ranged from 24 to 52 kb.

8. Mixed or Multiple Infections

Aizawa[16] has reviewed the early reports of double and mixed infections which commonly occurred in *Bombyx mori* and were also reported by Smith and Xeros[498] and Smith et al.[499] Single and double infections of NPV and CPV were observed in *Hyphantria drury* by Ryndovskaya et al.[500] Tchuchrity[32] has also reviewed mixed infections and the ultrastructural changes that occur in the nucleus and the cytoplasm of infected cells. Tanada's report[365] of an NPV and GV double infection in *Pseudaletia unipuncta* larvae indicating a synergistic effect led eventually to the discovery of the synergistic factor discussed in Section III.A and Chapter 7. Kurstak and Garzon[501] showed that multiple infections were successful when the second virus inoculum was diluted to a concentration that allowed the first virus to penetrate and initiate replication. Depending on the concentrations of inocula and timing sequences, interference or synergism could be produced with NPV, DNV (densonucleosis virus), and TIV (*Tipula* iridescent virus). Tchuchrity[502] noted interference in *M. brassicae* larvae simultaneously infected with MNPV and CPV (cytoplasmic polyhedrosis virus), as no ''virus assembly'' occurred. When CPV-infected gut epithelial cells were inoculated 24 h p.i. with NPV, double infection occurred. Double infections of NPV and CPV are very common. Hughes[226] found that some larvae of the Douglas fir tussock moth, *Orygia pseudotsugata*, succumbed to double infections.

The presence of single virions and multinucleocapsid virions in the same nucleus is common with many MNPVs,[239,242] and many MNPVs contain single virions.[199] Adams et al.[222] observed groups of virions infrequently in HzSNPV-infected tissues budding through plasma membranes to form multinucleocapsid virions. Ramoska and Hink[444] demonstrated that tissues of *T. ni* larvae inoculated with either of two plaque variants (MP, many polyhedra or FP, few polyhedra) always contained both variants, while cell cultures infected with one plaque variant always produced infections typical of that type. Falcon and Hess[503] compared the replication of AcMNPV and SeMNPV in *Spodoptera exigua* larvae. Interference was demonstrated by McIntosh et al.[504] in VO production of *T. ni* in *T. ni* cells infected with a high passage strain of AcMNPV. Viral interference was also demonstrated in LdMNPV-infected gypsy moth cells superinfected with AcMNPV. In that study, McClintock and Dougherty[505] demonstrated that AcMNPV was capable of interfering with the replication of LdMNPV and that a significant decrease in LdMNPV production occurred if AcMNPV superinfection occurred within the first 8 h after LdMNPV infection. However, AcMNPV replication occurred and appeared, at least qualitatively, to undergo normal replication if superinfection occurred by 6 h p.i. This may be due to the initiation of LdMNPV DNA synthesis at 12 h p.i., whereas AcMNPV DNA synthesis occurs at 8 h p.i. in the permissive or semipermissive systems.[281,456]

A mixed infection was observed in columnar and goblet cells of the midguts of *T. ni* and *A. californica* larvae by Hess et al.[506] Icosahedral nonenveloped viral particles of 40 nm diameter (SV_1) and 28 nm diameter (SV_2) as well as viral particles of CPV (63 nm diameter) and hexagonal nonenveloped particles of 80 nm diameter occurred along with AcMNPV infections in these species. The SV_2 often occurred in vacuoles in paracrystalline arrays.[506] These findings emphasize the importance of knowing the purity of the inoculum used for NPV studies. Sections of polyhedra contained SV$_1$ particles.[506] Polyhedra containing *T. ni* RNA virus particles and empty vesicles were also observed in later studies with AcMNPV.[507] Morris et al.[508,509] who characterized the virus particles physiochemically further established that this 35 nm RNA contaminant virus occurred in the preparations of AcMNPV examined. Further details may be found in Chapter 14.

Cotransfection experiments with *R. ou* MNPV (RoMNPV) and AcMNPV *Bam*HI or *Sma*I restrictions fragments in *G. mellonella* larvae produced recombinants primarily consisting of AcMNPV DNA or a mixture of AcMNPV and RoMNPV restriction fragments.[510] However, there was a relative lack of biochemical or genetic comparisons.

9. Tissue Specificity

Many SNPVs and MNPVs replicate in the nuclei of many susceptible tissues such as hypodermis, fat body, tracheal matrix, hemocytes, silk glands, and epithelial tissues of the nervous and reproductive systems, etc. while other NPVs are more tissue specific. MNPVs isolated from infected *Spodoptera* larvae in three different geographic areas were assayed in *S. littoralis* and *S. exigua* larvae by Merdan and Amargier.[511] They observed that fat body, hemocytes, and trachea were infected in *S. littoralis* while fat body was not infected in *S. exigua* larvae. Two strains of MNPV infect *Pseudaletia unipuncta*.[366] The hypertrophy strain (HNPV) develops more slowly than the typical strain (TNPV) producing a gradient of cells along the tracheal matrix showing three phases of development in the early infection process.[362,366,512] Differences were found in the location and quantity of the granular chromatin, the presence of dense compact interconnected fibrillar strands, and punctate densities in the HNPV-infected tissues.[512] Abnormal structures such as multilayered material and vacuoles lined with striations were observed in tissues infected with *Anticarsia gemmatalis* MNPV.[513] Abnormal cell proliferation in the hypodermis has been reported in *Hyphantrea cunea*.[360] Several studies on NPV infection in mosquitoes have established that only the gut columnar cells are infected.[363,364] Recent studies on additional species inoculated with field-collected NPVs from *Aedes sollicitans* propagated in laboratory-reared *A. epactius* confirmed earlier findings.[514] Once again there was a relative lack of biochemical or genetic comparisons.

10. Passage in Alternate Hosts

The effects of serial passage through specific host on the polyhedron morphology and through specific host on the polyhedron morphology and virulence of baculoviruses have been well documented. Tanada[345] showed a typical NPV development when *Colias eurytheme* larvae were infected with *Pieris rapae* MNPV. Tompkins et al.[162] demonstrated that TnMNPV infected *H. zea*. Vail and Jay[515] compared the histopathologies of larvae of *A. californica, T. ni, S. exigua, Estigmene acrea, H. zea, Bucculatrix thurberiella,* and *Plutella xylostella* following inoculation with AcMNPV. The nuclei of the hypodermis, fat body, and tracheal matrix were infected in all hosts. Other infected tissues are shown in Table 1.[515] Only *T. ni* showed viral infection of the Malpighian tubules; *E. acrea* and *T. ni* contained infected hindgut cells. High concentrations of AcMNPV were required to produce infection in *H. zea* with lower VO production in the tissues noted before than in other hosts which responded similarly to AcMNPV infection in its native host.[515]

Later studies by Vail and Collier[177] compared AcMNPV infections in *H. zea* and *H. virescens*. While *H. virescens* larval tissues showed typical NPV infection in fat body, hypodermis, tracheal matrix, midgut, hemocytes, and dorsal vessel, infection in *H. zea* larvae was limited to a few cells containing virogenic stroma and a few cells filled with VOs. Removal of specific tissues and organs of *H. zea* at specified times p.i. followed by homogenization and inoculation in TN-368 cells further confirmed that AcMNPV did not infect *H. zea* tissues.[179]

Comparisons of AcMNPV and SfMNPV infections in *S. exigua* larvae by Falcon and Hess[503] revealed that paracrystalline arrays were produced in goblet cells due to AcMNPV inoculation while similar structures were not observed in *S. exigua* infected with its native MNPV (see Reference 503, Figure 5). AcMNPV was also observed in all stages of development in the cytoplasm (see Reference 503, Figures 9 and 10) of tissues other than the fat body. When the black cutworm, *Agrotis ipsilon,* and the European corn borer, *Ostrinia nubilalis,* were inoculated with AcMNPV, RoMNPV, TnMNPV, and *Ephestia cautella* MNPV, only *A. ipsilon* inoculated with RoMNPV showed typical NPV infection in nuclei of hypodermis, fat body, tracheal matrix, and epithelial sheath of muscle, fore-, and hindgut.[187,516] *R. ou* MNPV and AcMNPV inoculations in *O. nubilalis* showed heavily infected fat body, hypodermis, and tracheal matrix, while TnMNPV and *E. cautella* MNPV produced light to moderate infections in the same tissues and generally gave less consistent results.[516] Electron microscopic examinations revealed that some abnormalities also occurred in viral replication similar to that reported earlier from continual *in vitro* passages in insect cell cultures.[434] Examples are shown in Figures 49 to 51 which include the following: (1) VOs formed with low numbers of occluded virions or none at all; (2) abnormal production of protein sheets in the nucleus and/or cytoplasm; (3) abnormal replication of the virion envelope; and (4) high budding rates from infected cells. The most significant observation concerning the abnormalities produced upon passage through these alternate hosts was that adjacent nuclei in the same tissue showed normal or abnormal viral replication, establishing that the tissue was not a factor in causing abnormalities but rather the inoculum was the source of these observed differences[242,516] (see Figures 55 and 56). This finding demonstrated that the abnormalities observed *in vitro* cultures were probably not due to the tissue culture cells or the media but were due to the inoculum used. Therefore, a knowledge of the history and biochemistry of the inoculum used for testing is imperative.

Hess and Falcon[517] studied the passage of AcMNPV in the midguts of *E. acrea* larvae, noting that viral replication proceeded normally to the stage of NC and virion formation; however, an abnormally high production of capsid-like profiles and paracrystalline protein inclusions in the cytoplasm were observed.

An excellent comparative cytopathological study was conducted by Hamm and Styer[518] using *S. frugiperda* and *S. exigua* larvae treated with SfMNPV isolates of known history and biochemistry (four original wild types [A, B, C, and D] and four plaque-purified viral isolates [B1, B2, D5, and D7]). Differences were noted in symptoms, percent mortality, numbers of VOs produced per larva, tissue specificity, and pathogenesis within susceptible tissues. The original isolates, particularly A, were generally more virulent to both species than the plaque-purified viruses. Following passage of the A and B2 isolates in *S. exigua,* a further passage in *S. frugiperda* or *S. exigua* indicated a great loss of infectivity compared to the virulence of these isolates in previous passages in *S. frugiperda.* Perhaps the inoculum produced in *S. exigua* had been altered by insertions of *S. exigua* host cell DNA. The A isolate caused rapid and extensive NPV infection with scattered heterochromatin at the periphery of hypertrophied nuclei and a clearing of the central area where virogenic stroma developed from scattered electron dense areas. Low numbers of VOs with abnormal morphology were observed. In *S. frugiperda* larvae inoculated with the other isolates, differences were noted with increasing types of abnormalities such as lower numbers of VOs produced per nucleus, lower numbers of virions or no virions occluded in VOs, abnormal virogenic stroma, abnormal paracrystalline aggregates of NCs, and "scalloped" polyhedra. In *S. exigua* larvae only a very few cells became infected following challenge with isolate A, and those hypertrophied cells in early stages of infection contained very scarce virogenic stroma and abnormal forms of NCs except in a few hypodermal nuclei where normal virogenic stroma and NCs were found. Other abnormalities observed were excessive production of NC envelope in the form of various sized vesicles, occlusion of empty vesicles within the few scattered VOs produced, and abnormally long NCs.[518] This morphological evidence of the expression of these viral isolates in *S. frugiperda* and *S. exigua* larvae greatly helps explain the differences in mortalities and symptoms, etc. This evidence also helps investigators toward a better understanding of "specificity" characteristic of NPVs.

Attempts by Maleki-Milani[519] to adapt AcMNPV for the control of *S. littoralis* larvae failed. However, repeated passages of AcMNPV in *S. littoralis* caused sufficient stress on the insect to stimulate the production of the natural host virus as confirmed by biochemical analyses of the isolated VOs. Groner[4] has also reported on many unsuccessful attempts to cross transmit NPVs to alternate hosts. An NPV from *Spodoptera littoralis* at 6.4 × 10[7] VOs/ml produced >90% mortality in termite castes of *Kalotermes flavicollis*.[520] All stages were equally susceptible; however, the investigations recommended that two applications of the virus would probably be required to eliminate the termites.[520] Hamelin et al.[521] have developed a simplified method for the characterization of nuclear polyhedrosis virus genomes using single larvae. This should greatly aid investigators in determining the stability of viruses after passages in alternate hosts.

The effects of chemical treatment and serial passage on virulence were previously described (see Section I.B). In order to evaluate the effects of *in vitro* serial passage on genotypic and phenotypic stability, recent investigations comparing genotypic variants (plaque-purified) of the wild type Hamden LdMNPV population demonstrated three distinguishable classes of variants (Ld-S, several polyhedra per cell; Ld-F, few polyhedra per cell; Ld-V, variable numbers of polyhedra per cell) with varying LD_{50} values.[447] In bioassay trials, two of the variants, Ld-V and Ld-S, exhibited lower LD_{50} values than the wild type virus, whereas the Ld-F variant was not infectious to larvae. When representative phenotypic variants were analyzed by restriction endonuclease cleavage and compared to the wild type digests, all variants displayed restriction site polymorphism. In order to identify the source of additional restriction fragments following serial passage, blots containing either viral DNA fragments or cell DNA digests were hybridized with cellular (IPLB-LD-652Y) DNA or viral DNA, respectively. No hybridization response was observed in either instance suggesting that the high passage variants arose by rearrangements involving solely viral DNA. It is interesting to note that by the 20th undiluted passage of Ld-S the *in vitro* CPE of this variant was altered; the majority of cells exhibited few polyhedra per nucleus while many cells showed typical CPE but contained no polyhedra.

There are several reports on *in vitro* baculovirus host ranges. Goodwin et al.[522] found that MNPVs could be successfully passed in alternate host cell lines while SNPVs infected only native host cell lines. Lynn and Hink[523] compared AcMNPV replication in the cell lines BR1-EEA from *E. acrea*, IPLB-LD64BA from *L. dispar*, I2D-MB0503 from *M. brassicae*, IPLB-SF1254 from *S. frugiperda*, and TN-368 from *T. ni*. They observed that the *T. ni* and *M. brassicae* cell lines produced the most polyhedra, but the AcMNPV (produced in TN-368 cells) gave the highest mortalities in the *T. ni* larvae tested. McIntosh et al.[524] studied the host range of five MNPVs in five lepidopteran cell lines showing differences in titers of VOs produced. AcMNPV was produced in *T. ni*, *S. frugiperda*, and *H. virescens* cell lines but not in *H. zea* and *H. armigera* cell lines. HzSNPV replicated in the *H. virescens* and *H. zea* cells but not in the *H. armigera* cell line. SfMNPV replicated only in the *S. frugiperda* cell line. Danyluk and Maruniak[525] tested AcMNPV-E2 and *S. frugiperda* (SfMNPV-2) in five insect cell lines (IAL-PID2, IAL-SFD1, IPLB-SF-21AE, TN-368, and IAL-TND1). AcMNPV-E2 replicated in all the cell lines tested but SfMNPV-2 did not replicate in IAL-TND1 and TN-368. The SfMNPV-2 was not infectious to *T. ni* larvae by *per os* or intrahemocoelic injection suggesting that a barrier to SfMNPV may exist at the cellular level in *T. ni* larvae. Lynn[526] has noted that over two dozen baculoviruses have been reported to replicate in cell cultures. Most of these are homologous infections (i.e., the same insect species was the source of virus and cells). Two notable exceptions are the *Spodoptera* sp. complex (in which a virus from one *Spodoptera* sp. will replicate in cell lines from other *Spodoptera*) and AcMNPV, where the latter has been shown to replicate in over 30 cell lines from 11 different species of Lepidoptera. The choice of insect cell line used is important in insect virus production, e.g., Lynn et al.[424] found that the fat body derived cell line of *L. dispar* produced the highest number of viral occlusion bodies of LdMNPV when compared with a cell line from embryos and an ovarian cell line. The choice of cell line for the highest production of AcMNPV-OBs was *Plutella xylostella* among the five cell lines tested by McIntosh and Ignoffo.[527]

Recent work by McClintock et al.[282] on AcMNPV infection of a gypsy moth cell line (IPLB-LD-652Y) has also demonstrated that some insect cell lines are semipermissive for replication of certain baculoviruses. In this study McClintock et al.[282] found that AcMNPV infection in IPLB-LD-652Y cells resulted in normal replication of the viral genome but only four viral or viral-induced proteins were observed after pulse labeling of infected cells with [^{35}S]-methionine followed by SDS-PAGE and autoradiography. These types of baculovirus host cell interactions, in which only a limited number of viral genes are correctly expressed and in which no viral progeny are produced, are of particular interest in that they may provide valuable insights into factors which regulate baculovirus host ranges.[528]

The interaction of AcMNPV with vertebrate cell lines, birds, mammals, and aquatic microorganisms has been reviewed by Groner.[4] Early studies by McIntosh and Shamy[529] with AcMNPV in a Chinese hamster cell (CHO) indicated restrictive replication of the virus. Volkman and Goldsmith[530] found that 35 vertebrate cell lines did not support AcMNPV replication according to peroxidase-antiperoxidase assays. Tjia et al.[531] showed that AcMNPV did not persist in the mammalian cell lines tested. Based on results from virus growth titrations, electron microscopy, DNA:DNA dot blot hybridizations, and synthesis of viral-induced proteins, Groner et al.[532] observed that although viral particles were absorbed and engulfed, AcMNPV replication did not occur in the Chinese hamster cell line CHO-K1 or a nonpermissive cell line (CP169). Recent studies by Brusca et al.[533] established that AcMNPV attached and entered but could not replicate in frog, turtle, trout, or codling moth cell lines. Neither viral DNA nor RNA synthesis was detected by nucleic acid probe hybridization. Free virions were observed in the cytoplasm, in cytoplasmic vacuoles, at nuclear pores, and within cell nuclei inside 2 h p.i. but viral replication was not observed. They proposed that the block in viral synthesis/replication probably occurred in the nucleus, due possibly to improper uncoating of the virion or inability of nonpermissive cells to transcribe AcMNPV DNA efficiently.

D. TRANSOVUM TRANSMISSION

Reviews of interest include those by Aizawa,[16] Tanada,[103,106] Hostetter and Bell,[9] Andreadis,[534] and Podgwaite and Mazzone.[144] Smith[24] suggested that transovum and transovarian transmission are very effective methods of dispersal of insect viruses among their hosts. Studies on sawfly populations led Bird[535] to propose that ovarial transmission occurred. Inoculation of older *Neodiprion swainei* larvae by Smirnoff[536] resulted in transovum transmission by adult sawflies who dispersed the virus without adversely affecting their fecundity. Tanada[103] observed that while infected sawflies dispersed viruses readily from the infected midguts which passed out in the feces to contaminate the tip of the abdomens and genitalia, less virus was shed by NPV-infected lepidopteran insects since the VOs are produced in other tissues than the midgut. Several studies have established that horizontal transmission occurs in laboratory[537] and field[9,538] populations of insects as well as transovum transmission by the contamination of female and male genitalia.[539-545] Virus-fed adults also contaminated the eggs.[546] NPV infections have also been observed in midguts of sawfly adults and pupae[547] but several laboratory studies designed to determine whether transovum or transovarian transmission occurs have concluded that the former accounts for most of the larval infection that occurs as hatching larvae consume the VOs on the contaminated egg shell.[540-545,548]

Studies on *Spodoptera ornithogalli* that survived SoNPV infection to postlarval stages showed reduced fecundity.[548]

VOs have been detected on the chorion of eggs by scanning electron microscopy.[542,543] Surface sterilization of eggs and pupae has been shown to prevent transovum transmission.[543,546,547,549-551] Histological studies of adults have revealed VOs on the surface of the cuticle near the tip of the abdomen.[543] Early studies on *Prodenia litura* in which egg homogenates were fed to larvae seemed to indicate that perhaps some transovarial transmission occurred.[552] A more recent report in which surface sterilized eggs and homogenates of surface sterilized eggs were fed to *S. mauritia* larvae indicated also that perhaps some transovarial transmission occurred.[549] Similar results were reported in tests with *Mythimna (Pseudaletia) separata*.[553] Nun moth virus antigens were detected in embryos by Larionov and Bakhvalov.[554] Doane's[555] early studies on transmission of LdMNPV in gypsy moth populations in New England forests indicated that the mode of viral transmission was predominantly transovum; however, recent studies by Wood et al.[556] using dot blot hybridization to probe DNA isolated from gypsy moth egg masses, found that 1 to 7% of the insect populations tested were positive. Positive samples contained at least one viral genome copy per cell. The larvae which hatched from positive samples contained LdMNPV but showed no symptoms of viral infection.[556] Studies on *T. ni* adults by Vail and Gough[557] found that high titers of *T. ni* NPV were obtained only when the virus was injected into the hemocoel. Examinations of eggs from diseased *T. ni* adults which were injected with *T. ni* NPV generally gave no indication of transovarial transmission. A low number of sectioned eggs revealed an occasional hypertrophied nucleus.[546] Histopathologic studies of *T. ni* pupae infected with *T. ni* NPV showed differences from typical larval infection in only the fat body which was not heavily infected as were all the other susceptible tissues.[557]

The follicular epithelium of many insect species is often infected during normal NPV pathogenesis but VOs are not observed in the developing oocytes. A recent report by Smith-Johannsen et al.[558] has shown that the thick basement membrane or basal lamella which isolates the follicular epithelium from the hemocoel appears to retard the spread of viral infection. Injections of collagenase in *Bombyx mori* pupae followed by injections of BmNPV virions removed the basement membrane thus allowing the more rapid spread of the virus into the follicular cells. Polyhedra were observed in immature oocytes, but no evidence, such as electron micrographs, was shown. These investigators proposed that chorion protein synthesis was adversely affected in the infected moths that were injected with collagenase and speculated that perhaps the eggs produced contained thin chorion which would desiccate easily and thus would not enable embryos to survive.[558]

E. SEROLOGY

Reviews of interest on the serology of invertebrate viruses include those of DiCaupua and Norton,[85] Mazzone and Tignor,[84] Mazzone,[83] Volkman,[64,65] Tinsley,[66] and McCarthy and Gettig.[73] Tinsley[66] has pointed out that since invertebrate animals do not possess an immune system similar to that of vertebrates our knowledge of the antigenic structure of insect viruses has been derived from reactions produced with antibodies produced by test animals in a hyperimmune state. What we are able to study is "the effects of insect virus antigens on the antibody responses in warm blooded animals".[66] Although animal virologists have serological techniques to rapidly identify animal viruses, invertebrate virologists have had to rely more on similarities in restriction enzyme profiles to identify specific baculoviruses. Confusion arose in early studies due to the impurity of preparations and the fact that baculoviruses share common antigens.[66,73] Several investigators showed that extreme precautions were necessary in order to obtain nucleocapsids free of VO proteins.[559-561] Harrap et al.[559] demonstrated that certain NC antigens were specific for three species of *Spodoptera* while proteins that were specific to the capsid envelope revealed antigens in common. The review by McCarthy and Gettig[73] has included the results from virus neutralization studies, fluorescence and immunoperoxidase assays, immunodiffusion and immunoelectrophoresis, Western blot analysis of BV polypeptides, enzyme linked immunosorbent assay (ELISA), radioimmunoassays (RIA), indirect immunoperoxidase techniques, and monoclonal antibodies.

Summers et al.[560] used the indirect immunoperoxidase technique to detect homologous antigens with antisera raised against virions at 6 to 8 h p.i. in TN-368 cells infected with AcMNPV LOVAL or PDV. Distinct regions of the cytoplasm were stained (near the nucleus) indicating the site of virion antigen synthesis and accumulation. At 12 h p.i. the staining was more intense in the nucleus than in the cytoplasm. At 15 h p.i. the staining was intense in the nucleus but less in the cytoplasm indicating a possible shutdown in synthesis of virion antigens. Using antiserum made against polyhedrin this protein was first detected at 12 h p.i. McCarthy et al.[561] used this technique to study the effect of inoculum concentration on the development of AcMNPV in TN-368 cells. NC antigens were detected 6 to 12 h p.i.; polyhedral proteins were detected 12 to 18 h p.i. At a tissue culture infective dose at 50% infection ($TCID_{50}$ per cell) or multiplicity of infection (m.o.i.) of 50, the extracellular virus titer increased between 6 and 12 h p.i.; while at m.o.i. of 25, 5, and 1, the ECV titer increased at 12 to 18 h p.i. The detection of baculovirus antigens and differences between NPV isolates was attempted by several investigators using ELISA,[389,560-565] RIA,[566] and protein blot immunoassay.[567] Brown et al.[564] found that *S. littoralis* MNPV was serologically distinct from NPVs of *S. frugiperda*, *S. exempta*, and *S. exigua*. McCarthy and Henchal[565] used ELISA to detect infectious BV at 72 h p.i. in fourth and fifth instar larvae inoculated with 2×10^4 VOs per milliliter. Smith and Summers[566] tested antisera against the virion structural polypeptides of 17 different baculovirus isolates which contained two MNPVs, one SNPV and a GV. They found that these isolates had similar antigenic determinants but these shared determinants may in fact be contaminating polyhedrin or granulin. Further studies using these isolates by Knell et al.[567] revealed a number of shared antigenic determinants but these shared determinants may in fact be contaminating polyhedrin or granulin.

Differences between the two phenotypes of NPV virions, i.e., virions (LOVAL or PDV) and virions with peplomers (BV or ECV) occur in morphology (see Figure 14) and infectivity. PDV was more infectious *per os* while ECV was more infectious by intrahemocoelic injection.[252,568-570] Neutralization studies revealed that antiserum raised to PDV neutralized PDV, while antiserum raised to ECV neutralized both phenotypes but the PDV was neutralized to a lesser extent.[568-572] These studies indicated that the differences in activity were associated with the viral envelopes and not the nucleocapsids of the virions. Following the production of a monoclonal antibody (AcV_1) reactive to the ECV phenotype of AcMNPV by Holman and Faulkner,[573] Keddie and Volkman[570] conducted *in vivo* neutralization studies with AcV_1 and ECV from AcMNPV in which ECV was incubated with AcV_1 before injection into the hemocoel of *T. ni* larvae. The infectivity of ECV was reduced approximately 240-fold while that of PDV *per os* was reduced approximately 46-fold. Another study by Volkman et al.[254] indicated that AcV_1 did not react with PDV or LOVAL in neutralization, ELISA, or indirect immunoperoxidase tests. Immunoelectron microscopic studies using peroxidase-antiperoxidase (PAP) procedures revealed staining in regions of heavy budding at the plasma membrane. The entire ECV envelope was stained but more intensively on the end with peplomers.[254]

Early studies indicated that the major protein of ECV was 67 to 70 K which was detected 6 to 12 h p.i. in infected cell cultures.[574-577] Volkman et al.[254] and Volkman and Goldsmith[578] using immunoprecipitation, SDS-PAGE, and autoradiography found that the target antigen of AcV_1 neutralization was a 64-K acidic phosphoglycoprotein. The question of whether this 64-K protein is similar to that of PDV is yet unresolved; however, studies to date indicate that differences exist because when the 64-K antigen of ECV is inactivated by AcV_1, it reacts in a manner similar to PDV.[64] Volkman and Goldsmith,[402] in virus neutralization studies, established that AcV blocked entry of ECVs into IPLB-SF-21 cells by the adsorptive endocytosis pathway. However, not all infectivity was neutralized by AcV₁. Further studies revealed that there is an alternate pathway for AcMNPV-ECV and AcV₁ linked ECV to gain entry into cells *in vitro*, i.e., fusion at the plasma membrane as viral envelope antigen was detected in the host cell plasma membrane by immunoelectron microscopy. When cells were treated with cytochalasin B, an inhibitor of phagocytosis, BV infectivity was not reduced, thus indicating that ECVs did not enter cells by phagocytosis.[403]

The effect of tunicamycin (TM), an inhibitor of N-linked glycosylation, was tested on the BV phenotype of AcMNPV by Charlton and Volkman.[579] They found that at 10 μg/ml glycosylation was completely inhibited, but the unglycosylated major envelope protein was still functional in normal viral entry. The lower infectivity obtained was attributed to a reduction in the major envelope protein and the number of extracellular particles produced.

Volkman et al.[410] showed that cytochalasin B and D prevented the production of infectious ECV by inhibiting synthesis of complete virions. BV particles lacking nucleocapsids budded from the plasma membrane in the AcMNPV-infected cells containing cytochalasin D while normal ECV budded from the plasma membranes of the control cells (see Figures 44A and B). Volkman[411] has found that cytochalasin D, an inhibitor of microfilament elongation, interfered with nucleocapsid assembly in the nucleus. Viral DNA and capsid antigen synthesis occurred where the latter was transported to the nucleus; however, NC assembly was blocked. Capsid protein was detected in long tubular structures by the positive association of anticapsid Mab39P10 with antibody anti-mouse IgG conjugated to 15-nm gold particles (Figure 45).

A monoclonal antibody (3D10) developed by Roberts and Naser[580] immunoprecipitated a polypeptide of 42 K from the BV of AcMNPV. Naser and Miltenberger[581,582] used a monoclonal antibody in a Western blotting-ELISA technique to determine the relatedness of a group of NPVs. The monoclonal antibody reacted only with AcMNPV and *Galleria mellonella* MNPV, a genotypic variant, and not with the other NPVs tested. Monoclonal antibodies against HzSNPV polyhedrin were developed by Huang et al.[583] who found that group and subgroup epitopes exist. Multiple copies of one epitope or a series of closely related epitopes existed in the polyhedrin molecule.[583] Quant et al.[584] developed monoclonal antibodies to the polyhedrin of two OpNPVs and later characterized p10 synthesis in *Orygia pseudotsugata* PDV using Mab210.[455] A cross reaction was detected to structures of the cytoskeletal system known to contain spindle fibers and intracellular bridges with Mabs 210, 211, and 209 in *L. dispar* cells. Anti-β-tubulin Mab produced immunofluorescent staining in insect cell cytoplasm and did not appear to stain p10 associated structures. Quant-Russell et al.[455] concluded that the p10 Mab210 was probably binding to a microtubule-associated protein in uninfected cells which was not tubulin. The epitope to which the Mab210 appeared to bind contained amino acids within the sequence (LPEIPDVP; site 77 to 84) coded between the *Alu*I and *Hpa*II sites.[455] Volkman and Falcon[585] used a monoclonal antibody in an ELISA to detect the presence of *T. ni* SNPV polyhedrin in *T. ni* larvae before the appearance of symptoms. The ELISA detected the presence of NPV 2 to 3 d in advance of a trained observer. The data of Volkman and Zaal[393] differ in finding that p10 was not involved in the depolymerization of microtubules but appeared to colocalize with microtubules in immunofluorescence and immunoelectronmicroscope studies.

F. VIRUS EXPRESSION VECTORS AND FUTURE PROSPECTS FOR NPVS

Miller[586] first proposed that insect baculoviruses could serve as virus expression vectors for invertebrates. The advantages of baculoviruses as virus expression vectors for cloned eukaryotic cells as presented by Smith et al.[587] include: (1) there is potential to encapsidate large pieces of foreign DNA into the viral genome; (2) safety is afforded since the recombinant viral vectors are not pathogenic to vertebrate cells; (3) the polyhedrin gene contains a nonessential region in which foreign DNA may be inserted; (4) a strong promoter (polyhedrin) directs transcription late in the infection cycle even after progeny virions have been produced; and (5) a genetic marker may be used to select for recombinants. Smith et al.[588] presented a method to introduce site-specific mutations into the genome of AcMNPV.

The potential of the baculovirus expression vector system is tremendous. Significant advances have resulted from applications in agricultural and medical research. Recent reviews include those by Miller,[34,586] Summers and Smith,[589] Miller et al.,[590] Doerfler,[69] Doerfler and Bohm,[70] Kuroda et al.,[62] Maeda,[63] Maramorosch,[57] Luckow and Summers,[333] and Atkinson et al.[591] A manual of methods for baculovirus expression vectors and insect cell culture procedures has been prepared by Summers and Smith.[58] The review by Luckow and Summers[333] lists 35 foreign genes expressed by baculovirus vectors, which include such proteins as human α- and β-interferons, influenza virus hemagglutinin, S RNA-coded genes of lymphocytic choriomeningitis arenavirus, HIV env, HIV gag, HTLV-I p40, and *Solanum tubersum* (potato) patatin. In a more recent study, the hemagglutinin-neuraminidase (HN) gene of Newcastle disease virus was cloned as a cDNA and inserted into the AcMNPV expression vector system to obtain high yields of HN proteins.[592] The baculovirus recombinant expressed HN, was electrophonetically and antigenically similar to the native HN, and displayed both hemagglutinating and neuraminidase activities. However, in another study, incorporation of a synthetic gene encoding an insect-specific paralytic scorpion neurotoxin in the AcMNPV expression vector resulted in low levels of toxin gene expression due possibly to toxin instability.[481]

Possibilities for the future development of genetically engineered baculoviruses with improved virulence and host range have been reviewed by Faulkner and Boucias,[128] Maramorosch,[159] Dougherty,[593] Payne,[150] and Kang.[594] The wide use of baculoviruses to construct expression vectors for bacteria and eukaryotic genes has already been noted.[58,68,69,159,333,586,588,590] As baculoviruses exist in nature it has been established that they are harmless to any other animals, bees,[595] or humans.[4-6] As Maramorosch[159] has noted "the potential of genetic engineering and cell fusion must be exploited without endangering the environment." Recent experiments by Bishop et al.[77] and Bishop[596] were aimed at determining the fate of genetically engineered NPVs by testing a "self destruct" NPV from which the polyhedrin gene was removed. Reviews by Dougherty[593] and Payne[150] have discussed the recent progress in genetic improvement of baculoviruses noting the great potential as new genes are introduced into specific locations of the BV genome. Two significant discoveries should contribute toward the genetic improvement of baculoviruses. O'Reilly and Miller[597] found that AcMNPV has a gene that blocks the molting of its insect host by producing an ecdysteroid UDP-glucosyl transferase. Hammock et al.[598] developed a technique to produce juvenile hormone esterase (JHE) in a baculovirus vector. The reduction in titer of juvenile hormone (JH) early in the last larval instar is associated with initiation of memtamorphosis and cessation of feeding. JHE hydrolyzes the methyl ester of JH to the biologically inactive JH acid. Elevated levels of JHE *in vivo* (with reduced levels of JH) at an early stage of development should cause insects to stop feeding. The JHE *in vivo* expression should contribute to the improvement of genetically engineered viral microbial control agents by reducing insect feeding.

Dougherty[593] has also noted that increased future use of insect viruses will come as improvements are made in formulation, application strategies, increasing virulence, and extending host range. More knowledge in virus epizootiology is needed.[593] A significant improvement in efficacy has been made by Tompkins et al.,[599] who have shown that encapsulation of BVs extended their efficacy in the control of cabbage loopers and imported cabbageworms on collards. In an effort to improve the effectiveness of baculoviruses as insecticides, Merryweather et al.[600] inserted the δ-endotoxin gene from *Bacillus thuringiensis* var. *kurstaki* HD-73 into the NPV of *Autographa californica*. Using two separate vector systems (polyhedrin-negative, Ac[PH⁻] Bt and polyhedrin-positive, Ac[PH⁺] Bt), recombinant virus-infected cells synthesized a large polypeptide (130 K), the cleaved active form (62 K), and a 44-K protein which reacted with antisera specific for the insecticidal toxin. To assess the biological activity of the two recombinant viruses, standard bioassays were performed using second instar *T. ni* larvae; when larvae were placed on a diet containing the Ac(PH⁻)Bt recombinant virus, feeding inhibition was observed. The Ac(PH⁺)Bt virus had an LD_{50} value approximately twofold higher than that observed for the wild type AcMNPV.

Payne[150] has advised that in order to make responsible use of the genetic engineered microorganisms, more detailed ecological studies of host pathogen interactions are needed as well as comparative laboratory and field studies of modified strains and naturally occurring viruses. Clearly, proper precautions will need to be taken before the introduction of biogenetically engineered microorganisms which are new into our ecosystems.

ADDENDUM

The use of baculovirus expression vectors to produce proteins for biomedical research and for the pharmaceutical industries has resulted in numerous publications such as those on proteins of human immunodeficiency disease,[601-606] amyloid proteins associated with Alzheimer's disease,[607] vaccines,[608] antibodies,[609] and plant virus proteins.[610-612] Recent reviews on the use of recombinant baculoviruses include those by Webb and Summers,[613] Page and Murphy,[614] and Christian and Oakeshott.[615] Processes for more efficient production of baculovirus (BVs) in insect cell cultures have been developed.[616-622] The effect of dissolution procedures on the infectivity of NPV derived virions for cell cultures has been reported by Vaughn et al.[623] and improved techniques for transfection of insect cells with BV DNA have been reported by Corsaro et al.[624] and Mann and King.[625] Tompkins et al.[626] showed that infectivity and virulence of NPVs were maintained during serial passage in noctuid cell lines by the addition of liposomes to TNFH-368 cells. IPLB-SF-21AE cells showed less effect of liposome addition since the cells utilized the exogenous sterols and lipids provided by the serum supplement in the standard medium. Tsuda et al.[627-629] have recently reported on three studies on infectivity of NPV in continuous cell cultures. Hink et al.[630] studied the expression of three recombinant proteins using BV vectors in 23 insect cell lines while Kitts et al.[631] improved the percentage of endonuclease obtained using digestion with a specific endonuclease but observed reduced infectivities compared to circular DNA.

Merryweather et al.[632] used the polyhedrin promoter to construct a polyhedrin-BV vector containing the delta endotoxin of *Bacillus thuringiensis* (HD-73 strain) and the p10 promoter to construct a polyhedrin + BV vector containing the same toxin. Martins et al.[633] introduced a large crystal protein gene into the genome of AcMNPV. The polyhedrin gene was replaced with the insecticidal crystal protein cryIA(b) of *Bacillus thuringiensis* subsp. *aizawai* 7.21. The *Bacillus thuringiensis* (*B. t.*) toxic crystals were expressed predominantly in the cytoplasm of *S. frugiperda* cell cultures infected with AcMNPV recombinant. Since a polyhedrin negative virus is impractical in the field, an AcMNPV recombinant is being constructed in which the *B. t.* crystal protein gene is inserted in the p10 gene locus under the crystal of the p10 promoter; this will yield a polyhedrin plus recombinant (Martins, unpublished). A baculovirus vector containing the gene encoding firefly luciferase has been constructed[634] and synthesized in large quantities *in vivo* in *T. ni* and *S. frugiperda* larvae.[635]

Genome maps have been prepared for *Spodoptera littoralis* MNPV[636] and *Bombyx mori* MNPV.[637] Sugimori et al.[638] analyzed the structural polypeptides of the N9 isolate of *B. mori* MNPV while Vialard et al.[639] identified and characterized a 37 K polypeptide of

AcMNPV which appears to be associated with the OB envelope (the DNA sequence of this 37 K polypeptide was reported by Wu and Miller[640]) and described the localization, time of synthesis and homology to spheroidin, the major component of entomopox viruses. The nucleotide sequence of the *Hin*dIII F region[641] and the *Hin*dIII P region[642] of AcMNPV have been reported.

Huh and Weaver[643] distinguished three classes of transcripts using cycloheximide and aphidicolin in *S. frugiperda* (IPLB-SF-21) cells infected with AcMNPV. They also identified the RNA polymerases that synthesize specific transcripts of AcMNPV, eg., transcripts from the p26 gene in the *Hin*dIII K/P region and the p35 gene in the *Hin*dIII K/Q region and the γ-transcripts in the *Hin*dIII K region are synthesized by the α-amanitin resistant virus-induced RNA polymerase late in infection.[644] Schetter et al.[645] identified an insect cell DNA insertion of 634 bp in the 81-map-unit segment of AcMNPV. The gene[646] encoding the envelope glycoprotein (gp64) of the budded virus (BV) of OpMNPV was mapped to the *Hin*dIII-E fragment of the viral genome and compared with the gp67 sequence of AcMNPV reported earlier by Whiteford et al.[480]

The time course of expression of three proteins (gp64, p39, and polyhedrin) were analyzed at 3 different m.o.i. (5, 10, and 100) in *L. dispar* cell cultures infected with OpMNPV by Bradford and Rohrmann.[647] The rate of BV production reached maximum levels at 24 to 36 h p.i. and continued at high levels. Polyhedra were first visible at 48 h p.i. The m.o.i. apparently influenced the magnitude but not the timing of early events in the viral replication cycle, the levels of BV production and percent of cells containing VOs but had little effect on the final rates of BV production and the time of detection of p39 and polyhedrin on western blots.[647] Russell et al.[648] used immunogold labeling and monospecific antibodies in immunoelectron microscopy to follow the temporal expression and localization of polyhedrin, p10, and p39 of OpMNPV in *L. dispar* cells. Using a polyclonal antibody, the presence of polyhedron envelope (PE) protein was followed. They found that p39, detected by other tests at 24 h p.i., was found in the nucleus at 24 h p.i. Polyhedrin was not found until 48 h p.i. where it was specifically localized to polyhedra; no accumulation occurred elsewhere in the nucleus or cytoplasm indicating that polyhedrin is synthesized in the cytoplasm and transported directly to the nucleus where it crystallizes around the virions to form polyhedra of viral occlusion bodies. A BV expression vector (BEV) was constructed (OplacZ) which expressed the β-galactosidase under the direction of the polyhedrin promoter. This BEV was used to study p10 and PE, which were labeled with 10 and 20 nm gold particles, respectively. The interior of the fibrillar structures showed the p10 label (10 nm gold) while the periphery was labeled with the 20 nm PE labeled gold particles. The PE label was not so intense in wt-virus-infected cells[473] perhaps due to the fact that polyhedrin served as a substrate for PE protein deposition. The OplacZ virus infected *L. dispar* cells, showed β-galactosidase staining (random distribution) in the nucleus and in the endoplasmic reticulum of the cytoplasm. It was detected at 36 h p.i. by the ONPG (*O*-nitriphenyl-βD-galactosidase) reaction but not until 48 h p.i. by immunoelectron microscopy or western blot analysis.[648] The IE-1 gene of OpMNPV has been identified and characterized[649] and compared with the IE-1 gene of AcMNPV.

Jarvis et al.[650] used the IE-1 promoter, to produce transformed insect cell clones which expressed a foreign gene product continuously. Sf9 cells were cotransfected with a mixture of a plasmid containing a neomycin-resistance gene and a plasmid containing a foreign gene, both under the control of the IE-1 promoter and assayed for the expression of the foreign gene. Many clones expressed the foreign gene but at much lower levels than usually produced in BV infected cells; however, the transformed cells were stable, expressed the foreign gene continuously and were able to process a complex glycoprotein more efficiently. This alternative approach was recommended for the expression of products that are relatively poorly expressed and/or processed in BV infected cells. The transcriptional regulation of an early gene encoding the 35 K protein was studied by Nissen and Friesen.[651] Lu and Carstens[652] determined the nucleotide sequence of the 60.1 to 65.5 m.u. region of AcMNPV which contains a gene essential for viral DNA replication, i.e., ORFI (p143) which is homologous to other proteins with DNA unwinding activity which could function to "unwind duplex DNA at the origin of replication and/or at the replication fork." Partington et al.[653] isolated and characterized two temperature-sensitive mutants of AcMNPV which were defective in very late gene expression, i.e., both synthesized viral DNA but were defective in ECV production at the nonpermissive temperature. All protein synthesis including host proteins was completely inhibited by 24 h.

Weyer et al.[654] used a new transfer vector pAcUWI to obtain recombinants of AcMNPV to produce high levels of polyhedrin of β-galactosidase under the control of the p10 gene promoter. Normal polyhedra were produced. This demonstrated that the p10 promoter can serve the function of the polyhedrin promoter. Another BEV, pAcUWZ (A) and (B) was constructed which allowed any foreign gene to be inserted upstream of the polyhedrin gene under the control of the p10 promoter.[654] Thiem and Miller[655] constructed a series of re-combinant viruses to study the regulation of expression of the promoter (Pcap) for the major capsid-protein encoding gene (p39) of AcMNPV which they had identified and sequenced earlier.[246] Hill-Perkins and Possee[656] constructed a BEV derived from the basic protein promoter of AcMNPV. The site chosen for the second copy of the basic protein promoter was the region of the polyhedrin coding sequences or promoter elements did not affect viral replication. The advantage of this BEV was that insect-specific toxins and hormones would be expressed at earlier times than with the p10 and polyhedrin expression systems, affecting the target insect earlier after ingestion.

A general analysis of receptor-mediated viral attachment to cell surfaces was recently presented by Wickham et al.[657] Three modes of receptor-mediated virus attachment were proposed; the third mode (spatial saturation with multivalent binding) includes the BVs. A mathematical model was presented to elucidate the virus attachment protein (VAP)/receptor affinity, receptor number, and density VAP number and virus concentration on the overall binding. Binding experiments with AcMNPV in Sf9 insect cells showed spatial saturation followed by virus adsorption in multilayers.

Cell specific proteins and viral DNA synthesis in permissive infections (SFIPLB-21) and nonpermissive infections (TN 368) of SfMNPV revealed 28 infected cell-specific polypeptides (ICSPs) ranging in M_r from 9,000 to 124,000 were detected in the infected SFIPLB-21 cells while only a 97 K and 29 K ICSP were detected in TN-368 cells infected with SfMNPV.[658] Thirty isolates of SeMNPV obtained from larvae of the beet armyworm, *S. exigua*, collected in viral epizootics in California appeared to be caused by a dominant genotype with closely related variants according to comparisons of REN patterns.[659] Maeda et al.[660] plaque-purified more than 100 isolates from 4 wild stocks of *Spodoptera litura* MNPV collected in Japan and compared them with *S. littoralis* collected in Egypt and BmMNPV. They classified them into four groups based on their *in vitro* host range, polyhedra characteristics, DNA restriction endonuclease patterns, and DNA sequence similarity. Two of the wild stocks were mixtures of the other groups. They proposed that the existence of different viruses in the wild stocks might account for propagation of progeny viruses when the original virus stock is ingested by other insect hosts.

Doyle et al.[661] tested 74 species of insect larvae from four orders for susceptibility to *M. brassicae* NPV(MbMNPV). None of the species outside of Lepidoptera were susceptible. Insect smears were made; if possible, then dot blots and RENs were carried out. Thirty-two species of Lepidoptera were susceptible, mostly from the family of Noctuidae, but also from the families of Geometridae, Yponomeutidae, and Nymphalidae. Another report on MbMNPV infection in *Aglais urticae* appeared recently.[662]

The co-occlusion and persistence of a baculovirus mutant was tested with AcMNPV in IPLB-SF-21 cells with wt and mutant (Ac-E10) AcMNPV by Hamblin et al.[663] To maintain the mutant, high doses of mutant and wt NPV were required in a ratio of 50:50.

An excellent study by Charlton and Volkman[664] traced the role of f-actin in the replication of AcMNPV in IPLB SF21 cells using fluorescence microscopy. TRITC-phalloidin labeling was used to compare uninfected and viral infected cells. While the actin microfilaments in uninfected cells appeared as a fine homogeneous network, infected cells contained coarse actin cables crossing the cells along the surface or in microspikes. At 1 to 3 h p.i. the thick actin microfilament cables formed (induced by a component of the viral inoculum) and disappeared by 6 h p.i. Ventral aggregates of f-actin appeared at 6 h p.i. (due to early viral gene activity) but were transient. At 12 h p.i. actin microfilaments began to polymerize around the virogenic stroma (due to viral DNA replication or a late gene product). At this time, with MAb 39 (39P10), p39 was localized in the virogenic stroma. As production of ECV increased, the colocalization of actin microfilaments with p39 increased in the nuclear ring zone. This evidence supported by earlier reports[393,410,411] showed that actin microfilaments may be involved in NC morphogenesis. Since p39 was retained by an f-actin binding protein. These observations led Charlton and Volkman[664] to propose that "f-actin attaches to the nuclear matrix in infected cells where it provides a scaffold for assembly and filling of capsids."

A proteinaceous viral derived factor(s) is secreted into gypsy moth cell lines for (IPLB-LD652Y and IPLB-LDFB) which are semipermissive replication of AcMNPV.[665] Guzo et al.[665] have designated it the macromolecular synthesis inhibition factor (MSIF) since it is produced and secreted from infected cells between 1 and 30 h p.i. and causes cytotoxic effects such as reduced levels of growth, mitosis, DNA and protein synthesis when uninfected cells are incubated in media containing MSIF. Preliminary tests indicate MSIF is heat-labile, proteinaceous and pH sensitive. Its activity was neutralized by three different monoclonal antibodies directed against the AcMNPV 64 K envelope glycoprotein. It was inhibited with leupeptin, a lysosome protease inhibitor. Additional studies on the nature of MSIF and its role *in vivo* are currently underway.[665] Six monoclonal antibodies against occluded virions of BmNPV were recently reported.[666] McIntosh[667] compared the *in vitro* host range of the celery looper, *Syngrapha falcifera* MNPV with AcMNPV and found that SfaMNPV also has a wide *in vitro* host range; the most productive cell line was BcIRL-HS-AM1.

Polyhedra without occluded virions were observed in the nuclei of columnar cells of the midguts of *H. zea* following inoculations with cytoplasmic polyhedrosis virus.[668] Keating et al.[669] determined the pH of the midguts of *L. dispar* larvae following feeding upon aspen and oak foliage and noted that larval susceptibility to LdMNPV was reduced when feeding on acid foliage and acidic diets due to a lowering of gut pH. Earlier studies had indicated that susceptibility varied according to the pH of the foliage consumed by the gypsy moth larvae.[670,671] Studies are continuing on the protection of microbials from the degradation of the ultraviolet rays of the sun. Variations in the biological activities of LdMNPV isolates from three geographical areas were found by Shapiro and Robertson.[672] They determined that the most logical selection criteria to be those isolates with the lowest LC_{50} vs. LC_{90}. In plots of LC_{50} vs. LC_{90} the most active isolates were those closest to the 0.0 intersection of the x and y axes. Recently, Ignoffo et al.[673] found that the addition of the uv protectants carbon and Congo red in starch encapsulations provided the best protection for HzSNPV.

Successful vertical transmission studies with HzSNPV and AcMNPV on *Heliothis virescens* male adults showed the effect of viral contamination on the scales of males' posterior segments, eggs laid by mated females (checked by scanning electron microscopy), and mortality of offspring.[674]

Hartig et al.[675] designed an *in vitro* testing scheme for viral pesticides and used it to test AcMNPV for potential health hazards. They concluded that since AcMNPV does not persist in primate cells, traditional plaque assays and nucleic acid blots are sufficient to check for viral persistence.

The safety of BVs to nontarget invertebrates was reviewed by Groner.[676] To date, no adverse effects of BVs to predators and parasites have been observed. Although penaeid shrimps have been observed with BV infections (see II this chapter), attempts to establish infections in aquatic invertebrates have generally been unsuccessful until the reports of Overstreet et al.[677] and LeBlanc and Overstreet.[678]

ACKNOWLEDGMENTS

The authors gratefully acknowledge the technical assistance of the following colleagues: T. A. Wilcox, A. V. Matthews, M. C. Fenton, M. A. Axum, L. B. Thaden, K. Riedl, M. E. Hartman, and J. M. Bradsher.

REFERENCES

1. **Matthews, R. E. F.,** Classification and nomenclature of viruses, *Intervirology,* 17, 1, 1982.
2. **Tinsley, T. W. and Kelly, D. C.,** Taxonomy and nomenclature of insect pathogenic viruses, in *Viral Insecticides for Biological Control,* Maramorosch, K. and Sherman, K. E., Eds., Academic Press, Orlando, FL, 1985, 3.
3. **Martignoni, M. E. and Iwai, P. J.,** A catalog of viral diseases of insects, mites and ticks, *U.S. Department of Agriculture Forest Service, Gen. Tech. Report,* PNW 195, 1, 1986.
4. **Groner, A.,** Specificity and safety of baculoviruses, in *The Biology of Baculoviruses,* Vol. 1, Granados, R. R. and Federici, B. A., Eds., CRC Press, Boca Raton, FL, 1986, 177.
5. **Doller, G.,** The safety of insect viruses as biological control agents, in *Viral Insecticides for Biological Control,* Maramorosch, K. and Sherman, K. E., Eds., Academic Press, Orlando, FL, 1985, 399.
6. **Laird, M., Lacey, L. A., and Davidson, E. W., Eds.,** *Safety of Microbial Insecticides,* CRC Press, Boca Raton, FL, 1990.
7. **Evans, H. F. and Entwistle, P. F.,** Epizootiology of the nuclear polyhedrosis virus of European spruce sawfly with emphasis on persistence of virus outside the host, in *Microbial and Viral Pesticides,* Kurstak, E., Ed., Marcel Dekker, New York, 1982, 449.
8. **Evans, H. F.,** Ecology and epizootiology of baculoviruses, in *The Biology of Baculoviruses,* Vol. 2, Granados, R. R. and Federici, B. A., Eds., CRC Press, Boca Raton, FL, 1986, 89.
9. **Hostetter, D. L. and Bell, M. R.,** Natural dispersal of baculoviruses in the environment, in *Viral Insecticides for Biological Control,* Maramorosch, K. and Sherman, K. E., Eds., Academic Press, Orlando, FL, 1985, 249.
10. **Benz, G.,** Introduction: Historical perspectives, in *The Biology of Baculoviruses,* Vol. 1, Granados, R. R. and Federici, B. A., Eds., CRC Press, Boca Raton, FL, 1986, 1.
11. **Steinhaus, E. A.,** *Insect Microbiology,* Cornell University, Press Hafner Publ., New York, 1949.
12. **Steinhaus, E. A.,** *Principles of Insect Pathology,* McGraw-Hill, New York, 1949.
13. **Bergold, G. H.,** Insect viruses, *Adv. Virus Res.,* 1, 91, 1953.
14. **Bergold, G. H.,** Viruses of insects, in *Handbuch der Virusforschung,* Vol. 4, Hallauer, C. and Meyer, K. F., Eds., Springer, Vienna, 1958, 60.
15. **Bergold, G. H.,** The nature of nuclear-polyhedrosis viruses, in *Insect Pathology, An Advanced Treatise,* Vol. 1, Steinhaus, E. A., Ed., Academic Press, New York, 1963, 413.
16. **Aizawa, K.,** The nature of infections caused by nuclear-polyhedrosis viruses, in *Insect Pathology-An Advanced Treatise,* Vol. 1, Steinhaus, E. A., Ed., Academic Press, New York, 1963, 381.
17. **Vago, C. and Bergoin, M.,** Viruses of invertebrates, *Adv. Virus Res.,* 13, 247, 1968.
18. **Tinsley, T. W. and Harrap, K. A.,** Moving frontiers in invertebrate virology, *Monographs in Virology,* Vol. 6, Tinsley, T. W. and Harrap, K. A., Eds., S. Karger, Basel, 1972.
19. **Bellett, A. J. D., Fenner, F., and Gibbs, A. J.,** The viruses, in *Viruses and Invertebrates,* Vol. 31, Gibbs, A. J., Ed., North Holland Res. Monographs-Frontiers in Biology, North Holland Publ., Amsterdam, 1973, 41.
20. **Vaughn, J. L.,** Viruses and rickettsial diseases, in *Insect Diseases,* Vol. 1, Cantwell, G. E., Ed., Marcel-Dekker, New York, 1974, 49.
21. **David, W. A. L.,** The status of viruses pathogenic for insects and mites, *Annu. Rev. Entomol.,* 20, 97, 1975.

22. **Paschke, J. D. and Summers, M. D.**, Early events in the infection of the arthropod gut by pathogenic insect viruses, in *Invertebrate Immunity*, Maramorosch, K. and Shope, R. E., Eds., Academic Press, New York, 1975, 75.

23. **Granados, R. R.**, Infection and replication of insect pathogenic viruses in tissue culture, *Adv. Virus Res.*, 20, 189, 1976.

24. **Smith, K. M.**, *Virus-Insect Relationships*, Longman, London, 1976.

25. **Summers, M. D.**, Baculoviruses (Baculoviridae), in *The Atlas of Insect and Plant Viruses*, Maramorosch, K., Ed., Academic Press, New York, 1977, 3.

26. **Weiser, J.**, *An Atlas of Insect Diseases*, Junk, W., B. V. Publ., The Hague, Netherlands, 1977.

27. **Tinsley, T. W. and Harrap, K. A.**, Viruses of Invertebrates, in *Comprehensive Virology*, Vol. 12, Fraenkel-Conrat, H. and Wagner, R. P., Eds., Plenum Press, New York, 1978, 1.

28. **Harrap, K. A. and Payne, C. C.**, The structural properties and identification of insect viruses, *Adv. Virus Res.*, 25, 273, 1979.

29. **Granados, R. R.**, Infectivity and mode of action of baculoviruses, *Biotechnol. Bioeng.*, 22, 65, 1980.

30. **Faulkner, P.**, Baculovirus, in *Pathogenesis of Invertebrate Microbial Diseases*, Davidson, E. W., Eds., Allanheld, Osmum & Co. Publ., Towata, NJ, 1981, 3.

31. **Kelly, D. C.**, Baculovirus replication, *J. Gen. Virol.*, 63, 1, 1982.

32. **Tchuchrity, M. G.**, *The Ultrastructure of the Viruses from the Lepidoptera-The Pests of Plants*, Shtiinksa Kishinev, 1982.

33. **Miller, L. K., Lingg, A. J., and Bulla, L. A., Jr.**, Bacterial, viral and fungal insecticides, *Science*, 219, 715, 1983.

34. **Miller, L. K.**, Exploring the gene organization of baculoviruses, *Methods in Virology*, Vol. 7, Academic Press, Orlando, FL, 1984, 227.

35. **Miltenburger, H. and Krieg, A., Eds.**, Bioinsecticides II. Baculoviridae in *Advances in Biotechnological Processes 3*, Mizrahi, A. and Van Wezel, A. L., Eds., Alan R. Liss, New York, 1984, 291.

36. **Tanada, Y. and Hess, R. T.**, The cytopathology of baculovirus infections in insects, in *Insect Ultrastructure*, Vol. 2, King, R. D. and Akai, H., Eds., Plenum Press, New York, 1984, 517.

37. **Kelly, D. C.**, The structure and physical characteristics of baculoviruses, in *Viral Insecticides for Biological Control*, Maramorosch, K. and Sherman, K. E., Eds., Academic Press, Orlando, FL, 1985, 469.

38. **Vaughn, J. L. and Dougherty, E. M.**, The replication of baculoviruses, in *Viral Insecticides for Biological Control*, Maramorosch, K. and Sherman, K. E., Eds., Academic Press, Orlando, FL, 1985, 569.

39. **Mazzone, H. M.**, Receptors in the infection process, in *Viral Insecticides for Biological Control*, Maramorosch, K. and Sherman, K. E., Eds., Academic Press, Orlando, FL, 1985, 695.

40. **Mazzone, H. M.**, Pathology associated with baculovirus infection, in *Viral Insecticides for Biological Control*, Maramorosch, K. and Sherman, K. E., Eds., Academic Press, Orlando, FL, 1985, 81.

41. **Padhi, S. B.**, Viral proteins for the identification of insect viruses, in *Viral Insecticides for Biological Control*, Maramorosch, K. and Sherman, K. E., Eds., Academic Press, Orlando, FL, 1985, 55.

42. **Vlak, J. M. and Rohrmann, G. F.**, The nature of polyhedrin, in *Viral Insecticides for Biological Control*, Maramorosch, K. and Sherman, K. E., Eds., Academic Press, Orlando, FL, 1985, 489.

43. **Arif, B. M.**, The structure of the viral genome, *Curr. Top. Microbiol. Immunol.*, 131, 21, 1986.

44. **Harrap, K. A. and Payne, C. C.**, The structural properties and identification of insect viruses, in *Adv. Virus Res.*, 25, 273, 1986.

45. **Rohrmann, G. F.**, Polyhedrin structure, *J. Gen. Virol.*, 67, 1499, 1986.

46. **Liang, D. G., Cai, Y. N., Lin, D. Y., Zhang, Q. L., Hu, Y. Y., He, H. J., and Zhao, K. B.**, *The Atlas of Insect Viruses in China*, Hunan Science and Technology Press, Hunan, China, 1986.

47. **Burand, J. P., Kawanishi, C. Y., and Huang, Y. S.**, Persistent baculovirus infections, in *The Biology of Baculoviruses*, Vol. 1, Granados, R. R. and Federici, B. A., Eds., CRC Press, Boca Raton, FL, 1986, 159.

48. **Bilimoria, S. L.**, Taxonomy and identification of baculoviruses, in *The Biology of Baculoviruses*, Vol. 1, Granados, R. R. and Federici, B. A., Eds., CRC Press, Boca Raton, FL, 1986, 37.

49. **Cochran, M. A., Brown, S. E., and Knudson, D. L.**, Organization and expression of the baculovirus genome, in *The Biology of Baculoviruses*, Vol. 1, Granados, R. R. and Federici, B. A., Eds., CRC Press, Boca Raton, FL, 1986, 239.

50. **Federici, B. A.**, Ultrastructure of baculoviruses, in *The Biology of Baculoviruses*, Vol. 1, Granados, R. R. and Federici, B. A., Eds., CRC Press, Boca Raton, FL, 1986, 61.

51. **Granados, R. R. and Williams, K. A.**, *In vivo* infection and replication of baculoviruses, in *The Biology of Baculoviruses*, Vol. 1, Granados, R. R. and Federici, B. A., Eds., CRC Press, Boca Raton, FL, 1986, 89.

52. **Volkman, L. E. and Knudson, D. L.**, *In vitro* replication of baculoviruses, in *The Biology of Baculoviruses*, Vol. 1, Granados, R. R. and Federici, B. A., Eds., CRC Press, Boca Raton, FL, 1986, 109.

53. **Rohrmann, G. R.**, Evolution of occluded baculoviruses, in *The Biology of Baculoviruses*, Vol. 1, Granados, R. R. and Federici, B. A., Eds., CRC Press, Boca Raton, FL, 1986, 203.

54. **Maruniak, J. E.,** Baculovirus structural proteins and protein synthesis, in *The Biology of Baculoviruses,* Vol. 1, Granados, R. R. and Federici, B. A., Eds., CRC Press, Boca Raton, FL, 1986, 129.

55. **Miller, L. K.,** The genetics of baculoviruses, in *The Biology of Baculoviruses,* Granados, R. R. and Federici, B. A., Eds., CRC Press, Boca Raton, FL, 1986, 217.

56. **Granados, R. R. and Federici, B. A., Eds.,** *The Biology of Baculoviruses,* Vol. 1, CRC Press, Boca Raton, FL, 1986.

57. **Maramorosch, K., Ed.,** *Biotechnology in Invertebrate Pathology and Cell Culture,* Academic Press, San Diego, CA, 1987.

58. **Summers, M. D. and Smith, G. E.,** A manual of methods for baculovirus vectors and insect cell culture procedures, *Tex. Agric. Exp. Stn. Bull.,* No. 1555, 1987.

59. **Fraser, M. J.,** FP mutation of nuclear polyhedrosis viruses: a novel system for the study of transposon mediated mutagenesis, in *Biotechnology in Invertebrate Pathology and Cell Culture,* Maramorosch, K., Ed., Academic Press, San Diego, CA, 1987, 265.

60. **Moore, N. F., King, L. A., and Possee, R. D.,** Viruses of insects, *Insects Sci. Appl.,* 8, 275, 1987.

61. **Granados, R. R., Dwyer, K. G., and Derksen, C. G.,** Production of viral agents in invertebrate cell cultures, in *Biotechnology in Invertebrate Pathology and Cell Culture,* Maramorosch, K., Ed., Academic Press, San Diego, CA, 1987, 167.

62. **Kuroda, K., Hansen, C., Rott, R., Klenk, H. G., and Doerfler, W.,** Biologically active influenza virus hemagglutinin expressed in insect cells by a baculovirus vector, in *Biotechnology in Invertebrate Pathology and Cell Culture,* Maramorosch, K., Ed., Academic Press, San Diego, CA, 1987, 236.

63. **Maeda, S.,** Expression of human interferon in silkworms with a baculovirus vector, in *Biotechnology in Invertebrate Pathology and Cell Culture,* Maramorosch, K., Ed., Academic Press, San Diego, CA, 1987, 222.

64. **Volkman, L. E.,** The 64K envelope protein of budded *Autographa californica* nuclear polyhedrosis virus, in *Curr. Top. Microbiol. Immunol.,* 131, 103, 1986.

65. **Volkman, L. E.,** Classification, identification, and detection of insect viruses by serologic techniques, in *Viral Insecticides for Biological Control,* Maramorosch, K. and Sherman, K. E., Eds., Academic Press, Orlando, FL, 1985, 27.

66. **Tinsley, T. W.,** Antigenic structure of insect viruses, *Immunochemistry of Viruses. The Basis for Serodiagnosis and Vaccines,* van Regenmortel, M. H. V. and Neurath, A. R., Eds., Elsevier, Amsterdam, 1985, 474.

67. **Faulkner, P. and Carstens, E. B.,** An overview of the structure and replication of baculoviruses, *Curr. Top. Microbiol. Immunol.,* 1, 131, 1986.

68. **Friesen, P. D. and Miller, L. K.,** The regulation of baculovirus gene expression, *Curr. Top. Microbiol. Immunol.,* 31, 131, 1986.

69. **Doerfler, W.,** Expression of the *Autographa californica* nuclear polyhedrosis virus genome in insect cells: homologous viral and heterologous vertebrate genes — the baculovirus vector system, in *Curr. Top. Microbiol. Immunol.,* 51, 131, 1986.

70. **Doerfler, W. and Bohm, P., Eds.,** The molecular biology of baculoviruses, *Curr. Top. Microbiol. Immunol.,* 1, 131, 1986.

71. **Wood, H. A. and Burand, J. P.,** Persistent and productive infections with the Hz-1 baculovirus, in *Curr. Top. Microbiol. Immunol.,* 119, 131, 1986.

72. **Kozlov, E. A., Levitina, T. L., and Gusak, M. M.,** The primary structure of baculovirus inclusion body proteins. Evolution and structure — function aspects, in *Curr. Top. Microbiol. Immunol.,* 135, 131, 1986.

73. **McCarthy, W. J. and Gettig, R. R.,** Current developments in baculovirus serology, in *The Biology of Baculoviruses,* Vol. 1, Granados, R. R. and Federici, B. A., Eds., CRC Press, Boca Raton, FL, 1986, 147.

74. **Miller, L. K.,** Expression of foreign genes in insect cells, in *Biotechnology in Invertebrate Pathology and Cell Culture,* Maramorosch, K., Ed., Academic Press, San Diego, CA, 1987, 295.

75. **Krell, P. J. and Beveridge, T. J.,** The structure of bacteria and molecular biology of viruses, *Int. Rev. Cytol.,* Suppl. 17, 15, 1987.

76. **Miller, L. K.,** Baculoviruses as gene expression vectors, *Annu. Rev. Microbiol.,* 42, 177, 1988.

77. **Bishop, D. H. L., Entwistle, P. F., Cameron, I. R., Allen, C. J., and Possee, R. D.,** Field tests of genetically-engineered baculovirus insecticides, in *The Release of Genetically-Engineered Micro-organisms,* Sussman, M., Collins, C. H., Skinner, F. A., and Stewart-Tull, D. E., Eds., Academic Press, London, 1988, 143.

78. **Oellig, C., Happ, T., Muller, T., and Doerfler, W.,** Expression of the *Autographa californica* nuclear polyhedrosis virus genome in insect cells, in *Invertebrate Cell Systems Applications,* Vol. 1, Mitisuhashi, J., Ed., CRC Press, Boca Raton, FL, 1990, 197.

79. **Maeda, S.,** Gene transfer vectors of a baculovirus, *Bombyx* mori nuclear polyhedrosis virus, and their use for expression of foreign genes in insect cells, in *Invertebrate Cell Systems Applications,* Vol. 1, Mitisuhashi, J., Ed., CRC Press, Boca Raton, FL, 1989, 167.

80. **Blissard, G. W. and Rohrmann, G. F.**, Baculovirus diversity and molecular biology, *Annu. Rev. Entomol.*, 35, 127, 1990.

81. **Kawanishi, C. Y. and Held, G. A.**, Viruses and bacteria as sources of insecticides, in *Safer Insecticides. Development and Use*, Hodgson, E. and Kuhr, R. J., Eds., Marcel Dekker, New York, 1990, 351.

82. **Volkman, L. E. and Keddie, B. A.**, Nuclear polyhedrosis virus pathogenesis, *Semin. Virol.*, 1, 249, 1990.

83. **Mazzone, H. M.**, Analysis of serological studies on the nucleopolyhedrosis and granulosis (capsule) viruses of insects, in *Baculoviruses for Insect Pest Control; Safety Considerations*, Summers, M., Engler, R., Falcon, L. A., and Vail, P., Eds., American Society of Microbiology, Washington, D.C., 1975, 33.

84. **Mazzone, H. M. and Tignor, G. H.**, Insect viruses: serological relationships, in *Advances in Virus Research*, Lauffer, M. A., Bang, F. B., Smith, K. M., and Maramorosch, K., Eds., Academic Press, New York, 1976, 237.

85. **DiCapua, R. A. and Norton, P. W.**, Immunochemical characterization of the baculoviruses: present status, in *Invertebrate Tissue Culture, Applications in Medicine, Biology and Agriculture*, Maramorosch, K., Ed., Academic Press, New York, 1976, 317.

86. **Mitsuhashi, J., Ed.**, *Invertebrate Cell Systems Applications*, Vols. 1 and 2, CRC Press, Boca Raton, FL, 1989.

87. **Tanada, Y.**, Microbial control of insect pests, *Annu. Rev. Entomol.*, 4, 277, 1959.

88. **Franz, J. M.**, Biological control of pest insects in Europe, *Annu. Rev. Entomol.*, 6, 183, 1961.

89. **Bird, F. T.**, The use of viruses in biological control, *Colloq. Int. Pathol. Insects*, Paris, 465, 1962.

90. **Cameron, J. W.**, Factors affecting the use of microbial pathogens in insect control, *Annu. Rev. Entomol.*, 8, 265, 1963.

91. **Rivers, C.**, Virus pesticides, *Discovery*, 25, 27, 1964.

92. **Franz, J. M.**, Bibliography on biological control, *Entomophaga*, 11, 11, 1966.

93. **Heimpel, A. M.**, Progress in developing insect viruses as microbial control agents, *Proc. Joint U.S.-Japan Semin. Microbial Control of Insect Pests*, Fukuoka, 1967, 51.

94. **Ignoffo, C. M.**, Virses living Insecticides, in *Curr. Top. Microbiol. Immunol.*, 42, 1968, 129.

95. **Ignoffo, C. M.**, Specificity of insect viruses, *Bull. Entomol. Soc. Am.*, 14, 265, 1968.

96. **Weiser, J.**, Recent advances in insect pathology, *Annu. Rev. Entomol.*, 15, 245, 1970.

97. **Burges, D. and Hussey, N. W., Eds.**, *Microbial Control of Insects and Mites*, Academic Press, London, 1971.

98. **Stairs, G. R.**, Use of viruses for microbial control of insects, in *Microbial Control of Insects and Mites*, Burges, H. D. and Hussey, N. W., Eds., Academic Press, London, 1971, 97.

99. **Tanada, Y.**, Recent advances in insect virology, *Proc. Hawaii. Entomol. Soc.*, 21, 113, 1971.

100. **Anon.**, The Use of Viruses for the Control of Insect Pests and Disease Vectors, World Health Organization Technical Report Series No. 531 FAO Agricultural Studies, No. 91, 3, 1973.

101. **Ignoffo, C. M.**, Development of a viral insecticide: concept to commercialization, *Exp. Parasitol.*, 33, 380, 1973.

102. **Smirnoff, W. A. and Juneau, A.**, Quinze Années de Recherches sur les Micro-organismes des Insectes Forestiers de la Province de Quèbec (1957—1972), *Ann. Soc. Entomol.*, 18, 147, 1973.

103. **Tanada, Y.**, Environmental factors external to the host, *Ann. N.Y. Acad. Sci.*, 217, 120, 1973.

104. **Ignoffo, C. M.**, Entomopathogens as insecticides, *Environ. Lett.*, 8, 23, 1975.

105. **Longworth, J. F.**, Insect viruses and pest control: the search for alternatives, *N.Z. J. Agric.*, 132, 16, 1976.

106. **Tanada, Y.**, Ecology of insect viruses, in *Perspectives in Forest Entomology*, Anderson, J. F. and Kaya, H. K., Eds., Academic Press, New York, 1976, 265.

107. **Heimpel, A. M.**, The use of viruses in plant protection, *Pontif. Accad. Sci., Scripta Varia*, 14, 275, 1977.

108. **Hostetter, D. L. and Ignoffo, C. M., Eds.**, Environmental Stability of Microbial Insecticides, in *Misc. Publ. Entomol. Soc. Am.*, 10, 1, 1977.

109. **Tinsley, T. W.**, Viruses and the biological control of insect pests, *Bioscience*, 27, 659, 1977.

110. **Dulmage, H. and Burgerjon, A.**, Industrial and international standardization of microbial pesticides-II Insect viruses, *Entomophaga*, 22, 131, 1977.

111. **Ignoffo, C. M. and Falcon, L. A.**, Formulation and application of microbial insecticides, *Misc. Publ. Entomol. Soc. Am.*, 10, 1, 1978.

112. **Tinsley, T. W.**, Use of insect pathogenic viruses as pesticidal agents, *Perspect. Virol.*, 10, 199, 1978.

113. **Kalmakoff, J. and Longworth, J. F.**, Microbial control of insect pests, *N.Z. Dep. Sci. Ind. Res., Bull.*, 228, 1980.

114. **Shieh, T. R. and Bohmfalk, G. T.**, Production and efficacy of baculoviruses, *Biotechnol. Bioeng.*, 22, 1357, 1980.

115. **Burges, H. D., Ed.,** *Microbial Control of Pests and Plant Disease (1970—1980),* Academic Press, London, 1981.

116. **Payne, C. C.,** Insect viruses as control agents, *Parasitology,* 84, 35, 1982.

117. **Kurstak, E., Ed.,** *Microbial and Viral Pesticides,* Marcel Dekker, New York, 720, 1982.

118. **Cunningham, J. C.,** Field trials with baculoviruses: control of forest pests, in *Microbial and Viral Pesticides,* Kurstak, E., Ed., Marcel Dekker, New York, 1982, 335.

119. **Longworth, J. F. and Kalmakoff, J.,** An ecological approach to the use of insect pathogens for pest control, in *Microbial and Viral Pesticides,* Kurstak, E., Ed., Marcel Dekker, New York, 1982, 425.

120. **Kalmakoff, J. and Crawford, A. M.,** Enzootic virus control of *Wiseana* sp. in the pasture environment, in *Microbial and Viral Pesticides,* Kurstak, E., Ed., Marcel Dekker, New York, 1982, 435.

121. **Shapiro, M.,** *In vivo* mass production of insect viruses for use as pesticides, in *Microbial and Viral Pesticides,* Kurstak, E., Ed., Marcel Dekker, New York, 1982, 463.

122. **Hink, W. F.,** Production of *Autographa californica* nuclear polyhedrosis virus in cells from large-scale suspension cultures, in *Microbial and Viral Pesticides,* Kurstak, E., Ed., Marcel Dekker, New York, 1982, 493.

123. **Yearian, W. C. and Young, S. Y.,** Control of insects pests of agricultural importance by viral insecticides, in *Microbial and Viral Pesticides,* Kurstak, E., Ed., Marcel Dekker, New York, 1982, 387.

124. **Deacon, J. W.,** *Microbial Control of Plant Pests and Diseases. Aspects of Microbiology No. 7,* Van Nostrand Reinhold, Berkshire, 1983.

125. **Sturrock, J. W. and Ulbricht, T. L. V., Eds.,** Special Issue: *Biological Control, Agriculture Ecosystems and Environment,* Vol. 10, Elsevier, Amsterdam, 1983, 99.

126. **Martignoni, M. E.,** Baculovirus: An attractive biological alternative, in *Chemical and Biological Control in Forestry,* American Chemical Society, Washington, D.C., 1984, 55.

127. **Hunter, F. R., Crook, N. E., and Entwistle, P. F.,** Viruses as pathogens for the control of insects, in *Microbial Methods for Environmental Biotechnology,* Grainger, J. M. and Lynch, J. M., Eds., Academic Press, New York, 1984, 323.

128. **Faulkner, P. and Boucias, D. G.,** Genetic improvement of insect pathogens: emphasis on the use of baculoviruses, in *Biological Control in Agricultural IPM Systems,* Hoy, M. A. and Herzog, D. C., Eds., Academic Press, Orlando, FL, 1985, 263.

129. **Entwistle, P. E. and Evans, H. F.,** Viral control, in *Comprehensive Insect Physiology, Biochemistry and Pharmacology,* Korkut, G. A. and Gilbert, L. I., Eds., Pergamon, Oxford, 1985, 347.

130. **Jaques, R. P.,** Stability of insect viruses in the environment, in *Viral Insecticides for Biological Control,* Maramorosch, K. and Sherman, K. E., Eds., Academic Press, Orlando, FL, 1985, 285.

131. **Kaupp, W. J. and Sohi, S. S.,** The role of viruses in the ecosystem, in *Viral Insecticides for Biological Control,* Maramorosch, K. and Sherman, K. E., Eds., Academic Press, Orlando, FL, 1985, 441.

132. **Tarasevich, L. M.,** *Insect Viruses of Use to Man,* ''Nauka'', Moscow, USSR, 1985.

133. **Morris, O. N., Cunningham, J. C., Finney-Crawley, J. R., Jaques, R. P., and Kinoshita, G.,** Microbial insecticides in Canada: their regulation and use in agriculture, forestry and public and animal health, *Bull. Entomol. Soc. Can.,* 18, 1986.

134. **Granados, R. R. and Federici, B. A., Eds.,** *The Biology of Baculoviruses,* Vol. 2, CRC Press, Boca Raton, FL, 1986.

135. **Hughes, P. R. and Wood, H. A.,** *In vivo* and *in vitro* bioassay methods for baculoviruses, in *The Biology of Baculoviruses,* Vol. 2, Granados, R. R. and Federici, B. A., Eds., CRC Press, Boca Raton, FL, 1986, 1.

136. **Shapiro, M.,** *In vivo* production of baculoviruses, in *The Biology of Baculoviruses,* Vol. 2, Granados, R. R. and Federici, B. A., Eds., CRC Press, Boca Raton, FL, 1986, 31.

137. **Weiss, S. A. and Vaughn, J. L.,** Cell culture methods for large-scale propagation of viruses, in *The Biology of Baculoviruses,* Vol. 2, Granados, R. R. and Federici, B. A., Eds., CRC Press, Boca Raton, FL, 1986, 63.

138. **Harper, J. D.,** Interactions between baculoviruses and other entomopathogens, chemical pesticides, and parasitoids, in *The Biology of Baculoviruses,* Vol. 2, Granados, R. R. and Federici, B. A., Eds., CRC Press, Boca Raton, FL, 1986, 133.

139. **Young, III, S. Y. and Yearian, W. C.,** Formulation and application of baculoviruses, in *The Biology of Baculoviruses,* Vol. 2, Granados, R. R. and Federici, B. A., Eds., CRC Press, Boca Raton, FL, 1986, 157.

140. **Huber, J.,** Use of baculoviruses in pest management programs, in *The Biology of Baculoviruses,* Vol. 2, Granados, R. R. and Federici, B. A., Eds., CRC Press, Boca Raton, FL, 1986, 181.

141. **Bohmfalk, G. T.,** Practical factors influencing the utilization of baculoviruses as pesticides, in *The Biology of Baculoviruses,* Vol. 2, Granados, R. R. and Federici, B. A., Eds., CRC Press, Boca Raton, FL, 1986, 223.

142. **Betz, F. S.,** Registration of baculoviruses as pesticides, in *The Biology of Baculoviruses,* Vol. 2, Granados, R. R. and Federici, B. A., Eds., CRC Press, Boca Raton, FL, 1986, 203.

143. **Briese, D. T.**, Insect resistance to baculoviruses, in *The Biology of Baculoviruses*, Vol. 2, Granados, R. R. and Federici, B. A., Eds., CRC Press, Boca Raton, FL, 1986, 237.

144. **Podgwaite, J. J. and Mazzone, H. M.**, Latency of insect viruses, *Adv. Virus Res.*, 31, 293, 1986.

145. **Fuxa, J. R.**, Ecological considerations for the use of entomopathogens in IPM, *Annu. Rev. Entomol.*, 32, 225, 1987.

146. **Aizawa, K.**, Strain improvement of insect pathogens, in *Biotechnology in Invertebrate Pathology and Cell Culture*, Maramorosch, K., Ed., Academic Press, San Diego, CA, 1987, 3.

147. **Harpaz, I.**, Improving the effectiveness of insect pathogens for pest control, in *Biotechnology in Invertebrate Pathology and Cell Culture*, Maramorosch, K., Ed., Academic Press, San Diego, CA, 1987, 451.

148. **McIntosh, A. H., Rice, W. C., and Ignoffo, C. M.**, Genotypic variants in wild-type populations of baculoviruses, in *Biotechnology in Invertebrate Pathology and Cell Culture*, Maramorosch, K., Ed., Academic Press, San Diego, CA, 1987, 305.

149. **Burges, H. D. and Pillai, J. S.**, Microbial bioinsecticides, in *Microbial Technology in the Developing World*, DaSilva, E. J., Dommergues, Y. R., Nyns, E. J., and Ratledge, C., Eds., Oxford University Press, Oxford, 1987, 121.

150. **Payne, C. C.**, Pathogens for the control of insects: biol control of pests, pathogens, and weeds: developments and prospects, *Philos. Trans. R. Soc. London, Ser. B*, 318, 225, 1988.

151. **Shieh, T. R.**, Industrial production of viral pesticides, *Adv. Virus Res.*, 36, 315, 1989.

152. **Heimpel, A. M.**, Safety of insect pathogens for man and vertebrates, in *Microbial Control of Insects and Mites*, Burges, H. D. and Hussey, N. W., Eds., Academic Press, London, 1971, 469.

153. **Summers, M., Engler, R., Falcon, L. A., and Vail, P. V.**, *Baculoviruses for Insect Pest Control: Safety Considerations*, American Society for Microbiology, Washington, D.C., 1975.

154. **Kurstak, E., Tijssen, P., and Maramorosch, K.**, Safety consideration and development problems make an ecological approach of biocontrol by viral insecticides imperative, in *Viruses and Environment*, Kurstak, E. and Maramorosch, K., Eds., Academic Press, New York, 1978, 571.

155. **Burges, H. D., Croizier, G., and Huber, J.**, A review of safety tests on baculoviruses, *Entomophaga*, 25, 329, 1980.

156. **Miltenburger, H. G.**, Safety Aspects of Baculoviruses as Biological Insecticides, in Symp. Proc., Miltenburger, H. G., Ed., Bundesministerium fur Forschung und Technologie, 1980.

157. **Burges, H. D.**, Safety, safety testing and quality control of microbial pesticides, in *Microbial Control of Pests and Plant Diseases*, Burges, H. D., Ed., Academic Press, London, 1981, 738.

158. **Harrap, K. A.**, Assessment of the human and ecological hazards of microbial insecticides, *Parasitology*, 84, 269, 1982.

159. **Maramorosch, K.**, Genetically engineered microbial and viral insecticides: safety considerations, in *Biotechnology in Invertebrate Pathology and Cell Culture*, Maramorosch, K., Ed., Academic Press, San Diego, CA, 1987, 485.

160. **Casida, J. E., Ed.**, *Pesticides and Alternatives: Innovative Chemical and Biological Approaches to Pest Control*, Elsevier, Amsterdam, 1990.

161. **Heimpel, A. M. and Adams, J. R.**, A new nuclear polyhedrosis virus of the cabbage looper, *Trichoplusia ni*, *J. Invertebr. Pathol.*, 8, 340, 1966.

162. **Tompkins, G. J., Adams, J. R., and Heimpel, A. M.**, Cross infection studies with *Heliothis zea* using nuclear polyhedrosis viruses from *Trichoplusia ni*, *J. Invertebr. Pathol.*, 14, 343, 1969.

163. **Vail, P. V., Jay, D. L., and Hunter, D. K.**, Cross infectivity of a nuclear polyhedrosis virus isolated from the alfalfa looper, *Autographa californica*, *Proc. IV Int. Colloquium on Insect Pathology*, College Park, MD, 1970, 297.

164. **Heimpel, A. M.**, Report to EPA, 1977.

165. **Hostetter, D. L.**, unpublished data, 1988.

166. **Vail, P. V., Sutter, G., Jay, D. L., and Gouch, D.**, Reciprocal infectivity of nuclear polyhedrosis virus of the cabbage looper and alfalfa looper, *J. Invertebr. Pathol.*, 17, 383, 1971.

167 **Vail, P. V., Jay, D. L., Hunter, D. K., and Staten, R. T.**, A nuclear polyhedrosis virus infective to the pink bollworm, *Pectinophora gossypiella*, *J. Invertebr. Pathol.*, 20, 124, 1972.

168. **Vail, P. V., Jay, D. L., and Hunter, D. K.**, Infectivity of a nuclear polyhedrosis virus from the alfalfa looper *Autographa californica*, after passage through alternate hosts, *J. Invertebr. Pathol.*, 21, 16, 1973.

169. **Stairs, G. R. and Lynn, D. E.**, Infection of the hickory tussock moth, *Halisidota caryae*, by a nuclear polyhedrosis virus of *Autographa californica* grown in *Trichoplusia ni* cell cultures, *J. Invertebr. Pathol.*, 24, 376, 1974.

170. **Harper, J. D.**, Cross-infectivity of six plusiine nuclear polyhedrosis virus isolates to plusiine hosts, *J. Invertebr. Pathol.*, 27, 275, 1976.

171. **Witt, D. J. and Janus, C. A.**, Aspects of the cross transmission to *Galleria mellonella* of a Baculovirus from the alfalfa looper *Autographa californica*, *J. Invertebr. Pathol.*, 27, 65, 1976.

172. **Kaya, H. K.**, Transmission of a nuclear polyhedrosis virus isolated from *Autographa californica* to *Alsophila pometaria*, *Hyphantria cunea*, and other forest defoliators, *J. Econ. Entomol.*, 70, 9, 1977.

173. **Lewis, L. C., Lynch, R. E., and Jackson, J. J.,** Pathology of a baculovirus of the alfalfa looper, *Autographa californica* in the European corn borer, *Environ. Entomol.,* 6, 535, 1977.

174. **Vail, P. V., Jay, D. L., Stewart, F. D., Martinez, A. J., and Dulmage, H. T.,** Comparative susceptibility of *Heliothis virescens* and *H. zea* to the nuclear polyhedrosis virus isolated from *Autographa californica, J. Econ. Entomol.,* 71, 293, 1978.

175. **Capinera, J. L. and Kanost, M. R.,** Susceptibility of the zebra caterpillar to *Autographa californica* nuclear polyhedrosis virus, *J. Econ. Entomol.,* 72, 570, 1979.

176. **Ignoffo, C. M. and Garcia, C.,** Susceptibility of larvae of the black cutworm to species of entomopathogenic bacteria, fungi, protozoa, and viruses, *J. Econ. Entomol.,* 72, 767, 1979.

177. **Vail, P. V. and Collier, S. S.,** Comparative replication, mortality and inclusion body production of the *Autographa californica* nuclear polyhedrosis virus in *Heliothis* sp., *Ann. Entomol. Soc. Am.,* 75, 376, 1982.

178. **Johnson, T. B. and Lewis, L. C.,** Pathogenicity of two nuclear polyhedrosis viruses in the black cutworm, *Agrotis ipsilon* (Lepidoptera:Noctuidae), *Can. Entomol.,* 114, 311, 1982.

179. **Vail, P. V. and Vail, S. S.,** Comparative replication of *Autographa californica* nuclear polyhedrosis virus in tissues of *Heliothis* spp., *Ann. Entomol. Soc. Am.,* 80, 734, 1987.

180. **Vail, P. V., Vail, S. S., and Summers, M. D.,** Responses of *Cactoblastis cactorum* (Lepidoptera:Phycitidae) to the nuclear polyhedrosis virus isolated from *Autographa californica* (Lepidoptera:Noctuidae), *Environ. Entomol.,* 13, 1241, 1984.

181. **Vail, P. V., Knell, J. D., Summers, M. D., and Cowan, P. K.,** In vivo infectivity of baculovirus isolates, variants, and natural recombinants in alternate hosts, *Environ. Entomol.,* 11, 1187, 1982.

182. **Burgerjon, A., Biache, G., and Chaufaux, J.,** Recherches sur la spécificité de trois virus a polyèdres nucléaires vis-à-vis de *Mamestra brassicae, Scotia segetum, Trichoplusia ni* et *Spodoptera exigua, Entomophaga,* 20, 153, 1975.

183. **Martignoni, M. E. and Iwai, P. J.,** Propagation of multicapsid nuclear polyhedrosis virus of *Orgyia pseudotsugata* in larvae of *Trichoplusia ni, J. Invertebr. Pathol.,* 47, 32, 1986.

184. **Hunter, D. K., Hoffman, D. F., and Collier, S. J.,** Cross infection of a nuclear polyhedrosis virus of the almond moth to the Indian meal moth, *J. Invertebr. Pathol.,* 22, 186, 1973.

185. **Carner, G. R., Hudson, J. S., and Barnett, O. W.,** The infectivity of a nuclear polyhedrosis virus of the velvet bean caterpillar for eight noctuid hosts, *J. Invertebr. Pathol.,* 33, 211, 1979.

186. **Adams, J. R., Wallis, R. L., Wilcox, T. A., and Faust, R. M.,** A previously undescribed polyhedrosis of the zebra caterpillar, *Ceramica picta, J. Invertebr. Pathol.,* 11, 45, 1968.

187. **Lewis, L. C. and Adams, J. R.,** Pathogenicity of a nuclear polyhedrosis virus from *Rachiplusia ou* to *Agrotis ipsilon, J. Invertebr. Pathol.,* 33, 253, 1979.

188. **Teakle, R. E.,** Relative pathogenicity of nuclear polyhedrosis viruses from *Heliothis punctigera* and *Heliothis zea* for larvae of *Heliothis armigera* and *Heliothis punctigera, J. Invertebr. Pathol.,* 34, 231, 1979.

189. **Allaway, G. P. and Payne, C. C.,** Host range and virulence of five baculoviruses from lepidopterous hosts, *Ann. Appl. Biol.,* 105, 29, 1984.

190. **Ignoffo, C. M., Heuttel, M. D., McIntosh, A. H., Garcia, C., and Wilkening, P.,** Genetics of resistance of *Heliothis subflexa* (Lepidoptera:Noctuidae) to *Baculovirus heliothis, Ann. Entomol. Soc. Am.,* 78, 468, 1985.

191. **Ignoffo, D. M., McIntosh, A. H., and Garcia, C.,** Susceptibility of larvae of *Heliothis zea, H. virescens,* and *H. armigera* (Lep:Noctuidae) to 3 baculoviruses, *Entomophaga,* 28, 1, 1983.

192. **Williams, C. F. and Payne, C. C.,** The susceptibility of *Heliothis armigera* larvae to three nuclear polyhedrosis viruses, *Ann. Appl. Biol.,* 104, 405, 1984.

193. **Stairs, G. R., Fraser, T., and Fraser, M.,** Changes in growth and virulence of a nuclear polyhedrosis virus from *Choristoneura fumiferana* after passage in *Trichoplusia ni* and *Galleria mellonella, J. Invertebr. Pathol.,* 38, 230, 1981.

194. **Belloncik, S., Lavallee, C., and Hamelin, C.,** Relative pathogenicity of nuclear polyhedrosis viruses from *Mamestra brassicae, Euxoa messoria, Agrotis segetum, Autographa californica,* and *Heliothis* spp. for larvae of *Euxoa scandens* (Lepidoptera:Noctuidae), *J. Invertebr. Pathol.,* 47, 8, 1986.

195. **Bensimon, A., Zinger, S., Gerassi, E., Hauschner, A., Harpaz, I., and Sela, I.,** "Dark cheeks" a lethal disease of locusts provoked by a lepidopterous baculovirus, *J. Invertebr. Pathol.,* 50, 254, 1987.

196. **Lobinger, G. and Skatulla, U.,** Cross infection test with two virus strains (NPV) against the nun moth *Lymantria monacha* L. (Lep., Lymantriidae) and the pine beauty *Panolis flammea* schiff (Lep. noctuidae), *Anz. Schaedlingskd., Pflanz. Umweltschutz,* 59, 147, 1986.

197. **McKinley, D. J., Brown, D. A., Payne, C. C., and Harrap, K. A.,** Cross-infectivity and activation studies with 4 baculoviruses, *Entomophaga,* 26, 79, 1981.

198. **Kelly, P. M. and Entwistle, P. F.,** In vivo mass production in the cabbage moth *(Mamestra brassicae)* of a heterologous *Panolis* and a homologous *(Mamestra)* nuclear polyhedrosis virus, *J. Virol. Methods,* 19, 249, 1988.

199. **Tompkins, G. J., Vaughn, J. L., Adams, J. R., and Reichelderfer, C. F.,** Effects of propagating *Autographa californica* nuclear polyhedrosis virus and its *Trichoplusia ni* variant in different hosts, *Environ. Entomol.,* 10, 801, 1981.

200. **Fraser, M. J. and Stairs, G. R.,** Susceptibility of *Trichoplusia ni, Heliothis zea* (Noctuidae), and *Manduca sexta* (Sphingidae) to a nuclear polyhedrosis virus from *Galleria mellonella* (Pyralidae), *J. Invertebr. Pathol.,* 40, 255, 1982.

201. **Tompkins, G. J., Dougherty, E. M., Adams, J. R., and Diggs, D.,** Changes in the virulence of nuclear polyhedrosis viruses when propagated in alternate noctuid (Lepidoptera:Noctuidae) cell lines and hosts, *J. Econ. Entomol.,* 81, 1027, 1988.

202. **McClintock, J. T. and Reichelderfer, C. F.,** *In vivo* treatment of a nuclear polyhedrosis virus of *Autographa californica* (Lepidoptera:Noctuidae) with chemical mutagens: determination of changes in virulence in four Lepidopteran hosts, *Environ. Entomol.,* 14, 691, 1985.

203. **Smith, G. E. and Summers, M. D.,** Restriction maps of five *Autographa californica* MNPV variants, *Trichoplusia ni* MNPV, and *Galleria mellonella* MNPV DNAs with endonucleases SmaI, KpnI, BamHI, SacI, XhoI and EcoRI, *J. Virol.,* 30, 828, 1979.

204. **Summers, M. D., Smith, G. E., Knell, J. D., and Burand, J. P.,** Physical maps of *Autographa californica* and *Rachoplusia ou* nuclear polyhedrosis virus recombinants, *J. Virol.,* 34, 693, 1980.

205. **Adams, J. R. and Wilcox, T. A.,** Scanning electron microscopical comparisons of insect virus occlusion bodies prepared by several techniques, *J. Invertebr. Pathol.,* 40, 12, 1982.

206. **Hughes, K. M. and Addison, R. B.,** Two nuclear polyhedrosis viruses of the douglas-fir tussock moth, *J. Invertebr. Pathol.,* 16, 196, 1970.

207. **Vago, C. and Amagier, A.,** Coloration histologique pour la différenciation des corps d'inclusion polyédriques de virus d'insectes, *Ann. Epiphyt.,* 14, 269, 1963.

208. **Wigley, P. J.,** Practical: diagnosis of virus infections-staining of insect inclusion body viruses, in *Microbial Control of Insect Pests,* Kalmakoff, J. and Longworth, J. F., Eds., N. Z. Dep. Sci. Ind. Res. Bull., 228, 1980, 35.

209. **Bergold, G. H.,** The molecular structure of some insect virus inclusion bodies, *J. Ultrastruct. Res.,* 8, 360, 1963.

210. **Bergold, G. H. and Brill, R.,** Spreitungsversuche mit insektenviren, *Kolloid-Z.* 99, 1, 1942.

211. **Harrap, K. A.,** The structure of nuclear polyhedrosis viruses. I. The inclusion body, *Virology,* 50, 114, 1972.

212. **Hughes, K. M.,** The macromolecular lattices of polyhedra, *J. Invertebr. Pathol.,* 31, 217, 1978.

213. **Minion, F. C., Coons, L. B., and Broome, J. R.,** Characterization of the polyhedral envelope of the nuclear polyhedrosis virus of *Heliothis virescens, J. Invertebr. Pathol.,* 34, 303, 1979.

214. **Whitt, M. A. and Manning, J. S.,** A phosphorylated 34-kDa protein and a subpopulation of polyhedrin are thiol linked to the carbohydrate layer surrounding a baculovirus occlusion body, *Virology,* 163, 33, 1988.

215. **Gombart, A. F., Pearson, M. N., Rohrmann, G. F., and Beaudreau, G. S.,** A baculovirus polyhedral envelope-associated protein: genetic location, nucleotide sequence, and immunocytochemical characterization, *Virology,* 169, 182, 1989.

216. **Huger, A. M. and Kreig, A.,** On spindle-shaped cytoplasmic inclusions associated with a nuclear polyhedrosis of *Choristoneura murinana, J. Invertebr. Pathol.,* 12, 461, 1968.

217. **Adams, J. R. and Wilcox, T. A.,** Histopathology of the almond moth, *Cadra cautella,* infected with a nuclear polyhedrosis virus, *J. Invertebr. Pathol.,* 12, 269, 1968.

218. **Francki, R. I. B.,** *Classification of Viruses,* in press, 1990.

219. **Injac, M., Vago, C., Veyrunes, J. C., and Duthoit, J. L.,** Libération (Release) des virions dans les polyédroses nucléaires, *C. R. Acad. Sci., Paris, Ser. D,* 273, 439, 1971.

220. **Kawamoto, F., Asayama, T., and Kobayashi, M.,** Acquisition of the envelope of nuclear polyhedrosis viruses in the Chinese oak silkworm, *Antheraea peryni* Guer-Min and the Japanese giant silkworm, *Appl. Entomol. Zool.,* 11, 59, 1976.

221. **Tanada, Y. and Hess, R. T.,** Development of a nuclear polyhedrosis virus in midgut cells and penetration of the virus into the hemocoel of the armyworm, *Pseudaletia unipuncta, J. Invertebr. Pathol.,* 28, 67, 1976.

222. **Adams, J. R., Goodwin, R. H., and Wilcox, T. A.,** Electron microscopic investigations on invasion and replication of insect baculoviruses *in vivo* and *in vitro, Biol. Cellulaire,* 28, 261, 1977.

223. **Kawamoto, F., Suto, C., Kumada, N., and Kobayashi, M.,** Cytoplasmic budding of a nuclear polyhedrosis virus and comparative ultrastructural studies of envelopes, *Microbiol. Immunol.,* 21, 255, 1977.

224. **Hess, R. T. and Falcon, L. A.,** Observations on the interaction of baculoviruses with the plasma membrane, *J. Gen. Virol.,* 36, 525, 1977.

225. **Granados, R. R. and Lawler, K. A.,** *In vivo* pathway of *Autographa californica* baculovirus invasion and infection, *Virology,* 108, 297, 1981.

226. **Hughes, K. M.,** Fine structure and development of two polyhedrosis viruses, *J. Invertebr. Pathol.,* 19, 198, 1972.

227. **Stoltz, D. S., Pavan, C., and Cunha, A. B.,** Nuclear polyhedrosis virus: a possible example of *de novo* intranuclear membrane morphogenesis, *J. Gen. Virol.,* 19, 145, 1973.

228. **Kawamoto, F. and Asayama, T.,** Studies on the arrangement patterns of nucleocapsids within the envelopes of nuclear-polyhedrosis virus in the fat-body cells of the brown tail moth, *Euproctis similis, J. Invertebr. Pathol.,* 26, 47, 1975.

229. **Skuratovskaya, I. N., Fodor, I., and Strokovskaya, L. I.,** Properties of the nuclear polyhedrosis virus of the great wax moth: oligomeric circular DNA and the characteristic of the genome, *Virology,* 12, 465, 1982.

230. **Entwistle, P. F. and Robertson, J. S.,** An unusual nuclear-polyhedrosis virus from larvae of a hepialid moth, *J. Invertebr. Pathol.,* 11, 487, 1968.

231. **Beaton, D. C. and Filshie, B. K.,** Comparative ultrastructural studies of insect granulosis and nuclear polyhedrosis viruses, *J. Gen. Virol.,* 31, 15, 1976.

232. **Burley, S. K., Miller, A., Harrap, K. A., and Kelly, D. C.,** Structure of the *Baculovirus* nucleocapsid, *Virology,* 120, 433, 1982.

233. **Bud, H. M. and Kelly, D. C.,** An electron microscope study of partially lysed *Baculovirus* nucleocapsids. The intra-nucleocapsid packaging of viral DNA, *J. Ultrastruct. Res.,* 73, 361, 1980.

234. **Revet, B. M. and Guelpa, B.,** The genome of a baculovirus infecting *Tipula* (Meig) (Diptera): a high molecular weight closed circular DNA of zero superhelix density, *Virology,* 96, 633, 1979.

235. **Tweeten, K. A., Bulla, L. A., and Consigli, R. A.,** Characterization of an extremely basic protein derived from granulosis virus nucleocapsids, *J. Virol.,* 33, 866, 1980.

236. **Elliott, R. M. and Kelly, D. C.,** Compartmentalization of the polyamines contained by a nuclear polyhedrosis virus from *Heliothis zea, Microbiologica,* 2, 409, 1979.

237. **Fraser, M. J.,** Ultrastructural observations of virion maturation in *Autographa californica* nuclear polyhedrosis virus infected *Spodoptera frugiperda* cell cultures, *J. Ultrastruct. Mol. Struct. Res.,* 95, 189, 1986.

238. **Summers, M. D.,** Electron microscopic observations on granulosis virus entry, uncoating and replication processes during infection of the midgut cells of *Trichoplusia ni, J. Ultrastruct. Res.,* 35, 606, 1971.

239. **Raghow, R. and Grace, T. D. C.,** Studies on a nuclear polyhedrosis virus in *Bombyx mori* cells *in vitro.* I. Multiplication kinetics and ultrastructural studies, *J. Ultrastruct. Res.,* 47, 384, 1974.

240. **Granados, R. R.,** Early events in the infection of *Heliothis zea* midgut cells by a baculovirus, *Virology,* 90, 170, 1978.

241. **Bassemir, U., Miltenburger, H. G., and David, P.,** Morphogenesis of nuclear polyhedrosis virus from *Autographa californica* in a cell line from *Mamestra brassicae* (cabbage moth), *Cell Tissue Res.,* 228, 587, 1983.

242. **Adams, J. R.,** unpublished data, 1988.

243. **Pearson, M. N., Russell, R. L. Q., Rohrmann, G. F., and Beaudreau, G. S.,** p39, A major baculovirus structural protein: immunocytochemical characterization and genetic location, *Virology,* 167, 407, 1988.

244. **Blissard, G. W., Quant-Russell, R. L., Rohrmann, G. E., and Beaudreau, G. S.,** Nucleotide sequence, transcriptional mapping, and temporal expression of the gene encoding p39, a major structural protein of the multicapsid nuclear polyhedrosis virus of *Orgyia pseudotsugata, Virology,* 168, 354, 1989.

245. **Guarino, L. A. and Smith, M. W.,** Nucleotide sequence and Characterization of the 39-K gene region of *Autographa californica* nuclear polyhedrosis virus, *Virology,* 179, 1, 1990.

246. **Thiem, S. M. and Miller, L. K.,** Identification, sequence, and transcriptional mapping of the capsid protein gene of the baculovirus, *Autographa californica* nuclear polyhedrosis virus, *Virology,* 63, 2008, 1989.

247. **Müller, R., Pearson, M. N., Russell, R. L. Q., and Rohrmann, G. F.,** A capsid-associated protein of the multicapsid nuclear polyhedrosis virus of *Orgyia pseudotusgata:* genetic location, sequence, transcription mapping, and immunocytochemical characterization, *Virology,* 176, 133, 1990.

248. **Wilson, M. E., Manprize, T. H., Friesen, P. D., and Miller, L. K.,** Location, transcription, and sequence of a baculovirus gene encoding a small arginine-rich polypeptide, *J. Virol.,* 61, 661, 1987.

249. **Wilson, M. E.,** A synthetic peptide to the predicted 6-9k translation product of the Hind III-H/EcoRI-D region of the AcNPV genome induces antibodies to the basic DNA-binding protein, *Virus Res.,* 9, 21, 1988.

250. **Wilson, M. E. and Miller, L. K.,** Changes in the nucleoprotein complexes of a baculovirus DNA during infection, *Virology,* 151, 315, 1986.

251. **Wilson, M. E. and Consigli, R. A.,** Functions of a protein kinase activity associated with purified capsids of the granulosis virus infecting *Plodia interpunctella, Virology,* 143, 526, 1985.

252. **Summers, M. D. and Volkman, L. E.,** Comparison of biophysical and morphological properties of occluded and extracellular nonoccluded baculovirus from *in vivo* and *in vitro* host systems, *J. Virol.,* 17, 962, 1976.

253. **Smith, G. E. and Summers, M. D.**, Analysis of baculovirus genomes with restriction endonucleases, *Virology*, 89, 517, 1978.

254. **Volkman, L. E., Goldsmith, P. A., Hess, R. T., and Faulkner, P.**, Neutralization of budded *Autographa californica* NPV by a monoclonal antibody: identification of the target antigen, *Virology*, 133, 354, 1984.

255. **Bergold, G. S.**, Fine structure of some insect viruses, *J. Insect Pathol.*, 5, 111, 1963.

256. **Ponsen, M. B., Henstra, S., and van der Scheer, C.**, Electron microscope observation on the structure of membranes from nuclear polyhedral viruses from *Malacosoma neustria Barathra brassicae* and *Adoxophyes reticulana*, *Neth. J. Plant Pathol.*, 71, 20, 1965.

257. **Amargier, A., Abol-Ela, S., Vergara, S., Meynadier, G., Mantouret, D., and Croizier, G.**, Histological and ultrastructural studies of the larvae *Pandemis heparana* (Lep. Tortricidae) during the advanced stages of a baculovirus due to a new virus inducing diapause, *Entomophaga*, 26, 319, 1981.

258. **Federici, B. A.**, Mosquito baculovirus: sequence of morphogenesis and ultrastructure of the virion, *Virology*, 100, 1, 1980.

259. **Bergold, G. H.**, Die isolierung des polyeder-virus und die natur der polyeder, *Z. Naturforsch. Teil B*, 2, 122, 1947.

260. **Bergold, G. H.**, Bundelformige ordnung in polyederviren, *Z. Naturforsch., Teil B*, 3, 35, 1948.

261. **Hughes, K. M.**, A demonstration of the nature of polyhedra using alkaline solutions, *J. Bacteriol.*, 59, 189, 1950.

262. **Smith, K. M. and Xeros, N.**, The development of virus in the cell nucleus, *Nature (London)*, 172, 670, 1953.

263. **Huger, A. M. and Krieg, A.**, Elektronenmikroskopische unfersuchungen zur virogenese bon bergoldia virus calypta steinhaus, *Naturwissenschaften*, 47, 546, 1960.

264. **Huger, A. M. and Kreig, A.**, Electron microscope investigations on the virogenesis of the granulosis of *Choristoneura murinana* (Hubner), *J. Insect Pathol.*, 3, 183, 1961.

265. **Smith, K. M. and Hills, G. H.**, Multiplication and ultrastructure of insect viruses, *Proc. 11th. Int. Congr. Entomol., Vienna*, 2, 1960, 823.

266. **Bird, F. T.**, On the development of insect polyhedrosis and granulosis virus particles, *Can. J. Microbiol.*, 10, 49, 1964.

267. **Summers, M. D.**, Biophysical and biochemical properties of baculoviruses, in *Baculoviruses for Insect Pest Control: Safety Considerations*, Summers, M. D., Engler, R., Falcon, L. A., and Vail, P. V., Eds., American Society of Microbiology, Washington, D.C., 1975, 17.

268. **Vlak, J. M. and Odink, K. G.**, Characterization of *Autographa californica* nuclear polyhedrosis virus deoxyribonucleic acid, *J. Gen. Virol.*, 44, 333, 1979.

269. **Summers, M. D. and Anderson, D. L.**, Characterization of nuclear polyhedrosis virus DNAs, *J. Virol.*, 12, 1336, 1973.

270. **Summers, M. D. and Anderson, D. L.**, Granulosis virus deoxyribonucleic acid: a closed, double-stranded molecule, *J. Virol.*, 9, 710, 1972.

271. **Summers, M. D. and Anderson, D. L.**, Characterization of deoxyribonucleic acid isolated from the granulosis viruses of the cabbage looper, *Trichoplusia ni* and the fall armyworm, *Spodoptera frugiperda*, *Virology*, 50, 459, 1972.

272. **McClintock, J. T. and Dougherty, E. M.**, Restriction mapping of *Lymantria dispar* nuclear polyhedrosis virus DNA: localization of the polyhedrin gene and identification of four homologous regions, *J. Gen. Virol.*, 69, 2302, 1988.

273. **Smith, I. R. L., Van Beek, N. A. M., Podgwaite, J. D., and Wood, H. A.**, Physical map and polyhedrin gene sequence of *Lymantria dispar* nuclear polyhedrosis virus, *Gene*, 71, 97, 1988.

274. **Johnson, D. W. and Maruniak, J. E.**, Physical map of *Anticarsia gemmatalis* nuclear polyhedrosis virus (AgMNPV-2) DNA, *J. Gen. Virol.*, 70, 1877, 1989.

275. **Miller, L. K. and Dawes, K. P.**, Physical map of the DNA genome of *Autographa californica* nuclear polyhedrosis virus, *J. Virol.*, 29, 1044, 1979.

276. **Smith, G. E. and Summers, M. D.**, Restriction map of *Rachiplusia ou* and *Rachiplusia ou-Autographa californica* baculovirus recombinants, *J. Virol.*, 33, 311, 1980.

277. **Vlak, J. M. and Smith, G. E.**, Orientation of the genome of *Autographa californica* nuclear polyhedrosis virus: a proposal, *J. Virol.*, 4, 1118, 1982.

278. **Lubbert, H., Kruczek, I., Tjia, S., and Doerfler, W.**, The cloned EcoRI fragments of *Autographa californica* nuclear polyhedrosis virus DNA, *Gene*, 16, 343, 1981.

279. **Cochran, M. A.**, Molecular cloning and physical mapping of restriction endonuclease fragments of *Autographa californica* nuclear polyhedrosis virus DNA, *J. Virol.*, 41, 940, 1982.

280. **Knebel, D., Lubbert, H., and Doerfler, W.**, The promoter of the late P-10 gene in the insect nuclear polyhedrosis virus *Autographa californica* activated by viral gene products and sensitivity to DNA methylation, *EMBO J.*, 4, 1301, 1985.

281. **Tjia, S. T., Carstens, E. B., and Doerfler, W.**, Infection of *Spodoptera frugiperda* cells with *Autographa californica* nuclear polyhedrosis virus. II. The viral DNA and the kinetics of replication, *Virology*, 99, 399, 1979.

282. **McClintock, J. T., Dougherty, E. M., and Weiner, R. M.,** Semipermissive replication of a nuclear polyhedrosis virus of *Autographa californica* in a gypsy moth cell line, *J. Virol.,* 57, 197, 1986.

283. **Gettig, R. R. and McCarthy, W. J.,** Genotypic variation among wild isolates of *Heliothis* spp. nuclear polyhedrosis viruses from different geographical regions, *Virology,* 117, 245, 1982.

284. **Monroe, J. E. and McCarthy, W. J.,** Polypeptide analysis of genotypic variants of occluded *Heliothis* spp. baculoviruses, *J. Invertebr. Pathol.,* 43, 32, 1984.

285. **Vlak, J. M. and Groner, A.,** Identification of two nuclear polyhedrosis viruses from the cabbage moth, *Manestra brassicae* (Lepidoptera:Noctuidae), *J. Invertebr. Pathol.,* 35, 269, 1980.

286. **Brown, D. A., Evans, H. F., Allen, C. J., and Kelly, D. C.,** Biological and biochemical investigations on five european isolates of *Mamestra brassicae* nuclear polyhedrosis virus, *Arch. Virol.,* 69, 209, 1981.

287. **Bilimoria, S. L.,** Genomic divergence among single-nucleocapsid nuclear polyhedrosis viruses of plusiine hosts, *Virology,* 127, 12, 1983.

288. **Kislev, N. and Edelman, M.,** DNA restriction-pattern differences from geographic isolates of *Spodoptera littoralis* nuclear polyhedrosis virus, *Virology,* 119, 219, 1982.

289. **Cherry, C. L. and Summer, M. D.,** Genotypic variation among wild isolates of two nuclear polyhedrosis viruses isolated from *Spodoptera littoralis, J. Invertebr. Pathol.,* 46, 289, 1985.

290. **Croizer, G., Quiot, J. M., Paradis, S., and Bouksoudmi-Amiri, K.,** Comparison de la compositon génétique de trois isolats du baculovirus de la polyédrose nucléaire du lépidoptère *Spodoptera littoralis, Entomophaga,* 31, 385, 1986.

291. **Allaway, G. P. and Payne, C. C.,** A biochemical and biological comparison of three European isolates of nuclear polyhedrosis viruses from *Agrotis segetum, Arch. Virol.,* 75, 43, 1982.

292. **Maruniak, J. E., Brown, S. E., and Knudson, D. L.,** Physical maps of SfMNPV baculovirus DNA and its genomic variants, *Virology,* 136, 221, 1984.

293. **Smith, I. R. L. and Crook, N. E.,** *In vivo* isolation of baculovirus genotypes, *Virology,* 166, 240, 1988.

294. **Jewell, J. E. and Miller, L. K.,** DNA sequence homology relationships among six Lepidopteran nuclear polyhedrosis viruses, *J. Gen. Virol.,* 48, 161, 1980.

295. **Smith, G. E. and Summers, M. D.,** DNA homology among subgroup A, B, and C Baculoviruses, *Virology,* 123, 393, 1982.

296. **Kislev, N.,** Homology relationship between *Spodoptera littoralis* nuclear polyhedrosis virus and other baculoviruses, *Intervirology,* 24, 50, 1985.

297. **Rohrmann, G. F., Martignoni, M. E., and Beaudreau, G. S.,** DNA sequence homology between *Autographa californica* and *Orgyia pseudotsugata* nuclear polyhedrosis viruses, *J. Gen. Virol.,* 62, 137, 1982.

298. **Arif, B. M., Kuzio, J., Faulkner, P., and Doerfler, W.,** The genome of *Choristoneura fumiferana* nuclear polyhedrosis virus: molecular cloning and mapping of the EcoRI, BamHI, SmaI, XbaI and BglII restriction sites, *Virus Res.,* 1, 605, 1984.

299. **Arif, B. M., Tjia, S. T., and Doerfler, W.,** DNA homologies between the genomes of *Choristoneura fumiferana* and *Autographa californica* nuclear polyhedrosis virus, *Virus Res.,* 2, 85, 1985.

300. **Arif, B. M. and Doerfler, W.,** Identification and localization of reiterated sequences in the *Choristoneura fumiferana* MNPV genome, *EMBO J.,* 3, 525, 1984.

301. **Kuzio, J. and Faulkner, P.,** Regions of repeated DNA in the genome of *Choristoneura fumiferana* nuclear polyhedrosis virus, *Virology,* 139, 185, 1984.

302. **Cochran, M. A. and Faulkner, P.,** Location of homologous DNA sequences interspersed at five regions in the baculovirus AcMNPV genome, *J. Virol.,* 45, 961, 1983.

303. **Kelly, D. C.,** The DNA contained by nuclear polyhedrosis viruses isolated from four *Spodoptera* sp. (Lepidoptera:Noctuidae). Genome size and homology assessed by DNA reassociated kinetics, *Virology,* 76, 468, 1977.

304. **Burand, J. P. and Summers, M. D.,** Alterations of *Autographa californica* nuclear polyhedrosis virus DNA upon serial passage in cell culture, *Virology,* 119, 223, 1982.

305. **Leisy, D. J., Rohrmann, G. F., and Beaudreau, G. S.,** Conservation of genome organization in two multicapsid nuclear polyhedrosis viruses, *J. Virol.,* 52, 699, 1984.

306. **Guarino, L. A., Gonzalez, M. A., and Summers, M. D.,** Complete sequence and enhancer function of the homologous DNA regions of *Autographa californica* nuclear polyhedrosis virus, *J. Virol.,* 60, 224, 1986.

307. **Guarino, L. A.,** Enhancers of early gene expression, in *Invertebrate Cell Systems Applications,* Vol. 1, Mitsuhashi, J., Ed., CRC Press, Boca Raton, FL, 1989, 211.

308. **Gombart, A. F., Blissard, G. W., and Rohrmann, G. F.,** Characterization of the genetic organization of the *Hind*III M-region of the multicapsid nuclear polyhedrosis virusis of *Orgyia pseudotsugata* reveals major differences among baculoviruses, *J. Gen. Virol.,* 70, 1815, 1989.

309. **Miller, D. W. and Miller, L. K.,** A virus mutant with an insertion of a copia-like transposable element, *Nature (London),* 299, 562, 1982.

310. **Fraser, M. J., Smith, G. E., and Summers, M. D.,** Acquisition of host cell DNA sequences by baculoviruses: Relationship between host DNA insertions and FP mutants of *Autographa californica* and *Galleria mellonella* nuclear polyhedrosis viruses, *J. Virol.,* 47, 287, 1983.

311. **Fraser, M. J., Brusca, J. S., Smith, G. E., and Summer, M. D.,** Transposon-mediated mutagenesis of a baculovirus, *Virology,* 145, 356, 1985.

312. **Friesen, P. D., Rice, W. C., Miller, D. W., and Miller, L. K.,** Bidirectional transcription from a solo long terminal repeat of the retrotransposon TED: symmetrical RNA start sites, *Mol. Cell. Biol.,* 6, 599, 1986.

313. **Friesen, P. D. and Miller, L. K.,** Divergent transcription of early 35- and 94 kilodalton protein genes encoded by the HindIII-K genome fragment of the baculovirus *Autographa californica* nuclear polyhedrosis virus, *J. Virol.,* 61, 2264, 1987.

314. **Carstens, E. B.,** Identification and nucleotide sequence of the regions of *Autographa californica* nuclear polyhedrosis virus genome carrying insertion elements derived from *Spodoptera frugiderda, Virology,* 161, 8, 1987.

315. **Friesen, P. D. and Nissen, M. S.,** Gene organization and transcription of TED, a lepidopteran retrotransposon integrated within the baculovirus genome, *Mol. Cell. Biol.,* 10, 3067, 1990.

316. **Fraser, M. J.,** Symposium: genetics in entomology. Transposon-mediated mutagenesis of baculoviruses: transposon shuttling and implications for speciation, *Ann. Entomol. Soc. Am.,* 79, 773, 1986.

317. **Beames, B. and Summers, M. D.,** Comparisons of host cell DNA insertions and altered transcription at the site of insertions in few polyhedra baculovirus mutants, *Virology,* 162, 206, 1988.

318. **Beames, B. and Summers, M. D.,** Sequence comparison of cellular and viral copies of host cell DNA insertions found in *Autographa californica* nuclear polyhedrosis virus, *Virology,* 174, 354, 1990.

319. **Cary, L. C., Goebel, M., Corsaro, B. G., Wang, H. G., Rosen, E., and Fraser, M. J.,** Transposon mutagenesis of baculoviruses: analysis of *Trichoplusia ni* transposon FP2 insertions within the FP-locus of nuclear polyhedrosis viruses, *Virology,* 172, 156, 1989.

320. **Wang, H. H., Fraser, M. J., and Cary, L. C.,** Transposon mutagenesis of baculoviruses: analysis of TFP3 lepidopteran transposon insertions at the locus of nuclear polyhedrosis virus, *Gene,* 81, 97, 1989.

321. **Van der Beek, C. O., Saaijer-Riep, J. D., and Vlak, J. M.,** On the origin of the polyhedrin protein of *Autographa californica* nuclear polyhedrosis virus. Isolation, characterization and translation of viral messenger RNA, *Virology,* 100, 326, 1980.

322. **Hofft van Iddekinge, B. J. L., Smith, G. E., and Summers, M. D.,** Nucleotide sequence of the polyhedrin gene of *Autographa californica* nuclear polyhedrosis virus, *Virology,* 131, 561, 1983.

323. **Carstens, E. B.,** Mapping the mutation site of an *Autographa californica* nuclear polyhedrosis virus polyhedron morphology mutant, *J. Virol.,* 43, 809, 1982.

324. **Carstens, E. B., Krebs, A., and Gallerneault, C. E.,** Identification of an amino acid essential to the normal assembly of *Autographa californica* nuclear polyhedrosis virus polyhedra, *J. Virol.,* 58, 684, 1986.

325. **Carstens, E. B., Lin-Bai, Y., and Faulkner, P.,** A point mutation in the polyhedrin gene of a baculovirus, *Autographa californica* MNPV, presents crystallization of occlusion bodies, *J. Gen. Virol.,* 68, 901, 1987.

326. **Howard, S. C., Ayres, M. D., and Possee, R. D.,** Mapping the 5' and 3' ends of *Autographa californica* nuclear polyhedrosis virus polyhedrin mRNA, *Virus Res.,* 5, 109, 1986.

327. **Ooi, B. G., Rankin, C., and Miller, L. K.,** Downstream sequences augment transcription from the essential initiation site of a baculovirus polyhedrin gene, *J. Mol. Biol.,* 210, 721, 1989.

328. **Gearing, K. L. and Possee, R. D.,** Functional analysis of a 603 nucleotide open reading frame upstream of the polyhedrin gene of *Autographa californica* nuclear polyhedrosis virus, *J. Gen. Virol.,* 71, 251, 1990.

329. **Rankin, C., Ooi, B. G., and Miller, L. K.,** Eight base pairs encompassing the transcriptional start point are the major determinant for baculovirus polyhedrin gene expression, *Gene,* 70, 39, 1988.

330. **Ooi, B. G. and Miller, L. K.,** Transcription of the baculovirus polyhedrin gene reduces the levels of an antisense transcript initiated downstream, *J. Virol.,* 64, 3126, 1990.

331. **Luckow, V. A. and Summers, M. D.,** High level expression of nonfused foreign genes with *Autographa californica* nuclear polyhedrosis virus expression vectors, *Virology,* 170, 31, 1989.

332. **Luckow, V. A. and Summers, M. D.,** Signals important for high level expression of foreign genes in *Autographa californica* nuclear polyhedrosis virus expression vectors, *Virology,* 167, 56, 1988.

333. **Luckow, V. A. and Summers, M. D.,** Trends in the development of baculovirus expression vectors, *Biol technol.,* 6, 47, 1988.

334. **Iatrou, K., Ito, K., and Witkiewicz, H.,** Polyhedrin gene of *Bombyx mori* nuclear polyhedrosis virus, *J. Virol.,* 54, 436, 1985.

335. **Rohrmann, G. M., Leisy, D. J., Chow, K. C., Pearson, G. D., and Beaudreau, G. S.,** Identification, cloning and R-Loop mapping of the polyhedrin gene from the multicapsid nuclear polyhedrosis virus of *Orgyia pseudotsugata, Virology,* 121, 51, 1982.

336. **Leisy, D., Rohrmann, G., and Beaudreau, G.,** The nucleotide sequence of the polyhedrin gene region from the multicapsid baculovirus of *Orgyia pseudotsugata, Virology,* 153, 280, 1986.

337. **Drake, E. L. and McEwen, F. L.,** Pathology of a nuclear polyhedrosis of the cabbage looper, *Trichoplusia ni* (Hubner), *J. Insect Pathol.,* 1, 281, 1959.

338. **Bird, R. T.,** On the multiplication of an insect virus, *Biochim. Biophys. Acta,* 8, 360, 1952.

339. **Steinhaus, E. A.,** New records of insect-virus diseases, *Hilgardia,* 26, 417, 1957.

340. **Benz, G.,** A nuclear polyhedrosis of *Malacosoma alpicola* (Staudinger), *J. Insect Pathol.,* 5, 215, 1963.

341. **Bergold, G. H.,** Fine structure of some insect viruses, *J. Insect Pathol.,* 5, 111, 1963.

342. **Bird, F. T. and Whalen, M. M.,** A virus disease of the European pine sawfly, *Neodiprion sertifer* (Geoffr.), *Can. Entomol.,* 85, 433, 1953.

343. **Hughes, K. M.,** The development of an insect virus within cells of its host, *Hilgardia,* 22, 391, 1953.

344. **Bird, F. T. and Whalen, M. M.,** Stages in the development of two insect viruses, *Proc. Int. Conf. Electron Microscopy,* London, 1954, 257.

345. **Tanada, Y.,** A polyhedrosis virus of the imported cabbage worm and its relation to a polyhedrosis virus of the alfalfa caterpillar, *Ann. Entomol. Soc. Am.,* 47, 553, 1954.

346. **Smith, K. M. and Xeros, M.,** Electron and light microscopic studies of the development of the virus rods of insect polyhedrosis, *Parasitology,* 44, 71, 1954.

347. **Injac, M., Vago, C., Veyrunes, J. C., Amargier, A., and Duthoit, J. L.,** Etude cytopathologique d'une prolifération cellulaire d'origine virale du tissue hypodermique chez un invertébré *Hyphantria cunea,* Drury, Lepidoptera, *Bull. Soc. Etud. Sci. Nat. Nimes,* 11, 47, 1971.

348. **Xeros, N.,** The virogenic stroma in nuclear and cytoplasmic polyhedrosis, *Nature (London),* 178, 412, 1956.

349. **Xeros, N.,** Origin of the virus-producing chromatic mass or net of the insect nuclear polyhedrosis, *Nature (London),* 175, 588, 1955.

350. **Bird, F. T.,** On the development of insect viruses, *Virology,* 3, 237, 1957.

351. **Day, M. F., Farrant, J. L., and Potter, C.,** The structure and development of a polyhedral virus affecting the moth larvae, *Pteroloccea amplicornis, J. Ultrastruct. Res.,* 2, 227, 1958.

352. **Bird, F. T.,** Polyhedrosis and granulosis viruses causing single and double infections in the spruce budworm, *Chroristoneura fumiferana* Clemens, *J. Insect Pathol.,* 1, 406, 1959.

353. **Drake, E. L. and McEwen, F. L.,** Pathology of a nuclear polyhedrosis of the cabbage looper, *Trichoplusia ni* (Hubner), *J. Insect Pathol.,* 1, 281, 1959.

354. **Laudeho, Y. and Amagier, A.,** Virose a polyedres nucléaires à localisation inhabituelle chez un lépidoptère, *Rev. Pathol. Végetale Entomol. Agr.,* 52, 207, 1963.

355. **Bird, F. T.,** The polyhedral disease of the spruce budworm *Choristoneura fumiferana* (Clem.), *For. Insect Invest.,* 5, 2, 1949.

356. **Hamm, J. J.,** Comparative histopathology of a granulosis and a nuclear polyhedrosis of *Spodoptera frugiperda, J. Invertebr. Pathol.,* 10, 320, 1968.

357. **Vail, P. V. and Hall, I. M.,** The histopathology of a nuclear polyhedrosis in larvae of the cabbage looper, *Trichoplusia ni,* related to symptoms and mortality, *J. Invertebr. Pathol.,* 13, 188, 1969.

358. **Hunter, D. K., Hoffman, D. F., and Collier, S. J.,** The histology and ultrastructure of a nuclear polyhedrosis virus of the webbing clothes moth, *Tineola bisselliella, J. Invertebr. Pathol.,* 21, 91, 1973.

359. **Mathad, S. B., Splittstoesser, C. M., and McEwen, F. L.,** Histopathology of the cabbage looper, *Trichoplusia ni,* infected with a nuclear polyhedrosis, *J. Invertebr. Pathol.,* 11, 456, 1968.

360. **Watanabe, H.,** Abnormal cell proliferation in the epidermis of the fall webworm, *Hyphantria cunea,* induced by the infection of a nuclear polyhedrosis, *J. Invertebr. Pathol.,* 12, 310, 1968.

361. **Baugher, D. G. and Yendol, W. G.,** Virulence of *Autographa californica* baculovirus preparations fed with different food sources to cabbage loopers, *J. Environ. Entomol.,* 74, 309, 1981.

362. **Tanada, Y., Hess, R. T., and Omi, E. M.,** Unique virus morphogenesis and cytopathology of a baculovirus (hypertrophy strain) in larva of the armyworm, *Pseudaletia unipuncta, J. Invertebr. Pathol.,* 40, 197, 1982.

363. **Clark, T. B., Chapman, H. C., and Fukuda, T.,** Nuclear-polyhedrosis and cytoplasmic-polyhedrosis virus infections in Louisiana mosquitoes, *J. Invertebr. Pathol.,* 14, 284, 1969.

364. **Federici, B. A. and Lowe, R. E.,** Studies on the pathology of a baculovirus in *Aedes triseriatus, J. Invertebr. Pathol.,* 20, 14, 1972.

365. **Tanada, Y.,** Synergism between two viruses of the armyworm, *Pseudaletia unipuncta* (Haworth) (Lepidoptera:Noctuidae), *J. Insect Pathol.,* 1, 215, 1959.

366. **Tanada, Y., Hukuhara, T., and Chang, G. Y.,** A strain of nuclear polyhedrosis virus causing extensive cellular hypertrophy, *J. Invertebr. Pathol.,* 13, 394, 1969.

367. **Tanada, Y. and Hukuhara, T.,** Enhanced infection of a nuclear-polyhedrosis virus in larvae of the armyworm, *Pseudaletia unipuncta,* by a factor in the capsule of a granulosis virus, *J. Invertebr. Pathol.,* 17, 116, 1971.

368. **Tanada, Y., Himeno, M., and Omi, E. M.,** Isolation of a factor from the capsule of a granulosis virus synergistic for a nuclear polyhedrosis virus of the armyworm, *J. Invertebr. Pathol.,* 21, 31, 1973.

369. **Tanada, Y., Hess, R. T., and Omi, E. M.,** Invasion of a nuclear polyhedrosis virus in midgut of the armyworm, *Pseudaletia unipuncta,* and the enhancement of a synergistic enzyme, *J. Invertebr. Pathol.,* 26, 99, 1975.

370. **Yamamoto, T. and Tanada, Y.,** Phospholipid, an enhancing component in the synergistic factor of a granulosis virus of the armyworm, *Pseudaletia unipuncta, J. Invertebr. Pathol.,* 31, 48, 1978.

371. **Yamamoto, T. and Tanada, Y.,** Physicochemical properties and location of capsule components, in particular the synergistic factor, in the occlusion body of a granulosis virus of the armyworm, *Pseudaletia unipuncta, Virology,* 107, 434, 1980.

372. **Tanada, Y., Inoue, H., Hess, R. T., and Omi, E. M.,** Site of action of a synergistic factor of a granulosis virus of the armyworm *Pseudaletia unipuncta, J. Invertebr. Pathol.,* 34, 249, 1980.

373. **Tanada, Y., Hess, R. T., Omi, E. M., and Yamamoto, T.,** Localization of a synergistic factor of a granulosis virus by its esterase activity in the larval midgut of the armyworm, *Pseudaletia unipuncta, Microbios,* 37, 87, 1983.

374. **Nagata, M. and Tanada, Y.,** Origin of an alkaline protease associated with the capsule of a granulosis virus of the armyworm, *Pseudaletia unipuncta* (Haworth), *Arch. Virol.,* 76, 245, 1983.

375. **Ohba, M. and Tanada, Y.,** A synergistic factor enhances the *in vitro* infection of an insect baculovirus, *Naturwissenschaften,* 70, 613, 1983.

376. **Ohba, M. and Tanada, Y.,** *In vitro* enhancement of nuclear polyhedrosis virus infection by the synergistic factor of a granulosis virus of the armyworm, *Pseudaletia unipuncta* (Lepidoptera:Noctuidae), *Ann. Virol.,* (Inst. Pasteur), 135, 167, 1984.

377. **Tanada, Y., Hess, R. T., and Omi, E. M.,** The movement and invasion of an insect baculvirus in tracheal cells of the armyworm, *Pseudaletia unipuncta, J. Invertebr. Pathol.,* 44, 198, 1984.

378. **Hukuhara, T., Tamura, K., Zhu, Y., Abe, H., and Tanada, Y.,** Synergistic factor shows specificity in enhancing nuclear polyhedrosis virus infection, *Appl. Ent. Zool.,* 22, 235, 1987.

379. **Hukuhara, T. and Zhu, Y.,** The effect of the synergistic factor on the *in vivo* infectivity of nuclear polyhedrosis viruses, in *Invertebrate and Fish Tissue Culture,* Kuruda, Y., Kurstak, E., and Maramorosch, K., Eds., Japan Scientific Soc. Press, Tokyo, 1988, 159.

380. **Tanada, Y.,** A synopsis of studies on the synergistic property of an insect baculovirus: a tribute to Edward A. Steinhaus, *J. Invertebr. Pathol.,* 45, 125, 1985.

381. **Tanada, Y.,** Synergistic interactions among insect viruses, with emphasis on the baculoviruses of the armyworm, *Pseudaletia unipuncta,* in *Invertebrate Cell System Applications,* Mitsuhashi, J., Ed., Vol. 2, CRC Press, Boca Raton, FL, 1989, 44.

382. **Vago, C. and Croissant, O.,** Recherches sur la pathogénèse des viroses d'insectes. La libération des virus dans le tube digestion de l'insecte à partir des corps d'inclusion ingérés, *Ann. Epiphyt.,* 10, 5, 1959.

383. **Derksen, A. C. G. and Granados, R. R.,** Alteration of a lepidopteran peritrophic membrane by baculoviruses and enhancement of viral activity, *Virology,* 167, 242, 1988.

384. **Granados, R. R. and Corsaro, B. G.,** Baculovirus enhancing proteins and their implications for insect control, *V Int. Collog. Invertebr. Microb. Control,* Adelaide, 1990, 174.

385. **Harrap, K. A. and Robertson, J. S.,** A possible infection pathway in the development of a nuclear polyhedrosis, *J. Gen. Virol.,* 3, 221, 1968.

386. **Harrap, K. A.,** Cell infection by a nuclear polyhedrosis virus, *Virology,* 42, 311, 1970.

387. **Summers, M. D.,** Apparent *in vivo* pathway of granulosis virus invasion and infection, *J. Virol.,* 4, 188, 1969.

388. **Tanada, Y. and Leutenegger, R.,** Multiplication of a granulosis in larval midgut cells of *Trichoplusia ni* and possible pathways of invasion into the hemocoel, *J. Ultrastruct. Res.,* 30, 589, 1970.

389. **Kelly, D. C., Edward, M. L., Evans, H. F., and Robertson, J. S.,** The use of the enzyme linked immunosorbent assay to detect a nuclear polyhedrosis virus in *Heliothis armigera* larvae, *J. Gen. Virol.,* 40, 465, 1978.

390. **Kawanishi, C. Y., Summers, M. D., Stoltz, D. B., and Arnott, H. J.,** Entry of an insect virus *in vivo* by fusion of viral envelope and microvillus membrane, *J. Invertebr. Pathol.,* 20, 104, 1972.

391. **Robertson, J. S., Harrap, K. A., and Longworth, J. F.,** Baculovirus morphogenesis: the acquisition of the virus envelope, *J. Invertebr. Pathol.,* 23, 248, 1974.

392. **Keddie, B. A., Aponte, G. W., and Volkman, L. E.,** The pathway of infection of *Autographa californica* nuclear polyhedrosis virus in an insect host, *Science,* 243, 1728, 1989.

393. **Volkman, L. E. and Zaal, K. J. M.,** *Autographa californica* M nuclear polyhedrosis virus: microtubules and replication, *Virology,* 175, 292, 1990.

394. **Vaughn, J. L., Zhu, G. K., and Stone, R. D.,** Release of polyhedra derived virions of *Autographa californica* NPV with the digestive fluids of a susceptible host insect *(Estigmene acrea)* in *Invertebrate Cell System Applications,* Vol. 2, Mitsuhashi, J., Ed., CRC Press, Boca Raton, FL, 1989, 15.

395. **Vaughn, J. L., Stone, R. D., and Zhu, G. K.,** The effect of dissolution procedures on the infectivity of the nuclear polyhedra derived virions for cell cultures, in *Invertebrate Cell System Applications,* Vol. 2, Mitsuhashi, J., Ed., CRC Press, Boca Raton, FL, 1989, 23.

396. **Elam, P., Vail, P. V., and Schreiber, F.,** Infectivity of *Autographa californica* nuclear polyhedrosis virus extracted with digestive fluids of *Heliothis zea, Estigmene acrea,* and carbonate solutions, *J. Invertebr. Pathol.,* 55, 278, 1990.

397. **Leland, J. and Miller, D. K.,** Uncoating of enveloped viruses, *Cell,* 28, 5, 1982.

398. **White, J., Kielian, M., and Helenius, A.,** Membrane fusion proteins of enveloped animal viruses, *Q. Rev. Biophys.,* 16, 151, 1983.

399. **Kislev, N., Harpaz, I., and Zelcer, A.,** Electron microscopic studies on hemocytes of the Egyptian cottonworm, *Spodoptera littoralis* (Boisduval) infected with a nuclear-polyhedrosis virus, as compared to noninfected hemocytes. II. Virus-infected hemocytes, *J. Invertebr. Pathol.,* 14, 245, 1969.

400. **Hirumi, H., Hirumi, K., and McIntosh, A. H.,** Morphogenesis of a nuclear polyhedrosis virus of the cabbage looper in a continuous cabbage looper cell line, *Ann. N.Y. Acad. Sci.,* 266, 302, 1975.

401. **Knudson, D. L. and Harrap, K. A.,** Replication of a nuclear polyhedrosis virus in a continuous culture of *Spodoptera frugiperda:* a microscopy study of the sequence of events of the virus infection, *J. Virol.,* 17, 254, 1976.

402. **Volkman, L. E. and Goldsmith, P. A.,** Mechanism of neutralization of budded *Autographa californica* nuclear polyhedrosis virus by a monoclonal antibody inhibition of entry by adsorptive endocytosis, *Virology,* 14, 185, 1985.

403. **Volkman, L. E., Goldsmith, P. A., and Hess, R. T.,** Alternate pathway of entry of budded *Autographa californica* nuclear polyhedrosis virus: fusion at the plasma membrane, *Virology,* 148, 288, 1986.

404. **Wang, X. and Kelly, D. C.,** Baculovirus replication-uptake of *Trichoplusia ni* nuclear polyhedrosis virus particles by insect cells, *J. Gen. Virol.,* 66, 541, 1985.

405. **Dougherty, E. M., Weiner, R. M., Vaughn, J. L., and Reichelderfer, C. F.,** Physical factors that affect *in vitro Autographa californica* nuclear polyhedrosis virus infection, *Appl. Environ. Microbiol.,* 41, 1166, 1981.

406. **Harrap, K. A.,** The structure of nuclear polyhedrosis viruses. III. Virus assembly, *Virology,* 50, 133, 1972.

407. **Kawamoto, F., Kumada, N., and Kobayashi, M.,** Envelope of the nuclear polyhedrosis virus of the oriental tussock moth, *Euproctis subflava, Virology,* 77, 867, 1977.

408. **Harrap, K. A.,** The structure of nuclear polyhedrosis viruses. II. The virus particle, *Virology,* 506, 124, 1972.

409. **Teakle, R. E.,** A nuclear-polyhedrosis virus of *Anthela varia* (Lepidoptera:Anthelidae), *J. Invertebr. Pathol.,* 14, 18, 1969.

410. **Volkman, L. E., Goldsmith, P. A., and Hess, R. T.,** Evidence for microfilament involvement in budded *Autographa californica* nuclear polyhedrosis virus production, *Virology,* 156, 32, 1987.

411. **Volkman, L. E.,** *Autographa californica* MNPV nucleocapsid assembly: inhibition by cytochalasin D, *Virology,* 163, 547, 1988.

412. **Hess, R. T., Goldsmith, P. A., and Volkman, L. E.,** Effect of cytochalasin D on cell morphology and AcMNPV replication in a *Spodoptera frugiperda* cell line, *J. Invertebr. Pathol.,* 53, 169, 1989.

413. **Allaway, G. P.,** Virus particle packing in baculovirus and cytoplasmic polyhedrosis virus inclusion bodies, *J. Invertebr. Pathol.,* 42, 357, 1983.

414. **Asayama, T. and Kawamoto, F.,** An electron microscope observation on the fat body cell of the brown tail moth, *Euproctis similis* Fuessly, infected with a nucleopolyhedrosis virus, *Jpn. J. Appl. Ent. Zool.,* 19, 1, 1975.

415. **Lynn, D. E., Boucias, D. G., and Pendlund, J. C.,** Nuclear polyhedrosis virus replication in epithelial cell cultures of Lepidoptera, *J. Invertebr. Pathol.,* 42, 424, 1983.

416. **Asayama, T., Inagaki, I., Kawamoto, F., and Suto, C.,** Electron microscope observations on the maturation process of the nucleopolyhedrosis viruses of the oriental tussock moth, *Euproctis subflava* Bremer, *Jpn. J. Appl. Ent. Zool.,* 18, 189, 1974.

417. **Granados, R. R., Lawler, K. A., and Burand, J. P.,** Replication of *Heliothis zea* baculovirus in an insect cell line, *Intervirology,* 16, 71, 1981.

418. **Kelly, D. C.,** Baculovirus replication: electron microscopy of the sequence of infection of *Trichoplusia ni* nucleocapsid in *Spodoptera frugiperda* cells, *J. Gen. Virol.,* 52, 209, 1981.

419. **MacKinnon, E. A., Henderson, J. F., Stoltz, D. B., and Faulkner, P.,** Morphogenesis of nuclear polyhedrosis virus under conditions of prolonged passage *in vitro, J. Ultrastruct. Res.,* 49, 419, 1974.

420. **Nappi, A. J. and Hammil, T. N.,** Viral release and membrane acquisition by budding through the nuclear envelope of hemocytes of the gypsy moth *Porthetria dispar, J. Invertebr. Pathol.,* 26, 387, 1975.

421. **Granados, R. R. and Hashimoto, Y.,** Infectivity of baculoviruses to cultured cells, in *Invertebrate Cell System Applications,* Vol. 2, Mitsuhashi, J., Ed., CRC Press, Boca Raton, FL, 1989, 3.

422. **Rubinstein, R., Lawler, K. A., and Granados, R. R.,** Use of primary fat body cultures for the study of baculovirus replication, *J. Invertebr. Pathol.,* 40, 266, 1982.

423. **Lynn, D. E., Dougherty, E. M., McClintock, J. T., and Loeb, M.,** Development of cell lines from various tissues of Lepidoptera, in *Invertebrate and Fish Tissue Culture,* Kuroda, Y., Kurstak, E., and Maramorosch, K., Eds., Japan Sci. Soc. Press, Tokyo, 1988, 239.

424. **Lynn, D. E., Dougherty, E. M., McClintock, J. T., and Shapiro, M.,** Comparative replication of *Lymantria dispar* nuclear polyhedrosis strains in three continuous-culture cell lines, *Appl. Environ. Microbiol.,* 55, 1049, 1989.

425. **Hink, W. F. and Hall, R. L.,** Recently established invertebrate cell lines, in *Invertebrate Cell System Applications,* Vol. 2, Mitsuhashi, J., Ed., CRC Press, Boca Raton, FL, 1989, 269.

426. **Lynn, D. E. and Oberlander, H.,** Characteristics of cell lines derived from imaginal disks of three species of lepidoptera, in *Invertebrate Cell System Applications,* Vol. 2, Mitsuhashi, J., Ed., CRC Press, Boca Raton, FL, 1989, 207.

427. **Vaughn, J. L.,** Insect cells for insect virus production, in *Advances in Cell Culture,* Vol. 1, Maramorosch, K., Ed., Academic Press, New York, 1981, 281.

428. **Kuroda, Y., Kurstak, E., and Maramorosch, K., Eds.,** *Invertebrate and Fish Tissue Culture,* Japan Sci. Soc. Press, Tokyo, 1988.

429. **Hink, W. F. and Hall, R. L.,** Recently established invertebrate cell lines, in *Invertebrate Cell System Applications,* Vol. 2, Mitsuhashi, J., Ed., CRC Press, Boca Raton, FL, 1989.

430. **Goodwin, R. H., Vaughn, J. L., Adams, J. R., and Louloudes, S. J.,** The influence of insect cell lines and tissue-culture media on baculovirus polyhedra production, *Entomol. Soc. Am.,* 9, 66, 1973.

431. **Hink, W. F. and Vail, P. V.,** A plaque assay for titration of alfalfa looper nuclear polyhedrosis virus in a cabbage looper (TN-368) cell line, *J. Invertebr. Pathol.,* 22, 168, 1973.

432. **Potter, K. N., Faulkner, P., and MacKinnon, E. A.,** Strain selection during serial passage of *Trichoplusia ni* nuclear polyhedrosis virus, *J. Virol.,* 18, 1040, 1976.

433. **Fraser, M. J. and Hink, W. F.,** The isolation and characterization of the MP and FP plague variants of *Galleria mellonella* nuclear polyhedrosis virus, *Virology,* 117, 366, 1982.

434. **Adams, J. R., Goodwin, R. H., and Wilcox, T. A.,** The specificity of *in vitro* and *in vivo* cellular responses in cells of insect species not susceptible to certain baculoviruses, *Proc. 1st Int. Colloq. Invertebr. Pathol., and IXth Annu. Meet. Soc. Invertebr. Pathol.,* Kingston, Canada, 1976, 332.

435. **Wood, H. A.,** Isolation and replication of an occlusion body-deficient mutant of the *Autographa californica* nuclear polyhedra, *Virology,* 105, 338, 1980.

436. **Brown, M., Faulkner, P., Cochran, M. A., and Chung, K. L.,** Characterization of two morphology mutants of *Autographa californica* nuclear polyhedrosis virus with large cuboidal inclusion bodies, *J. Gen. Virol.,* 50, 309, 1980.

437. **Croizier, L. G. and Quiot, J. M.,** Observation and analysis of 2 genetic recombinants of baculoviruses of lepidoptera, *Autographa californica* Speyer and *Galleria mellonella, Ann. Virol.,* 132, 3, 1981.

438. **Erlandson, M., Skepasts, P., Kuzio, J., and Carstens, E. B.,** Genomic variants of a temperature sensitive mutant of *Autographa californica* nuclear polyhedrosis virus containing specific reiterations of viral DNA, *Virus Res.,* 1, 565, 1984.

439. **Duncan, R., Chung, K. L., and Faulkner, P.,** Analysis of a mutant of *Autographa californica* nuclear polyhedrosis virus with a defect in the morphogenesis of the occlusion body macro-molecular lattice, *J. Gen. Virol.,* 64, 1531, 1983.

440. **Rohel, D. Z. and Cochran, M. A.,** Mapping of late transcripts of *Autographa californica* nuclear polyhedrosis virus by 'criss-cross' DNA-RNA hybridization, *J. Gen. Virol.,* 65, 809, 1984.

441. **Kumar, S. and Miller, L. K.,** Effects of serial passage of *Autographa californica* nuclear polyhedrosis virus in cell culture, *Virus Res.,* 7, 335, 1987.

442. **O'Reilly, D. R., Passarelli, A. L., Goldman, I. F., and Miller, L. K.,** Characterization of the Da26 Gene in a hypervariable region of the *Autographa californica* nuclear polyhedrosis virus genome, *J. Gen. Virol.,* 71, 1029, 1990.

443. **Miller, L. K. and Dawes, K. P.,** Restriction endonuclease analysis to distinguish two closely related nuclear polyhedrosis viruses, *Autographa californica* MNPV and *Trichoplusia ni* MNPV, *Appl. Environ. Microbiol.,* 35, 1206, 1978.

444. **Ramoska, W. A. and Hink, W. F.,** Electron microscope examination of two plaque variants from a nuclear polyhedrosis virus of the alfalfa looper, *Autographa californica, J. Invertebr. Pathol.,* 23, 197, 1974.

445. **Beames, B. and Summers, M. D.,** Location and nucleotide sequence of the 25K protein missing from baculovirus few polyhedra (FP) mutants, *Virology,* 168, 344, 1989.

446. **Croizier, G., Croizier, L., Bioche, G., and Chaufaux, J.,** Evolution de la composition genetique du pouvoir infectieux du baculovirus de *Mamestra brassica* L. au cours de 25 multiplications successives sur les larves de la noctuelle du chou, *Entomophaga,* 30, 365, 1985.

447. **Cusack, T. and McCarthy, W. J.,** Effect of serial passage on genetic homogeneity of a plaque variant of *Lymantria dispar* nuclear polyhedrosis virus, *J. Gen. Virol.,* 70, 2963, 1990.

448. **Adang, M. J. and Miller, L. K.**, Molecular cloning of DNA complementary to mRNA of the baculovirus *Autographa californica* nuclear polyhedrosis virus: location and gene products of RNA transcripts found late in infection, *J. Virol.*, 44, 782, 1982.

449. **Smith, G. E., Vlak, J. M., and Summers, M. D.**, *In vitro* translation of *Autographa californica* nuclear polyhedrosis virus early and late mRNAs, *J. Virol.*, 44, 199, 1982.

450. **Rohel, D. Z. and Cochran, M. A.**, Characterization of two abundant mRNAs of *Autographa californica* nuclear polyhedrosis virus present late in infection, *Virology*, 124, 357, 1982.

451. **Smith, G. E., Vlak, J. M., and Summers, M. D.**, Physical analysis of *Autographa californica* nuclear polyhedrosis virus transcripts for polyhedrin and 10,000 molecular weight protein, *J. Virol.*, 45, 215, 1983.

452. **Rankin, C., Ladin, B. F., and Weaver, R. F.**, Physical mapping of temporally regulated, overlapping transcripts in the region of the 10k protein gene in *Autographa californica* nuclear polyhedrosis virus, *J. Virol.*, 57, 18, 1986.

453. **Kuzio, J., Rohel, D. Z., Curry, D. J., Krebs, A., Carstens, E. B., and Faulkner, P.**, Nucleotide sequence of the p10 polypeptide gene of *Autographa californica* nuclear polyhedrosis virus, *Virology*, 139, 414, 1984.

454. **Leisy, D., Rohrmann, G., and Beaudreau, N. M.**, Nucleotide sequencing and transcription mapping of the Orgyia *pseudotsugata* multicapsid nuclear polyhedrosis virus p10 gene, *Virology*, 153, 157, 1986.

455. **Quant-Russell, R. L., Pearson, M. N., Rohrman, G. E., and Beaudreau, G. S.**, Characterization of baculovirus p10 synthesis using monoclonal antibodies, *Virology*, 160, 9, 1987.

456. **McClintock, J. T., Dougherty, E. M., and Weiner, R. M.**, Protein synthesis in gypsy moth cells infected with a nuclear polyhedrosis virus of *Lymantria dispar*, *Virus Res.*, 5, 307, 1986.

457. **Friesen, P. D. and Miller, L. K.**, Temporal regulation of baculovirus RNA: overlapping early and late transcripts, *J. Virol.*, 54, 392, 1985.

458. **Lubbert, H. and Doerfler, W.**, Mapping of early and late transcripts encoded by the *Autographa californica* nuclear polyhedrosis virus genome: is viral RNA spliced, *J. Virol.*, 50, 497, 1984.

459. **Summers, M. D. and Arnott, H. J.**, Ultrastructural studies on inclusion formation and virus occlusion in nuclear polyhedrosis and granulosis virus-infected cells of *Trichoplusia ni* (Hubner), *J. Ultrastruct. Res.*, 28, 462, 1969.

460. **Croizier, G. and Quiot, J. M.**, Etude en microscopie électronique des structures cellulaires réticultées induites chez lépidoptères par des baculovirus, *C. R. Acad. Sci., Ser. D*, 281, 1055, 1975.

461. **Croizier, G., Amargier, A., Godse, D. B., Jacquemard, P., and Duthoit, J. L.**, Un virus de polyédrose nucléaire decouvert chez le lépidoptère noctuidae *Diparopsis watersi* (Roth) nouveau variant du *Baculovirus* d' *Autographa californica* (Speyer), *Coton Fibers Trop. (Fr. Ed.)*, 35, 414, 1980.

462. **Chung, K. L., Brown, M., and Faulkner, P.**, Studies on the morphogenesis of polyhedral inclusion bodies of a baculovirus *Autographa californica* NPV, *J. Gen. Virol.*, 46, 335, 190.

463. **Croizier, G., Gonnet, P., and Devauchelle, G.**, Localisation cytologique de la protéine non structurale p10 du baculovirus de la polyédrose nucléaire du lépidoptère *Galleria mellonella* L., *C. R. Acad. Sci. Paris, Ser. III*, 305, 677, 1987.

464. **Van der Wilk, F., van Lent, J. W. M., and Vlak, J. M.**, Immunogold detection of polyhedrin, p10 and virion antigens in *Autographa californica* nuclear polyhedrosis virus infected *Spodoptera frugiperda* cells, *J. Virol.*, 68, 2615, 1987.

465. **Vlak, J. M., Klinkenberg, F. A., Zaal, K. J. M., Usmany, M., Klinge-Roode, E. C., Geervliet, J. B. F., Roosien, J., and Van Lent, J. W. M.**, Functional studies on the p10 gene of *Autographa californica* nuclear polyhedrosis virus using a recombinant expressing a p10-β-galactosidase fusion gene, *J. Gen. Virol.*, 69, 765, 1988.

466. **Weyer, U. and Possee, R. D.**, Functional analysis of the p10 gene 5′ leader sequence of the *Autographa californica* nuclear polyhedrosis virus, *Nucleic Acids Res.*, 16, 3635, 1988.

467. **Weyer, U. and Possee, R. D.**, Analysis of the promoter of the *Autographa californica* nuclear polyhedrosis virus p10 gene, *J. Gen. Virol.*, 70, 203, 1989.

468. **Qin, J., Liu, A., and Weaver, R. F.**, Studies on the control region of the p10 gene of the *Autographa californica* nuclear polyhedrosis virus, *J. Gen. Virol.*, 70, 1273, 1989.

469. **Williams, G. V., Rohel, D. Z., Kuzio, J., and Faulkner, P.**, A cytopathological investigation of *Autographa californica* nuclear polyhedrosis virus p10 gene function using insertion deletion mutants, *J. Gen. Virol.*, 70, 187, 1989.

470. **Gonnet, P. and Devauchelle, G.**, Obtention par recombinaison dans le gène du polypeptide p10 d'un baculovirus exprimant le gène de résistance à la néomycine dans les cellules d'insecte, *C. R. Acad. Sci. Paris, Ser. III*, 305, 111, 1987.

471. **Weyer, U., Knight, S., and Possee, R. D.**, Analysis of very late gene expression by *Autographa californica* nuclear polyhedrosis virus and the further development of multiple expression vectors, *J. Gen. Virol.*, 71, 1525, 1990.

472. **Hess, R. T. and Falcon, L. A.**, Electron microscope observations on the membrane surrounding polyhedral inclusion bodies of insects, *Arch. Virol.*, 56, 169, 1978.

473. **Russell, R. L. Q. and Rohrmann, G. F.**, A baculovirus polyhedron envelope protein: immunogold localization in infected cells and mature polyhedra, *Virology,* 174, 177, 1990.

474. **Zuidema, D., Klinge-Roode, E. C., Van Lent, J. W. M., and Vlak, J. M.**, Construction and analysis of an *Autographa californica* nuclear polyhedrosis virus mutant lacking the polyhedral envelope, *Virology,* 173, 98, 1989.

475. **van Lent, J. W. M., Groenen, J. T. M., Klinge-Roode, E. C., Rohrmann, G. F., Zuidema, D., and Vlak, J. M.**, Localization of the 34 kDa polyhedron envelope protein in *Spodoptera frugiperda* cells infected with *Autographa californica, Arch. Virol.,* 111, 103, 1990.

476. **Bicknell, J. N., Leisy, D. J., Rohrman, G. F., and Beaudreau, G. S.**, Comparison of the p26 gene region of two baculoviruses, *Virology,* 161, 589, 1987.

477. **Wu, J. and Miller, L. K.**, Sequence, transcription and translation of a late gene of the *Autographa californica* nuclear polyhedrosis virus encoding a 34.8 K polypeptide, *J. Gen. Virol.,* 70, 2449, 1989.

478. **Russell, R. L. Q. and Rohrmann, G. F.**, The p6.5 gene region of a nuclear polyhedrosis virus of *Orgyia pseudotsugata:* DNA sequence and transcriptional analysis of four late genes, *J. Gen. Virol.,* 71, 551, 1990.

479. **Kuzio, J., Jaques, R., and Faulkner, P.**, Identification of p74, a gene essential for virulence of baculovirus occlusion bodies, *Virology,* 173, 759, 1989.

480. **Whitford, M., Stewart, S., Kuzio, J., and Faulkner, P.**, Identification and sequence analysis of a gene encoding gp67, an abundant envelope glycoprotein of the baculovirus *Autographa californica* nuclear polyhedrosis virus, *J. Virol.,* 63, 1393, 1989.

481. **Carbonell, L. F., Hodge, M. R., Tomalski, M. D., and Miller, L. K.**, Synthesis of a gene encoding for an insect-specific scorpion neurotoxin and attempts to express it using baculovirus vectors, *Gene,* 73, 409, 1988.

482. **Roberts, T. E. and Faulkner, P.**, Fatty acid acylation of the 67k envelope glycoprotein of a baculovirus: *Autographa californica* nuclear polyhedrosis virus, *Virology,* 172, 377, 1989.

483. **Tomalski, M. D., Wu, J., and Miller, L. K.**, The Location, sequence, transcription, and regulation of a baculovirus DNA polymerase gene, *Virology,* 167, 591, 1988.

484. **Chisholm, G. E. and Hennen, D. J.**, Multiple early transcripts and splicing of the *Autographa californica* nuclear polyhedrosis virus IE-1 gene, *J. Virol.,* 62, 3193, 1988.

485. **Carstens, E. B., Tjia, S. T., and Doerfler, W.**, Infectious DNA from *Autographa californica* nuclear polyhedrosis virus, *Virology,* 101, 311, 1980.

486. **Dobos, P. and Cochran, M. A.**, Protein synthesis in cells infected by *Autographa californica* nuclear polyhedrosis virus (Ac-NPV): the effect of cytosine arabinoside, *Virology,* 103, 446, 1980.

487. **Maruniak, J. E. and Summers, M. D.**, *Autographa californica* polyhedrosis virus phosphoproteins and synthesis of intercellular proteins after virus infection, *Virology,* 109, 25, 1981.

488. **Granados, R. R., Nguyen, T., and Cato, B.**, An insect cell line persistently infected with a baculovirus-like particle, *Intervirology,* 10, 309, 1978.

489. **Ralston, A. L., Huang, Y. S., and Kawanishi, C. Y.**, Cell culture studies with the IMC-HZ-1 nonoccluded virus, *Virology,* 115, 33, 1981.

490. **Langridge, W. H. R.**, Biochemical properties of a persistent nonoccluded baculovirus isolated from *Heliothis zea* cells, *Virology,* 112, 770, 1981.

491. **Kelly, D. C., Lescott, T., Ayres, M. D., Carey, D., Coutts, A., and Harrap, K. A.**, Induction of nonoccluded baculovirus persistently infecting *Heliothis zea* cells by *Heliothis armigera* and *Trichoplusia ni* nuclear polyhedrosis viruses, *Virology,* 112, 174, 1981.

492. **Burand, J. P., Wood, H. A., and Summers, M. D.**, Defective particles from a persistent baculovirus infection, in *Trichoplusia ni* tissue culture cells, *J. Gen. Virol.,* 64, 391, 1983.

493. **McIntosh, A. H. and Ignoffo, C. M.**, Establishment of a persistent baculovirus infection in a lepidopteran cell line, *J. Invertebr. Pathol.,* 38, 395, 1981.

494. **Crawford, A. M. and Sheehan, C.**, Persistent baculovirus infections: *Spodoptera frugiperda* NPV and *Autographa californica* NPV in *Spodoptera frugiperda* cells, *Arch. Virol.,* 78, 65, 1983.

495. **Huang, Y. S., Hedberg, M., and Kawanishi, C. Y.**, Characterization of the DNA of a nonoccluded baculovirus, HZ-IV, *J. Virol.,* 43, 174, 1982.

496. **Burand, J. P. and Wood, H. A.**, Intracellular protein synthesis during standard and defective HZ-1 virus replication, *J. Gen. Virol.,* 67, 167, 1986.

497. **Chao, Y. C., Hamblin, M., and Wood, H. A.**, Physical map of HZ-1 baculovirus genome from standard and defective interfering particles, *J. Gen. Virol.,* 71, 1265, 1990.

498. **Smith, K. M. and Xeros, N.**, Cross-inoculation studies of polyhedral viruses. Experiments with a new virus from *Pyrameis cardui,* the painted lady butterfly, *Parasitology,* 43, 178, 1953.

499. **Smith, K. M., Wyckoff, R. W. G., and Xeros, N.**, Polyhedral virus diseases affecting the larvae of the privet hawk moth *(Sphinx ligustri), Parasitology,* 42, 287, 1953.

500. **Ryndovskaya, Y. L., Tarasevich, L. M., and Avakyan, A. A.**, Electron microscope study of a single and double infection in fall webworm moth, *Hyphantria cunea* durry, *Vopr. Virusol.,* 2, 200, 1972.

501. **Kurstak, E. and Garzon, S.,** Multiple infections of invertebrate cells by viruses, in *Pathobiology of Invertebrate Vectors of Disease,* Bulla, Jr., L. E., Ed., *Ann. N.Y. Acad. Sci.,* 266, 232, 1975.

502. **Tchukhrity, M. G.,** The process of virus assembly in insect virus mixed infections, *Acta Virol.,* 27, 412, 1983.

503. **Falcon, L. A. and Hess, R. T.,** Electron microscope study on the replication of *Autographa* nuclear polyhedrosis virus and *Spodoptera* nuclear polyhedrosis virus in *Spodoptera exigua, J. Invertebr. Pathol.,* 29, 36, 1977.

504. **McIntosh, A. H., Shamy, R., and Ilsley, C.,** Interference with polyhedral inclusion body (PIB) production in *Trichoplusia ni* cells infected with a high passage strain of *Autographa californica* nuclear polyhedrosis virus (NPV), *Arch. Virol.,* 60, 353, 1979.

505. **McClintock, J. T. and Dougherty, E. M.,** Superinfection of baculovirus-infected gypsy moth cells with the nuclear polyhedrosis viruses of *Autographa californica* and *Lymantria dispar, Virus Res.,* 7, 351, 1987.

506. **Hess, R. T., Summers, M. D., and Falcon, L. A.,** A mixed virus infection in midgut cells of *Autographa californica* and *Trichoplusia ni* larvae, *J. Ultrastruct. Res.,* 65, 253, 1978.

507. **Hess, R. T., Falcon, L. A., and Morris, T. J.,** Electron microscope observations on the specificity of inclusion bodies of insect polyhedrosis viruses, *J. Invertebr. Pathol.,* 43, 1, 1984.

508. **Morris, T. J., Vail, P. V., and Collier, S. S.,** An RNA virus in *Autographa californica* nuclear polyhedrosis preparations: detection and identification, *J. Invertebr. Pathol.,* 38, 201, 1981.

509. **Morris, T. J., Hess, R. T., and Pinnock, D. E.,** Physiochemical characterization of a small RNA virus associated with baculovirus infection in *Trichoplusia ni, Intervirology,* 11, 238, 1979.

510. **Croizier, G., Croizier, L., Quiot, J. M., and Lereclus, D.,** Recombination of *Autographa californica* and *Rachiplusia ou* nuclear polyhedrosis viruses in *Galleria mellonella* L., *J. Gen. Virol.,* 69, 177, 1988.

511. **Merdan, A. and Amargier, A.,** Actions cytopathogenes d'un meme baculovirus differentes vis à vis de deux especes de Lépitoptères appartement à un meme genre (*Spodoptera littoralis* et *Spodoptera exigua*), *Bull. Soc. Entomol. Egypte,* 63, 29, 1981.

512. **Ritter, K. S., Tanada, Y., Hess, R. T., and Omi, E. M.,** Eclipse period of baculovirus infection in larvae of the armyworm, *Pseudaletia unipuncta, J. Invertebr. Pathol.,* 39, 203, 1982.

513. **Hudson, S. S., Carner, G. R., and Barnett, O. W.,** Ultrastructure of fat body cells of the velvetbean caterpillar infected with a nuclear polyhedrosis virus, *J. Invertebr. Pathol.,* 33, 31, 1979.

514. **Stiles, B., Dunn, P. E., and Paschke, J. D.,** Histopathology of a nuclear polyhedrosis infection in *Aedes epactius* with observations in four additional mosquito species, *J. Invertebr. Pathol.,* 41, 191, 1983.

515. **Vail, P. V. and Jay, D. L.,** Pathology of a nuclear polyhedrosis virus of the alfalfa looper in alternate hosts, *J. Invertebr. Pathol.,* 21, 198, 1973.

516. **Adams, J. R., Lewis, L. C., and Goodwin, R. H.,** Bioassay and histopathological examinations in *Ostrinia nubilalis* and *Agrotis ipsilon* inoculated with four baculoviruses, *Soc. Invertebr. Pathol. XII Ann. Meet.,* Seattle, WA, 1980.

517. **Hess, R. T. and Falcon, L. A.,** Electron microscope observations of *Autographa californica* (Noctuidae) nuclear polyhedrosis virus replication in the midgut of the saltmarsh caterpillar *Estigmene acrea* (Arctiidae), *J. Invertebr. Pathol.,* 37, 86, 1981.

518. **Hamm, J. J. and Styer, E. L.,** Comparative pathology of isolates of *Spodoptera frugiperda* nuclear polyhedrosis virus in *S. frugiperda* and *S. exigua, J. Gen. Virol.,* 16, 1249, 1985.

519. **Maleki-Milani, H.,** Influence de passages répétés du virus de la polyédrose nucléaire de *Autographa californica* chez *Spodoptera littoralis* (Lep:Noctuidae), *Entomophaga,* 23, 217, 1978.

520. **Fazairy, A. A. A. and Hassan, F. A.,** Infection of termites by *Spodoptera littoralis* nuclear polyhedrosis virus, *Insect Sci. Appl.,* 9, 37, 1988.

521. **Hamelin, C., LaVallee, C., and Belloncik, S.,** A simplified method for the characterization of nuclear polyhedrosis virus genomes, *FEMS Microbiol. Lett.,* 60, 233, 1990.

522. **Goodwin, R. H., Adams, J. R., Vaughn, J. L., and Louloudes, S. J.,** Invertebrate tissue culture techniques and *in vitro* baculovirus host range, *Proc. 1st Int. Colloq. Invertebr. Pathol.,* (Abstr.), Kingston, Ontario, 1976, 94.

523. **Lynn, D. E. and Hink, W. F.,** Comparison of nuclear polyhedrosis virus replication in five lepidopteran cell lines, *J. Invertebr. Pathol.,* 35, 234, 1980.

524. **McIntosh, A. H., Ignoffo, C. M., and Andrews, P. L.,** *In vitro* host range of five baculoviruses in lepidopteran cell lines, *Intervirology,* 23, 150, 1985.

525. **Danyluk, G. M. and Maruniak, J. E.,** *In vivo* and *in vitro* host range of *Autographa californica* nuclear polyhedrosis virus and *Spodoptera frugiperda* nuclear polyhedrosis virus, *J. Invertebr. Pathol.,* 50, 207, 1987.

526. **Lynn, D. E.,** personal communication, 1988.

527. **McIntosh, A. H. and Ignoffo, C. M.,** Replication of *Autographa californica* nuclear polyhedrosis virus in five lepidopteran cell lines, *J. Invertebr. Pathol.,* 54, 97, 1989.

528. **Guzo, D.,** personal communication, 1988.

529. **McIntosh, A. H. and Shamy, R.,** Biological studies of a baculovirus in a mammalian cell line, *Intervirology,* 13, 331, 1980.

530. **Volkman, L. E. and Goldsmith, P. A.**, *In vitro* survey of *Autographa californica* nuclear polyhedrosis virus interaction with nontarget vertebrate host cells, *Appl. Environ. Microbiol.*, 45, 1085, 1983.

531. **Tjia, S. T., zu Altenschildesche, G. M., and Doerfler, W.**, *Autographa californica* nuclear polyhedrosis virus (AcNPV) DNA does not persist in mass cultures of mammalian cells, *Virology*, 125, 107, 1983.

532. **Groner, A., Granados, R. R., and Burand, J. P.**, Interaction of *Autographa californica* nuclear polyhedrosis virus with two non-permissive cell lines, *Intervirology*, 21, 203, 1984.

533. **Brusca, J., Summers, M., Couch, J., and Courtney, L.**, *Autographa californica* nuclear polyhedrosis virus efficiently enters but does not replicate in poikilothermic vertebrate cells, *Intervirology*, 26, 207, 1986.

534. **Andreadis, T. G.**, Transmission, in *Epizootiology of Insect Diseases*, Fuxa, J. R. and Tanada, Y., Eds., John Wiley & Sons, New York, 1987, 159.

535. **Bird, F. T.**, Transmission of some insect viruses with particular reference to ovarial transmission and its importance in the development of epizootics, *J. Insect Pathol.*, 3, 352, 1961.

536. **Smirnoff, W. A.**, Transovum transmission of virus of *Neodiprion swainei* Middleton (Hymenoptera:Tenthridinidae), *J. Insect Pathol.*, 4, 192, 1962.

537. **Jaques, R. P.**, The transmission of nuclear polyhedrosis virus in laboratory populations of *Trichoplusia ni* (Hubner), *J. Insect Pathol.*, 4, 433, 1962.

538. **Young, S. Y. and Yearian, W. C.**, Intra- and inter- colony transmission of a nuclear polyhedrosis virus of the loblolly pine sawfly, *Neodiprion taedae linearis* Ross on pine, *J. Entomol. Sci.*, 22, 29, 1987.

539. **Thompson, C. G. and Steinhaus, E. A.**, Further tests using a polyhedrosis virus to control alfalfa caterpillar, *Hilgardia*, 19, 412, 1950.

540. **Martignoni, M. E. and Milstead, J. E.**, Trans-ovum transmission of the nuclear polyhedrosis virus of *Colias eurytheme* (Boisduval) through contamination of the female genitalia, *J. Insect Pathol.*, 4, 113, 1962.

541. **Elmore, J. C. and Howland, H. E.**, Natural and artificial dissemination of nuclear polyhedrosis virus by contaminated adult cabbage loopers, *J. Insect Pathol.*, 6, 430, 1964.

542. **Neilson, M. M. and Elgee, D. E.**, The method and role of vertical transmission of a nucleopolyhedrosis virus in the european spruce sawfly *Diprion hercyniae*, *J. Invertebr. Pathol.*, 12, 132, 1968.

543. **Hamm, J. J. and Young, J. R.**, Mode of transmission of nuclear-polyhedrosis virus to progeny of adult *Heliothis zea*, *J. Invertebr. Pathol.*, 24, 70, 1974.

544. **Nordin, G. L.**, Transovum transmission of a nuclear polyhedrosis virus of the fall webworm, *Hyphantria cunea* (Lepidoptera:Arctiidae), *J. Kans. Entomol. Soc.*, 49, 589, 1976.

545. **Kunimi, Y.**, Transovum transmission of a nuclear polyhedrosis virus of the fall webworm, *Hyphantria cunea* Drury (Lepidoptera:Arctiidae), *Appl. Entomol. Zool.*, 17, 410, 1982.

546. **Vail, P. V. and Hall, I. M.**, The influence of infections of nuclear-polyhedrosis virus on adult cabbage loopers and their progeny, *J. Invertebr. Pathol.*, 13, 358, 1969.

547. **Bird, F. T.**, The effect of metamorphosis on the multiplication of an insect virus, *Can. J. Zool.*, 31, 300, 1953.

548. **Young, S. Y.**, Effect of nuclear polyhedrosis virus infection in *Spodoptera ornithogalli* larvae on post larval stages and dissemination by adults, *J. Invertebr. Pathol.*, 55, 69, 1990.

549. **Vasudevan Nair, K. P. and Jacob, A.**, Transmission of nuclear polyhedrosis virus of rice swarming caterpillar *Spodoptera maurita* (Boisduval) through egg, *Entomon*, 10, 1, 1985.

550. **Swaine, G.**, Generation-to-generation passage of the nuclear polyhedral virus of *Spodoptera exempta*, *Nature (London)*, 210, 1053, 1966.

551. **El Nagar, S., Tawtik, M. F. S., and Rahman, T. A.**, Transmission of nuclear polyhedrosis virus (NPV) disease of *Spodoptera littoralis* (Boisd.) via egg and pupal exposure to the virus, *Bull. Entomol. Soc. Egypt, Econ. Ser.*, No. 13, 31, 1985.

552. **Harpaz, I. and Shaked, Y. B.**, Generation-to-generation transmission of the nuclear-polyhedrosis virus of *Prodenia litura* (Fab), *J. Insect Pathol.*, 6, 127, 1964.

553. **Neelgund, Y. F. and Mathad, S. B.**, Transmission of nuclear polyhedrosis virus in laboratory population of the armyworm, *Mythimna (Pseudaletia) separata*, *J. Invertebr. Pathol.*, 3, 143, 1978.

554. **Larionov, G. V. and Bakhvalov, S. A.**, On the question of trans-ovum transmission of virus of nuclear polyhedrosis of nun moth, *Akad. Nauk. SSSR Ser. Biol. Nauk.*, 3, 60, 1974.

555. **Doane, C. C.**, Trans-ovum transmission of a nuclear-polyhedrosis virus in the gypsy moth and the inducement of virus susceptibility, *J. Invertebr. Pathol.*, 14, 199, 1969.

556. **Wood, H. A., Burand, J. P., Hughes, P. R., Flore, P. H., and Gettig, R. R.**, Transovarial transmission of *Lymantria dispar* nuclear polyhedrosis virus, *Fundamental and Applied Aspects of Invertebr. Pathol.*, (Abstr.), Foundation 4th Int. Colloq. Invertebr. Pathol., Wageningen, the Netherlands, 1986, 405.

557. **Vail, P. V. and Gough, D.**, Susceptibility of the pupal stage of the cabbage looper, *Trichoplusia ni* to nuclear-polyhedrosis virus. II. Histopathology, *J. Invertebr. Pathol.*, 15, 211, 1970.

558. **Smith-Johannsen, H., Witkiewicz, H., and IaTrou, K.**, Infection of silkmoth follicular cells with *Bombyx mori* nuclear polyhedrosis virus, *J. Invertebr. Pathol.*, 48, 74, 1986.

559. **Harrap, K. A., Payne, C. C., and Robertson, J. S.,** The properties of three baculoviruses from closely related hosts, *Virology*, 79, 14, 1977.

560. **Summers, M. D., Volkman, L. E., and Hsieh, C. H.,** Immunoperoxidase detection of baculovirus antigens in insect cells, *J. Gen. Virol.*, 40, 545, 1978.

561. **McCarthy, W. J., Lambiase, J. T., and Henchal, L. S.,** The effect of inoculum concentration on the development of the nuclear polyhedrosis virus of *Autographa californica* in TN-368 cells, *J. Invertebr. Pathol.*, 36, 48, 1980.

562. **Kelly, D. C., Brown, D. A., Robertson, J. S., and Harrap, K. A.,** Biochemical, biophysical and serological properties of two singly enveloped nuclear polyhedrosis viruses from *Heliothis armigera* and *Heliothis zea*, *Microbiologica*, 3, 319, 1980.

563. **Crook, N. E. and Payne, C. C.,** Comparison of three methods of ELISA for baculoviruses, *J. Gen. Virol.*, 46, 29, 1980.

564. **Brown, D. A., Allen, C. J., and Bignell, G. N.,** The use of protein A conjugate in an indirect enzyme-linked immunosorbent assay (ELISA) of four closely related baculoviruses from *Spodoptera* species, *J. Gen. Virol.*, 62, 375, 1982.

565. **McCarthy, W. J. and Henchal, S. L.,** Detection of *Autographa californica* baculovirus nonoccluded virions *in vitro* and *in vivo* by enzyme-linked immunosorbent assay, *J. Invertebr. Pathol.*, 41, 401, 1983.

566. **Smith, G. E. and Summers, M. D.,** Application of a novel radioimmunoassay to identify baculovirus structural proteins that share interspecies antigenic determinants, *J. Virol.*, 39, 125, 1981.

567. **Knell, J. D., Summers, M. D., and Smith, G. E.,** Serological analysis of 17 baculoviruses from subgroups A and B using protein blot immunoassay, *Virology*, 125, 381, 1983.

568. **Volkman, L. E., Summers, M. D., and Hsieh, C. H.,** Occluded and nonoccluded nucleopolyhedrosis virus grown in *Trichoplusia ni:* comparative neutralization, comparative infectivity, and *in vitro* growth studies, *J. Virol.*, 19, 820, 1976.

569. **Volkman, L. E. and Summers, M. D.,** *Autographa californica* nuclear polyhedrosis virus: comparative infectivity of the occluded, alkali-liberated, and nonoccluded forms, *J. Invertebr. Pathol.*, 30, 102, 1977.

570. **Keddie, B. A. and Volkman, L. E.,** Infectivity difference between the two phenotypes of *Autographa californica* nuclear polyhedrosis virus: importance of the 64k envelope glycoprotein, *J. Gen. Virol.*, 66, 1195, 1985.

571. **Roberts, P. L.,** Neutralization studies on the *Autographa californica* nuclear polyhedrosis virus, *Arch. Virol.*, 75, 149, 1983.

572. **Wang, X. and Kelly, D. C.,** Neutralization of *Trichoplusia ni* nuclear polyhedrosis virus with antisera of two forms of the virus, *Microbiologica*, 8, 141, 1985.

573. **Holmann, A. W. and Faulkner, P.,** Monoclonal antibodies to baculovirus structural proteins: determination of specificities by western blot analysis. *Virology*, 125, 432, 1983.

574. **Carstens, E. B., Tjia, S. T., and Doerfler, W.,** Infection of *Spodoptera frugiperda* cells with *Autographa californica* nuclear polyhedrosis virus. I. Synthesis of intracellular proteins after virus infection, *Virology*, 99, 386, 1979.

575. **Wood, H. A.,** *Autographa californica* nuclear polyhedrosis virus-induced proteins in tissue culture, *Virology*, 102, 21, 1980.

576. **Goldstein, N. I. and McIntosh, A. H.,** Glycoproteins of nuclear polyhedrosis viruses, *Arch. Virol.*, 64, 119, 1980.

577. **Stiles, B. and Wood, H. A.,** A study of the glycoproteins of *Autographa californica* nuclear polyhedrosis virus (AcNPV), *Virology*, 131, 230, 1983.

578. **Volkman, L. E. and Goldsmith, P. A.,** Budded *Autographa californica* NPV 64k protein: further biochemical analysis and effects of postimmunoprecipitation sample preparation conditions, *Virology*, 139, 295, 1984.

579. **Charlton, C. A. and Volkman, L. E.,** Effect of tunicamycin on the structural proteins and infectivity of budded *Autographa californica* nuclear polyhedrosis virus, *Virology*, 154, 214, 1986.

580. **Roberts, P. L. and Naser, W.,** Characterization of monoclonal antibodies to the *Autographa californica* nuclear polyhedrosis virus, *Virology*, 122, 424, 1982.

581. **Naser, W. L. and Miltenburger, H. G.,** A rapid method for the selective identification of *Autographa californica* nuclear polyhedrosis virus using a monoclonal antibody, *FEMS Microbiol. Lett.*, 15, 261, 1982.

582. **Naser, W. L. and Miltenburger, H. G.,** Rapid baculovirus detection, identification, and serological classification by western blotting-ELISA using a monoclonal antibody, *J. Gen. Virol.*, 64, 639, 1983.

583. **Huang, Y. S., Hu, P. C., and Kawanishi, C. Y.,** Monoclonal antibodies identify conserved epitopes on the polyhedrin of *Heliothis zea* nuclear polyhedrosis virus, *Virology*, 143, 380, 1985.

584. **Quant, R. L., Pearson, M. N., Rohrmann, G. F., and Beaudreau, G. S.,** Production of polyhedrin monoclonal antibodies for distinguishing two *Orgyia pseudotsugata* baculoviruses, *Appl. Environ. Microbiol.*, 48, 732, 1984.

585. **Volkman, L. E. and Falcon, L. A.,** Use of monoclonal antibody in an enzyme-linked immunosorbent assay to detect the presence of *Trichoplusia ni* (Lepidoptera:Noctuidae) S nuclear polyhedrosis virus polyhedrin in *T. ni* larvae, *J. Econ. Entomol.,* 75, 868, 1982.

586. **Miller, L. K.,** A virus vector for genetic engineering in invertebrates, *Genetic Engineering in the Plant Sciences,* Panopoulos, N. J., Ed., Praeger, New York, 1981, 203.

587. **Smith, G. E., Summers, M. D., and Fraser, M. J.,** Production of human beta interferon in insect cells infected with a baculovirus expression vector, *Mol. Cell. Biol.,* 3, 2156, 1983.

588. **Smith, G. E., Fraser, M. J., and Summers, M. D.,** Molecular engineering of the *Autographa californica* nuclear polyhedrosis virus genome: deletion mutations with the polyhedrin gene, *Virology,* 46, 584, 1983.

589. **Summers, M. D. and Smith, G. E.,** Genetic engineering of the genome of the *Autographa californica* nuclear polyhedrosis virus, in Banbury Report 22, *Genetically Altered Viruses and the Environment,* Fields, B., Martin, M. A., and Kamely, D., Eds., Cold Spring Harbor Laboratory, Cold Spring Harbor, New York, 1985.

590. **Miller, D. W., Safer, P., and Miller, L. K.,** An insect baculovirus host-vector system for high-level expression of foreign genes, in *Genetic Engineering 8. Principles and Methods,* Setlow, J. K. and Hollander, A., Eds., Plenum Press, New York, 1986, 277.

591. **Atkinson, A. E., Weitzman, M. D., Obosi, L., Beadle, D. J., and King, L. A.,** Baculoviruses as vectors for foreign gene expression in insect cells, *Pestic. Sci.,* 28, 215, 1990.

592. **Nagy, E., Derbyshire, B., Dobos, P., and Krell, P. J.,** Cloning and expression of NDV hemagglutinin-neuraminidase cDNA in a baculovirus expression vector system, *Virology,* 176, 426, 1990.

593. **Dougherty, E. M.,** Insect viral control agents, *Dev. Ind. Microbiol.,* 28, 63, 1987.

594. **Kang, C. Y.,** Baculoviruses vectors for expression of foreign genes, *Adv. Virus Res.,* 35, 171, 1988.

595. **Morton, H. L., Moffett, J. O., and Stewart, F. D.,** Effect of alfalfa looper nuclear polyhedrosis virus on honeybees, *J. Invertebr. Pathol.,* 26, 139, 1975.

596. **Bishop, D. H. L.,** UK release of genetically marked virus, *Nature (London),* 323, 496, 1986.

597. **O'Reilly, D. R. and Miller, L. K.,** A baculovirus blocks insect molting by producing ecdysteroid UDP-glucosyl transferase, *Science,* 245, 1110, 1989.

598. **Hammock, B. D., Bonning, B. C., Possee, R. D., Hanzlik, T. N., and Maeda, S.,** Expression and effects of the juvenile hormone esterase in a baculovirus vector, *Nature (London),* 344, 458, 1990.

599. **Tompkins, G. J., Linduska, J. J., Diggs, D., and Friend, D. R.,** Control of cabbage looper and imported cabbageworm on collards with encapsulated virus formulations, 1987, *Insecticide Acaracide Tests,* 13, 111, 1988.

600. **Merryweather, A. T., Weyer, U., Harris, M. P. G., Hirst, M., Booth, T., and Possee, R. D.,** Construction of genetically engineered baculovirus insecticides containing the *Bacillus thuringiensis* subsp. *kurstaki* HD-73 delta endotoxin, *J. Gen. Virol.,* 71, 1535, 1990.

601. **Farmer, J. L., Hampton, R. G., and Boots, E.,** Flow cytometric assays for monitoring production of recombinant HIV-1 Gp160 in insect cells infected with a baculovirus expression vector, *J. Virol. Methods,* 26, 279, 1989.

602. **Overton, H. A., Fujii, Y., Price, I. R., and Jones, I. M.,** The protease and gag gene products of the human immunodeficiency virus: authentic cleavage and post-translational modification in an insect cell expression system, *Virology,* 170, 107, 1989.

603. **Wells, D. E. and Compans, R. W.,** Expression and characterization of a functional human immunodeficiency virus envelope glycoprotein in insect cells, *Virology,* 176, 575, 1990.

604. **Mills, H. R. and Jones, I. M.,** Expression and purification of p24, the core protein of HIV, using a baculovirus insect cell expression system, *Aids,* 4, 1125, 1990.

605. **Morikawa, Y., Overton, H. A., Moore, J. P., Wilkinson, A. J., Brady, R. L., Lewis, S. J., and Jones, I. M.,** Expression of HIV-1 Gp120 and human soluble CD4 by recombinant baculoviruses and their interaction *in vitro, Aids Res. Human Retrovir.,* 6, 765, 1990.

606. **Murphy, C. I., Lennick, M., Lehar, DS. M., Beltz, G. A., and Young, E.,** Temporal expression of HIV-1 envelope proteins in baculovirus-infected insect cells-implications for glycosylation and CD4 binding, *Genet. Anal. Tech. Appl.,* 7, 160, 1990.

607. **Knops, J., Johnsonwood, K., Schenk, D. B., Sinha, S., Lieberburg, I., and McConlogue, L.,** Isolation of baculovirus-derived secreted and full-length beta-amyloid precursor protein, *J. Biol. Chem.,* 266, 7285, 1991.

608. **Vlak, J. M. and Keus, R. J. A.,** Baculovirus expression vector system for production of viral vaccines, in *Advances in Biotechnological Processes,* Vol. 14, Mizrahi, A., Ed., Wiley-Liss, New York, 1990, 91.

609. **Putlitz, J. Z., Kubasek, W. L., Duchene, M., Marget, M., Vonspecht, B. U., and Domdey, H.,** Antibody production in baculovirus-infected insect cells, *Bio-Technology,* 8, 651, 1990.

610. **Vlak, J. M., Schouten, A., Usmany, M., Belsham, G. J., Klinge-Roode, E. C., Maule, A. J., van Lent, J. W. M., and Zuidema, D.,** Expression of cauliflower mosaic virus gene-1 using a baculovirus vector based upon the p10 gene and a novel selection method, *Virology,* 179, 312, 1990.

611. **Zuidema, D., Schouten, A., Usmany, M., Belsham, A. J., Roosien, J., Klinge-Roode, E. C., van Lent, J. W. M., and Vlak, J. M.,** Expression of cauliflower mosaic virus gene-1 in insect cells using a novel polyhedrin-based baculovirus expression vector, *J. Gen. Virol.,* 71, 2201, 1990.

612. **Van Bokhoven, H. J., Wellink, M., Usmany, M., Vlak, J. M., Goldbach, R., and van Kammen, A.,** Expression of plant virus genes in animal cells — high level synthesis of cowpea mosaic virus β-RNA-encoded proteins with baculovirus expression vectors, *J. Gen. Virol.,* 71, 2509, 1990.

613. **Webb, N. R. and Summers, M. D.,** Expression of proteins using recombinant baculoviruses, *Technique (Phila.),* 2, 173, 1990.

614. **Page, M. J. and Murphy, V. F.,** Expression of foreign genes in cultured insect cells using a recombinant baculovirus vector, in *Methods in Molecular Biology, Animal Cell Culture,* Vol. 5, Pollard, J. W. and Walker, J. M., Eds., The Humana Press, Clifton, NJ, 1990, 573.

615. **Christian, P. D. and Oakeshott, J. G.,** The potential of genetically engineered baculoviruses for insect pest control, *Austral. J. Biotechnol.,* 3, 264, 1989.

616. **Van Lier, F. L. J., van den End, E. J., De Gooijer, C. D., Vlak, J. M., and Tramper, J.,** Continuous production of baculovirus in a cascade of insect-cell reactors, *Appl. Microbiol. Biotechnol.,* 33, 43, 1990.

617. **Kopier, R., Tramper, J., and Vlak, J. M.,** A continuous process for the production of baculovirus using insect-cell cultures, *Biotechnol. Lett.,* 10, 849, 1988.

618. **Wu, J., King, G., Daugulis, A. J., Faulkner, P., Bone, D. H., Goosen, M. F. A.,** Engineering aspects of insect cell suspension culture: a review, *Appl. Microbiol. Biotechnol.,* 32, 249, 1989.

619. **Kloeppinger, M., Fertig, G., Fraune, E., and Miltenburger, H. G.,** Multistage production of *Autographa-californica* nuclear polyhedrosis virus in insect cell cultures, *Cytotechnology,* 4, 271, 1990.

620. **Tramper, J., van den End, E. J., De Gooijer, C. D., Kopier, R., van Lier, F. L. J., Usmany, M., and Vlak, J. M.,** Production of baculovirus in a continuous insect-cell culture: bioreactor design, operation, and modeling, *Annu. N. Y. Acad. Sci.,* 589, 423, 1990.

621. **Caron, A. W., Archambault, J., and Massie, B.,** High-level recombinant protein production in bioreactors using the baculovirus insect cell expression system, *Biotechnol. Bioeng.,* 36, 1133, 1990.

622. **Vaughn, J. L., Fan, F., Dougherty, E. M., Adams, J. R., Guzo, D., and McClintock, J. T.,** The use of commerical serum replacements in media for the *in vitro* replication of nuclear polyhedrosis virus, *J. Invertebr. Pathol.,* 58, in press, 1991.

623. **Vaughn, J. L., Stone, R. D., and Zhu, G. K.,** The effect of dissolution procedures on the infectivity of the nuclear polyhedrosis polyhedra derived virions for cell culture, in *Invertebrate Cell System Applications,* Vol. II, Mitsuhashi, J., Ed., CRC Press, Boca Raton, FL, 1989, 23.

624. **Corsaro, B. G., DiRenzo, J., and Fraser, M. J.,** Transfection of cloned *Heliothis zea* cell lines with the DNA genome of the *Heliothis zea* nuclear polyhedrosis virus, *J. Virol. Methods,* 25, 283, 1989.

625. **Mann, S. G. and King, L. A.,** Efficient transfection of insect cells with baculovirus DNA using electroporation, *J. Gen. Virol.,* 70, 3501, 1989.

626. **Tompkins, G. J., Dougherty, E. M., Goodwin, R. H., and Adams, J. R.,** Maintenance of infectivity and virulence of nuclear polyhedrosis viruses during serial passage in noctuid (Lepidoptera, Noctuidae) cell lines, *J. Econ. Entomol.,* 84, 445, 1991.

627. **Tsuda, K., Mizuki, E., Kawarabata, T., and Aizawa, K.,** Infectivity of nuclear polyhedrosis virus in continuous cell cultures. II. Transfection with virus DNA to permissive and nonpermissive cells, *Jpn. J. Appl. Ent. Zool.,* 34, 303, 1990.

628. **Tsuda, K., Mizuki, E., Kawarabata, T., and Aizawa, K.,** Infectivity of nuclear polyhedrosis virus in continuous cell cultures. I. Adsorption of nonoccluded viruses to permissive and nonpermissive cells, *Jpn. J. Appl. Ent. Zool.,* 34, 297, 1990.

629. **Tsuda, K., Mizuki, E., Kawarabata, T., and Aizawa, K.,** Infectivity of nuclear polyhedrosis virus in continuous cell cultures. III. Comparative neutralization of nonoccluded and occluded virus of nuclear polyhedrosis viruses, *Jpn. J. Appl. Ent. Zool.,* 35, 31, 1991.

630. **Hink, W. F., Thomsen, D. R., Davidson, D. J., Meyer, A. L., and Castellino, F. J.,** Expression of three recombinant proteins using baculovirus vectors in twenty-three insect cell lines, *Biotechnol. Progr.,* 7, 9, 1991.

631. **Kitts, P. A., Ayres, M. D., and Possee, R. D.,** Linearization of baculovirus DNA enhances the recovery of recombinant virus expression vectors, *Nucleic Acids Res.,* 18, 5667, 1990.

632. **Merryweather, A. T., Weyer, U., Harris, M. P. G., Hirst, M., Booth, T., and Possee, R. D.,** Construction of genetically engineered baculovirus insecticides containing the *Bacillus-thuringiensis* subsp *kurstaki* HD-73 delta-endotoxin, *J. Gen. Virol.,* 71, 1535, 1990.

633. **Martens, J. W. M., Honee, G., Zuidema, D., van Lent, J. W. M., Visser, B., and Vlak, J. M.,** Insecticidal activity of a bacterial crystal protein expressed by a recombinant baculovirus in insect cells, *Appl. Environ. Microbiol.,* 56, 2764, 1990.

634. **Hasnain, S. E. and Nakhai, B.,** Expression of the gene encoding firefly *luciferase* in insect cells using a baculovirus vector, *Gene,* 91, 135, 1990.

635. **Jha, P. K., Nakhai, B., Sridhar, P., Talwar, G. P., and Hasnain, S. E.**, Firefly *luciferase,* synthesized to very high levels in caterpillars infected with a recombinant baculovirus, can also be used as an efficient reporter enzyme *in vivo, FEBS Lett.,* 274, 23, 1990.

636. **Croizier, G., Boukhoudmi-Amiri, K., and Croizier, L.**, A physical map of *Spodoptera littoralis* B-type nuclear polyhedrosis virus genome, *Arch. Virol.,* 104, 145, 1989.

637. **Maeda, S. and Majima, K.**, Molecular cloning and physical mapping of the genome of *Bombyx-mori* nuclear polyhedrosis virus, *J. Gen. Virol.,* 71, 1851, 1990.

638. **Sugimori, H., Nagamine, T., and Kobayashi, M.**, Analysis of structural polypeptide of *Bombyx-mori* (Lepidoptera: Bombycidae) nuclear polyhedrosis virus, *Appl. Entomol. Zool.,* 25, 67, 1990.

639. **Vialard, J. E., Yuen, L., and Richardson, C. D.**, Identification and characterization of a baculovirus occlusion body glycoprotein which resembles spheroidin, an entomopox virus protein, *J. Virol.,* 64, 5804, 1990.

640. **Wu, J. and Miller, L. K.**, Sequence transcription, and translation of a late gene of the *Autographa californica* nuclear polyhedrosis virus encoding a 34.8 K polypeptide, *J. Gen. Virol.,* 70, 2449, 1989.

641. **Tilakaratne, N., Hardin, S. E., and Weaver, R. F.**, Nucleotide sequence and transcription mapping of the Hind III F-region of the *Autographa californica* nuclear polyhedrosis virus genome, *J. Gen. Virol.,* 72, 285, 1991.

642. **Carstens, E. B. and Lu, A.**, Nucleotide sequence and transcription anlysis of the *Hind*III P region of a temperature-sensitive mutant of *Autographa californica* nuclear polyhedrosis virus, *J. Gen. Virol.,* 71, 3035, 1990.

643. **Huh, N. E. and Weaver, R. F.**, Categorizing some early and late transcripts directed by the *Autographa californica* nuclear polyhedrosis virus, *J. Gen. Virol.,* 71, 2195, 1990.

644. **Huh, N. E. and Weaver, R. F.**, Identifying the RNA polymerases that synthesize specific transcripts of the *Autographa californica* nuclear polyhedrosis virus, *J. Gen. Virol.,* 71, 195, 1990.

645. **Schetter, C., Oellig, C., and Doerfler, W.**, An insertion of insect cell DNA in 81-map-unit segment of *Autographa californica* nuclear polyhedrosis virus DNA, *J. Virol.,* 64, 1844, 1990.

646. **Blissard, G. W. and Rohrmann, G. F.**, Location, sequence, transcriptional mapping, and temporal expression of the gp64 envelope glycoprotein gene of the *Orgyia pseudotsugata* multicapsid nuclear polyhedrosis virus, *Virology,* 170, 537, 1989.

647. **Bradford, M. B., Blissard, G. W., and Rohrmann, G. F.**, Characterization of the infection cycle of the *Orgyia-pseudotsugata* multicapsid nuclear polyhedrosis virus in *Lymantria-dispar* cells, *J. Gen. Virol.,* 71, 2841, 1990.

648. **Russell, R. L. Q., Pearson, M. N., and Rohrmann, G. F.**, Immunoelectron microscopic examination of *Orgyia pseudotsugata* multicapsid nuclear polyhedrosis virus-infected *Lymantria dispar* cells-time course and localization of major polyhedron-associated proteins, *J. Gen. Virol.,* 72, 275, 1991.

649. **Theilmann, D. A. and Stewart, S.**, Identification and characterization of the IE-1 gene of *Orgyia pseudotsugata* multicapsid nuclear polyhedrosis virus, *Virology,* 180, 492, 1991.

650. **Jarvis, D. L., Fleming, J. A. G. W., Kovacs, G. R., Summers, M. D., and Guarino, L. A.**, Use of early baculovirus promoters for continuous expression and efficient processing of foreign gene products in stably transformed lepidopteran cells, *Bio-Technology,* 8, 950, 1990.

651. **Nissen, M. S. and Friesen, P. D.**, Molecular analysis of the transcriptional regulatory region of an early baculovirus gene, *J. Virol.,* 63, 493, 1989.

652. **Lu, A. and Carstens, E. B.**, Nucleotide sequence of a gene essential for viral DNA replication in the baculovirus *Autographa californica* nuclear polyhedrosis virus, *Virology,* 181, 336, 1991.

653. **Partington, S., Yu, H., Lu, A., and Carstens, E. B.**, Isolation of temperature sensitive mutants of *Autographa californica* nuclear polyhedrosis virus: phenotype characteriation of baculovirus mutants defective in very late gene expression, *Virology,* 175, 91, 1990.

654. **Weyer, U., Knight, S., and Possee, R. D.**, Analysis of very late gene expression by *Autographa californica* nuclear polyhedrosis virus and the further development of multiple expression vectors, *J. Gen. Virol.,* 71, 1525, 1990.

655. **Thiem, S. M. and Miller, L. K.**, Differential gene expression mediated by late very late and hybrid baculovirus promoters, *Gene,* 91, 87, 1990.

656. **Hill-Perkins, M. S. and Possee, R. D.**, A baculovirus expression vector derived from the basic protein promoter of *Autographa californica* nuclear polyhedrosis virus, *J. Gen. Virol.,* 71, 971, 1990.

657. **Wickham, T. J., Granados, R. R., Wood, H. A., Hammer, D. A., and Shuler, M. L.**, General analysis of receptor-mediated viral attachment to cell surfaces, *Biophysical J.,* 58, 1501, 1990.

658. **Liu, H. S. and Bilimoria, S. L.**, Infected cell specific protein and viral DNA synthesis in production and abortive infections of *Spodoptera frugiperda* nuclear polyhedrosis virus, *Arch. Virol.,* 115, 101, 1990.

659. **Gelernter, W. D. and Federici, B. A.**, Virus epizootics in Californian populations of *Spodoptera exigua*: dominance of a single viral genotype, *Biochem. System. Ecol.,* 18, 461, 1990.

660. **Maeda, S., Mukohara, Y., and Kondo, A.**, Characteristically distinct isolates of the nuclear polyhedrosis virus from *Spodoptera litura, J. Gen. Virol.,* 71, 2631, 1990.

661. **Doyle, C. J., Hirst, M. L., Cory, J. S., and Entwistle, P. F.,** Risk assessment studies-detailed host range testing of wild-type cabbage moth, *Mamestra brassicae* (Lepidoptera, Noctuidae), nuclear polyhedrosis virus, *Appl. Environ. Microbiol.,* 56, 2704, 1990.

662. **Doyle, C. J. and Hirst, M. L.,** Cross infection of nymphalid butterfly larvae with cabbage moth (*Mamestra brassicae*) nuclear polyhedrosis virus, *J. Invertebr. Pathol.,* 57, 131, 1991.

663. **Hamblin, M., van Beek, N. A. M., Hughes, P. R., and Wood, H. A.,** Co-occlusion and persistence of a baculovirus mutant lacking the polyhedrin gene, *Appl. Environ. Microbiol.,* 56, 3057, 1990.

664. **Charlton, C. A. and Volkman, L. E.,** Sequential rearrangement and nuclear polymerization of actin in baculovirus-infected *Spodoptera frugiperda* cells, *J. Virol.,* 65, 1219, 1991.

665. **Guzo, D., Dougherty, E. M., Braun, S. K., Lynn, D. E., and Weiner, R. M.,** Production of a cellular macromolecular synthesis inhibition factor(s) in gypsy moth cells infected with the *Autographa californica* nuclear polyhedrosis virus, *J. Invertebr. Pathol.,* 57, 413, 1991.

666. **Nagamine, T., Kobayashi, M., Saga, S., and Hoshino, M.,** Preparation and characterization of mono-clonal antibodies against occluded virions of *Bombyx mori* nuclear polyhedrosis virus, *J. Invertebr. Pathol.,* 57, 311, 1991.

667. **McIntosh, A. H.,** *In vitro* infectivity of a clonal isolate of *Syngrapha falcifera* (celery looper) multiple nuclear polyhedrosis virus, *J. Invertebr. Pathol.,* 57, 441, 1991.

668. **Bong, C. F. J. and Sikorowski, P. P.,** Presence of polyhedra in the midgut cell nuclei of *Heliothis zea* (Lepidoptera, Noctuidae) infected with cytoplasmic polyhedrosis virus, *J. Invertebr. Pathol.,* 57, 294, 1991.

669. **Keating, S. T., Schultz, J. C., and Yendol, W. G.,** The effect of diet on gypsy moth (*Lymantria dispar*) larval midgut pH, and its relationship with larval susceptibility to a baculovirus, *J. Invertebr. Pathol.,* 56, 317, 1990.

670. **Keating, S. T. and Yendol, W. G.,** Influence of selected host plants on gypsy moth (Lepidop-tera:Lymantriidae) larval mortality caused by a baculovirus, *Environ. Entomol.,* 16, 459, 1987.

671. **Keating, S. T., Yendol, W. G., and Schultz, J. C.,** Relationship between susceptibility of gypsy moth larvae (Lepidoptera:Lymantriidae) to a baculovirus and host plant foliage constituents, *Environ. Entomol.,* 17, 952, 1988.

672. **Shapiro, M. and Robertson, J. L.,** Natural variability of three geographic isolates of gypsy moth (Lep-idoptera: Lymantriidae) nuclear polyhedrosis virus, *J. Econ. Entomol.,* 84, 71, 1991.

673. **Ignoffo, C. M., Shasha, B. S., and Shapiro, M.,** Sunlight ultraviolet protection of the *Heliothis* nuclear polyhedrosis virus through starch encapsulation technology, *J. Invertebr. Pathol.,* 57, 134, 1991.

674. **Nordin, G. L., Brown, G. C., and Jackson, D. M.,** Vertical transmission of two baculoviruses infectious to the tobacco budworm, *Heliothis virescens* (F.) (Lepidoptera: Noctuidae) using an autodissemination technique, *J. Kansas Entomol. Soc.,* 63, 393, 1990.

675. **Hartig, P. C., Cardon, M. C., and Kawanishi, C. Y.,** Insect virus assays for viral replication and perisitence in mammalian cells, *J. Virol. Methods,* 31, 335, 1991.

676. **Groner, A.,** Safety to nontarget invertebrates of baculoviruses, in *Safety of Microbial Insecticides,* Marshall, L., Lacey, L. A., Davidson, E. W., Eds., CRC Press, Boca Raton, FL, 1990, 135.

677. **Overstreet, R. M., Stuck, K. C., Krol, R. A., and Hawkins, W. E.,** Experimental infections with *Baculovirus penai* in the white shrimp *Penaeus vannamei* (Crustacea: Decapoda) as a bioassay, *J. World Aquac. Soc.,* 19, 175, 1988.

678. **LeBlanc, B. D. and Overstreet, R. M.,** Prevalence of *Baculovirus penaei* in experimentally infected white shrimp (*Penaeus vannamei*) relative to age, *Aquaculture,* 87, 237, 1990.

PART 2.
NUCLEAR POLYHEDROSIS VIRUSES OF INVERTEBRATES OTHER THAN INSECTS

I. INTRODUCTION

Baculoviruses from noninsect hosts, mainly from Crustacea, have been reported since 1973 from sites around the world.[1-9] Shrimps, prawns, and crabs have been the hosts, other than insects, that most frequently harbor baculovirus infections.

Baculoviruses have been described or reported from eight species of penaeid shrimps from diverse locales such as the Gulf of Mexico, Hawaii, Japan, South America, and Australia. Species of these viruses also have been found to infect crabs from the Mediterranean and the east cost of the U.S. Therefore, in the last 14 years the widespread occurrence of new baculovirus forms in new host groups has been established worldwide. This recent chain of activities in baculovirus-crustacean research followed a particularly long period of research and study of baculoviruses in insects only, with little indication that other arthropod host groups might exist for baculoviruses.

Baculovirus infections in noninsect hosts, described to date, usually produce significant cytopathological changes and disease, but produce only a few gross signs in infected hosts. *Baculovirus penaei* (BP), the first to be described in a noninsect host, produces no gross signs of infection in adult or larval penaeid shrimps. However, characteristic tetrahedral occlusion bodies (OBs) in hepatopancreatic cells (see Figure 1) may occasionally be observed directly in infected, transparent larval shrimp (see Figure 2). Diagnosis of infection in most cases is dependent upon fresh squash examination of hepatopancreas (see Figure 3) or histological examination of tissue sections (see Figure 4). Heavy infections may result in appearance of characteristic occlusion bodies in feces (see Figure 5).

The baculovirus of *Penaeus monodon* (MBV) produces some obvious gross signs in infected hosts. According to Lightner et al.[7] heavily infected *P. monodon* are lethargic, do not feed normally, and appear to be more heavily fouled (shell commensals). Postlarvae that are infected may be smaller and darker in color than normal. Lightner et al.[7] also suggested that one other set of signs described as "white turbid liver" disease in cultured postlarval *P. monodon* by Liao[10] might be caused by MBV infections that usually occur in the hepatopancreas.

Other baculoviruses, mostly nonoccluded forms, reported from noninsect hosts have not yet been associated with grossly visible lesions in their hosts. These nonoccluded viruses will be considered in detail in Chapter 9. After more research is done on these viruses, particularly in larval as well as adult stages of their hosts, more may be known about disease signs or syndromes for them. Most of the insect baculoviruses cause lethal disease in the larvae of the insect host. Few studies have been done on larval stages of potential, baculoviral, crustacean hosts, and this should be the direction for future studies.

II. VIRUS STRUCTURE AND COMPOSITION

As pointed out by Johnson and Lightner,[8] *Baculovirus penaei* (BP) is the only noninsect baculovirus sufficiently characterized, to date, to be accepted by the *International Committee on Taxonomy of Viruses* (ICTV).[11]

Considerable work has been done, however, on the characterization of the *Penaeus monodon* baculovirus (MBV) by Lightner et al.[7] Therefore, remaining attention will be given to these two occluded crustacean viruses with emphasis on their morphology, morphogenesis, composition, and genome.

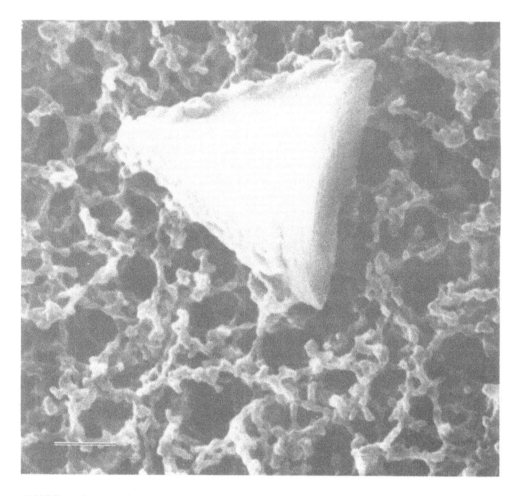

FIGURE 1.   Scanning electron micrograph of tetrahedral occlusion body of *Baculovirus penaei*. (Bar = 2 μm.)

A. *BACULOVIRUS PENAEI (BP)*
1. Morphology

This virus is an enveloped, rod-shaped, nuclear virus. Infections of host, crustacean digestive gland (hepatopancreas) cells result in the virus-controlled production of intranuclear, tetrahedral occlusion bodies containing multiple, embedded virions (Baculovirus, type A).

Virions are composed of rod-shaped nucleocapsids, surrounded by envelopes of trilaminar form (see Figures 6, 7, 8); nucleocapsids are approximately 260 nm long and 44.2 nm in diameter (see Figure 6). In certain planes of section, particularly median longitudinal, a knoblike protrusion may be observed extending from the base of the nucleocapsid (see Figure 6). This knob or shaft structure may be an extension of the nucleocapsid sheath as suggested by Fraser[12] in a description of *Autographa californica* baculovirus virion morphogenesis. As described by other authors for insect baculoviruses, BP virions also have a caplike structure opposite the base (see Figures 6, 8). Thus the crustacean BP virion has most of the morphological features of the other members of the family Baculoviridae.

2. Morphogenesis

The BP virion uncoats or injects its DNA into the host cell nucleus at the nuclear pore (see Figures 9, 10), following entry into the hepatopancreatic cell, presumably, via viropexis

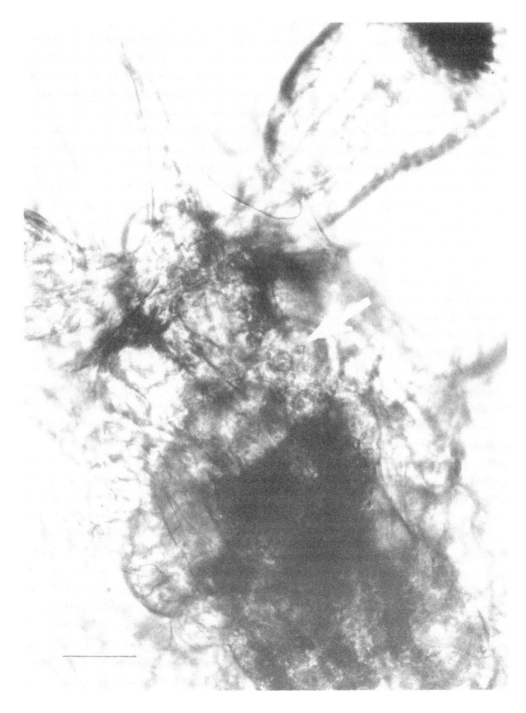

FIGURE 2.   Living larval penaeid shrimp, semitransparent, with small BP tetrahedral occlusion body visible in portion of hepatopancreas (arrow). (Bar = 100 μm.)

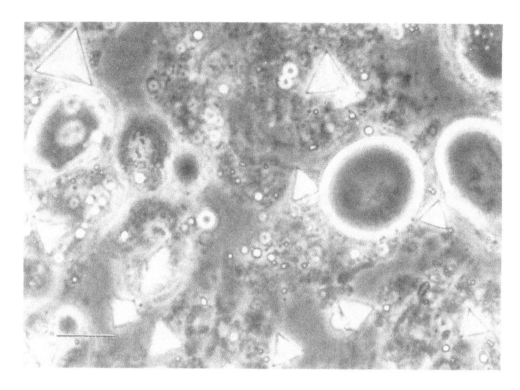

FIGURE 3. Fresh squash preparation of BP infected, hepatopancreas from pink shrimp; note tetrahedral occlusion bodies. (Bar = 10 μm.)

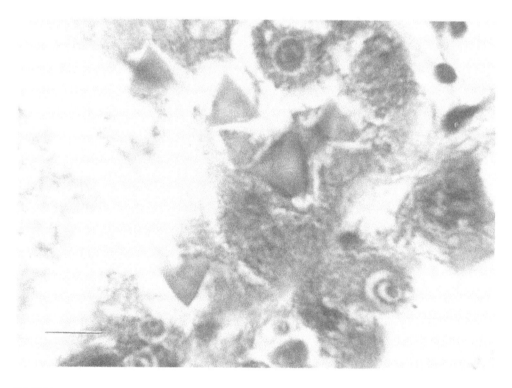

FIGURE 4. Histological section showing BP occlusion bodies in cells and free of cells in hepatopancreas of pink shrimp. (Bar = 10 μm.)

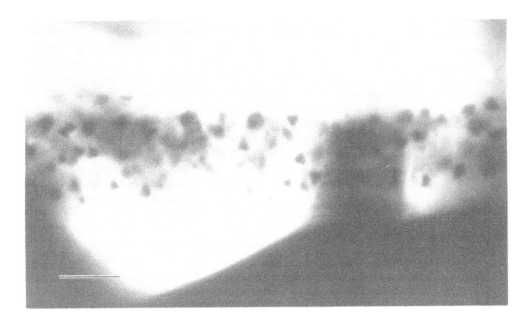

FIGURE 5. Protargol (silver) stained, larval shrimp hindgut, showing many BP occlusion bodies in fecal material to be excreted; the OBs stain dark purple with this method. (Bar = 50 μm.)

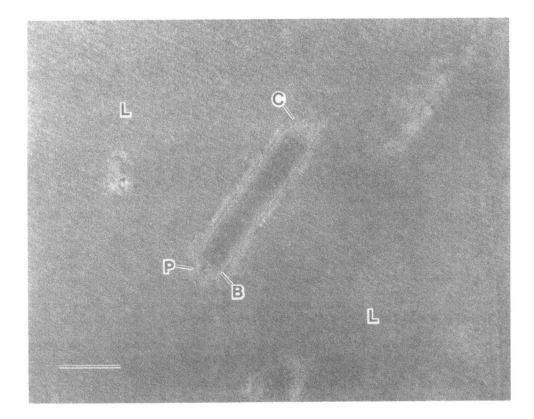

FIGURE 6. Occluded BP virion showing envelope and nucleocapsid; note cap (C) and base (B) of virion, and knoblike protrusion (P) from base; also note lattice (L) of OB paracrystalline structure. (Bar = 100 nm.)

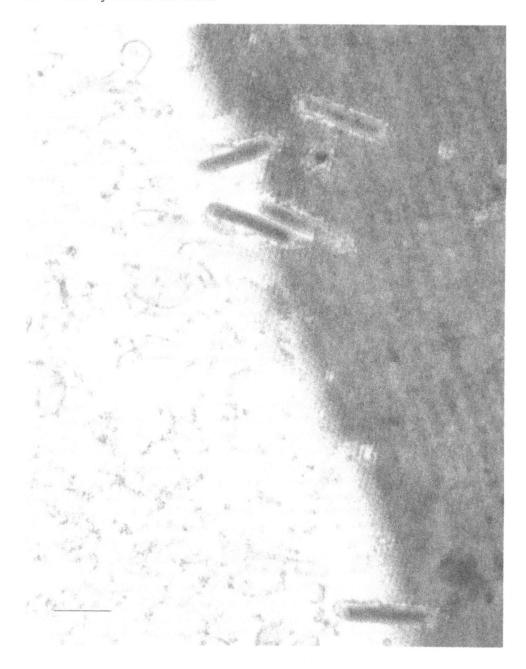

FIGURE 7.   BP virions partially and wholly occluded in OB; note trilaminar nature of virion envelopes, subunits of OB, and virogenic membranes in nucleoplasm of host cell. (Bar = 100 nm.)

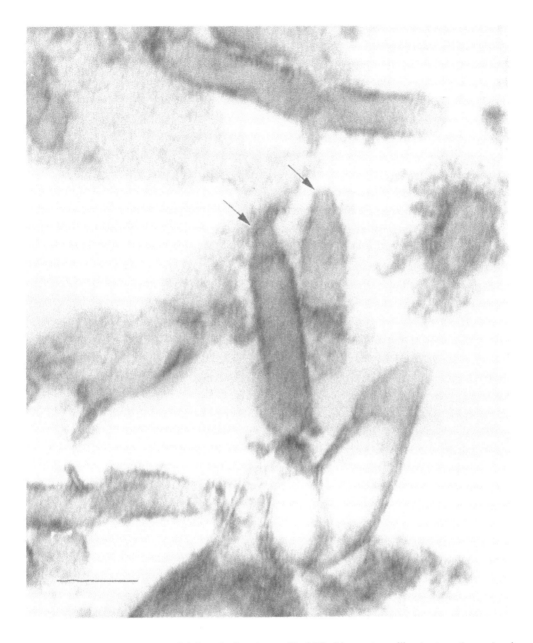

FIGURE 8. Empty appearing or lightly stained nucleocapsid of BP virion; note caplike structures (arrows) and base at opposite end. (Bar = 100 nm.)

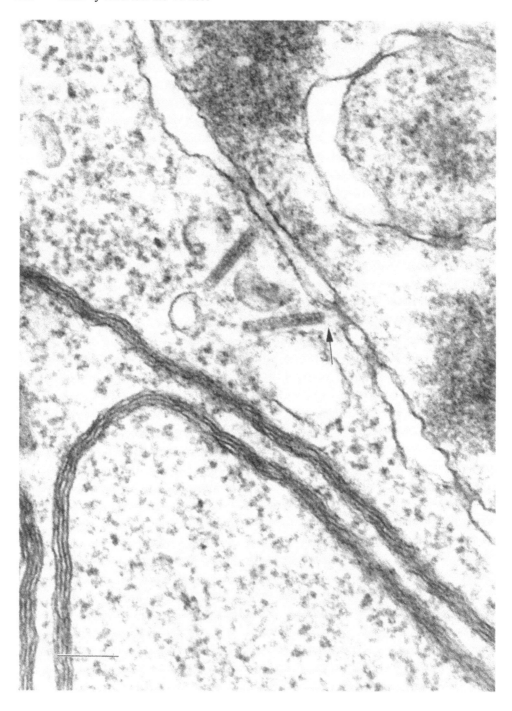

FIGURE 9.   Nucleocapsid of BP virion aligned at nuclear pore of shrimp cell nucleus (arrow); note fine filament, probably nucleic acid, and uneven density of nucleocapsid as it uncoats through nuclear pore. (Bar = 200 nm.)

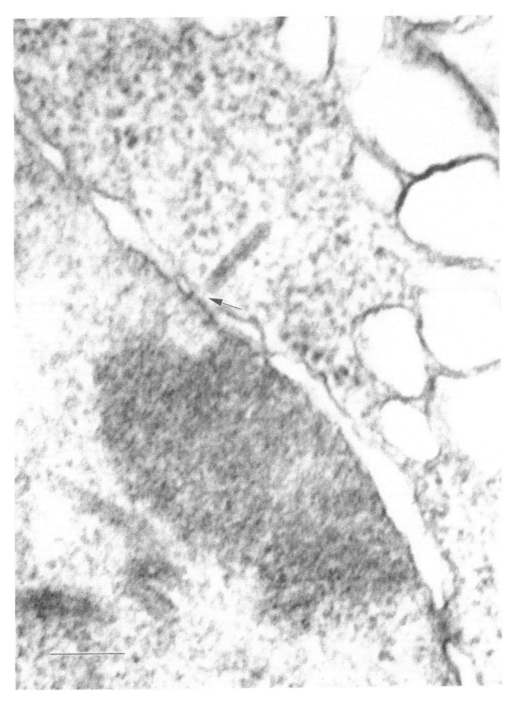

FIGURE 10.   Nucleocapsid of BP uncoating at nuclear pore; note probable nucleic acid filament entering nucleus (arrow). (Bar = 200 nm.)

or fusion of virion members with sites on host cell, microvillar membranes. Once integrated into the host cell genome, the virus directs a series of changes that can be morphologically recognized with electron microscopic studies. These changes conveniently may be described as a sequence of events that characterize the cellular infectious process to the production of mature virions and tetrahedral occlusion bodies (see Figures 11, 12, 13).[2,3] These stages are: (1) early and eclipsed infection characterized by nuclear hypertrophy, chromatin diminution, and segregation of the nucleoplasm into regions of granular and fibrillar stromata (see Figure 11); (2) intermediate infections characterized by appearance of nonoccluded, unenveloped, and enveloped virions in different stages of maturation, and by nuclear and cytoplasmic changes dominanted by proliferation of outer nuclear membranes and appearance of extensive membranous labyrinths (ML) contiguous with the endoplasmic reticulum (ER) (see Figure 12); and (3) advanced infections characterized by growth, coalescence, and paracrystallization of the tetrahedral OB that occludes mature virions as it develops in the nucleoplasm (see Figure 13).

Of singular significance is the nature and possible role(s) of the membranous labyrinth that first appears in all infected cells at the intermediate period during the development of the infectious virus (see Figure 14). Similar, but less dramatically developed membrane systems have been reported in certain insect baculovirus infections.[14,15] In BP infections of shrimp cells, the ML reaches extreme size and complexity (see Figures 15, 16), and is pathognomonic for BP infections. Though not demonstrated experimentally, it is possible that the ML may serve as a transport system or pathway for the entry of amino acids, peptides, proteins, or other precursors for *de novo* assembly of virion structures or occlusion bodies in the nucleus. This possibility is plausible when one realizes that there is no known protein synthetic machinery in the nucleus of cells, and precursor substances or preassembled structured components for virus reproduction may originate in the cytoplasm of cells at the ER/ribosomal level of organization. The timing of appearance of the ML, slightly preceding occlusion body formation and virion maturation, further suggests such a role for the ML. Many free ribosomes (monosomes) are found in the cytoplasm adjacent to the ML during and following its development (see Figures 12, 14).

The ML is an unusual elaboration of a portion of the cytocavitary system of the infected cells (i.e., components of the cytocavitary system of cells include nuclear envelope, ER, and Golgi, whose lumina are continuous with one another). Its appearance coincides with occurrence of many free ribosomes in the cell sap, a characteristic of undifferentiated or viral transformed cells.[16] Therefore, the infected shrimp cells are transformed for virus production and should possess such characteristics.

3. Composition and Genome

BP is the noninsect baculovirus that has been most extensively biologically, biochemically, and genetically characterized to date. Couch et al.[2,3,4,13,17,18] characterized BP morphologically, cytopathologically, and biologically. Morphological, cytopathological, and some biological features have already been discussed. Summers[19] completed partial characterization of the polyhedrin of BP occlusion bodies, DNA content, and conformation including Kleinschmidt preparation from BP virions, and determined relationships of the primary protein and DNA of BP to those of insect baculoviruses.

Summers[19] had to use a strong denaturing agent sodium dodecyl sulfate (SDS) in order to open the conformation of the BP polyhedrin (protein) to determine its relatedness to insect baculovirus polyhedrin. He found some crossreaction with *Autographa californica* baculovirus polyhedrin and *Trichoplusia ni* granulin antisera, which confirmed presence of similar primary sequences or antigenic sites in BP and insect polyhedrins.

The BP DNA (double standard, circular DNA) possessed size and structural characteristics similar to those of insect baculovirus DNAs. The ratio of ccDNA relative to rcDNA

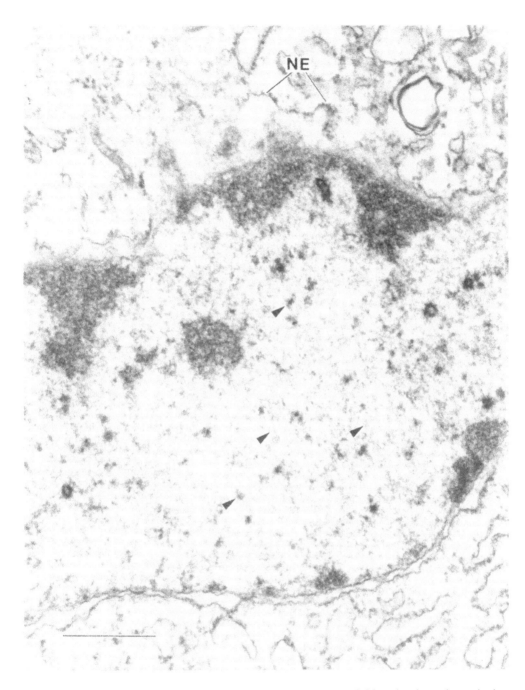

FIGURE 11. Early or eclipsed infection of BP in pink shrimp; note absence of virions, but chromatin margination, and early appearance of granular and fibrillar stroma (arrowheads); note very early alteration in outer nuclear envelope (NE) with ribosomes still attached. (Bar = 1 μm.)

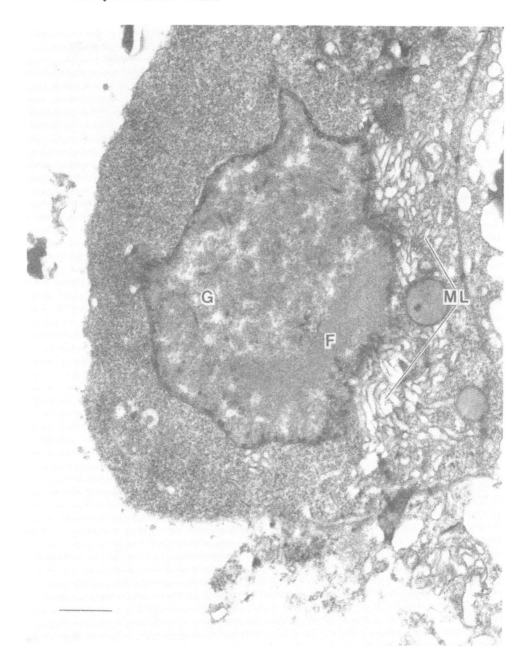

FIGURE 12.   Intermediate infection of BP in pink shrimp; note many free ribosomes in cytoplasm, early proliferation of membranous labyrinth (ML) adjacent to nucleus, and nuclear changes including fibrillar (F) and granular
(G) stroma plus virion development. (Bar = 1 μm.)

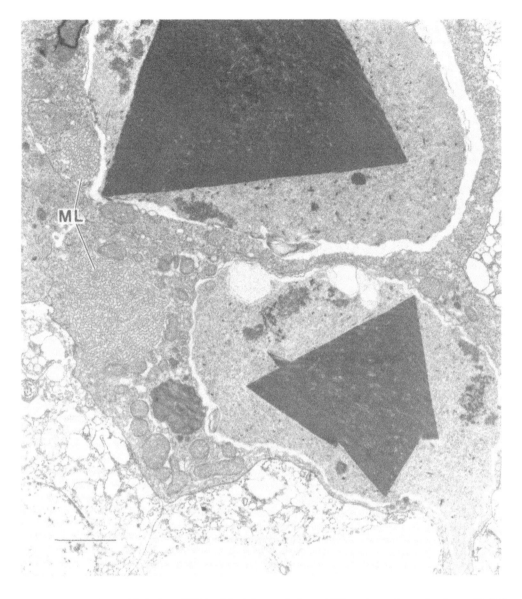

FIGURE 13. Advanced infection of BP in pink shrimp; note formed OBs and granular nucleoplasm with both free and occluded virions; note the well-formed membranous labyrinth (ML) in two cells with OBs and enlarged, atypical mitochondria. (Bar = 2 μm.)

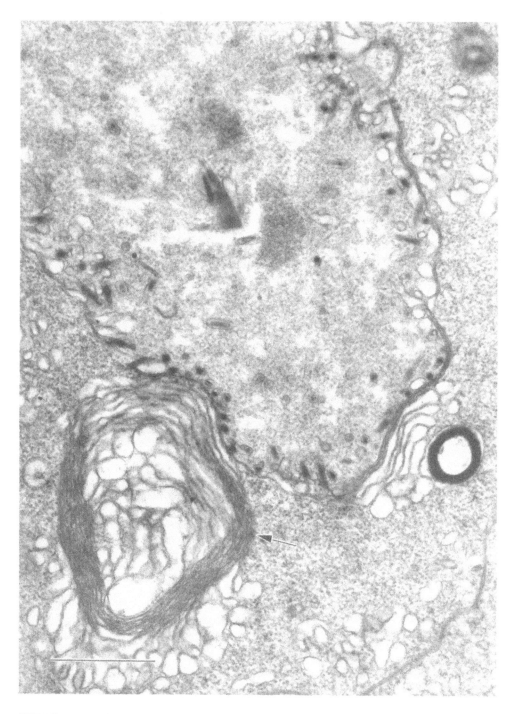

FIGURE 14.   Membranous labyrinth in BP-infected cell in intermediate stage of the infectious cycle; note the unusual, concentric collapse (arrow) of cytocavitary membranes (myelin-like) that have their origin at the outer membrane of the nuclear envelope; note the fine granular matrix in the nucleoplasm; these granules are probably polyhedrin precursors that have entered the nucleus via the ML preparatory to coalescing to form the occlusion body of BP. (Bar = 1 μm.)

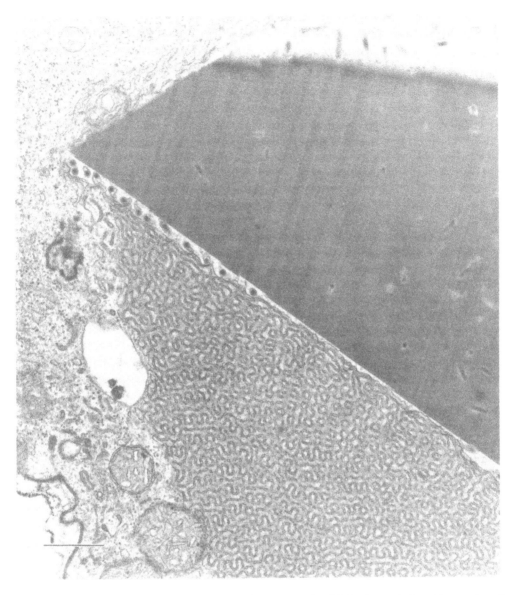

FIGURE 15. Well-formed, highly organized ML adjacent to growing OB that has occluding virions at time of fixation; note fine granular material in cisternae of ML interconnected with remains of nuclear envelope and surrounded by many free ribosomes in the cytoplasm. (Bar = 500 nm.)

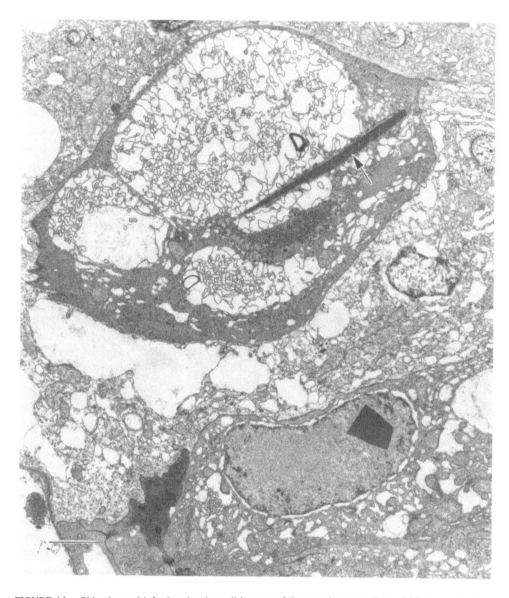

FIGURE 16.   Old, advanced infection showing cell in stage of degeneration; note sliver of OB (arrow) in plane of section, and note particularly the bizarre remnants of the ML; this cell is at point of collapse because of loss of structural integrity. (Bar = 2 μm.)

and dlDNA are very similar for the BP DNA and insect baculovirus DNA. BP DNA has a size of 75 × 10⁶ Da, which places it close to, and in the range of sizes of many insect baculovirus DNAs.[19,20]

B. *PENAEUS MONODON* BACULOVIRUS (MBV)

1. Morphology

MBV is the second described, occluded baculovirus (type A) from noninsect hosts (i.e., penaeid shrimps). Lightner and Redman,[6] first reported MBV from *Penaeus monodon* from Mexico. Since then it has been found in other Pacific shrimps in Taiwan, Tahiti, and the Philippines. Virions of MBV are rod-shaped and membrane-enveloped and are occluded in spherical occlusion bodies (see Figures 17, 18). Nucleocapsids were approximately 42 nm

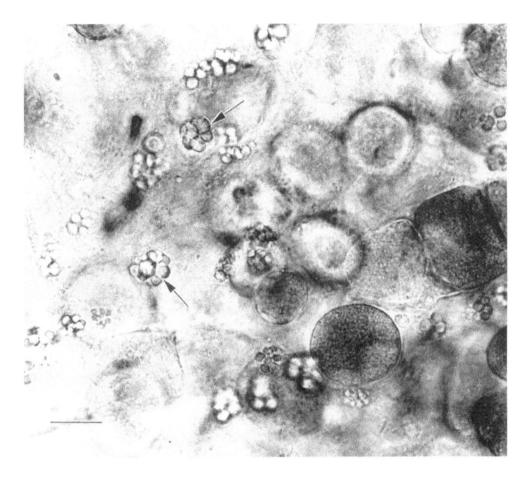

FIGURE 17. Light micrograph of spherical OBs (arrows) of MBV in hepatopancreatic squash of *Penaeus monodon*, the Pacific shrimp; note multiple numbers of OBs in single nuclei (Bar = 20 μm.) (Micrograph provided by Dr. Don Lightner.)

in diameter and 246 nm long. Entire virions including envelope membranes were approximately 75 nm by 324 nm.[8] As with BP and most insect baculoviruses, the MBV virion also has base and caplike structures (see Figure 16). The mature nucleocapsid of MBV is not always evenly electron dense, and it is uncertain from study of available electron micrographs if knoblike structures or projections are found at the base of MBV virion as they are in BP and certain insect baculoviruses.

2. Morphogenesis

Serendipitous findings of empty or partially uncoated virions of MBV at nuclear pores have been reported; therefore, the mode of infection for MBV apparently agrees with that of other baculoviruses, i.e., uncoating via extrusion of virion DNA through nuclear pore into nuclei of shrimp host cells. Early infections resemble those of BP to a large extent: hypertrophy of nucleus, margination of chromatin, formation of granular and fibrillar virogenic stroma in the nucleus, and loss of nucleolar structures. In intermediate to advanced infections (see Figures 19, 20), cytoplasm revealed increased numbers of free ribosomes (monosomes), proliferations of membranous labyrinth membranes (ML) as in BP, and formation of polyhedra at or near the ER. Nuclear changes included increased hypertrophy, appearance of many enveloped virions (see Figures 18, 19, 20), and appearance (see Figure 20) and growth of occlusion bodies (see Figures 18, 20) to occlude many mature virions.[7]

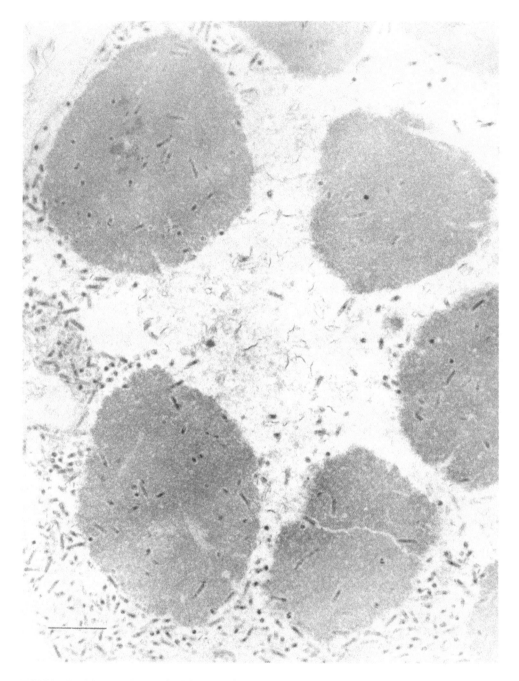

FIGURE 18.   Electron micrograph of OBs and virions of MBV from Pacific shrimp; note lattice fine structure and round form of OBs (Bar = 1 μm.) (Micrograph provided by Dr. Don Lightner.)

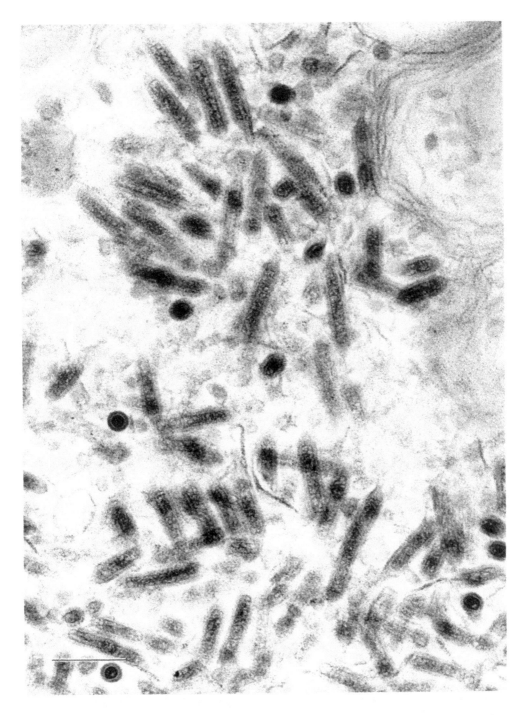

FIGURE 19.   Fine structure of MBV virions in nucleus (Bar = 200 nm.) (Micrograph provided by Dr. Don Lightner.)

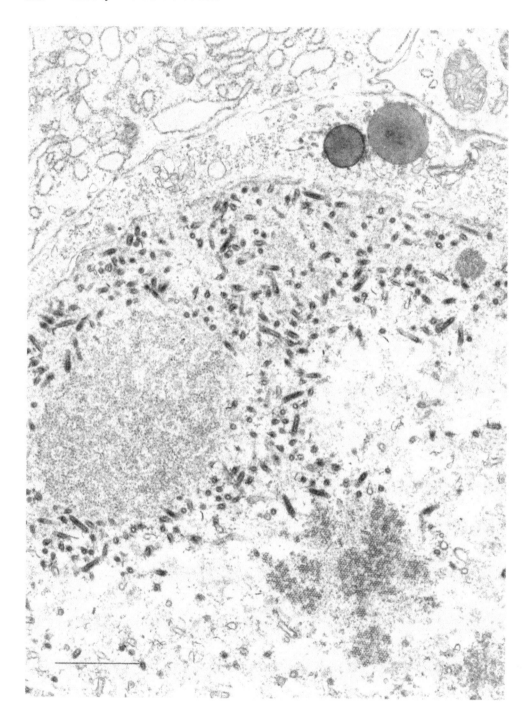

FIGURE 20.   Early formation of OBs in MBV-infected nucleus; note the virogenic stroma and many completed virions (Bar = 1 μm.) (Micrograph provided by Dr. Don Lightner.)

Lightner et al.[7] report a difference in the formation of the ML for MBV infections and BP infections. They point out that proliferation of the ML in MBV infections begins at the Golgi. Upon examination of their electron micrographs, we are convinved that the proliferation of ML of MBV is essentially the same as in BP infections and that the chief points of origin of the ML proliferation is between the outer nuclear envelope membrane and the nearby Golgi, both having continuity as part of the interconnected cytocavitary system or membrane system of the host cell. Lightner's et al.[7] finding of apparent early polyhedrin paracrystalline units in the cytoplasm near the ER of the cytocavitary system further supplies supporting evidence that the proliferated membranes of the ML (cytocavitary system) may function to provide a conduit for transport of the precursors of polyhedrin occlusion bodies to the nucleus for incorporation into the growing occlusion bodies.

3. Composition and Genome

MBV is assumed to contain DNA, but little biochemical or genetic characterization of this virus has been completed, and we must refer to the excellent morphological, cytopathological, and biological characterization of Lightner et al.[7] to arrive at probable affinities to the Baculoviridae.

III. CYTOPATHOLOGY

Many of the cytopathic changes associated with these viruses (BP and MBV) have been illustrated in figures already presented in this section, particularly in the electron micrographs of tissues and cells infected with BP or MBV. It remains here to reemphasize and reiterate some points about these cytopathic effects.

Cytopathic alterations involving the nucleus in both BP and MBV infections include: nuclear hypertrophy, chromatin diminution, loss of nucleolus, replacement of nuclear components by fibrillar and granular virogenic stromata, growth of occlusion bodies, and eventual rupture of nucleus. Cytoplasmic changes include: swollen mitochondria, increase of free ribosomes, loss of ribosomes attached to ER, and most strikingly, proliferation of membranes of the cytocavitary system to form massive membraneous labyrinths that eventually lead to complete loss of cellular integrity and structure, and eventual lysis or fragmentation of infected cells (see Figures 12, 13, 15, 16).

It is probable that the ML plays at least a dual role in the virus infectious cycle. It may play the role of conduit or transport system for viral structural precursors from the cytoplasm to the nucleoplasm, and after this function is accomplished, the ML provides the basis for the major or ultimate cytopathic events that lead to the collapse and/or fragmentation of the infected cell and release of virions and OBs.

Recent methods for predictable experimental transmission of BP have been reported in penaeid shrimp by Overstreet et al.[21] This development permits the further study of these crustacean baculoviruses at a level heretofore not permitted in the laboratory, and should provide a source and quality of viral DNA not previously available.

REFERENCES

1. **Couch, J. A. and Nimmo, D. R.,** Cytopathology, ultra-structure, and virus infection in pink shrimp exposed to the PCB, Aroclor 1254, in *Program Jt. Meet. Soc. Invertebr. Pathol. Int. Colloq. Insect Pathol. Microbial Control,* (Abstr.), Oxford University, England, 1973, 105.
2. **Couch, J. A.,** Free and occluded virus, similar to Baculovirus, in hepatopancreas of pink shrimp, *Nature (London),* 247, 229, 1974.

3. **Couch, J. A.,** An enzootic nuclear polyhedrosis virus of pink shrimp: ultrastructure, prevalence, and enhancement, *J. Invertebr. Pathol.,* 24, 311, 1974.

4. **Couch, J. A.,** Viral diseases of invertebrates other than insects, in *Pathogenesis of Invertebrate Microbial Diseases,* Davidson, E. W., Ed., Allanheld, Osmun Publ., Totowa, NJ, 1981, chap. 5.

5. **Lightner, D. V.,** A review of the diseases of cultured Penaeid shrimps and prawns with emphasis on recent discoveries and developments, in *Proc. Imp. Int. Conf. the Culture of Penaeid Prawns/Shrimps,* Iloilo City, Philippines, 1986, 79 (© 1985 SEAFDE, Aquaculture Department).

6. **Lightner, D. V. and Redman, R. M.,** A baculovirus-caused disease of the penaeid shrimp, *Penaeus monodon, J. Invertebr. Pathol.,* 38, 299, 1981.

7. **Lightner, D. V., Redman, R. M., and Bell, T. A.,** Observations on the geographic distribution, pathogenesis and morphology of the baculovirus from *Penaeus monodon* Fabricius, *Aquaculture,* 32, 209, 1983.

8. **Johnson, P. T. and Lightner, D. V.,** The rod-shaped nuclear viruses of crustaceans: gut-infecting species, *Dis. Aquatic Organisms,* 5, 123, 1988.

9. **Johnson, P. T.,** A baculovirus from the blue crab, *Proc. 1st Int. Coll. Invertebr. Pathol.,* 1976, 24.

10. **Liao, I.,** A culture study on grass prawn, *Penaeus monodon,* in Taiwan — the patterns, the problems and the prospects, *J. Fish. Soc. Taiwan,* 5, 11, 1977.

11. **Matthews, R. E. F.,** Classification and nomenclature of viruses, *Intervirology,* 17, 1, 1982.

12. **Frazer, M. J.,** Ultrastructural observations of virion maturation in *Autographa californica* nuclear polyhedrosis virus infected *Spodoptera frugiperda* cell cultures, *J. Ultrastruct. Mol. Struct. Res.,* 95, 189, 1986.

13. **Couch, J. A., Summers, M. D., and Courtney, L.,** Environmental significance of Baculovirus infections in estuarine and marine shrimp, *Ann. N.Y. Acad. Sci.,* 266, 528, 1975.

14. **Summers, M. D.,** Electron microscopic observations on granulosis virus, entry, uncoating and replication processes during infection of the midgut cells of *Trichoplusia ni, J. Ultrastruct. Res.,* 35, 606, 1971.

15. **Gouranton, J.,** Development of an intranuclear nonoccluded rod-shaped virus in some midgut cells of an adult insect, *Gyrinus natator* L. (Coleoptera), *J. Ultrastruct. Res.,* 39, 281, 1972.

16. **Trump, B. F. and Arstila, A. J.,** Cell injury and cell death, in *Principles of Pathobiology,* La Via, M. F. and Hill, Jr., R. B., Eds., Oxford University Press, New York, 1971, chap. 2.

17. **Couch, J. A. and Courtney, L.,** Interaction of chemical pollutants and virus in a crustacean: a novel bioassay system, *Ann. N.Y. Acad. Sci.,* 298, 497, 1977.

18. **Couch, J. A.,** Attempts to increase Baculovirus prevalence in shrimp by chemical exposure, *Prog. Exp. Tumor Res.,* 20, 304, 1976.

19. **Summers, M. D.,** Characterization of shrimp Baculovirus, Ecological Research Series, EPA publication EPA-600/3-77-130, 1977.

20. **Summers, M. D. and Anderson, D. L.,** Characterization of nuclear polyhedrosis virus DNAs, *J. Virol.,* 12, 1336, 1973.

21. **Overstreet, R. M., Stuck, K. C., Krol, R. A., and Hawkins, W. E.,** Experimental infections with *Baculovirus penaei* in the white shrimp *Penaeus vannamei* (Crustacea:Decapoda) as a bioassay, *J. World Aquac. Soc.,* 19, 175, 1988.

Chapter 7

BACULOVIRIDAE. GRANULOSIS VIRUSES

Yoshinori Tanada and Roberta T. Hess

TABLE OF CONTENTS

I. INTRODUCTION

A. CLASSIFICATION

The granulosis virus infection was first named pseudograsserie by Paillot[1] and later classified as granulosis by Steinhaus[2] because of the presence of minute granules in the hemocoel of infected larva. The generic name *Bergoldia*[2] was applied to the granulosis virus (GV) in honor of Bergold who first demonstrated the presence of rod-shaped virions in the capsules. Weiser[3] separated the GVs into the cytoplasmic *(Bergoldia)* and nuclear *(Steinhausia)* forms, but the committee on insect virus nomenclature retained a single genus, *Bergoldiavirus*.[4] At present, a scientific name is not applied to the GVs which are known as members of subgroup B of family Baculoviridae.[5] The subgroup A includes the nuclear polyhedrosis viruses (NPVs). In this family, the rod-shaped nucleocapsid consists of a DNA core surrounded by a protein capsid.

B. HISTORICAL ACCOUNT

The first granulosis was reported in the European cabbageworm, *Pieris brassicae,* by Paillot[1] who later described three others in *Agrotis segetum* (= *Euxoa segetum*).[6-8] Similar infections were observed in *Natada nararia* in 1931,[9] the variegated cutworm, *Peridroma saucia* (= *P. margaritosa*) in 1947,[10] and in the imported cabbageworm, *Pieris rapae,* possibly as early as 1886 to 1941.[11] In 1948, Bergold[12] described the rod-shaped virus particles in the occlusion bodies (capsules) obtained from infected larvae of the pine shoot roller, *Cacoecia murinana;* and in 1949, Steinhaus[2] observed virions and occlusion bodies in the variegated cutworm. By means of nucleic acid and nucleotide analyses, Wyatt[13,14] showed that the GV contained DNA. Later workers confirmed the DNA nature of the GV and furthermore showed that the DNA was double-stranded, supercoiled, covalently closed, and circular.[15-19]

II. SUSCEPTIBLE INSECT HOSTS AND TISSUES

A. INSECT HOSTS

The GVs infect only insects, mainly species in the order Lepidoptera and possibly a few in the order Hymenoptera.[20] The approximately 150 susceptible species in Lepidoptera are found mostly in the families Noctuidae (nearly one third of the host species) and Tortricidae (about one fifth of the host species); with the remainder in 20 other families. In the order Hymenoptera, Martignoni and Iwai[20] listed two host species with one case in the family Argidae and another in the family Pamphilidae. In general, the GVs have a relatively specific host range.[21,22] Examples of the cross transmission to two or more different insect species are the GVs of the imported cabbageworm, *Pieris rapae;*[23-26,165] codling moth, *Laspeyresia pomonella;*[27] cutworm, *Agrotis segetum;*[28] *Cadra cautella;*[29] armyworm, *Pseudaletia unipuncta;*[30] clover cutworm, *Scotogramma trifolii,*[31] and *Heliothis armigera.*[164]

B. *IN VITRO* REPLICATION

At present, there are only a few cases of successful *in vitro* replication of GVs.[32] This is the major factor for the limited investigation of these viruses in biochemistry, molecular biology, and genetic engineering as compared to the NPVs. Vago and Bergoin[33] were the first to report the replication of *Pieris brassicae* GV in the ovarian cell line of the gypsy moth, *Lymantria dispar;* however, for nearly two decades their work could not be duplicated with other GVs and cell lines except that initial virus structural proteins were observed but not virus replication.[34] Recently, the GVs of the codling moth[35,36] and the cabbage looper, *Trichoplusia ni,*[37] have been propagated in homologous cell lines. The properties of the cell line populations are most important for the *in vitro* replication of baculoviruses[36] and may be the cause for the limited success with GVs.

III. VIRUS STRUCTURE AND COMPOSITION

A. VIRUS PARTICLE

The genomes of GVs, like those of NPVs, are supercoiled, double-stranded DNAs that are assembled in a proteinaceous rod-shaped structure to form the nucleocapsid. The genome size of GVs is large, usually greater than 100 kilobase pairs (kbp). The sizes and shapes of the nucleocapsids are similar to those of the NPVs. Their sizes range from 30 to 60 by 260 to 360 nm.

The nucleocapsid is a cylindrical core made of DNA and protein, and capped at both ends[38,39] (see Figure 1A). The capsid is composed of a stacked series of rings of subunits with the rings aligned perpendicular to the longitudinal axis. The capsid symmetry, therefore, is not helical. Burley et al.[39] report that each ring in the capsid of *Spodoptera frugiperda* GV has approximately 12 subunits. The capsid is made up of two main polypeptides: (1) a major structural protein of high molecular weight and (2) an extremely basic arginine-rich protein of low molecular weight.[39-41] The basic protein may be involved in the condensation and packaging of the large supercoiled DNA genome within the capsid. The gene coding for the basic arginine-rich protein in the NPV has been isolated,[42] and a similar or identical gene may occur in the GV. This gene is abundantly transcribed late in infection[42] and is closely associated with the viral DNA.[40]

The basic arginine-rich protein serves as the acceptor for the kinase activity located in the GV capsid.[43,44] The kinase catalyzes the transfer of phosphate to both serine and arginine residues of acceptor proteins. The activation of the protein kinase results in the release of the DNA from the nucleocapsid.[44] Wilson and Consigli[44] speculate that the protein kinase is activated when the nucleocapsid aligns with the nucleopore, possibly from the high concentration of Mn^{2+} within the nuclear membrane of the insect cell. The kinase activation results in the phosphorylation of the basic protein and causes the release of DNA into the nucleoplasm. This may be the case for the uncoating of the nucleocapsid at the nuclear pore, but when the nucleocapsid enters the nucleus, the mechanism for uncoating may not be the same.

The nucleocapsids are naked when formed; also during certain periods, e.g., prior to emerging from the host cell, upon entering or infecting the cell; and during the passage into the nucleus. The ends of the GV nucleocapsid differ, as in the case of those of NPVs, with one end being conical or hemispherical and the other end being blunt[45] (see Figure 1A). The nucleocapsid has polarity in that only the conical end contacts the plasma membrane in the process of budding out of the cell.

The GV nucleocapsids acquire envelopes in a manner similar to those of NPVs. An amorphous but definite layer exists between the nucleocapsid and the envelope[38] (see Figure 1B). The enveloped nucleocapsids measure 40 to 100 by 250 to 360 nm. Robertson et al.[46] report three ways of envelopment: (1) by morphogenesis or possibly *de novo* synthesis in the nucleus, though usually after the disintegration of nuclear membrane; (2) by morphogenesis or *de novo* synthesis in the cytoplasm; or (3) by budding through the cell plasma membrane (synhymenosis) (see Figure 2A). Aside from synhymenosis, the origin of the envelope is still in doubt. Envelopes obtained by synhymenosis possess peplomers only on the portion of the envelope around the conical end that has initially contacted the cell plasma membrane during emergence from the cell (see Figure 2A). Each envelope generally contains one nucleocapsid and rarely two or more[47,48] (see Figure 2B).

Subviral, low molecular weight components of GVs have been reported to be infectious as in the case of NPVs. The subviral particles of GVs have been isolated by centrifugation

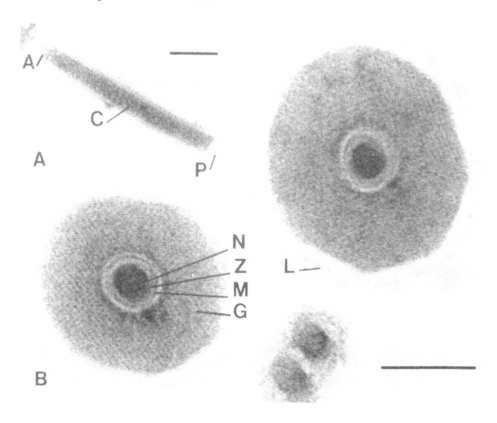

FIGURE 1.   Electron micrographs of nucleocapsid and capsules of codling moth GV. (A) Longitudinal section of a nucleocapsid shows structural differentiations, most frequently seen in negatively stained preparations, at the ends of the capsid; the end, which appears to preferentially orient against the plasma membrane during the budding process to become associated with peplomers, is conical with a dense cap and a clear zone behind it (A); the opposite end, which has been likened to a claw, is more blunt and striated (P); the nucleocapsid core (C) is seen as an elongated structure which contains the nucleic acid as well as protein. (B) Cross sections of the capsules show an electron dense nucleocapsid (N) consisting of the core complex and capsid, a surrounding amorphous zone (Z), a unit membrane (M), capsule matrix of granulin (G), a clear zone, and an outer epicapsular layer of membrane (L); an envelope containing two nucleocapsids is seen in the lower center of the figure. (Both bars = 100 nm.)

techniques and inoculated into host larvae to demonstrate their infectivity.[49,50] Whether centrifugation will eliminate all nucleocapsids is questionable. Pinnock and Hess[51] have observed in electron micrographs, subviral particles (30 to 80 nm) arising in the nucleus and appearing to be involved in virogenesis. They speculate that the subviral particles are a product of blocked GV infection or may represent a new virus infective only to the midgut. This subject of subviral particles is yet to be resolved.[52]

B. VIRUS GENOME

The DNAs of a number of GVs have been characterized by restriction enzyme digestions, e.g., the GVs of *Trichoplusia ni* and *Spodoptera frugiperda*,[53,54] *Plodia interpunctella*,[55] *Pieris* (= *Artogeia*) *rapae*,[25,26,55,56,166,167] *Pieris brassicae*,[25,26,57] *Heliothis armigera*,[54,58] *Phthorimaea operculella*,[58] *Laspeyresia pomonella*,[58-60] *Choristoneura* spp.,[61] *Lacanobia oleracea*,[62] *Pseudaletia unipuncta, Scotogramma trifolii*,[63] and *Chilo* spp.[168] The DNAs of each GV have unique restriction enzyme patterns. In some GV isolates, submolar fragments have been detected indicating the presence of genetic heterogeneity.[25,53,55,56,60]

Physical maps have been prepared for the DNAs of *L. pomonella*,[60] *P. rapae*,[56,166,167] and *P. brassicae*.[57] The gene of granulin, the occlusion body matrix protein, has been cloned

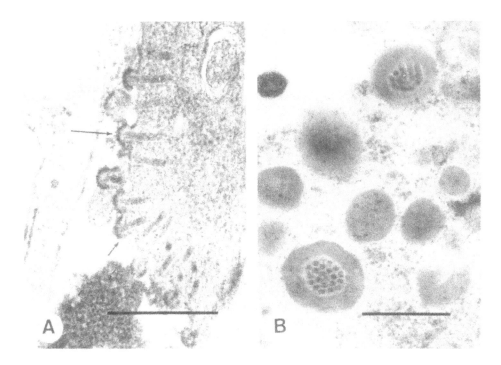

FIGURE 2. (A) Electron micrograph of a codling moth larval midgut cell showing the budding of nucleocapsids from the thickened basal plasma membrane; during synhymenosis, the virions acquire peplomers (arrows). (From Hess, R. T. and Falcon, L. A., *J. Invertebr. Pathol.*, 50, 85, 1987. With permission.) (B) Electron micrograph of a larval midgut cell showing the packaging of one or more nucleocapsids in an envelope of a capsule; note one capsule contains more than 20 nucleocapsids. (From Falcon, L. A. and Hess, R. T., *J. Invertebr. Pathol.*, 45, 356, 1985. With permission.) (Both bars = 0.5 μm.)

and sequenced.[56,57,64,166] The nucleotide and amino acid sequences of the granulin gene are over 50% homologous with those of the polyhedrin (polyhedra matrix protein) gene sequences.[56,57,64] Dwyer and Granados,[56] after detecting areas of noncolinearity and nonhomology, speculate that some genome changes have occurred since the evolutionary divergence of these two subgroups of baculoviruses.

Transcriptional studies of the 106-kbp circular DNA of *P. rapae* GV (PrGV) have been conducted with infected larvae as sources of viral mRNAs.[169] Over 100 viral transcripts have been isolated during the early and late periods of larval infection and mapped on the PrGV genome. A translational map of 36 viral polypeptides of this virus was established by *in vitro* translation of hybrid-selected transcripts.

Early study had indicated that glycoproteins were absent in the GVs,[65] but subsequently such proteins have been detected both in the nucleocapsid and in the envelope.[66] The glycoproteins are essential for infection (fusion) in the NPVs[67] and may play a similar role with the GVs.[66]

C. CAPSULE

Occlusion bodies of GVs usually are called capsules and less frequently granules (see Figure 1B). They are generally ovocylindrical and measure 120 to 350 by 300 to 500 nm, but some are ovoidal or ellipsoidal. However, their shapes and sizes may vary widely.[21] In some cases, the shapes vary with insect host, especially with an unusual host, and bizarre forms may occur with many enveloped nucleocapsids.[29] There are cuboidal capsules resembling the NPV polyhedra but with one or rarely more enveloped nucleocapsids.[47,62,68-71] A GV strain with cuboidal capsule has been isolated from the codling moth.[70] Although a capsule generally has a single nucleocapsid, up to 22 virions per capsule may be occluded in the codling moth GV depending on the density and distribution of the virions[72] (see Figure 2B). The matrix of the GV capsule consists of a paracrystalline protein lattice (see Figure 1B) as in the case of the NPV polyhedra.[9,73,74] The matrix protein is called granulin with molecular weights ranging from 276,000 to 378,000 Da depending on virus species.[75] Granulin, a phosphoglycoprotein,[41] is composed of subunits of approximately 26 to 28 kDa and is similar to the polyhedrin of the NPV polyhedra. The gene coding for granulin has been isolated (see Section III.A). Current evidence indicates that the capsule matrix protein is composed of a single protein species, the molecules of which are shaped like a six-armed nodal unit and are oriented in a face-centered cubic lattice.[76]

Granulin generally occludes (encapsulates) the enveloped nucleocapsid at one end and progresses to the other end;[71,77-79] infrequently occlusion starts from the middle or from both ends.[80] Yamamoto et al.[81] incubated a suspension of granulin with enveloped nucleocapsids and observed granulin molecules attached to the envelopes, apparently at receptor sites (see Figure 3A). Earlier, Arnott and Smith[80] speculated that nucleation sites occurred on the GV envelope.

The controversy on the existence of specific structures (envelopes or membranes) surrounding the GV capsule and the NPV polyhedra has been resolved.[76] In the GV capsule, the convincing evidence is the presence of an envelope layer of repeating structures that is different from the capsule protein lattice.[82] The surface layer, viewed with the electron microscope, reveals a structure with hexagonal arrangement of holes 6 nm in diameter, typical of membranous surface. Similar membrane-like profiles (see Figure 1B) or epicapsular layers have been observed in other GVs.[45,47,48] After considering the ultrastructural and biochemical data, Federici[76] concludes that the mature capsules and polyhedra are delineated by an envelope that is composed primarily of carbohydrates.

The capsules may occlude, in addition to the nucleocapsids, extraneous materials, e.g., membranes and vacuoles from infected cells.[48,83] A protease, considered to be endogenous by some workers, has been detected in capsules as well as in NPV polyhedra, and it assists in the dissolution of occlusion bodies.[84-86] The protease, however, appears to be a contaminant from the midgut juices since the capsules are collected often by triturating whole infected larvae. When capsules are collected from larvae with their digestive tracts removed, they lack the protease.[87] Moreover, the protease is readily adsorbed by protease-free capsules, and it can be extracted from the capsules with sodium chloride.[88]

Several GVs interact synergistically with NPVs and enhance their infectivity.[31,93,94,170] The enhancement is associated with lipoproteins (mol wt 93,000 to 126,000)[89,90] that occur in capsule matrices of GVs of the armyworm[89,91,92] and of the clover cutworm, *Scotogramma trifolii*.[31] Certain lipoproteins act as a synergistic or enhancing factor (SF) in the infection of baculoviruses in lepidopterous larvae.[31,93-97] Since the SF occurs in only one of two strains of the GV infecting the armyworm,[98] it appears to be virus coded. Enhancement by the SF also occurs *in vitro*.[99,100] The SF agglutinates susceptible insect cells indicating that the SF acts as an attachment molecule for enveloped virions to receptor sites on the cell plasma membrane.[101,163] The SF acts *in vivo* on a number of different baculoviruses, but does not enhance some NPVs.[102] On the other hand, Derkson and Granados[171] report that there are enhancing factors in the matrices of the GV capsule and the NPV polyhedron. These factors enhance virus infections by degrading the peritrophic membrane and exposing the larval midgut epithelium to the virions.

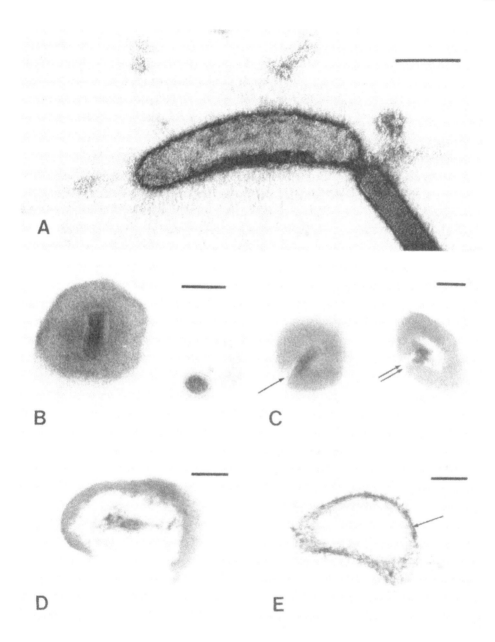

FIGURE 3. Electron micrographs of enveloped virions and capsules of the armyworm GV. (A) Reattachment of granulin in orderly arrays on the viral envelope indicates the presence of receptor or nucleation sites on the envelope surface. (From Yamamoto, Y., Hess, R. T., and Tanada, Y., *J. Ultrastruct. Res.*, 75, 127, 1981. With permission.) (B) to (E) Process of GV capsule dissolution in the larval midgut of *Pseudaletia unipuncta:* (B) capsules and enveloped virions can be obtained from the midgut lumen immediately after virus ingestion and by 15 min, all stages of capsule dissolution are found; (C) the initial stage of dissolution appears to be a "cracking" of the capsule surface (arrow) followed by a loosening of the capsule matrix around the virion (double arrow); (D) the capsule dissolution results in a virion which may separate from the capsule remains; (E) the capsule finally may be represented by a ghost in which the outer, electron-dense layer possibly represents a capsule envelope (arrow). (Bars = 100 nm.)

IV. GROSS PATHOLOGY

A. TYPES OF SYNDROMES

The pathology of GVs varies with a number of factors, such as the type, virulence, and dosage of the GV, host larval age, and, in particular, the type of susceptible host tissues. In general, the young larval stages are more susceptible to GVs than the older stages. In the codling moth, the larvae show an increase in tolerance to the GV when they are induced for pupation and for diapause.[103]

The granuloses were separated, at one time, into two types, the monorganotropic diseases affecting tissues only of mesoderm origin (fat body) and polyorganotropic diseases which also involve tissues of ectodermal origin (epidermis, tracheal matrix).[21,104] Subsequent studies have shown that, although the first type involves mainly the fat body, it also affects the midgut and therefore would not be monorganotropic. There appears to be a monorganotropic GV that infects the midgut of the grape leaf skeletonizer, *Harrisinia brillians*[105] (see Section IV.B).

B. PATHOLOGY IN DIVERSE SYNDROMES

The pathogenesis of GVs falls into three types of syndromes: type 1 in which the hypodermis (integument) is uninfected and more than one organ is infected, type 2 in which the hypodermis and other organs are infected, and type 3 in which the principal organ infected is the midgut.[106] Most GVs, including the type species infecting *Trichoplusia ni*, produce type 1 syndrome with the fat body being the major target. In this syndrome, the infected larva grows more slowly and may become larger than an uninfected larvae.[107-110] In some cases, the life of the infected larva is prolonged beyond the normal pupal stage and even up to adult emergence.[107,110,111] These pathophysiological effects may be associated with a malfunction of the hormone system.[109] The period of lethal infection, however, depends largely on the virus dosage, virulence of virus strains, and host larval age. For example, infection in the first instar armyworm larvae may cause death in a few days, whereas larvae infected in third or older instars may die in the last instar.[107] The integuments of moribund and dead larvae are firm and leathery.

Larvae with the type 1 syndrome show the first indication of GV infection with loss of appetite and progressive color changes from the normal color to a pale whitish or milky yellow appearance, particularly on the ventral side.[21] At an advanced stage of infection, they become severely mottled due to the hypertrophied fat bodies especially in larvae with lightly pigmented integuments. With the initiation of color change, the larvae become progressively weaker, sluggish, and flaccid, and their appetites are greatly reduced. A few days prior to death, the larval color may turn brownish and rapidly darken at death, very likely due to the invasion and multiplication of bacteria. The hemolymph of larvae at an advanced stage of infection is milky white and turbid, owing to the presence of large numbers of virus capsules from disintegrated infected tissues. The infected fat body cells under the light microscope are opaque and have a yellowish to light brown coloration.[112] Spherical vesicles containing capsules often are expelled from heavily, GV-infected cells. The capsules in the vesicles are agitated by Brownian movement and are liberated when the vesicles eventually burst. These vesicles, characteristic for granuloses, are called "boules hyalines".[8,12]

The type 2 syndrome closely resembles the syndrome of nuclear polyhedroses in lepidopterous larvae, and multiple organs, in particular the integument, are infected. The lethal period of infection in the type 2 syndrome also depends on the type of viruses, virus dosage, and host age, but it is usually much shorter (4 to 7 d) than that of the type 1 syndrome. Infected larvae turn chlorotic much more rapidly than in the type 1 syndrome, lose their appetites, turn flaccid, and die in a wilted condition with the integument becoming very fragile due to the disintegration of the infected hypodermis. The codling moth typically exhibits type 2 syndrome, but when fed low dosages of GV, has a higher and longer rate of weight increase than uninfected larvae.[113]

The third type of syndrome is represented, at present, by only the granulosis of the grape leaf skeletonizer, *Harrisinia brillians,* and the infection is in the midgut epithelium.[105] Whether the GV infects other larval organs is still unknown. Infected larvae initially exhibit loss of appetite and abnormal feeding behavior resulting in a "peppered" appearance rather than a uniform skeletonizing of the grape leaves. The infected larval midgut turns opaque and flaccid. Infected midgut cells display the characteristic brown coloration and ruptured cells expel spherical vesicles. The larvae become flaccid and shrunken in size and change in color from lemon yellow through shades of grayish yellow to brown or black.[105,114] They are inclined to wander and usually become diarrheic with brownish discharge. Sometimes moribund larvae are fastened to the leaves by the dried discharge, but most often fall off and dry into brittle black cadavers. The integument remains firm and intact at larval death.

C. PHOTOPOSITIVE BEHAVIOR

Some lepidopterous larvae infected with NPV exhibit photopositive behavior or "Wipfelnkranheit" (tree top disease).[115] Similar behavior is exhibited in granuloses, e.g., by GV-infected larvae of the armyworm on grass[93] and of the clover cutworm.[172]

V. VIRUS REPLICATION

When Paillot[1] described pseudograsserie in *P. brassicae,* he stated that the virus multiplied in the cytoplasm of fat and hypodermal cells. In 1936,[116] he reported that the virus of one of the pseudograsserie of *A. segetum* appeared first in the cytoplasm, while in another the virus multiplied especially in the nucleus and also in the cytoplasm. In a third pseudograsserie of this insect, he cited the nucleus as the site of virus replication. Lipa and Ziemnicka[28] reported that the three granuloses might be caused by the same virus. Later workers with the light microscope reached similar conclusions that the nucleus or the cytoplasm was the site of GV multiplication depending on the GV. Subsequent electron microscopic investigations indicated that the GV virogenesis was initiated in the nucleus and, with the fragmentation of the nuclear envelope, extended into the cytoplasm.[78,79,117] However, more recent electron microscopic studies revealed that some GV replicated in the cytoplasm,[48,110] and the following is a treatment of nuclear and cytoplasmic GV replications.

A. REPLICATION IN NUCLEUS (NUCLEAR GV)

GVs most commonly infect the host larvae through the midgut. The infected GV capsules are dissolved by the alkaline digestive juice to liberate the occluded enveloped virions[162] (see Figures 3B—E). The process of invasion into the midgut takes place by fusion with the microvillus (see Figure 4A—D) as in the case of NPVs.[52,118,156] The movement and replication of the virions in the midgut cell apparently follow the functional aspects of the cell.[45] The GV nucleocapsids upon entering the cytoplasm may pass directly into the hemocoel, as in the case of the NPVs, without replicating in the midgut cell.[52,119,120] In general, the nucleocapsids within the midgut cell get to the nuclear pores, possibly in association with microtubules as proposed for NPVs by Granados.[121] Uncoating of the nucleocapsids occurs either at the nuclear pores[122] (see Figures 5 A, B) or after the nucleocapsids enter the nucleus[45,123] (see Figures 6A, B). The two sites of uncoating would negate the suggestion by Granados[52] that the site of virus uncoating differentiated GVs from NPVs. Virogenesis in the nucleus produces the cytopathology described in Section VI.A. A virogenic stroma is formed and tubular elements (40 nm), presumably capsids, appear associated with the stroma. Subsequently, the tubular elements are transformed into nucleocapsids. As virus

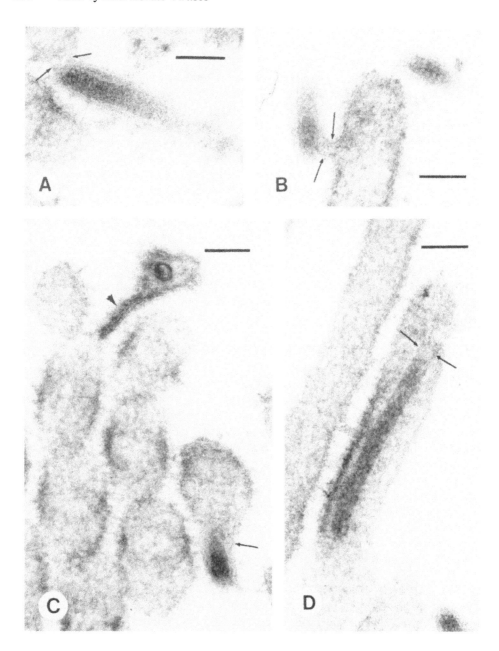

FIGURE 4.   Electron micrographs of larval midgut cells of *P. unipuncta* showing the uptake of GV. (A) Virions are frequently found in close proximity to the microvilli where fine filaments form bridges between the two (arrows). (B) and (C) Invasion of the microvillus is achieved by the attachment and fusion of the virion envelope with the microvillar membrane ([B], arrows); the fused membranes often appear as electron-dense plaques ([C], arrowhead). (D) A nucleocapsid within the microvillus has fine filaments (arrows) associated with it; whether the filaments are comparable to those commonly found in microvilli or to others that are associated with virus entry needs to be investigated. (Bars = 100 nm.)

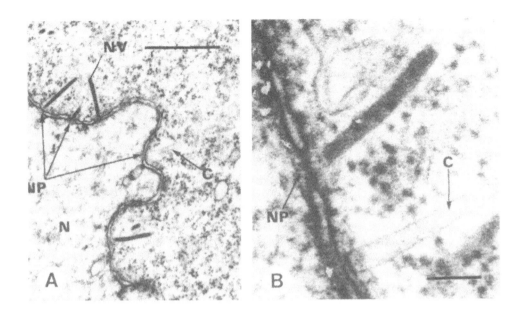

FIGURE 5. (A) and (B) Electron micrographs of GV nucleocapsids (NV) on the exterior of the cell nucleus (N) of *Trichoplusia ni* illustrating their alignment at the nucleopores (NP); various stages of nucleic acid release into the nucleoplasm are suggested by the presence of full (electron dense), partially full, and empty capsids (C); nucleic acid strands are observed in the capsids and nucleus further suggesting that uncoating is occurring. (Bars in [A] = 1.0 μm, in [B] = 100 nm.) (From Summers, M. D., *J. Ultrastruct. Res.*, 35, 606, 1971. With permission.)

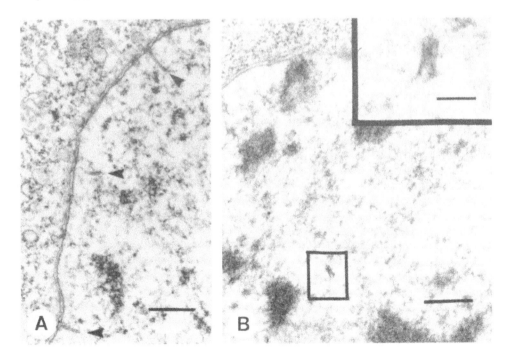

FIGURE 6. Electron micrographs of portions of cell nuclei shortly after GV ingestion showing no overt cytological signs of infection. (A) Nucleocapsids (arrowheads) are seen within the cell nucleus of *Spodoptera frugiperda* at 1 d posttreatment; (B) a nucleocapsid of codling moth GV (boxes) appears to be uncoating in a fat body cell nucleus at 24 h p.i. (Bars = 0.5 μm; bar in insert B = 100 nm.) [A] is from Walker, S., Kawanishi, C. Y., and Hamm, J. J., *J. Ultrastruct. Res.*, 80, 163, 1982. With permission.)

replication progresses, the nuclear envelope fragments and the nucleoplasm together with the nucleocapsids pass into the cytoplasm.

The first-formed nucleocapsids pass to the cell periphery, and become aligned perpendicular to the plasma membrane with their conical ends in contact with the membrane. The area of the plasma membrane at the point of contact with the nucleocapsid is thickened, apparently by the insertion of viral material as in the case of NPV.[124] Fibers in certain instances connect the conical end of the nucleocapsid with the thickened membrane. The nucleocapsid, as it buds through the plasma membrane, becomes surrounded by the altered plasma membrane which forms the envelope. The portion of the envelope surrounding the initial contact point (conical end) of the nucleocapsid has peplomers that function in the attachment and invasion of susceptible cells by viropexis. Whether all budded viruses acquire peplomers is still unknown. The budded virions pass into the basal space between the plasma membrane and the basal lamina and into the basolateral intercellular space.[45,125] The virions eventually pass through the basal lamina into the hemocoel to infect other susceptible tissues. When the cell lyses, various stages of the virus may occur in the hemocoel. The pathway of infection of hemocoelic tissues has not been observed and is presumed to occur by viropexis and fusion.

In the codling moth larva, virus replication continues in the virogenic area of the midgut nucleus and the nucleocapsids are budded out of the cell until about 56 hours post infection (hpi) when some of them are occluded to form capsules.[45] In fat body cells, the encapsulation occurs within 32 hpi, much earlier than in midgut cells, and takes place in the nucleus and cytoplasm. The capsules are noninfectious until ingested by another host larva.[9,111]

B. REPLICATION IN CYTOPLASM (CYTOPLASMIC GV)

The process of the initial infection of the GV which replicates in the cytoplasm has not been described, and it is presumed to be the same as that of the GV replicating in the nucleus. In the fruit tree leaf roller, *Archips argyrospila,* Pinnock and Hess[48] have observed the GV in all stages of replication in the cytoplasm of some hypodermal and fat body cells with intact nuclei showing no signs of virogenesis. The replication processes in the cytoplasm appear morphologically similar to those in the nucleus. They have also reported presumably phagocytic cells with GV capsules enclosed in large numbers within membrane-bound vacuoles.

In the fruit tortrix, *Adoxophyes orana,* Schmid et al.[110] report two types of GV infection in fat body cells. One GV develops typically in the nucleus and may infect 60% of the cells while the other is found in the cytoplasm infecting 40% of the cells. The nuclear type causes larval death much faster than the cytoplasmic type in which the larva dies usually in the last instar. The cytoplasmic GV is found in cytoplasmic enclaves (vesicles) surrounded by a membrane. The nuclei of such infected cells are intact.

The first indication of cytoplasmic GV infection in cells of *A. orana* is a modification within the vesicles which begin to swell at about 10 days postinfection (dpi). Two days later, the first naked and encapsulated nucleocapsids appear in the vesicles. On 19 dpi, all developmental stages up to clusters of mature capsules occur in the membrane enclosed vesicles. At a late stage, large masses of capsules occur in the completely disordered cytoplasm of the cell in which the nucleus is still distinct.

Drolet and Benz[126] have isolated the cytoplasmic GV of *A. orana* with a micromanipulator from single cells infected only with this GV. They have concluded that the cytoplasmic and nuclear GVs are two different types of viruses. If this is substantiated by biochemical analysis, the proposal by Weiser[3] to separate nuclear and cytoplasmic GVs needs to be reconsidered.

VI. CYTOPATHOLOGY

A. MAJOR TARGET ORGANS

The following description is concerned largely with the temporal development of cytopathology in cells of the major target organ, the fat body. Walker et al.[123] report that the first overt sign of granulosis in the cells of *Spodoptera frugiperda* is the formation of intranuclear membranous blebs followed by the appearance of annulate lamellae at the interior of the nuclear envelope (see Figure 7A). Such blebs have not been observed in the nuclei of cells of other GV-infected larvae, in which the first sign of infection is the clearing of the nucleoplasm (see Figure 7B) concomitant with the margination of the chromatin and nucleolar complex.[45,156] The nucleus becomes hypertrophied, and the virogenic stroma appears as a network of fine filaments distributed as strands throughout the nucleoplasm (see Figure 8A, B). The granular components of the nucleolus are very prominent in the cleared nucleus, suggesting an increase in the overall size of the components (see Figure 8B). The nuclear pores increase in numbers.[45,123] In infected cells of some granuloses, there are intranuclear annulate lamellae and crystalline deposits (see Figure 9A) on the nuclear membrane,[48,123] but in others such structures and deposits are absent.[45]

Clear tubules (40 nm) appear near the virogenic stroma located in the central area of the nucleoplasm.[45] These tubules are presumably empty capsids since their interiors increase in electron density to form nucleocapsids which are found throughout the central region of the nucleus (see Figure 9B). At this point, the nuclear envelope is composed predominantly of interconnected nuclear pores, and some areas are almost devoid of ribosomes. The nuclear envelope in certain areas is closely associated with groups of cytoplasmic microtubules. Subsequently, the nuclear envelope fragments into multiple pieces[11,51,78,79,112,117,122,123,125,127-130] that remain in the nuclear area (see Figure 9A) and appear similar to sections of annulate lamellae or nuclear pore complexes[45] (see Figure 10A, B). Wäger and Benz[131] call the original area of the nucleus, the nuclear field. The rapid and drastic disassembly of the nuclear envelope is a major difference in the cytopathology between granuloses and nuclear polyhedrosis; in the latter the nuclear envelope remains intact until polyhedra formation and up to cell lysis.

The nucleolus disperses into fibrillar and granular components[45] (see Figure 10A, B). At a later stage of infection, membrane-like complexes, which may be the pseudonuclei reported by Wäger and Benz,[131] are associated with the nucleolus remnants (see Figure 10B). Since the RNA transcripts are believed to be located in the nucleolus, the disassociation of this structure would affect RNA synthesis (see Section VII). The nature of the membrane-like complexes, however, has yet to be resolved. They have been regarded as structures similar to the paired cisternae of rough endoplasmic reticulum (RER),[45,109] as coiled membranous structures surrounded by electron-dense material that may represent nuclear fragments,[128] as remnants of nuclear membrane which appear as myelin-like whorls,[130] as stacks of annulate lamellae,[48] and as arrays of tubular membranes that may originate from paired cisternae or from the excessive production of cellular membranes.[123] Benz[109] has suggested that the complexes formed from RER fragments become filamentous loops within which viral nucleoprotein and granulin are produced.

In the codling moth larvae, the whorls of RER are very obvious in midgut cells which had produced an excess of budded virions over a long period that persisted throughout infection, indicating the importance of RER in envelope production.[45] An abundance of RER also occurs in midgut cells of *Trichoplusia ni*.[125]

Prior to the occlusion of the enveloped nucleocapsids, tubular elements (20 to 30 nm) are formed close to the enveloped nucleocapsids or scattered in areas throughout the cytoplasm (see Figure 11A).[45] These tubular elements differ from microtubules in that they are branched, vary in thickness, and are not hollow (see Figure 11B). They may be associated

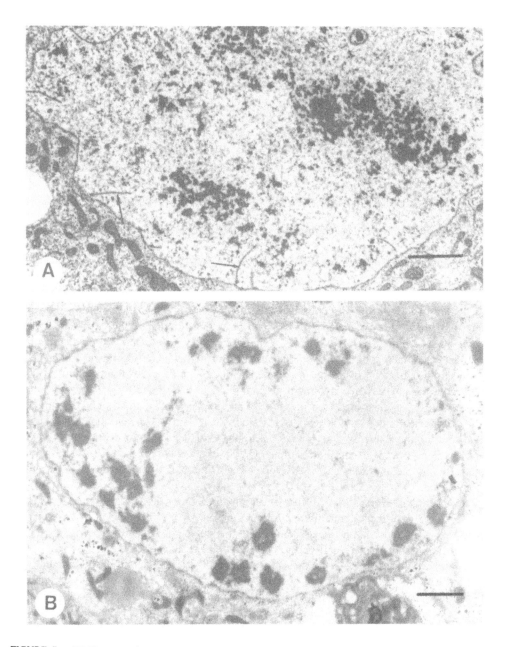

FIGURE 7. (A) Electron micrograph of a portion of a *S. frugiperda* cell, following treatment with GV, showing first cellular alteration as small intranuclear blebs and subsequently, intranuclear annulate lamellae (arrows). (B) Nucleus of a GV-infected midgut cell of *P. unipuncta* showing the clearing of the nucleus with margination of the nuclear components; this cytopathology is the first reported sign of cellular infection in most granuloses. (Bars = 1.0 μm.) ([A] is from Walker, S., Kawanishi, C. Y., and Hamm, J. J., *J. Ultrastruct. Res.*, 80, 163, 1982. With permission.)

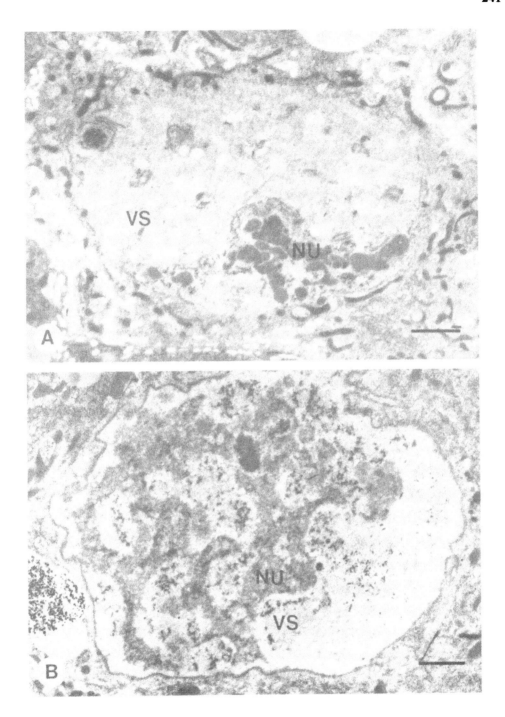

FIGURE 8. (A) Electron micrograph of a nucleus of a GV-infected *P. unipuncta* cell after the nucleus had cleared, showing a network of fine filaments which appear as dense strands to form the virogenic stroma (VS) in which virus assembly occurs; granular and fibrillar elements of the nucleolus (NU) are very apparent and appear to be larger than those in the nucleolus of an uninfected nucleus. (B) Nucleus showing a later stage of infection than that in (A); as infection progresses, the nucleolus fragments into many pieces or forms an extensive network (NU) throughout the nucleus and is intermixed with the virogenic area (VS); the nuclear membrane appears to contain increased numbers of nucleopores. (Bars = 1.0 μm.)

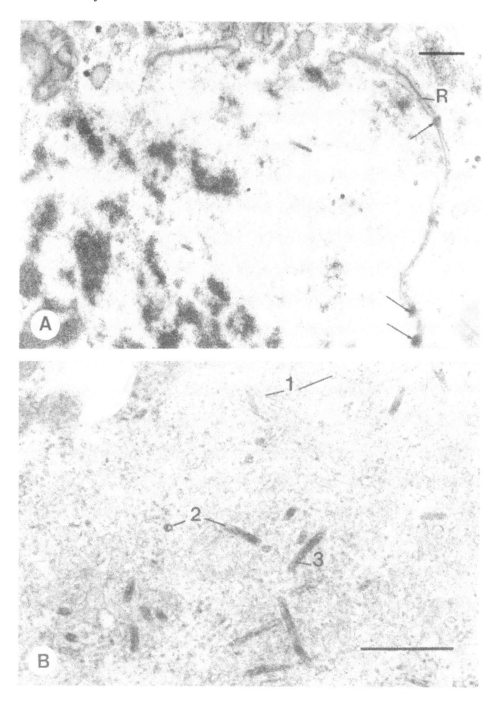

FIGURE 9. (A) Electron micrograph of a portion of a GV-infected cell of *Archips argyrospila* showing electron-dense crystalline deposits (arrows) on the fragmented nuclear envelope; such deposits occur in granuloses when the nuclear membrane is disrupted; in addition, an area (R) of the nuclear envelope appears to be structurally similar to the cisternae of the rough endoplasmic reticulum shown in Figure 12B; virus assembly within the virogenic area occurs before or during nuclear envelope disassembly and continues thereafter within the general nuclear area. (B) Portion of a GV-infected codling moth cell showing various stages of viral encapsidation; encapsidation is indicated by the appearance of empty capsid profiles (1), capsids with partial electron-dense deposits (2), and almost completely dense, mature nucleocapsids (3). (Bars = 0.5 μm.)

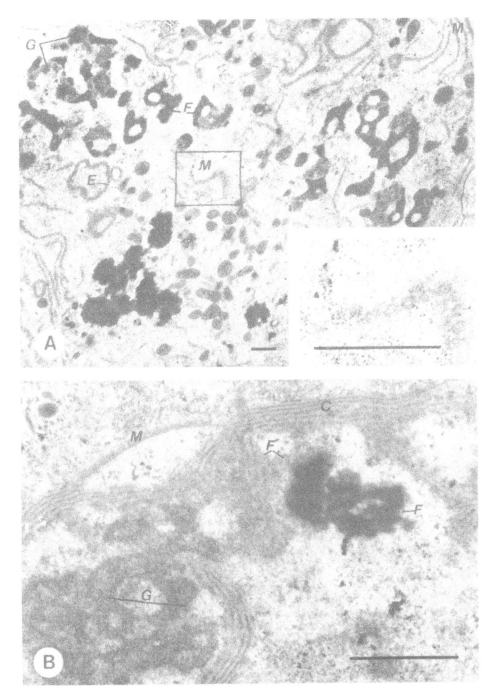

FIGURE 10.   Electron micrographs of portions of GV-infected codling moth cells. (A) Fibrillar (F) and granular (G) components of the nucleolus are found during the virus replicative phase; these components, as well as the nuclear membrane fragments, are dispersed among cellular organelles; boxed area (M) shows a fragment of the nuclear envelope on which high numbers of nucleopores can be seen; rough endoplasmic reticulum forms cisternae with membranes in the center (E) (also shown in Figure 12C). (B) Membrane-like complexes (C), that are formed possibly from the fibrillar elements of the nucleolus (F), are found associated with these elements; electron micrographs suggest that the fibrillar elements become less electron dense, form into linear sheets, and eventually condense into stacked membranes of the complex; remnants of the nuclear membrane (M) without ribosomes, but with nucleopores (or annulate lamellae) are associated with these membrane-like complexes; granular components of nucleolus (G). (Bars = 0.5 μm.) (From Hess, R. T. and Falcon, L. A., *J. Invertebr. Pathol.*, 50, 25, 1987. With permission.)

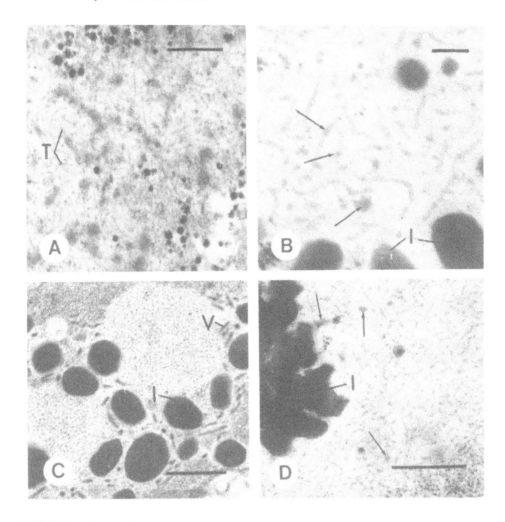

FIGURE 11. Electron micrographs of coiled tubular elements (T) present in the granulosis of the codling moth. These elements also occur in other granuloses. (A) The elements are first dispersed in the peripheral cell area prior to capsule formation. (B) Coiled tubular elements, initially with variable thickness, appear to contain a material similar to granulin (arrows) and are closely associated with incomplete and complete capsules (I). (C) During capsule formation, islands of coiled tubular elements are surrounded by enveloped nucleocapsids (V) as well as by capsules (I). (D) Coiled tubular elements are most evident in cells in which aberrant capsule (I) formation has occurred or in which capsule formation appears inhibited or reduced; arrows indicate possible granulin association with the tubules and capsule. (Bars in [A], [C], and [D] = 0.5 μm and in [B] = 100 nm.) (From Hess, R. T. and Falcon, L. A., *J. Invertebr. Pathol.*, 50, 85, 1987. With permission.)

with granulin formation or with transport. In some sections, the occluded nucleocapsids are associated with islands of coiled tubular elements of varying diameters (10 to 25 μm) (see Figure 11C), which occur more frequently in infected cells producing only a few large aberrant capsules or in cells with little or no capsule formation (see Figure 11D). Some capsules have an outer epicapsular layer (8 to 12 nm wide) (see Figure 1B) that is separated from a fairly uniform distance from the capsule surface.[45,47,48] The epicapsular layer is not always continuous. It has been considered equivalent to the envelope surrounding the polyhedra of NPV (see Section III.C).

When the cell ruptures, usually at late infection, and releases its contents into the hemocoel, masses of coiled tubular elements are evident among cellular debris.[45] The diameters of the coiled elements are largest at the early stage of infection and decrease in cells in which capsule formation has been completed. Residual components of the nucleus are scattered throughout the cytoplasm and appear to have undergone some structural changes. Fragments of the nuclear envelope occur singly or in pairs (see Figure 10A). The nucleo-capsids sometimes are associated with remnants of the nuclear envelope, and they are aligned perpendicular to the nuclear pores with the conical ends opposed to the pore lumina. Con-tinuous sheets of parallel membrane-like structures (10 nm wide) form arrays of complexes. A fibrous material is deposited between the membrane-like complexes. In cross section, the complexes appear as interconnected circular elements and the fibrous material, as hollow tubules (see Figure 12A). The surfaces of the stacks of the membrane-like complexes are often associated with ribosomes (see Figure 12C), and the structure appears similar to RER. In the lumen of the paired RER cisternae inner membranes are found (see Figure 12B). The endoplasmic reticulum cisternae are clearly associated with pieces of nuclear membrane suggesting a connection or continuation between these elements. The stacks of membrane-like complexes are frequently associated with portions of granular and dense fiber-like elements of the nucleolus (see Figure 12A), and the complexes may be formed from nuclear remains. The capsules are found at this time mainly in the cell periphery and may occur in ordered arrays[45] (see Figure 13A).

In addition to the cellular dissociation described before, the following cytopathic changes occur in GV-infected fat body cells.[45] The RER decreases in numbers, and there is a loss of normal cellular organelles such as lipid droplets, glycogen, electron-dense granules, multivesicular bodies, and pinocytic vesicles. The cellular junctions, especially the hemi-desmosomes, are no longer apparent. In some cases, the mitochondria are swollen. Even-tually, breaks occur in the plasma membrane through which the cell contents spill out into the hemocoel.

Paillot's observation[7,116] that GV caused cellular proliferation has been confirmed by others.[21,28,79,104,108,123,132] The cellular proliferation in GV-infected larvae of *E. griseana* occurs within 16 hpi.[104] In some cases, extensive cellular proliferation has not been reported after GV inoculation,[11,112,133] but proliferation may be apparent when new cells replace injured or lysed cells.[45,120]

Some investigators have speculated that the GV-infected cell exhibits processes that parallel those of a normal dividing cell during mitosis.[45,133] At the prophase stage in a typical cell, the nuclear membrane breaks down and the nucleolus dissociates into fragments. This is mimicked in a GV-infected cell and provides the virus unrestricted access to the cytoplasm and nuclear components of the cell for the synthesis and assembly of virus progeny. The association of GV infection with mitosis is an intriguing concept.

B. TEMPORAL DEVELOPMENT OF GV

The most thorough and detailed studies on the temporal development of GVs in various target organs have been conducted with the codling moth larva.[45,131] For a comprehensive description, the readers are referred to the report of Hess and Falcon.[45] The virus replication times in the various organs differ and may be associated with differences in metabolism, the functional cell types, and the age of the cells.[45] There are differences in the polarity of the midgut, fat body, and hypodermal cells concerning virus invasion and subsequent release of virus progeny. These differences result from the location, morphology and/or attachment of the cells with respect to the hemocoel environment. For example, infection in the hy-podermis occurs later than in the fat body.[45,134]

In the midgut epithelium, the GV invades via the microvillus[156] (see Figures 4A—D), and the progeny virus exists from the basal-lateral and basal portions of the cells (see Figure 13B). The process of initial infection in fat body and hypodermal cells has not been established[118,120] and probably occurs by viropexis and fusion as in the case of NPVs.[135,136] The virions in fat body cells bud from any cell surface, mostly in areas of the plasma

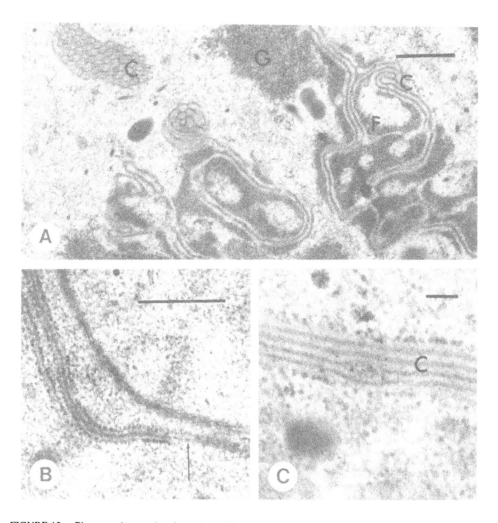

FIGURE 12.   Electron micrographs of membrane-like complexes and rough endoplasmic reticulum from GV-infected codling moth larval cells. (A) Membrane-like complexes (C) are formed in close proximity to and possibly from the fibrillar components (F) of the nucleolus in codling moth GV replication; granular components (G) of the nucleolus are closely associated with membrane-like complexes which are stacks of continuous linear parallel sheets extending for variable distances throughout the cell; in cross section they appear interconnected and tubular and contain a fibrillar material in their center. (B) Cisternae of rough endoplasmic reticulum in GV-infected cells contain a membrane-like structure within their lumena; cisternae may represent single units of membrane-like complexes to which they are similar in appearance (compare [B] and [C]); arrow indicates a putative continuation between outer and inner membranes of the cisternae and remnants of the nuclear envelope. (C) Membrane-like complexes (C) in their final or "mature" form are seen as stacks of parallel membranes, ribosomes are associated with the outer surfaces of the stacks. (Bars in [A] and [B] = 0.5 μm and in [C] = 100 nm.) (From Hess, R. T. and Falcon, L. A., *J. Invertebr. Pathol.*, 50, 85, 1987. With permission.)

membrane open to the hemocoel. Some virions, however, exit into intercellular spaces that have no junctions. In hypodermal cells, virion release occurs along the basal regions of the cells (see Figure 14) and not along the entire lateral plasma membrane until late in infection when the cellular junctions are disrupted. Apparently, no virions bud through the apical portion of the cell beneath the cuticle. The selected areas on the plasma membrane for virus emergence suggest that in only these areas can the viral glycoproteins needed for budding be inserted as proposed by Volkman et al.[124] The virions accumulate in spaces (lateral intercellular and basal) in all organs, indicating a high proliferation and exit of virions from

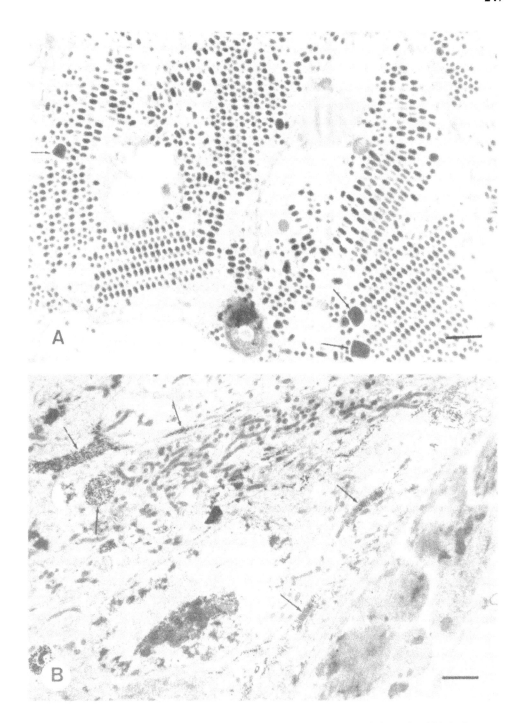

FIGURE 13. (A) Electron micrograph of a cell of GV-infected codling moth fat body which is the most productive organ for GV capsules; in this cell, some capsules are aligned in an orderly array to form crystalline-like deposit and a few are aberrant cuboidal capsules (arrows); this micrograph illustrates the extent of abnormal capsule formation. (B) A GV-infected codling moth midgut cell showing the virus pathway; the virus enters the midgut cell from the gut lumen, replicates in the nucleus, and exits the cell from sites on the basal and basal-lateral plasma membrane; after budding, the virus accumulates in the intercellular spaces (arrows) as well as in the basal region of the midgut between the plasma membrane and basal lamina. (Bars = 1 μm.) (From Hess, R. T. and Falcon, L. A., *J. Invertebr. Pathol.*, 50, 85, 1987. With permission.)

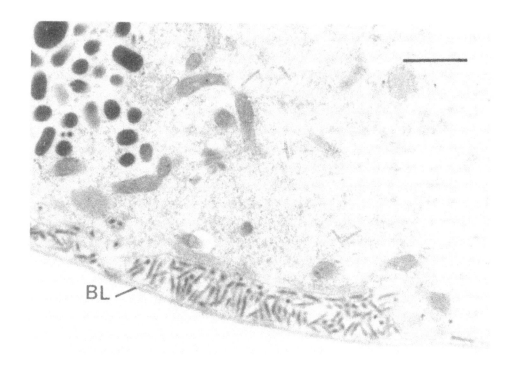

FIGURE 14.   Electron micrograph of a portion of a GV-infected codling moth, epidermal cell showing the emergence of the virus along the basal plasma membrane. Virions appear not to exit from the cell apical portion which is bound by the cuticle, nor from the lateral region where the entire surface appears connected by cellular junctions to the adjacent cell. Virions, which have budded out of the cell, collect in the space between basal plasma membrane and basal lamina (BL). (Bar = 1 μm.)

the cell and the resistance to penetration through the basal lamina (see Figures 13B, 14). In all tissues, there are residual cells that are not infected. The virions appear incapable of invading such cells even though some are attached to the cell plasma membrane.[45]

VII. PATHOPHYSIOLOGY

Pathophysiology covers a broad subject involving the "mechanisms of disease".[115,137] Such studies on the pathophysiology of granuloses are limited. The effect of GVs on weight distribution and period of lethal infection has been discussed in Section IV.B. In general, the granuloses with the type 1 syndrome caused marked increase in larval weight and greatly extend the period of lethal infection in infected larvae.

A. HEMATOLOGY

In GV-infected larvae of *P. unipuncta* with the type 1 syndrome, the differential hemocyte counts closely approximate those of uninfected larvae except for the 6-d period immediately following infection when the microplasmatocyte (granulocyte according to Shapiro[138]) counts significantly increase, whereas the macroplasmatocyte (plasmatocyte by Shapiro[138]) counts decrease.[111,139] On the other hand, the total hemocyte counts in infected larvae drop sharply at 6 dpi with the appearance of GV capsules, while the counts increase in uninfected larvae and reach a maximum at 10 to 12 dpi.[11]

Dendrolimus sibiricus larvae exhibit granulosis with a greatly prolonged period of infection[140] as in the case of *P. unipuncta*. Within 2 to 4 dpi, the number of granulocytes increases, and up to 15 to 20 dpi, the numbers of proleukocytes (prohemocytes by Shapiro[138]) and micronucleocytes (plasmatocytes by Shapiro[138]) increase rapidly to as high as 1000 times that of uninfected larvae. At 6 to 8 dpi, the hemocytes contain vesicles filled with vibrating capsules. Some hemocytes are infected, and their nuclei undergo pycnosis or karyorhexis. Later, the number of affected cells increases to 30 to 40% of the hemocyte population. Approximately 20 to 30 dpi, there is a mass proliferation of virus in hemocytes of all types. At the terminal stage of infection, the granulocytes form 50 to 80% of the total number of cells. All hemocytes are destroyed during the period preceding death of the larva which may live, however, for 7 to 14 d or more even though it is devoid of hemocytes. The changes in the hemolymph of GV-infected larva are intimately related to the pathologies in the target organs, e.g., when the intestinal epithelium is infected, granulocyte numbers increase (1 to 5%); with infection in the fat body, proleukocytes increase; the appearance of pyriform, tadpole-shaped, spindle-shaped, and differently shaped cells coincides with the infection in the hypodermis.

Insect hemocytes treat the virus capsules as inert particles and phagocytize them. Microplasmatocytes, and to a lesser extent with macroplasmatocytes and podocytes[111,139] (variant of plasmatocyte by Gupta[141]), are responsible for phagocytosis; these cells eventually degenerate. The phagocytes often are completely filled with vibrating capsules and become luminescent under phase microscopy, a characteristic for granuloses.[111,139,142]

Hemocytes may infiltrate GV-infected fat body and form nodules in this organ.[8,116,134] In some cases, the hemocytes form pseudoepithelium to replace degenerating epidermis in the GV-infected larva.[134]

Martignoni and Mistead[143] have investigated the transaminase activity of the hemolymph of the variegated cutworm, *Peridroma saucia,* affected with granulosis (Type 1 syndrome) and with nuclear polyhedrosis. The glutamate-aspartate transaminase activity in the hemolymph of GV-infected larvae rises slowly by 5 dpi and increases considerably only after 7 dpi. In nuclear polyhedrosis, the hemolymph transaminase activity is significantly elevated at 4 dpi and attains a much higher level than in granulosis. This difference in hyperenzynemia is the result of dissimilar cytocidal activities of the GV and the NPV.[143]

B. PROTEIN METABOLISM

Most studies on pathophysiology of granuloses are directed at protein metabolism. Hypoproteinemia developing in the hemolymph of GV-infected larvae of *P. saucia* is very significant by 4 dpi, whereas it is insignificant at this period of infection in nuclear polyhedrosis of the same insect.[144]

The relationships of DNA, RNA, and protein metabolism have been investigated with radioactive isotopes. When [^{3}H]uridine or [^{3}H]thymidine is inoculated into codling moth larvae at different periods of GV infection, the RNA synthesis (mostly ribosomal RNA) associated with protein synthesis increases sharply in fat body cells.[109,133] With the degeneration of the nucleolus and chromatin, there is a concomitant decrease in RNA and DNA syntheses to normal and subnormal levels. A second resurgence of RNA synthesis occurs with the appearance of the virogenic stroma. The increase lasts for a relatively long duration (three times normal) with a tremendous increase of DNA synthesis and a pronounced maximum (30 times normal level) at 60 to 70 hpi. In larval fat body cells of the fall webworm, *Hyphantria cunea,* the granulosis exhibits similar RNA, DNA, and protein syntheses.[79] At the early stage of infection, the active synthesis is restricted to the nucleus and subsequently to the nuclear strands after the disruption of the nuclear envelope. In the armyworm larva, the host cellular protein synthesis, as detected by the sequential labeling with [^{35}S]methionine, is not significantly altered in GV-infected fat body cell but the expression of the specific viral proteins is sequential.[145]

The hemolymph proteins in GV-infected armyworm larvae show electrophoretic patterns that are similar to those of nuclear polyhedrosis.[146] As granulosis progresses, the number and concentration of protein bands in the gel are reduced. The similarity of the protein bands in granulosis and nuclear polyhedrosis is due to the infection of both viruses in the major target organ, the fat body which is the center of hemolymph protein synthesis.

In virus infections in vertebrates, damage of certain tissues occurs not from intracellular virus replication and release but during subsequent inflammatory and immune responses initiated by the appearance of viral (or host) antigens on affected cells.[147] This aspect has not been investigated with insect baculoviruses. An analogous, if not similar, response occurs in the granulosis and nuclear polyhedrosis of the armyworm in which toxic proteinaceous substances appear in the hemolymph.[148-152] The toxic substances affect the tissues of the internal insect parasitoid *(Glyptapanteles militaris)*, but not those of the host armyworm larvae. The toxins upset the permeability of the integuments of the embryo and larvae of the parasitoid and cause developmental aberration and death. The toxins in granulosis may not be the same as that in nuclear polyhedrosis. The GV toxin, with a molecular weight of about 64,000,[152] appears to be specific for certain parasitoid species since not all braconids and ichneumonids are affected.[153] Moreover, other strains of GV and NPV which infect the armyworm do not produce these toxic substances.

VIII. COMPARISON OF GVS AND NPVS

The GVs and NPVs are morphologically, biophysically, and biochemically similar but they are distinct and have been assigned to two different subgroups in the family Baculoviridae. There are speculations on the molecular phylogeny of the two subgroups. Smith and Summers[54] concluded from their study on ten multiply enveloped NPVs, two singly enveloped NPVs, and four GVs that the homologous DNA sequences of these viruses present good evidence for evolution from a common ancestor. Rohrmann et al.[154] from their genome study speculated that the GVs evolved from a unicapsid NPV presumably arising in a stock infecting dipterous hosts and before the divergence of the virus of Hymenoptera which gave rise to the Lepidoptera NPV. On the other hand, Evans and Entwistle[155] considered that the restriction of the GVs to Lepidoptera suggested divergence of the GVs much later in Lepidoptera of the Cretaceous period or possibly in a pre-Cretaceous ancestral Trichopteran-Lepidopteran stock. However, the report of two GVs on hymenopterous hosts[20] may negate the latter speculation.

Briefly, the GVs and NPVs differ in: (1) biochemical composition, (2) number of nucleocapsids in an envelope, (3) size and shape of the occlusion bodies, (4) occluding process of enveloped virions, (5) number of enveloped nucleocapsids in an occlusion body, (6) cytopathology, (7) host range, and (8) tissue specificity. Biochemical analyses of the nucleotide and amino acid sequences of the viral genomes show that the GVs and NPVs are distinct.[57,64,156] In the GVs, a single nucleocapsid occurs usually within an envelope, and no strain or variant with two or more nucleocapsids per envelope has been isolated; on the other hand, in NPVs, the multiply enveloped nucleocapsids are common but strains with singly enveloped nucleocapsids have been isolated.[157-161] The GV capsules are much smaller than the NPV polyhedra and are generally ovocylindrical, whereas the polyhedra are cubic or many sided. Although GV strains with polyhedral-shaped capsules are known, no NPV strains have been isolated with ovocylindrical polyhedra. In GVs, the enveloped nucleocapsids are occluded with granulin from one end and rarely from both ends or from the middle, while the polyhedrin of NPV appears to be deposited at certain undefined foci in the nucleus in which enveloped nucleocapsids accumulate and are occluded by the additional deposition of polyhedrin. There is usually one and rarely more enveloped nucleocapsids in a GV capsule, whereas exceedingly large numbers occur commonly in a polyhedra, and no NPV has been isolated with polyhedra containing only a single enveloped nucleocapsid. The cytopathologies of GVs and NPVs show some similarity, but a major difference is the dissociation of the nuclear envelope during virogenesis in a GV-infested cell nucleus, whereas in nuclear polyhedrosis, the envelope of the infected nucleus remains intact until after polyhedra formation and cell lysis. The host range of GV is restricted mainly to Lepidoptera with two possible cases in Hymenoptera. The NPVs infect a large number of species in Lepidoptera and Hymenoptera, some Diptera, and other insect orders. The GVs also are more host specific as compared to NPVs. Aside from the NPVs of Hymenoptera in which infection is confined to the midgut, the NPVs infect a number of tissues of Lepidoptera, in particular, the hypodermis. Some GVs infect multiple organs of Lepidoptera, but others infect primarily the fat body, to a lesser extent the midgut, and not the hypodermis.

ACKNOWLEDGMENTS

We wish to thank Dr. Harry K. Kaya, Department of Nematology, University of California, Davis for critically reading the manuscript. Portions of this work were supported by the National Science Foundation under grant number DCB-8517580. Any opinions, findings, and conclusions or recommendations expressed in this publication are those of the authors and do not necessarily reflect the views of the National Science Foundation.

REFERENCES

1. **Paillot, A.,** Sur une nouvelle maladie du noyau ou grasserie des chenilles de *Pieris brassicae* et un nouveau groupe de micro-organismes parasites, *C. R. Acad. Sci., Paris,* 182, 180, 1926.
2. **Steinhaus, E. A.,** *Principles of Insect Pathology,* McGraw-Hill, New York, 1949.
3. **Weiser, J.,** Zur taxonomie der insektenviren, *Cesk. Parazitol.,* 5, 203, 1958.
4. **Bergold, G. H., Aizawa, K., Smith, K., Steinhaus, E. A., and Vago, C.,** The present status of insect virus nomenclature and classification, *Int. Bull. Bacteriol. Nomencl. Taxon.,* 10, 259, 1960.
5. **Matthews, R. E. F.,** Classification and nomenclature of viruses, *Intervirology,* 17, 1, 1982.
6. **Paillot, A.,** Un nouveau type de maladie à ultravirus chez les insectes, *C. R. Acad. Sci., Paris,* 198, 204, 1934.
7. **Paillot, A.,** Nouvel ultravirus parasite d'*Agrotis segetum* provoquant une prolifération des tissus infectés, *C. R. Acad. Sci., Paris,* 201, 1062, 1935.
8. **Paillot, A.,** Nouveau type de pseudo-grasserie observé chez les chenilles d'*Euxoa segetum, C. R. Acad. Sci., Paris,* 205, 1264, 1937.
9. **Bergold, G. H.,** Virus of insects, in *Handbuch der Virusforschung,* Vol. 4, Hallauer, C. and Meyer, K. F., Eds., Springer, Vienna, 1958, 60.
10. **Steinhaus, E. A.,** A new disease of the variegated cutworm, *Peridroma margaritosa* (Haw.), *Science,* 106, 323, 1947.
11. **Tanada, Y.,** Description and characteristics of a granulosis virus of the imported cabbageworm, *Proc. Hawaii. Entomol. Soc.,* 15, 235, 1953.
12. **Bergold, G.,** Über die Kapselvirus-Krankheit, *Z. Naturforsch.,* 3b, 338, 1948.
13. **Wyatt, G. R.,** The nucleic acids of some insect viruses, *J. Gen. Physiol.,* 36, 201, 1952.
14. **Wyatt, G. R.,** Specificity in the composition of nucleic acids, *Exp. Cell Res., Suppl.,* 2, 201, 1952.
15. **Shvedchikova, N. G., Ulanov, V. P., and Tarasevich, L. M.,** Structure of the granulosis virus of Siberian silkworm *Dendrolimus sibiricus* Tschetw., *Mol. Biol. (USSR),* 3, 283, 1969.
16. **Shvedchikova, N. G. and Tarasevich, L. M.,** Electron microscope investigation of granulosis viruses of *Dendrolimus sibiricus* and *Agrotis segetum, J. Invertebr. Pathol.,* 18, 25, 1971.
17. **Summers, M. D. and Anderson, D. L.,** Characterization of deoxyribonucleic acid isolated from the granulosis viruses of the cabbage looper, *Trichoplusia ni* and the fall armyworm, *Spodoptera frugiperda, Virology,* 50, 459, 1972.
18. **Summers, M. D. and Anderson, D. L.,** Granulosis virus deoxyribonucleic acid: a closed, double-stranded molecule, *J. Virol.,* 9, 710, 1972.
19. **Tweeten, K. A., Bulla, L. A., Jr., and Consigli, R. A.,** Supercoiled circular DNA of an insect granulosis virus, *Proc. Natl. Acad. Sci. U.S.A.,* 74, 3574, 1977.

20. **Martignoni, M. E. and Iwai, P. J.,** A catalogue of viral diseases of insects, mites, and ticks, *U.S. Department of Agriculture, For. Serv. Pac. Northwest Res. Stn., Gen. Tech. Rep. PNW-195,* 4th ed. revised, 1986, 50 pp.

21. **Huger, A.,** Granuloses of insects, in *Insect Pathology, An Advanced Treatise,* Vol. 1, Steinhaus, E. A., Academic Press, New York, 1963, 531.

22. **Ignoffo, C. M.,** Specificity of insect viruses, *Bull. Entomol. Soc. Am.,* 14, 265, 1968.

23. **Smith, K. M. and Rivers, C. F.,** Some viruses affecting insects of economic importance, *Parasitology,* 46, 235, 1956.

24. **David, W. A. L.,** The granulosis virus of *Pieris brassicae* L. in relation to natural limitation and biological control, *Ann. Appl. Biol.,* 56, 331, 1965.

25. **Crook, N. E.,** A comparison of the granulosis viruses form *Pieris brassicae* and *Pieris rapae, Virology,* 115, 173, 1981.

26. **Crook, N. E.,** Restriction enzyme analysis of granulosis viruses isolated from *Artogeia rapae* and *Pieris brassicae, J. Gen. Virol.,* 67, 781, 1986.

27. **Falcon, L. A., Kane, W. R., and Bethell, R. J.,** Preliminary evaluation of a granulosis virus for control of the codling moth, *J. Econ. Entomol.,* 61, 1208, 1968.

28. **Lipa, J. J. and Ziemnicka, J.,** Studies on the granulosis virus of cutworms, *Agrotis* spp. (Lepidoptera, Noctuidae), *Acta Microbiol. Pol.,* Ser B, 3, 155, 1971.

29. **Hunter, D. K. and Hoffman, D. F.,** Cross infection of a granulosis virus of *Cadra cautella,* with observations on its ultrastructure in infected cells of *Plodia interpunctella, J. Invertebr. Pathol.,* 20, 4, 1972.

30. **Tanada, Y. and Omi, E. M.,** Persistence of insect viruses in field populations of alfalfa insects, *J. Invertebr. Pathol.,* 23, 360, 1974.

31. **Stoddard, P. J.,** Persistence and transmission of baculoviruses in insect populations in alfalfa, Ph.D. thesis, University of California, Berkeley, 1980, 101 pp.

32. **Volkman, L. E. and Knudson, D. L.,** *In vitro* replication of baculoviruses, in *The Biology of Baculoviruses: Biological Properties and Molecular Biology,* Vol. 1, Granados, R. R. and Federici, B. A., Eds., CRC Press, Boca Raton, FL, 1987, 109.

33. **Vago, C. and Bergoin, M.,** Développement des virus à corps d'inclusion du Lépidoptère *Lymantria dispar* en cultures cellulaires, *Entomophaga,* 8, 253, 1963.

34. **Rubinstein, R., Lawler, K. A., and Granados, R. R.,** Use of primary fat body cultures for the study of baculovirus replication, *J. Invertebr. Pathol.,* 40, 266, 1982.

35. **Naser, W. L., Miltenburger, H. G., Harvey, J. P., Huber, J., and Huger, A. M.,** *In vitro* replication of the *Cydia pomonella* (codling moth) granulosis virus, *FEMS Microbiol. Lett.,* 24, 117, 1984.

36. **Miltenburger, H. G., Naser, W. L., Harvey, J. P., Huber, J., and Huger, A. M.,** The cellular substrate: a very important requirement for baculovirus *in vitro* replication, *Z. Naturforsch.,* 39, 993, 1984.

37. **Granados, R. R., Derksen, A. C. G., and Dwyer, K. G.,** Replication of the *Trichoplusia ni* granulosis and nuclear polyhedrosis viruses in cell cultures, *Virology,* 152, 472, 1986.

38. **Beaton, C. D. and Filshie, B. K.,** Comparative ultrastructural studies of insect granulosis and nuclear polyhedrosis viruses, *J. Gen. Virol.,* 31, 151, 1976.

39. **Burley, S. K., Miller, A., Harrap, K. A., and Kelly, D. C.,** Structure of the *Baculovirus* nucleocapsid, *Virology,* 120, 433, 1982.

40. **Tweeten, K. A., Bulla, L. A., Jr., and Consigli, R. A.,** Characterization of an extremely basic protein derived from granulosis virus nucleocapsids, *J. Virol.,* 33, 866, 1980.

41. **Russell, D. L. and Consigli, R. A.,** Two-dimensional polyacrylamide gel analysis of *Plodia interpunctella* granulosis virus, *J. Virol.,* 60, 82, 1986.

42. **Wilson, M. E., Mainprize, T. H., Friesen, P. D., and Miller, L. K.,** Location, transcription, and sequence of a baculovirus gene encoding a small arginine-rich polypeptide, *J. Virol.,* 61, 661, 1987.

43. **Wilson, M. E. and Consigli, R. A.,** Characterization of a protein kinase activity associated with purified capsids of the granulosis virus infecting *Plodia interpunctella, Virology,* 143, 516, 1985.

44. **Wilson, M. E. and Consigli, R. A.,** Functions of a protein kinase activity associated with purified capsids of the granulosis virus infecting *Plodia interpunctella, Virology,* 143, 526, 1985.

45. **Hess, R. T. and Falcon, L. A.,** Temporal events in the invasion of the codling moth, *Cydia pomonella,* by a granulosis virus: an electron microscope study, *J. Invertebr. Pathol.,* 50, 85, 1987.

46. **Robertson, J. S., Harrap, K. A., and Longworth, J. F.,** Baculovirus morphogenesis: the acquisition of the virus envelope, *J. Invertebr. Pathol.,* 23, 248, 1974.

47. **Arnott, H. J. and Smith, K. M.,** Ultrastructure and formation of abnormal capsules in a granulosis virus of the moth *Plodia interpunctella* (Hbn.), *J. Ultrastruct. Res.,* 22, 136, 1968a.

48. **Pinnock, D. E. and Hess, R. T.,** Morphological variations in the cytopathology associated with granulosis virus in the fruit-tree leaf roller, *Archips argyrospila, J. Ultrastruct. Res.,* 63, 252, 1978.

49. **Barefield, K. P. and Stairs, G. R.,** Infectious components of granulosis virus of the codling moth, *Carpocapsa pomonella, J. Invertebr. Pathol.,* 15, 401, 1969.

50. **Summers, M. D. and Paschke, J. D.**, Alkali-liberated granulosis virus of *Trichoplusia ni.* I. Density gradient purification of virus components and some of their *in vitro* chemical and physical properties, *J. Invertebr. Pathol.*, 16, 227, 1970.

51. **Pinnock, D. E. and Hess, R. T.**, Electron microscope observations on granulosis virus replication in the fruit tree leaf roller, *Archips argyrospila:* infection of the midgut, *J. Invertebr. Pathol.*, 30, 354, 1977.

52. **Granados, R. R.**, Infectivity and mode of action of baculoviruses, *Biotechnol. Bioeng.*, 22, 1377, 1980.

53. **Smith, G. E. and Summers, M. D.**, Analysis of baculovirus genomes with restriction endonucleases, *Virology*, 89, 517, 1978.

54. **Smith, G. E. and Summers, M. D.**, DNA homology among subgroup A, B, and C baculoviruses, *Virology*, 123, 393, 1982.

55. **Tweeten, K. A., Bulla, L. A., Jr., and Consigli, R. A.**, Restriction enzyme analysis of the genomes of *Plodia interpunctella* and *Pieris rapae* granulosis viruses, *Virology*, 104, 514, 1980.

56. **Dwyer, K. G. and Granados, R. R.**, A physical map of the *Pieris rapae* granulosis virus genome, *J. Gen. Virol.*, 68, 1471, 1987.

57. **Chakerian, R., Rohrmann, G. F., Nesson, M. H., Leisy, D. J., and Beaudreau, G. S.**, The nucleotide sequence of the *Pieris brassicae,* granulosis virus granulin gene, *J. Gen. Virol.*, 66, 1263, 1985.

58. **Burges, S.**, EcoRI restriction endonuclease fragment patterns of eight lepidopteran baculoviruses, *J. Invertebr. Pathol.*, 42, 401, 1983.

59. **Harvey, J. P. and Volkman, L. E.**, Biochemical and biological variation of *Cydia pomonella* (codling moth) granulosis virus, *Virology*, 124, 21, 1983.

60. **Crook, N. E., Spencer, R. A., Payne, C. C., and Leisy, D. J.**, Variation in *Cydia pomonella* granulosis virus isolates and physical maps of the DNA from three variants, *J. Gen. Virol.*, 66, 2423, 1985.

61. **Arif, B. M., Guangyu, Z., and Jamieson, P.**, A comparison of three granulosis viruses isolated from *Choristoneura* spp., *J. Invertebr. Pathol.*, 48, 180, 1986.

62. **Crook, N. E. and Brown, J. D.**, Isolation and characterization of a granulosis virus from the tomato moth, *Lacanobia oleracea,* and its potential as a control agent, *J. Invertebr. Pathol.*, 40, 221, 1982.

63. **Harvey, J. and Tanada, Y.**, Characterization of the DNAs of five baculoviruses pathogenic for the armyworm, *Pseudaletia unipuncta, J. Invertebr. Pathol.*, 46, 174, 1985.

64. **Akiyoshi, C., Chakerian, R., Rohrmann, G. F., Nesson, M. H., and Beaudreau, G. S.**, Cloning and sequencing of the granulin gene from the *Trichoplusia ni* granulosis virus, *Virology*, 141, 328, 1985.

65. **Tweeten, K. A., Bulla, L. A., Jr., and Consigli, R. A.**, Structural polypeptides of the granulosis virus of *Plodia interpunctella, J. Virol.*, 33, 877, 1980.

66. **Russell, D. L. and Consigli, R. A.**, Glycosylation of purified enveloped nucleocapsids of the granulosis virus infecting *Plodia interpunctella* as determined by lectin, *Virus Res.*, 4, 83, 1985.

67. **Stiles, B., Wood, H. A., and Hughes, P. R.**, Effect of tunicamycin on the infectivity of *Autographa californica* nuclear polyhedrosis virus, *J. Invertebr. Pathol.*, 41, 405, 1983.

68. **Bird, F. T.**, Polyhedrosis and granulosis viruses causing single and double infections in the spruce budworm, *Choristoneura fumiferana* Clemens, *J. Insect Pathol.*, 1, 406, 1959.

69. **Bird, F. T.**, Effects of mixed infections of two strains of granulosis virus of the spruce budworm, *Choristoneura fumiferana* (Lepidoptera:Tortricidae), on the formation of viral inclusion bodies, *Can. Entomol.*, 108, 865, 1976.

70. **Stairs, G. R.**, Selection of a strain of insect granulosis virus producing only cubic inclusion bodies, *Virology*, 24, 520, 1964.

71. **Stairs, G. R., Parrish, W. B., Briggs, J. D., and Allietta, M.**, Fine structure of a granulosis virus of the codling moth, *Virology*, 30, 583, 1966.

72. **Falcon, L. A. and Hess, R. T.**, Electron microscopic observations of multiple occluded virions in the granulosis virus of the codling moth, *Cydia pomonella, J. Invertebr. Pathol.*, 45, 356, 1985.

73. **Bergold, G. H.**, Biochemistry of insect viruses, in *The Viruses,* Burnett, F. M. and Stanley, W. M., Eds., Academic Press, New York, 1959, 505.

74. **Bergold, G. H.**, The molecular structure of some insect virus inclusion bodies, *J. Ultrastruct. Res.*, 8, 360, 1963.

75. **Bergold, G. H.**, The nature of nuclear-polyhedrosis viruses, in *Insect Pathology, An Advanced Treatise,* Vol. 1, Steinhaus, E. A., Ed., Academic Press, New York, 1963, 413.

76. **Federici, B. A.**, Ultrastructure of baculoviruses, in *The Biology of Baculoviruses, Practical Application for Insect Control,* Vol. 1, Granados, R. R. and Federici, B. A., Eds., CRC Press, Boca Raton, FL, 1987, 61.

77. **Hughes, K. M.**, Development of the inclusion bodies of a granulosis virus, *J. Bacteriol.*, 64, 375, 1952.

78. **Huger, A. and Krieg, A.**, Electron microscope investigations on the virogenesis of the granulosis of *Choristoneura murinana* (Hübner), *J. Insect Pathol.*, 3, 183, 1961.

79. **Watanabe, H. and Kobayashi, M.**, Histopathology of a granulosis in the larva of the fall webworm, *Hyphantria cunea, J. Invertebr. Pathol.*, 16, 71, 1970.

80. **Arnott, H. J. and Smith, K. M.,** An ultrastructural study of the development of a granulosis virus in the cells of the moth *Plodia interpunctella* (Hbn.), *J. Ultrastruct. Res.,* 21, 251, 1968b.

81. **Yamamoto, T., Hess, R. T., and Tanada, Y.,** Assembly of occlusion-body proteins around the enveloped virion of an insect baculovirus, *J. Ultrastruct. Res.,* 75, 127, 1981.

82. **Longworth, J. F., Robertson, J. S., and Payne, C. C.,** The purification and properties of inclusion body protein of the granulosis virus of *Pieris brassicae, J. Invertebr. Pathol.,* 19, 42, 1972.

83. **Hess, R. T., Falcon, L. A., and Morris, T. J.,** Electron microscope observations on the specificity of inclusion bodies of insect polyhedrosis viruses, *J. Invertebr. Pathol.,* 43,. 7, 1984.

84. **Summers, M. D. and Smith, G.,** *Trichoplusia ni* granulosis virus granulin: a phenol-soluble, phosphorylated protein, *J. Virol.,* 16, 1108, 1975.

85. **Tweeten, K. A., Bulla, L. A., Jr., and Consigli, R. A.,** Characterization of an alkaline protease associated with a granulosis virus of *Plodia interpunctella, J. Virol.,* 26, 702, 1978.

86. **Langridge, W. H. R. and Balter, K.,** Protease activity associated with the capsule protein of *Estigmene acraea* granulosis virus, *Virology,* 114, 595, 1981.

87. **Nagata, M. and Tanada, Y.,** Origin of an alkaline protease associated with the capsule of granulosis virus of the armyworm, *Pseudaletia unipuncta* (Haworth), *Arch. Virol.,* 76, 245, 1983.

88. **Maeda, S., Nagata, M., and Tanada, Y.,** Ionic conditions affecting the release and adsorption of an alkaline protease associated with the occlusion bodies of insect baculoviruses, *J. Invertebr. Pathol.,* 42, 376, 1983.

89. **Yamamoto, T. and Tanada, Y.,** Protein components of two strains of granulosis virus of the armyworm, *Pseudaletia unipuncta* (Lepidoptera, Noctuidae), *J. Invertebr. Pathol.,* 32, 158, 1978.

90. **Hotchkin, P. G.,** Comparison of virion proteins and granulin from a granulosis virus produced in two host species, *J. Invertebr. Pathol.,* 38, 303, 1981.

91. **Yamamoto, T. and Tanada, Y.,** Phospholipid, an enhancing component in the synergistic factor of a granulosis virus of the armyworm, *Pseudaletia unipuncta, J. Invertebr. Pathol.,* 31, 48, 1978.

92. **Yamamoto, T., Kita, H., and Tanada, Y.,** Capsule components of two strains of a granulosis virus of the armyworm *(Pseudaletia unipuncta):* purification by affinity chromatography from proteinase-free capsules and analysis by peptide mapping, *J. Gen. Virol.,* 45, 371, 1979.

93. **Tanada, Y.,** Some factors affecting the susceptibility of the armyworm to virus infections, *J. Econ. Entomol.,* 49, 52, 1956.

94. **Tanada, Y.,** Synergism between two viruses of the armyworm, *Pseudaletia unipuncta* (Haworth) (Lepidoptera, Noctuidae), *J. Insect Pathol.,* 1, 215, 1959.

95. **Tanada, Y. and Hukuhara, T.,** Enhanced infection of a nuclear-polyhedrosis virus in larvae of the armyworm, *Pseudaletia unipuncta,* by a factor in the capsule of a granulosis virus, *J. Invertebr. Pathol.,* 17, 116, 1971.

96. **Tanada, Y., Hess, R. T., and Omi, E. M.,** Invasion of a nuclear polyhedrosis virus in midgut of the armyworm, *Pseudaletia unipuncta,* and the enhancement of a synergistic enzyme, *J. Invertebr. Pathol.,* 26, 99, 1975.

97. **Tanada, Y., Hess, R., and Omi, E. M.,** Localization of esterase activity in the larval midgut of the armyworm *(Pseudaletia unipuncta), Insect Biochem.,* 10, 125, 1980.

98. **Tanada, Y. and Hukuhara, T.,** A nonsynergistic strain of a granulosis virus of the armyworm, *Pseudaletia unipuncta, J. Invertebr. Pathol.,* 12, 263, 1968.

99. **Ohba, M. and Tanada, Y.,** A synergistic factor enhances the *in vitro* infection of an insect baculovirus, *Naturwissenschaften,* 70, 613, 1983.

100. **Ohba, M. and Tanada, Y.,** *In vitro* enhancement of nuclear polyhedrosis virus infection by the synergistic factor of a granulosis virus of the armyworm, *Pseudaletia unipuncta (Lepidoptera:Noctuidae), Ann. Virol. (Inst. Pasteur),* 135E, 167, 1984.

101. **Ohba, M. and Tanada, Y.,** A synergistic factor of an insect granulosis virus agglutinates insect cells, *Experientia,* 40, 742, 1984.

102. **Hukuhara, T., Tamura, K., Zhu, Y., Abe, H., and Tanada, Y.,** Synergistic factor shows specificity in enhancing nuclear polyhedrosis virus infections, *Appl. Entomol. Zool.,* 22, 235, 1987.

103. **Camponovo, F. and Benz, G.,** Age-dependent tolerance to *Baculovirus* in last larval instars of the codling moth, *Cydia pomonella* L., induced either for pupation or for diapause, *Experientia,* 40, 938, 1984.

104. **Martignoni, M. E.,** Contributo alla conoscenza di una granulosi di *Eucosma griseana* (Hübner) (Tortricidae, Lepidoptera) quale fattore limitante il pullulamento dell'insetto nella Engadina alta, *Mitt. Schweiz. Anst. Forstl. Versuchswes.,* 32, 371, 1957.

105. **Smith, O. J., Hughes, K. M., Dunn, P. H., and Hall, I. M.,** A granulosis virus disease of the western grape leaf skeletonizer and its transmission, *Can. Entomol.,* 88, 507, 1956.

106. **Tanada, Y. and Hess, R. T.,** The cytopathology of baculovirus infections in insects, in *Insect Ultrastructure,* Vol. 2, King, R. C. and Akai, H., Eds., Plenum, New York, 1984, 517.

107. **Tanada, Y.,** Descriptions and characteristics of a nuclear polyhedrosis virus and a granulosis virus of the armyworm, *Pseudaletia unipuncta* (Haworth) (Lepidoptera, Noctuidae), *J. Insect Pathol.,* 1, 197, 1959.

108. **Hamm, J. J. and Paschke, J. D.,** On the pathology of a granulosis of the cabbage looper, *Trichoplusia ni* (Hubner), *J. Insect Pathol.,* 5, 187, 1963.

109. **Benz, G.,** Pathophysiology of insect viruses, in Progress in Invertebrate Pathology, *Proc. Int. Coll. Invertebr. Pathol. XIth Annu. Meet. Soc. Invertebr. Pathol.,* Sept. 11-17, 1978, Weiser, J., Ed., 1979, 15.

110. **Schmid, A., Cazelles, O., and Benz, G.,** A granulosis virus of the fruit tortrix, *Adoxophyes orana* F.v.R. (Lep., Tortricidae), *Mitt. Schweiz. Entomol. Ges.,* 56, 225, 1983.

111. **Wittig, G.,** Phagocytosis by blood cells in healthy and diseased caterpillars. III. Some observations concerning virus inclusion bodies, *J. Invertebr. Pathol.,* 10, 211, 1968.

112. **Hughes, K. M. and Thompson, C. G.,** A granulosis of the omnivorous looper, *Sabulodes caberata* Guenée, *J. Infect. Dis.,* 89, 173, 1951.

113. **Jans, P. and Benz, G.,** Weight increase of granulosis virus infected allatectomized larvae of the codling moth, *Cydia pomonella* (L.) (Lep., Tortricidae), *Mitt. Schweiz. Entomol. Ges.,* 58, 341, 1985.

114. **Steinhaus, E. A. and Hughes, K. M.,** A granulosis of the western grape leaf skeletonizer, *J. Econ. Entomol.,* 45, 744, 1952.

115. **Martignoni, M. E.,** Pathophysiology in the insect, *Annu. Rev. Entomol.,* 9, 179, 1964.

116. **Paillot, A.,** Contribution à l'étude des maladies à ultravirus des insectes, *Ann. Épiphyt. Phytogénét.,* 2, 341, 1936.

117. **Bird, F. T.,** On the development of granulosis viruses, *J. Insect Pathol.,* 5, 368, 1963.

118. **Mazzone, H. M.,** Receptors in the infection process, in *Viral Insecticides for Biological Control,* Maramorosch, K. and Sherman, K. E., Eds., Academic Press, New York, 1985, 695.

119. **Blumer-Wolf, A.,** Observations on early stages of granulosis virus development *in vivo* and *in vitro, Abstr. 3rd Int. Colloq. Invertebr. Pathol.,* University of Sussex, Brighton, U.K., 1982, 171.

120. **Granados, R. R. and Williams, K. A.,** *In vivo* infection and replication of baculoviruses, in *The Biology of Baculoviruses, Biological and Molecular Biology,* Vol. 1, Granados, R. R. and Federici, B. A., Eds., CRC Press, Boca Raton, FL, 1987, 89.

121. **Granados, R. R.,** Early events in the infection of *Heliothis zea* midgut cells by a baculovirus, *Virology,* 90, 170, 1978.

122. **Summers, M. D.,** Electron microscopic observations on granulosis virus entry, uncoating and replication processes during infection of the midgut cells of *Trichoplusia ni, J. Ultrastruct. Res.,* 35, 606, 1971.

123. **Walker, S., Kawanishi, C. Y., and Hamm, J. J.,** Cellular pathology of a granulosis virus infection, *J. Ultrastruct. Res.,* 80, 163, 1982.

124. **Volkman, L. E., Goldsmith, P. A., Hess, R. T., and Faulkner, P.,** Neutralization of budded *Autographa californica* NPV by a monoclonal antibody: identification of the target antigen, *Virology,* 133, 354, 1984.

125. **Tanada, Y. and Leutenegger, R.,** Multiplication of a granulosis virus in larval midgut cells of *Trichoplusia ni* and possible pathways of invasion into the hemocoel, *J. Ultrastruct. Res.,* 30, 589, 1970.

126. **Drolet, J. and Benz, G.,** Separation of two types of granulosis viruses infecting the fat body of *Adoxophyes orana* F.v.R. (Lep., Tortricidae), in *Fundamental and Applied Aspects of Invertebrate Pathology,* Vlak, J. M. and Peters, D., Eds., Found. 4th Int. Colloq. Invertebr. Pathol., Wagenigen, Netherlands, 1986, 131.

127. **Huger, A. and Krieg, A.,** Elektronenmikroskopische Untersuchungen zur Virogenese von *Bergoldiavirus calypta* Steinhaus, *Naturwissenschaften,* 47, 546, 1960.

128. **Tanada, Y. and Leutenegger, R.,** Histopathology of a granulosis-virus disease of the codling moth, *Carpocapsa pomonella, J. Invertebr. Pathol.,* 10, 39, 1968.

129. **Summers, M. D.,** Apparent *in vivo* pathway of granulosis virus invasion and infection, *J. Virol.,* 4, 188, 1969.

130. **Hunter, D. K., Hoffman, D. F., and Collier, S. J.,** Observations on a granulosis virus of the potato tuberworm, *Phthorimaea operculella, J. Invertebr. Pathol.,* 26, 397, 1975.

131. **Wäger, R. and Benz, G.,** Histochemical studies on nucleic acid metabolism in granulosis-infected *Carpocapsa pomonella, J. Invertebr. Pathol.,* 17, 107, 1971.

132. **Hamm, J. J.,** Comparative histopathology of a granulosis and a nuclear polyhedrosis of *Spodoptera frugiperda, J. Invertebr. Pathol.,* 10, 320, 1968.

133. **Benz, G. and Wäger, R.,** Autoradiographic studies on nucleic acid metabolism in granulosis-infected fat body of larvae of *Carpocapsa, J. Invertebr. Pathol.,* 18, 70, 1971.

134. **Wittig, G.,** Untersuchungen über den Verlauf der Granulose bei Raupen von *Choristoneura murinana* (Hb.) (Lepidopt., Tortricidae), *Arch. Gesamte Virusforsch.,* 9, 365, 1959.

135. **Adams, J. R., Goodwin, R. H., and Wilcox, T. A.,** Electron microscopic investigations on invasion and replication of insect baculoviruses *in vivo* and *in vitro, Biol. Cell,* 28, 261, 1977.

136. **Tanada, Y., Hess, R. T., and Omi, E. M.,** The movement and invasion of an insect baculovirus in tracheal cells of the armyworm, *Pseudaletia unipuncta, J. Invertebr. Pathol.,* 44, 198, 1984.

137. **Benz, G.,** Physiopathology and histochemistry, in *Insect Pathology, An Advanced Treatise,* Vol. 1, Steinhaus, E. A., Ed., Academic Press, New York, 1963, 299.

138. **Shapiro, M.,** Changes in hemocyte populations, in *Insect Hemocytes: Development, Forms, Functions, and Techniques,* Gupta, A. P., Ed., Cambridge University Press, Cambridge, 1979, 475.

139. **Wittig, G.,** A study of the role of blood cells in insect disease, in *Proc. 12th Int. Congr. Entomol. London, 1964,* 1965, 743.

140. **Luk'yanchikov, V. P.,** Changes in the hemolymph of *Dendrolimus sibiricus* Tschet. (Lepidoptera, Lasiocampidae) in granulosis, *Entomol. Obozr.,* 43, 297, 1964.

141. **Gupta, A. P.,** Hemocyte types: their structures, synonymies, interrelationships, and taxonomic significance, in *Insect Hemocytes: Development, Forms, Functions, and Techniques,* Gupta, A. P., Ed., Cambridge University Press, Cambridge, 1979, 85.

142. **Wittig, G.,** Untersuchungen am Blut gesunder und granulosekranker Raupen von *Choristoneura murinana* (HB.) (Lepidopt. Tortricidae), *Z. Angew. Entomol.,* 46, 385, 1960.

143. **Martignoni, M. E. and Milstead, J. E.,** Glutamate-aspartate transaminase activity in the blood plasma of an insect during the course of two viral diseases, *Ann. Entomol. Soc. Am.,* 60, 428, 1967.

144. **Martignoni, M. E. and Milstead, J. E.,** Hypoproteinemia in a noctuid larva during the course of a granulosis, *J. Invertebr. Pathol.,* 8, 261, 1966.

145. **Maeda, S. and Tanada, Y.,** Protein synthesis of a granulosis virus in the larval fat body of the armyworm, *Pseudaletia unipuncta* (Noctuidae, Lepidoptera), *J. Invertebr. Pathol.,* 41, 265, 1983.

146. **Tanada, Y. and Watanabe, H.,** Disc electrophoretic patterns of hemolymph proteins of larvae of the armyworm, *Pseudaletia unipuncta,* infected with a nuclear-polyhedrosis and a granulosis virus, *J. Invertebr. Pathol.,* 17, 127, 1971.

147. **Mims, C. A. and White, D. A.,** *Viral Pathogenesis and Immunology,* Blackwell Scientific, Oxford, 1984.

148. **Kaya, H. K.,** Toxic factor produced by a granulosis virus in armyworm larva: effect on *Apanteles militaris, Science,* 168, 251, 1970.

149. **Kaya, H. K. and Tanada, Y.,** Response of *Apanteles militaris* to a toxin produced in a granulosis-virus-infected host, *J. Invertebr. Pathol.,* 19, 1, 1972.

150. **Kaya, H. K. and Tanada, Y.,** Pathology caused by a viral toxin in the parasitoid *Apanteles militaris, J. Invertebr. Pathol.,* 19, 262, 1972.

151. **Kaya, H. K. and Tanada, Y.,** Hemolymph factor in armyworm larvae infected with a nuclear-polyhedrosis virus toxic to *Apanteles militaris, J. Invertebr. Pathol.,* 21, 211, 1973.

152. **Hotchkin, P. G. and Kaya, H. K.,** Isolation of an agent affecting the development of an internal parasitoid, *Arch. Biochem. Physiol.,* 2, 375, 1985.

153. **Hotchkin, P. G. and Kaya, H. K.,** Interactions between two baculoviruses and several insect parasites, *Can. Entomol.,* 115, 841, 1983.

154. **Rohrmann, G. F., Pearson, M. N., Bailey, T. J., Becker, R. R., and Beaudreau, G. S.,** N-terminal polyhedrin sequences and occluded *Baculovirus* evolution, *J. Mol. Evol.,* 17, 329, 1981.

155. **Evans, H. F. and Entwistle, P. E.,** Viral Diseases, in *Epizootiology of Insect Diseases,* Fuxa, J. R. and Tanada, Y., Eds., John Wiley & Sons, New York, 1987, 257.

156. **Tanada, Y. and Hess, R. T.,** unpublished data, 1987.

157. **Hughes, K. M. and Addison, R. B.,** Two nuclear polyhedrosis viruses of the Douglas-fir tussock moth, *J. Invertebr. Pathol.,* 16, 196, 1970.

158. **Ignoffo, C. M., Shapiro, M., and Hink, W. F.,** Replication and serial passage of infectious *Heliothis* nucleopolyhedrosis virus in an established line of *Heliothis zea* cells, *J. Invertebr. Pathol.,* 18, 131, 1971.

159. **Goodwin, R. H., Vaughn, J. L., Adams, J. R., and Louloudes, S. J.,** The influence of insect cell lines and tissue-culture media on baculovirus polyhedra production, *Misc. Publ. Entomol. Soc. Am.,* 9, 66, 1973.

160. **Ramoska, W. A. and Hink, W. F.,** Electron microscope examination of two plaque variants from a nuclear polyhedrosis virus of the alfalfa looper, *Autographa californica, J. Invertebr. Pathol.,* 23, 197, 1974.

161. **Sohi, S. S., Percy, J., Arif, B. M., and Cunningham, J. C.,** Replication and serial passage of singly enveloped baculovirus of *Orgyia leucostigma* in homologous cell lines, *Intervirology,* 21, 50, 1984.

162. **Kawanishi, C. Y., Egawa, K., and Summers, M. D.,** Solubilization of *Trichoplusia ni* granulosis virus proteinic crystal. II. Ultrastructure, *J. Invertebr. Pathol.,* 20, 95, 1972.

163. **Nakagaki, M., Ohba, M., and Tanada, Y.,** Specificity of receptor sites on insect cells for the synergistic factor of an insect baculovirus, *J. Invertebr. Pathol.,* 50, 169, 1987.

164. **Hamm, J. J.,** Extension of the host range for a granulosis virus from *Heliothis armiger* from South Africa, *Environ. Entomol.,* 11, 159, 1982.

165. **Ripa, R., Tatchell, G. M., and Payne, C. C.,** The susceptibilities of *Pieris brassicae* and *P. rapae* to a granulosis virus, in *Progress in Invertebrate Pathology,* Proc. Int. Colloq. Invertebr. Pathol., September 11-17, 1978, Weiser, J., Ed., Agricultural College Campus, Prague, CSSR, 1979, 165.

166. **Smith, I. R. L. and Crook, N. E.,** Physical maps of the genomes of four variants of *Artogeia rapae* granulosis virus, *J. Gen. Virol.,* 69, 1741, 1988.

167. **Belloncik, S., Lavallee, C., Mailloux, G., and Arella, M.,** Characterization of granulosis virus infection in imported cabbageworm (*Artogeia rapae*) in the province of Quebec, *Phytoprotection,* 69, 93, 1988.

168. **Easwaramoorthy, S. and Cory, J. S.,** Characterization of the DNA of granulosis viruses isolated from two closely related moths, *Chilo infuscatellus* and *C. saccariphagus indicus, Arch. Virol.,* 110, 113, 1990.
169. **Dwyer, K. G. and Granados, R. R.,** Mapping *Pieris rapae* granulosis virus transcripts and their *in vitro* translation products, *J. Virol.,* 64, 1533, 1988.
170. **Goto, C.,** Enhancement of a nuclear polyhedrosis virus (NPV) infection by a granulosis virus (GV) isolated from the spotted cutworm, *Xestia c-nigrum) L. (Lepidoptera:Noctuidae), Appl. Entomol. Zool.,* 25, 135, 1990.
171. **Derkson, A. C. G. and Granados, R. R.,** Alteration of a lepidopteran peritrophic membrane by baculoviruses and enhancement of viral infectivity, *Virology,* 167, 242, 1988.
172. **Tanada, Y. and Hess, R. T.,** unpublished data.

Chapter 8

ENTOMOPOXVIRINAE

R. H. Goodwin, R. J. Milner, and C. D. Beaton

TABLE OF CONTENTS

I. MEMBERS OF THE SUBFAMILY AND THEIR RELATIONSHIPS

The entomopoxviruses, now the subfamily Entomopoxvirinae of the family Poxviridae, comprise a poxvirus subgroup apparently restricted to insects.[1] The subfamily is presently represented by three probable genera; Genus A (type species from *Melolontha melolontha*) primarily infecting the Coleoptera; Genus B (type species from *Amsacta moorei*) primarily infecting the Lepidoptera and Orthoptera; and Genus C (type species from *Chironomus luridus*) possibly restricted to the Diptera. The subfamily is characterized by virions that contain a single molecule of double-stranded DNA in the molecular size range of 123 to 240 × 10⁶ (see References 1 and 2) and a guanine plus cytosine content of 18 to 26%.[2] The known entomopoxviruses are summarized alphabetically in Table 1 according to the probable genera. Earlier reviews have dealt with viral components,[2] viral morphology and development,[3-7] disease symptomatology and pathology,[4-6] tissue culture studies,[2,5] and diagnosis.[3,8,9]

II. GENUS A

The first entomopoxvirus (EPV) to be described was discovered infecting larvae of *Melolontha melolontha* (Coleoptera:Melolonthinae) in France.[10] This virus was intensively studied during the next 10 years resulting in publications on the epizootiology,[11-13] host range,[14] ultrastructure,[15,16-18] developmental cycle in the hemocytes,[19] amino acid composition,[20] spindles,[21] immunology,[22] and histochemistry.[23]

At least ten similar viruses have since been discovered and are now classified as Genus A (see Table 1). These are all similar to the *Melolontha* EPV in that they cause slow infections of the larval fat body of soil inhabiting beetle larvae. An exception is the EPV described from a soil inhabiting caterpillar, *Oncopera alboguttata*.[24] This led to the hypothesis that the three genera of EPVs represented ecological rather than host-related categories, with Genus A being from soil-inhabiting insects, Genus B from aboveground foliage feeders, and Genus C from aquatic insects.[24] Another EPV has been descibed from *Wiseana umbraculata* (a caterpillar which has a very similar ecology to *Oncopera*) that successfully cross infected another soil-inhabiting hepialid, *W. cervinata*.[25] Unfortunately, this virus has not yet been characterized to the genus level.

Genus A viruses are characterized by large spheroids in which up to 100 virus particles are embedded, each particle being large and unilaterally concave. Usually virus-free spindles are also formed but may be embedded within the spheroids. Rarely, spindles are absent altogether. These diseases develop slowly in the fat body, which can be severely affected, often growing over the rectal sac (in coleopteran larvae), and becoming very white, enlarged, and amorphous. Diagnosis is usually made with a phase contrast microscope and may be aided by staining in lactophenol cotton blue.[26] Such diagnosis should be confirmed with the electron microscope, which will enable the virus to be placed in the correct genus.

In nature EPVs are very widespread but rarely cause epizootics. Undoubtedly there are a large number of undescribed EPVs from scarabaeid hosts and probably from other soil-inhabiting beetle larvae such as wireworms and weevils. EPVs have been shown to play a role in the natural control of some insect pests,[27] but are generally not sufficiently pathogenic to warrant development as microbial pesticides.[28]

A. VIRUS PARTICLE

The virus particles mature only after they are embedded within the proteinaceous spheroids. This maturation process involves the formation of the outer beaded coat as well as the lateral body and viral core (see Figures 1A, B). In mature particles, the beaded coat is

TABLE 1
Characteristics of the Entomopoxviruses

Host	Virus MW (×10)	Virus size (nm)	Virus shape	Virus core shape (vertical section)	Spheroid size (μm)	Virus-free spindles	Tissues affected	Ref.
Genus A								
Anomala cuprea (Scarabaeidae:Col.)		440 × 250	Ellipsoid	Unilaterally concave	5 × 8	Present, NO, large	Fat body, hemocytes	64
Aphodius tasmaniae (Scarabaeidae:Col.)		380 × 430 × 250—300	Ellipsoid	Unilaterally concave	5—12	Present, NO	Fat body	65
Demodena boranensis (Scarabaeidae:Col.)		420 × 230	Ellipsoid	Unilaterally concave	7.8—11.0	Present, NO, large	Fat body	66
Dermolepida albohirtum (Scarabaeidae:Col.)		420—450 × 220—140	Ellipsoid	Unilaterally concave	3—5	Absent	Fat body	65
Figulus sublaevis (Lucanidae:Col.)		330 × 290	Ellipsoid	Unilaterally concave	1—5	Present, NO, large	Fat body, hypodermis	29
Geotrupes silvaticus (Scarabaeidae:Col.)		366—416 × 255—286	Ellipsoid	Unilaterally concave	3.5—11 × 2.0—7.5	Absent	Fat body	67
Melolontha melolontha (Scarabaeidae:Col.)		450 × 250	Ellipsoid	Unilaterally concave	10—24	Present, NO, macrospindle	Fat body, hemocytes	10, 15, 30

TABLE 1 (continued)
Characteristics of the Entomopoxviruses

Host	Virus MW (×10)	Virus size (nm)	Virus shape	Virus core shape (vertical section)	Spheroid size (μm)	Virus-free spindles	Tissues affected	Ref.
Oncopera alboguttata (Hepialidae:Lep.)		390 × 270 × 230	Ellipsoid	Unilaterally concave	5.7 × 8.2	Present, O	Fat body	24
Onthnonius batesi (Scarabaeidae:Col.)	200.4	470 × 265	Ellipsoid	Unilaterally concave	5—10	Present, NO, large	Fat body, hypodermis	33, 35
Phyllopertha horticola (Rutelidae:Col.)		400 × 240	Ellipsoid	Unilaterally concave	6—25	Present, NO, large	Fat body	68
Proagopertha lucidula (Scarabaeidae:col.)		500 × 250	Ellipsoid	Unilaterally concave	8—10	Present, NO, large	Fat body: hemocytes	69
Rhopaea verreauxi (Scarabaeidae:Col.)		500 × 300	Ellipsoid	Unilaterally concave	16 × 13	Present, NO large	Fat body	70
Genus B								
Adoxophyes sp. (Tortricidae:Lep.)		290—240	Ellipsoid			Present, O	Polytropic	71
Amsacta moorei (Arctiidae:Lep.)	134.7	350 × 250	Ellipsoid	Rectangular	1—4	Rare	Polytropic	33, 72
Arphia conspersa (Acrididae:Orth.)	129	320 × 250	Ellipsoid	Rectangular	2—11	Absent	Fat body	43, 48

Species								
Choristoneura biennis (Tortricidae:Lep.)	132—142	400 × 300	Ellipsoid	Rectangular	2.2—3.2	Present, O	Fat body, midgut	42, 51
Choristoneura conflictana (Tortricidae:Lep.)		273 × 235	Ellipsoid	Rectangular	4.3 × 7	Present, O	Fat body	38
Choristoneura diversana (Tortricidae:Lep.)		280 × 220	Ellipsoid	Rectangular	3.1 × 4.6	Present, O	Fat body	73
Elasmopalpus lignosellus (Pyralidae:Lep.)	140.8	270 × 200	Ellipsoid	Rectangular	1.5	Present, O	Fat body, hemocytes	32, 37
Euxoa auxiliaris (Noctuidae:Lep.)	135.6	260 × 165	Ellipsoid	Rectangular	3.7 × 4.7	Absent	Fat body, hemocytes	33, 56, 74
Heliothis spp. (Noctuidae:Lep.)		340 × 270	Ellipsoid	Rectangular	4 × 7	Present, NO	Fat body	70
Melanoplus sanguinipes (Acrididae:Orth.)	124.4	320 × 250	Ellipsoid	Rectangular	2—11	Absent	Fat body, hemocytes	36, 43, 75
Operophtera brumata (Geometridae:Lep.)		400 × 350	Ellipsoid	Rectangular	3 × 15	Present, large	Fat body	76, 83
Oreopsyche angustella (Psychidae:Lep.)		360 × 260	Ellipsoid	Rectangular	2—7 × 3—10	Present	Fat body	77
Phoetaliotes nebrascensis (Acrididae:Orth.)	125	320 × 250	Ellipsoid	Rectangular	2—11	Absent	Fat body	43, 48

TABLE 1 (continued)
Characteristics of the Entomopoxviruses

Genus C

Host	Virus MW (×10)	Virus size (nm)	Virus shape	Virus core shape (vertical section)	Spheroid size (μm)	Virus-free spindles	Tissues affected	Ref.
Aedes aegypti (Culicidae:Dip.)		320 × 230	Flattened rectangular cushion	Dumbell				60
Camptochironomus tentans (Chironomidae:Dip.)		200—250 × 270—300 × 130—150	Flattened rectangular cushion	Dumbell	2.16 × 8.10	Absent	Fat body (hemocytes?)	57
Chironomus attenuatus (Chironomidae:Dip.)		300 × 250 × 130	Flattened rectangular cushion	Dumbell	1—6	Absent	Hemocytes	61
Chironomus decorus (Chironomidae:Dip.)		300 × 260 × 160	Flattened rectangular cushion	Dumbell	5 × 8	Absent	Hemocytes	78, 79, 80
Chironomus luridus (Chironomidae:Dip.)		320 × 230 × 110	Flattened rectangular cushion	Dumbell	4 × 7	Absent	Polytropic	58, 59
Chironomus salinarius (Chironomidae:Dip.)		320 × 140	Flattened rectangular cushion	Dumbell	2.47—9.26 × 3.4—13.23	Absent	Hemocytes, fat body	63
Goeldichironomus holoprasinus (Chironomidae:Dip.)	199.2	346 × 300 × 160	Flattened rectangular cushion	Dumbell	3 × 5	Absent	Hemocytes	33, 81

Possible Genus D

Bombus impatiens	260 × 155	Ellipsoid	Vaned	Absent	Absent	Salivary gland, hypodermis	62
Bombus pennsyl-						
vanicus						
Bombus fervidis						
(Api-						
dae:Hymen.)						

Note: Abbreviations: O = occluded; NO = nonoccluded.

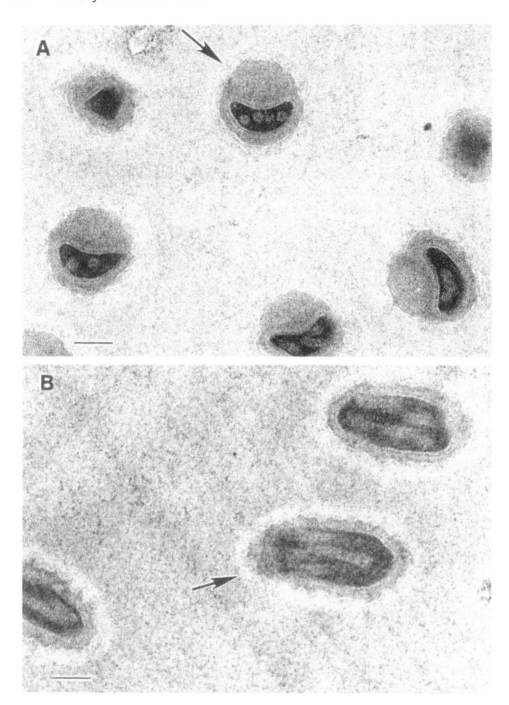

FIGURE 1.   Section of mature spheroids of *Othnonius batesi* EPV, showing mature virions. (A) Cross sections showing four profiles of the viral core (arrow); (B) a favorably oriented longitudinal section (arrow) suggesting four profiles of the viral core. (Bars = 0.1 μm.)

clearly visible both in sections surrounded by a halo (see Figure 1B) and after dissolution of the spheroid in negatively stained preparations. The internal structure has been best elucidated for the *Melolontha* EPV and shown in schematic diagrams.[15] The virus core is unilaterally concave, with a single lateral body. Within the core there is a ropelike structure which shows up to five profiles in cross sections and is thought to be a single length of

DNA some 1.2 to 1.5 nm long. This structure is not always clearly visible; for example, in the *Oncopera* EPV it has not proven possible to determine the number of coils. In section this core measures about 20 nm in diameter.

The size of the particles range from 330 × 290 nm in *Figulus*[29] to 470 × 265 nm in *Othnonius*. In *Oncopera* the whole virus particle measures 390 × 270 × 230 nm while the unilaterally concave core is 320 × 80 nm.[24]

In Genus A the spheroids are usually very large. In *Melolontha* EPV they measure 10 to 25 μm in diameter;[30] while more usually they measure 5 to 10 μm as in *Oncopera*, and in *Figulus* the spheroids are only 1 to 5 μm in diameter.[29] As a general rule the virus size and spheroid size are correlated with small particles in small spheroids and vice versa. Virus-free spindles are also characteristic for each species and occasionally, as in *Dermolepida* EPV, are absent altogether.[65] In *Oncopera* EPV, as in some members of Genus B, micro-spindles are formed (see Figure 2C) which are often embedded within the spheroids (see Figure 2A). In fact, they often form the core of a spheroid, suggesting that the spheroid is formed around them.[24] In all cases, the protein lattice of the two inclusions is different (see Figure 2D); in *Oncopera* the line-to-line distance is 11 nm for spheroids and 6 nm in spindles. The two proteins have been shown to be chemically and antigenically distinct.[20-22] The origin and function of the spindles are unknown.

Purification of spindles and spheroids from diseased insect tissue is often facilitated by the removal of lipids using a mixture of alcohol and ether (3:1,v:v). The two types of inclusion bodies can then be separated and washed by centrifugation.[20,31] Dissolution of the spheroid to release the virus particles is normally achieved in an alkaline medium with a reducing agent such as thioglycolate,[26] mercaptoacetic acid,[31] or mercaptoethanol.[32] Normally this requires a pH of 10 to 11; however, for the *Oncopera* EPV, the pH must be increased to 12.[24] Mitchell[32] described a procedure for use with *Elasmopalpus* EPV (Genus B) in which dissolution occurs at pH 7.5. The released particles can then be negatively stained and examined with the transmission electron microscope. The highly beaded appearance (see Figure 1B) is analogous to the M (mulberry)-form of the vertebrate vaccinia virus. These released particles are normally infectious on injection into the host insect hemocoel. The amino acid compositions of spindles and spheroids of the *Melolontha* EPV were described by Bergoin.[20]

Relatively little work has been done on the biochemistry of this genus. Langridge and Roberts[33,34] studied the nucleic acid of *Othnonius* EPV. They reported that the double-stranded DNA had a molecular weight of 200.4 × 10^6 with an average G + C ratio of 16.3%. Thus, this genus has the largest genome of all entomopoxviruses, and it is about 32% larger than that of the vertebrate vaccinia group.

B. REPLICATION CYCLE

Studies on replication of Genus A viruses has been inhibited by the lack of tissue culture systems to support growth and development of the viruses *in vitro*. Detailed *in vivo* observations have been made for *Melolontha* EPV in adipose tissue[18] and hemocytes.[19] More limited observations have been made of replication in *Othnonius* EPV[35] and in *Oncopera* EPV.[24]

Two types of viroplasm which apparently develop independently are commonly recognized often within the same host. Viroplasmic Types I and II as used here follow the system originally designated by Granados.[4] Type I, which predominates in *Melolontha* EPV and *Othnonius* EPV, is characterized by a well-defined area of dense granular material (see Figure 3A) from which semicircular pieces of double membrane filled with dense material are budded off. These eventually become circular in cross section and separate off completely from the granular area (see Figure 3B; also Reference 4, Figure 34). Type II viroplasms, which predominate in *Oncopera* EPV, are more diffuse areas which generate short curved

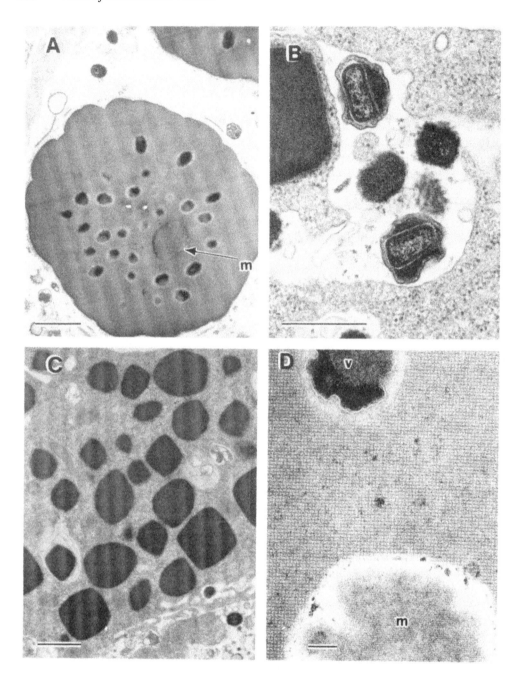

FIGURE 2.   *Oncopera* EPV. (A) Section of a complete spheroid showing its rounded-off form, the relatively random placement of the virions within it, and a microspindle (m). (Bar = 1 μm.) (B) An early stage in the development of the spheroid showing complete virions just prior to incorporation. (Bar = 0.5 μm.) (C) Survey field showing a large number of microspindles. (Bar = 1 μm.) (D) High magnification of part of a spheroid showing the different lattices of the spheroid and the microspindle (m); mature virion (v). (Bar = 0.1 μm.)

pieces of double membrane *de novo* (see Figure 3B). These membranes eventually grow to become circular in cross section, and the dense granular material gradually fills the spherical particle so that it resembles those produced by Type I viroplasms (see Figure 4E). A postulated sequence is shown in Figures 4A to F. The subsequent development is common to both types. The virus particle forms internally with the outer membrane eventually disappearing

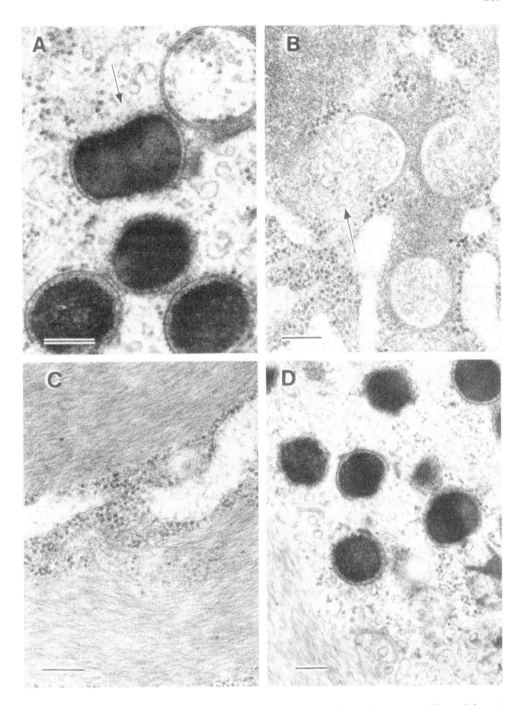

FIGURE 3. Early stages in the development of *Oncopera* EPV. (A) Type I viroplasm (arrow). (Bar = 0.2 μm.) (B) Type II viroplasm (arrow). (Bar = 0.2 μm.) (C) Cytoplasmic fibrils. (Bar = 0.2 μm.) (D) Fibrils associated with viroplasms. (Bar = 0.2 μm.)

(see Figures 2B and 4E) leaving one lateral body around the ropelike membrane-bound core. The outer beaded coat forms from the outside prior to occlusion within the spheroid.[18] The number of viral particles occluded within the same spheroid varies within and between viruses. For *Oncopera* EPV, it has been estimated as up to 180 particles per spheroid (see Figure 2A).

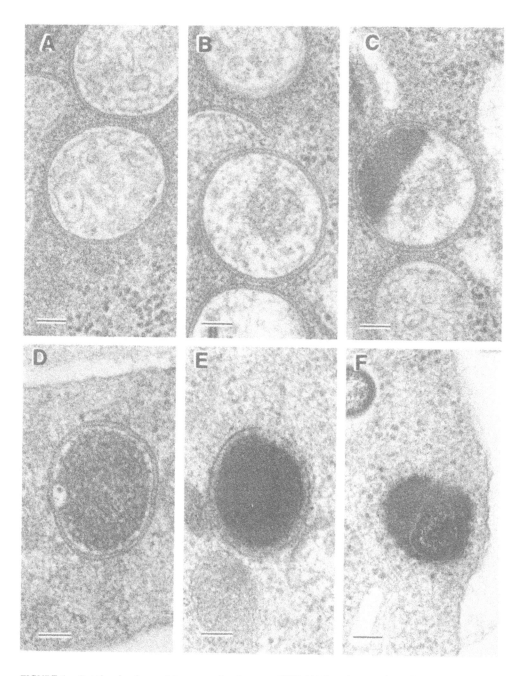

FIGURE 4.   Putative developmental sequence for *Oncopera* EPV. (A) Complete membrane-bound bodies similar to those formed by Type II viroplasms are seen. (B) to (D) Membrane-bound bodies with increasing amounts of electron-dense material. (E) and (F) After the particle has completely filled with electron-dense material, the outer membrane disposition alters, perhaps by invagination, producing the viral core. (Bars = 0.1 μm.)

In *Oncopera* EPV, fibrils of unknown function associated with the infection appear in the cell cytoplasm (see Figure 3C, D). These fibrils are also seen in Lepidoptera infected with Genus B viruses and may be adjacent to viroplasms and developing viral particles (see Figure 5C).

C. CYTOPATHOLOGY

The only reports of cytopathological changes associated with infections by Genus A are those of Devauchelle[19] for *Melolontha* EPV infecting hemocytes. The nuclei became more rounded with wide lacunae in the nuclear membranes and marked condensation of the chromatin, especially around the nuclear periphery. In the cytoplasm, ribosomes were less numerous. The infected cells became rounded as the infection progressed.

III. GENUS B

Wieser and Vago[83] were the first to describe an entomopoxvirus from a lepidopteran, the winter moth, *Operophtera brumata*. Shortly afterward, Roberts and Granados[72] described a similar virus from *Amsacta moorei*. This is now the best studied entomopoxvirus and the first one to have been grown in tissue culture.[84] It is the type species of Genus B.[1] There are now at least nine viruses in this genus known from Lepidoptera (see Table 1). A similar group of viruses are now known to infect grasshoppers. Henry[36] described an "inclusion body virus" from *Melanoplus sanguinipes,* and subsequent studies led to its placement in Genus B. A total of six grasshopper entomopoxviruses are known, three of which have been reported in the literature (see Table 1).

Genera A and B are distinct, both morphologically and biologically. In Genus B the virus particle is smaller than in Genus A. Hosts of Genus A viruses are very slow to develop frank infections (taking several months to show symptoms) while hosts of Genus B viruses develop symptoms in 10 to 12 d and are often killed within 21 d. These viruses can be highly infectious; Mitchell and Smith[37] reported that the LC_{50} for first instar *Elasmopalpus lignosellus* was just 9 spheroids, while for the fifth and sixth instars this increased to about 700 spheroids.

The main tissue affected is the fat body, with the cytoplasm of the cells becoming packed with virus-containing spheroids, and sometimes virus-free spindles (see Figure 1C), while the nucleus is relatively unaffected. The *Amsacta* EPV is unusual in attacking other tissues such as nerve, hemocyte, muscle, and midgut.[5] The spheroids vary in size from about 1.5 μm in diameter for *Elasmopalpus* EPV to up to 7 × 11 μm for *Oreopsyche* EPV. In *Heliothis* EPV the spheroids measure about 4 × 7 μm and are ellipsoidal in shape with rounded ends (see Figures 6A, C, D) while the rarer virus-free spindles are similar in size but with sharply pointed ends (see Figure 6B). Some Genus B viruses form microspindles which may be occluded within the spheroids, for example, *Choristoneura* EPV[38] and *Elasmopalpus* EPV.[32]

A. VIRUS PARTICLE

The virions of this genus are characteristically smaller than those in Genus A, measuring 300 to 400 × 200 to 300 nm. The ropelike core is less well-defined and shows three or four profiles in cross section (see Figure 5D). The shape of the core is rectangular with a single sleevelike lateral body (see Figure 6D). The ultrastructure of the *Amsacta* EPV has been described in detail by Granados and Roberts.[39]

The biochemistry of Genus B has been extensively studied, particularly in relation to the morphologically similar vaccinia virus of vertebrates. The ease of growing the host caterpillars and the relatively fast disease course make viruses from this genus attractive for biochemical studies. Initial studies[40] showed that the *Amsacta* EPV shared three of the four enzymes with vaccinia. The fourth enzyme, an RNA polymerase, was subsequently detected by using a gentler preparative procedure.[41] In contrast, major differences were found in terms of the proportion of DNA, the buoyant density, molecular weight, and G + C ratio.[33,41,42]

More recently, Langridge[31] used restriction endonucleases to compare five viruses from Lepidoptera and Orthoptera with vaccinia. He concluded that the five invertebrate viruses

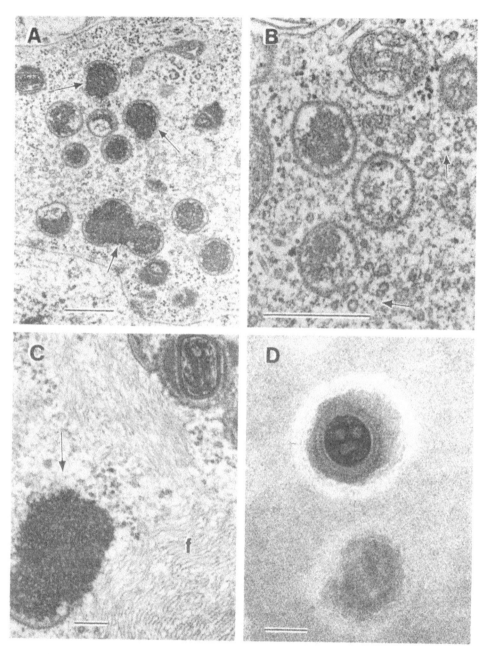

FIGURE 5.   Stages in the development of *Heliothis sp.* EPV. (A) Areas of Type I viroplasm (arrows). (Bar = 0.5 μm.) (B) Type II viroplasm (arrows). (Bar = 0.5 μm.) (C) Fibrils (f) which are seen here associated with Type I viroplasm (arrow), (Bar = 0.1 μm.) (D) Mature virions occluded within the spheroid; the lattice of the spheroid is observable at this magnification, as are the beaded outer layer of the virion and the three cross sections of its core. (Bar = 0.1 μm.)

were distinct and genetically unrelated to each other or the vaccinia virus. Subsequently, Langridge[43] reported on DNA base sequence homology studies. He was unable to detect homology either between the vertebrate and invertebrate viruses or between viruses from different insect orders. Within the Lepidoptera and Orthoptera, however, there was considerable homology. Interestingly, the *Oncopera* EPV, which has been placed in Genus A,

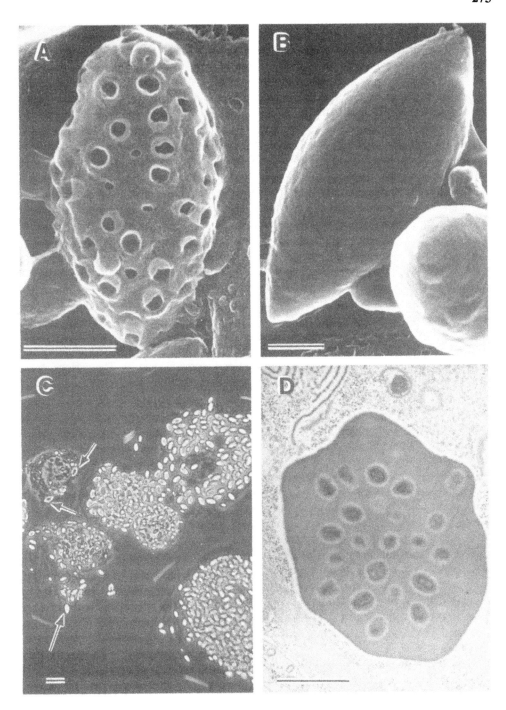

FIGURE 6. Spheroids and virus-free spindles of *Heliothis* sp. EPV. (A) Scanning electron micrograph of an incompletely synthesized spheroid; the pits on the surface indicate sites from which virions were removed during preparation. (Bar = 1 μm.) (B) Scanning electron micrograph of a virus-free spindle. (Bar = 1 μm.) (C) Light micrograph of a squash preparation of infected fat body showing the relative numbers of spindles (arrows) and spheroids. (Bar = 10 μm.) (D) Thin section of a spheroid showing its irregular and rounded-off polygonal shape and the random placement of the virions within it. (Bar = 1 μm.)

showed only a very slight homology with the other lepidopteran viruses. An ELISA technique has been developed which can detect low levels of entomopoxvirus in live hosts or tissue culture.[44,92] Attempts to grow the *Amsacta* EPV in vertebrate cells were unsuccessful, though some large molecular weight protein, equivalent to the spheroid protein, was detected.[44] In this connection, it is worth noting that a *Choristoneura* EPV was extensively safety tested prior to field tests as a biological control agent and found to be harmless to mammals and birds.[45]

The structural proteins of the spheroids and virions of the *Choristoneura* EPVs were investigated by Bilimoria and Arif.[46,47] The spheroids are composed of a single subunit acidic protein, termed spheroidin, with a molecular weight of 120,000 Da. Langridge[48] carried out some similar research on three orthopteran entomopoxviruses and reported a slightly lower molecular weight (100,000) for the spheroidin from those viruses. For the *Choristoneura* EPV, it was further shown that the protein could be broken down into discrete polypeptides by means of its endogenous alkaline proteases to give fragments of 3,000 to 52,000 Da, or by detergents to give 16,000 to 86,000 Da fragments. Studies on the virions showed at least 40 polypeptides ranging in weight from 12,000 to 250,000; however, 12 of these components made up 95% of the total virus protein.

B. REPLICATION CYCLE

The basic scheme of replication involving the two types of viroplasm is very similar to that found in Genus A. In *Heliothis* EPV the Type II viroplasm (see Figure 5B) predominates and may be associated with characteristic fibrils (see Figures 5C and 7). The various stages of viral maturation and occlusion are shown in Figure 7. Small Type I viroplasms are also present (see Figure 5A). The beaded outer membrane is present prior to virus occlusion as can be seen in Figure 7. When the partially filled spheroids are observed with the scanning electron microscope, the outer virions are lost revealing striking holes (see Figure 6A). A scanning electron micrograph of *Euxoa* EPV spheroids was shown by Adams and Wilcox.[82]

Replication of the *Amsacta* EPV has been observed in hemocyte cell cultures of a related host, *Estigmene acrea*.[84] The virus developed normally, replicating in Type I viroplasms and forming virus-containing spheroids. However, these spheroids contained fewer virions than is normal in the caterpillar host. Virus particles gained entry into cells by means of phagocytosis, resulting in the viral core being inserted into the cell cytoplasm.[5] Interestingly, viral replication was inhibited by rifampicin, an antibiotic which also inhibits vertebrate poxvirus growth by binding to viral DNA-dependent RNA polymerase. DNA and viral protein synthesis in this cell culture system were followed by Langridge[49] using hybridization of polymerase ^{32}P-labeled material. Viral specific DNA was detected just 6 to 12 h after inoculation with a rapid increase after 12 to 24 h. Virus-containing spheroids were seen 18 h after inoculation, although the 110,000 Da spheroid protein was not detected until after 24 h. Some 37 virus-related structural proteins with molecular weights ranging from 13,000 to 208,000 were reported.

Very little is known about the host specificity of entomopoxviruses. The *Amsacta* EPV readily infects the related host caterpillar *E. acrea*[50] and cell lines from a variety of lepidopteran hosts but not mosquito or mammalian cells.[5] Similarly, the *Choristoneura biennis* EPV will infect other members of the same genus.[51] McCarthy[52] reported that the *Euxoa* EPV would not infect other larvae from the Noctuidae, Arctiidae, Galleriidae, or Pyralidae. The ability of the *Melanoplus* EPV to infect and kill *Locusta migratoria* nymphs was cited as evidence that this virus might be a suitable choice as a wide host-range viral pesticide.[53]

C. CYTOPATHOLOGY

There have been very few observations of cytopathological changes in insect cells infected with Genus B entomopoxviruses. Bird[51,54] found that larvae of *Choristoneura fu-*

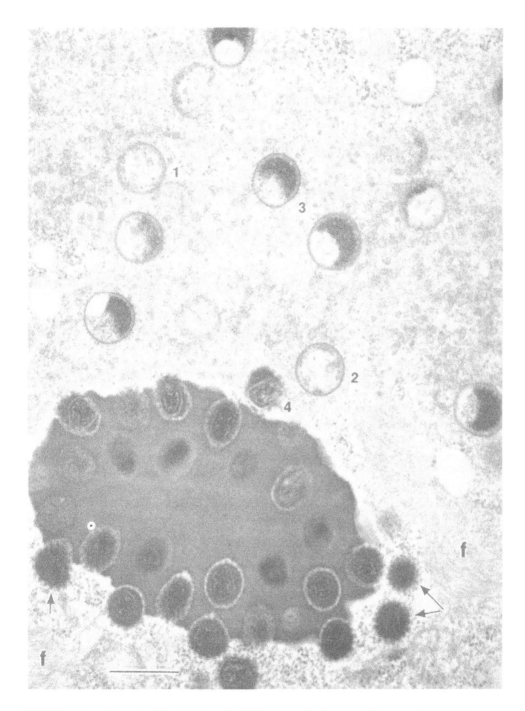

FIGURE 7.   Incomplete spheroid in a fat body cell of *Heliothis* sp., showing some of the stages of viral development indicated by the numerals 1 to 4. Arrows point to the beaded outer membrane which is visible on the virions before occlusion; fibrils (f). (Bar = 0.5 μm.).

miferana developed abnormally if infected with the entomopoxvirus from *C. biennis*. The symptoms could be mimicked by treatment with a juvenile hormone analog, suggesting that an imbalance of this hormone caused the teratological effect.[55]

An unusual and unexplained phenomenon was described by Sutter,[56] who found that the mitochondria of the infected cells in *Euxoa* contained electron-dense granules. The nuclei of fat body cells infected with *Amsacta* EPV contain irregular electron-dense masses and also some crystalline inclusions.[4]

IV. GENUS C

Since Weiser's,[57] Gotz's,[58] and Huger's[59] early descriptions in 1969 of entomopoxviruses from the Diptera, there have been relatively few Genus C descriptions in the literature. It is remarkable that these viruses have been isolated from only one family (Chironomidae) of aquatic dipterans, except one possible but rather unconvincing report of a poxvirus in the Culicidae.[60] While the Genus C virions are the smallest (320 × 230 nm maximum) of the three genera, their DNA molecular weights may be among the highest (see Table 1) when more measurements are published. DNA properties (see Reference 2, Table 2) and other characteristics of these viruses are practically unknown. As with the other genera, these viruses are largely tissue specific in their hosts. The reported Genus C viruses are, with one polytrophic exception in *Chironomus luridus*,[58,59] restricted to the cytoplasm of the hemocytes and/or fat body of aquatic Chironomid host larvae.

A. VIRUS PARTICLE

The virus particles of Genus C most closely resemble the vertebrate poxviruses, being flattened rectangular cushion shaped before and after occlusion (see Table 1). The immature particles are apparently bounded initially by either single or double trilaminar unit membranes (see Reference 61, Figures 10A, B; and Reference 39) depending upon the stage in the replication process or the interpretation of the outer spicule or subunit layer.[39,61] The outer spicule layer of the shell is progressively lost during maturation (see Reference 61, Figures 14B—G). The development of the viral core begins with the appearance of a small electron-dense nucleoid at one end of the membrane-contained viroplasmic material (see Reference 61, Figures 11 to 13). The nucleoid enlarges, becoming approximately rectangular in outline in one dimension, and somewhat flattened in the other (see Reference 61, Figure 14H; Reference 59, Figure 3D). The nucleoid becomes somewhat biconcave or dumbbell shaped in cross section, apparently due to pressure from the lateral bodies (see Reference 61, Figure 17), or the lateral body matrix material,[62] depending upon interpretation. The coiled filament or tube noted in nucleoids of other genera is apparently lacking in Genus C.

B. REPLICATION CYCLE

Genus C virus particles are generally assumed to form exclusively within Type II viroplasms,[4] although the virus from *Chironomus luridus* may also generate Type I viroplasms (see Reference 59, Figure 2A—Type II, and Figures 2B, D—Type I?). As the virus particle contents differentiate, the particles progressively condense and become embedded or occluded within the enlarging protein matrix of the developing spheroids.[59,61] The fully embedded virus particles are the most flattened and the most markedly biconcave stage, (see Figure 8).[59,61] Uniquely, the virus particles of this genus are very haphazardly arranged within the spheroids rather than radially arrayed as in Genus A and B. Spheroids may fuse during growth, resulting in lobate spheroids as in the other genera (see Figure 8). The completed spheroids are surrounded by an outer coat in which regularly arrayed subunits may be visible (see Reference 59, Figure 5; Reference 61, Figures 6, 7). Spread of the virus within the tissues of the infected larva occurs via budding of virus particles at the plasma membrane of the virus-containing cells (see Reference 61, Figure 26).

C. CYTOPATHOLOGY

The "concentric lamellae" appearing in whorls in the cytoplasm of infected larval cells seem to be an excess production of spheroid outer coat material (see Reference 59, Figure

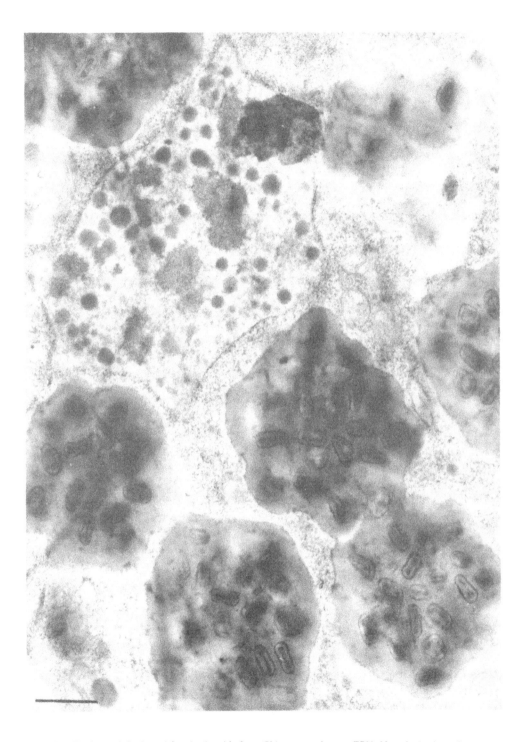

FIGURE 8.   Sections of single and fused spheroids from *Chironomus decorus* EPV. Note the haphazard arrangement of virions within the spheroids that is characteristic of Genus C viruses. Host cell nucleus shows chromatin fragmentation. (Bar = 0.5 μm.) (Courtesy of Dr. J. R. Harkrider, unpublished.)

4A to F; Reference 61, Figure 5). Other cytological changes include nuclear margination (see Reference 61, Figure 3; Reference 63, Figure 1) or chromatin fragmentation into small bits and pieces (see Figure 8; also Reference 63, Figure 1 and Reference 59, Figure 2C). Solid inclusions may also form within the nucleus (see Reference 59, Figure 4F).

V. HYMENOPTERAN VIRUS (POSSIBLE GENUS D)

A. VIRUS PARTICLE

Virus particles were initially observed within the hemolymph of adult bumblebees collected at Beltsville, MD. The particles were easily seen with a $100\times$ phase or dark field optical microscope objective. They gave the host hemolymph a slightly cloudy to milky white appearance depending upon the extent of infection. Thick sections of virus particle-containing salivary glands showed a Feulgen-positive staining where the virus particles were concentrated. This indicates that the virus particles contain DNA, which suggests their relationship to the entomopoxviruses. Thin sections of these particles were examined in an electron microscope.[62] The particles are enclosed within a trilaminar unit membrane (see Reference 62, Figure 3), and contain a dark lobate or vaned central core separated from the outer membrane by a proximal clear zone and a more distal zone of intermediate density. The particles were most commonly noted within the cytoplasm of adult salivary glands, but also occurred freely in the hemolymph and within the cytoplasm of hypodermal cells and tracheolar cells of the hypodermis.[93] This distinctive morphology and their cytoplasmic location further indicates their close similarity to the described entomopoxviruses. In addition, the structures described in Section V.B seem to leave no alternative to their inclusion among the entomopoxviruses, in spite of the absence of further biochemical or serological data.

B. REPLICATION CYCLE

The earliest probable stages of virus particle assembly are similar to the viroplasm Type I and II stages that occur in both Genus A and B of the entomopoxviruses.[4] The Type I assembly occurring in a salivary gland cell of an adult *Bombus impatiens* is depicted in Figure 3 of Clark's article.[62] This early virus assembly stage is contained within a cytoplasmic vacuole. Virus particles being assembled in hypodermal and tracheolar cells (see Reference 62, Figures 1, 4) also may occur within cytoplasmic vacuoles. Viroplasm Type II assembly in a salivary gland cell is depicted in Figure 9. Concentric arched membranes are present as well as entire membranes containing a lighter granular core material than that usually present in the Type I core contents. Note that these early particles are nearly spherical whereas the subsequent stages are progressively more ellipsoidal.

The particle contents first condense into a brick-shaped fibrillar prenucleoid (see Reference 62, Figure 2). Subsequently, the nucleoid (core) further condenses, developing four lateral concavities along its length (see Figure 10). Clark[62] described the particle contents between the nucleoid and the outer trilaminar virus membrane as a "lateral body matrix" since it appeared to be continuous around the nucleoid (see also virions of Genus A and B) and is indeterminate in shape (see Figure 10). This seems to be a more accurate term than "lateral body" except for Genus C, where the lateral bodies apparently have a more discrete structure.[61] Cross sections of mature virus particles most clearly show the final nucleoid with its lateral concavities (see Figure 10). Within cross sections of the consequent four nucleoid lobes or vanes, lighter toned circles are often visible (see Reference 62, Figure 3) that may be homologous to cross sections of the filamentous rod visible within the nucleoid of some entomopoxvirus virions.[35] There were no indications in any of Clark's observations that virions were being occluded within proteinaceous matrix bodies (spheroids). This is the most distinctive departure in the replication cycle between this virus and other entomopox-

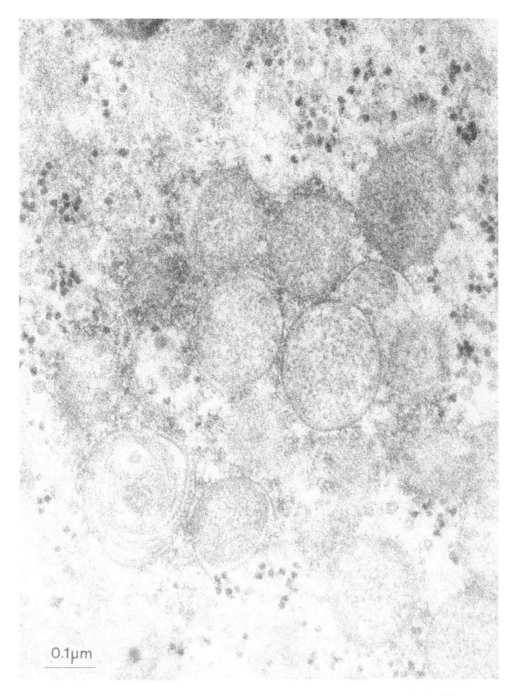

FIGURE 9.   Type II viroplasm of *Bombus impatiens* EPV, within an adult salivary gland cell. The diffuse contents of these early virus particles are characteristic of the Type II viroplasm source. (Courtesy of U.S. Department of Agriculture Insect Pathology Lab. Taken by Dr. Truman B. Clark.)

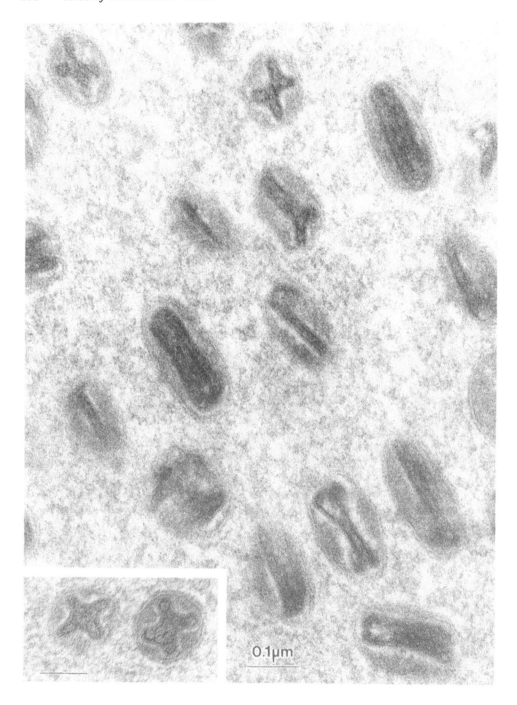

FIGURE 10.   Mature EPV particles in an adult salivary gland cell from *Bombus impatiens*. The various eccentric, longitudinal, and cross sections of the depicted virions clarify the unique vaned nucleoid present in this entomopoxvirus. (Bar = 0.1 μm.) (Courtesy of U.S. Department of Agriculture Insect Pathology Lab. Taken by Dr. Truman B. Clark.)

viruses. Together with the relatively small virus size (260 × 155 nm), the vaned nucleoid, the unusual tissue tropism, and the absence of both spheroids and spindles, the nonoccluded final replicaiton stage suggests that this virus should eventually be placed in a separate entomopoxvirus genus.

C. PATHOLOGY AND CYTOPATHOLOGY

Infected adult bumblebees (*Bombus impatiens, Bombus pennsylvanicus,* and *Bombus fervidus*) were field collected while foraging in an apparently normal manner. Infected and uninfected bumblebees survived equally well in captivity. There were no outward signs of infection. Clark suggested that the virus may be transmitted to bumblebee larvae during feeding by adults with salivary juice-contaminated pollen balls and/or to other adults through shared floral nectar sources,[93] but he was unable to demonstrate this transmission mode or transmission by intrahemocoelic injection in preliminary transmission tests.[62] Infected hemolymph was cloudy to milky white rather than clear in appearance. Infected salivary glands demonstrated a slight opalescence of a few lobes to a complete iridescent green opacity of the entire organ. In some infected bees, the cytoplasm of infected salivary gland cells was often densely filled with virus particles.

VI. NEW ISOLATES, DNA STUDIES, AND TISSUE CULTURE REPLICATIONS

An additional nonoccluded entomopoxvirus has been isolated from the venom apparatus (accessory gland) of the parasitic wasp, *Biosteres longicaudatus* (Hymenoptera:Braconidae).[85,86] This virus also infects the hemocytes within the fruit fly host (*Anastrepha suspensa;* Diptera:Tephritidae) of the parasite. This is the first report of an entomopoxvirus apparently serving as a parasitic wasp symbiote (but see the previous studies by Edson, et al., cited in [85]); the previously reported virus symbiotes of parasitic wasps are known as polydnaviruses (DNA viruses related to the baculoviruses). A polytrophic entomopoxvirus was isolated from *Pseudaletia separata* (Lepidoptera:Noctuidae) in China. This virus was infectious for one cell line from the same species and for one line from *Bombyx mori* (Lepidoptera:Bombycidae).[87] While the *P. separata* virus produces macrospindles *in vivo* and *in vitro*, a newly described polytrophic and cross-infectious lepidopteran entomopoxvirus isolated in Japan from a tortricid moth[88] produces microspindles that are occluded within the spheroids in a similar manner to the previously reported tortricid entomopoxviruses (see Table 1). The new tortricid virus can also be replicated in continuous cell lines, which means that there are now three (lepidopteran) entomopoxviruses that are active in cultured insect cells. The original cell line replicable virus from *A. moorei* has recently been serially passaged in serum-free cultures from the gypsy moth, *Lymantria dispar* (Lepidoptera:Lymantriidae).[89] A special complex basal medium supplemented with a nutrient sterol-phospholipid peptoliposome preparation was required to obtain *in vitro* replication under serum-free conditions. The virus successfully replicated through six sequential passages in the serum-free cells with no apparent morphological aberration and while retaining the capacity to infect larvae of the salt marsh caterpillar, *E. acrea* (Lepidoptera:Arctiidae), p.o. and larvae of the gypsy moth by intrahemocoelic injection.[89]

A physical map was recently constructed from the DNA of the *A. moorei* entomopoxvirus.[90] The genome size estimate in this study based on field inversion gel electrophoresis was somewhat higher than that based on restriction enzymes. Sequencing of the spheroidin gene of the *Choristoneura biennis* entomopoxvirus revealed substantial homology between its 5′ flanking region and the vaccinia major core protein precursor gene P4b.[91] The described spheroidin gene promoter may be one of the strongest known eukaryotic promoters.

REFERENCES

1. **Matthews, R. E. F.,** Classification and nomenclature of viruses. Fourth report of the International Committee on Taxonomy of Viruses, *Intervirology,* 17, 1, 1982.
2. **Arif, B. M.,** The Entomopoxviruses, *Adv. Virus Res.,* 29, 195, 1984.
3. **Bergoin, M. and Dales, S.,** Comparative observations on poxviruses of invertebrates and vertebrates, in *Comparative Virology,* Maramorosch, K. and Kurstak, E., Eds., Academic Press, New York, 1971, 169.
4. **Granados, R. R.,** Insect Poxviruses: pathology, morphology, and development, *Misc. Publ. Entomol. Soc. Am.,* 9, 73, 1973.
5. **Granados, R. R.,** Entomopoxvirus infections in insects, in *Pathogenesis of Invertebrate Microbial Disease,* Davidson, E. W., Ed., Allenheld, Osmun, Totowa, NJ, 1981, 101.
6. **Smith, K. M.,** Virus-insect relationships, Longman, London, 1976, 291 pp.
7. **Kurstak, E. and Garzon, S.,** Entomopoxviruses (poxviruses of invertebrates), in *The Atlas of Insect and Plant Viruses, Including Mycoplasma Viruses and Viroids,* Maramorosch, K., Ed., Academic Press, New York, 1977, 29.
8. **Goodwin, R. H. and Roberts, R. J.,** Diagnosis and infectivity of entomopoxviruses from three Australian scarab beetle larvae (Coleoptera:Scarabaeidae), *J. Invertebr. Pathol.,* 25, 47, 1975.
9. **Goodwin, R. H.,** Recognition and diagnosis of diseases in insectaries and the effects of disease agents on insect biology, in *Advances and Challenges in Insect Rearing,* King, E. G. and Leppla, N. C., Eds., U.S. Department of Agriculture Handbook, 1984.
10. **Vago, C.,** A new type of insect virus, *J. Insect Pathol.,* 5, 275, 1963.
11. **Hurpin, B.,** The influence of temperature and larval stage on certain diseases of *Melolontha melolontha, J. Invertebr. Pathol.,* 10, 252, 1968.
12. **Robert, P. H.,** Laboratory experiments on the persistence of the virulence of *Vagoiavirus melolonthae, Entomophaga,* 14, 159, 1969.
13. **Hurpin, B. and Robert, P. H.,** Conservation in the soil of three germs pathogenic to *Melolontha melolontha* larvae, Coleoptera, Scarabaeidae, *Entomophaga,* 21, 73, 1976.
14. **Hurpin, B. and Robert, P.,** Sur la specificité de *Vagoiavirus melolonthae, Entomophaga,* 12, 175, 1967.
15. **Bergoin, M., Devauchelle, G., and Vago, C.,** Electron microscopy study of *Melolontha melolontha* poxvirus: the fine structure of occluded virions, *Virology,* 43, 453, 1971.
16. **Bergoin, M., Devauchelle, G., Duthoit, J. L., and Vago, C.,** Étude au microscope electronique des inclusions de la virose à fuseaux des coléoptères, *C. R. Acad. Sci., Paris, Ser. D.,* 266, 2126, 1968.
17. **Bergoin, M., Devauchelle, G., and Vago, C.,** Observations au microscope électronique sur le developpement du virus de la ''maladie à fuseaux'' du coléoptère *Melolontha melolontha* L., *C. R. Acad. Sci., Paris, Ser. D.,* 267, 382, 1968.
18. **Bergoin, M., Devauchelle, G., and Vago, C.,** Electron microscopy study of the pox-like virus of *Melolontha melolontha* L. (Coleoptera:Scarabaeidae). Virus morphogenesis. *Arch. Gesamte Virusforsch.,* 28, 285, 1969.
19. **Devauchelle, G., Bergoin, M., and Vago, C.,** Étude ultra-structurale du cycle de replication d'un entomopoxvirus dans les hemocytes de son hote, *J. Ultrastruct. Res.,* 37, 301, 1971.
20. **Bergoin, M., Veyrunes, J. C., and Scalla, R.,** Isolation and amino acid composition of the inclusions of *Melolontha melolontha* poxvirus, *Virology,* 40, 760, 1970.
21. **Bergoin, M., Devauchelle, G., and Vago, C.,** Les inclusions fusiformes associées à l'entomopoxvirus du coléoptère *Melolontha melolontha, J. Ultrastruct. Res.,* 55, 17, 1976.
22. **Croizier, G. and Veyrunes, J. C.,** Analyse immunochimique des inclusions de la ''maladie à fuseaux'' du coléoptère *Melolontha melolontha* L., *Ann. Inst. Pasteur,* 120, 709, 1971.
23. **Morris, O. H.,** Histochemical changes in *Melolontha melolontha* (Linnaeus) (Coleoptera) infected by a new virus, *Can. J. Microbiol.,* 12, 965, 1966.
24. **Milner, R. J. and Beaton, C. D.,** An entomopoxvirus from *Oncopera alboguttata* (Lepidoptera, Hepialidae) in Australia, *Intervirology,* 11, 341, 1979.
25. **Moore, S. G., Kalmakoff, J., and Miles, J. A. R.,** Virus disease of porina (*Wiseana spp.* Lepidoptera:Hepialidae). I. Field isolations, *N.Z. J. Sci.,* 16, 139, 1973.
26. **Moore, S. and Milner, R. J.,** Quick stains for differentiating entomopoxvirus inclusion bodies, *J. Invertebr. Pathol.,* 22, 467, 1973.
27. **Milner, R. J.,** The role of disease during an outbreak of *Oncopera alboguttata* and *Oncopera rufobrunnea* (Lepidoptera:Hepialidae) in the Ebor-Dorrigo region of New South Wales, Australia, *J. Aust. Entomol. Soc.,* 16, 21, 1977.
28. **Milner, R. J. and Lutton, G. G.,** The pathogenicity of an entomopoxvirus from *Othnonius batesi* (Coleoptera:Scarabaeidae) and its possible use as a control agent, *Entomophaga,* 20, 213, 1975.
29. **Vago, C., Monsarrat, P., Duthoit, J. L., Amargier, A., Meynadier, G., and Van Waerebeke, D.,** Nouvelle virose à fuseaux observée chez un Lucanide (Coleoptera) de Madagascar, *C. R. Acad. Sci., Paris, Ser. D.,* 266, 1621, 1968.

30. **Hurpin, B. and Vago, C.,** Une Maladie à inclusions cytoplasmiques fusiformes chez le coléoptère *Melolontha melolontha,* (L.), *Rev. Pathol. Veg. Entomol. Agric. Fr.,* 42, 115, 1963.
31. **Langridge, W. H. R.,** Partial characterization of DNA from 5 entomopoxviruses, *J. Invertebr. Pathol.,* 42, 369, 1983.
32. **Mitchell, F. L., Smith, G. E., and Smith, Jr., J. W.,** Characterization of an entomopoxvirus of the lesser cornstalk borer *Elasmopalpus lignosellus, J. Invertebr. Pathol.,* 42, 299, 1983.
33. **Langridge, W. H. R. and Roberts, D. W.,** Molecular weight of DNA from 4 entomopoxviruses determined by electron microscopy, *J. Virol.,* 21, 301, 1977.
34. **Langridge, W. H. R., Bozarth, R. F., and Roberts, D. W.,** The base composition of entomopoxvirus DNA, *Virology,* 76, 616, 1977.
35. **Goodwin, R. H. and Filshie, B. K.,** Morphology and development of an occluded virus from the blacksoil scarab, *Othonius batesi, J. Invertebr. Pathol.,* 13, 317, 1969.
36. **Henry, J. E., Nelson, B. P., and Jutila, J. W.,** Pathology and development of the grasshopper inclusion body virus in *Melanoplus sanguinipes, J. Virol.,* 3, 605, 1969.
37. **Mitchell, F. L. and Smith, Jr., J. W.,** Pathology and bioassays of the lesser cornstalk borer *Elasmopalpus lignosellus* entomopoxvirus, *J. Invertebr. Pathol.,* 45, 75, 1985.
38. **Cunningham, J. C., Burke, J. M., and Arif, B. M.,** An entomopoxvirus found in populations of the large aspen tortrix, *Choristoneura conflictana* Wlk. (Lepidoptera:Tortricidae) in Ontario, *Can. Entomol.,* 105, 767, 1973.
39. **Granados, R. R. and Roberts, D. W.,** Electron microscopy of a pox-like virus infecting an invertebrate host, *Virology,* 40, 230, 1970.
40. **Pogo, B. G. T., Dales, S., Bergoin, M., and Roberts, D. W.,** Enzymes associated with an insect poxvirus, *Virology,* 43, 306, 1971.
41. **McCarthy, W. J., Granados, R. R., and Roberts, D. W.,** Isolation and characterization of entomopox virions from virus-containing inclusions of *Amsacta moorei* (Lepidoptera:Arctiidae), *Virology,* 59, 59, 1974.
42. **Arif, B. M.,** Isolation of an entomopoxvirus and characterization of its DNA, *Virology,* 69, 626, 1976.
43. **Langridge, W. H. R.,** Detection of DNA base sequence homology between entomopoxviruses isolated from Lepidoptera and Orthoptera, *J. Invertebr. Pathol.,* 43, 41, 1984.
44. **Langridge, W. H. R.,** Detection of *Amsacta moorei* entomopoxvirus and vaccinia virus proteins in cell cultures restrictive for poxvirus multiplication, *J. Invertebr. Pathol.,* 42, 77, 1983.
45. **Buckner, C. H. and Cunningham, J. C.,** The effect of the poxvirus of the spruce budworm, *Choristoneura fumiferana,* (Lepidoptera:Tortricidae) on mammals and birds, *Can. Entomol.,* 104, 1333, 1972.
46. **Bilimoria, S. L. and Arif, B. M.,** Subunit protein and alkaline protease of entomopoxvirus spheroids, *Virology,* 96, 596, 1979.
47. **Arif, B. M. and Brown, K. W.,** Two entomopoxvirus strains isolated from the spruce budworm, *Choristoneura fumiferana, For. Pest Manage. Inst., Can. For. Serv. Bi-mon. Res. Notes,* 36, 1, 1980.
48. **Langridge, W. H. R., Oma, E., and Henry, J. E.,** Characterization of the DNA and structural proteins of entomopoxviruses from *Melanoplus sanguinipes, Arphia conspersa,* and *Phoetaliotes nebrascensis,* Orthoptera, *J. Invertebr. Pathol.,* 42, 327, 1983.
49. **Langridge, W. H. R.,** Virus DNA replication and protein synthesis in *Amsacta moorei* entomopoxvirus infected *Estigmene acrea* cells, *J. Invertebr. Pathol.,* 41, 341, 1983.
50. **Granados, R. R. and Roberts, D. W.,** Electron microscopy of a pox-like virus infecting an invertebrate host, *Virology,* 40, 230, 1970.
51. **Bird, F. T., Sanders, C. J., and Burke, J. M.,** A newly discovered virus disease of the spruce budworm, *Choristoneura biennis* (Lepidoptera:Tortricidae), *J. Invertebr. Pathol.,* 18, 159, 1971.
52. **McCarthy, W. J., Granados, R. R., Sutter, G. R., and Roberts, D. W.,** Characterization of entomopox virus of the army cutworm *Euxoa auxiliaris* (Lepidoptera:Noctuidae), *J. Invertebr. Pathol.,* 25, 215, 1975.
53. **Jaeger, B. and Langridge, W. H. R.,** Infection of *Locusta migratoria* with entomopoxviruses from *Arphia conspersa* and *Melanoplus sanguinipes* grasshoppers, *J. Invertebr. Pathol.,* 43, 374, 1984.
54. **Bird, F. T.,** Cell proliferation in the spruce budworm, *Choristoneura fumiferana* (Lepidoptera:Tortricidae) infected with entomopoxvirus, *Can. Entomol.,* 108, 859, 1976.
55. **Retnakaran, A. and Bird, F. T.,** Apparent hormone imbalance syndrome caused by an insect virus, *J. Invertebr. Pathol.,* 20, 358, 1972.
56. **Sutter, G. R.,** A poxvirus of the army cutworm, *J. Invertebr. Pathol.,* 19, 375, 1972.
57. **Weiser, J. A.,** A pox-like virus in the midge *Camptochironomus tentans, Acta Virol.,* 13, 549, 1969.
58. **Gotz, P., Huger, A. M., and Kreig, A.,** Uber ein insektenpathogenes virus aus der gruppe der pockenviren, *Naturwissenschaften,* 56, 145, 1969.
59. **Huger, A. M., Kreig, A., Emschermann, P., and Gotz, P.,** Further studies on *Polypoxvirus chironomi,* an insect virus of the pox group isolated from the midge *Chironomus luridus, J. Invertebr. Pathol.,* 15, 253, 1970.
60. **Buchatsky, L. P.,** Poxvirus of *Aedes aegypti, Mikrobiol. Zh. (Kiev),* 36, 797, 1974.
61. **Stoltz, D. B. and Summers, M. D.,** Observations on the morphogenesis and structure of a hemocytic poxvirus in the midge *Chironomus attenuatus, J. Ultrastruct. Res.,* 40, 581, 1972.

62. **Clark, T. B.,** Entomopoxvirus like particles in three species of bumble bees, *J. Invertebr. Pathol.,* 39, 119, 1982.

63. **Majori, G., Ali, A., Donelli, G., Tangucci, F., and Harkrider, R.,** The occurrence of a virus of the pox group in a field population of *Chironomus salinarius* Kieffer (Diptera:Chironomidae) in Italy, *Fla. Entomol.,* 69, 418, 1986.

64. **Katagiri, K., Kushida, T., Kasuga, S., and Ohba, M.,** An entomopoxvirus disease of the cupreous chafer, *Anomala cuprea, Jpn. J. Appl. Entomol. Zool.,* 19, 243, 1975.

65. **Goodwin, R. H. and Filshie, B. K.,** Morphology and development of entomopoxviruses from two Australian scarab beetle larvae (Coleoptera:Scarabaeidae), *J. Invertebr. Pathol.,* 25, 35, 1975.

66. **Vago, C., Amargier, A., Hurpin, B., Meynadier, G., and Duthoit, J. L.,** Virose à fuseaux d'un scarabaéide d'amérique de sud, *Entomophaga,* 13, 373, 1968.

67. **Lipa, J. J. and Bartkowski, J.,** A newly discovered poxlike virus disease of dung beetles, *Geotrupes silvaticus* (Coleoptera:Scarabaeidae), *J. Invertebr. Pathol.,* 20, 218, 1972.

68. **Vago, C., Robert, P., Amargier, A., and Duthoit, J. L.,** Nouvelle virose à spheroides et à fuseaux observée chez le coléoptère *Phyllopertha horticola* L., *Mikroskopie,* 25, 378, 1969.

69. **Ding Cui and Cai Xiu Yu,** Light and electron microscopical study of entomopoxvirus in *Proagopertha lucidula* (Coleoptera:Scarabaeidae), *Acta Entomol. Sinica,* 28, 30, 1985.

70. **Milner, R. J. and Beaton, C. D.,** unpublished data, 1987.

71. **Ishikawa, I., Shimamura, A., and Watanabe, H.,** A new entomopoxvirus disease of the smaller tea tortrix *Adoxophyes sp.,* Lepidoptera, Tortricidae, *Jpn. J. Appl. Entomol. Zool.,* 27, 300, 1983.

72. **Roberts, D. W. and Granados, R. R.,** A pox-like virus from *Amsacta moorei* (Lepidoptera:Arctiidae), *J. Invertebr. Pathol.,* 12, 141, 1968.

73. **Katagiri, K.,** A newly discovered entomopoxvirus of *Choristoneura diversana* (Lepidoptera:Tortricidae), *J. Invertebr. Pathol.,* 22, 300, 1973.

74. **McCarthy, W. J., Granados, R. R., Sutter, G. R., and Roberts, D. W.,** Characterization of entomopox virions of the army cutworm *Euxoa auxiliaris* (Lepidoptera Noctuidae), *J. Invertebr. Pathol.,* 25, 215, 1975.

75. **Langridge, W. H. R. and Henry, J. E.,** Molecular weight and base composition of DNA isolated from *Melanoplus sanguinipes* entomopoxvirus, *J. Invertebr. Pathol.,* 37, 34, 1981.

76. **Weiser, J., Tchubianishvilli, C., and Zizka, Z.,** Ultrastructure of the spindle-shaped virus, *Vagoiavirus operophterae,* in the winter moth, *Operophtera brumata* L., *Acta Virol.,* 14, 314, 1970.

77. **Meynadier, G., Fosset, J., Vago, C., Duthoit, J. L., and Bres, N.,** Une virose à inclusions ovoides chez un Lépidoptère, *Ann. Epiphyt. (Paris),* 19, 703, 1968.

78. **Harkrider, J. R. and Hall, I. M.,** The occurrence of an entomopoxvirus in a field population of *Chironomus* sp., *Proc. Calif. Mosq. Control Assoc.,* 43, 103, 1975.

79. **Harkrider, J. R. and Hall, I. M.,** The effect of an entomopoxvirus on larval populations of an undescribed species in the *Chironomus decorus* complex under laboratory conditions, *Environ. Entomol.,* 8, 631, 1979.

80. **Harkrider, J. R. and Hall, I. M.,** Evidence for the transovum transmission of a chironomid entomopoxvirus, *Mosq. News,* 40, 116, 1980.

81. **Federici, B. A., Granados, R. R., Anthony, D. W., and Hazard, E. I.,** An entomopoxvirus and nonoccluded virus-like particles in larvae of the chironomid *Goeldichironomus holoprasinus, J. Invertebr. Pathol.,* 23, 117, 1974.

82. **Adams, J. R. and Wilcox, T. A.,** Scanning electron microscopical comparisons of insect virus occlusion bodies prepared by several techniques, *J. Invertebr. Pathol.,* 40, 12, 1982.

83. **Weiser, J. and Vago, C.,** A newly described virus of the winter moth, *Operophtera brumata* Hubner (Lepidoptera:Geometridae), *J. Invertebr. Pathol.,* 8, 314, 1966.

84. **Granados, R. R. and Naughton, M.,** Replication of *Amsacta moorei* entomopoxvirus and *Autographa californica* nuclear polyhedrosis virus in hemocyte cell lines from *Estigmene acrea,* in *Invertebrate Tissue Culture Applications in Medicine, Biology, and Agriculture,* 4th Int. Conf., Mont. Gabriel, Quebec, Canada, Kurstak, E. and Maramorosch, K., Eds., Academic Press, New York, 1975, 379.

85. **Lawrence, P. O. and Akin, D.,** Virus-like particles from the poison glands of the parasitic wasp *Biosteres longicaudatus* (Hymenoptera:Braconidae), *Can. J. Zool.,* 68, 539, 1990.

86. **Lawrence, P. O.,** Replication of an insect parasite-transmitted entomopoxvirus in hemocytes of a tephritid fruit fly host, 5th Int. Colloq. Invertebr. Pathol. Microb. Control, Adelaide, Australia, 20—24 August 1990, *Proc. Abstr.,* (abstr. no. 75), Society for Invertebrate Pathology, 205.

87. **Hukuhara, T., Xu, J., and Yano, K.,** Replication of an entomopoxvirus in two lepidopteran cell lines, *J. Invertebr. Pathol.,* 56, 222, 1990.

88. **Sato, T.,** Establishment of eight cell lines from neonate larvae of tortricids (Lepidoptera), and their several characteristics including susceptibility to insect viruses, in *Invertebrate Cell System Applications, Vol. 2,* Mitsuhashi, J., Ed., CRC Press, Boca Raton, FL, 1989, chap. 23.

89. **Goodwin, R. H., Adams, J. R., and Shapiro, M.,** Replication of the entomopoxvirus from *Amsacta moorei* in serum-free cultures of a gypsy moth cell line, *J. Invertebr. Pathol.,* 56, 190, 1990.

90. **Hall, R. L. and Hink, W. F.,** Physical mapping and field inversion gel electrophoresis of *Amsacta moorei* entomopoxvirus DNA, *Arch. Virol.,* 110, 77, 1990.

91. **Yuen, L., Dionne, J., Arif, B., and Richardson, D.,** Identification and sequencing of the spheroidin gene of *Choristoneura biennis* entomopoxvirus, *Virology,* 175, 427, 1990.

92. **McGuire, M.,** personal communication.

93. **Clark, T. B.,** personal communication.

Chapter 9

BACULOVIRIDAE. NONOCCLUDED BACULOVIRUSES

A. M. Huger and A. Krieg

TABLE OF CONTENTS

I. INTRODUCTION

The family Baculoviridae is characterized by rod-shaped virions with a probable relative molecular weight (M_r) of 10^9 for enveloped single nucleocapsids. The genome consists of monopartite, circular supercoiled dsDNA of M_r 58 to 119 $\times$ 10^6.[1,2] Two major proteins are associated with the nucleocapsids, several others with the envelope of unit-membrane type. Thus far only one genus, *Baculovirus,* has been described. Based on morphological properties, members of this genus are divided into three subgroups: A, nuclear polyhedrosis viruses; B, granulosis viruses; and C, nonoccluded rod-shaped nuclear viruses.[1] The proposed subgroup D comprising "symbiotic viruses" of parasitic hymenoptera (Ichneumonidae and Braconidae) was found to possess multipartite, double-stranded, circular DNA genomes and was therefore removed from the family Baculoviridae to be assigned to the new family Polydnaviridae[3] (see Chapter 10).

The type species of subgroup C is the so-called *Oryctes* baculovirus which was detected and first described in 1966 by Huger.[4] Contrary to the members of subgroups A and B, virions of subgroup C are not occluded in paracrystalline proteinaceous bodies such as polyhedra or granules. The cryptogram for *Oryctes* baculovirus in its updated form is $D/2:80—87/C:U_e/E:I/I$.[2] True and possible members of subgroup C have been reported from the arthropod classes Insecta, Arachnoidea, and Crustacea.

II. MEMBERS OF THE GROUP

During past decades, many rod-shaped viruses or virus-like particles of arthropods have been reported in literature that have general properties or structural and other similarities to nonoccluded baculoviruses (NOBV) of subgroup C. Pertinent hosts belonging to Coleoptera, Lepidoptera, Orthoptera, Diptera, Siphonaptera, Hymenoptera, Homoptera, Acarina, Araneina, and Crustacea either suffer from acute infections or are only carriers of the virus without overt signs and symptoms. However, the majority of the viruses described have not been investigated in detail, and have been categorized as probable or possible members of the group mainly on the basis of structural features displayed in thin sections. Only results of future biochemical and biophysical studies will decide on their final assignment to subgroup C or even to new taxonomic groups within, or separate from, the family Baculoviridae.

As yet there are only three NOBVs of insects that have been studied in sufficient detail to warrant their formal inclusion in subgroup C: the NOBV of the palm rhinoceros beetle, *Oryctes rhinoceros,* being the type species of this group;[1,4] the so-called Hz-1 virus persistently established in the lepidopteran cell line IMC-Hz-1;[5] and the NOBV most recently discovered in crickets.[6,7] Therefore, only these three members will be presented in greater detail in Section III dealing with NOBVs of insects. As they differ considerably in their biology, they will be treated separately. Reports on the occurrence of NOBV-like particles in insects and main properties known so far are listed in Tables 1 and 2 (see Section III).

III. NONOCCLUDED BACULOVIRUSES OF INSECTS

A. BACULOVIRUS OF *ORYCTES RHINOCEROS*
1. Occurrence and History

In 1963 extensive field studies were carried out by Huger[4] in Southeast Asia to find any efficient natural factor that could be used in the control of the palm rhinoceros beetle, *O. rhinoceros,* then a real problem pest of palm plantations in vast tropical areas. In the course of this survey a virus disease was discovered with larvae of this pest in Malaysia. Diagnostic studies revealed that it was caused by a new type of insect virus structurally resembling

TABLE 1
Rod-Shaped Virus and Virus-Like Particles of Insects with Structural and Other Similarities to Baculovirus Subgroup C

Host	Host stage and site of infection	Size of virus particles (nm)	Size of nucleocapsids (nm)	Biochemical data	Ref.
Chaoborus crystallinus (Diptera:Chaoboridae)	Larvae Midgut epithelium (nucleus)	260—300 × 75—106 (thin sections)	210—226 × 38—43 (thin sections)	—	30
Chaoborus astictopus (Diptera:Chaoboridae)	Larvae Midgut epithelium (nucleus)	220 × 90 (thin sections)	150 × 60 (thin sections)	—	27
Gyrinus natator (Coleoptera:Gyrinidae)	Adults Midgut epithelium (nucleus)	ca. 160 × 75 (thin sections)	150 × 35 (thin sections)	DNA ([^{3}H] thymidine incorporation)	49
Diabrotica undecimpunctata (Coleoptera:Chrysomelidae)	Adults Hemocytes (nucleus)	ca. 295 × 95 (thin sections)	230 × 52 (thin sections)	—	50
Bacillus rossius (Orthoptera:Phasmidae)	Midgut epithelium (nucleus)	250—300 × 100 (thin sections)	210 × 50—60 (thin sections)	DNA ([^{3}H] thymidine incorporation)	49
Acheta domesticus (Orthoptera:Gryllidae)	Hemocytes (?)	—	223 × 57 (isolations)	—	52
Aphis spec.	Fat body, muscle cells (nucleus)	200—300 × 50—60 (thin sections)	180—200 × 40—50 (thin sections)	—	53
Pentalonia nigronervosa (Homoptera:Aphidae)					
Pulex simulans (Siphonaptera:Pulicidae)	Midgut epithelium (nucleus)	120 × 50 (thin sections)	90 × 30 (thin sections)	—	53a

TABLE 2
Virus and Virus-Like Particles of Insects with Filamentous or Elongated Nucleocapsids

Host	Host stage and site of infection	Size (nm) and shape of virus particles	Size of nucleocapsids (nm)	Biochemical data	Ref.
Apis mellifera (Hymenoptera:Apidae)	Adults (worker bees) Fat body (nucleus)	400—450 × 100—150 Ellipsoidal (isolations)	ca. 3000 × 40—60 folded in virus particles (isolations)	dsDNA, $M_r \sim 30 \times 10^6$; 12 Polypeptides M_r 13 × 10^3 . . . 70 × 10^3	54—56
Tenebrio molitor (Coleoptera:Tenebrionidae)	Adults Midgut epithelium (nucleus)	ca. 450 × 200 Ovoidal (thin sections)	500 × 25 folded in virus particles (thin sections)	Probably DNA ([^{3}H]thymidine incorporation)	57, 58
Scolytus scolytus (Coleoptera:Scolytidae)	Larvae Midgut epithelium (nucleus)	ca. 314 × 173 Ovoidal (thin sections)	485 × 22 folded in virus particles (thin sections)	—	59
Diabrotica undecimlineata (Coleoptera:Chrysomelidae)	Adults Midgut epithelium (nucleus)	>1000 × 60 Filamentous (thin sections)	2000 × 25 sometimes hairpin-like folded in virus particles (thin sections)	—	50
Glossina pallidipes (Diptera:Muscidae)	Adults Salivary gland (nucleus)	Two size classes: 869 × 57 Elongated; 1175 × 57 Filamentous (isolations)	—	dsDNA; 12 Polypeptides, major component: M_r 39 × 10^3	60, 61
Merodon equestris (Diptera:Syrphidae)	Pupae, adults Salivary gland (nucleus, cytoplasm)	650—700 × 65 Elongated (thin sections) 700 × 60 Elongated (isolations)	650 × 35 (thin sections) 650 × 47 (isolations)	DNA	62

nuclear polyhedrosis and granulosis viruses, but not being occluded in proteinaceous bodies. Because insect viruses had not been officially classified at that time, the *Oryctes* virus was assigned to a new genus and named *Rhabdionvirus oryctes*.[4] After classification and nomenclature of insect viruses had been completely revised by the ICTV, and after the *Oryctes* virus had been characterized sufficiently, it has become the type species of subgroup C within the family Baculoviridae.[1]

In further field studies, autochthonous occurrence of the *Oryctes* virus was stated throughout the Philippines, Sumatra, West Kalimantan (Borneo), Thailand, and India. It did not exist in any South Pacific countries which until recently suffered from severe palm damage by *O. rhinoceros* due to its accidental introduction. As the biology of the pest was thwarting efficient conventional control measures, means of biological control received primary option in the mid-1960s. By initiative of the senior author, trials for colonization of the *Oryctes* virus were started in 1967 in the frame of the UNDP/SPC-Rhinoceros Beetle Project in Western Samoa. Surprisingly, from a few local breeding sites artificially contaminated with virus, the disease became established in the populations and spread autonomously throughout the Western Samoan islands. Concomitantly a drastic decline of the beetle populations took place and the coconut plantations conspicuously recovered from damage.[8,9] The striking autodissemination of the virus in the pest populations only became understandable when it was disclosed that the adult beetles themselves are very efficient natural vectors of the disease, responsible for its spread and transmission.[9-11] The unique mechanism of pathogenicity in virus-vectoring adults is described in Section III.A.4.c.

Though colonization of the disease in field populations via virus-contaminated breeding sites proved to be very effective, this method was finally superseded by the simple, efficient, and most economical way of releasing laboratory-infected adult beetles. In one of the most ambitious biocontrol projects in the South Pacific, effective control of *Oryctes* populations as a consequence of deliberate virus release was also achieved in other countries of this region (e.g., Fiji, Tonga, Wallis Island, Tokelau Islands, American Samoa) and in Mauritius, with a final decline in palm damage by up to 95%.[12,13] Follow-up field surveys during past decades proved that subsequent to the primary epizootic wave induced by virus introduction, the disease stays permanently in the decimated populations, thus exerting effective long-term suppression and regulation of this pest. The incidence of disease in persistently infected populations usually fluctuates between 30 to 50%.[10,14] Yet to preserve and enhance the control capacity of the virus, and to prevent local *Oryctes* outbreaks usually resulting from accumulations of breeding sites, an integrated control program has to be practiced. In this connection plantation hygiene deserves primary attention.[13,15] It is agreed that the spectacular success achieved with the elegant and economical long-term control of the serious palm pest *O. rhinoceros* by its homologous virus is a milestone in microbial pest control.[16] This is especially true in view of the outstanding importance of coconut as subsistence and export crop in large tropical areas.

2. Gross Pathology and Host Range

Both larval and adult stages of *O. rhinoceros* are highly susceptible to the virus via the peroral or parenteral route. While the virus causes lethal infections in larvae, adults undergo a chronic disease process. After larvae have ingested virus with contaminated food, the infection begins in the midgut epithelium and spreads to nearly all organs and tissues. Depending on the larval stage and virus dose administered, the period of lethal infection is from about 6 to 30 d.[4,17] Acutely infected larvae display striking signs and symptoms. Usually, some days postinfection they cease feeding, develop diarrhea, and their body gradually becomes turgid and assumes a glassy, beige-waxen or pearly appearance. Extensive cytolysis of the fat body results in the formation of a highly viscous mass. Moreover, the amount of turbid hemolymph is considerably increased, so that the larvae appear translucent when viewed against light. Prolapse of the rectum is frequently observed. At the terminal phase of disease the abdomen often attains a whitish mottled pattern due to chalky white deposits formed beneath the integument. Finally, the totally lethargic larvae, mostly lying on the surface of the medium, die from secondary septicemia. Now they are flaccid and their color changes to brownish, metallic blue, and bluish black.[4]

Adults can be readily infected either by applying a drop of virus suspension to the mouthparts or by forcing them to swim for 10 min in a 10% suspension of macerated, freshly virus-killed larvae.[18] Chronically infected adults generally show no external symptoms. Soon after infection they stop feeding and egg laying.[10,18] Especially the latter fact plays a major role in the reduction and long-term limitation of populations below the economic injury level. In the first phase of disease the infection is restricted to the midgut which in the course of massive cell proliferation becomes densely packed with heavily virus-infected cells.[9] As seen by dissection, such midguts are characteristically swollen and whitish in color, a diagnostic feature being particularly helpful in monitoring the incidence of disease in field populations.[19] Chronically infected adults are very productive virus reservoirs which void enormous quantities of virus with their feces. It is estimated that up to 0.3 mg of virus per day may be produced by each individual.[20] As the beetles are actively flying around for many weeks, they virtually represent mobile virus factories, spreading infective virus into the species-specific habitats (feeding burrows in palm crowns and widely distributed breeding substrates) hardly accessible to conventional control measures.[9] In this way they provide for rapid horizontal virus transmission, most frequently to other adults via oral contact with fecal virus during mating or just by staying in joint habitats. Larvae and adults inhabiting virus-contaminated breeding sites may also become infected. Larvae hatching from eggs laid by infected females are only rarely infected.[11]

A series of other dynastine agricultural pests also proved to be susceptible to the *Oryctes* virus, though its pathogenic effect usually is less severe as compared to the homologous host. Apart from other species, larvae and/or adults of *Oryctes monoceros*, *Oryctes boas*, *Oryctes nasicornis*, *Scapanes australis grossepunctatus*, and *Strategus aloeus* were shown to be capable of supporting virus replication.[4,13,21,22] After successful introduction of the virus into populations of *O. monoceros* in the Seychelles, the percentage of infected beetles fluctuated between 20 and 50 and a modest population reduction (ca. 30%) was recorded.[23] Pilot virus release trials against this palm pest were recently conducted in Tanzania; their outcome is still pending.

3. Virus Structure and Composition

a. Particle Structure

Virions of the *Oryctes* baculovirus are basically rod-shaped, singly enveloped nucleocapsids (see Figures 1A to D). From thin sections it is suggested that mature virions have a thin intermediate layer of moderate electron density between nucleocapsid and envelope (see Figure 1D). Depending on isolation procedure, negatively stained virions range in size from about 200 to 235 nm in length and 100 to 120 nm in width[24-26] as compared to 195 × 70 nm in thin sections.[4] Treatment of purified virus samples with a detergent (e.g., 1% [v/v] NP 40 in 0.05 *M* Tris buffer, pH 7.5, 30 min) causes the virus envelope to detach, so that homogeneous preparations of nucleocapsids can be obtained by subsequent sucrose gradient (10 to 50% w/v) centrifugation. The density in CsCl gradients and the sedimentation coefficient of nucleocapsids were determined to be 1.47 g/ml and 806 S, respectively, compared with 1.28 g/ml and 1640 S for complete virus particles.[26]

Negatively stained nucleocapsids are cylindrical in shape and have slightly rounded ends. In various preparations they averaged 160 × 50 nm[26] and 180 × 65 nm[7] in size. Properly stained nucleocapsids reveal a thin-walled cylindrical capsid tightly comprising the electron-dense DNA protein core. The ends of the nucleocapsids are clearly thickened or "capped"

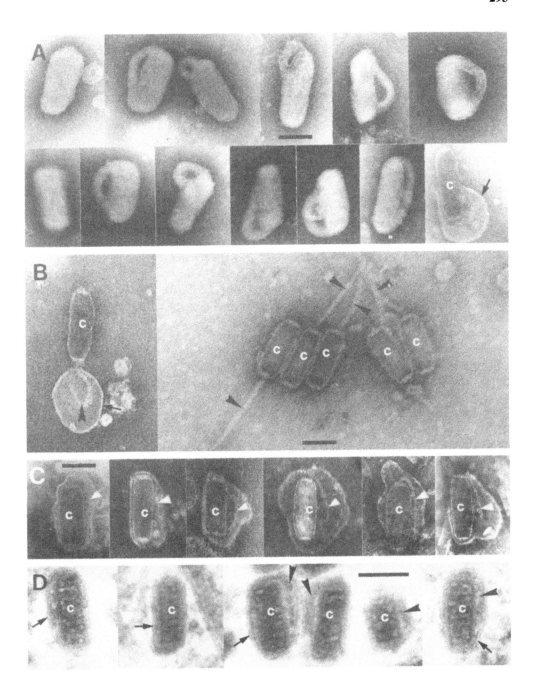

FIGURE 1. Electron micrographs illustrating major components of the baculovirus of *Oryctes rhinoceros:* (A) to (C) isolated from adult midgut and negatively stained with phosphotungstic acid. (A) Virions unpenetrated by stain, most of them with the typical mug-shaped distortion of the envelope; below right is a nucleocapsid (c) in the stage of envelope detachment (arrow). (B) Free nucleocapsids (c) displaying capped ends and the characteristic tail-like appendage (arrowheads) which in one case (left) is still within the detached envelope (arrow). (C) Virions penetrated by stain, showing enveloped capped nucleocapsids (c) with the tail-like appendage (white arrowheads). (D) Longitudinal, thin sections of virions from larval fat body nuclei, showing the unit-membrane envelope (arrows), the helicoid structure of the nucleoprotein core (c) in the capsid, and, in suitable orientation, the presence of the tail-like appendage (arrowheads) of the nucleocapsid within a unilateral dilatation of the envelope. (Bars [A] to [D] = 100 nm.) (Photographed by A. M. Huger.)

with two layers being discernible[24,25] (see Figures 1B and C). In this way they resemble baculoviruses of subgroups A and B.[27] A most typical structure regularly observed in preparations of the *Oryctes* virus is the tail-like appendage at one end of the nucleocapsid[26,28] (see Figures 1B to D). These flexous tails are relatively consistent in size, measuring approximately 270 × 10 nm,[26] and apparently have a twisted or helicoidal substructure.[28] They are suggested to consist of nucleoprotein and to be originally an internal component of the nucleocapsid that is released when the viral envelope is removed.[26,28] Apical protuberances and mug-shaped distortions of the envelope commonly occurring in purified preparations of the relatively fragile virions might be stages of this process[24,26] (see Figure 1A). In negatively stained preparations there is always a certain number of distorted and undistorted virions that display this tail between nucleocapsid and envelope (see Figure 1C). Nevertheless, the question of whether the tail normally is an internal or external component of the nucleocapsid can only be clarified by further studies. Some evidence for its external presence comes: (1) from well-preserved, undistorted virions which in suitable orientation also show a tail-like structure outside the nucleocapsid and (2) from transversely and longitudinally sectioned virons inside and outside host cells displaying a bulge in one side of the envelope which holds an electron-dense structure resembling the tail.[7,24,29] (see Figure 1D). The latter condition is also clearly exhibited by the NOBV-like particles described from midgut nuclei of the phantom midge, *Chaoborus crystallinus*.[30]

From purified preparations of virions defecated by *Oryctes* adults, twelve structural proteins were identified by PAGE with relative molecular weights ranging from 10 × 10³ to 76 × 10³.[26] The major proportion of the virions is composed of the two smallest polypeptides (M_r 10 × 10³ and 13 × 10³) which may amount to as much as 50% of the total protein.[25] Nucleocapsids contained 8 of the 12 structural polypeptides identified for virions.

A total of 27 structural proteins were detected with virions propagated in the cell line DSIR-HA-1179 and analyzed by aid of silver-stained polyacrylamide gels.[29] By pulse-labeling infected cells with L-[³⁵S]methionine at 2 h intervals during infection, the synthesis of only eight of the major structural proteins was observed. Late, new protein synthesis analogous to the occlusion body protein synthesis of subgroup A baculoviruses was not detected.[29]

Tube precipitation and gel diffusion tests revealed that the nucleocapsids share a common antigen with nucleocapsids of the subgroup A nuclear polyhedrosis virus of *Spodoptera littoralis*.[26] Also antigenic similarities between the *Oryctes* virus and the subgroup B granulosis virus of *Mamestra oleracea* were noticed in immunofluorescence studies used for detection of antigens in virus-infected cells.[31] Apparently the three subgroups of baculoviruses share a common group-specific internal antigen.

b. Genome Structure

The genome of the *Oryctes* virus is a single molecule of supercoiled, circular, double-stranded DNA, with a density in CsCl of 1.70 g/ml. DNA molecules could be isolated and examined by electron micrographs; they measured 47 ± 3 μm in length.[32] Estimates of the M_r of virus DNA amounted to 87 × 10⁶ [25] and 92 × 10⁶.[32] Later on, the latter value was corrected to 80 × 10⁶ (cited in Reference 33). The percentage of C + G content was estimated to 43%[25] or 44%.[32]

The electrophoresis profiles of *Oryctes* virus DNA digested with *Bam*HI, *Eco*RI, *Hind*III, and *Pst*I restriction endonucleases showed 21, 43, 23, and 7 fragments, respectively. After cloning of DNA fragments (representative of 96% of the virus genome) into bacterial plasmids, the restriction sites were mapped using further digestions of cloned fragments as well as hybridization of labeled fragments. Within the physical map six regions containing re-iterated DNA sequences were demonstrated. The DNA appears to lack the conserved sequences associated with the polyhedrin gene of subgroup A and B baculoviruses.[34]

The genotypic variation among 12 geographical virus isolates cloned in DSIR-HA-1179 cells was analyzed by endonuclease digestion. It was found that only slight differences exist in the electrophoretic fragment profiles. Most of the changes were due to small insertions or deletions within genome regions that contain reiterated sequences.[35]

To overcome difficulties in assessing the effectiveness of virus release programs in host populations where the virus is endemic, a fragment of foreign DNA was inserted into the genome of the *Oryctes* virus.[36] Thus, the recombinant virus can be easily discriminated from wild type virus strains.

4. Cytopathology

a. *Larvae*

Since virus-infected nuclei exhibit characteristic changes, infected tissues usually can be recognized in squash preparations, stained paraffin sections or Giemsa-stained smears. In each case nuclei at the advanced stage of infection appear greatly hypertrophied, have a dense and dark ring-zone adjacent to the nuclear membrane and a lighter, more or less homogeneous center[4,19] (see Figures 4B, 6D, 7). These typical features also proved to be a useful practical tool in monitoring the incidence of virus disease in adult field populations by examination of feces.[37]

Though larvae develop systemic infections, the nuclei of the midgut epithelium (see Figure 4) and fat body (see Figures 2 and 3) are the principal sites of virus reproduction. As is best seen in light microscope sections, large areas of the fat body, especially in the abdomen, disintegrate completely to a loose network of cell membranes and connective tissues often embedded in a highly viscous mass. Other areas of the fat body display extensive cell proliferation. The proliferated cells also become infected, separate from each other, and accumulate in the hemolymph intermingled with infected blood cells.[4]

The pathway of cell infection by the *Oryctes* virus is still poorly understood. In cell cultures the virus apparently enters the cell by pinocytosis.[29] It is not yet clear where uncoating takes place. Preliminary studies with primary cardiac and hemocyte cultures from *O. rhinoceros* larvae suggest that the nucleocapsids uncoat at nuclear pores[38] as reported for granulosis viruses (subgroup B of Baculoviridae). Virions leaving cultured cells acquire a second envelope as they bud through the plasma membrane.[29]

Early cytopathic changes associated with virus infection include increasing hypertrophy of the nucleus, margination and diminution of nuclear chromatin, and formation of a central electron-lucent area of more or less dense fibrillar and granular virogenic stroma. Concomitant with the appearance of the first virus particles, or even earlier, an abundance of vesicular virus envelope material is produced *de novo,* a most characteristic cytopathic feature in the reproduction of the *Oryctes* virus[4] (see Figure 2A). In thin sections these vesicular structures, averaging 160 nm in diameter, appear as circular, cup-shaped or spirally coiled profiles of the unit-membrane type. During their formation they enclose or engulf fibrillar virogenic stroma (see Figure 3A). Successive stages of virus morphogenesis in thin sections suggest that the virus envelope and nucleocapsid shell assemble first, followed by condensation of the electron-dense nucleoprotein core.[7,29] To a relatively low extent complete virions may be assembled in mature vesicles and thus obtain one or two additional envelopes[4] (see Figure 3A).

Together with virus particles and excess envelope material, long unenveloped and enveloped tubular filaments with diameters of capsids and complete virions, respectively, are produced in varying amounts in many nuclei. Most of these tubular structures finally have an electron-dense core, thus representing filamentous virus forms. They may break across to form virus rods. There are always nuclei where virus filaments may even prevail[4] (see Figure 3B).

At later stages of infection enormous accumulations of rod-shaped virions are usually

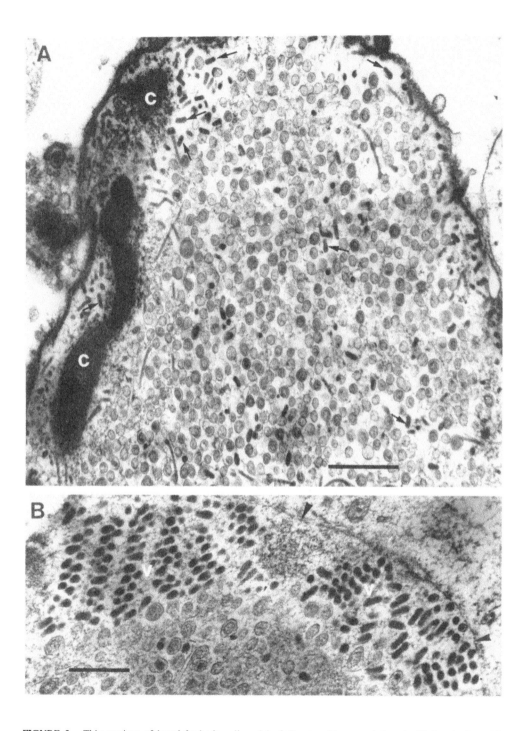

FIGURE 2. Thin sections of larval fat body cell nuclei of *Oryctes rhinoceros* infected with baculovirus. (A) Nucleus packed with vesicular virus envelope material; note rod-shaped virus particles (arrows) preferably gathering in the marginal area where also remnants of disintegrating chromatin (c) are visible. (Bar = 1 μm.) (B) Virus particles (v) accumulated in the marginal nuclear space adjacent to the nuclear membrane (arrowheads), thus forming the so-called ring-zone of infected nuclei; note the paracrystalline arrangement of virus rods. (Bar = 500 nm.) (Photographed by A. M. Huger.)

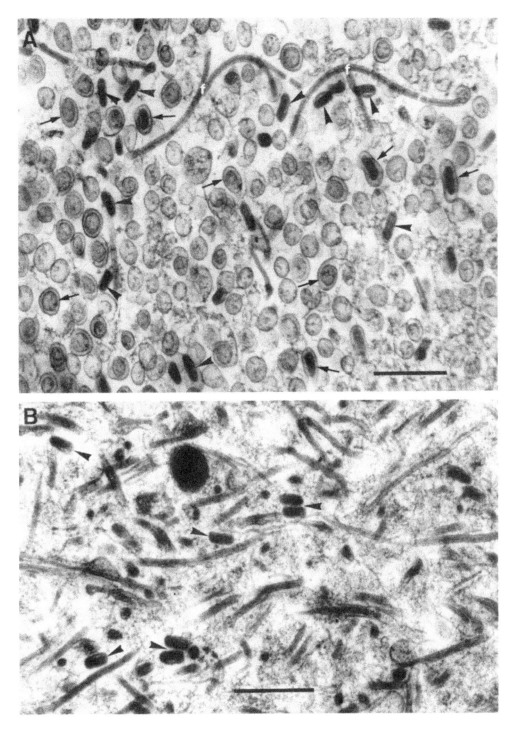

FIGURE 3. Thin sections of larval cell nuclei of *Oryctes rhinoceros* infected with baculovirus. (A) Nucleus from fat body showing accumulation of vesicular virus envelope material in various stages of assembly; note enclosure or engulfment of fibrillar virogenic stroma; virions assembling within discrete vesicles may have an additional envelope (arrows); common virions are marked by arrowheads; filamentous virus structures (f). (B) Nucleus from midgut with prevailing filamentous virus assembly stages embedded in fibrillar virogenic stroma; some rod-shaped virions are marked by arrowheads. (Bars = 500 nm.) (Photographed by A. M. Huger.)

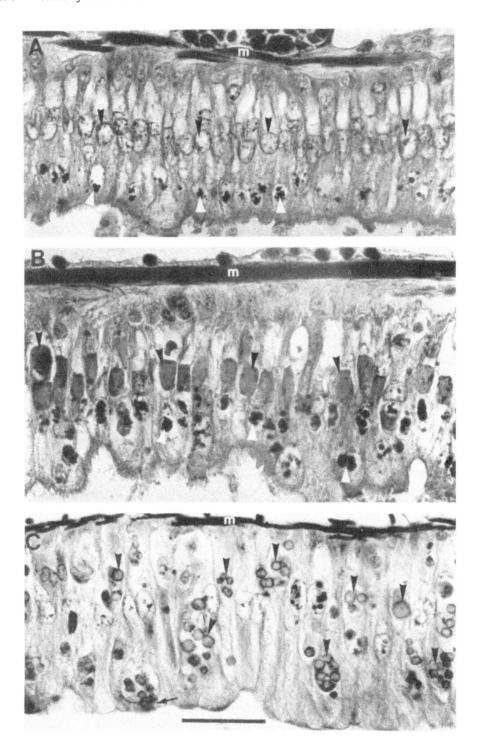

FIGURE 4.   Light micrographs of longitudinally sectioned larval (L₃) midgut epithelia of *Oryctes rhinoc-eros*. (Stained with Heidenhain's hematoxylin.) (A) Healthy. (B) and (C) Infected with baculovirus; note the light healthy nuclei in (A) (black arrowheads) strikingly contrasting with the dense, virus-infected, hypertrophied nuclei in B (black arrowheads); in both cases the cells contain many secretory granules (white arrowheads); midgut muscularis (m). (C) The virus-infected nuclei contain varying numbers of polyhedral virus occlusion bodies (arrowheads) which are gradually discharged into the midgut lumen (arrow); midgut muscularis (m). For details see Figure 5. (Bar for [A] to [C] = 50 μm.) (Photographed by A. M. Huger.)

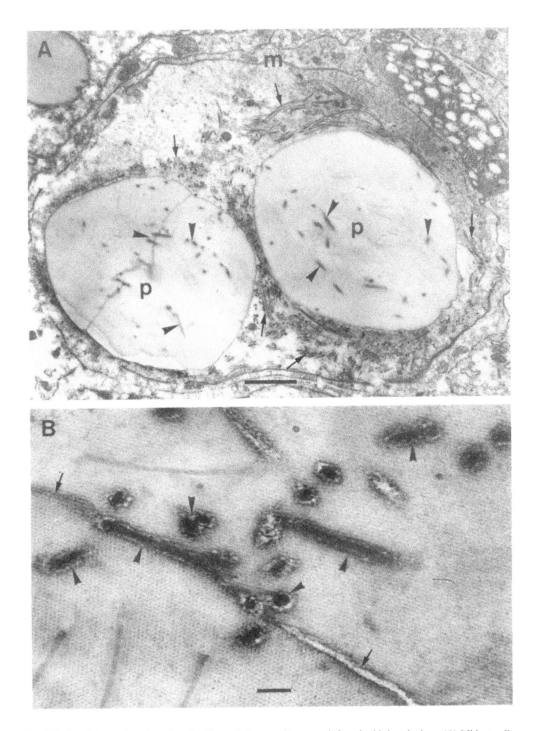

FIGURE 5. Thin sections from larval midgut of *Oryctes rhinoceros* infected with baculovirus. (A) Midgut cell showing two large polyhedra (p) with occluded virus particles (arrowheads) in the nucleus and many unoccluded virus particles and -filaments (arrows) in the nuclear space; a crack is seen in the polyhedron on the left; nuclear membrane (m). (Bar = 1 μm.) (B) Detail of a polyhedron in higher magnification showing virus particles and filaments (arrowheads) embedded in a paracrystalline proteinaceous lattice; note a crack (arrows) traversing the polyhedron. (Bar = 100 nm.)

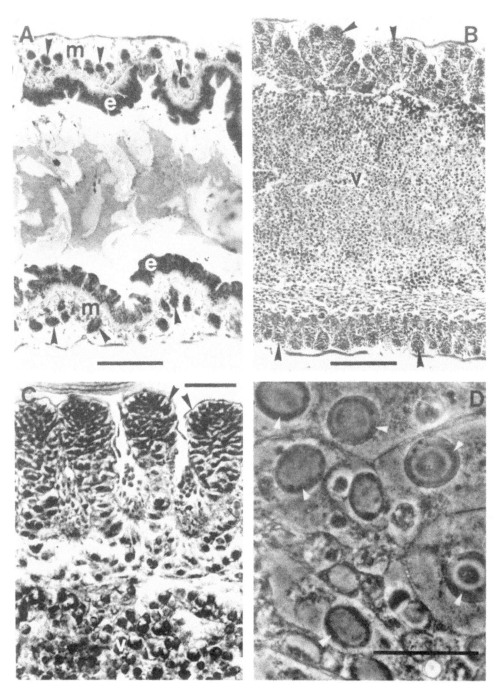

FIGURE 6.   Light micrographs showing the unique pathological changes in the midgut of *Oryctes rhinoceros* adults infected with baculovirus. (A) Longitudinally sectioned midgut of a healthy beetle; the midgut epithelium (e) is provided with an apical layer of regenerative crypts (arrowheads) embedded in the midgut muscularis (m). (B) Same type of section as (A), but 7 d after peroral infection with baculovirus; note that the midgut lumen is completely filled up with virus-infected cells (v) proliferating from the strikingly enlarged regenerative crypts (arrowheads). (Bars in [A] and [B] = 200 μm.) (C) Detail from (B) in higher magnification, showing regenerative crypts (arrowheads) proliferating cells from their apical region into the midgut lumen; note in the latter the virus-infected hypertrophied nuclei (v). ([A] to [C] stained with Heidenhain's hematoxylin.) (D) Wet mount preparation in phase contrast, showing greatly hypertrophied virus-infected midgut nuclei with the characteristic virus-harboring ring-zone (white arrowheads). For details see Figure 7. (Bars in [C] and [D] = 50 μm.) (Photographed by A. M. Huger.)

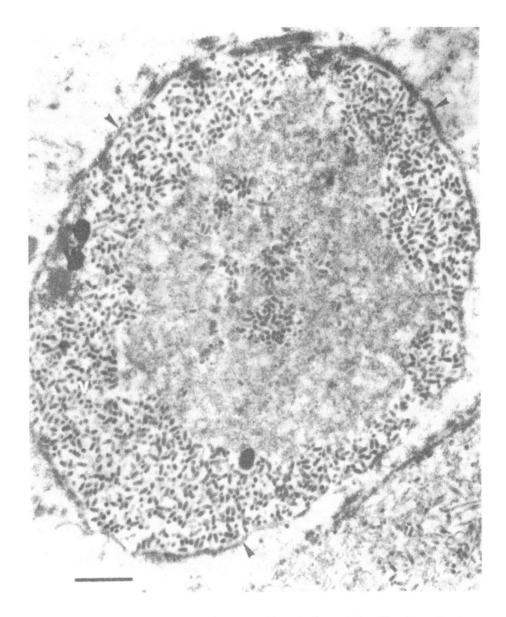

FIGURE 7.   Thin section through nucleus from the proliferated cell mass in the midgut lumen of a virus-vectoring *Oryctes rhinoceros* adult. Note the characteristic accumulation of virus particles (v) in the marginal ring-zone adjacent to the nuclear membrane (arrowheads). While virus assembly is still progressing in the nuclear center, the cytoplasm is already disintegrated. (Bar = 1 μm.) (Photographed by A. M. Huger.)

present in the marginal area of the nucleus, accomplishing the ring-zone cited before. Especially here clusters of virus particles may be densely packed in a two- or three-dimensional pseudocrystalline pattern[4] (see Figure 2B). Often viruses are closely attached or aligned to the inner layer of the nuclear membrane. Also virus particles budding through the nuclear and cytoplasmic membrane are frequently seen. It is, therefore, not surprising that they are repeatedly encountered in the cytoplasm, either singly or in groups. However, small virogenic foci or electron-dense bodies (approximately 10 μm in diameter) that harbor virions may also occur in the cytoplasm.[4,31]

b. Larval Virus Occlusion Bodies

While many investigators confirmed the nonoccluded nature of the *Oryctes* baculovirus, the facultative occurrence of proteinaceous, polyhedral virus occlusion bodies (VOBs) in nuclei of the larval midgut epithelium was documented by Huger[39] in earlier histopathologic studies (see Figure 4C). The factors responsible for this phenomenon are not yet clear. It is noteworthy that VOBs were preferably found in field-collected larvae that had been reared for a longer period of time, largely on a rearing substrate of inferior nutritive value. Moreover, larvae generating VOBs commonly carried subacute infections. Accidental infections in the laboratory with low virus doses cannot be excluded. Yet also larvae artificially infected with virus repeatedly developed VOBs. It is further interesting to note that VOBs were only apparent in nuclei of the larval midgut, not in other tissues liable to infection or in adults at all. Similar observations were also made with larvae of *O. monoceros*.[7]

Electron microscope studies disclosed that nuclei generating VOBs usually supported only slight or moderate virus replication. On the other hand, VOBs were lacking in nuclei with intensive virus reproduction as well as in larvae dying of acute infections. The number of VOBs per nucleus varied greatly from one or two voluminous occlusions up to several or many smaller ones (see Figures 4C and 5A). Accordingly, their size ranges considerably from 2 μm up to about 10 to 12 μm for single giant forms. VOBs readily dissolve in weak alkaline solutions.

Thin sections revealed that the proteinaceous VOBs containing both virus rods and virus filaments have a distinct paracrystalline substructure comparable to that of VOBs of subgroups A and B[39] (see Figure 5B). The center-to-center distance of the parallel lattice lines averages 12.5 nm and is thus about twice that of the classical polyhedral VOBs.[7] Varying amounts of virus available in the nuclei remain unoccluded. VOBs repeatedly show cracks (see Figure 5); they may agglomerate to large clusters. There are also nuclei with no visible morphological virus forms but harboring one or some paracrystalline polyhedral bodies also devoid of virus. Similar midgut nuclei were also observed in obviously healthy larvae. This could imply that in infected larvae, virus particles become occluded by chance in nuclear proteinaceous inclusions or deposits formed under yet unknown conditions. It seems less likely that a virus strain with genes coding for polyhedrin synthesis is involved in this phenomenon.

c. Adults

The mode of virus replication and the induced cytopathic changes as described for larvae also apply to adults. However, the chronically infected adults undergo a unique histopathologic transformation of their midgut.[9] At the early stage of disease, virus reproduction only takes place in the hypertrophied nuclei of the midgut epithelium. As the midgut cells get out of function, a severe cell proliferation is induced in the adjacent layer of regenerative crypts. By this "repair mechanism" the badly affected midgut epithelium is pushed off into the gut lumen. Due to continued vigorous cell proliferation the midgut lumen is completely filled up with a dense aggregation of cells within 1 to 2 weeks (see Figures 6A to C). Since the proliferating cells already have become infected while leaving the crypts, their hypertrophied nuclei finally are packed with virus (see Figures 6D and 7). While the cytoplasm disintegrates, the virus-laden nuclei usually show a large cisterna formed by distention of the outer layer of the nuclear membrane.[9] Enormous amounts of virus are produced and excreted for many weeks by the mobile adults, thus effecting self-control as described before.

Subsequent to the peculiar pathological changes in the adult midgut, the virus infection gradually spreads to other organs and tissues such as fat body, tracheal matrix, muscles, Malpighian tubules, and blood cells.[9] Moreover, virions have been found in the nucleus and cytoplasm of spermatids, in the cells and lumens of accessory glands, in the ejaculatory canal, as well as in the chorionated oocytes, follicle cells, and spermatheca.[31,40]

B. HZ-1 BACULOVIRUS

1. Origin and Host Range

So far only one nonoccluded baculovirus is known from Lepidoptera. This so-called Hz-1 virus was first described in 1978 by Granados et al.[5] as a persistent agent in an established cell line (IMC-Hz-1) derived from adult ovarian tissue of *Heliothis zea*. In earlier studies this cell line appeared to be nonsusceptible to most insect viruses (HzNPV, TnNPV, AcNPV) or displayed unexpected cytopathic effects. It was possible to transmit the Hz-1 virus from the persistently infected carrier culture to several lepidopterous tissue cultures, e.g., IPLB-1075 *(H. zea)*, TN-368 *(Trichoplusia ni)*, IPLB-SF-21 *(Spodoptera frugiperda)*, and IPLB-65Z *(Lymantria dispar)*.[5,41,42] Once infected with the Hz-1 virus, these cell cultures were refractory to NPV infections with one exemption: persistently infected *T. ni* cells were still susceptible to AcNPV without activation of Hz-1 virus. Retrospectively, it became clear that the nonsusceptibility of the IMC-Hz-1 cells to NPVs was due to virus interference (cross protection) and that spontaneous cytopathic effects were caused by induction of a productive Hz-1 virus replication.[43] The latter condition was affected not only by HzNPV, AcNPV, SfNPV, SlNPV and HaGV, but also by the homologous Hz-1 virus and by active as well as UV- or heat-inactivated TnNPV. No activation of Hz-1 virus occurred with *Iridovirus* (Type 22) or with mitomycin C.[42]

Hz-1 virus apparently did not replicate in cultured cells from *Estigmene acrea*[5] and *Manduca sexta*.[41] Also no clear infections were obtained after inoculation of Hz-1 virus into larvae of *H. zea, H. armigera, E. acrea, S. frugiperda,* and *S. littoralis*.[5,42]

2. Virus Structure and Composition

As a typical baculovirus, the Hz-1 virus has a rod-shaped enveloped nucleocapsid.[5] In connection with cloning of this virus two types of virus populations could be isolated: a standard virus and, in a low percentage, a defective virus. The bacilliform standard virus particles measure $414 \pm 30 \times 80 \pm 3$ nm (see Figure 8B); they contain a superhelical, circular dsDNA with a mean relative molecular weight about 156×10^6.[45,46] Thus, the length of the virion and its genome size are approximately twice that of the *Oryctes* virus. By centrifugation in a neutral CsCl gradient, the buoyant density of the Hz-1 virus DNA was determined to be 1.7024 g/ml, which corresponds to a G + C content of 43%. Altogether 28 structural proteins could be identified ranging from M_r 14 to 153×10^3; 14 of them were found to be glycosylated.[47] The defective Hz-1 virus particles, however, are more heterogeneous in length (370 ± 76 nm) (see Figure 8C) and contain genomes with deletions of up to 100 kb. However, the structural proteins of the defective particles are the same as in the standard virus.[43]

3. Replication and Cytopathology

Infections of permissive cell lines with standard virus induce cell lysis within 12 to 24 h. Inoculations preferably containing defective virus particles usually result in the survival of less than 10% of the cells carrying persistent infections. Defective Hz-1 virus populations contain small amounts of standard virus particles which are probably required for the replication of the defective particles. Since the latter do not only induce persistent infections but also are able to interfere with the infection of standard virus, they are denoted as defective interfering (DI) particles.[43] Inoculation of TN-368 cells with DI particles results in the synthesis of all 37 virus-induced proteins promoted by the standard virus.[44] However, there are differences in the rate of protein synthesis. In the presence of DI particles the yield of nine structural and nine nonstructural virus proteins is reduced. Contrary to occluded baculoviruses, in Hz-1 virus infections the shutdown of host protein synthesis occurs early in the infection cycle. In IMC Hz-1 cell lines the Hz-1 virus causes cytopathic effects in less than 1% of the cells as compared to 90 to 100% in TN-368 cell cultures.[5]

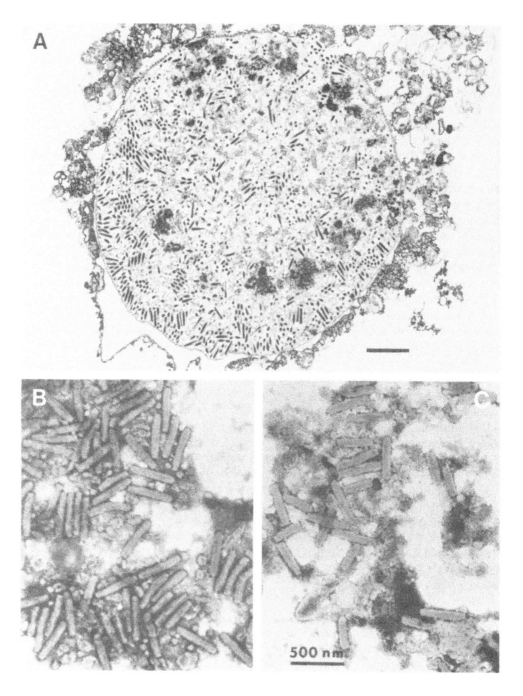

FIGURE 8.   (A) Thin section of TN-368 cell showing reproduction of Hz-1 virus in the nucleus. (Bar = 1 μm.) (From Ralston, A. L., Huang, Y., and Kawanishi, C. Y., *Virology,* 115, 33, 1981. With permission.) (B) Standard and (C) defective Hz-1 virus particles (negatively stained with phosphotungstic acid). (Bars [B] and [C] = 500 nm.) (From Burand, J. P., Wood, H. A., and Summers, M. D., *J. Gen. Virol.,* 64, 391, 1983. With permission.)

With TN-368 cells, entry of Hz-1 virus occurred by fusion of the virus envelope with the cellular membrane and insertion of the nucleocapsid into the cell cytoplasm. However, enveloped virus particles within phagocytic vacuoles were occasionally observed. The nucleocapsids became attached to nuclear pores where uncoating took place.[43] First morphogenic changes at 8 h p.i. were: rounding of the cells, hypertrophy of the nucleus, and margination of its chromatin. In the nuclear virogenic stroma, nucleocapsids assembled within accumulations of open-ended vesicular membrane structures apparently formed *de novo*. Finally, the enlarged nucleus contained numerous enveloped rod-shaped virus particles, partly enveloped nucleocapsids, nonenveloped nucleocapsids, and also empty capsids (see Figure 8A). By 16 h p.i. openings in the nuclear membrane allowed the virus particles to enter the cytoplasm. At 18 to 24 h p.i. virus particles were released into the culture medium by lysis of the cells.[5,43] This stage is displayed by plaques in permissive cell monolayers.

While in persistently infected cell cultures only few cells lyse, the others seem to be unaffected as the expression of virus genes is suppressed. However, derepression and induction of virogenesis followed by cell lysis is possible, if the latent virus infection is triggered by superinfection. From this it is suggested that in persistently infected cells like IMC-Hz-1 the DI particles induce self-limitation of the pathogenic process. As Hz-1 virus can give rise to both productive and suppressed replication in a number of cell lines, useful systems for studying persistent virus infections are available.

C. BACULOVIRUS OF CRICKETS

1. Host Range

The cricket baculovirus was isolated by Huger[6] from heavily diseased nymphs and adults of the field cricket, *Gryllus campestris*, derived from rearings in southern Germany. Laboratory studies confirmed that subsequent losses in colonies of other cricket species, such as *Gryllus bimaculatus*, *Teleogryllus oceanicus*, and *Teleogryllus commodus*, were also caused by this virus. Experiments to transmit the virus perorally to various developmental stages of the house cricket, *Acheta domesticus*, were not successful.

2. Gross Pathology

Virus infections in rearings primarily occurred during nymphal development, especially by cannibalistic feeding on moribund or dead specimens. In peroral infection experiments with early nymphal stages of *G. bimaculatus* the period of lethal infection was between 3 to 12 weeks. Affected crickets are smaller in size; they may molt repeatedly while becoming progressively uncoordinated and lethargic until they finally die. Sometimes infected specimens get crippled. In the advanced stage of disease, the crickets are often strikingly swollen and harbor an enormous amount of viscous and milky opalescent hemolymph.[6]

3. Virus Structure and Composition

Pure virus isolates were obtained by centrifuging prepurified homogenates of diseased crickets in 30 to 63% (w/v) sucrose gradients. The virus particles formed a distinct major band at a density of about 1.23 g/ml. Negatively stained virions appear irregularly ellipsoidal to rod-shaped, and mostly have a longitudinal fold (see Figure 9B). Their size ranges from 145 to 240 nm in length and from 80 to 100 nm in width with an average of 177 × 87 nm. Virus particles penetrated by stain reveal a single rod-shaped, capped nucleocapsid (average size 162 × 66 nm) comprised by an envelope[6] (see Figure 9C). Thin sections display the latter to be of the unit-membrane type (see Figure 10B). Moreover, a possibly filamentous structure is repeatedly seen between one side of the nucleocapsid and the dilated envelope. This structure is also evident in thin sections (see Figures 9C and 10B).

The molecular weight of purified DNA as calculated from restriction fragments amounted to about 90 kb. Preliminary analysis of virus proteins separated by SDS-PAGE showed profiles being distinctly different from those of the *Oryctes* virus. There was very little cross reaction between the proteins of both viruses in Western blot-ELISA tests.[48]

4. Cytopathology

Nuclei of the fat body are the principal sites of virus reproduction. They undergo extreme hypertrophy reaching diameters up to 115 μm. At an advanced stage of infection they exhibit

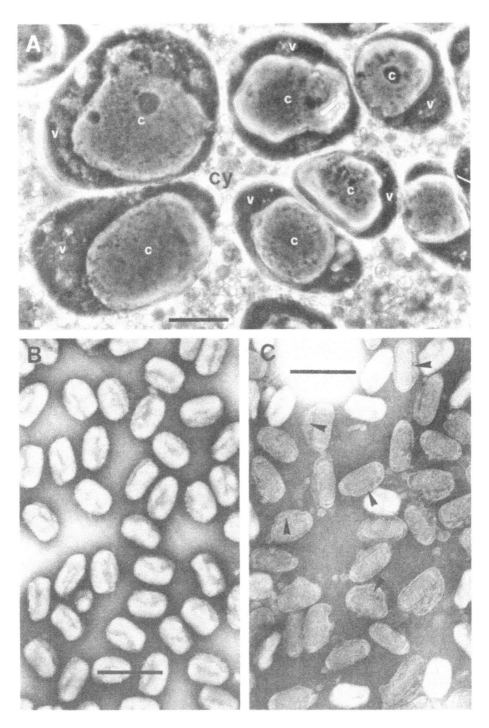

FIGURE 9.   (A) Phase contrast wet mount preparation of nymphal fat body cells of *Gryllus campestris* infected
with baculovirus; note the enormously hypertrophied nuclei with a dense, virus-harboring, peripheral ring-zone
(v) comprising the lighter central area (c); compare with Figure 10A; cytoplasm (cy). (Bar = 25 μm.) (B) and
(C) Purified *Gryllus* baculovirus (negatively stained with phosphotungstic acid) in (C) with penetration of stain;
virions in (B) show a longitudinal fold; (C) capped nucleocapsids within envelopes which show a unilateral
dilatation holding a possibly filamentous structure (arrowheads); compare with Figure 10B. (Bars [B] and [C]
= 250 nm.) (Photographed by A. M. Huger.)

a very characteristic ring-zone in their marginal area accomplished by myriads of virions being in rapid Brownian movement under phase contrast and dark field microscopy (see Figure 9A). As seen in thin sections, this viral ring-zone comprises a dense accumulation of *de novo* formed, rounded vesicles in the nuclear center comparable to *Oryctes* virus infections.[4,6] Nucleocapsids assemble within these vesicles, thus acquiring their envelope[7] (see Figures 10A and B). Even at the final stage of virus reproduction the nuclei still harbor plenty of vesicular envelope material.

D. ROD-SHAPED VIRUS AND VIRUS-LIKE PARTICLES OF INSECTS WITH AFFINITIES TO THE GROUP

As already mentioned, there is a number of preliminary descriptions of nonoccluded baculovirus-like particles from various insect orders, mostly based on studies of thin sections. The main results known so far are listed in Table 1. Therefore, only some additional information will be presented here.

Two possible baculoviruses were found in larval midgut nuclei of phantom midges (Diptera). As shown in Figures 11C and D, the virions of *Chaoborus crystallinus* have a unilateral evagination which apparently contains a tail-like structure similar to that of the *Oryctes* virus. Especially at the nuclear periphery, virions accumulate to paracrystalline clusters (see Figure 11A). They are often arranged in chainlike aggregates[30] (see Figure 11B). A similar virus infection was reported from *Chaoborus astictopus*.[27]

Apart from *O. rhinoceros*, nonoccluded rod-shaped viruses were briefly described from two other coleopteran species, *Gyrinus natator* and *Diabrotica undecimpunctata*. In symptomless adults of the whirligig beetle, *G. natator*, virus replication in nuclei of the midgut epithelium is similar to that of the *Oryctes* virus. Viral DNA synthesis was demonstrated by autoradiographic studies using [³H]thymidine.[49] In *D. undecimlineata* rod-shaped virus-like particles were reported from nuclei of hemocytes, whereas a filamentous virus type occurred in nuclei of the midgut epithelium.[50]

Also from two orthopteran species, the stick insect *(Bacillus rossius)* and the house cricket *(Acheta domesticus)*, rod-shaped baculovirus-like particles were described. In *B. rossius* virus particles in midgut nuclei were often packed in a paracrystalline array. Infected hosts showed no gross symptoms of disease.[51] From symptomless crickets *(A. domesticus)* virus-like particles were isolated from the hemolymph. It was suggested that the virus was derived from infected hemocytes.[52]

Finally, the detection of baculovirus-like particles in fat body sections of two homopterous species (*Aphis* sp. and *Pentalonia nigronervosa*)[53] and in the midgut epithelium of the flea *(Pulex simulans)*[53a] indicate that many more nonoccluded baculovirus infections will be found in the future, provided that adequate diagnostic techniques are applied.

E. VIRUS AND VIRUS-LIKE PARTICLES OF INSECTS WITH FILAMENTOUS OR ELONGATED NUCLEOCAPSIDS

Review of literature still reveals a small group of viruses with extremely elongated rod-shaped or filamentous and folded nucleocapsids. Though as yet only some preliminary analyses on virus genomes and proteins have been done, it appears that for most or all of them there is currently no obvious taxonomic group of DNA viruses in which they could be placed. Since these nonoccluded viruses have some morphological similarities to baculoviruses, they should be briefly referred to in this chapter.

Table 2 presents known data on virus infections of *Apis mellifera* and three coleopterous species which are characterized by a filamentous, flexuous nucleocapsid often folded within the virus envelope. Different substructures of complete virus particles have been reported:

1. The ellipsoidal virus particles of *A. mellifera* contain a folded filamentous nucleocapsid

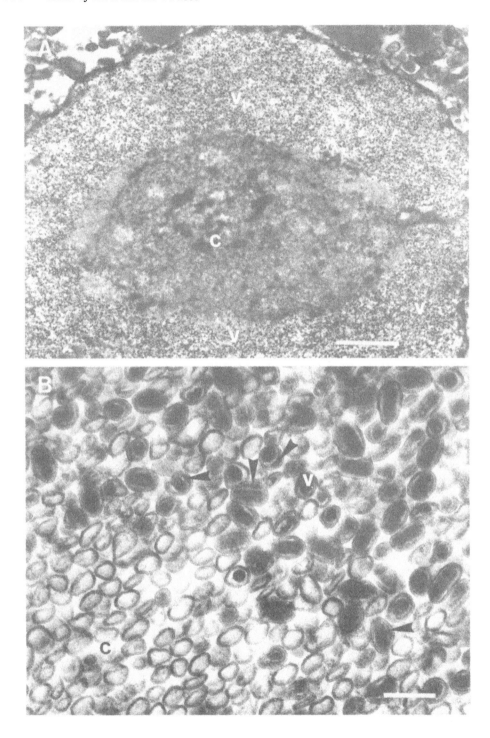

FIGURE 10.   Thin sections of nuclei of *Gryllus campestris* fat body cells infected with baculovirus. (A)
Nucleus densely packed with virus particles (v) in the marginal ring-zone comprising the accumulation of
virus envelope material in the center (c). (Bar = 2 μm.) (B) Transitional region between nuclear ring-zone
(v) and -center (c) showing rod-shaped virions and vesicular virus envelope material, respectively; virions
in suitable orientation exhibit a unilateral dilatation holding a possibly filamentous structure (arrowheads).
(Bar = 200 nm.) (Photographed by A. M. Huger.)

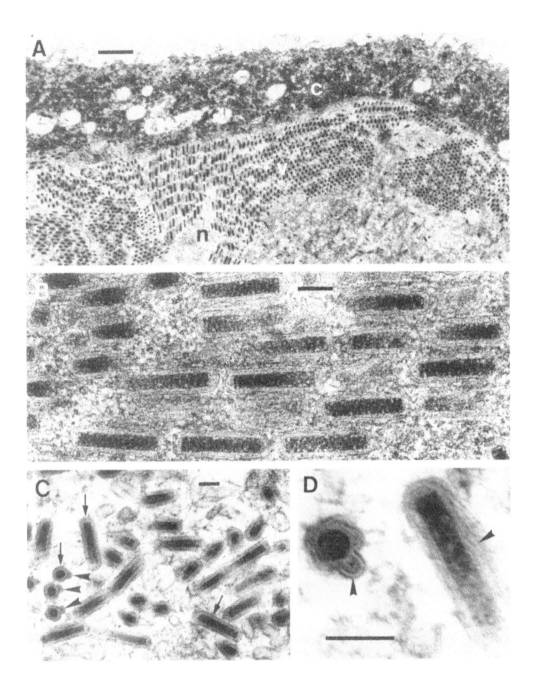

FIGURE 11. Thin sections of larval midgut cells of *Chaoborus crystallinus* infected with a possible baculovirus. (A) Part of infected cell nucleus (n) showing paracrystalline clusters of virus particles (v) in the marginal zone; cytoplasm (c). (Bar = 1 μm.) (B) Chainlike aggregates of longitudinally sectioned enveloped virus rods; note the blunt ends of the electron-dense nucleocapsids. (C) Free virus particles (arrows) in the gut lumen; note the typical unilateral bulge (arrowheads) of three transversely sectioned virions at the left. (D) Higher magnification of two virus particles in transverse (left) and longitudinal (right) section showing details of the unilateral bulge (arrowheads) which holds an electron-dense, possibly tail-like structure. (Bars [B] to [D] = 100 nm.) ([A] to [D] courtesy of Dr. R. Larsson. [B] to [D] from Larsson, R., *J. Invertebr. Pathol.*, 44, 178, 1984. With permission.)

enveloped by a single trilaminar unit membrane (see Figure 12). Updated results suggest the dsDNA to be linear with a M_r about 30×10^6.[54-56]

2. The virus-like particles of *Tenebrio molitor*[57,58] and *Scolytus scolytus*[59] are ovoidal in shape. Their flexuous, filamentous nucleocapsid is also folded within an envelope of unit-membrane structure. In addition, most of the particles become further enclosed by a characteristic coat or shell of fibrillar or lamellar material and by an outer membrane.

3. Thin sections of nuclei in the midgut of adult *Diabrotica undecimlineata* displayed filamentous virus-like particles of variable length. Their nucleocapsid is sometimes folded like a hairpin within the envelope.[50]

Infection experiments have been conducted only with the virus of *A. mellifera*. It was transmissible to bees *per os* and by intrahemocoelic injection. Virus infections induced weakening of worker bees and caused some mortality. Nuclei of fat body cells were the primary and principal sites of infection. Subsequently virus particles were also found in other tissues except midgut cells. Finally virus particles accumulated in the hemolymph which attained an opalescent, milky white appearance. No infections were obtained after peroral or parenteral applications of the virus to larvae and queens, respectively.[54,57]

Table 2 further lists two extremely elongated rod-shaped insect viruses described from flies. In the tsetse fly *(Glossina pallidipes)* the virus replicates in nuclei of the salivary glands. The virus particles were of two size classes with average lengths of 869 and 1175 nm. Unlike baculoviruses, no fully enveloped virions were found in purified preparations. There was some evidence for a double-stranded, linear virus DNA.[60] The virus was infectious for adults when fed or injected. Its transmission to offspring probably allows a low incidence of virus persistence in host populations. Besides salivary gland hyperplasia the virus also induces reduced life span and pathological changes in gonads resulting in sex ratio distortion. Moreover, insemination rates and fecundity are reduced.[61]

A similar virus-induced hyperplasia of salivery glands and atrophy of gonads was found in the syrphid fly, *Merodon equestris*. Long, rod-shaped nucleocapsids assemble in the nucleus, pass through the nuclear membrane, and become enveloped in the cytoplasm.[62]

This mode of virus morphogenesis recently was also reported from the ichneumonid wasp, *Diadegma terebrans*, where reproduction of long virus-like particles ($>1000 \times 65$ nm) was encountered in the apical region of the calyx. It is noteworthy that this virus coexists with a typical polydnavirus.[63]

IV. POSSIBLE NONOCCLUDED BACULOVIRUSES OF ARTHROPODS OTHER THAN INSECTS

A. VIRUSES OF ACARINA
1. Members of the Group

As yet, two nonoccluded possible baculoviruses have been reported from phytophagous mites. The virus infection of the European red mite, *Panonychus ulmi,* collected in Ontario was at first briefly described by Bird.[64] Reed and Hall[65] identified the virus disease of the citrus red mite, *Panonychus citri,* in California.

2. Symptoms, Host Range, and Transmission

Virus-diseased mites of *P. ulmi* turn from olive green to reddish brown apparently due to reduced ingestion of plant cell contents. They usually accumulate birefringent spheroidal crystals in their midgut which facilitate diagnosis of infected specimens. Mites became infected by ingestion of virus-contaminated feces and residues of diseased individuals. Suspensions of infected mites triturated in water were rather inefficient inocula. No evidence

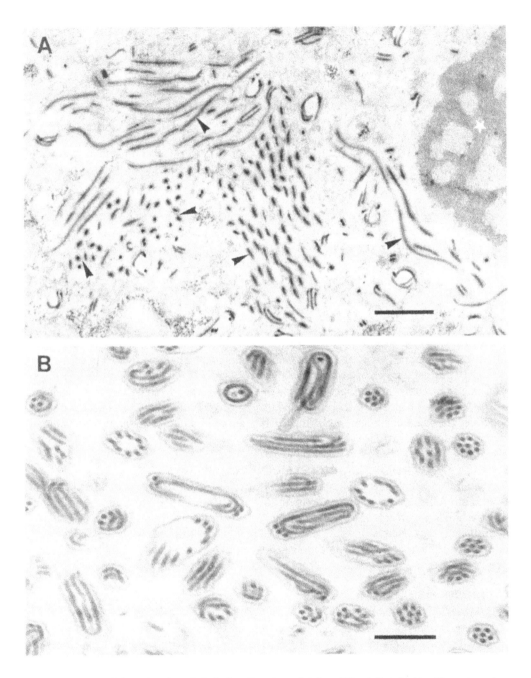

FIGURE 12. (A) Thin section through fat body cell nucleus of *Apis mellifera* infected with a filamentous virus; note virus filaments (arrowheads) and virogenic stroma (asterik). (Bar = 500 nm.) (B) Thin section through pellet of the filamentous bee virus isolated from hemolymph; note folded filamentous nucleocapsids within a trilaminate envelope. (Bar = 250 nm.) (From Clark, T. B., *J. Invertebr. Pathol.*, 32, 332, 1978. With permission.)

of transovarial transmission was obtained. Introduction of the virus into orchard populations induced epizootics which rapidly reduced population density. Natural epizootics were found only in dense populations. Attempts to infect *Tetranychus urticae* failed.[66]

Also the virus of *P. citri* could not be transmitted to most other phytophagous mites such as *Eotetranychus sexmaculus*, *Oligonychus punicae*, *Tetranychus urticae*, *T. pacificus*, and *T. evansi.* Only in *T. cinnabarinus* a low percentage of virus infections was obtained.[67] Refractoriness against the virus was also demonstrated with predatory mites such as *Amblyseius hibisci, A. limonicus,* and *Typhlodromus occidentalis.*[68] In virus transmission experiments, mites of *P. citri* did not acquire the virus when probing plant cells previously fed on by virus-infected hosts, yet they encountered the virus in feces and debris of infected mites on the plant surface.[69] In laboratory studies, dried deposits of aqueous virus suspensions sprayed on lemons lost most of their infectivity within some hours, whereas natural deposits of diseased mites remained infective for at least 8 d.[70] Field studies showed that the virus is an important mortality factor in the suppression of *P. citri* populations where it is also easily recognized by the presence of birefringent crystals in the midgut of infected mites.[68]

3. Virus Structure

Both viruses from phytophagous mites are rod-shaped. In thin sections the enveloped virions of *P. ulmi* measured about 200 × 90 nm with an electron-dense nucleocapsid of about 150 × 38 nm.[64] Average dimensions of the virions and nucleocapsids of *P. citri* were 266 × 111 nm and 194 × 58 nm, respectively.[65] Isolated virus particles of *P. citri* measured 180 to 233 × 71 to 86 nm. Their structure is very similar to that of the *Oryctes* virus[71] (see Figure 13).

4. Cytopathology

As shown by thin sections, virus replication in *P. ulmi* takes place in the nuclei of fat body cells,[64] while in *P. citri,* it occurs in nuclei of midgut and hindgut cells.[65] In each case enveloped virus particles escape from the nucleus by a process of budding. In this way they may either be singly or multiply enclosed by an envelope derived from the nuclear membrane[64,65] (see Figure 13). It is suggested that in *P. citri* virions move from the infected midgut into the adjacent hindgut. From infected hindgut cells, cytoplasmic strands containing virus are continually sloughed into the gut lumen to be defecated.[72]

B. VIRUS OF *PISAURA MIRABILIS* (ARANEINA)

The virus infection of the spider *Pisaura mirabilis* was detected by Morel et al.[73] Virus particles are rod-shaped with dimensions of 160 × 110 nm. Their structure differs from other baculoviruses in that the nucleocapsid (300 × 40 nm) is always bent in a U-shape within the envelope. Virus replication takes place in nuclei of hepatopancreatic cells. Experimental infections *per os* or by injection do not induce overt signs or symptoms of disease. After oral infection, virions become excreted only 7 weeks later.

C. VIRUSES OF CRUSTACEA
1. Members of the Group

Among a variety of viruses described from marine crustaceans, nonoccluded rod-shaped viruses with structural similarities to members of baculovirus subgroup C have been reported from the European shore crabs *Carcinus maenas*[74] at the French Atlantic coast and *Carcinus mediterraneus*[75,76] at the French Mediterranean coast; from the blue crab, *Callinectes sapidus,*[77-79] collected in bays at the Atlantic coast of the U.S. and from cultured Kuruma shrimps, *Penaeus japonicus,*[80] in Japan. While baculovirus infections of *C. sapidus* and *C. maenas* have not yet been associated with fatal disease, the so-called τ-virus may induce lethal infections in *C. mediterraneus.*

2. Virus Structure

The nonoccluded baculovirus of *C. maenas* has a rod-shaped enveloped nucleocapsid and approximate dimensions of 300 to 320 × 90 to 100 nm.[74] Two similar types of baculovirus A and B were reported from *C. sapidus.*[79] In thin sections the nucleocapsids of

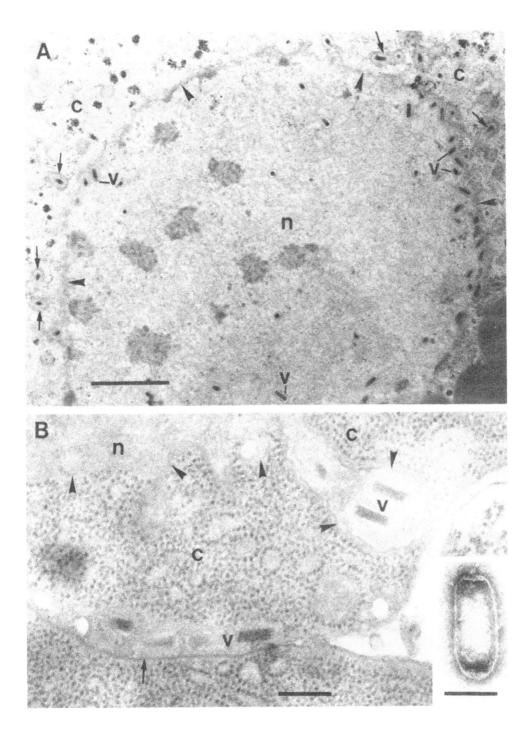

FIGURE 13. Thin sections of midgut epithelial cells of *Panonychus citri* infected with baculovirus. (A) Nucleus (n) with virus particles (v) budding through the nuclear membrane (arrowheads) into the cytoplasm (c), thus acquiring an additional membrane (arrows). (Bar = 1 μm.) (B) Detail of this budding process; symbols as in (A); some virus particles are attached to the cell membrane (arrow) below. (Bar = 200 nm.) Inset, nucleocapsid of *P. citri* virus (negatively stained with uranyl acetate), showing capped ends. (Bar = 100 nm.) ([A] and [B] from Reed, D. K. and Hall, I. M., *J. Invertebr. Pathol.*, 20, 272, 1972. With permission.)

baculovirus A are approximately 300 nm in length and 50 nm in width.[78] The τ-virus of *C. mediterraneus* shares the general characteristics of nonoccluded baculoviruses, but differs by the occurrence of both rod-shaped and bow-shaped enveloped nucleocapsids in infected cell nuclei. The virions average 350 × 80 nm in size.[76] During purification their envelope undergoes mug-shaped distortions similar to that of the *Oryctes* virus (see Section III). Negatively stained nucleocapsids (300 to 320 × 65 to 70 nm) also appear capped and furthermore display a regularly striated surface structure. In addition, a tail-like structure (13 to 15 nm in diameter and 100 to 200 nm in length) resembling that of the *Oryctes* virus may protrude from one end of the nucleocapsid; it is suggested that it is nucleoproteinic in nature[75] (see Figure 14B).

3. Cytopathology

The baculovirus of *C. maenas* replicates in the nuclei of hemocytes and connective tissue cells.[74] While the baculovirus B of *C. sapidus* also affects hemocytes (see Figures 15A and B), the baculovirus A replicates in nuclei of epithelial cells of the hepatopancreas[79] (see Figure 15C). Both secretory and reserve cells are infected, but not the apical, actively dividing "embryonic" cells. Overtly infected nuclei are strongly hypertrophied. Infected cells are expelled into the lumen of the tubules, while cellular regeneration and tissue repair take place. Thin sections of nuclei infected with baculovirus A or B display the joint occurrence of mature virus particles and virus envelope material in which nucleocapsids may be assembled. Both virus types often form paracrystalline aggregations[77,78] (see Figures 15A to C).

The τ-baculovirus of *C. mediterraneus* primarily affects epithelial cells of the hepatopancreas (see Figures 14A, C, D) with subsequent focal infections in the midgut. Infected cells are characterized by nuclear hypertrophy and cytoplasmic vacuolization. Nuclei in the hepatopancreas often exhibit marginal paracrystalline virus accumulations (see Figure 14A) and migrate to the apical part of the cell. Complete cellular breakdown is followed by release of cell contents into the lumen of the tubule, thus contributing to virus dissemination in the digestive tract and finally into the external environment.[76]

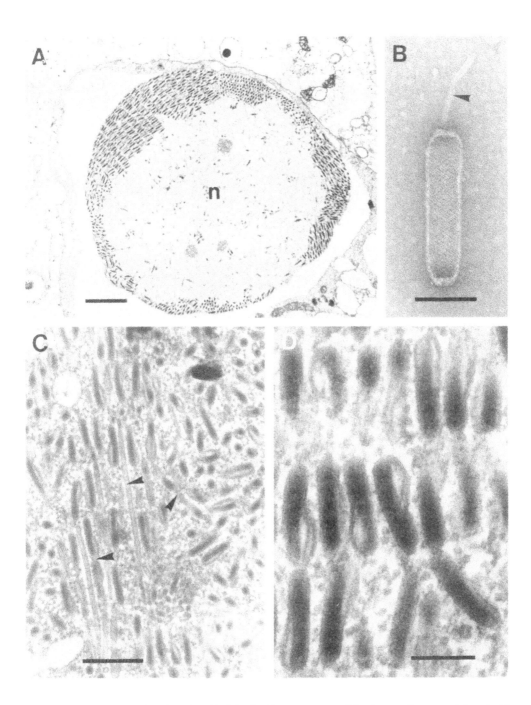

FIGURE 14. (A) Thin section through epithelial cell of hepatopancreas of *Carcinus mediterraneus* infected with τ-baculovirus; note in the nucleus (n) the paracrystalline aggregation of virus particles in a marginal ring-zone. (Bar = 2 μm.) (B) Nucleocapsid of τ-baculovirus (negatively stained with phosphotungstic acid), showing regular surface striations of the capsid shell and release of a nucleoproteinic strand (arrowhead). (Bar = 100 nm.) (C) Thin section of nucleus showing stages of morphogenesis of τ-virus particles with apparent involvement of filamentous structures (arrowheads). (Bar = 500 nm.) (D) Thin section of mature τ-virus particles. (Bar = 200 nm.) (Courtesy of Drs. J.-R. Bonami and J. Mari, Université des Sciences et Techniques du Languedoc, Laboratoire de Pathologie Comparée, Montpellier, France.)

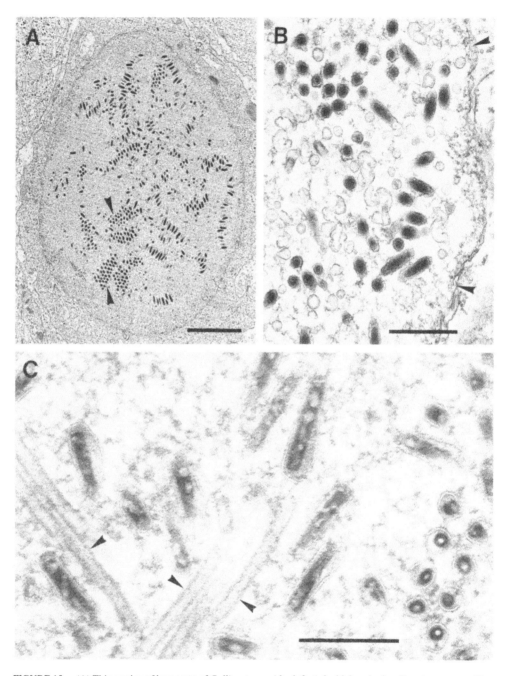

FIGURE 15.   (A) Thin section of hemocyte of *Callinectes sapidus* infected with baculovirus B; note paracrystalline arrays of virus particles (arrowheads) in the nucleus. (Bar = 2 μm.) (B) Same as (A), showing virus rods and vesicular virus envelope material in the nucleus; arrowheads at nuclear membrane. (Bar = 500 nm.) (C) Thin section of hepatopancreatic cell nucleus of *C. sapidus* infected with baculovirus A; note virus rods in longitudinal and cross section as well as filamentous virus envelope structures (arrowheads). (Bar = 250 nm.) (Courtesy of Dr. P. T. Johnson, National Marine Fisheries Service, Northeast Fisheries Center, Oxford, MD.)

REFERENCES

1. **Matthews, R. E. F.,** Classification and nomenclature of viruses. Fourth report of the International Committee on Taxonomy of Viruses, *Intervirology*, 17, 1, 1982.
2. **Fenner, F. and Gibbs, A.,** Cryptograms — 1982, *Intervirology*, 19, 121, 1983.
3. **Brown, F.,** The classification and nomenclature of viruses: summary of results of meetings of the International Committee on Taxonomy of Viruses in Sendai, September 1984, *Intervirology*, 25, 141, 1986.
4. **Huger, A. M.,** A virus disease of the Indian rhinoceros beetle, *Oryctes rhinoceros* (Linnaeus), caused by a new type of insect virus, *Rhabdionvirus oryctes* gen. n., sp. n., *J. Invertebr. Pathol.*, 8, 38, 1966.
5. **Granados, R. R., Nguyen, T., and Cato, B.,** An insect cell line persistently infected with a baculovirus-like particle, *Intervirology*, 10, 309, 1978.
6. **Huger, A. M.,** A new virus disease of crickets (Orthoptera:Gryllidae) causing macronucleosis of fatbody, *J. Invertebr. Pathol.*, 45, 108, 1985.
7. **Huger, A. M.,** unpublished data.
8. **Marschall, K. J.,** Introduction of a new virus disease of the coconut rhinoceros beetle in Western Samoa, *Nature (London)*, 225, 288, 1970.
9. **Huger, A. M.,** Grundlagen zur biologischen Bekämpfung des Indischen Nashornkäfers, *Oryctes rhinoceros* (L.), mit *Rhabdionvirus oryctes:* Histopathologie der Virose bei Käfern, *Z. Angew. Entomol.*, 72, 309, 1972/73.
10. **Zelazny, B.,** Studies on *Rhabdionvirus oryctes*. II. Effect on adults of *Oryctes rhinoceros, J. Invertebr. Pathol.*, 22, 122, 1973.
11. **Zelazny, B.,** Transmission of a baculovirus in populations of *Oryctes rhinoceros, J. Invertebr. Pathol.*, 27, 221, 1976.
12. **Bedford, G. O.,** Biology, ecology, and control of palm rhinoceros beetles, *Annu. Rev. Entomol.*, 25, 309, 1980.
13. **Bedford, G. O.,** Control of the rhinoceros beetle by baculovirus, in *Microbial Control of Pests and Plant Diseases 1970—1980*, Burges, H. D., Ed., Academic Press, London, 1981, 409.
14. **Kaske, R. and Huger, A. M.,** Ein Beispiel für biologische Schädlingsbekämpfung: Erfahrungen mit dem Indischen Nashornkäfer *(Oryctes rhinoceros)* im pazifischen Raum, in *Alternative Konzepte*, Vol. 50, Vogtmann, H., Boehncke, E., and Fricke, I., Eds., Verlag C. F. Müller GmbH, Karlsruhe, Federal Republic of Germany, 1986, 167.
15. **Huger, A. M.,** Virusanwendung als Komponente eines integrierten Programms zur Bekämpfung des Indischen Nashornkäfers, *Oryctes rhinoceros* (L.) (Col.), *Mitt. Dtsch. Ges. Angew. Entomol.*, 1, 246, 1978.
16. **Caltagirone, L. E.,** Landmark examples in classical biological control, *Annu. Rev. Entomol.*, 26, 213, 1981.
17. **Zelazny, B.,** Studies on *Rhabdionvirus oryctes*. I. Effect on larvae of *Oryctes rhinoceros* and inactivation of the virus, *J. Invertebr. Pathol.*, 20, 235, 1972.
18. **Zelazny, B.,** *Oryctes rhinoceros* populations and behavior influenced by a baculovirus, *J. Invertebr. Pathol.*, 29, 210, 1977.
19. **Zelazny, B.,** Methods of inoculating and diagnosing the baculovirus disease of *Oryctes rhinoceros, FAO Plant Prot. Bull.*, 26, 163, 1978.
20. **Monsarrat, P. and Veyrunes, J. C.,** Evidence of *Oryctes* virus in adult feces and new data for virus characterization, *J. Invertebr. Pathol.*, 27, 387, 1976.
21. **Paul, W.-D.,** Integrierte Bekämpfung von Palmenschädlingen in Tanzania, in *Monographs on Agriculture and Ecology of Warmer Climates*, Vol. 1, Verlag J. Margraf, Aichtal, Federal Republic of Germany, 1985.
22. **Lomer, C. J.,** Infection of *Strategus aloeus* (L.) (Coleoptera:Scarabaeidae) and other Dynastinae with *Baculovirus oryctes, Bull. Entomol. Res.*, 77, 45, 1987.
23. **Lomer, C. J.,** Release of *Baculovirus oryctes* into *Oryctes monoceros* populations in the Seychelles, *J. Invertebr. Pathol.*, 47, 237, 1986.
24. **Monsarrat, P., Veyrunes, J.-C., Meynadier, G., Croizier, G., and Vago, C.,** Purification et étude structurale du virus du Coléoptère *Oryctes rhinoceros* L., *C. R. Acad. Sci., Ser. D.*, 277, 1413, 1973.
25. **Payne, C. C.,** The isolation and characterization of a virus from *Oryctes rhinoceros, J. Gen. Virol.*, 25, 105, 1974.
26. **Payne C. C., Compson, D., and De Looze, S. M.,** Properties of the nucleocapsids of a virus isolated from *Oryctes rhinoceros, Virology*, 77, 269, 1977.
27. **Federici, B. A.,** Ultrastructure of baculoviruses, in *The Biology of Baculoviruses*, Vol. 1, Granados, R. R. and Federici, B. A., Eds., CRC Press, Boca Raton, FL, 1986, 61.
28. **Monsarrat, P., Revet, B., and Gourevitch, I.,** Mise en évidence, stabilisation et purification d'une structure nucléoprotéique intracapsidaire chez le *Baculovirus* d'*Oryctes rhinoceros* L., *C. R. Acad. Sci., Ser. D*, 281, 1439, 1975.
29. **Crawford, A. M. and Sheehan, C.,** Replication of *Oryctes* baculovirus in cell culture: viral morphogenesis, infectivity and protein synthesis, *J. Gen. Virol.*, 66, 529, 1985.

30. **Larsson, R.,** Baculovirus-like particles in the midgut epithelium of the phantom midge, *Chaoborus crystallinus* (Diptera, Chaoboridae), *J. Invertebr. Pathol.,* 44, 178, 1984.

31. **Monsarrat, P., Meynadier, G., Croizier, G., and Vago, C.,** Recherches cytopathologiques sur une maladie virale du Coléoptère *Oryctes rhinoceros* L., *C. R. Acad. Sci., Ser. D,* 276, 2077, 1973.

32. **Révet, B. and Monsarrat, P.,** L'acide nucléique du virus du Coléoptère *Oryctes rhinoceros* L.: un ADN superhélicoïdal de haut poids moléculaire, *C. R. Acad. Sci., Ser. D,* 278, 331, 1974.

33. **Révet, B. M. J. and Guelpa, B.,** The genome of a baculovirus infecting *Tipula paludosa* (Meig.) (Diptera): a high molecular weight closed circular DNA of zero superhelix density, *Virology,* 96, 633, 1979.

34. **Crawford, A. M., Ashbridge, K., Sheehan, C., and Faulkner, P.,** A physical map of the *Oryctes* baculovirus genome, *J. Gen. Virol.,* 66, 2649, 1985.

35. **Crawford, A. M., Zelazny, B., and Alfiler, A. R.,** Genotypic variation in geographical isolates of *Oryctes* baculovirus, *J. Gen. Virol.,* 67, 949, 1986.

36. **Crawford, A. M. and Zelazny, B.,** Recent progress with *Oryctes* baculovirus control of the coconut palm rhinoceros beetle, in *Program and Abstracts, SIP XX Annu. Meet., University of Florida,* Gainesville, FL, July 20—24, 1987, 110.

37. **Mohan, K. S., Jayapal, S. P., and Pillai, G. B.,** Diagnosis of baculovirus infection in coconut rhinoceros beetles by examination of excreta, *Z. Pflanzenkrankh. Pflanzenschutz,* 93, 379, 1986.

38. **Quiot, J.-M., Monsarrat, P., Meynadier, G., Croizier, G., and Vago, C.,** Infection des cultures cellulaires de Coléoptères par le ''virus *Oryctes*'', *C. R. Acad. Sci., Ser. D,* 276, 3229, 1973.

39. **Huger, A. M.,** Studies on pathological changes of the midgut of *Oryctes rhinoceros* infected with *Rhabdionvirus oryctes,* in *South Pacific Commission, UNDP(SF)/SPC Rhinoceros Beetle Project, Report June 1970—May 1971,* Noumea, New Caledonia, 1971, 201.

40. **Monsarrat, P., Duithoit, J.-L., and Vago, C.,** Mise en évidence de virions de type Baculovirus dans l'appareil génital du Coléoptère *Oryctes rhinoceros* L., *C. R. Acad. Sci., Ser. D,* 278, 3259, 1974.

41. **Ralston, A. L., Huang, Y., and Kawanishi, C. Y.,** Cell culture studies with the IMC-Hz-1 nonoccluded virus, *Virology,* 115, 33, 1981.

42. **Kelly, D. C., Lescott, T., Ayres, M. D., Carey, D., Coutts, A., and Harrap, K. A.,** Induction of a nonoccluded baculovirus persistently infecting *Heliothis zea* cells by *Heliothis armigera* nuclear polyhedrosis virus and *Trichoplusia ni* nuclear polyhedrosis virus, *Virology,* 112, 174, 1981.

43. **Wood, H. A. and Burand, J. P.,** Persistent and productive infections with the Hz-1 Baculovirus, in *Current Topics in Microbiology and Immunology,* Vol. 131, Doerfler, W. and Böhm, P., Eds., Springer-Verlag, Berlin-Heidelberg, 1986, 119.

44. **Burand, J. P. and Wood, H. A.,** Intracellular protein synthesis during standard and defective Hz-1 virus replication, *J. Gen. Virol.,* 67, 167, 1986.

45. **Huang, Y.-S., Hedberg, M., and Kawanishi, C. Y.,** Characterization of the DNA of a nonoccluded baculovirus, Hz-1V, *J. Virol.,* 43, 174, 1982.

46. **Burand, J. P., Wood, H. A., and Summers, M. D.,** Defective particles from a persistent baculovirus infection in *Trichoplusia ni* tissue culture cells, *J. Gen. Virol.,* 64, 391, 1983.

47. **Burand, J. P., Stiles, B., and Wood, H. A.,** Structural and intracellular proteins of the nonoccluded baculovirus Hz-1, *J. Virol.,* 46, 137, 1983.

48. **Harvey, J. P., Naser, W. L., and Huger, A. M.,** unpublished data, 1986.

49. **Gouranton, J.,** Development of an intranuclear nonoccluded rod-shaped virus in some midgut cells of an adult insect, *Gyrinus natator* L. (Coleoptera), *J. Ultrastruct. Res.,* 39, 281, 1972.

50. **Kim, K. S. and Kitajima, E. W.,** Nonoccluded baculovirus- and filamentous virus-like particles in the spotted cucumber beetle, *Diabrotica undecimpunctata* (Coleoptera:Chrysomelid), *J. Invertebr. Pathol.,* 43, 234, 1984.

51. **Scali, V., Montanelli, E., Lanfranchi, A., and Bedini, C.,** Nuclear alterations in a baculovirus-like infection of midgut epithelial cells in the stick insect, *Bacillus rossius, J. Invertebr. Pathol.,* 35, 109, 1980.

52. **Grégoire, C.,** Virus-like bodies in the blood of the house cricket, *J. Gen. Microbiol.,* 5, 121, 1951.

53. **Kitajima, E. W., Costa, C. L., and Sá, C. M.,** Baculovirus-like particles in two aphid species, *J. Invertebr. Pathol.,* 31, 123, 1978.

53a. **Beard, C. B., Butler, J. F., and Maruniak, J. E.,** A Baculovirus in the flea, *Pulex simulans, J. Invertebr. Pathol.,* 54, 128, 1989.

54. **Clark, T. B.,** A filamentous virus of the honey bee, *J. Invertebr. Pathol.,* 32, 332, 1978.

55. **Bailey, L., Carpenter, J. M., and Woods, R. D.,** Properties of a filamentous virus of the honey bee *(Apis mellifera), Virology,* 114, 1, 1981.

56. **Bailey, L., Ball, B. V., and Overton, H.,** Unusual bee viruses, in *Proc. III. Int. Coll. Invertebr. Pathol.,* Brighton, Sept. 1982, 146.

57. **Devauchelle, G. and Vago, C.,** Présence de particules d'allure virale dans les noyaux des cellules de l'intestin moyen du Coléoptère *Tenebrio molitor* (Linné), *C. R. Acad. Sci., Ser. D,* 269, 1142, 1969.

58. **Thomas, D. and Gouranton, J.,** Development of viruslike particles in the crystal-containing nuclei of the midgut cells of *Tenebrio molitor, J. Invertebr. Pathol.,* 25, 159, 1975.

59. **Arnold, M. K. and Barson, G.,** Occurrence of viruslike particles in midgut epithelial cells of the large elm bark beetle, *Scolytus scolytus, J. Invertebr. Pathol.,* 29, 373, 1977.

60. **Odindo, M. O., Payne, C. C., Crook, N. E., and Jarrett, P.,** Properties of a novel DNA virus from the tsetse fly *Glossina pallidipes, J. Gen. Virol.,* 67, 527, 1986.

61. **Jaenson, T. G. T.,** Sex ratio distortion and reduced lifespan of *Glossina pallidipes* infected with the virus causing salivary gland hyperplasia, *Entomol. Exp. Appl.,* 41, 265, 1986.

62. **Armargier, A., Lyon, J.-P., Vago, C., Meynadier, G., and Veyrunes, J.-C.,** Mise en évidence et purification d'un virus dans la prolifération monstrueuse glandulaire d'insectes. Étude sur *Merodon equestris* F. (Diptère, Syrphidae), *C. R. Acad. Sci., Ser. D,* 289, 481, 1979.

63. **Krell, P. J.,** Replication of long virus-like particles in the reproductive tract of the ichneumonid wasp *Diadegma terebrans, J. Gen. Virol.,* 68, 1477, 1987.

64. **Bird, F. T.,** A virus disease of the European red mite *Panonychus ulmi* (Koch), *Can. J. Microbiol.,* 13, 1131, 1967.

65. **Reed, D. K. and Hall, I. M.,** Electron microscopy of a rod-shaped noninclusion virus infecting the citrus red mite, *J. Invertebr. Pathol.,* 20, 272, 1972.

66. **Putman, W. L.,** Occurrence and transmission of a virus disease of the European red mite, *Panonychus ulmi, Can. Entomol.,* 102, 305, 1970.

67. **Beavers, J. B. and Reed, D. K.,** Susceptibility of seven tetranychids to the nonoccluded virus of the citrus red mite and the correlation of the carmine spider mite as a vector, *J. Invertebr. Pathol.,* 20, 279, 1972.

68. **Reed, D. K.,** Control of mites by non-occluded viruses, in *Microbial Control of Pests and Plant Diseases 1970—1980,* Burges, H. D., Ed., Academic Press, London, 1981, 427.

69. **Tashiro, H., Beavers, J. B., Groza, M., and Moffitt, C.,** Persistence of a nonoccluded virus of the citrus red mite on lemons and in intact dead mites, *J. Invertebr. Pathol.,* 16, 63, 1970.

70. **Gilmore, J. E. and Munger, F.,** Stability and transmissibility of a viruslike pathogen of the citrus red mite, *J. Invertebr. Pathol.,* 5, 141, 1963.

71. **Reed, D. K. and Desjardins, P. R.,** Morphology of a non-occluded virus isolated from citrus red mite, *Panonychus citri, Experientia,* 38, 468, 1982.

72. **Reed, D. K., Tashiro, H., and Beavers, J. B.,** Determination of mode of transmission of the citrus red mite virus, *J. Invertebr. Pathol.,* 26, 239, 1975.

73. **Morel, G., Bergoin, M., and Vago, C.,** Mise en évidence chez l'Araignée *Pisaura mirabilis* Cl. d'un nouveau type de virus apparenté aux *Baculovirus, C. R. Acad. Sci., Ser. D,* 285, 933, 1977.

74. **Bazin, F., Monsarrat, P., Bonami, J.-R., Croizier, G., Meynadier, G., Quiot, J. M., and Vago, C.,** Particules virales de type baculovirus observées chez le crabe *Carcinus maenas, Rev. Trav. Inst. Pêches Marit.,* 38, 205, 1974.

75. **Pappalardo, R. and Bonami, J.-R.,** Infection des crustacés marins due à un virus de type nouveau apparenté aux baculovirus, *C. R. Acad. Sci., Ser. D,* 288, 535, 1979.

76. **Pappalardo, R., Mari, J., and Bonami, J.-R.,** τ(tau)Virus infection of *Carcinus mediterraneus:* histology, cytopathology, and experimental transmission of the disease, *J. Invertebr. Pathol.,* 47, 361, 1986.

77. **Johnson, P. T.,** Viral diseases of the blue crab, *Callinectes sapidus, Mar. Fish. Rev.,* 40, 13, 1978.

78. **Johnson, P. T.,** A baculovirus from the blue crab, *Callinectes sapidus,* in *Proc. 1st Int. Coll. Invertebr. Pathol.,* Kingston, Ont., Canada, 1976, 24.

79. **Johnson, P. T.,** New information on viral diseases of the blue crab, *Callinectes sapidus,* in *Abstr. Int. Coll. Invertebr. Pathol.,* Prague, September 1978, 55.

80. **Sano, T., Nishimura, T., Oguma, K., Momoyana, K., and Takeno, N.,** Baculovirus infection of cultured Kuruma shrimp, *Penaeus japonicus,* in Japan, *Fish Pathol.,* 15, 185, 1981.

Chapter 10

POLYDNAVIRIDAE

Peter John Krell

TABLE OF CONTENTS

I. INTRODUCTION

The Polydnaviridae is a family consisting of certain viruses which replicate in the ovaries of some species of parasitic hymenoptera.[1-3] They are distinguished by their polydisperse (or multipartite) superhelical DNA genome, a characteristic which is the basis for the etymology of the family name (namely, poly-DNA-viridae). Although originally classified in subgroup D of the Baculoviridae, they are clearly unrelated to the insect baculoviruses and are now properly in their own Polydnaviridae family.[1] This family constitutes two major subgroups (probably at the level of subfamilies) of viruses, one representing polydnaviruses derived exclusively from the Ichneumonidae family of parasitic wasps and the second group representing polydnaviruses derived exclusively from the Braconidae family. Hence, these polydnaviruses will be referred to as either ichneumonid polydnaviruses or braconid polydnaviruses, respectively. The differentiation of these two groups is based on more than just the family status of the parasite from which the viruses are derived. This designation of the two groups based on the family name does not preclude the possibility that an ichneumonid polydnaviruses will be referred to as either ichneumonid polydnaviruses or braconid polydnaviruses, respectively. The differentiation of these two groups is based on more than just since this basis of classification is unambiguous, and this is preferred by at least this author.

In addition to polydnaviruses derived from some species of parasitic hymenoptera, several other viruses, or virus-like particles, of unknown etiology have also been observed in some of the same species of wasps. Although these are clearly not members of the Polydnaviridae, they will be described briefly at the end of this chapter.

A. MEMBERS OF THE GROUP

Under the current classification scheme only one genus, Polydnavirus, has been defined.[1] For largely historical reasons the species chosen originally as the type species is polydnavirus Type 1, originally called *Hyposoter exiguae* virus (HeV) since it was isolated from the ichneumonid wasp, *Hyposoter exiguae*.[4] Although this was the first ichneumonid polydnavirus which was purified and characterized, a polydnavirus from the ichneumonid wasp, *Campoletis sonorensis,* and therefore named *Campoletis sonorensis* virus (CsV) has been studied much more extensively and should more properly be considered the "type virus" for this family.[5-13]

There is as yet no consensus on an appropriate nomenclature for polydnaviruses. Therefore, the approach already used in the literature for naming HeV[4] and CsV[5] will be used here: a polydnavirus is named after the parasite in which it replicates. In the abbreviation of the virus name, the first letter is uppercase and is the first letter of the genus of the parasite, while the second is lowercase and is the first letter of the species of the parasite; V stands for virus.

The multipartite nature of the genomes of CsV and HeV has been demonstrated by Krell and Stoltz[4] and Krell et al.[5] (see Figure 1A). Although some of the different sized circles of the genomes of either of these viruses show some sequence homology, many of the circles are heterologous and unrelated by sequence homology to each other. Thus, the polydisperse genomes of at least two polydnaviruses are multipartite not only in size but also, to a large extent, by sequence heterogeneity. However, only in rare cases has the multipartite nature of the genome been investigated or established. Nevertheless, many parasitic hymenoptera have been shown by electron microscopy to contain polydnaviruses. In all cases where the genomes of viruses with a morphology similar to ichneumonid or braconid polydanviruses have been analyzed, they have been shown to be polydisperse in nature. Thus, it is likely that viruses which at least appear morphologically like polydnaviruses also have a multipartite genome and can be considered as genuine polydnaviruses.

The braconid and ichneumonid parasite species in which particles with the morphology

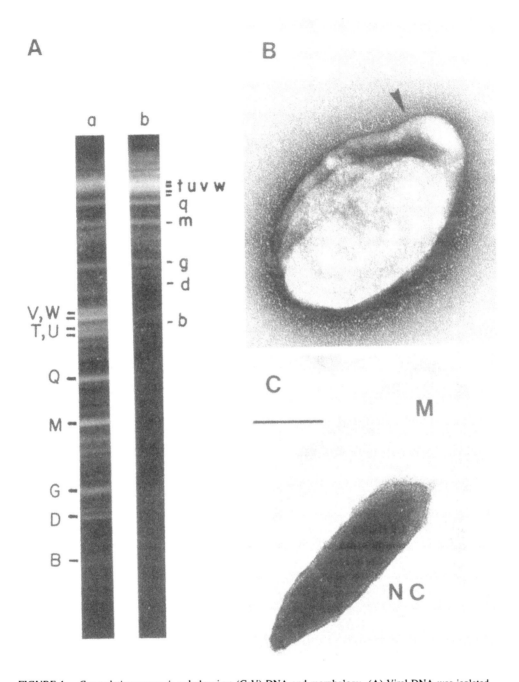

FIGURE 1. *Campoletis sonorensis* polydnavirus (CsV) DNA and morphology. (A) Viral DNA was isolated from purified CsV by equilibrium sedimentation in CsCl/EtBr into a superhelical fraction and a relaxed circular fraction and analyzed by agarose gel electrophoresis (lanes [a] and [b], respectively); some of the major superhelical bands are identified by uppercase letters in lane (a) while the corresponding relaxed circular bands are identified by the corresponding lowercase letters next to lane (b). (B) An intact CsV virion is negatively stained and shows the enveloped hook structure at one end (arrowhead). (C) The uranyl acetate stained nucleocapsid (NC) of CsV is shown and was released by partial trypsin digestion prior to staining; part of the inner membrane (M) is still attached. (Figures 1B and C are at the same magnification; bar = 200 nm.) (Figure 1B is from Krell, P. J. and Beveridge, T. J., *Int. Rev. Cytol.*, Suppl. 17, 15, 1987. With permission.)

of either group of polydnaviruses have been found are summarized in Tables 1 and 2, respectively. These tables also summarize data on the nature of the genomes (minimum number of different sized DNA circles and estimated molecular size range of the DNA circles) of viruses for which this information is available.

Particles observed in the parasite species listed in Table 1 have a morphology which, for the most part, is similar to braconid polydnaviruses such as *Cotesia* (formerly *Apanteles*) *melanoscela* virus (CmV) and *Cardiochiles nigriceps* virus (CnV),[14,15] while those in Table 2 are morphologically similar to ichneumonid polydnaviruses such as CsV and HeV.[16,17] Needless to say, not all the particles listed in these tables can be categorically classified as polydnaviruses. Only those which are shown to contain a polydisperse circular dsDNA genome can be considered as true members of the Polydnaviridae family.

There has been some effort devoted to determining the relatedness of polydnaviruses from different ichneumonid parasites. This includes both nucleic acid hybridization and serological analysis. When total DNA from *Hyposoter fugitivus* polydnavirus (HfV) was used as probe against DNA from several ichneumonid polydnaviruses, hybridization was observed against polydnaviruses from *Diadegma terebrans*, *Diadegma acromyctae*, *Diadegma interruptum*, *Olesicampe* spp., *Hyposoter annulipes*, *Hyposoter rivalis*, and *Hyposoter* spp., but not against polydnaviruses from a *Campoplex* sp., a *Glypta* sp., and two *Campoletis* sp. (*C. aprilis* and *C. sonorensis*).[51] In a more detailed study[52] two cloned *Eco*RI fragments from the genome of HfV hybridized to certain bands of DNA from several polydnaviruses. In one case a 1.8 kb fragment from HfV DNA hybridized to a supercoiled circle from each of four polydnaviruses from *Olesicampe* spp., *Hyposoter rivalis*, *Hyposoter lymantriae*, and *Casinaria arjuna*. A 3.8 kb *Eco*RI fragment from HfV also hybridized to certain superhelical DNA bands from each of two polydnaviruses from *Hyposoter lymantriae* and *Casinaria arjuna*. In both these experiments, however, the hybridization reaction to DNAs of heterologous polydnavirus was not as strong as for DNA of the homologous polydnavirus. These preliminary data suggest that at least some ichneumonid polydnaviruses are somewhat related to each other but not to other polydnaviruses. Thus, it should be possible to further classify some of the ichneumonid polydnaviruses on the basis of genome relatedness.

In an extensive Western blot analysis of ichneumonid polydnaviruses from 14 species representing 6 genera of ichneumonid wasps, Cook and Stoltz[18] demonstrated that antisera to HeV reacted either strongly or moderately to polydnaviruses from other *Hyposoter* spp. Polydnaviruses from species of unrelated genera of wasps reacted only weakly with the same antisera. These observations parallel closely the nucleic acid hybridization analysis described before.

Both the nucleic acid hybridization and serological comparisons of different polydnaviruses show that the degree of relatedness of the polydnaviruses tested seems to be similar to the degree of phylogenetic relatedness of the parasite species from which they are derived. Thus, as noted before, polydnaviruses from the *Hyposoter* genus are more closely related to each other than to polydnaviruses from other genera of parasitic wasps.

B. PATHOLOGY

There does not appear to be any pathology associated with polydnavirus replication. Adult parasitic wasps, which are the hosts of polydnaviruses, live only long enough to mate and to parasitize suitable host larvae. In general, adults of many species of hymenoptera survive only a few days (or sometimes only a few weeks) postemergence. For any given species, the male, in which virus apparently does not replicate, survives as long as the female in which it does. Thus, it is unlikely that the polydnavirus "infection" in females of the species contributes to their short life span. If it did then females would be expected to die earlier than their male counterpart.

As will be described in more detail later, polydnavirus replicates solely in the ovary

TABLE 1
Braconid Polydnaviruses

Parasite source	Virus	Minimum number of circles	Size range estimate (kb)	Ref.
Apanteles crassicornis				14
Apanteles fumiferanae				14
Ascogaster argentifrons				2
Ascogaster quardridentata				2
Cardiochiles nigriceps		15	4.5—20	48, 49
Chelonus altitudinis				2
Chelonus blackburni				2
Chelonus near *curvimaculatus*		15	8.4—27	32
Chelonus insularis		12	9.7—38.5	32
Chelonus texanus				15
Cotesia congregatus[a]				14
Cotesia flavipes				14
Cotesia glomeratus				50
Cotesia hyphantriae				2
Cotesia marginiventris				14
Costesia melanoscela	CmV	16	3—27	33
Cotesia rubecula				2
Cotesia schaeferi				35
Glyptapanteles liparidis[a]				14
Hypomicrogaster ecdytdophae				2
Microgaster canadensis				2
Microplitis croceipes		10	6—27	2, 32
Phanerotoma flavitestacea				46
Pholetesor ornigis[a]				14
Protapanteles paleacritae[a]				14
Protomicroplitis facetosa				2

[a] Formerly genus *Apanteles*.

epithelial cells in a specialized region of the ovary known as the calyx. Infected cells in the calyx can be easily identified cytologically. These cells secrete copious amounts of virus into the ovary lumen to produce a bluish calyx fluid. Although these virus-infected cells are the major component of the ovary, the structural integrity of this tissue does not seem to be adversely affected by polydnavirus replication. Thus, polydnavirus replication in the parasitoid host does not result in any obvious pathology.

The polydnavirus is transmitted during oviposition to larvae of the habitual host of the parasitoid. In these lepidopteran host species, several effects attributed to the presence of polydnavirus have been suggested. It is known that at least for the ichneumonid polydnavirus CsV and the braconid polydnavirus CmV that viral-specific transcription occurs in parasitized host larvae.[6,7,10,21] Also new proteins are synthesized in parasitized hosts.[19-22,23] In one case the synthesis of a new glycoprotein in parasitized host larvae was related directly to the presence of CsV injected by the parasite.[19]

Some additional physiological changes have also been attributed to polydnavirus expression. These include decreased levels of active phenol oxidase,[24] hemocytic transformation and immunosuppression,[25,26] decreased host weight gain,[13] decreased hemocyte counts, decreased spreading ability of plasmatocytes,[27] and changes in ecdysteroid levels.[20]

Some or all of these effects could contribute to gross changes in the health and development of the parasitized larva. Although it may be difficult to differentiate between changes due directly to the polydnavirus and those due to the parasite egg and other injected secretions,

TABLE 2
Ichneumonid Polydnaviruses

Parasite source	Virus	Minimum number of circles	Size range estimate (kb)	Ref.
Compoletis aprilis		5	6—20	TR
Campoletis flavicinta				2
Campoletis sonorensis	CsV	28	6—20	5, 6
Campoplex sp.				TR
Casinaria forcipata				18
Casinaria infesta				2
Casinaria arjuna		10	5—20	47
Casinaria sp.		10	6.9—9.3	30
Diadegma acromycta		15	2—6	18, TR
Diadegma interruptum		15	2—6	30, TR
Diadegma terebrans	DtV	10	2.3—5.5	28
Glypta sp.		10	6—20	30
Hyposoter annulipes		15	3.3—5.7	30, TR
Hyposoter fugitivus		10	2—6	2
Hyposoter exiguae	HeV	15	3.0—6	4
Hyposoter lymantriae		10	206	47
Hyposoter pilosulus				2
Hyposoter rivalis		15	2.5—6	TR
Olesicampe benefactor		10	3.3—6.4	30
Olesicampe geniculatae		10	2—6	47
Tranosema sp.		10	2.5—4.4	30

Note: TR = this report.

some polydnaviruses are, at least in part, responsible for some alterations in physiology which adversely affect the parasitized host larva; thus, they may be considered to have a pathological effect on the host larva.

II. VIRUS STRUCTURE AND COMPOSITION

A. PARTICLE STRUCTURE
1. Ichneumonid Polydnaviruses

The morphology of three polydnaviruses (HeV, CsV, and DtV) have been described in some detail.[2,4,16,17,28] All three are morphologically similar and are members of one of perhaps two morphological groups of the ichneumonid polydnaviruses. The large ovoid virion measures at least 130 × 350 nm, and each contains a partially lenticular-shaped cylindrical nucleocapsid measuring 85 nm at the widest diameter and at least 330 nm in length (see Figures 1B, C; 2A). There is a distinct substructure visible on the nucleocapsid surface, and the conical ends are not identical. The nucleocapsid is ensheathed in two membranes which appear to be attached to each other and one end of the nucleocapsid. A hooklike structure consisting of several distinct globular subunits appears to be attached to one end of the virion of CsV and is enveloped by at least one of the viral membranes.[2,17,29]

All virions from *C. sonorensis* appeared identical in morphology and formed a single band by sucrose density gradient centrifugation or by CsCl equilibrium centrifugation.[5] On sucrose gradients, CsV bands at 1.18 g/cc while in CsCl isopycnic gradients it bands at a buoyant density of 1.20 g/cc.

A second morphological subgroup of the ichneumonid polydnaviruses has been described by Stoltz et al.[30] (see Figure 2B). However, there are less detailed morphological data available for this group. One of the viruses in this second group is an ichneumonid poly-

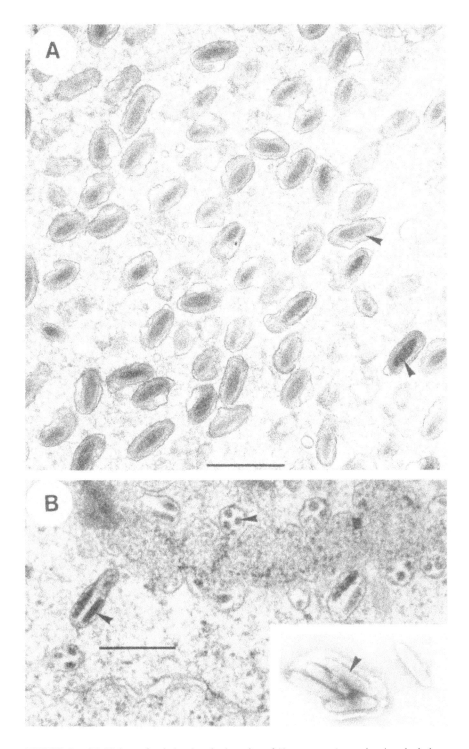

FIGURE 2. (A) Virions of polydnavirus in the calyx of *Hyposoter exiguae* showing single large nucleocapsids surrounded by a double membrane. (B) Virions in a calyx cell nucleus of *Glypta* sp. with a slightly smaller nucleocapsid and with 2 to 4 nucleocapsids surrounded by a membrane. Inset of (B) is a negatively stained preparation of the *Glypta* virus showing the surface substructure of the nucleocapsids; a viral membrane encloses four nucleocapsids. Arrowheads indicate the individual nucleocapsids. (Bars = 500 nm in [A] and 250 nm in [B].) (Figure 2A is from Stoltz, D. B. and Vinson, S. B., *Adv. Virus Res.*, 24, 125, 1979. With permission.) (Both [A] and [B] are courtesy of Dr. D. B. Stoltz.)

dnavirus from a *Glypta* species and has several (two or more) nucleocapsids per envelope. In this one respect they are similar to some of the braconid polydnaviruses which also have several nucleocapsids per envelope. The nucleocapsids from the *Glypta* polydnaviruses are much smaller than those of the ichneumonid polydnaviruses described before, but are otherwise morphologically similar.

The morphology of the particles from the ichneumonid wasp *Bathyplectes curculionis* (BcV) is somewhat indeterminate.[30,31] These particles appear to have several small nucleocapsids per envelope but their morphology is unlike that of any ichneumonid polydnaviruses. Although BcV is derived from an ichneumonid wasp and may represent a third morphological subgroup of the ichneumonid polydnaviruses, their relationship, if any, to the Polydnaviridae is tenuous. The virus is described in more detail in Chapter 22.

2. Braconid Polydnaviruses

Braconid polydnaviruses are morphologically similar, in some respects, to the insect baculoviruses. Hence, the first reference to their morphology describes them as "baculovirus-like particles".[15] They were also originally placed in the baculovirus family due to this similarity, but are now placed in their own polydnavirus family because of other characteristics not shared with the baculoviruses. The cylindrical nucleocapsids have a uniform diameter of approximately 40 nm (see Figure 3). The length of the nucleocapsid is variable even within a viral species. One of the shortest nucleocapsids is 25 nm from *Cardiochiles nigriceps* virus, and some are longer than 100 nm as in *Cotesia melanoscela* virus.[2] The nucleocapsids of this latter polydnavirus range in length from 30 to 105 nm. The nucleocapsids can be enveloped as single nucleocapsids per envelope as in braconid polydnaviruses from *Cardiochiles nigriceps*, *Chelonus texanus*, *Microplitis croceipes*, *Apanteles crassicornis*, *Pholetesor* (formerly *Apanteles*) *ornigis*, and *Apanteles fumiferanae*. In other species previously identified as *Apanteles* species, nucleocapsids are enveloped in multiples of several nucleocapsids, often of different lengths, in one virion.[2,14,15]

The braconid nucleocapsids are asymmetric and have easily identifiable end structures (see Figure 3). A long tail structure, 8.5 nm × 2 to 300 nm, can be seen on one end of each nucleocapsid in polydnavirus from *Cotesia melanoscela* and *Chelonus insularis*.[14,32] This tail may be important for orientation to the nuclear pore and subsequent uncoating since in several micrographs they are seen penetrating the pore while the main nucleocapsid component remains in the cytoplasm[14] (see Figure 3C).

An electron-dense amorphous matrix is observed between the nucleocapsid(s) and the enveloping membrane. The membrane appears by electron microscopy as a trilaminar structure typical of unit membranes. This envelope in some polydnaviruses such as CmV, also contains a tail-like projection which may function in the penetration of virus through the basement membrane of parasitized host larvae coinjected with both the parasite egg and the corresponding polydnavirus during parasitization.

B. PROTEIN COMPOSITION

As might be expected for viruses with such a complex morphology their protein composition is likewise complex. Polyacrylamide gel electrophoresis analysis of the braconid polydnavirus from *Cotesia melanoscela*[33] and the two ichneumonid polydnaviruses from *Hyposoter exiguae*[4] and *Campoletis sonorensis*[5] has been reported. In each case at least 20 polypeptides ranging in size from approximately 11 to 200 kDa could be resolved. In the braconid polydnavirus, CmV, five nucleocapsid polypeptides have been tentatively identified and include two major proteins of 32 and 38 kDa.

The aggregate molecular weight of all structural polypeptides in each polydnavirus examined is just in excess of 10^3 kDa. These many proteins would thus require a coding capacity in excess of 20 megadaltons (MDa) (30 kb) for their synthesis. In each of the

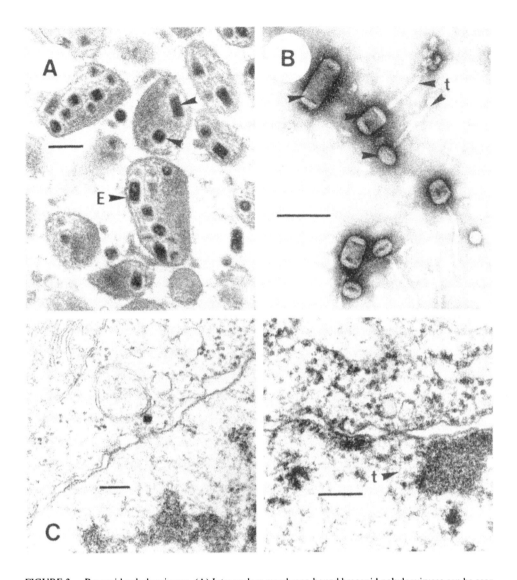

FIGURE 3. Braconid polydnaviruses. (A) Intranuclear membrane-bound braconid polydnaviruses can be seen in sections of calyx epithelial cells of the braconid wasp *Protapanteles* (formerly *Apanteles) paleacritae;* the rod-shaped nucleocapsids (arrowheads) can be seen in both cross and longitudinal sections; the variable length of the nucleocapsids, the distinctive end structures, and the tails on some nucleocapsids are readily seen; several nucleocapsids are surrounded by a common trilaminar envelope (E). (B) Negatively stained CmV, a braconid polydnavirus from *Cotesia* (formerly *Apanteles) melanoscelus* showing some liberated nucleocapsids showing the variable length, distinctive end structures and a tail (t) attached to one end of each nucleocapsid. (C) Two profiles of braconid polydnavirus nucleocapsids of CmV uncoating at the nuclear pores of fat body cells of *Orgyia leucostigma* larvae following parasitization by *Cotesia melanoscelus;* the tail (t) of an empty nucleocapsid can readily be seen penetrating into the nucleus through the nuclear pore while the capsid remains in the cytoplasm. (Bar = 100 nm.) (Courtesy of Dr. D. B. Stoltz, and Stoltz, D. B. and Vinson, S. B., *Can. J. Microbiol.*, 23, 28, 1977. With permission.)

polydnaviruses examined, the estimated aggregate genome size exceeds this requirement suggesting that each polydnavirus can at least code for the synthesis of its own structural proteins. However, rarely is this much DNA present on only one of the circular DNAs which collectively comprise the viral genome. Thus, the DNA coding for all the viral polypeptides is probably distributed among several circles of DNA for each polydnavirus.

C. GENOME STRUCTURE

By far the most intriguing characteristic of all polydnaviruses is the multicomponent nature of the genome. It is this criterion alone which defines the polydnavirus family.[3] No other family of double-stranded DNA viruses has this characteristic. That the genomes of CmV and HeV consisted of circular DNA was first established by electron microscopy and agarose gel electrophoresis.[4,33] Moreover, that these genomes consisted of several different sizes of DNA was also shown by these techniques.

The range in size of viral DNA and aggregate genome size varies from species to species. In some species the range is fairly narrow as in the ichneumonid polydnaviruses HeV (estimated range, 3.4 to 6.0 kilobase pairs [kbp]), HaV and HrV (both estimated to range from 3.3 to 5.7 kb) from the *Hyposoter* genus, and DaV and DiV (estimated to range from 2 to 6 kb) from the *Diadegma* genus. The aggregate genome size has been calculated for one of these, HeV, to be approximately 50 kb. In some ichneumonid polydnaviruses the range in size can be broader than just described and some circles can be quite large. For example, in CsV, CaV, and in some polydnaviruses from the genus *Glypta* the range can be from 6 kb to in excess of 20 kb. The aggregate size of the genome of one of these, CsV, has been calculated to be in excess of 200 kb, at least four times that of HeV.[4,5] Whether the ichneumonid polydnaviruses can be segregated into at least two groups based on the aggregate size of the genome and on the size range of the circular viral DNAs is worth considering and might provide additional criteria for subdividing the family of ichneumonid polydnaviruses.

The viral DNAs of CmV, HeV, and CsV have mean buoyant densities in CsCl of 1.694 g/cc (35% G + C), 1.689 g/cc (29% G + C), and 1.71 g/cc (50% G + C), respectively.[4,5,33] Unfortunately density data for other polydnaviruses are unavailable. It would be interesting to determine if polydnaviruses whose DNAs have similar G + C content are closely related by other criteria such as virus morphology, genome size, and range of sizes of DNA circles and ultimately by sequence homology.

When polydnaviruses were initially discovered to have a polydisperse circular DNA genome, it was felt that the smaller circles may have been derived from larger circular genomic DNA and generated defective interfering viruses. However, Southern blot hybridization using only a few select major supercoiled DNAs as probes showed that there was only limited DNA homology among these major DNAs and other DNA circles.[4,5] These preliminary experiments demonstrated that the genome was in fact multicomponent and not a collection of defective deletions of a larger DNA.

A more detailed study of the genome of CsV demonstrated that there were some regions with homologous sequences among some of the circles. Using cloned CsV DNA as probes, Blissard et al.[6,7] showed that certain probes hybridized to several (but not all) other superhelical CsV DNAs. A similar observation was noted using a 3.8 kb cloned *Eco*RI fragment of HfV as a probe against HfV DNA. This probe hybridized to two or three superhelical DNAs including the original 3.8 kb circle.[34]

In a very extensive analysis, Theilman and Summers[12] demonstrated that the cross hybridization among some circular DNAs was attributed to a family of imperfectly conserved repeat elements with an average length of 540 base pairs. Although there is some cross hybridization among DNA circles of individual polydnaviruses, this is not universal and appears to be restricted to certain families of SH DNAs. Despite this limited cross homology, the genome of at least HeV and CsV can still be considered as multipartite as initially suggested by Krell et al.[5] In addition, the genome of CsV has been shown to be transcriptionally multipartite. Blissard et al.[6,7] showed that several different cloned cDNA fragments specific to CsV hybridized to different SH DNA circles.

The sequence heterogeneity in the polydisperse genome of any of the braconid polydnaviruses has not yet been studied in any detail. However, in an unrelated study on polydnavirus transmission, Stoltz et al.[35] showed that one cloned fragment of CmV hybridized to only one CmV SH DNA circle (and its relaxed circular equivalent) while a second cloned fragment hybridized to two different SH DNA circles (and their relaxed circular equivalents). Thus, in the braconid polydnaviruses, at least for CmV, the genomes are likewise multipartite but with a limited degree of cross hybridization.

The significance of the cross hybridization observed among some circular DNAs of both ichneumonid and braconid polydnaviruses is unknown. They may help to identify families of related circles. They may also represent common regulatory sequences for transcription, replication, or recombination. Clearly, the function of these shared sequences is fertile ground for conjecture and further research.

III. CYTOPATHOLOGY

A. ICHNEUMONID POLYDNAVIRUSES
1. Cytopathology in the Parasite

Although not properly called "cytopathology", the cells in which the virus replicates are quite distinctive from other cells of the parasite in which the virus is found. The cytopathology of only two ichneumonid polydnaviruses (CsV and DtV) has been described in any detail. That of CsV was described by Norton et al.[16] before it was known to be a polydnavirus. They described cells of the parasitoid *Campoletis sonorensis* in which they found "DNA-containing nuclear secretory particles" (viz the polydnavirus, CsV). CsV is seen only in the epithelial cells of the calyx, a specialized region of the ovary in female *C. sonorensis* adults. CsV nucleocapsids like those of DtV assemble in the enlarged nuclei of these cells in areas which consist of a fibrogranular stroma. Nucleocapsid core material is condensed from the periphery of the stroma and enveloped by a membrane which appears to be derived *de novo* within the stroma. The completely enveloped nucleocapsids exit the nucleus by budding through the abundant serrated nuclear envelope. Once in the cytoplasm, the nuclear membrane component of CsV is lost, leaving the nucleocapsids surrounded by the limiting membrane derived during nucleocapsid morphogenesis in the nucleus. The cytoplasmic form of CsV is then "secreted" through the epithelial cell membrane into the calyx lumen. This region of the cell periphery is lined with numerous microvilli. The enveloped cytoplasmic nucleocapsids bud through the microvillar membrane and thus acquire a second envelope. In the calyx lumen the mature CsV then consists of the nucleocapsids surrounded by an inner membrane derived *de novo* from within the nucleocapsid and the outer envelope derived from the microvillar membrane.

DtV replicates in a similar manner, also exclusively in the calyx epithelial cells of the host parasitoid, *Diadegma terebrans* (see Figure 4).[28] In this species, biogenesis of the intranuclear nucleocapsid inner envelope appears to be coincident with condensation of nuclear material from within the fibrogranular stroma to form the enveloped nucleocapsid. The DtV nucleocapsids bud through the nuclear envelope and then through the cell membrane into the ovary lumen where they form the major component of the thick "calyx fluid".

Initially the cytopathology of CsV (and DtV) was observed only in adult parasites. However, Norton and Vinson[36] noted that initiation of CsV morphogenesis occurred in the stage II pupae (only eyes, head, and thorax are pigmented) of *C. sonorensis*. Evidence of earlier stages of CsV replication was seen in the intranuclear fibrogranular stroma observed in stage I pupae (only eyes are pigmented).

A detailed study of the distribution of polydnavirus infected cells was used to map the spatial distribution and limits of virus infection in the ovary of adult *D. terebrans* females.[28] DtV infected cells were found to completely encircle the ovary and were found along the full length of the calyx region (see Figure 4). Very few epithelial cells in this region were virus free. DtV was absent from the ovarioles at one end of the ovary and the lateral oviducts at the other. DtV was also absent from any other tissues of male or female parasitoids.

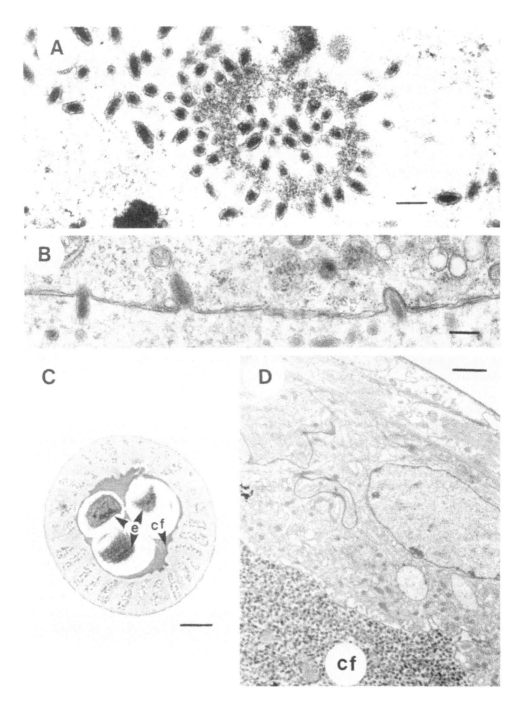

FIGURE 4.   Replication of DtV an ichneumonid polydnavirus in calyx epithelial cells of *Diadegma terebrans*. (A) Replication begins by condensation and envelopment of nuclear material by *de novo* formed membranes in the granular regions of the nucleus. (B) The enveloped nucleocapsids bud through the nuclear envelope. (C) Cytoplasmic DtV buds through the plasma membrane to accumulate as the major component of calyx fluid (cf) shown as a darkly stained fluid in a cross section of the whole ovary; also seen in (C) is the surrounding layer of DtV-infected epithelial cells with hypertrophied nuclei having a granular appearance; three eggs (e) can be seen within the oviduct lumen. (D) The high concentration of DtV in the calyx fluid (cf) of the lateral oviduct is surrounded by a limiting cellular sheath, the cells of which are not infected by DtV. (Bars = 200 nm in [A] and [B], 50 μm in [C], and 1 μm in [D].) (Figures 4[C] and [D] are from Krell, P. J., *Can. J. Microbiol.*, 33, 176, 1987. With permission.)

The DtV-infected cells are directly adjacent to the calyx lumen into which DtV is secreted. These cells also constitute the major structural component of this tissue. The infected cells are easily identified even by light microscopy. They are very large and wedge-shaped with large distinctive nuclei. The nuclei have an extensive serrated nuclear envelope and have a granular appearance. The intranuclear granules are coincident with the intranuclear virogenic stroma seen by electron microscopy. Despite the extensive distribution of polydnavirus-infected cells, the overall structural integrity of the ovary remains intact.

2. Cytopathology in Parasitized Larvae

In addition to being found in parasitic wasps, polydnaviruses are also seen in parasitized larvae.[17,37] The ichneumonid polydnavirus CsV, for example, is injected into the hemocoel of larvae of *Heliothis virescens* during oviposition of the parasitoid egg. The polydnavirus penetrates the larval basement membranes, perhaps with the aid of the outer envelope protrusion (and associated hook-structure). The outer envelope is lost at the basement membrane. CsV subsequently enters larval cells presumably by membrane fusion between the inner viral membrane and the cytoplasmic membrane. Only the nucleocapsid (devoid of either of the two membranes) of CsV is observed in the cytoplasm of infected larval cells. The CsV nucleocapsid enters the nucleus intact, presumably through the nuclear pores. These nucleocapsids become uncoated within the nucleus allowing limited transcription from only certain viral circular DNAs.[6-9] Viral replication, per se, has not been observed in parasitized larvae. The viral DNA of CsV does persist, however, allowing for continued viral transcription.[12]

B. BRACONID POLYDNAVIRUSES

1. Cytopathology in the Parasite

Braconid polydnavirus replication also appears to be restricted to calyx epithelial cells. Morphogenesis starts with the formation, *de novo,* of "open-ended" membrane profiles within affected nuclei of *C. melanoscela*.[2,14] Viral nucleocapsids form within the intranuclear virogenic stroma where they also become enveloped with either one or more nucleocapsids per envelope, depending on the species. The mechanism of extracellular release of these mature intranuclear virions is unknown. Sloughing and lysis of infected cells may account for the release of braconid polydnaviruses from the ovary of, for example, *Cotesia congregatus*. Other less radical routes of release may also exist since in the calyx fluid of some braconid species there is little evidence of cellular debris which would be expected following cell lysis. More detailed analysis of morphogenesis and release of braconid polydnavirus is an area which would require additional effort.

2. Cytopathology in Parasitized Larvae

Braconid polydnaviruses are also transmitted to parasitized larvae during oviposition by the braconid parasite.[2,14] For CmV, the outer (and only) viral envelope is lost during penetration through the basement membranes and cell membranes of parasitized larvae. Naked nucleocapsids of CmV move through the cytoplasm toward nuclear pores. The nucleocapsid tail penetrates and remains in the nuclear pore, and the nucleocapsid core remains on the cytoplasmic side of the nuclear envelope (see Figure 3). Viral DNA then appears to be injected, presumably through the nucleocapsid tail into the nucleus during uncoating. Thus full, partially filled, and empty nucleocapsids can often be seen sitting on nuclear pores of cells in parasitized larvae. Using reconstruction experiments, Stoltz and colleagues[21] demonstrated that venom was required for this uncoating process in host larvae infected with CmV. Braconid polydnaviruses such as CmV do not appear to replicate in cells of parasitized larvae since no new viral particles have been reported in such cells. However, viral DNA persists in CmV-infected cells of *Orgyia leucostigma* or *Lymantria dispar* after parasitization by *C. melanoscela*.[21] In the same studies the authors showed that viral-specific transcription followed uncoating, suggesting that some viral-specific translation was also possible in parasitized larvae.

Several physiological studies of immunosuppression and encapsulation in larvae parasitized by braconid wasps such as *A. glomeratus*[38,39] and *C. melanoscela*[40,41] implicated components of both the calyx fluid (presumably the resident polydnavirus) and the venom. Neither fluid alone was effective in immunosuppression. Stoltz et al.[21] suggested that the venom was required for uncoating of the viral genome which could then lead to viral specific transcription in parasitized larvae. Thus, unlike the situation for the ichneumonid polydnaviruses the braconid polydnaviruses cannot effect "host regulation" independently. Research in this area should lead to some interesting mechanisms of how braconid wasps can influence host regulation postparasitism.

C. POLYDNAVIRUS TRANSMISSION

Polydnaviruses are very efficiently transmitted to progeny wasps so that all females of an affected species can be shown to carry the corresponding polydnavirus in their ovaries. Since viral DNA from CsV has been found to occur in integrated form in parasite tissues, this raised the possibility of germ line transmission of the viral genome.[9] Nevertheless, in some parasite species the viral DNA can also occur in extrachromosomal form.[9,35] If transmission was through the latter route, this could give rise to maternal inheritance of the viral genome. In two related studies, Stoltz and colleagues[34,35] used genomic circular DNA length polymorphism and restriction fragment length polymorphism of polydnavirus DNA to follow the fate of certain "marked" DNAs after different mating experiments. These studies demonstrated that DNA from both the braconid polydnavirus CmV (from *C. melanoscela*) and from the ichneumonid polydnavirus HfV (from *H. fugitivus*) was transmitted to progeny wasps in chromosomal form (i.e., integrated into the wasp genome). Also the DNA length polymorphism markers could be inherited from either parent in a Mendelian fashion.[34,35]

IV. OTHER VIRUSES IN PARASITIC HYMENOPTERA

Several virus-like particles which do not appear to be related to any of the polydnaviruses have been seen in some braconid and ichneumonid species. Some of these are also covered in Chapter 22, but it is worthwhile introducing some of them here because of their coexistence with polydnaviruses. While this coexistence may be entirely fortuitous, there is a possibility of a functional interdependence which cannot be ignored.

Long viruses or virus-like particles have been observed in the ovary of the braconid parasites *Cotesia congregatus*, *Cotesia hyphantriae*, *Microplitis croceipes*, and *Mesoleius tenthredinis* and in the ichneumonid parasite, *Diadegma terebrans*[2,14,42,43] (see Figure 5).

The virus-like particles from *M. tenthredinis* are morphologically similar to baculoviruses and probably contain a circular, high molecular weight genome.[42] The virus infection cycle and morphogenesis are also like those of baculoviruses suggesting that this virus may belong to the Baculoviridae.

The long particles observed in the braconid wasps *(C. congregatus, M. croceipes, C. hyphantriae)* and in the ichneumonid wasp *(D. terebrans)* appear to be enveloped and are morphologically very similar to each other. However, their viral nature has not been established and they are unlike any virus yet described. Therefore, they remain unclassified. The braconid long particles are found to coexist in the ovary of braconid parasites with braconid polydnavirus, even within the same cell.[2] In the ichneumonid wasp ovary of *D. terebrans*, the long virus-like particles are seen to replicate independently of the polydnavirus, DtV, in different cells of the ovary and only in ovary cells near the junction with the ovarioles.[43] These particles are often more than 1 µm long, having a diameter of 65 nm, and the enveloped nucleocapsids are 20 nm in diameter.

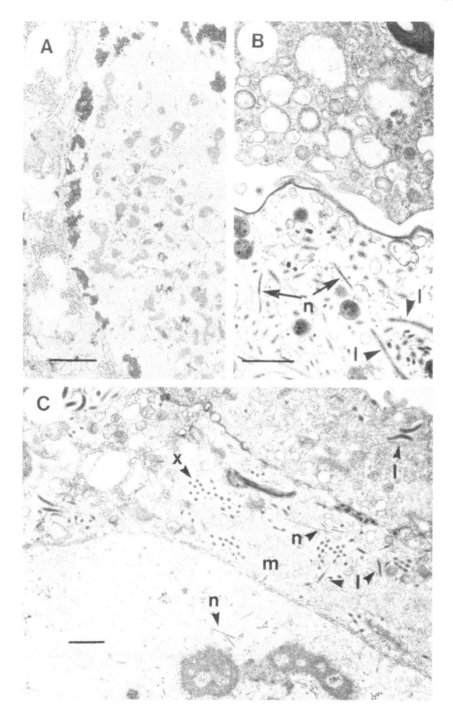

FIGURE 5.   Long virus-like particles of parasitoid hymenoptera. (A) Long virus-like particles are seen in the calyx cells of the parasitoid *Microplitis croceipes*. (B) Long virus-like particles (l) are seen along with regular braconid polydnaviruses (arrows) in the calyx fluid of a braconid wasp, *Cotesia* (formerly *Apanteles*) *congregatus*. (C) Long nucleocapids (n) of long particles are seen in the nucleus and cytoplasm of ovary cells of an ichneumonid wasp, *Diadegma terebrans;* membrane-bound nucleocapsids can be seen in longitudinal (l) and cross (x) section along with many membrane-bound vesicles (m) in the cytoplasm of infected cells. (Bars in [A] = 2 μm, in [B], and [C] = 0.5 μm.) ([A] and [B] are from Stoltz, D. B. and Vinson, S. B., *Adv. Virus Res.*, 24, 125, 1979. With permission. Courtesy of Dr. D. B. Stoltz. [C] is from Krell, P. J., *J. Gen. Virol.*, 68, 1477, 1987. With permission.)

An intriguing virus-like particle has been discovered in *C. melanoscela*. This particle (CmV$_2$) replicates in hemocytes and fat body of parasitized tussock moth *(Orgyia leucostigma)* larvae. CmV$_2$ replicates in the nucleus and has a large genome of approximately 120 kb.[44,45] The virus has two envelopes and has a long, 800 nm lenticular nucleocapsid with a maximum diameter of 100 nm. These particles are morphologically similar to the ichneumonid polydnaviruses, but differ in two respects. Unlike polydnaviruses, the *C. melanoscela* particles contain a single size genome and can replicate in the parasitized host. Although these latter particles may be extremely unique and interesting, their coexistence with a braconid polydnavirus in *C. melanoscela* may be entirely coincidental.

Some of these virus-like particles and others[46] found in parasitic wasps may play a complementary role for the resident polydnavirus. However, this possible beneficial interaction has not been established for any of the parasite virus-like particles just described. Thus, these particles may, in fact, exist quite independently of the corresponding cohabiting polydnavirus and may represent members of yet another new virus family.

ACKNOWLEDGMENTS

Much of the work covered by this review involves collaboration with Dr. Don Stoltz of Dalhousie University, Halifax, Nova Scotia, and Dr. Max Summers and Dr. Brad Vinson both at Texas A and M University, College Station, TX. These three labs and my own have been the major contributors to the research on the Polydnaviridae. I owe a special debt of gratitude to all three and their co-workers for their continuing interest and collaboration on these most unusual entities now called polydnaviruses. I am particularly grateful for the contribution of the excellent micrographs provided by Dr. D. B. Stoltz as indicated in the figure legends. I would like to acknowledge receipt of operating grants from the Natural Sciences and Engineering Research Council of Canada and from Agriculture Canada for supporting my work on the Polydnaviridae. I would also like to thank Lesley Burke and Frances Newcombe for typing the manuscript.

REFERENCES

1. **Brown, F.,** The classification and nomenclature of viruses: summary of results of meetings of the International Committee on Taxonomy of Viruses in Sendai, *Intervirology*, 25, 141, 1986.
2. **Stoltz, D. B. and Vinson, S. B.,** Viruses and parasitism in insects, *Adv. Virus Res.*, 24, 125, 1979.
3. **Stoltz, D. B., Krell, P., Summers, M. D., and Vinson, S. B.,** Polydnaviridae — a proposed family of insect viruses with segmented, double-stranded, circular DNA genomes, *Intervirology*, 21, 1, 1984.
4. **Krell, P. J. and Stoltz, D. B.,** Virus-like particles in the ovary of an ichneumonid wasp: purification and preliminary characterization, *Virology*, 101, 408, 1980.
5. **Krell, P. J., Summers, M. D., and Vinson, S. B.,** Virus with a multipartite superhelical DNA genome from the ichneumonid parasitoid *Campoletis sonorensis*, *J. Virology*, 859, 1982.
6. **Blissard, G. W., Fleming, J. G. W., Vinson, S. B., and Summers, M. D.,** *Campoletis sonorensis* virus: expression in *Heliothis virescens* and identification of expressed sequences, *J. Insect Physiol.*, 32, 351, 1986.
7. **Blissard, G. W., Vinson, S. B., and Summers, M. D.,** Identification, mapping, and *in vitro* translation of *Campoletis sonorensis* virus mRNAs from parasitized *Heliothis virescens* larvae, *J. Virol.*, 318, 1986.
8. **Blissard, G. W., Smith, O. P., and Summers, M. D.,** Two related viral genes are located on a single superhelical DNA segment of the multipartite *Campoletis sonorensis* virus genome, *Virology*, 160, 120, 1987.
9. **Fleming, J. G. W. and Summers, M. D.,** *Campoletis sonorensis* endoparasitic wasps contain forms of *C. sonorensis* virus DNA suggestive of integrated and extrachromosomal polydnavirus DNAs, *J. Virol.*, 57, 552, 1986.

10. **Fleming, J. G. W., Blissard, G. W., Summers, M. D., and Vinson, S. B.,** Expression of *Campoletis sonorensis* virus in the parasitized host, *Heliothis virescens, J. Virol.,* 48, 74, 1983.
11. **Theilmann, D. A. and Summers, M. D.,** Molecular analysis of *Campoletis sonorensis* virus DNA in the lepidopteran host *Heliothis virescens, J. Gen. Virol.,* 67, 1961, 1986.
12. **Theilmann, D. A. and Summers, M. D.,** Physical analysis of the *Campoletis sonorensis* virus multipartite genome and identification of a family of tandemly repeated elements, *J. Virol.,* 61, 2589, 1987.
13. **Vinson, S. B., Edson, K. M., and Stoltz, D. B.,** Effect of a virus associated with the reproductive system of the parasitoid wasp, *Campoletis sonorensis,* on host weight gain, *J. Invertebr. Pathol.,* 34, 133, 1979.
14. **Stoltz, D. B. and Vinson, S. B.,** Baculovirus-like particles in the reproductive tracts of female parasitoid wasps II: the genus *Apanteles, Can. J. Microbiol.,* 23, 28, 1977.
15. **Stoltz, D. B., Vinson, S. B., and MacKinnon, E. A.,** Baculovirus-like particles in the reproductive tracts of female parasitoid wasps, *Can. J. Microbiol.,* 22, 1013, 1976.
16. **Norton, W. N., Vinson, S. B., and Stoltz, D. B.,** Nuclear secretory particles associated with the calyx cells of the ichneumonid parasitoid *Campoletis sonorensis* (Cameron), *Cell Tissue Res.,* 162, 195, 1975.
17. **Stoltz, D. B. and Vinson, S. B.,** Penetration into caterpillar cells of virus-like particles injected during oviposition by parasitoid ichneumonid wasps, *Can. J. Microbiol.,* 25, 207, 1979.
18. **Cook, D. and Stoltz, D. B.,** Comparative serology of viruses isolated from ichneumonid parasitoids, *Virology,* 130, 215, 1983.
19. **Cook, D. I., Stoltz, D. B., and Vinson, S. B.,** Induction of new haemolymph glycoprotein in larvae of permissive hosts parasitized by *Campoletis sonorensis, Insect Biochem.,* 14, 45, 1984.
20. **Beckage, N. E. and Templeton, T. J.,** Physiological effects of parasitism by *Apanteles congregatus* in terminal-stage tobacco hornworm larvae, *J. Insect Physiol.,* 32, 299, 1986.
21. **Stoltz, D. B., Guzo, D., Belland, E. R., Lucarotti, C. J., and MacKinnon, E. A.,** Venom promotes uncoating *in vitro* and persistence *in vivo* of DNA from a braconid polydnavirus, *J. Gen. Virol.,* 69, 903, 1988.
22. **Ferkovich, S. M., Greany, P. D., and Dillard, C.,** Changes in haemolymph proteins of the fall armyworm, *Spodoptera frugiperda* (J. E. Smith), associated with parasitism by the braconid parasitoid *Cotesia marginiventris* (Cresson), *J. Insect Physiol.,* 29, 933, 1983.
23. **Jones D., Jones, G., Rudnicka, M., and Click, A.,** Precocious expression of the final larval instar developmental pattern in larvae of *Trichoplusia ni* pseudoparasitized by *Chelonus* spp., *Comp. Biochem. Physiol.,* 83B, 339, 1986.
24. **Stoltz, D. B. and Cook, D.I.,** Inhibition of host phenoloxidase activity by parasitoid hymenoptera, *Experientia,* 39, 1021, 1983.
25. **Edson, K. M., Vinson, S. B., Stoltz, D. B., and Summers, M. D.,** Virus in a parasitoid wasp: suppression of the cellular immune response in the parasitoid's host, *Science,* 211, 582, 1981.
26. **Stoltz, D. B. and Guzo, D.,** Apparent haemocytic transformations associated with parasitoid-induced inhibition of immunity in *Malacosoma disstria* larvae, *J. Insect Physiol.,* 32, 377, 1986.
27. **Vinson, S. B., Davies, D. H., and Dover, B. A.,** Effects of the polydnaviridae in select parasitoid host insect systems, in *Fundamental and Applied Aspects of Invertebrate Pathology,* Samson, R. A., Vlak, J. M., and Peters, D., Eds., 4th Int. Colloq. Invertebr. Pathol., The Netherlands, 1986, 74.
28. **Krell, P. J.,** Polydnavirus replication and tissue organization of infected cells in the parasitic wasp *Diadegma terebrans, Can. J. Microbiol.,* 33, 176, 1987.
29. **Krell, P. J. and Beveridge, T. J.,** The structure of bacteria and molecular biology of viruses, in *International Review of Cytology, Supplement 17,* Bourne, G. H., Ed., Academic Press, NY, 1987, 15.
30. **Stoltz, D. B., Krell, P. J., and Vinson, S. B.,** Polydisperse viral DNA's in ichneumonid ovaries: a survey, *Can. J. Microbiol.,* 27, 123, 1981.
31. **Hess, R. T., Poinar, Jr., G. O., Etzel, L., and Merritt, C. C.,** Calyx particle morphology of *Bathyplectes anurus* and *B. curculionis* (Hymenoptera:Ichneumonidae), *Acta Zool. (Stockholm),* 61, 111, 1980.
32. **Jones, D., Sreekrishna, S., Iwaya, M., Yang, J. N., and Eberely, M.,** Comparison of viral ultrastructure and DNA banding patterns from the reproductive tracts of eastern and western hemisphere *Chelonus* spp. (Braconidae:Hymenoptera), *J. Invertebr. Pathol.,* 47, 105, 1986.
33. **Krell, P. J. and Stoltz, D. B.,** Unusual baculovirus of the parasitoid wasp *Apanteles melanoscelus:* isolation and preliminary characterization, *J. Virol.,* 29, 1118, 1979.
34. **Stoltz, D. B.,** Evidence for chromosomal transmission of polydnavirus DNA, *J. Gen. Virol.,* 71, 1051, 1990.
35. **Stoltz, D. B., Guzo, D., and Cook, D.,** Studies on polydnavirus transmission, *Virology,* 155, 120, 1986.
36. **Norton, W. N. and Vinson, S. B.,** Correlating the initiation of virus replication with a specific pupal developmental phase of an ichneumonid parasitoid, *Cell Tissue Res.,* 231, 387, 1983.
37. **Stoltz, D. B.,** Interactions between parasitoid-derived products and host insects: an overview, *J. Insect Physiol.,* 32, 347, 1986.
38. **Kitano, H.,** Effect of the venom of the gregarious parasitoid, *Apanteles glomeratus* on its hemocytic encapsulation by the host, *Pieris, J. Invertebr. Pathol.,* 40, 61, 1982.

39. **Kitano, H.,** The role of *Apanteles glomeratus* venom in the defense response of its host, *Pieris rapae crucivora, J. Insect Physiol.,* 32, 369, 1986.
40. **Guzo, D. and Stoltz, D. B.,** Observations on cellular immunity and parasitism in the tussock moth, *J. Insect Physiology,* 33, 19, 1987.
41. **Guzo, D. and Stoltz, D. B.,** Obligatory multiparasitism in the tussock moth, *Orgyia leucostigma, Parasitology,* 90, 1, 1985.
42. **Stoltz, D. B.,** A putative baculovirus in the ichneumonid parasitoid, *Mesoleius tenthredinis, Can. J. Microbiol.,* 27, 116, 1981.
43. **Krell, P. J.,** Replication of long virus-like particles in the reproductive tract of the ichneumonid wasp *Diadegma terebrans, J. Gen. Virol.,* 68, 1477, 1987.
44. **Stoltz, D. B. and Faulkner, G.,** Apparent replication of an unusual virus-like particle in both a parasitoid wasp and its host, *Can. J. Microbiol.,* 24, 1509, 1978.
45. **Stoltz, D. B., Krell, P., Cook, D., MacKinnon, E. A., and Lucarotti, C. J.,** An unusual virus from the parasitic wasp *Cotesia melanoscela, Virology,* 162, 311, 1988.
46. **Poinar, Jr., G. O., Hess, R., and Caltagirone, L. E.,** Virus-like particles in the calyx of *Phanerotoma flavitestacea* (Hymenoptera:Braconidae) and their transfer into host tissues, *Acta Zool. (Stockholm),* 57, 161, 1976.
47. **Stoltz, D. B.,** Viruses of parasitic hymenoptera in *Fundamental and Applied Aspects of Invertebrate Pathology,* Samson, R. A., Vlak, J. M., and Peters, D., Eds., 4th Int. Colloq. Invertebr. Pathol., The Netherlands, 1986, 81.
48. **Vinson, S. B. and Scott, J. R.,** Particles containing DNA associated with the oocyte of an insect parasitoid, *J. Invertebr. Pathol.,* 25, 375, 1975.
49. **Krell, P. J.,** Viruses in parasitoid hymenoptera: a preliminary characterization, Ph.D. thesis, Dalhousie University, Halifax, Nova Scotia, Canada, 1980.
50. **Kitano, H. and Nakatsuji, N.,** Resistance of *Apanteles* eggs to the haemocytic encapsulation by their habitual host *Pieris, J. Insect Physiol.,* 24, 261, 1978.
51. **Krell, P. J.,** unpublished data.
52. **Krell, P. J. and Stoltz, D. B.,** unpublished data.

Chapter 11

ASCOVIRIDAE

Brian A. Federici, John J. Hamm, and Eloise L. Styer

TABLE OF CONTENTS

I. INTRODUCTION

The name of this recently discovered group of viruses, Ascoviruses,[1] is derived from the Greek word *askós,* meaning sac, and is used to indicate the most distinctive feature of these viruses — the routine occurrence of virions in large, membrane-bound vesicles. Viruses belonging to this group are currently only known from larvae of species in the lepidopteran family Noctuidae, in which they cause a chronic, fatal disease.[1-9] During the course of the disease, large numbers ($>10^8$/ml) of virion-containing vesicles accumulate in the hemolymph. The virion-containing vesicles are formed by a unique developmental sequence in which each infected host cell cleaves into a cluster of vesicles as virion assembly proceeds.[1] The virions of ascoviruses are large (130 × 400 nm), enveloped, allantoid to bacilliform in shape, with complex symmetry, and contain a linear, double-stranded DNA genome of about 170 kb.[1]

II. MEMBERS OF THE GROUP AND THEIR RELATIONSHIPS

The first ascoviruses were discovered in 1977 in larvae of two different noctuid species, the cotton bollworm, *Heliothis zea,*[2] and the clover cutworm, *Scotogramma trifolii.*[3] Since then, ascoviruses have also been isolated from larvae of *Autographa precationis,*[6] *Heliothis virescens,*[6] *Spodoptera frugiperda,*[6] and *Trichoplusia ni*[4] (see Table 1, Figures 1 to 6). To date, all of the reported isolations of ascoviruses are from populations of noctuid larvae occurring within the continental U.S. (see Table 1). However, the relatively broad host range of this viral type among the noctuid species reported before indicates the ascovirus group likely occurs worldwide. That this viral type was discovered only recently is probably due to a combination of factors, including the lack of significant gross pathology in diseased larvae, the chronic nature of the disease, and its low prevalence in larval populations.[4,6]

Because ascoviruses are a newly discovered group, little is known about the relationships of the various isolates. In a recent study of genetic relatedness based on DNA homologies, it was shown that the viruses isolated from *Heliothis zea* in South Carolina and *Trichoplusia ni* in California, are variants of the same viral species.[7] In the same study, no homology was detected between the DNA of the ascovirus isolated from *Spodoptera frugiperda* and the DNAs of the isolates from *H. zea* or *T. ni.* Thus, we know that there are at least two distinct viral biotypes or species within the ascovirus group.

III. PATHOLOGY OF ASCOVIRUS DISEASE

A. GROSS PATHOLOGY

A notable feature of ascoviruses is that there is little gross pathology associated with the disease. Infected larvae, particularly those collected in the field, differ only slightly in appearance from healthy larvae. In advanced stages of disease, a white or cream-colored hue develops throughout the abdomen, particularly along the ventrum, and may be obvious in light-colored larvae. However, this discoloration is often inapparent, especially in larvae that are normally dark or have heavily pigmented cuticle. In some larvae, the disease can be recognized by an inability to completely cast the exuvium after molting. In these, the old cuticle remains around the body, usually in the abdominal region.

Another feature of the disease caused by ascoviruses is the inability of the larvae to progress in development after acquiring infection.[1,5,6,8] Larvae will typically survive for from 2 to 6 weeks after infection, but feed only intermittently, gain little weight, and eventually die.

B. HISTOPATHOLOGY

The histopathology of ascoviruses varies among the different isolates. The isolate from

TABLE 1
Isolation of Ascoviruses from Larvae of the
Lepidopteran Family, Noctuidae

Species	Year of isolation	Location (U.S.)	Ref.
Scotogramma trifolii	1977	California	3
Heliothis zea	1977	Mississippi	2
Trichoplusia ni	1979	California	4
Spodoptera frugiperda	1982	Florida	6
Heliothis spp.	1982	South Carolina	5
Autographa precationis	1983	Georgia	6
Heliothis virescens	1985	Georgia	6

T. ni[1,9] infects a range of tissues and is most commonly observed in the fat body, tracheal matrix, and epidermis. The closely related isolate from *H. zea*[9] is found primarily in the epidermis and tracheal matrix, and although there is a distinctive pathology observed in the fat body of infected larvae, viral replication and progeny virions have not been observed in this tissue. The ascovirus isolated from *Spodoptera frugiperda*,[9] however, apparently replicates only in fat body cells. Because ascoviruses can be transmitted *per os* though only at a low rate, they may be capable of infecting the midgut epithelium.

The most remarkable histological aspect of ascovirus infection is the enormous accumulation of virion-containing vesicles that occurs in the hemolymph (see Figure 1). Vesicles 2 to 10 μm in diameter can be found in the hemolymph 2 to 3 d after infection, and continue to accumulate there as the disease advances, reaching concentrations as high as 10^8/ml. The high concentrations of vesicles found in the hemolymph make it milky white in color. When examined in wet mounts with phase microscopy, the virion-containing vesicles appear highly refractile and typically contain several refractile viral inclusions.

C. CYTOPATHOLOGY

The cytopathology caused by ascoviruses is their most unique feature and clearly sets them apart from all other known viruses including bacteriophages, and plant and other animal viruses. The cytopathology begins with hypertrophy of the nucleus in infected cells and occasionally invagination of the nuclear membrane (see Figure 3A). Hypertrophy of the nucleus continues, leading to cellular hypertrophy. The nuclear membrane eventually ruptures and fragments (see Figure 3B). After fragmentation of the nuclear membrane, sheets of cytoplasmic membrane assemble throughout the cell and coalesce, partitioning the cell into a cluster of vesicles. In the ascoviruses isolated from *Spodoptera frugiperda* and *Autographa precationis*, the formation of the vesicles is initiated at the periphery of the cell by invagination and proliferation of the plasma membrane (see Figures 4A, 5). Once vesicle formation is complete, the vesicles dissociate from each other (see Figure 6A) and accumulate within the tissue in which they formed. As the disease progresses, the basement membrane of infected tissues is disrupted and vesicles are released into the hemolymph (see Figure 3C). Completely formed vesicles often have long filamentous extensions of the plasma membrane projecting from the surface (see Figure 6A).

IV. VIRION MORPHOGENESIS AND ULTRASTRUCTURE

A. VIRION MORPHOGENESIS

Virion morphogenesis is initiated after the nucleus ruptures, and occurs prior to and during the cleavage of the cell into vesicles.[1] The first recognizable structural component of the virion to form is the multilaminar layer of the inner particle[3] (see Figure 2A, B).

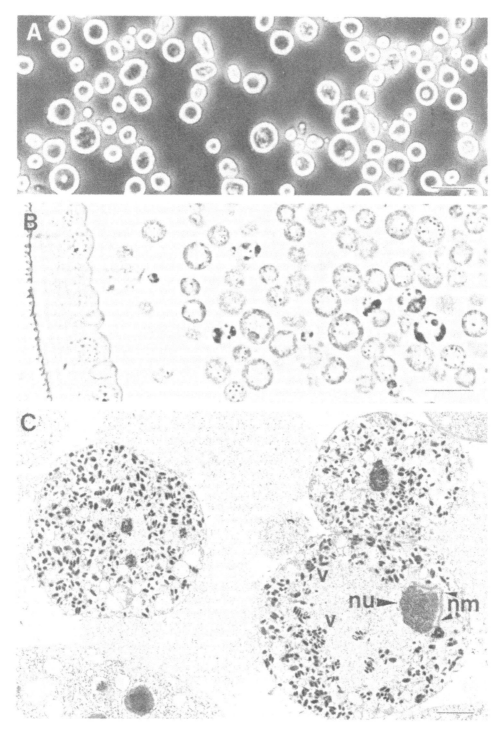

FIGURE 1.   Micrographs of vesicles induced by an ascovirus in larvae of the clover cutworm, *Scotogramma trifolii*. (A) Wet mount of hemolymph illustrating numerous refractile vesicles characteristic of the disease caused by all ascoviruses (phase contrast). (B) Micrograph of a section through a portion of the hemocoel and epidermis of a diseased fourth instar embedded in plastic; note the vesicles in the hemolymph; the dense regions along the periphery of the vesicles are areas of virion maturation (phase contrast). (C) Electron micrograph illustrating the internal structure of vesicles such as those in (A) and (B); note the masses of virions (v) and remnant of the nuclear membrane (nm) and nucleolus (nu) from the host cell. (Bars in [A] and [B] = 10 μm; in [C] = 1 μm.) (From Federici, B. A., *J. Invertebr. Pathol.*, 40, 44, 1982. With permission.)

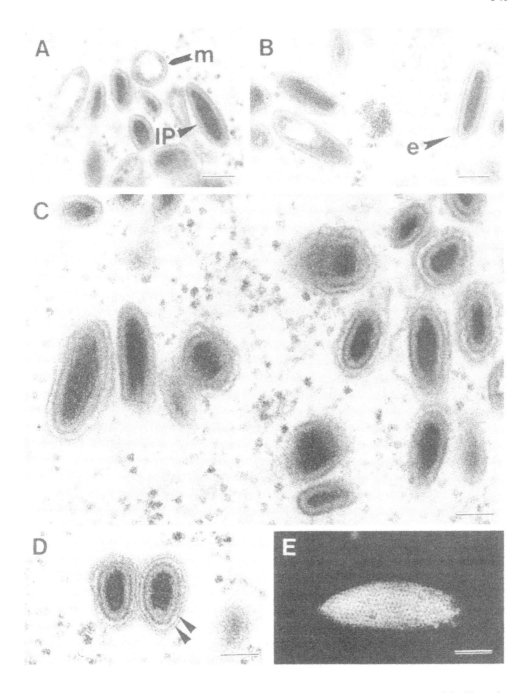

FIGURE 2. Electron micrographs of an ascovirus from the clover cutworm, *Scotogramma trifolii*, illustrating virion ultrastructure and stages of development. (A) and (B) Illustrate stages of inner particle formation; in the sequence of virion assembly, the first clearly recognizable virion component is the multilaminar layer (m) of the inner particle (IP); this layer apparently consists of a unit membrane bounded externally by a distinctive layer of protein subunits, both of which are more apparent in fully formed virions (see Figure 3G); the inner particle forms by the accumulation and condensation of viral nucleoprotein on the inner surface of the multilaminar layer; subsequently, the inner particle obtains an envelope (e). (C) and (D) Ultrathin sections through inner particles and virions; in (D) note the inner and outer layers (arrowheads) of the virion bilayer envelope. (E) Negatively stained virion illustrating the reticulate appearance characteristic of mature viral particles. (Bars in [A] through [E] = 100 nm.) (From Federici, B. A., *J. Invertebr. Pathol.*, 40, 50, 1982. With permission.)

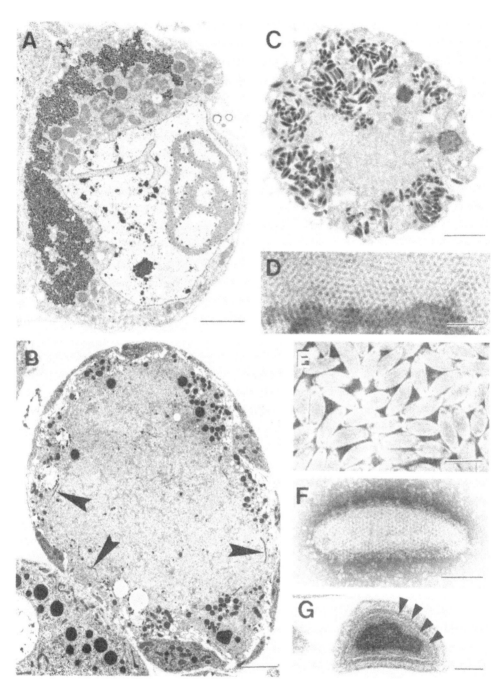

FIGURE 3. Electron micrographs illustrating aspects of the cytopathology and virion structure typical for the ascovirus isolated from the cabbage looper, *Trichoplusia ni*. (A) Fat body cell in an early stage of infection; the nucleus is slightly hypertrophied and the nuclear membrane is distorted and invaginated. (B) Infected fat body cell prior to virion assembly and vesicle formation; cell is greatly enlarged and nuclear membrane has fragmented (arrows). (C) Fully developed, virion-containing vesicle from larval hemolymph. (D) Portion of a negatively stained membrane from a virion-containing vesicle; note the similarity of the pattern in the membrane to that which occurs in the virions (see Figures 2E, 3F, 4D, 5B). (E) Negatively stained virions purified from vesicles using sucrose density gradient centrifugation; note the allantoid shape of the particles. (F) Negatively stained virion from larval hemolymph. (G) Cross section through a virion within a vesicle; note inner electron-translucent layer and bounding protein layer of inner particle (arrowheads). (Bars in [A] = 2 μm, in [B] = 3 μm, in [C] = 1 μm, in [D] = 70 nm, in [E] = 350 nm, in [F] = 100 nm, and in [G] = 60 nm.) (Adapted from Federici, B. A., *Proc. Natl. Acad. Sci. U.S.A.*, p. 7765, 1983.)

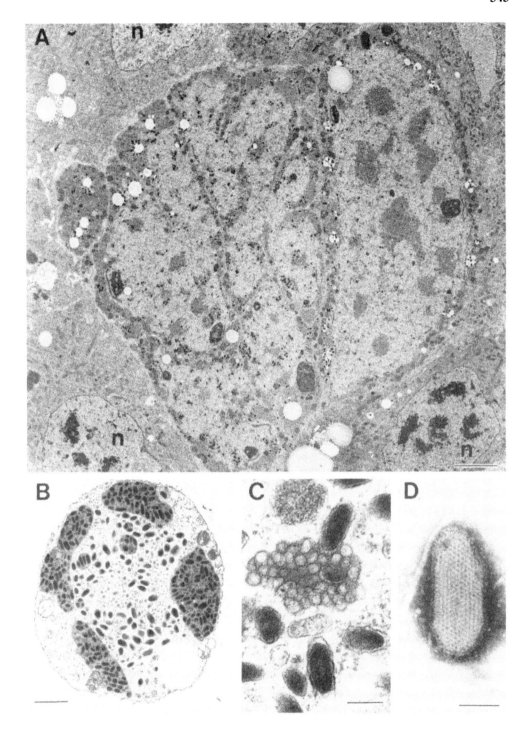

FIGURE 4. Electron micrographs illustrating aspects of the cytopathology and virion structure characteristic for the ascovirus isolated from the fall armyworm, *Spodoptera frugiperda*. (A) Three infected, greatly hypertrophied fat body cells at early to mid-stages in the process of vesicle formation; the nuclear membrane has fragmented, and vesicles are being formed through invagination of the cell's plasma membrane; note the apparently uninfected nuclei (n) adjacent to the infected cells. (B) Virion-containing vesicle in larval hemolymph; note the inclusions containing virions near the vesicle periphery. (C) Developing viral inclusion body illustrating the vesicular matrix in which the virions are occluded. (D) Negatively stained virion from larval hemolymph. (Bars in [A] = 3 μm, in [B] = 1 μm, in [C] = 200 nm, and in [D] = 100 nm.) (Adapted from Federici, B. A., *J. Gen. Virol.*, p. 1664, 1990.)

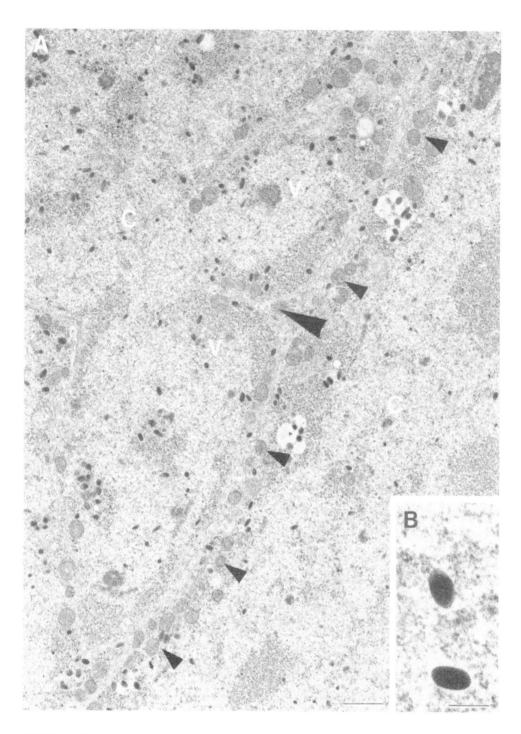

FIGURE 5.   Enlargement of Figure 4A, illustrating developing and mature virus particles in infected cells. (A) Two cells (c) in different stages of vesicle formation; the small arrowheads point to mitochondria along the proliferating plasma membrane of the cell on the lower right half of the figure just prior to the initiation of cleavage; in the upper left half of the figure, cleavage has been initiated as indicated by the invagination of the proliferating plasma membrane (large arrowhead). Note that virions and inner particles (B) are more numerous in the cell in which cleavage has been initiated. (Bar in [A] = 830 nm, in [B] = 200 nm.)

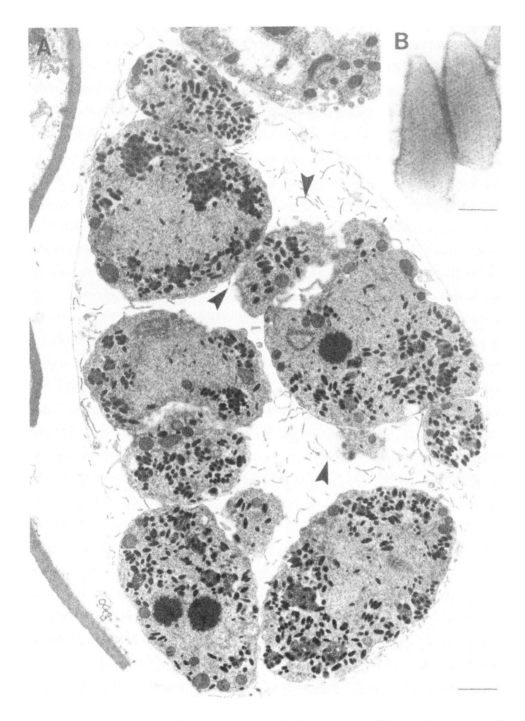

FIGURE 6. Electron micrographs illustrating an ascovirus isolated from larvae of *Autographa precationis*. (A) Completely formed virion-containing vesicles within the remnants of an epidermal cell; note the thin, membrane-bound filaments that extend from the vesicles (arrows). (B) Negatively stained virions from larval hemolymph. (Bar in [A] = 1 μm, in [B] = 100 nm.)

Based on its ultrastructure, this layer consists of a unit membrane and an exterior layer of protein subunits. As the multilaminar layer assembles, a dense nucleoprotein core forms internally. This process continues and the allantoid or bacilliform shape of the inner particle gradually becomes apparent. Once the inner particle is formed, it is enveloped (see Figures 2B, C; 4C).

In the ascoviruses isolated from *Trichoplusia ni*, *Heliothis zea*,[5,7] and *Scotogramma trifolii*,[3] after formation the virions accumulate toward the periphery of the vesicle, often forming inclusion bodies (see Figure 3C). In the ascovirus isolated from *Spodoptera frugiperda*,[6,7] virions are actually occluded in a "foamy" or vesicular matrix (see Figure 4C).

B. VIRION ULTRASTRUCTURE

The typical ascovirus virion is allantoid in shape (see Figures 2E; 3E, F), although the virion of the isolate from *Spodoptera frugiperda* is essentially bacilliform (see Figures 4C, D; 5B). Virions of all isolates are in the range of 400 nm in length by 130 nm in diameter. Evidence obtained from ultrastructural studies indicates the virion contains two unit-membrane structures, an outer envelope, and a membrane surrounding the DNA nucleoid within the inner particle. The electron translucent lipid layer of the latter membrane is most easily observed in virions packed in aggregations, such as the one illustrated in Figure 3G. Currently, there is no biochemical evidence for an inner membrane. The implication that the electron translucent layer of the inner particle is that of a membrane is derived from its similarity in structure to the nucleoid membrane found in the core of virions of poxviruses.[10]

Both the inner and outer membranes bear a distinctive external layer of subunits (see Figures 2D, 3G). In negatively stained preparations, the virions exhibit a characteristic reticular appearance (see Figures 2E, 3F, 4D, 6B). This appearance is imparted to the virions by the subunits of the virion envelope or by a combination of these subunits superimposed on those of the inner membrane. A similar reticular appearance has also been observed in large membrane fragments found in negatively stained preparations of virion-containing vesicles (see Figure 3D). The latter membranes are thought to be fragments of the membrane which delimits the vesicle, but could also be derived from the membrane precursors of virion envelopes from inside the vesicle.

In summary, the ascovirus virion as currently understood consists of an internal DNA/protein core surrounded by a unit-membrane structure, which in turn is surrounded by an envelope.

REFERENCES

1. **Federici, B. A.,** Enveloped double-stranded DNA insect virus with novel structure and cytopathology, *Proc. Natl. Acad. Sci. U.S.A.*, 80, 7664, 1983.
2. **Adams, J. R., Stadelbacher, E. A., and Tompkins, G. J.,** A new virus-like particle from the cotton bollworm, *Heliothis zea*, in *37th Annu. Proc. Elect. Micros. Soc. Am.*, Bailey, G. W., Ed., 1979, 52.
3. **Federici, B. A.,** A new type of insect pathogen in larvae of the clover cutworm, *Scotogramma trifolii*, *J. Invertebr. Pathol.*, 40, 41, 1982.
4. **Browning, H. W., Federici, B. A., and Oatman, E. R.,** Occurrence of a disease caused by a rickettsialike organism in a larval population of the cabbage looper, *Trichoplusia ni* in southern California, *Environ. Entomol.*, 11, 550, 1982.
5. **Carner, G. R. and Hudson, J. S.,** Histopathology of virus-like particles in *Heliothis* spp., *J. Invertebr. Pathol.*, 41, 238, 1983.
6. **Hamm, J. J., Pair, S. D., and Marti, Jr., O. G.,** Incidence and host range of a new ascovirus isolated from fall armyworm, *Spodoptera frugiperda* (Lepidoptera: Noctuidae), *Fla. Entomol.*, 69, 524, 1986.
7. **Federici, B. A., Vlak, J. M., and Hamm, J. J.,** Comparison of virion structure, protein composition, and genomic DNA of three Ascovirus isolates. *J. Gen. Virol.*, 71, 1661, 1990.

8. **Govindarajan, R. and Federici, B. A.,** Ascovirus infectivity and the effects of infection on the growth and development of Noctuid larvae, *J. Invertebr. Pathol.,* 56, 291, 199.

9. **Federici, B. A. and Govindarajan, R.,** Comparative histopathology of three ascovirus isolates in larval noctuids, *J. Invertebr. Pathol.,* 56, 300, 1990.

10. **Kurstak, E. and Garzon, S.,** Entomopoxviruses, in *The Atlas of Insect and Plant Viruses,* Maramorosch, K., Ed., Academic Press, New York, 1977, 29.

Chapter 12

NODAVIRIDAE

Simon Garzon and Guy Charpentier

TABLE OF CONTENTS

I. INTRODUCTION

The family Nodaviridae is composed of insect small icosahedral riboviruses which contain a bipartite genome. The two single-stranded (ss) RNA molecules with molecular weight of 1 and 0.5×10^6 for RNA1 and RNA2, respectively, are packaged in the same virus particle. Both ssRNAs are required for infectivity.[1,2] This bipartite ssRNA genome differentiates the Nodaviridae from the other small animal RNA viruses (Picornaviridae, Caliciviridae, *Nudaurelia-β* virus group)[3] or from the known bipartite plant viruses in which each RNA molecule is encapsidated in different particles.[4] The Nodaviridae are roughly spherical particles with a diameter of 29 to 30 nm; they contain a major coat protein of 40 kDa and two minor coat proteins with molecular weights of approximately 43 and 33 to 38 kDa. The type species, Nodamura virus differs from other Nodavirus with respect to its wide host range including mammals.

A. MEMBERS OF THE GROUP

Four viruses have been recognized as members of the Nodaviridae: Nodamura virus (NV, the type species) isolated from female mosquitoes *Culex tritaeniorhyncus;*[5] the black beetle virus (BBV) isolated from the New Zealand black beetle *Heteronychus arator* (Coleoptera:Scarabaeidae) (F.);[6] Boolarra virus (BoV) from the grass grub *Oncopera intricoides* (Lepidoptera:Hepialidae) (Walker);[7] and the flock house virus (FHV) isolated from the New Zealand grass grub, *Costelytra zealandica* (Coleoptera:Scarabaeidae) (White).[8]

Among possible members, an endogenous virus of *Drosophila* cells which could be a mutant of BBV;[9] a small RNA virus from *Lymantria ninayi* (Lepidoptera:Lymantriidae) (LNV_G);[10,11] and another virus from *Sesamia calamistis* (Lepidoptera:Agrotidae) (Hampson)[12] were also shown to contain a bipartite genome. Recently, a second distinctive virus with a bipartite RNA genome, Manawatu virus, from the New Zealand grass grub, *Costelytra zealandica*, was shown to be serologically related but not identical to FHV or any of the other members of Nodaviruses.[13]

Many other small icosahedral riboviruses have been isolated or visualized by electron microscopy from different insects in latent or mixed infections but since their physico-biochemical properties are not available, they are still unclassified. Some of these viruses are difficult to characterize because of their association in nature with numerous other picorna-like viruses, which could give the illusion of a bipartite genome. The Arkansas bee virus (ABV) proposed as a possible member of the Nodaviridae was in fact contaminated by a distinct and not related virus, the Berkeley bee picornavirus (BBPV).[3,14] ABV shares some physicochemical properties with a number of plant viruses and contains only one species of ssRNA (mol wt 1.8×10^6) which excludes further consideration of ABV as a Nodavirus.[15]

B. HOST RANGE

Although very similar by their physicochemical properties and antigenically related but distinct from each other, the Nodaviruses differ in their biological properties. Significant differences in host range occur within this group of viruses. Normally indigenous to insects, the Nodaviruses can extend their host range to other taxonomic kingdoms. Indeed NV replicates well in both insect and mammalian cells whereas FHV recently was shown to replicate in plant cells.[16]

NV multiplies without apparent effect in swine, mosquitoes, ticks, and the moth *Plodia interpunctella*, and can be transmitted to mice by *Aedes aegypti*.[5] Tesh[17] showed that NV kills the mosquitoes *Aedes albopictus* and *Toxorhynchites amboinensis* but not *Culex quinquefasciatus* after intrathoracic infection, whereas oral infection was shown to induce only an inapparent infection.[18] When inoculated into suckling mice or hamsters, NV produces flaccid paralysis and death.[18,19] NV also kills adult honey bees, *Apis mellifera*, and wax

TABLE 1
Host Range of Nodaviridae

Virus	Natural host	Cross-transmission hosts	Cell cultures
Nodamura virus (NV)	*Culex tritaeniorhyncus* (Diptera)[5]	Diptera: *Culex tarsalis, Aedes aegypti, A. albopictus, Toxorhynchites amboinensis* Ticks: *Ornithodoros savignyi* Hymenoptera: *Apis mellifera* Lepidoptera: *Plodia interpunctella, Galleria mellonella* Coleoptera: *Attagenus piceus* Mammals: Suckling mice and hamsters	*A. aegypti, A. albopictus, A. pseudoscutellaris,* (AP-61), baby hamster kidney (BHK-21)
Black beetle virus (BBV)	*Heteronychus arator* (Coleoptera)[48]	Coleoptera: *Aphodius tasmaniae, Pericoptus truncatus* Lepidoptera: *Galleria mellonella, Pseudalotia separata*	*Drosophila* D1[9]
Boolarra virus (BoV)	*Oncopera intricoides* (Lepidoptera)[7]	Lepidoptera: *Galleria mellonella*	*Drosophila* D1
Flock house virus (FHV)	*Costelytra zealandica* (Coleoptera)[8]	Lepidoptera: *Galleria mellonella* Plants: barley, cowpea, chenopodium, tobacco, *Nicotiana benthamiana*[16]	*Drosophila* D1, *Drosophila* GM1, *Coleoptera* Ha1179[27]

Note: No CPE or virus production could be detected in suckling mice or BHK-21 cell line infected with BoV, FHV, or BBV.

moth larvae, *Galleria mellonella,* after parenteral infection.[20] FHV replication was detected in plant cells inoculated with FHV genomic RNA or in plant protoplasts infected with FHV whole virions.[16] BBV and BoV show a more restricted host range to Lepidoptera and/or Coleoptera (see Table 1). The wax moth larvae, *G. mellonella,* are permissive to the replication of the four nodaviruses (see Table 1) and also to the bipartite ribovirus isolated from *Lymantria ninayi.*[10]

In permissive cell cultures some of these viruses induce widespread cytopathic effect (CPE) like BBV and FHV in D1 *Drosophila* cell line, whereas NV-infected *Aedes* or baby hamster kidney (BHK-21) cell lines, or BoV-infected D1 *Drosophila* cells produce viruses without apparent CPE.[7-9,21] RNAs of NV and BBV can be translated *in vitro* in cell-free lysates from wheat germ embryo or HeLa cells,[22] or from rabbit reticulocyte or *D. melanogaster* for NV and BBV, respectively.[23-25]

C. GROSS PATHOLOGY

Most of the Nodaviruses and, in particular, NV and FHV produce inapparent or persistent infection in nature. These infections result in a slower development of the insects or a slight reduction in the total population due to a reduction of egg viability. Some other Nodaviruses, BoV or BBV, for some ecological reasons like overcrowding or mixed infections give an

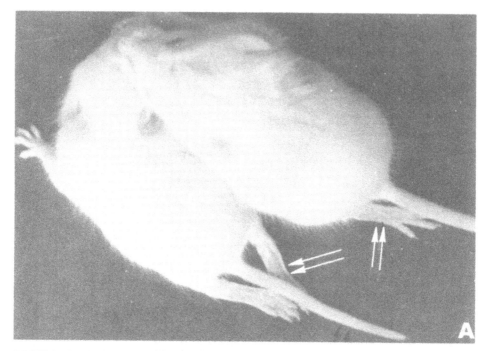

FIGURE 1.   Suckling mice (A) and suckling hamster (B) intracerebrally infected by Nodamura virus, and presenting, after 8 days postinfection the typical paralysis of the hind limbs (arrows).

epizootic, and dead or moribund larvae can be collected. NV when injected to *G. mellonella*, suckling mice or hamsters, mosquitoes, and honey bees, induced a fatal paralysis of the hind legs after 7 to 20 d (see Figure 1). The BBV-infected larvae of *Heteronychus arator* and LNV$_\text{G}$-infected *G. mellonella* larvae become flaccid and translucent whereas FHV-infected grubs appear only retarded in their development.[6-8,17]

II. VIRUS STRUCTURE AND COMPOSITION

Biophysical and biochemical properties of members of the *Nodaviridae* described in this section will be mainly based on the properties of two members of the family, NV and BBV, which have been extensively studied. These biophysical and biochemical properties are summarized in Table 2.

A. PARTICLE STRUCTURE

The size of the *Nodaviridae* particles range from 29 to 32 nm. Regarding the diameters, *Nodaviridae* are undistinguishable from the insect picornaviruses but distinct from *Nudaurelia*-β (38 nm in diameter). An extensive study of BBV-structure by Hosur et al.[26] with negative staining and X-ray diffraction of virus crystals indicates that the particles are icosahedral in shape. In negative staining electron micrography of nonpurified preparations, many of the virus particles remain associated with cellular membrane debris (see Figure 2B). The type member of this family, Nodamura virus, has a sedimentation coefficient of 135 S and a particle density of 1.34 g/ml (for the other members see Table 2). FHV and BBV particles show different electrophoretic mobility.[27] NV is resistant to diethyl ether and chloroform[5] and displays variable resistance to sodium deoxycholate.[28] This indicates that NV does not contain lipid as arboviruses do. These viruses seem to be stable through pH ranges of 3 to 9 and seem to be moderately sensitive to temperature.[27,29] Unlike FHV[25] some other members of this family are unstable in chloride ions,[1] or in CsCl gradients,[27] or both.[1] The two ssRNA fragments of *Nodaviridae* have sedimentation coefficients of, respectively, 22 S and 15 to 17 S and molecular weights of 0.95 to 1.15 × 10^6 and 0.46 to 0.47 × 10^6.

NV capsid is composed of a single major polypeptide of molecular weight 40 kDa and two minor polypeptides of molecular weight 38 and 43 kDa. Except for the FHV for which no minor polypeptide was detected,[27] all the other *Nodaviridae* show in SDS-PAGE various minor bands of polypeptides. The BBV seems to have different features since it has a polypeptide of low molecular weight (5 kDa) as a minor component of unknown function.[30] It has been suggested by Friesen and Rueckert[30] that a coat protein precursor of 47 kDa (α-protein) could be cleft during the assembly process in two polypeptides; one of 43 kDa (β-protein) and the other of 5 kDa (γ-protein). In FHV, the cleavage process occurs only after assembly of the protein into provirion, and this cleavage, probably autocatalytic, increases the stability of the virus particles.[31] A similar maturation process analogous to the picornavirus maturation has been observed for the NV[22] and LNV_G,[11] and seems to be a general characteristic of the Nodaviridae family.

Hosur et al.[26] recently proposed that a BBV particle is constituted from 180 protein subunits (protomers) of 44 kDa with T = 3 quasi-symmetry. This was confirmed by Hosur et al.[32] by an atomic resolution study of BBV. The protomers are constituted of three visible domains, a β-barrel as in other animal and plant virus protomers, an outer domain mainly composed of a β-sheet formed by three large insertions between strands of the barrel, and a carboxy terminal domain composed of two distorted helices lying inside the shell. The outer domains of quasi-threefold related protomers form pyramidal surface protrusions. Both termini, 64 residues at the amino end and 62 at the carboxy end, lie inside the shell but are crystallographically disordered and therefore invisible.

B. GENOME STRUCTURE

Both virion RNAs are necessary for infection[33] and they possess distinct nucleotide sequences.[34] However, they can be translated independently.[2] The *Nodaviridae* RNAs lack a 3′ terminal polyadenylated sequence.[27,35,36] The 5′ terminal sequences of BBV RNAs are capped.[36,37] The 3′ terminal sequence of BBV RNA2 had an unidentified moiety that confers its resistance to poly A addition and to ligation.[38] Apart from the discovery that BBV

TABLE 2
Biophysical and Biochemical Properties of Members of the Nodaviridae[a]

Virus		Nodamura	BBV (black beetle virus)	FHV (flock house virus)	BoV (Boolarra virus)	LNV$_G$[b] (Lymantria ninayi virus)
Diameter of virion (nm)		29	30	29	30	32
Sedimentation coefficient of virion (S)		135	137	142	140	ND
Buoyant density in CsCl (g/ml)		1.34	1.335	1.351	1.34	1.30
Molecular weight of major capsid proteins (kDa)		40	43	43	38	35
Molecular weight of minor capsid proteins (kDa)		43	47	No minor species	40.6	2 Minor species
		38	5			
RNA in virion (%)		20.5%	28.2%	ND	21.2%	ND
Sedimentation coefficient of RNAs (S)	RNA1	22	22	22	22	ND
	RNA2	15	15	15	17	ND
Molecular weight of RNAs (×10^6)	RNA1	1.15	1.12	1.1	1.03	0.95
	RNA2	0.46	0.46	0.46	0.47	0.47
G + C content of RNAs (%)	RNA1	55%	47%	ND	ND	ND
	RNA2	51%	49%	ND	ND	ND
Tested for the absence of poly A on RNAs		Yes	Yes	Yes	ND	ND
Capping of RNAs 5' termini		ND[c]	RNA1 and RNA2 capped	ND	ND	ND

[a] The main sources of information for this table are from: Dasmahapatra et al.,[36] Friesen and Rueckert,[30,43] Greenwood and Moore,[10] Longworth and Carey,[48] Newman and Brown,[1,33,35] Reavy et al.,[11] Reinganum et al.,[7] Scotti et al.,[27] Moore and Tinsley,[70] and Moore et al.[1,7,10,11,27,30,33,35,36,43,48,70,71]

[b] The virus of *Lymantria ninayi* seems similar to Nodaviridae in spite of the fact that the molecular weight of the two species of RNA and of the capsid proteins are lower.[11]

[c] ND = not determined.

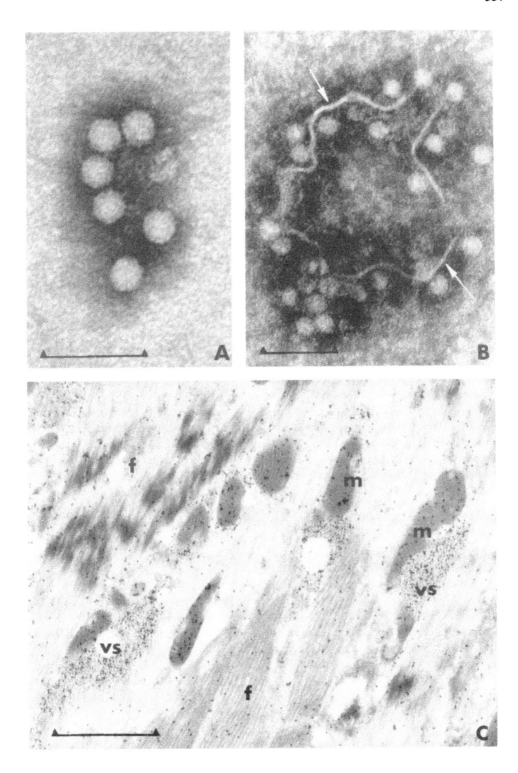

FIGURE 2. (A) and (B) Negative staining of partially purified Nodamura virus; many viral particles are membrane-associated (arrows). (Bars in [A] = 0.1 μm, in [B] = 0.3 μm.) (C) Localization of RNA by the RNase gold technique;[72] labeling is intense in the assembly site of viruses (VS) and around and inside the mitochondria. (Bar = 1 μm.)

replicates to high titers in *Drosophila* line 1, only few studies *in vitro* have been performed. The study of Nodaviridae replication cycle was performed either on intact *Drosophila* line 1 cells together with the use of translation cell-free system,[24] or on heat shock- and actinomycin D-treated cells.[39] These experiments demonstrated that RNA1 was the messenger for a 110 kDa protein (protein A) presumably a RNA-dependent RNA polymerase,[23] the properties of which were extensively examined.[40,41] RNA2 is the messenger for protein-α (47 kDa), a precursor of the coat protein.[23] In cultured *Drosophila* cells, during the course of the infection another nonstructural protein of low molecular weight (8 to 10 kDa) also appears.[30,42] The role of this protein, named protein B, has to be elucidated. The 9 S RNA messenger for this protein B is found in the infected cells but not in the virion and must be a subgenomic viral RNA (RNA3).[43] RNA3 originates from RNA1 and acts as a possible regulator of protein A translation.[2] There is also evidence for a translational control in BBV replication through competition between viral mRNAs. This translational control allows early and late functions to be temporarily segregated during viral replication.[25] Early functions are mainly expressed by maximum synthesis of proteins A and B from 4 to 10 h postinfection (hpi) and late functions by maximum synthesis of protein-α at 24 h.[30] These events and the replication cycle are summarized in Figure 9. Regarding NV RNAs translated in cell-free extracts, RNA1 directed the synthesis of a 105-kDa protein, and RNA2 directed the synthesis of a 43-kDa protein. The tryptic fingerprint of vp 43 was similar to that of the major coat protein vp 40.[22]

The three RNAs of BBV have been sequenced and their secondary structure determined.[36-38] The BBV RNA1 sequence (3,105 bases) contains, after a 5′ noncoding region of 38 nucleotides, a coding region for a protein of a predicted molecular weight of 101,873 Da. BBV RNA3 has a sequence (389 bases) homologous to that of the 3′ terminal region of virion RNA1. This RNA3 has two overlapping coding regions for proteins of 10,768 and 11,633 Da. One of these reading frames probably codes for a 10-kDa protein in infected *Drosophila* cells, more probably the 11,633-Da region. The RNA2 sequence (1399 bases) has two coding regions. The first one, starting at the base 23, codes for a protein 407 amino acids long corresponding to the coat protein precursor. The second reading frame beginning at base 1110 codes for a putative 72 amino acids long protein. Comparison of the nucleotide sequences of BBV, NV, FHV, and BoV RNA2 shows 80% homology between FHV and BBV, whereas BoV and NV are distantly related to BBV (50% homology).[44]

The advantage of the split genome is to separate the genes involved in early functions (replicative) from those involved in late functions (viral assembly).[25] Sequence regions favorable to bonding between RNA1 and RNA2 have been pointed out and may tie the two RNA species together for their joint encapsidation.[37] Preliminary attempts to make hybrid viruses by coinfection of cells by BBV RNA1 with NV RNA2 resulted in synthesis of BBV RNA1 and RNA3 but not of NV RNA2, suggesting the specificity of the BBV RNA polymerase, unable to recognize the heterologous NV RNA2 as template.[2] Infectious reassortant viruses can be made, however, by infecting D1 cells with BBV RNA1 and FHV RNA2 and vice versa.[45] It should also be possible to replicate suitable engineered foreign mRNAs with BBV RNA1.[46]

III. CYTOPATHOLOGY

Only fragmentary information on the *in vivo* histopathology or morphogenesis of the Nodaviridae are available, even though the virus produces visible pathology. In this case, the difficulty of following the infection is due to the fact that lesions are focal and sparse in the infected host and that different cells of a same tissue present a variable susceptibility to the virus. Another limitation is the small diameter of the isolated viral particles (28 to 30 nm) which can hardly be differentiated from ribosomes on thin sections. Direct evidence of viral replication is found only in specimens from hosts with severe paralysis or flaccidity. Early events in the progression of the disease are then difficult to resolve, and this will require the application of *in situ* hybridization method for the Nodaviridae RNA localization or the immunocytolocalization of viral proteins using poly- or monoclonal antibodies.

A. PATHOGENESIS

The pathogenesis and morphogenesis of NV are quite similar in mice or in the lepidopteran *G. mellonella* infected tissues. The cellular lesions are localized mainly in the cytoplasm of muscle cells and are characterized by the disorganization of muscle fibrils and the formation of dense viral inclusions which gradually replace the sarcolemma. These inclusions show a red fluorescence on staining with acridine orange. In the suckling mice, NV causes a myositis of hind limbs and paravertebral muscles, a necrosis of brown fat, and the degeneration of spinal cord neurons.[18,29] Whereas in *G. mellonella* larvae, the adipose tissue, the salivary and molting glands, the hypodermis, the trachea, and the hemocytes are also sensitive to the infection.[47] No data are available on the pathogenesis of Nodamura when injected to honey bees *Apis mellifera* or to *Aedes albopictus* and *Toxorhynchites amboinensis* mosquitoes; it is only known that head squash preparation of the paralyzed mosquitoes shows the replication of the virus by immunofluorescence or by titration on mosquito cell culture.[17] Similar NV histopathologic lesions are reported in the infected muscles of *Sesamia calamistis*.[12] BBV replication takes place in the cytoplasm of gut and fat body cells of the black beetle larvae, whereas *G. mellonella* midgut columnar cells and to a lesser extent fat body cells are susceptible to BBV.[48]

In vitro, NV replication is detected in the cytoplasm of *A. albopictus* or BHK-21 cell cultures by immunofluorescence in a proportion of 25 to 30% of the cells but no cytopathic effect (CPE) is produced.[17,21] No CPE could be detected in suckling mice or BHK-21 cell culture when infected by BBV, BoV, or FHV; and although no CPE appeared in BBV, BoV, or FHV infected D1 cells of *D. melanogaster,* viral particles are produced, the majority of which remain associated with cells or cellular debris.[7-9] A selected strain of BBV which grows more rapidly and produces cytolysis of D1 cells has been produced by repeated passages of noncytolytic BBV wild type at low multiplicity through *Drosophila* cells and is used for BBV plaque titration.[45] FHV replicates also in the coleopteran cell line Ha1179 or the *Drosophila* GM1 cells. The infectivity of FHV is unaffected by heating for 10 min at 40°C. Most infectivity is lost at 53 or 58°C.[27] The BBV has little effect on the host-cell protein synthesis during the early stages of infection of D1 cells of *D. melanogaster,* shut-off of host synthesis does not occur in infected cells until 8 to 10 hpi.[49]

B. MORPHOGENESIS

The following data are summarized from the morphogenesis study of the only two Nodaviridae, NV and BoV, which have been observed, respectively, *in vivo*[29,47] and *in vitro*[50] by electron microscopy. Both viruses present a replication very similar to the one described for the Picornaviridae of invertebrates[51-53] or vertebrates, particularly when referring to the role of cytoplasmic membranes for the virus biosynthesis.[54-59] Nodaviridae replication takes place exclusively in the cytoplasm of the host cells, in close association with the intracytoplasmic cell membranes: endoplasmic reticulum or mitochondria, which support the development of the virogenic stroma. The process of viral self-assembly occurs within these foci which are thought to contain viral RNA and proteins. Mature particles are gradually released free in the cytoplasm or are packed inside the endoplasmic reticulum cisternae leading to vesicles filled by viral particles. The viruses will be later extruded into the cytoplasm by the disruption of the vesicle membrane. Cytoplasmic viruses at high concentration form paracrystalline arrays which gradually displace the cellular organelles or fibrils. In BoV-infected D1 or D2 *Drosophila* cells, particles with electron-lucent centers about 16 nm in diameter, possibly precursor of the electron-dense mature particles (30 nm in diameter) are observed.[50]

The conclusion that intracellular membranes could support Nodaviridae replication is suggested by different observations. First, as already mentioned the Nodaviridae particles remain largely membrane-associated when released from infected cells[9,27,33] or after mechanical disruption of the infected cells (see Figure 2B). Second, it has been found that the RNA-dependent RNA polymerase of the BBV is bound to particulate material and forms a complex with its template BBV RNA.[40] Such membrane-enzyme-template complexes have also been described for picornaviruses[60] and several plant viruses.[61,62] Third, BoV and NV are often found intracellularly in several arrangements, either associated with membranes of the endoplasmic reticulum or vesicles, or located in between membranes of the mitochondria or of vesicular bodies probably derived from the transformation of mitochondria of the infected cells. Such vesicular bodies reported in BoV-infected *Drosophilia* cells[50] are similar to those seen in some virus-infected plants[63] or insects.[64] The modified mitochondria of infected cells seem to contain ribonucleic acids involved in virus replication as revealed by the RNase gold labeling (see Figure 2C). Similar structures in other virus-infected cells are also found to contain nucleic acids.[65,66] Fourth, in a recent ultrastructural study some earlier events of the cycle have been observed.[67] In cells with well-developed endoplasmic reticulum as the hemocytes or the adipose tissue cells, the sequence of replication events inside the reticulum can be established: first, stringlike arrays between two smooth unit membranes; then different arrangements in two, three, or more rows of bounded-vesicles virions; and last completely filled vesicles (see Figure 3). In muscle cells, the endoplasmic reticulum is not so developed, and although virus could be found in stringlike or vesicular arrangements (see Figure 4), numerous clusters of mitochondria appear scattered between the fibrils. Whether this particular accumulation of mitochondria results from the migration of cell mitochondria to the sites of viral replication or if a multiplication of mitochondria is induced by the virus is still unknown (see Figure 5A). A higher magnification reveals the presence, between the mitrochondrial membranes, of a dense material from which viral particles (28 nm in diameter) are assembled (see Figure 5B). Following the infection and during the virogenic stroma development, accumulation of viral particles will gradually induce the degeneration of the mitochondria (see Figure 6), and the cytoplasm is completely filled by viral particles which crystallize occasionally (see Figure 7).[47]

Thin sections of limb muscle of NV-infected suckling mice reveal a large number of virus particles, either dispersed or gathered in a crystalline array within the sarcoplasm (see Figure 8). Severe destruction of skeletal muscle cell organization is associated with the accumulation of virus particles. Fibroblasts in the skeletal muscles are also infected whereas macrophages and Kupffer cells in liver contain large masses of virus particles in membrane-bound structures[29] which could be the result of a phagocytosis process. Similar results are obtained in the muscles of the NV-infected larvae of *G. mellonella*.[47] In NV-infected larvae, helical filaments of 110 Å in diameter of unknown origin and function have been observed in the cytoplasm and inside the basal membrane of the infected muscle or adipose tissue.

The propagation and dispersion of Nodaviridae can be brought about by the lysis of the cells or by the migration of the virus along the endoplasmic reticulum to the intercellular spaces and basal membrane.[47,50]

C. ANTIGENIC RELATIONSHIPS OF THE NODAVIRIDAE

Serological relationship between NV, BBV, and BoV was demonstrated by immuno-electron microscopy,[7] though with another method, no cross reactions were observed between BBV and NV.[48] However, it is clear that BoV shares antigens with both viruses. The endogenous virus isolated from *Drosophila* line 1 cells (DL1V) was related but not identical to these viruses by immunodiffusion test.[9] By ELISA double antibody sandwich method

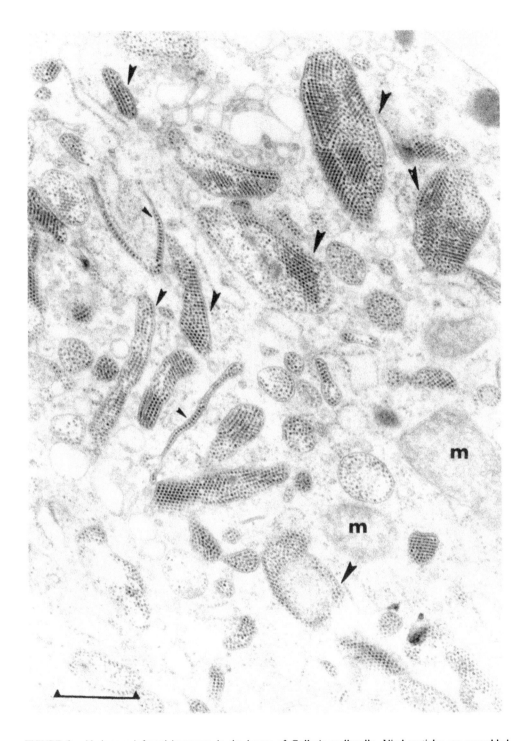

FIGURE 3.   Nodamura-infected hemocyte in the larvae of *Galleria mellonella*. Viral particles are assembled between cellular membranes probably from the endoplasmic reticulum. Different arrangements from stringlike array (small arrowheads) to two, three, or more lines of viruses (medium arrowheads) to vesicles filled with viruses sometimes packed in paracrystalline array (large arrowheads), can be observed in the cytoplasm, suggesting a sequence of events in virus production inside membranes or vesicles. The mitochondria (m) are still recognizable. (Bar = 0.5 μm.)

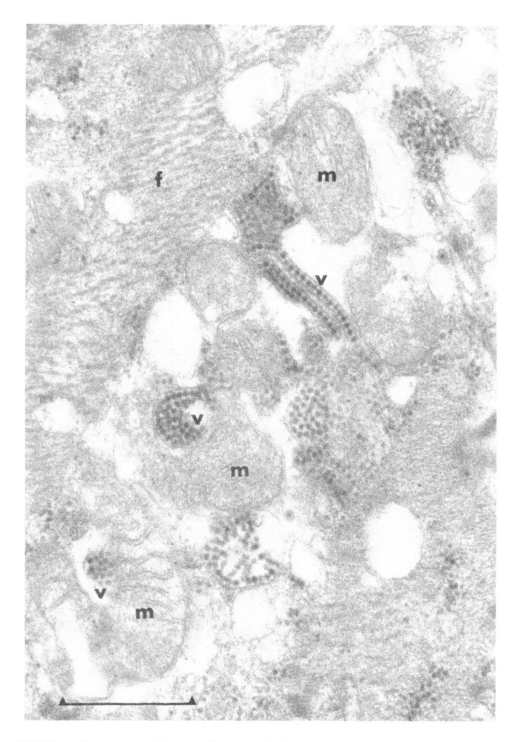

FIGURE 4. Nodamura infected *G. mellonella* muscle cell, illustrating the production and accumulation of viral particles (v) between two unit membranes, or in vesicles in close association with mitochondria (m), in an oedematous interfibrillar space; muscle fibrils (f). (Bar = 0.5 μm.)

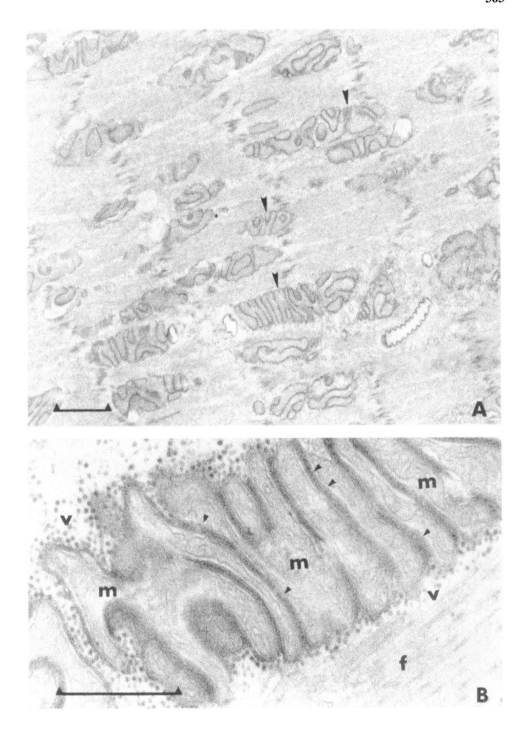

FIGURE 5.   Nodamura infected muscle cell in an early stage of infection of *G. mellonella* larvae. (A) The fibrils are not very disorganized and the Z bands are still distinguishable; however, notice the tightly agglutinated mitochondria (arrowheads) scattered between the muscle fibrils. (Bar = 1 μm.) (B) Higher magnification of one of this agglutinated mitochondria (m); the inner structure of the individual mitochondria is distinct; however, a dense material (arrowheads) is produced in between the external unit membranes of the adjacent mitochondria; viral particles (v) seem to be assembled from this dense material. (Bar = 0.5 μm.)

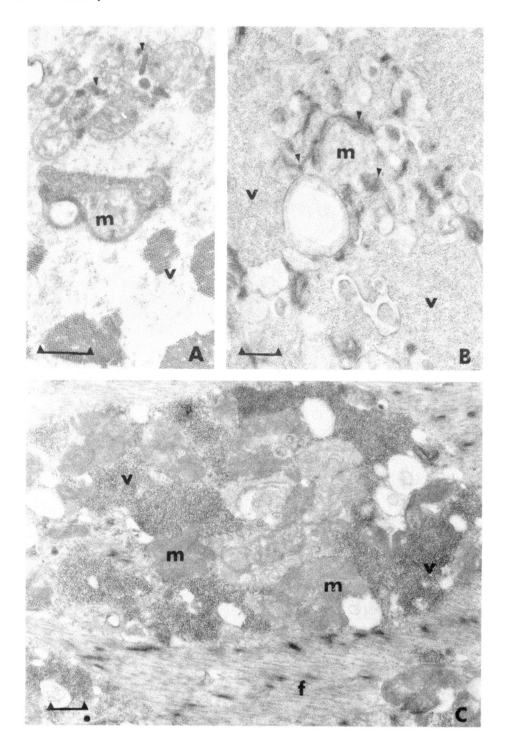

FIGURE 6.   Nodamura replication in association with mitochondria presenting the gradual accumulation of viral particles (v) and the concomitant degradation of the mitochondria (m). Arrowheads point out some dense virogenic stroma localized around the mitochondria. (Bar = 0.5 μm.)

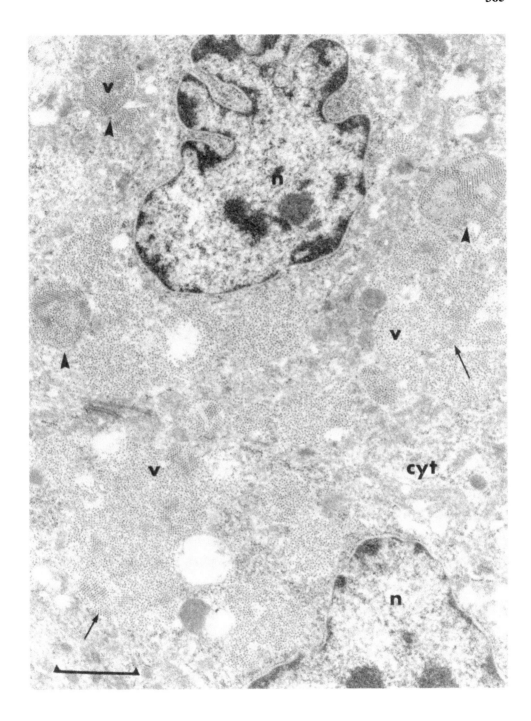

FIGURE 7. Late stage in the Nodamura virus infection of hypodermal cells of *G. mellonella* larvae. The cytoplasm (cyt) of the cells is almost replaced by the accumulation of viral particles (v) often in paracrystalline arrays (arrows) or in vesiculated bodies (arrowheads). The cytoplasmic membrane between the two cells is almost "eclipsed". (Bar = 1 μm.)

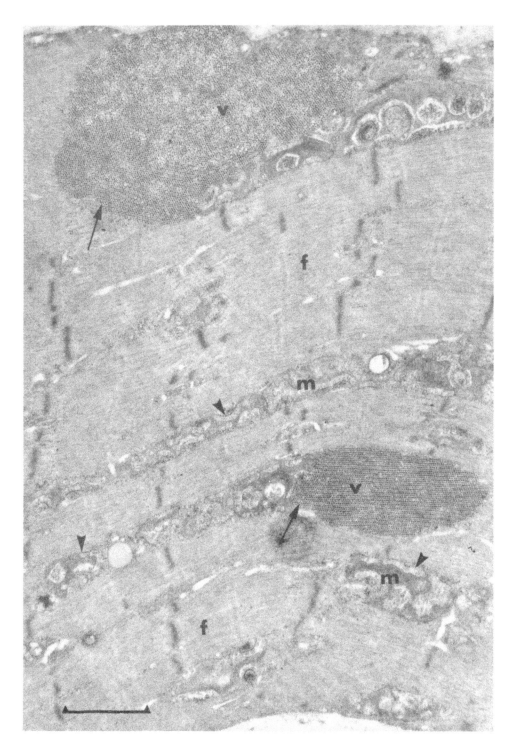

FIGURE 8.   Nodamura infected muscle cell from suckling mice. Notice the two viral inclusions packed with mature viruses, most of them in paracrystalline arrangements (arrows). Mitochondria (m) are modified by the infection and appear as multivesiculated bodies (arrowheads). (Bar = 1 μm.)

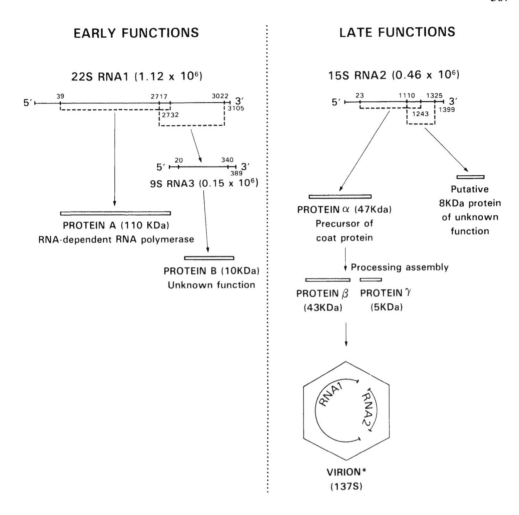

FIGURE 9.   Scheme for the replication and particle assembly of BBV. Note: This scheme is a modification of a scheme proposed by Gallagher et al. (1983), based on some new information given by Dasgupta et al. (1984), Guarino et al. (1984), and Dasmahapatra et al. (1985). *RNA1 and RNA2 may be tied by homologous sequences.

and by crossover electrophoresis, LNV$_G$ showed a weak serological relationship with BBV; however, partial proteolysis fragment studies failed to reveal any similarities between these two viruses.[11] By immunodiffusion test FHV was antigenically related to, but distinct from, BBV and DL1 virus. Using crossover electrophoresis, Greenwood and Moore[10] failed to demonstrate any relationship between LNV$_G$ and members of the *Nudaurelia*-β virus group. No serological relationships were found between NV and 11 small RNA viruses from *Nudaurella*-β and insect picornaviruses group.[68] Finally, with the seroneutralization test, complement fixation, and hemagglutination inhibition, no relationship could be detected between NV and a great number of arboviruses, enteroviruses from human, and some other human or animal viruses like herpes simplex virus and vesicular stomatitis virus.[69]

ACKNOWLEDGMENTS

We are grateful for the secretarial assistance of Mrs. Andrée B. Desrochers and Mrs. M. H. Colas de la Noue for the English revision of the manuscript. We also thank C. Alarie and H. Strykowski for technical assistance. This work was partly supported by grant no. 83-EQ-2098 from F.C.A.R. (Québec, Canada).

REFERENCES

1. **Newman, J. F. E. and Brown, F.**, Further physicochemical characterization of Nodamura virus. Evidence that the divided genome occurs in a single component, *J. Gen. Virol.*, 38, 83, 1977.
2. **Gallagher, T. M., Friesen, P. D., and Rueckert, R. R.**, Autonomous replication and expression of RNA1 from black beetle virus, *J. Virol.*, 46, 481, 1983.
3. **Matthews, R. E. F.**, Classification and nomenclature of viruses. Fourth report of the International Committee on Taxonomy of Viruses, *Intervirology*, 17, 1, 1982.
4. **Bruenning, G.**, Plant covirus systems: two-component systems, in *Comprehensive Virology*, Vol. 11, Fraenkel-Conrat, H. and Wagner, R. R., Eds., Plenum Press, New York, 1977, 55.
5. **Scherer, W. F. and Hurlbut, H. S.**, Nodamura virus from Japan: a new and unusual arbovirus resistant to diethyl ether and chloroform, *Am. J. Epidemiol.*, 86, 271, 1967.
6. **Longworth, J. F. and Archibald, R. D.**, A virus of black beetle *Heteronychus arator*, F. (Coleoptera:Scarabaeidae), *N.Z. J. Zool.*, 2, 233, 1975.
7. **Reinganum, C., Bashiruddin, J. B., and Cross, G. F.**, Boolarra virus: a member of the *Nodaviridae* isolated from *Oncopera intricoides* (Lepidoptera:Hepialidae), *Intervirology*, 24, 10, 1985.
8. **Dearing, S. C., Scotti, P. D., Wigley, P. J., and Dhana, S. D.**, A small RNA virus isolated from the grass grub, *Costelytra zealandica* (Coleoptera:Scarabaeidae), *N.Z. J. Zool.*, 7, 267, 1980.
9. **Friesen, P., Scotti, P., Longworth, J., and Rueckert, R. R.**, Black beetle virus: propagation in *Drosophila* line 1 cells and an infection-resistant subline carrying endogenous black beetle virus-related particles, *J. Virol.*, 35, 741, 1980.
10. **Greenwood, L. K. and Moore, N. F.**, The purification and partial characterization of a small RNA-virus from *Lymantria* — the identification of a Nodamura-like virus, *Microbiologica*, 5, 49, 1982.
11. **Reavy, B., Crump, W. A. L., Greenwood, L. K., and Moore, N. F.**, Characterization of a bipartite genome ribovirus isolated from *Lymantria ninayi*, *Microbiologica*, 5, 305, 1982.
12. **Jacquemard, P., Croizier, G., Amargier, A., Veyrunes, J.-C., Croizier, L., Bordat, D., and Vercambre, B.**, Présence de trois virus chez *Sesamia calamistis* Hampson (lépidoptère *Agrotidae)* á l'île de la Réunion, *L'Agronomie Tropicale*, 40, 66, 1985.
13. **Scotti, P. D. and Fredericksen, S.**, Manawatu virus: a nodavirus isolated from *Costelytra zealandica* (White) (Coleoptera:Scarabaeidae), *Arch. Virol.*, 97, 85, 1987.
14. **Bailey, L. and Woods, R. D.**, Three previously undescribed viruses from the honeybee, *J. Gen. Virol.*, 25, 175, 1974.
15. **Lommel, S. A., Morris, T. J., and Pinnock, D. E.**, Characterization of nucleic acids associated with Arkansas bee virus, *Intervirology*, 23, 199, 1985.
16. **Selling, B. H., Allison, R. F., and Kaesberg, P.**, Genomic RNA of an insect virus directs synthesis of infectious virions in plants, *Proc. Natl. Acad. Sci. U.S.A.*, 87, 434, 1990.
17. **Tesh, R. B.**, Infectivity and pathogenicity of Nodamura virus for mosquitoes, *J. Gen. Virol.*, 48, 177, 1980.
18. **Scherer, W. F., Verna, J. E., and Richter, G. W.**, Nodamura virus, an ether- and chloroform-resistant arbovirus from Japan, *Am. J. Trop. Med. Hyg.*, 17, 120, 1968.
19. **Charpentier, G.**, personal communication.
20. **Bailey, L. and Scott, H. A.**, The pathogenicity of Nodamura virus for insects, *Nature (London)*, 241, 545, 1973.
21. **Bailey, L., Newman, J. F. E., and Porterfield, J. S.**, The multiplication of Nodamura virus in insect and mammalian cell cultures, *J. Gen. Virol.*, 26, 15, 1975.
22. **Newman, J. F. E., Matthews, T., Omilianowski, D. R., Salerno, T., Kaesberg, P., and Rueckert, R.**, *In vitro* translation of the two RNAs of *Nodamura* virus, a novel mammalian virus with a divided genome, *J. Virol.*, 25, 78, 1978.

23. **Guarino, L. A., Hruby, D. E., Ball, L. A., and Kaesberg, P.,** Translation of black beetle virus RNA and heterologous viral RNAs in cell-free lysates derived from *Drosophila melanogaster*, *J. Virol.*, 37, 500, 1981.

24. **Crump, W. A. L. and Moore, N. F.,** *In vivo* and *in vitro* synthesis of the proteins expressed by the RNA of black beetle virus, *Arch. Virol.*, 69, 131, 1981.

25. **Friesen, P. D. and Rueckert, R. R.,** Early and late functions in a bipartite RNA virus: evidence for translational control by competition between viral mRNAs, *J. Virol.*, 49, 116, 1984.

26. **Hosur, M. V., Schmidt, T., Tucker, R. C., Johnson, J. E., Selling, B. H., and Rueckert, R. R.,** Black beetle virus — crystallization and particle symmetry, *Virology*, 133, 119, 1984.

27. **Scotti, P. D., Dearing, S., and Mossop, D. W.,** Flock house virus: a nodavirus isolated from *Costelytra zealandica* (White) (Coleoptera:Scarabaeidae), *Arch. Virol.*, 75, 181, 1983.

28. **Scherer, W. F.,** Variable results of sodium deoxycholate tests of Nodamura virus, an ether and chloroform resistant arbovirus, *Proc. Soc. Expt. Biol. Med.*, 129, 194, 1968.

29. **Murphy, F. A., Scherer, W. F., Harrison, A. K., Dunne, H. W., and Gary, G. W., Jr.,** Characterization of Nodamura virus, an arthropod transmissible picornavirus, *Virology*, 40, 1008, 1970.

30. **Friesen, P. D. and Rueckert, R. R.,** Synthesis of black beetle virus proteins in cultured *Drosophila* cells: differential expression of RNAs 1 and 2, *J. Virol.*, 37, 876, 1981.

31. **Gallagher, T. M. and Rueckert, R. R.,** Assembly-dependent maturation cleavage in provirions of a small icosahedral insect ribovirus, *J. Virol.*, 62, 3399, 1988.

32. **Hosur, M. V., Schmidt, T., Tucker, R. C., Johnson, J. E., Gallagher, T. M., Selling, B. H., and Rueckert, R. R.,** Structure of an insect virus at 3.0 Å resolution, *Proteins: Struct., Funct., Genet.*, 2, 167, 1987.

33. **Newman, J. F. E. and Brown, F.,** Evidence for a divided genome in Nodamura virus, an arthropod-borne picornavirus, *J. Gen. Virol.*, 21, 371, 1973.

34. **Clewley, J. P., Crump, W. A. L., Avery, R. J., and Moore, N. F.,** Two unique RNA species of the Nodavirus black beetle virus, *J. Virol.*, 44, 767, 1982.

35. **Newman, J. F. E. and Brown, F.,** Absence of poly(A) from the infective RNA of Nodamura virus, *J. Gen. Virol.*, 30, 137, 1976.

36. **Dasmahapatra, B., Dasgupta, R., Ghosh, A., and Kaesberg, P.,** Structure of the black beetle virus genome and its functional implications, *J. Mol. Biol.*, 182, 183, 1985.

37. **Guarino, L. A., Ghosh, A., Dasmahapatra, B., Dasgupta, R., and Kaesberg, P.,** Sequence of the black beetle virus subgenomic RNA and its location in the viral genome, *Virology*, 139, 199, 1984.

38. **Dasgupta, R., Ghosh, A., Dasmahapatra, B., Guarino, L. A., and Kaesberg, P.,** Primary and secondary structure of black beetle virus RNA2, the genomic messenger for BBV coat protein precursor, *Nucleic Acids Res.*, 12, 7215, 1984.

39. **Crump, W. A. L. and Moore, N. F.,** The proteins induced in black beetle virus-infected *Drosophila* cells at elevated temperature, *Ann. Virol. (Inst. Pasteur)*, 132E, 495, 1981.

40. **Guarino, L. A. and Kaesberg, P.,** Isolation and characterization of an RNA-dependent RNA polymerase from black beetle virus-infected *Drosophila melanogaster* cells, *J. Virol.*, 40, 379, 1981.

41. **Saunders, K. and Kaesberg, P.,** Template-dependent RNA polymerase from black beetle virus-infected *Drosophila melanogaster* cells, *Virology*, 147, 373, 1985.

42. **Crump, W. A. L. and Moore, N. F.,** The polypeptides induced in *Drosophila* cells by a virus of *Heteronychus arator*, *J. Gen. Virol.*, 52, 173, 1981.

43. **Friesen, P. D. and Rueckert, R. R.,** Black beetle virus: messenger for protein B is a subgenomic viral RNA, *J. Virol.*, 42, 986, 1982.

44. **Dasgupta, R. and Sgro, J.-Y.,** Nucleotide sequences of three Nodavirus RNA2s: the messengers for their coat protein precursors, *Nucleic Acids Res.*, 17, 7525, 1989.

45. **Selling, B. H. and Rueckert, R. R.,** Plaque assay for black beetle virus, *J. Virol.*, 51, 251, 1984.

46. **Dasmahapatra, B., Dasgupta, R., Saunders, K., Selling, B., Gallagher, T., and Kaesberg, P.,** Infectious RNA derived by transcription from cloned cDNA copies of the genomic RNA of an insect virus, *Proc. Natl. Acad. Sci. U.S.A.*, 83, 63, 1986.

47. **Garzon, S., Charpentier, G., and Kurstak, E.,** Morphogenesis of the Nodamura virus in the larvae of the lepidopteran *Galleria mellonella* (L.), *Arch. Virol.*, 56, 61, 1978.

48. **Longworth, J. F. and Carey, G. P.,** A small RNA virus with a divided genome from *Heteronychus arator* (F.) [Coleoptera:Scarabaeidae], *J. Gen. Virol.*, 33, 31, 1976.

49. **Crump, W. A. L., Reavy, B., and Moore, N. F.,** Intracellular RNA expressed in black beetle virus-infected *Drosophila* cells, *J. Gen. Virol.*, 64, 717, 1983.

50. **Bashiruddin, J. B. and Cross, G. F.,** Boolarra virus: ultrastructure of intracytoplasmic virus formation in cultured *Drosophila* cells, *J. Invertebr. Pathol.*, 49, 303, 1987.

51. **Teninges, D. and Plus, N.,** P virus of *Drosophila melanogaster* as a new picornavirus, *J. Gen. Virol.*, 16, 103, 1972.

52. **Bailey, L.,** Viruses attacking the honey bee, *Adv. Virus Res.*, 20, 271, 1976.

53. **Bailey, L.,** *Honey Bee Pathology,* Academic Press, New York, 1981.
54. **Bienz, K., Bienz-Isler, G., Egger., D., Weiss, M., and Loeffler, H.,** Coxsackie-virus infection in skeletal muscles of mice. An electron microscopic study. II. Appearance and fate of virus progeny, *Arch. Ges. Virusforsch.,* 31, 257, 1970.
55. **Bienz, K., Egger, D., and Pasamontes, L.,** Association of polioviral proteins of the P2 genomic region with the viral replication complex and virus-induced membrane synthesis as visualized by electron microscopic immunocytochemistry and autoradiography, *Virology,* 160, 220, 1987.
56. **Bienz-Isler, G., Bienz, K., Weiss, M., and Loeffler, H.,** Coxsackie-virus infection in skeletal muscles of mice. An electron microscopic study. I. Cell and nucleus alterations, *Arch. Ges. Virusforsch.,* 31, 247, 1970.
57. **Caliguiri, L. A. and Tamm, I.,** The role of cytoplasmic membranes in poliovirus biosynthesis, *Virology,* 42, 100, 1970.
58. **Caliguiri, L. A. and Tamm, I.,** Characterization of poliovirus specific structures associated with cytoplasmic membranes, *Virology,* 42, 112, 1970.
59. **Godman, G. C.,** Picornaviruses, in *Ultrastructure of Animal Viruses. Ultrastructure in Biological Systems,* Vol. 5, Dalton, A. J. and Haguenau, F., Eds., Academic Press, New York, 1973, 133.
60. **Traub, A., Diskin, B., Rosenberg, H., and Kalmar, E.,** Isolation and properties of the replicase of encephalomyocarditis virus, *J. Virol.,* 18, 375, 1976.
61. **Clerx, C. M. and Bol., J. F.,** Properties of solubilized RNA-dependent RNA polymerase from alfalfa mosaic virus-infected and healthy tobacco plants, *Virology,* 91, 453, 1978.
62. **Zabel, P. H., Weenen-Swaans, H., and van Kammen, A.,** *In vitro* replication of cowpea mosaic virus RNA. I. Isolation and properties of the membrane-bound replicase, *J. Virol.,* 14, 1049, 1974.
63. **Binnington, K. C.,** Structural transformation of blowfly mitochondria by a putative virus; similarities with virus-induced changes in plant mitochondria, *J. Gen. Virol.,* 68, 201, 1987.
64. **Binnington, K. C., Lockie, E., Hines, E., and van Gerwen, C. M.,** Fine structure and distribution of three types of virus-like particles in the sheep blowfly, *Lucilia cuprina* and associated cytopathic effects, *J. Invertebr. Pathol.,* 49, 175, 1987.
65. **Kim, K. S.,** Cytopathology of spotted cucumber beetle hemocytes containing virus-like particles, *J. Invertebr. Pathol.,* 36, 292, 1980.
66. **Chao, Y. C., Young, S. Y., and Kim, K. S.,** Cytopathology of the soybean looper, *Pseudoplusia includens,* infected with the *Pseudoplusia includens* icosahedral virus, *J. Invertebr. Pathol.,* 45, 16, 1985.
67. **Garzon, S., Strykowski, H., and Charpentier, G.,** Implication of mitochondria in the replication of Nodamura virus in larvae of the Lepidoptera, *Galleria mellonella* (L) and in suckling mice, *Arch. Virol.,* 113, 163, 1990.
68. **Reinganum, C. and Scotti, P. D.,** Serological relations between twelve small RNA viruses of insects, *J. Gen. Virol.,* 31, 131, 1976.
69. **Taylor, R. M.,** Nodamura, in *Catalogue of Arthropod Borne Viruses of the World,* Public Health Service Publ. No. 1760, U.S. Department of Health, Education and Welfare, Besthesda, MD, 1967, 145.
70. **Moore, N. F. and Tinsley, T. W.,** The small RNA-viruses of insects, *Arch. Virol.,* 72, 229, 1982.
71. **Moore, N. F., Reavy, B., and King, L. A.,** General characteristics, gene organization and expression of small RNA viruses of insects, *J. Gen. Virol.,* 66, 647, 1985.
72. **Bendayan, M.,** Enzyme-gold electron microscopic cytochemistry: a new affinity approach for the ultrastructural localization of macromolecules, *J. Electron Microsc. Tech.,* 1, 349, 1984.

Chapter 13

PICORNAVIRIDAE: PICORNAVIRUSES OF INVERTEBRATES

N. F. Moore and S. M. Eley

TABLE OF CONTENTS

I. INTRODUCTION

Picornaviruses appear, by electron microscopy, to have no distinctive surface structure, and they have an apparent diameter of ≈27 nm. Each virus contains a single molecule of positive sense single-stranded RNA with a molecular weight of 2.5 to 3×10^6. The RNA is polyadenylated at the 3′ end and has a polypeptide, the genome-linked protein (VpG), attached to the 5′ end. Each virus particle contains 60 molecules of each of four proteins, three with molecular weight of ≈30×10^3 and one of molecular weight of ≈10×10^3. RNA replication occurs via two types of replicative intermediate (RI) which are partially double-stranded. One type of RI uses positive strand RNA as a template, and the other uses negative strand RNA to make full length copies. Viral structural proteins, an RNA-dependent RNA polymerase, and other functional proteins are made by a series of post-translational cleavages from a polyprotein. The family *Picornaviridae* is divided into four genera: the Enteroviruses, the Cardioviruses, the Rhinoviruses, and the Aphthoviruses. There have been numerous reports of small viruses which may be picornaviruses, associated with marine and fresh-water invertebrates. It is known that ticks, mosquitoes, and biting flies transmit many viruses which cause human or animal diseases such as reoviruses and bunyaviruses. There have been, however, very few reports of well-characterized picornaviruses from invertebrates other than insects; hence, this chapter is focused on the insect viruses which are possible picornaviruses.

Relatively few viruses of insects have been studied in sufficient detail to be classified as picornaviruses by the International Committee on Taxonomy of Viruses.[1] The criteria for classification cannot be based merely on an electron microscopic examination of the viruses and demonstration of the correct macromolecular complement. Few of the small RNA viruses of insects replicate sufficiently well in continuous tissue culture cells to permit studies of the function of viral RNA and the synthesis of virus structural proteins.[2-4] Hence, this chapter will describe a number of viruses which are possibly picornaviruses on the basis of an appropriate complement of proteins and RNA, size and sedimentation coefficient.

A. MEMBERS OF THE GROUP

Three probable picornaviruses of insects have been studied in considerable detail. Cricket paralysis virus has been studied because it grows to high titers in *Drosophila melanogaster* cells; thus, its mechanism of replication can be studied in detail.[5-7] *Drosophila* C virus has also been studied because it replicates in tissue culture cells, and because of its potential impact on laboratory colonies of *Drosophila* and the fact that several different isolates have been made.[8-10] Flacherie virus[11,12] of the silkworm *Bombyx mori* has also been examined in detail because this virus causes significant losses to the silkworm industry in Japan. Matthews,[1] however, mentions only three viruses of insects which are members of the family *Picornaviridae* not yet assigned to genera; these are cricket paralysis, *Drosophila* C, and *Gonometa* viruses.[13] It is likely that the latter virus will be excluded as replicating isolates do not exist. In addition to the viruses mentioned before, Bailey and colleagues[14-17] isolated several viruses from bees with physiochemical characteristics which may result in their eventual inclusion in the group. Plus and colleagues[18,19] have isolated viruses from *Drosophila* which are distinctly different from *Drosophila* C virus on the basis of serology and/or physiochemical properties. However, the information available on the size and structure of the RNA and replicative mechanisms is insufficient to classify them.

B. VIRUS IDENTIFICATION AND PATHOLOGY

In 1965 in the Kigezi district of Uganda many specimens of the commercially forested pine *Pinus patula* were defoliated by larvae of *Gonometa podocarpi*. Numerous dead and moribund larvae were seen; larvae were found in a limp condition hanging by the last pair or two pairs of prolegs. Subjecting larvae and pupae to extraction and differential centrifugation resulted in large amounts of a small virus which had the structural protein content and physiochemical properties associated with mammalian picornaviruses. Large numbers of virus particles were found associated with the cytoplasm of columnar cells, and cellular structures showed pronounced degeneration.[13] The virions, however, were not maintained in a suitable laboratory host insect or continuous tissue culture cells, and while this virus strain undoubtedly exists in the field, there are no viable isolates available and further studies are currently impossible.

A disease of the Australian field crickets, *Teleogryllus oceanicus* and *T. commodus*, was recognized when early instar nymphs showed evidence of paralysis in their hind legs resulting in uncoordinated movement and mortality.[5,6] Extraction of the virus present and its injection into young crickets resulted in the same symptoms and mortality occurring within 3 d. The virus was called cricket paralysis virus (CrPV); Reinganum and colleagues[5,6] showed that the same virus occurred in the black field cricket, *T. commodus*, collected from a large number of locations in the western district of Victoria, Australia. CrPV is one of the few putative picornaviruses whose host range has been examined. The virus infects several species of Orthoptera and Lepidoptera. Reinganum[20] also identified CrPV in larvae of the emperor gum moth, *Antheraea eucalypti*, and studied cross infection of the virus to crickets, showing that they acquired lethal doses of the virus. Hence, it would appear that the moths, in forming a food source for the cricket could pass on a lethal infection.

Several other viruses similar to the original isolate of CrPV have been identified.[21,22] The biological and serological properties of eight different isolates of CrPV from Australia, New Zealand, and the U.S. were compared, and a combination of tests allowed the isolates to be divided into several distinct groups.[23] There have been numerous isolates of a very similar but serologically distinct virus from *Drosophila* spp. *Drosophila* C virus (DCV) has been identified in natural populations of *D. melanogaster* from Morocco, France, and the French Antilles.[18,21,24] The original isolate of the virus was from dead flies in a laboratory colony; flies injected with extracts of these insects died within 3 d. The normal routes of infection must be presumed to be by contamination of the eggs by an infected female and/ or by feeding on infected material. Both routes were demonstrated to be possible by a reduction in the lifespan and survival rates of adults fed with infected material.

CrPV and DCV isolates have been compared by serology, complementary DNA hybridization analysis, oligonucleotide mapping, and expressed intracellular proteins.[25-28] CrPV was found to be the most distinct, though the different isolates of DCV showed marked differences from each other. The DCV isolates, however, showed similarities related to their geographical origin.

II. VIRUS STRUCTURE AND COMPOSITION

A. PARTICLE STRUCTURE

There is now a considerable amount of extremely sophisticated information available on the 3D structure of mammalian picornaviruses; the exact relationship of the structural proteins on the surface of the virion has been elucidated using a combination of methods including X-ray crystallography and complete sequence analysis. An understanding of the structure of the mammalian viruses, e.g., poliovirus or foot and mouth disease virus will assist in the design of improved antiviral compounds or immunogens. The aims of insect virology, however, are either to eradicate the disease from useful insects such as cross pollinators, bees, or silkworms, or to use the viruses as biological field control agents of pest species. As insects do not have an immunologic system it is unlikely that eradication of viruses will occur by this route; hence, a detailed knowledge of structure is unnecessary. There is a strong case, however, for deriving a detailed comparative knowledge of the insect and mammalian picornaviruses in order to eliminate the possibility of vertebrate disease resulting from the use of picornaviruses as biological control agents.

Electron microscopic examination of CrPV and DCV showed spherical particles with diameters of between 27 and 30 nm (see Figures 1A, B). Both viruses have bouyant densities of 1.34 g/cm^3 and sedimentation coefficients of between 153 S and 167 S.[2,3] The virions contain a single-stranded RNA of molecular weight between 2.6×10^6 and 3.0×10^6.[29,30] Both CrPV and DCV contain three major structural proteins of molecular weight about 30×10^3. DCV has a fourth protein of molecular weight about 10×10^3 which is not readily detectable in CrPV. As seen in Figure 1C it is sometimes difficult to separate the structural proteins of the insect picornaviruses. This figure shows the [^{35}S] methionine labeled proteins of DCV. Another probable picornavirus was identified in field-collected larvae of the tussock moth, *Lymantria ninayi* from Papua New Guinea;[31] a polyacrylamide gel of the structural proteins (see Figure 2) shows an improved separation of the three major proteins. As with many of the mammalian picornaviruses, the purified insect viruses contain minor amounts of an additional protein (VPO) the cleavage of which as the immediate precursor of two of the viral structural proteins (VP2 and VP4) is the end point of the series of proteolytic cleavages from the polyprotein. The *L. ninayi* protein of molecular weight 43×10^3 is probably VPO. The VPO of CrPV and DCV (see Figure 1C) have been positively identified by a series of experiments showing the intracellular processing of the precursor proteins. The original studies on the *G. podocarpi* virus demonstrated a virus with the protein complement of mammalian picornaviruses. While we were able to isolate large amounts of the virus in a pure form from *G. podocarpi* (see Figure 3), we were unable to perform any significant studies on the virus in the absence of a susceptible laboratory host insect.

Two picorna-like viruses have been purified from the aphid *Rhopalosiphum padi* (see Figures 4 to 6). The most recent isolate of aphid lethal paralysis virus caused behavioral abnormalities in the colony members. The affected aphids move away from the food source and become uncoordinated before paralysis and speedy death. This virus contained four structural proteins, one with a molecular weight of 11×10^3 and polyadenylated RNA with a size of 3.4×10^6 Da.[33] The classification of the first small virus isolated from *R. padi* continues to be evaluated but there is evidence from the replicative events that it is not a picornavirus.[52]

B. GENOME STRUCTURE

CrPV RNA has a 5′ genome-linked protein (VpG) and is polyadenylated at the 3′ end.[34,35] Infectious flacherie virus and DCV appear to have similar genome structures to CrPV.[27,36-38] The majority of the genome of CrPV has been cloned into bacteria as complementary DNA, and the sequence data for the polymerase demonstrates that the genome is significantly different from that of mammalian picornavirus.[39,40] One region of homology was found between CrPV, mammalian, and plant polymerases, and this consisted of a conserved triplet, GDD, surrounded by hydrophobic amino acids, and, 20 to 30 bases upstream, a region containing several conserved amino acids. It is presumed that this region is the active portion of the nucleic acid binding site of the polymerase.[40]

The arrangement of the coding sequence of the viral protein of CrPV appears to be similar to that found with mammalian picornaviruses as has been demonstrated by pactamycin mapping.[41] In addition the virion genome appears to encode a polyprotein which by a series of proteolytic cleavages gives rise to the viral structural proteins, a protease and a polymerase. A similar coding strategy appears to be used by flacherie virus as demonstrated by cell-free translation of the viral RNA[36] and by DCV as seen by *in vitro* and *in vivo* studies.[42-44]

III. VIRUS REPLICATION

CrPV and DCV have been demonstrated in crystalline arrays in the cytoplasm of cultured

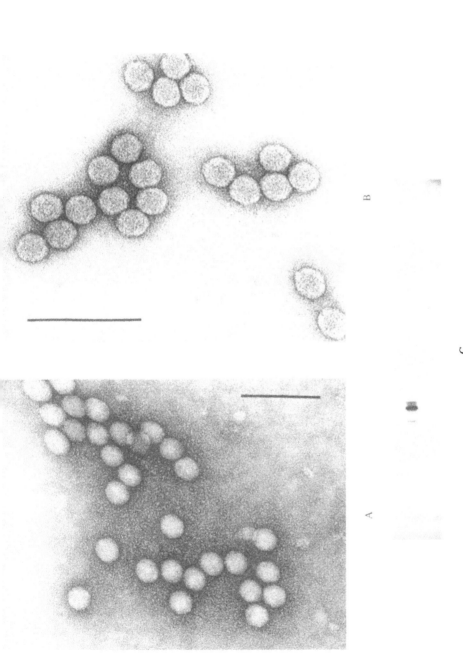

FIGURE 1. Electronmicrograph of (A) purified cricket paralysis virus and (B) *Drosophila* C virus. (Bars = 100 nm.) (C) [^{35}S]methionine-labeled protein of purified *Drosophila* C virus.

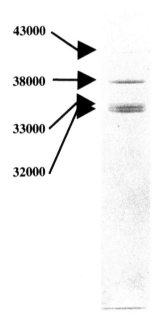

43000

38000

33000

32000

FIGURE 2.   Silver-stained protein of purified *Lymantria ninayi* putative picornavirus.

cells. In insect hosts, CrPV is found in the alimentary canal cells and in nerve and epidermal cells while DCV has been located mainly in tracheal cells and cells surrounding the cerebral ganglia.[21,22] CrPV gave a very marked and distinctive cytopathic effect in Schneider's *Drosophila* cells. After an input inoculum of greater than 1 plaque forming unit (pfu)/cell, the cells rounded up after a few hours and complete cell lysis occurred within 24 h. The cells are all fragmented, and yields in excess of 10^9 pfu/ml are often obtained with the growth, infection, and overlay conditions described by Moore and Pullin.[45] The excellent cytopathic effect made a plaque assay system easy to develop in *D. melanogaster* cells (see Figure 7A). In the same cell line, DCV gave a much slower cytopathic effect, and if a high multiplicity of infection was used, lysis (without the marked fragmentation noted with CrPV) occurred within 48 h. *Drosophila* C virus can persistently infect *D. melanogaster* cells without apparent cytopathic effect.

Viral RNAs are detected as early as 3 hpi and induced RNAs are either of genome size or high molecular weight dsRNA (replicative forms and intermediates).[30] The marked inhibition of cell-coded protein is normally greater with CrPV than DCV, using the same multiplicity of infection.[27] CrPV and DCV RNAs function as polycistronic mRNAs as evidenced by the large number of proteins identified in infected cells (see Figure 7B) whose presence can be reduced or increased by the use of various protease inhibitors, amino acid analogs, and conditions such as elevated temperature.[42,46,47]

The multiplication of several of the probable picornaviruses in the tissues of their host insects has been examined by electron microscopy. For CrPV and DCV, the location of the virus within different tissues has been reviewed in detail. While the viruses are found in different cells in different tissues, all the evidence from insect studies and tissue culture has indicated that the replication is cytoplasmic. In a recent study we examined the replication of CrPV in a major pest of olives *Dacus oleae*, the olive fly.[48-51] Figures 8 and 9 show examples of the type of assembly of virus particles normally found. Large numbers of virus particles could be visualized in the hindgut and cells lining the tracheae but no virus could be detected in the foregut. The most heavily infected tissue was the fat body which contained

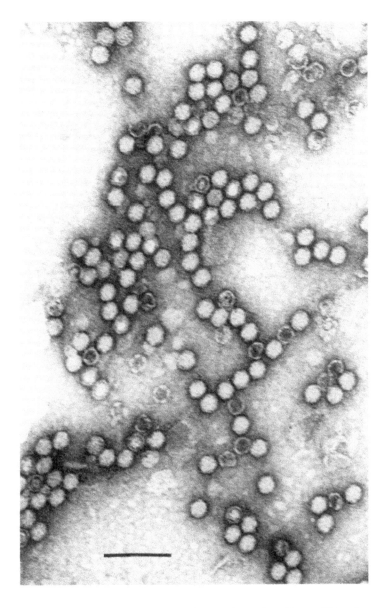

FIGURE 3. Purified *Gonometa podocarpi* picornavirus. (Bar = 100 nm.)

numerous scattered virions in addition to some excellent crystalline arrays. Muscle tissues also contained some individual virions but no arrays could be seen. The presence of crystalline arrays allows the infection to be visualized; random distribution of virus particles and the associated cell damage is not readily seen. Histopathologic examination of silkworm larvae infected with flacherie virus showed two types of inclusion body formed by: (1) the endogenous reaction of the host cell and (2) the degeneration and denaturation of the goblet cell. Again flacherie virus is located within the cytoplasm of infected cells (see Figure 10). Further information on flacherie virus is presented in Chapter 20 of this volume.

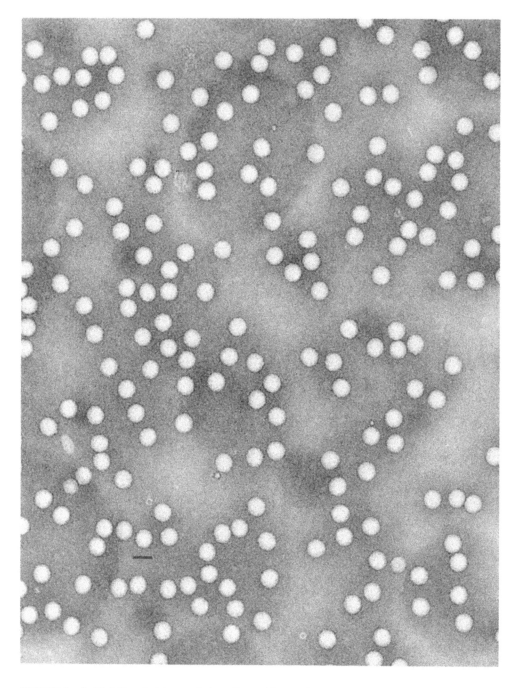

FIGURE 4.   Purified *Rhopalosiphum padi* virus. (Bar = 25 nm.) (Courtesy of C. Williamson and G. F. Kasdorf.)

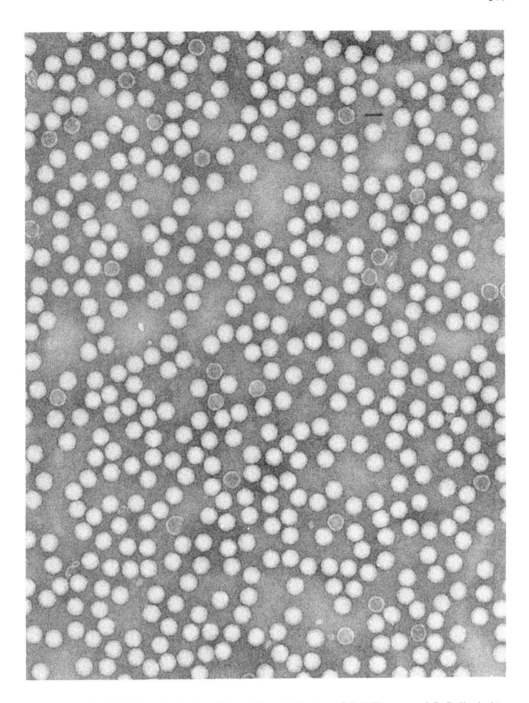

FIGURE 5.   Aphid lethal paralysis virus. (Bar = 26 nm.) (Courtesy of C. Williamson and G. F. Kasdorf.)

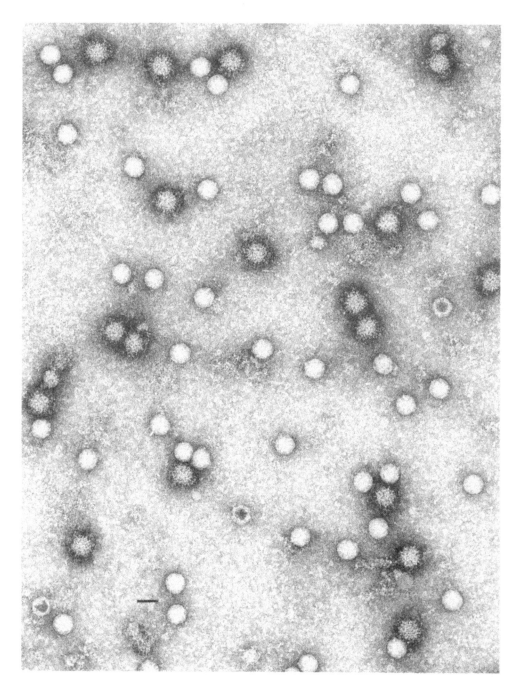

FIGURE 6. Distinguishing the viruses using antiserum against *Rhopalosiphum padi* virus. (Bar = 26 nm.) (Courtesy of C. Williamson and G. F. Kasdorf.)

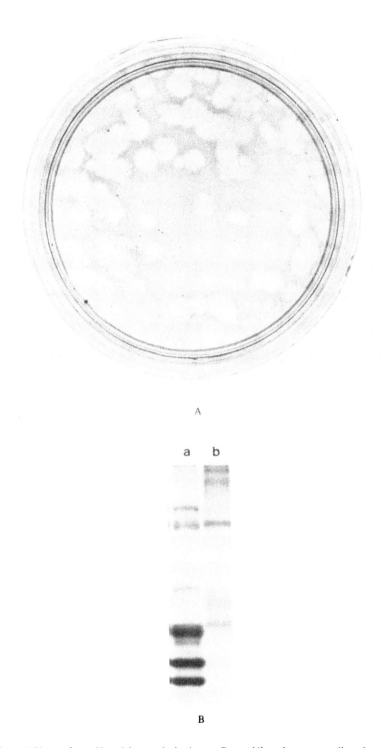

A

a   b

B

FIGURE 7. (A) Plaques formed by cricket paralysis virus on *Drosophila melanogaster* cells under an agar overlay; (B) [³⁵S]methionine intracellular proteins expressed by CrPV (a) infected cells and (b) noninfected cells.

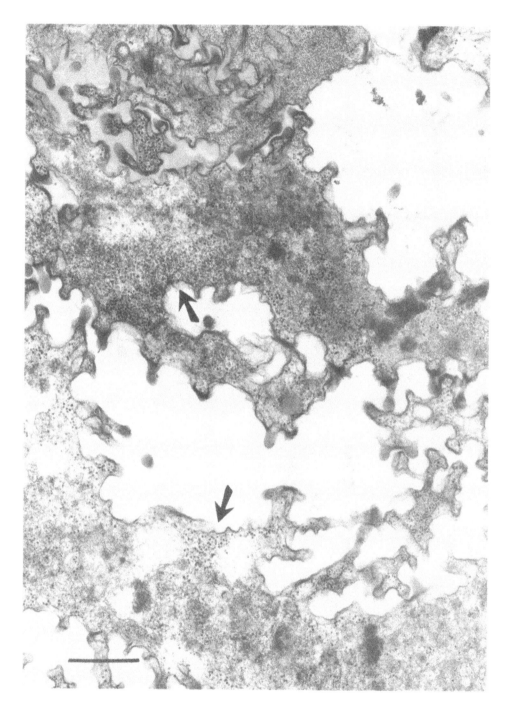

FIGURE 8.   Cricket paralysis virus particles in tissues of adult *Dacus oleae* (arrows) tracheolar tissue. (Bar = 1 μm.)

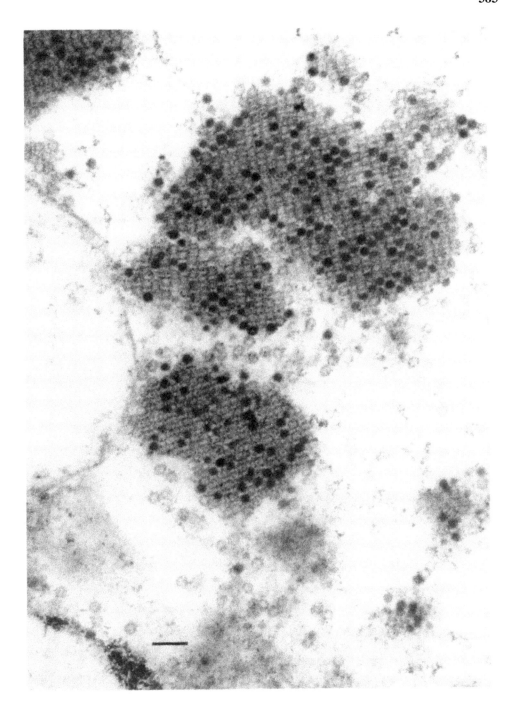

FIGURE 9.   Cricket paralysis virus particles in tissue of adult *Dacus oleae*. Cytoplasmic array in the fat body. (Bar = 100 nm.)

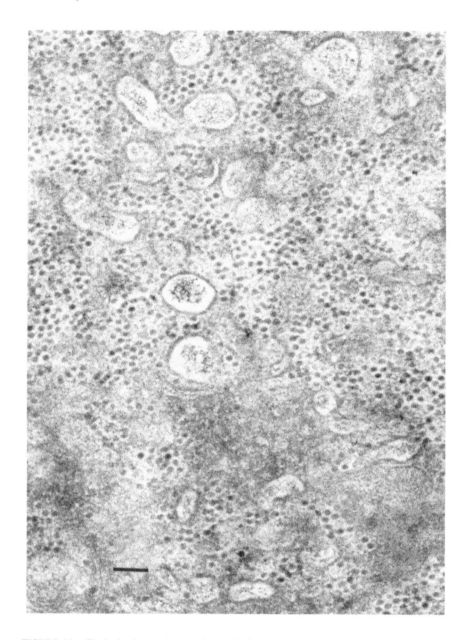

FIGURE 10.   Flacherie virus in the cytoplasm of columnar cell of the midgut of *Bombyx mori*. (Bar = 100 nm.) (Courtesy of Dr. Yoshimitsu Iwashita.)

REFERENCES

1. **Matthews, R. E. F.,** Classification and nomenclature of viruses, *Intervirology,* 17, 131, 1982.
2. **Moore, N. F. and Tinsley, T. W.,** The small RNA-viruses of insects, *Arch. Virol.,* 72, 229, 1982.
3. **Moore, N. F., Reavy, B., and King, L. A.,** General characteristics, gene organization and expression of small RNA viruses of insects, *J. Gen. Virol.,* 66, 647, 1985.
4. **Reavy, B. and Moore, N. F.,** The replication of small RNA-containing viruses of insects, *Microbiologica,* 5, 63, 1982.
5. **Reinganum, C.,** Studies on a non-occluded virus of the field cricket *Teleogryllus* spp., Masters thesis, Monash University, Melbourne, 1973.
6. **Reinganum, C., O'Loughlin, G. T., and Hogan, T. W.,** A non-occluded virus of the field crickets *Teleogryllus oceanicus* and *T. commodus* (Orthoptera:Gryllidae), *J. Invertebr. Pathol.,* 16, 214, 1970.
7. **Moore, N. F., Kearns, A., and Pullin, J. S. K.,** Characterisation of cricket paralysis virus-induced polypeptides in *Drosophila* cells, *J. Virol.,* 33, 1, 1980.
8. **Jousset, F-X., Bergoin, M., and Revet, B.,** Characterisation of the *Drosophila* C virus, *J. Gen. Virol.,* 34, 269, 1977.
9. **Moore, N. F., Reavy, B., Pullin, J. S. K., and Plus, N.,** The polypeptides induced in *Drosophila* cells by *Drosophila* C virus (strain Ouarzazate), *Virology,* 112, 411, 1981.
10. **Moore, N. F., Pullin, J. S. K., Crump, W. A. L., and Plus, N.,** The proteins expressed by different isolates of *Drosophila* C virus, *Arch. Virol.,* 74, 21, 1982.
11. **Kawase, S., Hashimoto, Y., and Nakagaki, M.,** Characterisation of flacherie virus of the silkworm, *Bombyx mori, J. Sericul. Sci. Jpn.,* 49, 477, 1980.
12. **Iwashita, Y. and Kanke, E.,** Histopathological diagnosis of diseased larvae infected with flacherie virus of the silkworm *Bombyx mori* Linnaeus, *J. Sericul. Sci. Jpn.,* 38, 64, 1969.
13. **Longworth, J. F., Payne, C. C., and Macleod, R.,** Studies on a virus isolated from *Gonometa podocarpi* (Lepidoptera:Lasiocampidae), *J. Gen. Virol.,* 18, 119, 1973.
14. **Bailey, L.,** Viruses attacking the honey bee, *Adv. Virus Res.,* 20, 271, 1976.
15. **Bailey, L. and Woods, R. D.,** Bee viruses, in *The Atlas of Insect and Plant Viruses,* Maramorosch, K., Ed., Academic Press, New York, 1977, 141.
16. **Bailey, L., Carpenter, J. M., and Woods, R. D.,** Egypt bee virus and Australian isolates of Kashmir bee virus, *J. Gen. Virol.,* 43, 641, 1979.
17. **Bailey, L., Carpenter, J. M., and Woods, R. D.,** A strain of sacbrood virus from *Apis cerana, J. Invert. Pathol.,* 39, 264, 1982.
18. **Plus, N., Croizier, G., Duthoit, J. L., David, J., Anxolabehere, D., and Periquet, G.,** Découverte chez la Drosophile, de virus appartenant à trois nouveaux groupes, *C. R. Hebd. Seances Acad. Sci. (Ser. D),* 280, 1501, 1975.
19. **Plus, N., Croizier, G., Veyrunes, J-C., and David, J.,** A comparison of buoyant density and polypeptides of Drosophila P, C and A viruses, *Intervirology,* 7, 346, 1976.
20. **Reinganum, C.,** The isolation of cricket paralysis virus from the emperor gum moth, *Antheraea eucalypti* Scott and its infectivity towards a range of insect species, *Intervirology,* 5, 97, 1975.
21. **Scotti, P. D., Longworth, J. F., Plus, N., Croizier, G., and Reinganum, C.,** The biology and ecology of strains of an insect small RNA virus complex, *Adv. Virus Res.,* 26, 117, 1981.
22. **Moore, N. F.,** Pathology associated with small RNA viruses of insects, in *Viral Insecticides for Biological Control,* Maramorosch, K. and Sherman, K. E., Eds., Academic Press, London, 195, 1987, 233.
23. **Plus, N. and Scotti, P. D.,** The biological properties of eight different isolates of cricket paralysis virus, *Ann. Virol. (Inst. Pasteur),* 135E, 257, 1984.
24. **Plus, N. Croizier, G., Reinganum, C., and Scotti, P. D.,** Cricket paralysis virus and *Drosophila* C virus: serological analysis and comparison of capsid polypeptides and host range, *J. Inverteb. Pathol.,* 31, 296, 1978.
25. **King, L. A., Massalski, P. R., Cooper, J. I., and Moore, N. F.,** Comparison of the genome RNA sequence between cricket paralysis virus and *Drosophila* C virus by complementary DNA hybridization analysis, *J. Gen. Virol.,* 65, 1193, 1984.
26. **Clewley, J. P., Pullin, J. S. K., Avery, R. J., and Moore, N. F.,** Oligonucleotide fingerprinting of the RNA species obtained from six *Drosophila* C virus isolates, *J. Gen. Virol.,* 64, 503, 1983.
27. **Moore, N. F., Pullin, J. S. K., Crump, W. A. L., Reavy, B., and Greenwood, L. K.,** Comparison of two insect picornaviruses, *Drosophila* C and cricket paralysis viruses, *Microbiologica,* 4, 359, 1981.
28. **Pullin, J. S. K., Moore, N. F., Clewley, J. P., and Avery, R. J.,** Comparison of the genome of two insect picornaviruses, cricket paralysis virus and *Drosophila* C virus, by ribonuclease T1 oligonucleotide fingerprinting, *FEMS Micro. Lett.,* 15, 215, 1982.
29. **Reavy, B.,** Replication of small RNA-viruses of insects, Ph.D. thesis, University of Oxford, 1982.

30. **Reavy, B., Crump, W. A. L., and Moore, N. F.,** Characterisation of cricket paralysis virus and *Drosophila* C induced RNA species synthesized in infected *Drosophila melanogaster* cells, *J. Invertebr. Pathol.*, 41, 397, 1983.

31. **Pullin, J. S. K., Black, F., King, L. A., Entwistle, P. F., and Moore, N. F.,** Characterization of a small RNA containing virus in field collected larvae of the tussock moth, *Lymantria ninayi* from Papua New Guinea, *App. and Environ. Microbiol.*, 48, 504, 1984.

32. **D'Arcy, C. J., Burnett, P. A., and Hewings, A. D.,** Detection, biological effects, and transmission of a virus of the aphid, *Rhopalosiphum padi.*, *Virology*, 114, 268, 1981.

33. **Williamson, C., Rybicki, E. P., Kasdorf, C. G. F., and von Wechmar, M. B.,** Characterization of a new picorna-like virus isolated from aphids, *J. Gen. Virol.*, 69, 787, 1988.

34. **King, L. A. and Moore, N. F.,** Evidence for the presence of a genome-linked protein in two insect picornaviruses, cricket paralysis and Drosophila C viruses, *FEMS Lett.*, 50, 41, 1988.

35. **Eaton, B. T. and Steacie, A. D.,** Cricket paralysis virus RNA has 3' terminal poly (A), *J. Gen. Virol.*, 50, 167, 1980.

36. **Hashimoto, Y., Watanabe, A., and Kawase, S.,** *In vitro* translation of infectious flacherie virus RNA in a wheat germ and a rabbit reticulocyte system, *Biochim. Biophys. Acta*, 781, 76, 1984.

37. **Hashimoto, Y., Watanabe, A., and Kawase, S.,** Evidence for the presence of a genome-linked protein in infectious flacherie virus, *Arch. Virol.*, 90, 301, 1986.

38. **Hashimoto, Y. and Kawase, S.,** Characteristics of structural proteins of infectious flacherie virus from the silkworm, *Bombyx mori*, *J. Invertebr. Pathol.*, 41, 68, 1983.

39. **King, L. A.,** *Molecular Biology of Insect Picornaviruses*, Ph.D. thesis, University of Oxford, 1985.

40. **King, L. A., Pullin, J. S. K., Stanway, G., Almond, J. W., and Moore, N. F.,** Cloning of the genome of cricket paralysis virus: sequence of the 3' end, *Virus Res.*, 6, 331, 1987.

41. **Reavy, B. and Moore, N. F.,** The genenome organisation of a small RNA-containing insect virus: comparison with that of mammalian picornaviruses, *Virology*, 131, 551, 1983.

42. **Moore, N. F. and Pullin, J. S. K.,** Heat shock used in combination with amino acid analogues and protease inhibitors to demonstrate the processing of the proteins of an insect picornavirus *(Drosophila C* virus) in *Drosophila melanogaster* cells, *Ann. Virol.*, 134E, 285, 1983.

43. **Reavy, B. and Moore, N. F.,** Cell-free translation of *Drosophila* C virus RNA: identification of a virus protease activity involved in capsid protein synthesis and further studies on *in vitro* processing of cricket paralysis virus specified proteins, *Arch. Virol.*, 76, 101, 1983.

44. **Moore, N. F., King, L. A., and Pullin, J. S. K.,** Insect picornaviruses, in *The Molecular Biology of the Positive Strand RNA Viruses*, Rowlands, D. J., Mayo, M. A., and Mahy, B. W. J., Eds., Academic Press, London, 1987, chap. 5.

45. **Moore, N. F. and Pullin, J. S. K.,** Plaque purification of cricket paralysis virus using a agar overlay on *Drosophila* cells, *J. Invertebr. Pathol.*, 39, 10, 1982.

46. **Moore, N. F., Pullin, J. S. K., and Reavy, B.,** The intracellular proteins induced by cricket paralysis virus in *Drosophila* cells: the effect of protease inhibitors and amino acid analogues, *Arch. Virol.*, 70, 1, 1981.

47. **Moore, N. F., Reavy, B., and Pullin, J. S. K.,** Processing of cricket paralysis virus induced polypeptides in *Drosophila* cells: production of high molecular weight polypeptides by treatment with iodoacetomide, *Arch. Virol.*, 68, 1, 1981.

48. **Manousis, T. and Moore, N. F.,** Control of *Dacus oleae*, a major pest of olives, *Insect Sci. Appl.*, 8, 1, 1987.

49. **Manousis, T., Arnold, M. K., and Moore, N. F.,** Electron microscopical examination of tissues and organs of *Dacus oleae* flies infected with cricket paralysis virus (CrPV), *J. Invertebr. Pathol.*, 51, 119, 1988.

50. **Manousis, T., Koliais, S. I., and Moore, N. F.,** An inapparent infection with a probable picornavirus in several stocks of laboratory reared and naturally occuring populations of *Dacus oleae* Gmel. pupae in Greece, *Microbios*, 51, 81, 1987.

51. **Manousis, T. and Moore, N. F.,** Cricket paralysis virus, a potential control agent for the olive fruit fly, *Dacus oleae* Gmel., *Appl. Environ. Microbiol.*, 53, 142, 1987.

52. **Gildow, F. E.,** personal communication.

Chapter 14

TETRAVIRIDAE

Carl Reinganum*

TABLE OF CONTENTS

*   Unfortunately, Dr. Reinganum passed away suddenly in 1989 due to a very sudden and severe brain hemorrhage. We salute his highly significant contributions to invertebrate virology.

I. INTRODUCTION

There are a few groups of viruses which are confined exclusively to insects and which have no vertebrate counterparts; the Tetraviridae is one such group. Although sufficient information has been amassed to warrant its recognition by the International Committee on Taxonomy of Viruses (ICTV),[1] the tetraviruses, unlike the baculoviruses have not been studied in any great depth. This is partly due to the unlikelihood that any of them will be used as biocontrol agents and to the failure, so far, to develop a permissive cell line in which the *in vitro* synthesis of the viruses can be studied.

II. MEMBERS OF THE GROUP

A. NATURAL HOST RANGE

The group presently recognizes seven members, together with nine whose distinctiveness is yet to be established. These are presented in Table 1.

Individual members will be referred to as "β-viruses", in conformity with the type species. This is done for the sake of convenience and is not to suggest that the names are those which may ultimately be adopted.

B. ARTIFICIAL HOST RANGE

Almost nothing is known about the host range of the viruses. Attempts to transmit viruses to another species must take into account the possibility that crossinfection will activate a different but related virus in that host. However, Tripconey[5] succeeded in transmitting *Nudaurelia-β* virus (NdβV) to *Bombyx mori* (Bombycidae), *Metahepialus xenoctenus* (Helialidae) and *Dionycopus amasis* (Arctiidae), although it required at least one serial passage before the virus became adapted to the alternate host and could be detected serologically.

III. GROSS PATHOLOGY

The response of larvae to virus infection depends upon the particular host-virus combination, larval age, and virus dose. Neonate larvae of *Trichoplusia ni* fed doses of *Drichoplusia-β* virus (TrβV) will support a chronic infection where the only gross symptom of disease is a marked reduction in larval and pupal size.[3] On a weight-for-weight basis, such larvae may contain appreciable quantities of virus. Since the virus multiplies primarily in the cells of the midgut (see Figure 1), the size reduction in chronically infected larvae is most probably due to starvation. On the other hand, virus administered to late instar larvae of *T. ni* has little or no effect and the virus can persist as an inapparent infection in succeeding generations. The ability to support limited multiplication through succeeding generations points to an efficient vertical transmission of the virus. Whether such transmission is transovum or transovarial is unknown. Larvae of *Nudaurelia cytheraea capensis* infected with NdβV may pupate, but the adults which emerge are usually underdeveloped and have malformed wings.[5]

Larvae of *N. cytheraea capensis* and *Antheraea eucalypti* which have actually been killed by virus either drop from the food plant or remain hanging by the prolegs and are flaccid. Because the integrity of the internal tissues is largely unaffected, dead larvae do not rupture on handling, such as occurs with larvae killed by a baculovirus.

IV. VIRUS COMPOSITION AND STRUCTURE

A. THE VIRION

The virions are icosahedral and have a diameter of 35 to 38 nm (see Figure 2). The sedimentation coefficient ranges between 194 and 210 S and the buoyant density in CsCl

TABLE 1
Members of the Group

Virus (abbreviation)	Host
Nudaurelia-β virus (NdβV) (type species)	*N. cytheraea capensis* (Saturniidae)
Antheraea-β virus (AnβV)[a]	*A. eucalypti* (Saturniidae)
Darna-β virus (DrβV)	*D. trima* (Limacodidae)
Thosea-β virus (ThβV)	*T. asigna* (Limacodidae)
Philosamia-β virus (PhβV)	*P. cynthia × ricini* (Saturniidae)
Dasychira-β virus (DsβV)	*D. pudibunda* (Lymantriidae)
Trichoplusia-β virus (TrβV)	*T. ni* (Noctuidae)

Provisional members have been isolated from the following:

Saturnia pavonia (Saturniidae)
Acherontia atropas (Sphingidae)
Setora nitens (Limacodidae)
Lymantria ninayi (Lymantriidae)
Eucytia meeki (Cocytiidae)
Euploea corea (Nymphalidae)
Agraulis vanillae (Nymphalidae)
Hypocrita jacobeae (Arctiidae)
Callimorpha quadripunctata (Arctiidae)

[a] There is evidence that this virus is the same as NdβV, since the two are antigenically indistinguishable by gel double diffusion tests.

between 1.27 and 1.31 g/ml. The virions are unusually stable, being unaffected by 67% acetic acid or mild alkali. Resistance to ether indicates the probable absence of lipids while carbohydrates have not been detected.

B. THE CAPSID

The capsid consists of 240 copies of a single species of polypeptide, arranged in T = 4 surface symmetry[7] (see Figure 3). It is this property which is unique among icosahedral viruses, and distinguishes them from, for example, the picornaviruses where 180 polypeptide chains are arranged in T = 3 symmetry. Estimates of the molecular weight of the polypeptide depend upon whether electrophoresis is conducted in a continuous or discontinuous SDS-PAGE system. The former yields a value of about 62,000 and the latter about 66,000.* Whatever the case, there are small differences in polypeptide molecular weights between species.

Amino acid analyses of the polypeptides of TrβV, NdβV, and DsβV (*Dasychira-β* Virus) indicate that the latter two have a more similar composition than has either to the former, which contains higher amounts of glutamic acid and glycine.[8]

C. THE GENOME

The RNA is a single-stranded, positive-sense molecule with a molecular weight of about 1.8×10^6 and a sedimentation coefficient value of 32, as determined for NdβV.[6] It comprises 10 to 11% by weight of the virion. Analyses of the RNA of DsβV and TrβV indicate the absence of a terminal poly(A) tract.[9] The base ratio of NdβV RNA is guanine:adenine:cytosine:uracil = 28:24:25:23.[6]

---

* The value of 62,000 is more consistent with the other physicochemical data. Thus, since the capsid contains 240 copies of the polypeptide, then its molecular weight would be $62,000 \times 240 = 14.9 \times 10^6$. Adding to this the molecular weight of the RNA (1.8×10^6), gives a value of 16.7×10^6 for the virion.

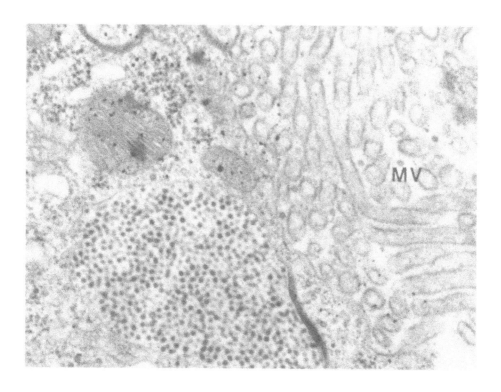

FIGURE 1.   Section of midgut cell of *Trichoplusia ni* showing membrane-bound array of TrβV in cytoplasm MV = microvilli. (Magnifications ca. × 50,000.) (Courtesy of Dr. N. F. Moore.)

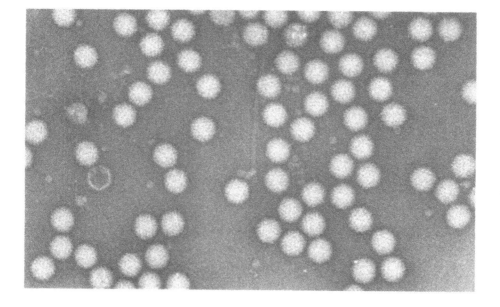

FIGURE 2.   Virions of NdβV stained with 1% uranyl acetate and photographed over a hole in the carbon film. (Magnification × 210,000.) (Figures 2 and 3 from Finch, J. T., Crowthers, R. A., Hendry, D. A., and Struthers, J. K., *J. Gen. Virol.*, 24, 191, 1974. With permission.)

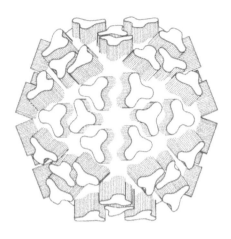

FIGURE 3.   Illustration of the surface morphology
of NdβV.

TABLE 2
Physicochemical Properties of Group Members[a]

	DrβV	NdβV	PhβV	ThβV	TrβV	DsβV
Size (nm)	35	35	35	35	35	38
Sedimentation coefficient ($S_{20,w}$)	199	210	206	194	200	ND[b]
Density in CsCl (g/ml)	1.289	1.298	1.275	1.275	1.3	1.31
%RNA	11.5	11	9.5	10.2	10—15	ND
Mol wt of polypeptide						
Continuous[c]	62,100	62,600	62,400	60,800		
Discontinuous	66,000	66,000			68,000	66,000
Mol wt of RNA ($\times 10^6$)	ND	1.8	ND	ND	1.9	ND
Mol wt of virion ($\times 10^6$)	ND	16.3	ND	ND	ND	ND

[a]   Data from: Morris et al.,[4] Tripconey,[5] Struthers and Hendry,[6] Reinganum et al.,[12] Greenwood and Moore.[13]
[b]   ND = not determined.
[c]   Upper row of figures are for gels run by SDS-PAGE in a continuous buffer system, the lower in a discontinuous system.

The base sequence of the RNA is unknown. cDNA-RNA liquid hybridization analyses showed that DsβV and TrβV share little sequence homology, which is reflected in their distant antigenic relationship.[9]

In the absence of a permissive cell line, attempts to investigate the genome strategy have been limited to cell-free translation of the RNA using rabbit reticulocytes.[10,11] The genome actived as mRNA and stimulated protein synthesis five- to tenfold above that of endogenous RNA which, in comparison with other viruses, is a low level of efficiency. A number of proteins were detected, including one of similar size to that of the capsid polypeptide. Evidence for the production of this polypeptide by postranslational cleavage from a high molecular weight precursor was equivocal.

The physicochemical properties of individual group members are presented in Table 2.

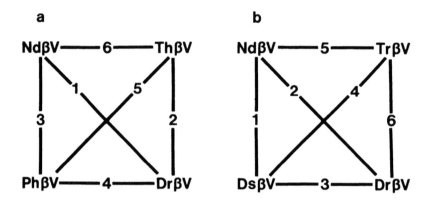

FIGURE 4.   Diagram illustrating hierarchy of antigenic relationships between group members. (a) Data derived from reciprocal complementation tests.[12] (b) Data derived from ELISA tests.[14]

V. ANTIGENIC RELATIONSHIPS

All group members are antigenically related to each other. TrβV, however, stands apart because of its distant relationship to the others.[14] The hierarchy of antigenic relationships between group members is shown diagrammatically in Figure 4.

REFERENCES

1. **Matthews, R. E. F.,** Classification and nomenclature of viruses. Fourth report of the International Committee on Taxonomy of Viruses, *Intervirology,* 17, 1, 1982.
2. **Juckes, I. R. M., Longworth, J. F., and Reinganum, C. A.,** Serological comparison of some nonoccluded insect viruses, *J. Invertebr. Pathol.,* 21, 119, 1973.
3. **Vail, P. V., Morris, T. J., and Collier, S. S.,** An RNA virus in *Autographa californica* nuclear polyhedrosis virus preparations: gross pathology and symptoms, *J. Invertebr. Pathol.,* 41, 179, 1983.
4. **Morris, T. J., Hess, R. T., and Pinnock, D. E.,** Physicochemical characterisation of a small RNA virus associated with baculovirus infection in *Trichoplusia ni, Intervirology,* 11, 238, 1979.
5. **Tripconey, D.,** Studies on a nonoccluded virus of the pine emperor moth, *J. Invertebr. Pathol.,* 15, 268, 1970.
6. **Struthers, J. K. and Hendry, D. A.,** Studies of the protein and nucleic acid components of *Nudaurelia capensis* β virus, *J. Gen. Virol.,* 22, 355, 1974.
7. **Finch, J. T., Crowther, R. A., Hendry, D. A., and Struthers, J. K.,** The structure of *Nudaurelia capensis* β virus: the first example of a capsid with icosahedral surface symmetry T = 4, *J. Gen. Virol.,* 24, 191, 1974.
8. **Moore, N. F., Greenwood, L. K., and Rixon, K. R.,** Studies on the capsid protein of several members of the *Nudaurelia* β group of small RNA viruses of insects, *Microbiologica,* 4, 59, 1981.
9. **King, L. A. and Moore, N. F.,** The RNAs of two viruses of the *Nudaurelia* β family share little homology and have no terminal poly (A) tracts, *FEMS Microbiol. Lett.,* 26, 41, 1985.
10. **King, L. A., Merryweather, A. T., and Moore, N. F.,** Proteins expressed *in vitro* by two members of the *Nudaurelia* β family of viruses, *Trichoplusia ni* and *Dasychira pudibunda* viruses, *Annu. Virol. (Inst. Pasteur),* 135E, 335, 1984.
11. **Reavy, B. and Moore, N. F.,** *In vitro* translation of a small RNA virus from *Trichoplusia ni, J. Invertebr. Pathol.,* 44, 244, 1984.
12. **Reinganum, C., Robertson, J. S., and Tinsley, T. W.,** A new group of RNA viruses from insects, *J. Gen. Virol.,* 40, 195, 1978.
13. **Greenwood, L. K. and Moore, N. F.,** A single protein Nudaurelia-β-like virus of the pale tussock moth, *Dasychira pudibunda, J. Invertebr. Pathol.,* 38, 305, 1981.
14. **Greenwood, L. K. and Moore, N. F.,** The *Nudaurelia* -β group of small RNA viruses of insects: serological comparison of four members, *Microbiologica,* 4, 271, 1981.

Chapter 15

REOVIRIDAE

Tosihiko Hukuhara and Jean-Robert Bonami

TABLE OF CONTENTS

Common features of Reoviridae are (1) a quasi-spherical capsid 60 to 80 nm in diameter, exhibiting icosahedral symmetry elements; (2) viral genome consisting of several (10 to 12) segments of double-stranded RNA (dsRNA) with molecular weights ranging from 0.2 to 3.0 × 10[6]; and (3) characteristic cytopathic changes, namely, "viroplasm" formed in the cytoplasm of infected cells.[1] Within this family there are six genera. Invertebrate viruses have been included in the cytoplasmic polyhedrosis virus (CPV) group. Reovirus-like particles which do not fit properly into this group have been described in several invertebrate animals. They are dealt with in this chapter as possible members of Reoviridae.

Some vertebrate and plant viruses of Reoviridae (*Orbivirus, Phytoreovirus,* and *Fijivirus*) multiply in arthropod vectors. The arthropods are really alternate hosts, and hence the viruses infecting them can also be considered as invertebrate viruses. However, no attempt will be made to collate them since such information is available in other publications.[2]

I. VIRUSES OF INSECTS

A. INTRODUCTION
1. Cytoplasmic Polyhedrosis Viruses

Cytoplasmic polyhedroses have been recorded in 201 Lepidoptera, 39 Diptera, 7 Hymenoptera, and 2 Coleoptera.[3,4] The CPV of the silkworm, *Bombyx mori,* is the most thoroughly studied virus of this group and serves as a typical representative. In the absence of suitable tissue culture systems for the development of serum neutralization tests for virus identification, CPVs are provisonally classified on the basis of variations in the electrophoretic migration patterns of the genome segments.[5] Twelve distinct electropherotypes have been defined, which differ from one another in the migration of at least three genome segments.[6] The antigenic variation of the virus structural proteins and the RNA sequence variation correlate well with this classification. No RNA sequence homology[7] and relatively low levels of serological cross-reactivity[8] have been detected between different electropherotypes.

Cytoplasmic polyhedroses are characterized by the presence of polyhedral inclusion bodies, or polyhedra, in the cytoplasm of infected cells. The site of virus multiplication is restricted to the midgut epithelial cells. Infection usually takes place in the larval stage. CPV infections have little or no adverse effect on most dipterous insects, and infected larvae usually pupate and emerge as apparently healthy adults. Larvae of other insect orders suffer from lethal infections, although some larvae survive when infected late in the larval stage or with a sublethal dose. Diseased larvae lag behind uninfected insects in their development, become sluggish, are sometimes diarrheic, and emit white fecal pellets. Their midguts assume an opaque yellow to white appearance due to the presence of large numbers of polyhedra.

2. Reovirus-Like Particles

Viruses or virus-like particles similar to reovirus morphologically have been detected in several insects (see Table 1). Many of them are imperfectly characterized, but their presence seems to be restricted to insects. They differ from the typical CPVs in that no polyhedra are produced. Viruses from two species of Hemiptera and five species of Diptera have 10 to 11 segments of dsRNA and fulfill the primary prerequisite for inclusion in the Reoviridae family.

Leafhopper A virus (LAV) multiplies in the maize leafhopper, *Cicadulina bimaculata,* but not in maize plants colonized by the insect. The virus is transmitted vertically through insect eggs or horizontally in insects feeding on plants in which it circulates transiently after being injected by coinfesting virus-infected maize leafhoppers. Thus, the maize plant is a transitory virus reservoir and can be looked upon as a circulative but nonpersistent vector of an insect virus. Leafhoppers infected with LAV appear very similar to uninfected ones in their development and behavior although the virus infection shortens their life span by

TABLE 1

List of Insects in which Reovirus-Like Particles Have Been Detected

Order:family	Scientific name	Vernacular name	Tissue tropism	Particle diam(nm)	Infectivity test[a]	No. dsRNA segments	Ref.
Thysanoptera:Thripidae	*Frankliniella fusca*	Tobacco thrips	Polytropic	64	−	—	9
Hemiptera:Cicadellidae	*Cicadulina bimaculata*	Maize leafhopper	?	70	+	10	10, 11
Hemiptera:Cimicidae	*Cimex lectularius*	Bedbug	Gut	56	−	11	12
Hemiptera:Delphacidae	*Peregrinus maidis*	Delphacid planthopper	Polytropic	60—65	+	—	13
Coleoptera:Chrysomelidae	*Cerotoma trifurcata*	Bean leaf beetle	Hemocytes	70—75	−	—	14
Coleoptera:Chrysomelidae	*Diabrotica undecimpunctata*	Spotted cucumber beetle	Polytropic	70	−	—	15
Coleoptera:Coccinellidae	*Epilachna varivestis*	Mexican bean beetle	Hemocytes	70—75	−	—	16
Diptera:Calliphoridae	*Lucilia cuprina*	Sheep blowfly	Gut	60	−	—	17
Diptera:Drosophilidae	*Drosophila melanogaster*	Common vinegar fly	Midgut	60—69	+	10	18
Diptera:Drosophilidae	*Drosophila simulans*	—	Epiderm	60—69	+	10	19
Diptera:Muscidae	*Musca domestica*	Housefly	Hemocytes	63—80	+	10	20—22
Diptera:Trypetidae	*Ceratitis capitata*	—	?	60—69	−	10	18, 23
Diptera:Trypetidae	*Dacus oleae*	Olive fly	Midgut	60—70	+	10	24
Diptera:Trypetidae	*Dacus trvoni*	Queensland fruitfly	Midgut	40—65	+	—	25

[a] (+) Infectivity established; (−) infectivity test not done or unsuccessful.

20%. No information is available concerning the infectivity and the pathogenicity of a virus derived from the bedbug, *Cimex lectularius*.

Reo-like viruses have been isolated repeatedly in both laboratory and wild populations of *Drosophila melanogaster*. All isolates are almost nonpathogenic and give only low virus yields. The prototype of these viruses are designated as *Drosophila* F virus (DFV). *Ceratitis* I virus derived from *Ceratitis capitata* and a virus from *Dacus oleae* are mildly pathogenic, while the housefly virus (HFV) is highly pathogenic. *Drosophila* S virus (DSV) is responsible for the phenotype (malformation and/or suppression of thoracic bristles) of a *D. simulans* strain. The virus is maternally transmitted, slightly pathogenic and mildly infectious to larvae having permissive genotypes.

Several established cell lines from *Drosophila melanogaster* are persistently infected with reo-like viruses. Two virus isolates from malignant blood cell lines are very similar to DFV and presumably derived from the flies themselves.[18,26,27] Whether the other isolates[28-30] are also of endogenous nature or introduced by external contamination has yet to be determined.

B. VIRUS STRUCTURE AND COMPOSITION
1. Polyhedra

The polyhedra consist of a number of virions and a crystalline matrix protein. The virions comprise only about 5% by weight of a polyhedron.[31] They are distributed in a random fashion and never in closely packed clusters (see Figure 1). Their density is 259 virions per μm^3 of the matrix protein.[32] The polyhedra vary greatly in size (0.1 to 10 μm). Polyhedra of certain shapes can be cloned to give rise to polyhedra of that particular shape in subsequent infections.[33] The surface of the polyhedra as observed by scanning electron microscopy has many cavities of the size and shape of virions, which suggests that the virions once may have been impregnated in them but were removed during the preparation procedure (see Figure 2).[34]

The polyhedron matrix protein has a crystalline lattice structure. Two lattice patterns are commonly observed: a dot and a line pattern (see Figures 1 and 14). The center-to-center spacing between the rows of dots or lines is in the range of 50 to 65 Å.[35] The overall pattern of the lattice is undisturbed by the occluded virions.

Interpretation of the lattice structure of matrix protein is controversial. Bergold and Suter[36] considered that the molecular packing was a simple cubic as in nuclear polyhedrosis virus (NPV) polyhedra and granulosis virus (GV) capsules. Bergold[37] later changed his view and proposed a face-centered cubic arrangement of the protein molecules of NPV polyhedra and GV capsules. Unfortunately, the model he used to demonstrate this (see Reference 37, Figure 12 and Table 1) was a simple cubic rather than a face-centered cubic structure. X-ray diffraction diagrams of CPV[38] and NPV[39] polyhedra revealed the body-centered type of arrangement of the polyhedron protein (see Figure 3).

The polyhedron protein is composed of one major polypeptide (polyhedrin) with molecular weights ranging from 25 to 30 kDa.[5] The polyhedrin has no serological relationship with the virion structural proteins[40] and is coded for by the smallest genome segment (segment 10).[41,42] The complete nucleotide sequencing of segment 10 has revealed the absence of amino acid sequence homology between CPV and NPV polyhedrins, although they form a morphologically similar crystalline structure.[43-45] Some strains of *B. mori* CPV are deficient in forming polyhedra but produce crystalline or amorphous inclusions in the nucleus.[35]

2. Particle Structure
a. Cytoplasmic Polyhedrosis Viruses

Solubilization of polyhedra under certain alkaline conditions releases the occluded virions without destroying their infectivity.[31] Extensive structural studies have concentrated almost

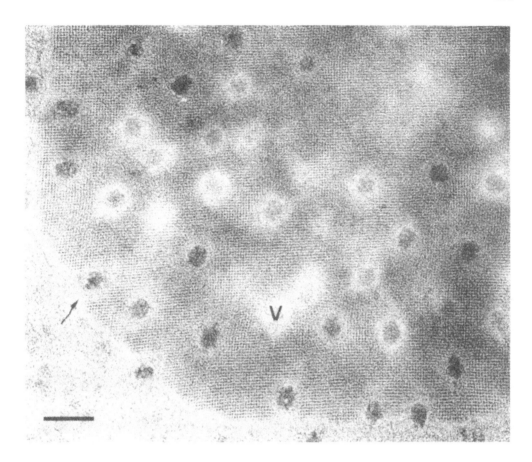

FIGURE 1.   Thin section of a polyhedron of *Bombyx mori* CPV (I strain) showing a lattice structure of the matrix protein and a number of occluded virions (V). The overall pattern of the lattice is undisturbed by the presence of virions. Note the incomplete fusing of the crystal (arrow) resulting from the recent incorporation of the virions. (Bar = 200 nm).

exclusively on *Bombyx mori* CPV particles purified from polyhedra by differential or density-gradient centrifugation. Therefore, the morphology of this virus is presented in this and the following sections as a model for the CPV group.

The virus contains dsRNA inside an icosahedrally constructed protein shell (55 to 69 nm in diameter) which has twelve projections localized at twelve vertices of the icosahedron (see Figure 4). According to an early model (see Figure 5), the projections are hollow and built of four pentagonal prisms, 7.5 nm in mean length.[46] The proximal two prisms are often covered by globules, 12 nm in diameter.[47,48] A recent model proposes that the projection consists of two spikes: an external spike (9 nm long, 10.7 nm wide) and a larger spike (9 nm long, 14.7 nm wide) attached to the shell (see Figure 6, Table 2).[49] Each projection rests on a base plate which forms part of the shell.[50]

Although a double-shelled structure of the reovirus type was proposed in the early model,[46] the electron micrographs show no evidence for it.[49] The number of capsomeres in the capsid has not been clearly established in spite of efforts to determine the details of the shell structure, including attempts to enhance the images by the rotation technique. The capsid is composed of many subunits which contain a hollow space about 10 nm in diameter in their centers.[50] Negatively stained particles often show an appearance of capsomeres around certain sectors (but never the whole) of their periphery. Counts indicated 20 per complete periphery.[51] However, these are almost certainly artifacts caused by partial superimposition of two or more real capsomeres.

FIGURE 2.   Polyhedra of *Pectinophora* gossypiella CPV; scanning electron microscopy. (Bar = 1 μm.) (From Adams, J. R. and Wilcox, T. A., *J. Invertebr. Pathol.*, 40, 12, 1982. With permission.)

FIGURE 3.   X-ray powder photograph of polyhedra of *Bombyx mori* CPV (H strain) showing a number of reflections. The edge of the unit cell is 105 Å long. (From Fujiwara, T., Yukibuchi, E., Tanaka, Y., Yamamoto, Y., Tomita, K., and Hukuhara, T., *Appl. Entomol. Zool.*, 19, 402, 1984. With permission.)

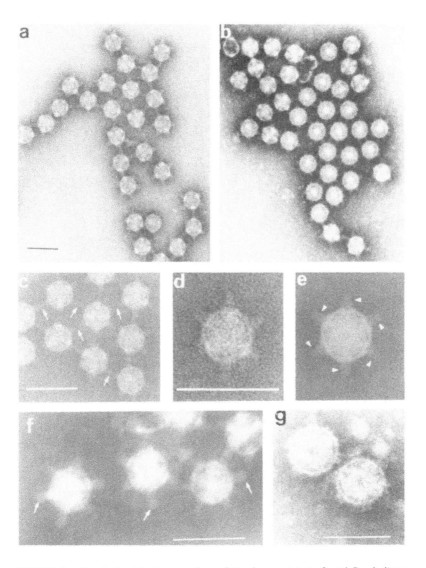

FIGURE 4. Negatively stained preparations of *Bombyx mori* (a to f) and *Dendrolimus punctatus wenshanensis* (g) CPV particles. ([a] and [b] preparations were stained with uranyl acetate [UA]; [c] to [e], with ammonium molybdate [AM]; and [f] and [g], with potassium phosphotungstate [PTA].) (b) The CPV particles are arranged so that their fivefold axes of symmetry are perpendicular to the supporting film. Particles in other preparations are arranged with their threefold axes perpendicular to the supporting films. The projections on the particles are adjacent to each other (arrows in [c]). Two spikes and their junctions (arrowheads) are clearly resolved in (e), which is the same particle as in (d) after photographic enhancement by three-times rotation. Globules are associated with the projections (arrows in [f]). (Bars = 100 nm.) ([a] to [e] from Hatta, T. and Francki, R. I. B., *Intervirology*, 18, 203, 1982; [f] from Asai, J., Kawamoto, F., and Kawase, S., *J. Intevertebr. Pathol.*, 19, 279, 1972; [g] from Liang, D. and Hu, Y., *Atlas of Insect Viruses in China*, Sci. Technol. Publ., Hunan, 1986. With permission.)

The majority of virions in the infected midguts are free in the cell cytoplasm and not occluded within polyhedra even at the late stages of infection.[47] These virions purified by sucrose gradient centrifugation are not different in morphology from virions derived from polyhedra.[48,49]

It now seems that *B. mori* CPV particles are composed of at least four morphological

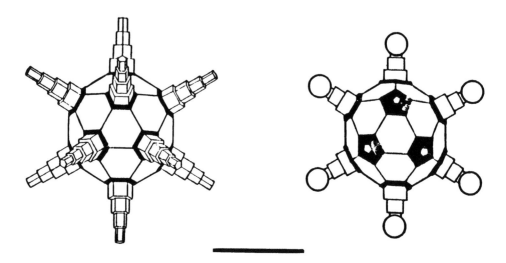

FIGURE 5.   Structural model of the *Bombyx mori* CPV particle showing projections (left) and globules associated with them (right). The front three projections of the right model are omitted to show the pentagonal disks on the vertices. (Bar = 50 nm.) (Left model, from Aizawa, K., *Protecting the Silkworm from Diseases,* Iwanami, Tokyo, 1975. Right model, from Kawamoto, F., Kawase, S., and Asai, J., *Jpn. J. Appl. Entomol. Zool.,* 19, 133, 1975. With permission.)

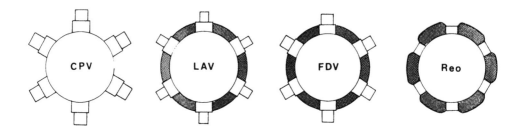

FIGURE 6.   Diagrammatic representation of sections through four different viruses belonging to the family Reoviridae. Hatched areas represent the outer capsid shells which can be removed from LAV and reovirus (reo) particles by digestion with chymotrypsin but in Fiji disease virus (FDV) are lost spontaneously. (From Hatta, T. and Francki, R. I. B., *Intervirology* 1, 203, 1982. With permission.)

TABLE 2
Dimensions of Structures of Some Reoviridae Particles in Uranyl Acetate Stain[a]

Structure	Dimension (nm)	*B. mori* CPV	LAV	FDV	Reovirus
Intact particles	Diameter	56.5	71.6	70.8	76.4
A spikes	Width	10.7	7.5	10.6	—
	Height	9.0	8.5	9.3	—
Core particles	Diameter	—	57.6	54.0	52.1
B spikes	Width	14.7	14.1	13.6	10.0
	Height	9.0	8.5	8.0	5.7

[a]   Modified from Hatta and Francki.[49]

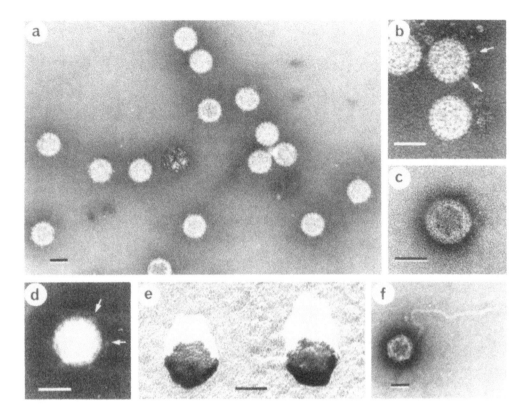

FIGURE 7. The appearance of LAV particles. ([a] and [b] preparations were stained with AM; [c] and [f], with UA, and [d], with PTA.) The preparation (e) was freeze-dried and shadowed. A-spikes on the particles can be seen on some of the stained particles (arrows in [b] and [d] and on those which were shadowed [e]). A helical filament associated with LAV is seen in (f). (Bars = 50 nm.) (From Boccardo, G., Hatta, T., Francki, R. I. B., and Grivel, C. J., *Virology*, 100, 300, 1980. With permission.)

components, each likely to be made of a different protein: two types of spikes, base plates, and the rest of the capsid. The virion structural proteins are coded for by larger segments of the genomic RNA.[41,42,53] It would be of interest to assign seven virion proteins[54] to specific viral structures.

The diameters of CPV particles isolated from insects other than *B. mori* are between 50 and 70 nm. The presence of surface projections appears to be a common feature of this group, with the possible exception of a CPV of the larch sawfly, *Anoplonyx destructor*.[55]

b. Reovirus-Like Particles

The LAV particles have a double-shelled icosahedral structure with external projections 7.5 nm wide and 8.5 nm long located at the 12 vertices (see Figure 7). The projections have been termed A-spikes to distinguish them from the B-spikes on the inner shell. The A-spikes are attached to the B-spikes which are 14 nm wide and 8.5 nm long. When the outer shells are removed enzymatically, most of the resulting 57-nm cores have only B-spikes (see Figure 8). Some cores are provided with long projections (17 nm) consisting of A- and B-spikes. LAV are similar in particle structure to plant viruses of the genus *Fijivirus* (see Figure 6). Both viruses replicate in insect hosts without causing any indication of disease. This suggests a long association and may be indicative of evolution from common insect-infecting ancestors.[56] CPV particles are remarkably similar to cores of LAV and FDV. The three also show some similarities to the reovirus particle which, however, lacks A-spikes. Furthermore, the outer shell of reovirus particles is thicker than the height of the B-spikes.

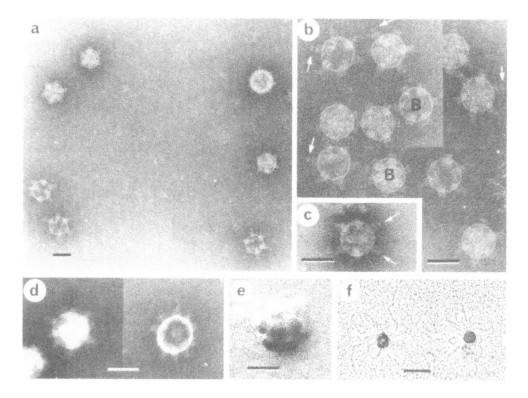

FIGURE 8.   The appearance of LAV cores. ([a] and [b] preparations were stained with AM; [c], with UA; and [d], with PTA.) Both A- and B-spikes are attached to some of the particles (arrows in [b] and [c]), whereas only B-spikes to others (B in [b]). (e) The preparation was freeze-dried and shadowed showing the prominent B-spikes whereas (f) was spread on cytochrome c and shadowed. (Bars = 50 nm in [a] to [e] and 200 nm in [f].) (From Boccardo, G., Hatta, T., Francki, R. I. B., and Grivell, C. J., *Virology*, 100, 300, 1980. With permission.)

The reovirus-like particles from the bedbug and five species of Diptera have a double-shelled structure. Their diameters are between 60 and 80 nm (see Table 1). The dipteran viruses vary in their sensitivity to physical and chemical treatments. DFV and *Ceratitis* I virus are resistant to low pH, to pelleting, to high CsCl concentration, to lipid solvents, and even to chymotrypsin as found with CPVs.[18] On the contrary, HFV and DSV are fragile *in vitro* and difficult to purify and concentrate.[22,57]

3. Genome Structure
a. Cytoplasmic Polyhedrosis Viruses
The genomic RNA of *B. mori* CPV exists in the virion as ten equimolar segments with a total molecular mass of 20.3 MDa (see Table 3).[58-60] An extra subgenomic segment occurs by internal deletion of segment 10.[43] Each segment is composed of an mRNA (plus strand) and its complement (minus strand) in an end-to-end base-paired configuration, except for the protruding 5' cap in the plus strand.[61,62]

The virus synthesizes and releases its own mRNA *in vitro* without particle degradation.[63,64] The virus-associated RNA polymerases and methyltransferases catalyze the synthesis of capped mRNAs.[65,66] Each genome segment is coupled with its own transcription enzymes and acts as an independent transcription unit which undergoes multiple cycles of transcription.[64,67] This picture, built up mainly on biochemical data, has structural basis as described here.

The *B. mori* CPV RNA extracted from virions with the phenol method has been visualized

TABLE 3

Approximate Estimation of the Molecular Weights ($\times 10^6$) of Double-Stranded RNA Segments from *B. mori* CPV and Several Reovirus-Like Particles

B. mori CPV[58]	*B. mori* CPV[59]	LAV[10]	*Cimex* virus[12]	DFV[18]	DSV[57]	HFV[21]	*Ceratitis* I virus[23]	*D. oleae* virus[75]
3.61	2.55	3.15	2.75	2.24	2.50	2.75	2.24	2.47
3.28	2.42	2.64	2.70	2.20	2.25	2.66	2.20	2.25
3.28	2.32	2.62	2.51	2.20	1.85	2.60	2.20	2.25
2.76	2.03	2.51	2.40	1.85	1.85	2.27	1.85	1.82
2.45	1.82	2.30	2.25	1.74	0.99	1.84	1.74	1.82
1.38	1.12	1.75	1.70[a]	1.56	0.77	1.50	1.56	1.40
1.11	0.84	1.26	1.65	1.24	0.72	1.43	1.24	1.16
0.96	0.62	1.05	1.33	1.06	0.68	1.38	1.06	0.96
0.83	0.52	0.98	1.16	0.62	0.55	0.75	0.62	0.64
0.66	0.35	0.78	0.90	0.52	0.35	0.56	0.52	0.50
			0.75					
			0.15					
20.32	14.63	19.04	18.55	13.03	12.51	17.74	13.03	15.27

[a] Minor band excluded in calculation of total molecular weight.

in the electron microscope. The contour lengths of the extracted RNA filaments are distributed bimodally at 0.4 and 1.3 μm, which correspond to the molecular weights of 1 and 3 MDa, respectively.[60] Extraction with urea or sodium perchlorate results in some long RNA filaments (6.8 μm, mol wt = 13 to 15 MDa).[68] Disruption of CPV polyhedra with a detergent, Sarkosyl, in the presence of pronase yields open circular RNA filaments (15 μm, mol wt = 35 MDa), which is interpreted to indicate that the CPV genome is a single molecule containing a number of sites sensitive to contaminating RNase.[69] Although this possibility cannot be completely excluded, the proposed tandem linkages of genome segments are not compatible with the biochemical data on RNA transcription.

Genome strands extruding from the virions have been demonstrated with an improved preparative method.[70] The projections on CPV particles appear as spherical blobs, 23 nm in diameter, with a slight constriction at the attachment site to the shell (see Figure 9). It is probable that the tips of the projections are removed under this preparative procedure. Disruption of virions directly on a grid with EDTA results in the extrusion of genome strands from the projections. If the virions are disrupted after glutaraldehyde fixation, the released genome strands often carry a protein particle 23 nm in diameter usually at their proximal ends. The particle is presumed to be the basal part of the projection. Virions in the process of mRNA synthesis have swollen projections and the strands extruding from them frequently assume a loop structure with the protein particles attached at various positions along the strands.

Mild destruction of virions followed by dispersion with detergents has yielded genome segments which are separated into ten bands in polyacrylamide gel electrophoresis (see Figure 10).[71] Their mobilities are less than those of the corresponding dsRNA species. The strands are bent and wavy and 8 to 12 nm in thickness. They are about twice as thick and half as long as dsRNA filaments purified by phenol treatment (see Figure 11). RNA in the genome strands may be compacted as a result of mutual strand coiling and protein binding. Some of the strands possess partially loosened regions in their thick configurations. RNA filaments obtained by the phenol treatment of the genome segments are linear. If the segments are first treated with an RNA-protein cross-linker, diepoxybutane, and then with phenol, the extracted dsRNA assumes a circular form, which suggests the binding of the two ends of linear genome dsRNA by protein(s). Each genome segment is a complex of dsRNA and protein(s) which are essential to the construction and retainment of the circular and supercoiled structure of linear genome RNA.

The following hypothesis has been proposed on the basis of these observations.[70,71] A part of a genome segment may be attached to the basal part of a projection where enzymes for mRNA synthesis integrate in the virus structure. In a virion synthesizing mRNA, unwinding of the supercoil would occur and the genome strand may move from within the shell to the projection or vice versa through a pore or "gate", which is reported to be present in the base plate.[50] During this movement mRNA is synthesized and released through the hollow projection. The circular genome segment is capable of returning to the initial site after one cycle of transcription.

CPV particles disrupted by ultraviolet light illumination release a genome-enzyme complex in a filamentous form.[72] Individual genome segments purified from the complex retain the activities of RNA polymerase and methyltransferase. The complex contains three protein components, which are serologically related to the virus structural proteins.[73] The RNA polymerase and methyltransferase extracted from the complex are enhanced in their activities when reconstituted with CPV dsRNA.[74] It is probable that the enzymes for mRNA synthesis are tightly bound to the genomic RNA and that there is at least one complex of enzymes for each of the ten genome segments.

b. Reovirus-Like Particles

Reovirus-like particles listed in Table 3 differ from each other in their genome profiles

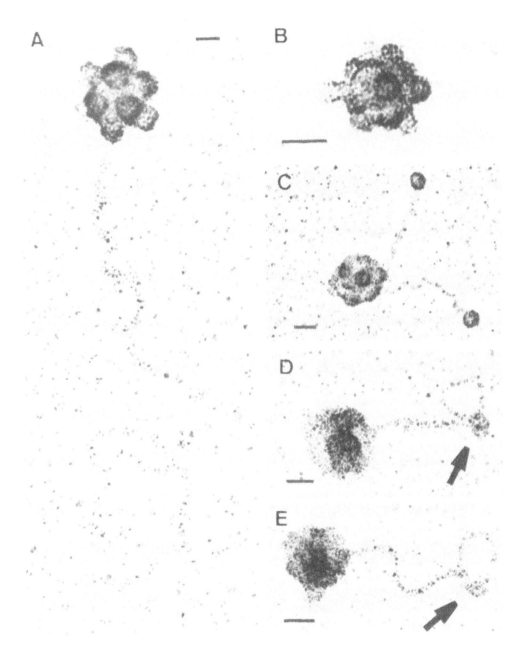

FIGURE 9. *Bombyx mori* CPV particles (stained with UA and shadowed with platinum-palladium). (A) Disrupted virion showed a genome strand extruding from a spherical projection. (B) Virion synthesizing mRNA. (C) Fixed and disrupted virion showing two extruding genome strands with attached protein particles. (D) and (E) Fixed and disrupted virion synthesizing mRNA. Arrows point to protein particles. (Bar = 30 nm.) (From Yazaki, K. and Miura, K., *Virology*, 105, 467, 1980. With permission.)

except for DFV and *Ceratitis* I virus which exhibit exactly the same distribution of RNA segments.[23,57]

The genomic dsRNA of LAV is composed of ten segments with a total molecular mass of 19.0 MDa. The genome profile closely resembles the profiles observed for members of *Fijivirus* although the molecular weights of the individual segments are different. Examination of LAV core preparations by the cytochrome c spreading technique discloses many

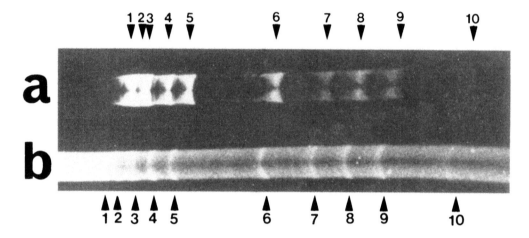

FIGURE 10.   Electrophoretic separation of dsRNA (a lane) and genome (b lane) of *Bombyx mori* CPV. Numerals represent the band numbers. (From Yazaki, K., Mizuno, A., Sano, T., Fujii, H., and Miura, K., *J. Virol. Methods*, 14, 275, 1986. With permission.)

strands of dsRNA extruding from the cores (see Figure 8f). No such strands have been observed in preparations of intact particles. Helical filaments (12 nm thick) associated with LAV (see Figure 7f) may be derived from the material that makes up the viroplasm.[56]

C. CYTOPATHOLOGY

1. Cytoplasmic Polyhedrosis Viruses

The CPV virions are always found in the cytoplasm. In ultrathin sections they appear to consist of a central dense core surrounded by a thin shell-like structure which is separated from the core by an electron-lucent zone. The core usually has a hexagonal outline and is 50 to 70% of the diameter of the virion. The surface projections are observed when the virions are sectioned in an appropriate plane. Those of a monarch butterfly CPV are 17 nm in length and appear to consist of two components.[76]

The cytoplasms of infected cells contain unique structures, the virogenic stromata or viroplasms that are composed of massive accumulations of finely textured material (see Figures 12 and 13). They have no regular shape and lack a surrounding membrane. At an early stage of infection, they increase in size and fuse, and occupy a large area of the cytoplasm. They contain empty or partly filled viral capsids. Their size decreases at a late stage of infection. At this stage, complete virions move away from the viroplasmic matrix into the surrounding cytoplasm, where they usually remain scattered at random, occasionally aggregate in crystals, or are incorporated into lysosomes (see Figure 13).

The virons scattered in the cytoplasm become embedded in polyhedron protein. The granular and fibrous protein diffuses into relatively large accumulations of virions and crystalizes to form polyhedra. The growing polyhedra are often associated with a matrix of closely packed fibrils. This matrix, designated as the crystallogenic matrix,[76] is less dense than the polyhedra but more dense than the viroplasms (see Figures 12 to 14). It is presumed on the basis of circumstantial evidence that the synthesis and/or assembly of polyhedron protein occurs in the crystallogenic matrix and those of virus (capsid) proteins occur in the viroplasm.[76] The crystallogenic matrix contains many complete virions, which are difficult to observe owing to the superimposition of fibrils. This matrix regresses as the crystal formation proceeds. The incorporation of a virion begins with the close association of the virion to the surface of a polyhedron (see Figure 14C). The crystal grows around the virion by a process of deposition and reorganization of polyhedron protein. The parameter of crystal

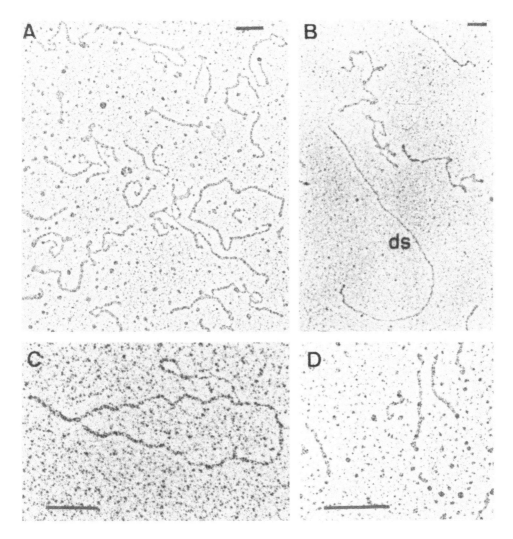

FIGURE 11. Genome strands extracted from *Bombyx mori* CPV particles. (A) Genome strands showing different lengths. (B) Comparison of a dsRNA filament (ds) with genome strands. (C) Genome segment obtained by treatment with diepoxybutane and phenol. (D) Genome strands recovered from band 9 of the polyacrylamide gel. (Bars = 100 nm.) (From Yazaki, K., Mizuno, A., Sano, T., Fujii, H., and Miura, K., *J. Virol. Methods*, 14, 275, 1986. With permission.)

growth is coded for by the virus.[35] With some CPVs, such as those that infect mosquitoes and chironomids, the individual virions are occluded in protein crystals that may coalesce to form larger crystals with more than a single particle.[77]

In the typical CPV infection, the nuclei are relatively unaffected and only rarely contain inclusions. Some strains of *B. mori* CPV, however, induce the formation of nuclear inclusions and associated cytological changes in most infected cells.[33] The changes include the enormous hypertrophy of the nucleoli and the aggregation of dense reticulum at multiple foci within the nuclei (see Figures 12 and 15). The dense reticulum lacks the characteristic granules of nucleolonemata and are composed of small granules linked together by fine filaments.[78] The reticulum is reduced in amount as the inclusion increases in size. Nuclear inclusions appear within the areas where many masses of dense reticulum conglomerate. The nuclear inclusions may be crystalline or amorphous depending on the virus strain that induces them. The similarity of the protein of nuclear inclusions to the polyhedron protein has been considered

FIGURE 12.   Thin section of portions of two *Bombyx mori* cells infected with CPV (B_2 strain). Viroplasm (Vp) in the upper cell is surrounded by several clusters of rod-shaped polyhedra. Some of the clusters are still embedded within a crystallogenic matrix (arrow). Cytoplasm also contains masses of vesiculate rough endoplasmic reticulum, oval mitochondria, and few vacuoles. In the nucleus (N) of the lower cell, three rod-shaped inclusions (I) are visible, associated with several hypertrophied nucleoli (No) and many masses of dense reticulum (arrowhead). (Bar = 1 μm).

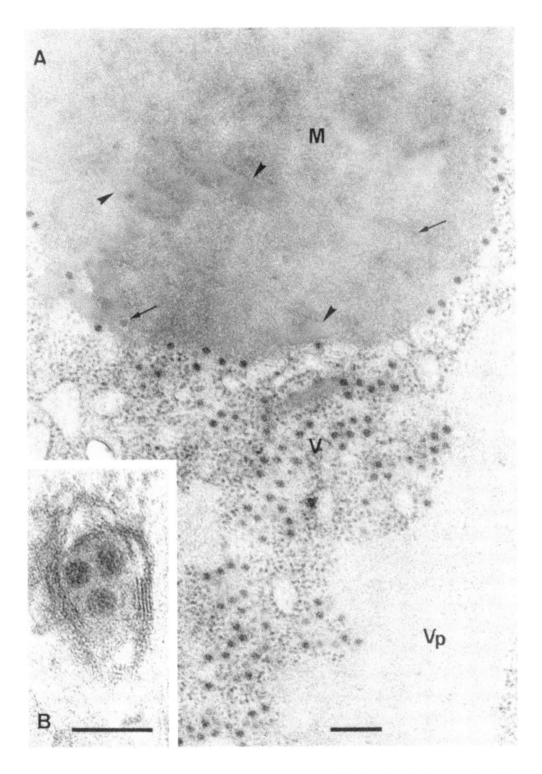

FIGURE 13. (A) Thin section of a *Bombyx mori* cell infected with CPV (B$_2$ strain) showing a viroplasm (Vp), a crystallogenic matrix (M), and accumulation of virions (V) in the surrounding cytoplasm. Rod-shaped polyhedra (arrowheads) and virions in the matrix (arrows) are difficult to recognize owing to the presence of tightly packed fibrils. (Bar = 300 nm.) (B) Thin section of a lysosome enclosing apparently intact virions. (Bar = 100 nm.)

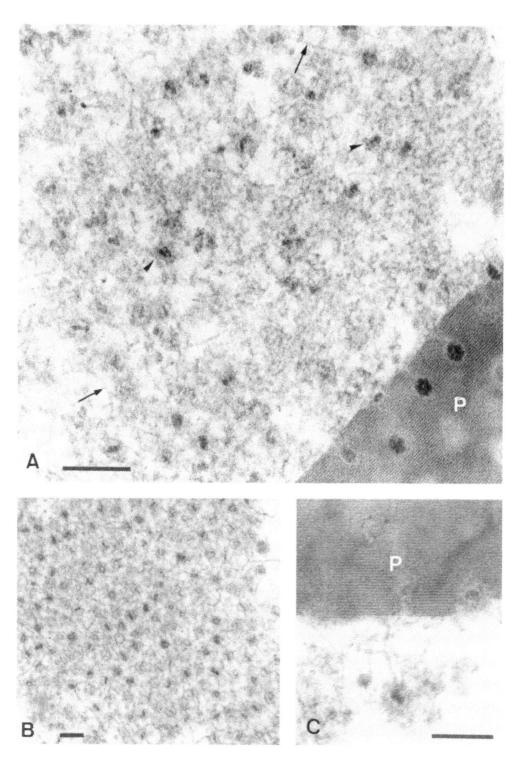

FIGURE 14.   Thin sections of a *Bombyx mori* cell infected with CPV (A strain). (A) Margin of a polyhedron and the associated crystallogenic matrix. (B) Crystallogenic matrix. (C) Margin of a polyhedron illustrating the incorporation of virions and accumulation of fibrils of polyhedron protein. The matrix consists of a granular and fibrous mass associated with many complete virions (arrowheads). Long filaments (arrows) extend from the surface of many virions connecting them between each other or to the surface of the polyhedron (P). (Bar = 100 nm.)

FIGURE 15. (A) Thin section of a *Bombyx mori* cell infected with CPV (A strain) showing many masses of dense reticulum (arrowhead) and two inclusions (I) in the nucleus (N). (Bar = 1 μm.) (B) Higher resolution micrograph of the inclusion showing the crystalline lattice structure. (Bar = 20 nm.)

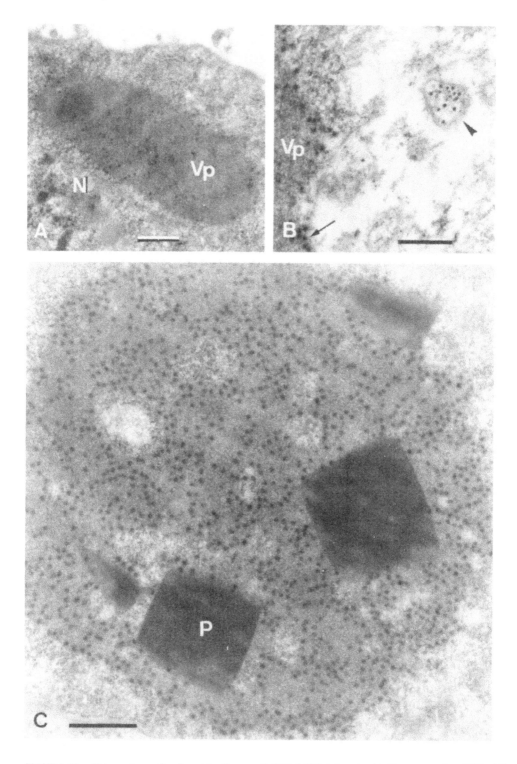

FIGURE 16.   Thin sections of a *Lymatria dispar* cell (SCLd-135) infected with *Euxoa scandens* CPV. (A) Viroplasm (Vp) in the cytoplasm; nucleus (N). (B) Portion of a viroplasm with associated virions (arrow) and a lysosome (arrowhead) enclosing virions. (C) Crystallogenic matrix containing many virions and several polyhedra (P). (Bars = 500 nm.)

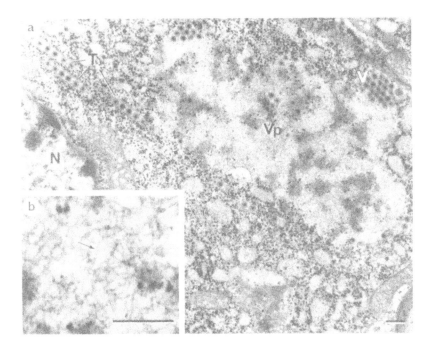

FIGURE 17.   Thin sections of cells of *Cicadulina bimaculata* infected with LAV. (a) The cell contains some LAV particles in an aggregate (V) and others in tubules (T) near a viroplasm (Vp). (b) Higher resolution micrograph of the viroplasm shows typical fibrils 7 nm in diameter (arrow). (From Baccardo, G., Hatta, T. Francki, R. I. B., and Grivell, C. J., *Virology*, 100, 300, 1980. With permission.)

as evidence that the nucleus may be the site of assembly of the polyhedrin transported from the cytoplasm.[35,45,79]

Several insect cell lines are susceptible to CPV infection.[6,80] Many of the polyhedra produced in infected cell cultures are cubic and differ from the shape of polyhedra produced in larvae by the same virus isolate. The change in polyhedron shape may be attributed to the different crystallization process of polyhedron protein in two distinct cellular systems rather than to the selection of different virus strains from a heterologous virus population.[33,81] Otherwise, the cytopathology *in vitro* is typical of CPV development *in vivo* (see Figure 16).[82-85] Virion development is restricted to viroplasms in the cytoplasm. Mature virus particles are associated with crystallogenic matrices and are incorporated into polyhedra. Lysosomes occasionally contain complete virions at a late stage of infection.

2. Reovirus-Like Particles

The LAV particles usually occur in the cytoplasm of infected leafhopper cells either singly or in crystalline arrays, often in association with viroplasms, unbranched tubules, and residual bodies or phagosomes (see Figure 17).[10] The viroplasms are similar to those formed in CPV infection. However, the crystallogenic matrix and polyhedra are never formed in LAV infection.

The HFV particles measure 60 to 76 nm in diameter and have pentagonal and hexagonal outlines in ultrathin sections.[20] Viroplasms have irregular shapes and vary in size from 0.5 to 1.5 μm in diameter. Virions are scattered over their surface and on their periphery. Dense lysosomes contain virions.

The *Cimex* virus and DFV particles occur in the cytoplasm of infected cells in crystalline arrays.[12,18] The cytopathology of other reovirus-like particles is still to be determined.

II. VIRUSES OF INVERTEBRATES OTHER THAN INSECTS

A. INTRODUCTION

Contrary to insect reoviruses, which for the majority belong to the group of CPVs, almost all of the reoviruses or reo-like viruses known in invertebrates other than insects are nonoccluded viruses; one exception is found in a daphnid[86] and considered as a true member of this group (see Table 4).

All the others which did not fit properly into this group are described in this chapter as possible members of Reoviridae, although some were determined as possessing a genome consisting of several dsRNA segments and have not yet been accepted by the International Committee on Taxonomy of Viruses (ICTV) as members of the family.[1]

1. Cytoplasmic Polyhedrosis Viruses

The only CPV recorded was found in an Iridovirus-infected fresh-water daphnid, *Simocephalus expinosus* (Crustacea), collected in Florida.[86]

2. Nonoccluded Viruses

The nonoccluded viruses have been reported in helminths, mollusks, crustaceans, and arachnids. In Arachnida, only one reo-like virus was reported in the scorpion *Buthus occitanus*,[87] but it will not be mentioned here, since this will be discussed in the chapter "Unclassified Viruses of Arachnida" (see Chapter 25).

In mollusks, a virus named 13P2 was observed in juvenile American oysters *Crassostrea virginica*,[88] by using fish cell lines, particularly BF2 cells, where typical cytopathic effects (CPE) were produced. However, no evidence of its development in oyster tissue was given, and its presence in oysters could be interpreted as a result of the filtration potential of this mollusk. There seemed to be a similar phenomenon reported for some birnaviruses isolated from mollusks[89] or crustaceans.[90] Because this agent is more probably a fish virus than an oyster virus,[91] we will describe it as a special case in this chapter.

On the basis of morphological considerations — essentially morphogenesis, tissue location, and cytoplasmic arrangement — the other nonoccluded viruses may be separated into three types.

a. Nonoccluded Type I Viruses

Type I represents those viruses which are characterized during their development by the formation of a paracrystalline arrangement of particles associated with filamentous or tubular structures and developing in connective or parenchymatous tissues or in blood. Such viruses were recorded in one helminth[92] and two crustaceans.[93-95]

The helminth virus has never been isolated and has only been observed in a Plathelminth, *Diplectanum aequans* (Monogena) parasitizing the gills of sea bass, *Dicentrarchus labrax* (Teleost) (see Figure 18), from the French Mediterranean coasts.[92]

In Crustacea, the RLV (reo-like virus) has been described in the blue crab *Callinectes sapidus* from both Chincoteague and Chesapeake Bays (U.S.).[95,96]

In France, the P virus isolated in the swimming crab *Macropipus depurator* from the Mediterranean area,[93,94] could represent the model of this group because it was the most investigated at the level of its characterization as well as in its biological and pathological properties.[97,98]

By its characteristics, the RLV of *C. sapidus* seemed to be closely related, if not identical, to the P virus of *M. depurator*.

b. Nonoccluded Type II Viruses

Type II includes viruses forming "rosettes" in the course of their morphogenesis, in

TABLE 4
List of Reovirus or Reo-Like Virus Particles Reported in Invertebrates Other than Insects

	Name	Hosts	Infectivity tests[a]	Tissue tropism	Size (nm)	No. ds-RNA segments	Ref.
CPV	—	*Simocephalus expinosus*	—	Midgut epithelial cells	60[s]	—	86
NO[b]	13P2	*Crassostrea virginica*	—	(c)	79	11	88, 91, 109, 111
NO Type I	—	*Diplectanum aequans*	—	Subtegumental parenchyme	66—70[s]	—	92
	RLV	*Callinectes sapidus*	+	Connective cells hemocytes	55[s]	—	95, 96
	P	*Macropipus depurator*	+	Hematopoietic organ	59—61	At least 11	93, 94, 97, 98, 102
NO Type II	W	*Carcinus maenas*	+	Connective cells	65—70	At least 9	99
	W2	*Carcinus mediterraneus*	+	Hemocytes	55—60	—	100, 101, 110
	GENV	*Carcinus mediterraneus*	—	Gill epithelium	85—95	—	98, 102
NO Type III 3	RC84	*Carcinus mediterraneus*	+	B cells of hepatopancreas	85—95	—	100, 101, 103
	GNSV	*Penaeus japonicus*	+	R cells of hepatopancreas	60	—	104, 105, 107
	—	*Sepia officinalis*	—	Stomach epithelium	75[s]	—	108

Note: Exception is the reo-like virus of scorpion (Arachnida); see Chapter 25.

a Infectivity established (+); not established or not done (−).
b Nonoccluded (NO) viruses.
c Virus development does not occur in oyster tissues.
s Diameter of the particle as seen in section. The other values have been determined after negative staining (PTA).

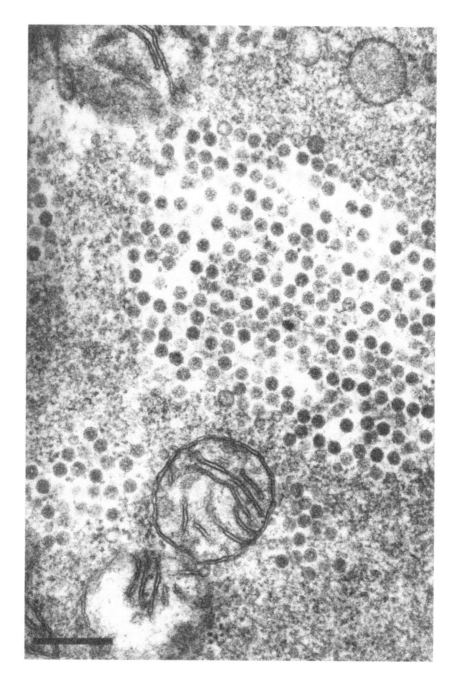

FIGURE 18.   Viral area of an infected plathelminth, *Diplectanum aequans* (Monogena) ectoparasite of the gills of sea bass, *Dicentrarchus labrax*. (TEM; bar = 0.5 µm.) (Courtesy of A. Lambert and C. Maillard.)

association with curved filamentous or tubular structures. In infected cells, the rosettes are typical structures made by the association of six particles as seen in section. To our knowledge, such rosette structures are a unique event in the group of cytoplasmic double-stranded and multisegmented RNA viruses.

They include the W virus of the green crab *C. maenas* from the Channel area (U.K. coasts),[98] the W2 virus,[100,101] and the gill epithelium necrosis (GEN) virus,[98,102] both of the shore crab *(C. mediterraneus)* from the Languedoc coasts (France). Although the W and W2 viruses look very much alike owing to their general properties and location in connective tissues and blood cells, only the GEN virus develops in epithelial cells even though its general morphological characteristics are similar to those of the W and W2 viruses.

c. Nonoccluded Type III Viruses

Viruses of the third type develop in digestive cells where they form nonstructured viral areas, and only occasionally paracrystalline arrays. They have been reported in crustaceans and cephalopods.

In Crustacea, the RC84 virus was found in *C. mediterraneus* from the French Mediterranean coasts[100,101,103] and the reo-like virus of the tiger shrimp, *P. japonicus*,[104,105] was found associated with the "gut and nerve syndrome" (GNS)[106] in experimental farms both in the Languedoc area (France) and in Hawaii (U.S.).[107]

The last virus belonging to this third type was described in stomach cells of the cuttlefish *Sepia officinalis*,[108] in the course of electron microscopic investigations of this organ. Authors emphasized ultrastructural analogies with members of the Reoviridae family, although this agent was never isolated and furthermore investigated.

B. VIRUS STRUCTURE AND COMPOSITION

Very little is known about the CPV of the daphnid *S. expinosus*.[86] Authors reported the presence of pleomorphic inclusion bodies, some brick-shaped, others (up to 2 μm), were spherical. Icosahedral in shape, virions were about 60 nm in diameter (see Figure 19a and b).

1. Particle Structure

Because the cytoplasmic polyhedrosis virus of *S. expinosus* has never been isolated, data on the particle structure of Reoviridae from invertebrates other than insects only concern the nonoccluded viruses.

The surface structure of virions has been well investigated in the case of the 13P2 virus isolated from *C. virginica*, using the technique of rotational enhancement images.[109] When observed in negative staining, the particles, 79 nm in diameter, are icosahedral in shape, and exhibit a hexamer-pentamer pattern of capsomer arrangement. Most of the capsomers of disrupted virions were hexagonal and composed of six subunits, but only five subunits were visualized in the others. In spontaneously degraded particles, six symmetrical pheripheral projections could be evidenced, apparently protruding from the inner shell. There may be a total of 32 capsomers, divided into 12 pentamers and 20 hexamers.

a. Nonoccluded Type I Viruses

In the case of the P virus of *M. depurator,* particles had a paraspherical structure of 60 ± 1 nm in diameter. Empty particles exhibited an electron-dense center, 40 nm in diameter, when observed in negative staining (see Figure 20a). In such PTA-penetrated particles, two-, three-, and five-fold symmetrical axes were determined indicating an icosahedral structure of particles.[97] The surface of intact particles was formed with subunits or capsomers, 8 to 9 nm in diameter, which presented a hollow tubular structure (see Figure 20b). In some hexagonally shaped particles, four capsomers were clearly evidenced on each side. As for the integrity of the particles, ethyl ether treatment did not induce any modifications, but led to evidence the double-shelled structure of virions (see Figure 20c); each shell was about 6 nm thick. The inner shell seemed to be formed with smaller subunits than those of the outer shell. In section, the inner shell appeared electron-lucent whereas the denser outer shell exhibited a striated structure. The center, highly electron dense and generally paraspherical in shape, sometimes presented a star-shaped structure.[98]

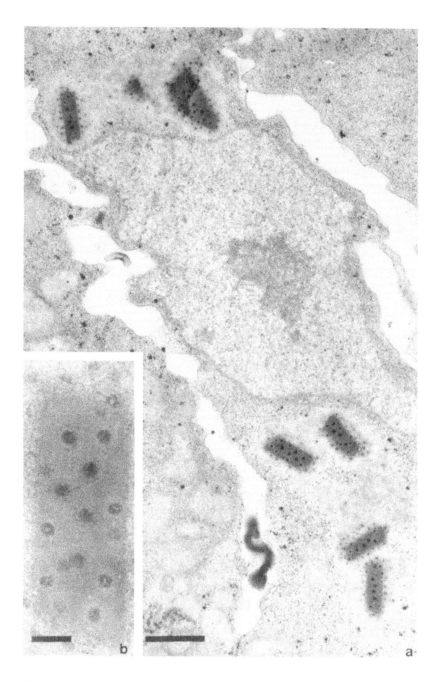

FIGURE 19. Cytoplasmic polyhedrosis virus of the daphnid *Simocephalus expinosus.* (a) The cytoplasm of an infected midgut epithelial cell contains brick-shaped inclusion bodies. (TEM; bar = is 1 μm.) (b) Higher magnification of an inclusion body containing occluded virions. (TEM; bar = 0.1 μm.) (Courtesy of B. A. Federicci.)

b. Nonoccluded Type II Viruses

Negatively stained W2 virus particles from *C. mediterraneus*[101] were grossly paraspherical, 65 to 70 nm in diameter (see Figure 21a). Their surface was formed with 8- to 9-nm subunits leading to a notched look on the periphery of PTA-penetrated particles. These capsomers were 8 to 9 nm in diameter, with an electron-dense central area about 5 nm in diameter, indicating a prismatic or hollow tubular structure (see Figure 21a). The capsid

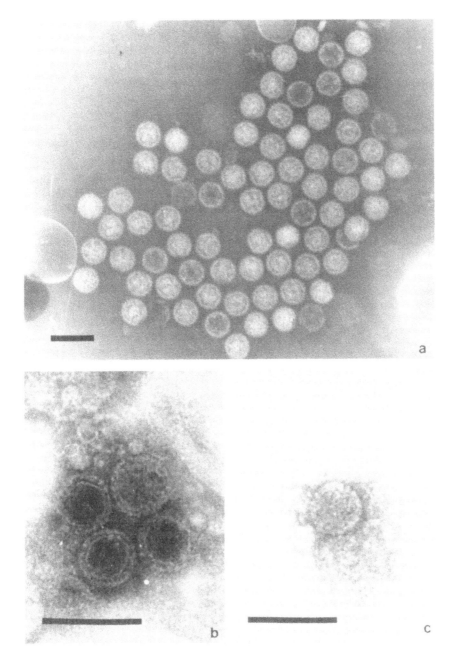

FIGURE 20. Negatively stained (PTA) P virus of *Macropipus depurator* (Crustacea). (a) Full and empty particles of purified virions; (b) the double-shelled structure of virions is evidenced after ether treatment; (c) isolated particles showing the hollow tubular structure of capsomers. (Bars = 0.1 μm.)

was formed with two layers of identical thickness, 4 to 5 nm each, delimiting an electron-dense center, 50 to 55 nm in diameter, which sometimes appeared hexagonal in shape.

In older preparations, damaged particles showed a very fragile outer shell releasing an inner component, 55 to 60 nm in diameter (see Figure 21b), which exhibited small projections at each angle of the particle. Rearrangement in paracrystalline arrays of subunits released from damaged particles has been reported; such structures were formed by the juxtaposition

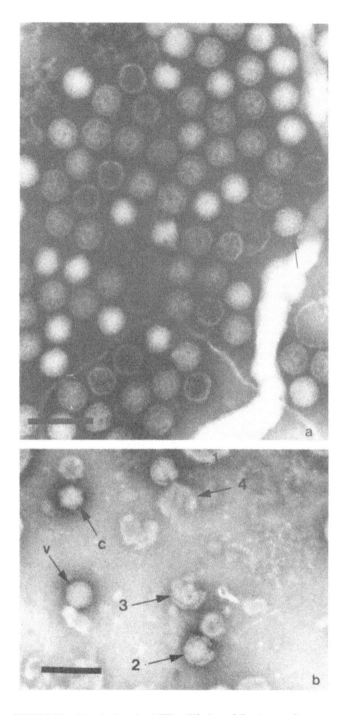

FIGURE 21.   Negatively stained (PTA) W2 virus of *Carcinus mediterraneus*
(Crustacea). (a) Full and empty particles of purified virions; capsomers exhibit
a hollow tubular or prismatic structure (arrow). (b) damaged W2 virions showing the different degrees of disruption; unaltered particle (v). (c) The inner
component or core is released by disruption of the outer shell (arrows 1, 2,
3, and 4). (Bars = 150 nm.)

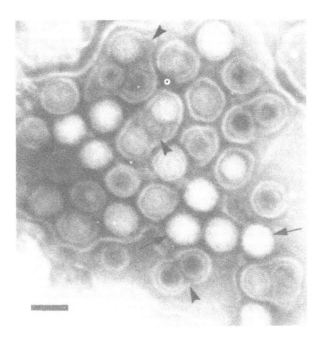

FIGURE 22. Negatively stained (PTA) RC84 virus of *Carcinus mediterraneus* (Crustacea). Note nearby unaltered particles (arrows), the different releasing stages of the outer shell; full or empty cores are surrounded by the outer shell. In some cases this external shell exhibits the aspect of an envelope-like structure containing two or three cores (arrowheads). (Bar = 0.1 μm). (Courtesy of J. Mari.)

of small hollow prismatic elements identical to the capsomers described forming the outer shell of virions.[101,110]

c. Nonoccluded Type III Viruses

The negatively stained preparations of the RC84 virus of *C. mediterraneus*,[103] usually exhibited small groups of virions which, because of their fragility, showed all the different stages of damaged particles allowing us to distinguish the different components (see Figure 22). So, beside intact particles, 85 to 95 nm in diameter, without noticeable superficial ornamentation, some others displayed an external layer containing a smaller particle or core of 65 to 70 nm. If the inner particle was penetrated by PTA, the virions seemed to be formed with two concentric structures delimiting an electron-dense center, 50 nm in diameter. These two concentric layers were separated with a space variable in thickness depending on the degree of damage of the particle. The inner shell, 6 to 6.5 nm thick, exhibited small projections or spikes which gave the core a notched look.[101,103]

The external shell, 8.5 to 9.5 nm thick, did not show any peculiar arrangement and appeared relatively fragile because disruptions could be observed in PTA-penetrated virions; sometimes such external shells seemed to be coalescing with neighboring particles (see Figure 22), giving the aspect of an envelope-like structure surrounding two or three particles.[101,103] To our knowledge, such fragile morphological features have never been described in Reoviridae and constitute a characteristic of the RC84 virus structure.

2. Particle Composition

a. Genome Structure

To date, results on genome structure are heterogeneous; as a matter of fact, authors have investigated the genome structure using essentially different electrophoresis gel types, such as polyacrylamide,[111] agarose,[101,110] and agarose-acrylamide.[101] These techniques were used to demonstrate that all the genomes studied were of a multisegmented dsRNA nature, but most of the results must be interpreted as temporary since the same 10% polyacrylamide gel electrophoresis technique was not used in genome-reovirus studies; indeed, considerable variations in segment mobilities were reported according to the different gel type or gel concentration used.[62]

Using a 9% polyacrylamide gel, the genome of the 13P2 virus was determined to be composed of 11 segments of double-stranded RNA, which fell into three size classes, S, M, and L (small, medium, and large, respectively), with a total molecular weight of 15.04 × 10^6.[111]

Genomes of the W, W2, and P viruses were investigated using successively composite agarose-acrylamide gels and agarose gels.

The first comparative studies using composite gels emphasized the similarities between W and W2, giving an identical electrophoretype characterized by seven segments, whereas the P virus genome electrophoretype was different, exhibiting nine segments.[101] Using more accurate gels, i.e., 1% agarose gels, electrophoresis has evidenced that the W2 virus genome was composed of at least 9 segments, whereas the P virus genome was composed of at least 11 segments (see Figure 23a and b);[101,110] the total molecular weight was estimated in these cases to be 13.43 and 13.11 × 10^6, respectively. Table 5 summarizes these results.

Whereas the presence of RNA in 13P2 particles was indirectly demonstrated using DNA inhibitor 5-iodo-2-desoxyuridine (IdUrd),[88] the type of nucleic acid was determined by colorimetric methods in the cases of the P,[98] W,[99] and W2 viruses.[101,110] Moreover, they were found to be double stranded after formaldehyde treatment and melting temperature determination.[98,101,110]

b. Capsid Composition

Analysis by SDS-PAGE (12% gels) revealed that the 13P2 virus possesses five structural proteins:[111] two large polypeptides of approximately 135 and 125 × 10^3 mol wt, one medium 70 × 10^3 mol wt and two small polypeptides of 45 and 34 × 10^3 mol wt.

Using 8% gels, the P, W, and W2 viruses were analyzed by SDS-PAGE revealing that the W type virus contains six structural proteins with a molecular weight range from 120 to 24 × 10^3 Da,[101,110] while the P virus had four polypeptides also ranging from 120 to 24 × 10^3 mol wt.[101,110] These results are summarized in Table 6.

It is very interesting to emphasize similarities observed in molecular weight between the four polypeptides of the P virus and the first, second, fifth, and sixth polypeptides of the W and W2 viruses. This must be considered in parallel with the similarities of the genome structure of the P and W2 viruses, which exhibit close relationships showing similar facies but not exactly the same electrophoretype.

C. CYTOPATHOLOGY

Morphogenesis of the CPV of *S. expinosus* occurred in the cytoplasm of midgut epithelial cells showing particles which exhibit electron translucent centers.[86] They matured apparently in close vicinity of inclusion bodies by acquiring an electron-dense core of about 38 nm in diameter during their inclusion process.

The 13P2 virus, isolated from juvenile oysters, *C. virginica*, seemed to be unable to develop and produce lesions in experimentally contaminated oysters, but appeared pathogenic to fingerling bluegills, *Lepomis macrochirus,* in which it produced a focal necrotic hepatitis leading to a 44% mortality.[91] Affected hepatocytes produced typical cytoplasmic arrays of particles.

1. Nonoccluded Type I Viruses

The RLV of *C. sapidus* and the P virus of *M. depurator* developed in the cytoplasm of

P   W2

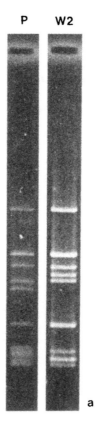

a

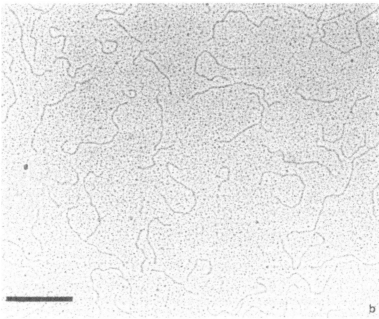

b

FIGURE 23. (a) 1% agarose gel coelectrophoresis of dsRNA of two crab viruses (P virus of *Macropipus depurator* and W2 virus of *Carcinus mediterraneus*); ethidium bromide staining. (b) Shadowed (platinum-palladium) dsRNA segments of W2 virus after phenol extraction, evidencing the different size classes of the genome segments. (Bar = 0.5 μm.)

TABLE 5
Estimates of Molecular Weight
($\times$ 10^{-6}) of dsRNA Segments from
Four Reo-Like Viruses of
Invertebrates Other than Insects

13P2[111]	W and W2[101,110]	P[101]
—	3.06	3.11
2.40	—	—
2.30	—	—
2.30	—	—
1.90	1.88	1.88
1.70	1.69	1.70
—	1.57	1.54
—	1.51	—
1.40	—	1.42
—	1.10	1.11
—	0.92	0.94
—	—	0.91
0.85	0.87	0.87
—	0.83	0.84
0.80	—	0.82
0.57	—	—
0.47	—	—
0.35	—	—
Total 15.04	13.43	13.11

TABLE 6
Estimates of Molecular Weights
($\times$ 10^{-3}) of the Polypeptides
from Four Reo-Like Viruses of
Invertebrates Other than Insects

13P2[111]	W[99]	W2[101,110]	P[101]
135	—	—	—
128	122	120	120
—	95	94	94
70	77	76	—
45	44	44	—
35	32	32	32
—	25	24	24

connective cells, hemocytes, cells of the hematopoietic organ, where they produced numerous large fuchsinophilic and crystalline arrays 6 to 10 μm long, 1 to 3 μm wide (see Figure 24a and b).[94-98] In the case of P virus, these crystalline arrays formed by the association of virions in regular and parallel lines showed a periodicity of about 80 nm (see Figure 25a and b). In such formations, the capsid exhibited an average thickness of 15 nm with an electron-dense center of 30 nm in diameter.[98]

Associated with these lesions, membranous or tubular structures, often branched, 20 to 25 nm thick, were located in high ribosomal concentration areas (see Figure 25a).[94,96,98] In less infected cells, electron-dense cytoplasmic arrays contained some isolated viral particles showing no peculiar arrangement; such areas could be interpreted as viroplasms. Here, the

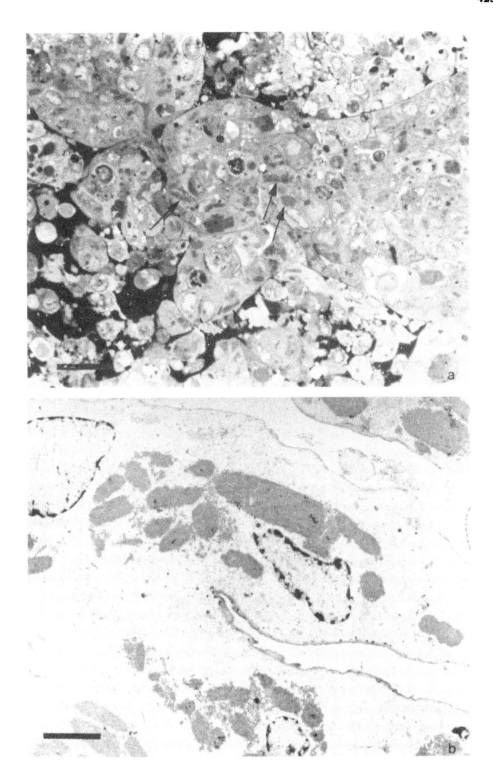

FIGURE 24.   (a) RLV of *Callinectes sapidus* (Crustacea); large crystalline inclusions of virions (arrows) are located in the cytoplasm of infected hemopoietic cells. Semithin section (toluidine blue). (Bar = ca. 20 μm.) (Courtesy of P. T. Johnson.) (b) Electron micrograph of P virus-infected connective cells of the crab *Macropipus depurator*; note the large cytoplasmic inclusions of virions forming crystalline clusters. (TEM; bar = 5 μm.)

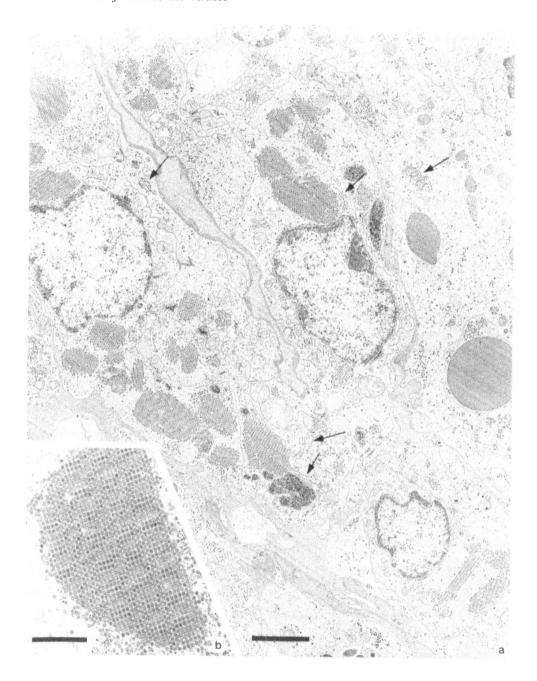

FIGURE 25. RLV-infected cells in a blood vessel of *Callinectes sapidus* (Crustacea). (a) Note the presence of membranous structures (arrows) associated with virus development. (TEM; bar = 3 μm.) (b) Higher magnification of crystalline array of virions in an hemocyte. (TEM; bar is 1 μm.) (Courtesy of P. T. Johnson.)

virion structure appeared more clearly than in crystals; the capsid was formed of two layers equal in thickness; i.e., an electron-lucent inner shell, and an electron-dense outer shell which exhibited a striated structure. The center of the particles, highly electron-dense, usually paraspherical in shape, sometimes exhibited a starlike shaped structure.[98]

Infected cells often exhibited an enlargement of the nuclear cisternae, a decrease in the chromatin content and swelling mitochondria.

In heavily infected animals, the hemolymph carried numerous cell debris and virions. The importance of cytoplasmic lesions, i.e., large clusters of virus, seemed to be at the origin of the rupture of the hyperinfected cell, which released a large quantity of virions. This stage corresponded to the viremic aspect of the infection.[98]

2. Nonoccluded Type II Viruses

Except for the GEN virus, which developed in epithelial cells of the gills of *C. mediterraneus*,[98,102] the W virus of *C. maenas*[99] and the W2 virus of *C. mediterraneus*[101,110] infected the same location as nonoccluded Type I viruses, i.e., the cytoplasm of connective cells and hemocytes.

Such tissues infected with the W2 virus were degenerative, associated with an important cellular degeneration.[101,110] The mitochondria exhibited a disorganized inner structure characterized by the disappearance of cristae. Some cells were highly vacuolized with a cytoplasm devoid of organelles; others exhibited inclusions and abnormal formations in a cytoplasm where free ribosomes were exceptionally numerous. The nuclear cisternae were enlarged and the chromatin was diminished. In the cytoplasm, abnormal formations corresponded to degenerating concentric structures or electron-dense material, variable in size, associated or not with fibrillar structures 25 to 30 nm thick. An unusual feature was the presence of ring-shaped structures in the cytoplasm of such cells (see Figure 26a, b).

These ring-shaped structures or rosettes were formed, in sections, with the association of 5 to 7 paraspherical particles, 55 to 60 nm in diameter each (see Figure 26b). They corresponded to small spherical-shaped clusters of virions with an empty center. Bow-shaped and nonbranched tubular or membranous structures were always associated in the cytoplasm with virus development (see Figure 26c).[101,110]

When virions were isolated, either in the cytoplasm or in extracellular space after release by disruption of infected cells, they exhibited a capsid, 15 to 20 nm thick and an electron-dense center, 27 to 35 nm in diameter, usually paraspherical or ovoid, but sometimes starlike shaped.[101,110]

Viral morphogenesis occurred in a poorly defined viroplasm generally extended to the whole cytoplasm. This viroplasm was characterized by a granular electron-dense appearance, showing some filamentous areas. At this stage, it seems that virions were formed already placed in their own position in the rosette. Rosette formation occurs at the same time and parallel to virion morphogenesis. At a more advanced stage, the cytoplasm exhibited a more uniform aspect.[101,110]

Virions were released in intercellular spaces and in the hemolymph after disruption of heavily infected cells. At this stage, they had lost their rosette arrangement and appeared isolated from each other. Such a massive release of virions was at the origin of a truly viremic stage of this disease.[101,110]

3. Nonoccluded Type III Viruses

The reo-like virus of *S. officinalis* develops in the cytoplasm of stomach cells where it induces an important lysis of infected cells.[108] Morphogenesis occurred at the periphery of large viroplasms showing a heterogeneous structure formed with electron-dense granular areas inside a clear matrix; ribosomes and polysomes were numerous around and sometimes inside these areas. Before complete cell degeneration, cytoplasmic organelles could not be recognized, but a paracrystalline arrangement of virions occurred. Bundles of tubules, often enclosed in a unit-membrane structure, were present in some cells; these long tubules had a diameter close to those of the viral particles.

The two other viruses of this type, i.e., RC84 of *C. mediterraneus* (see Figure 27) and the GNS-virus (see Figure 28) of *P. japonicus*, seemed to have identical cytopathologies

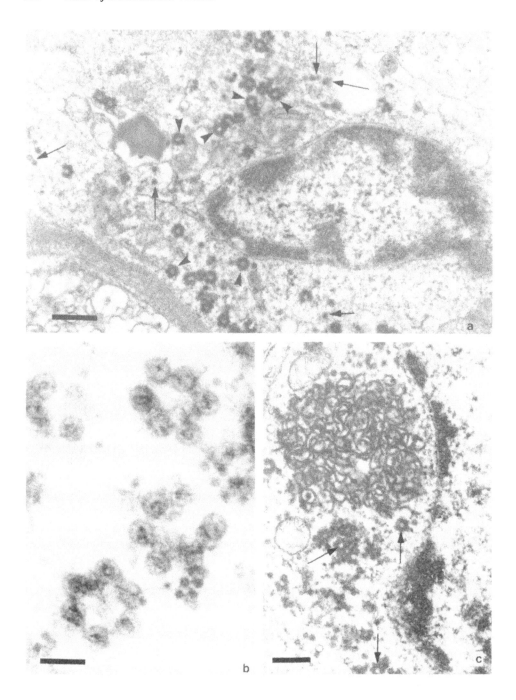

FIGURE 26. W2 virus infection of *Carcinus mediterraneus*. (a) infected connective cell showing "rosette" formations in the cytoplasm (arrowheads) and isolated virions (arrows). (TEM; bar = 0.6 μm.) (b) Detail of rosettes showing virion arrangement. (TEM; bar = 100 nm.) (c) Bow-shaped membranous structures associated in cytoplasm with virus development; rosettes are arrowed. (TEM; bar = 0.5 μm.)

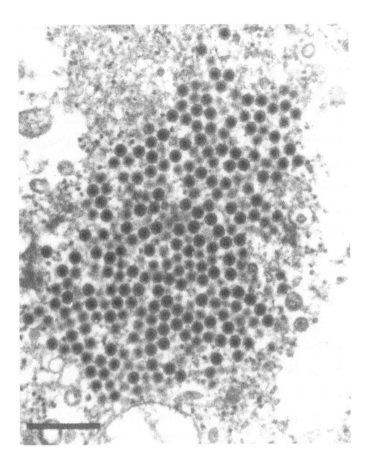

FIGURE 27. RC84 virions in the cytoplasm of an infected epithelial cell of hepatopancreas of the crab *Carcinus mediterraneus*. (TEM; bar = 0.5 μm.) (Courtesy of J. Mari.)

except for their cell location which was strictly the cytoplasm of B cells of the hepatopancreas for the RC84 virus[101,103] and the cytoplasm of R cells for the GNS virus.[104,105,107]

The RC84 virus of *C. mediterraneus* formed large viral areas in the B cells, variable in size and shape, and scattered over the whole cytoplasm (see Figure 27).[101,103] Such areas could reach 8 μm in length. Some are near the nucleus which seemed to have a normal aspect. Numerous small empty vacuoles were often observed around viral areas.

Some diffuse and filamentous electron-dense material, associated with numerous ribosomes gave a highly electron-dense aspect to the cytoplasm; they were located near or inside the viral area and looked like a viroplasm.

Paracrystalline arrays of virions were very infrequent; they showed a period of 60 nm inconsistent with the virion diameter, revealing a tridimensional organization.

The disruption of degenerating infected cells released viral particles into hepatopancreatic tubules, the digestive tract, then with feces into the seawater.[101,103]

The GNS virus seemed to have the same releasing and dissemination modes[105,106] as the RC84 virus.

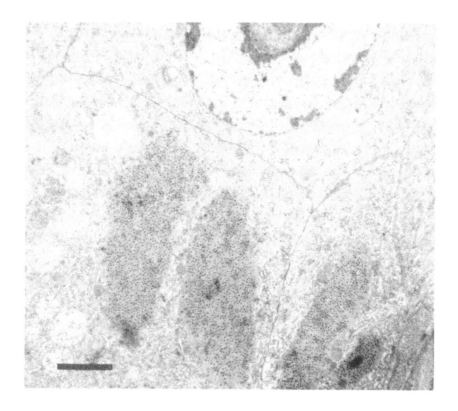

FIGURE 28.   GNS-associated virus of the tiger shrimp, *Penaeus japonicus*. Large viroplasms are located in the cytoplasm of R cells of the hepatopancreas. (TEM; bar = 2 μm.)

ACKNOWLEDGMENTS

Tosihiko Hukuhara would like to thank Dr. Y. Tanada for critically reading the manuscript; and Drs. J. R. Adams, K. Aizawa, R. I. B. Francki, Y. Hu, S. Kawase, D. Liang, and K. Yazaki for supplying material for the illustrations.

Jean-Robert Bonami would like to thank Mr. J. Luciani for his technical assistance.

REFERENCES

1. **Matthews, R. E. F.**, Classification and nomenclature of viruses. IVth Report of the International Committee on Taxonomy of Viruses, *Intervirology*, 17, 1, 1982.
2. **Joklik, W. K.**, *The Reoviridae*, Plenum Press, New York, 1983, Chaps. 7 and 10.
3. **Martignoni, M. E. and Iwai, P. J.**, A catalog of viral diseases of insects, mites, and ticks, in *Gen. Tech. Rep. PNW-195*, 4th ed., U.S. Department of Agriculture, Forest Service, Pacific Northwest Research Station, Portland, OR, 1986.
4. **Liang, D. and Hu, Y.**, *Atlas of Insect Viruses in China*, Sci. Technol. Publ., Hunan, 1986.
5. **Payne, C. C. and Rivers, C. F.**, A provisional classification of cytoplasmic polyhedrosis viruses based on the sizes of the RNA genome segments, *J. Gen. Virol.*, 33, 71, 1976.
6. **Payne, C. C. and Mertens, P. P. C.**, Cytoplasmic polyhedrosis viruses, in *Reoviridae*, Joklik, W. K., Ed., Plenum Press, New York, 1983, chap. 9.
7. **Gallinski, M. S., Stanik, V. H., Rohrmann, G. F., and Beaudreau, G. S.**, Comparison of sequence diversity in several cytoplasmic polyhedrosis viruses, *Virology*, 130, 372, 1983.

8. **Mertens, P. P. C., Crook, N., Rubinstein, R., Pedley, S., and Payne, C.**, Cytoplasmic polyhedrosis virus classification by electropherotype; validation by serological analyses and agarose gel electrophoresis, *J. Gen. Virol.*, 70, 173, 1989.

9. **Paliwal, Y. C.**, Occurrence and localization of spherical viruslike particles in tissues of apparently healthy tobacco thrips, *Frankliniella fusca*, a vector of tomato spotted wilt virus, *J. Invertebr. Pathol.*, 33, 307, 1979.

10. **Boccardo, G., Hatta, T., Francki, R. I. B., and Grivell, C. J.**, Purification and some properties of reovirus-like particles from leafhoppers and their possible involvement in Wallaby ear diseases of maize, *Virology*, 100, 300, 1980.

11. **Ofori, F. A. and Francki, R. I. B.**, Transmission of leafhopper A virus, vertically through eggs and horizontally through maize in which it does not multiply, *Virology*, 144, 152, 1985.

12. **Eley, S. M., Gardner, R., Molyneax, D. H., and Moore, N. F.**, A reovirus from the bedbug, *Cimex lectularis*, *J. Gen. Virol.*, 68, 195, 1987.

13. **Herold, F. and Munz, K.**, Virus particles in apparently healthy *Peregrinus maidis*, *J. Virol.*, 1, 1028, 1967.

14. **Kim, K. S. and Scott, H. A.**, Two types of virus-like particles in the bean leaf beetle, *Cerotoma trifurcata*, *J. Invertebr. Pathol.*, 31, 77, 1978.

15. **Kim, K. S.**, Cytopathology of spotted cucumber beetle hemocytes containing virus-like particles, *J. Invertebr. Pathol.*, 36, 292, 1980.

16. **Kitajima, E. W., Kim, K. S., Scott, H. A., and Gergerich, R. C.**, Reovirus-like particles and their vertical transmission in the Mexican bean beetle, *Epilachna varivestis* (Coleoptera:Coccinellidae), *J. Invertebr. Pathol.*, 46, 83, 1985.

17. **Binnington, K. C., Lockie, E., Hines, E., and van Gerwen, A. C. M.**, Fine structure and distribution of three types of virus-like particles in the sheep blowfly, *Lucilia cuprina*, and associated cytopathic effects, *J. Invertebr. Pathol.*, 49, 175, 1987.

18. **Plus, N., Gissman, L., Veyrunes, J. C., Pfister, H., and Gateff, E.**, Reoviruses of *Drosophila* and *Ceratitis* populations and of *Drosophila* cell lines: a possible new genus of the *Reoviridae* family, *Ann. Virol. (Inst. Pasteur)*, 132E, 261, 1981.

19. **Louis, C., Lopez-Ferber, M., Comendador, M., Plus, N., Kuhl, G., and Baker, S.**, *Drosophila* S virus, a hereditary reolike-virus, probale agent of the morphological S character in *Drosophila simulans*, *J. Virol.*, 62, 1266, 1988.

20. **Moussa, A. Y.**, A new virus disease in the housefly, *Musca domestica* (Diptera), *J. Invertebr. Pathol.*, 31, 204, 1978.

21. **Moussa, A. Y.**, The housefly virus contains double-stranded RNA, *Virology*, 106, 173, 1980.

22. **Moussa, A. Y.**, Studies of the housefly virus: structure and disruption during purification procedures, *Micron*, 12, 131, 1981.

23. **Plus, N. and Croizier, G.**, Further studies on the genome of *Ceratitis capitata* I virus *(Reoviridae)*, *Ann. Virol. (Inst. Pasteur)*, 133E, 489, 1982.

24. **Anagnou-Veroniki, M., Bergoin, M., and Veyrunes, J. C.**, Research on the viral infections of *Dacus oleae* (Gmel.), *CEE Programme on Integrated and Biological Control*, Final Report 1979—1983, 201, 1984.

25. **Moussa, A. Y.**, A new cytoplasmic inclusion virus from Diptera in the Queensland fruitfly, *Dacus tryoni* (Frogg) (Diptera:Tephritidae), *J. Invertebr. Pathol.*, 32, 77, 1978.

26. **Haars, R., Zentgraf, H., Gateff, E., and Bautz, F. A.**, Evidence for endogenous reovirus-like particles in a tissue culture cell line from *Drosophila melanogaster*, *Virology*, 101, 124, 1980.

27. **Gateff, E., Gissmann, L., Shrestha, R., Plus, N., Pfister, H., Schröder, J., and Zur Hausen, H.**, Characterization of two tumorous blood cell lines of *Drosophila melanogaster* and the viruses they contain, in *Invertebrate Systems in vitro*, Kurstak, E., Maramorosch, M., and Dübendorfer, A., Eds., Elsevier/North-Holland, Amsterdam, 1980, 517.

28. **Alatortsev, V. E., Ananiev, E. V., Gushchina, E. A., Grigoriev, V. B., and Guchchin, B. V.**, A virus of the *Reoviridae* in established cell lines of *Drosophila melanogaster*, *J. Gen. Virol.*, 54, 23, 1981.

29. **Hsu, J. T. and Sanders, M. M.**, Characterization of a segmented double-stranded RNA virus in *Drosophila* K_c cells, *Nucleic Acids Res.*, 11, 3665, 1983.

30. **Teninges, D., Ohanessian, A., Richard-Molard, C., and Contamine, D.**, Contamination and persistent infection of *Drosophila* cell lines by reovirus type particles, *In Vitro*, 15, 425, 1979.

31. **Hukuhara, T. and Hashimoto, Y.**, Serological studies of the cytoplasmic- and nuclear-polyhedrosis viruses of the silkworm, *Bombyx mori*, *J. Invertebr. Pathol.*, 8, 234, 1966.

32. **Allaway, G. P.**, Virus particle packaging in baculovirus and cytoplasmic polyhedrosis virus inclusion bodies, *J. Invertebr. Pathol.*, 42, 357, 1983.

33. **Hukuhara, T.**, Pathology associated with cytoplasmic polyhedrosis viruses, in *Viral Insecticides for Biological Control*, Maramorosch, K. and Sherman, K. E., Eds., Academic Press, New York, 1985, 121.

34. **Adams, J. R. and Wilcox, T. A.,** Scanning electron microscopical comparisons of insect virus occlusion bodies prepared by several techniques, *J. Invertebr. Pathol.,* 40, 12, 1982.

35. **Hukuhara, T. and Midorikawa, M.,** Pathogenesis of cytoplasmic polyhedrosis in the silkworm, in *Double-Stranded RNA Viruses,* Compans, R. W. and Bishop, D. H. L., Eds., Elsevier, New York, 1983, 405.

36. **Bergold, G. H. and Suter, J.,** On the structure of cytoplasmic polyhedra of some Lepidoptera, *J. Insect Pathol.,* 1, 1, 1959.

37. **Bergold, G. H.,** The molecular structure of some insect virus inclusion bodies, *J. Ultrastr. Res.,* 8, 360, 1963.

38. **Fujiwara, T., Yukibuchi, E., Tanaka, Y., Yamamoto, Y., Tomita, K., and Hukuhara, T.,** X-ray diffraction studies of polyhedral inclusion bodies of nuclear- and cytoplasmic·polyhedrosis viruses, *Appl. Entomol. Zool.,* 19, 402, 1984.

39. **Engström, A. and Kilkson, R.,** Molecular organization in the polyhedra of *Porthetria dispar* nuclear-polyhedrosis, *Exp. Cell Res.,* 53, 305, 1968.

40. **Rubinstein, R.,** Characterization of the proteins and serological relationships of cytoplasmic polyhedrosis virus of *Heliothis armigera, J. Invertebr. Pathol.,* 42, 292, 1983.

41. **McCrae, M. A. and Mertens, P. P. C.,** *In vitro* translation studies on and RNA coding assignments for cytoplasmic polyhedrosis viruses, in *Double-Stranded RNA Viruses,* Compans, R. W. and Bishop, D. H. L., Eds., Elsevier, New York, 1983, 35.

42. **Pullin, J. S. K. and Moore, N. F.,** Gene assignments of a cytoplasmic polyhedrosis virus (type 2) from *Nymphalis io, Microbiology,* 18, 131, 1985.

43. **Arella, M., Lavallée, C., Belloncik, S., and Furuichi, Y.,** Molecular cloning and characterization of cytoplasmic polyhedrosis virus polyherdin and a viable deletion mutant gene, *J. Virol.,* 62, 211, 1988.

44. **Fossiez, F., Belloncik, S., and Arella, M.,** Nucleotide sequence of the polyhedrin gene of *Euxoa scandens* cytoplasmic polyhedrosis virus (EsCPV), *Virology,* 169, 462, 1989.

45. **Mori, H., Minobe, Y., Sasaki, T., and Kawase, S.,** Nucleotide sequence of the polyhedrin gene of *Bombyx mori* cytoplasmic polyhedrosis virus A strain with nuclear localization of polyhedra, *J. Gen. Virol.,* 70, 1885, 1989.

46. **Hosaka, Y. and Aizawa, K.,** The fine structure of the cytoplasmic polyhedrosis virus of the silkworm *(Bombyx mori* L.), *J. Insect Pathol.,* 6, 53, 1964.

47. **Asai, J., Kawamoto, F., and Kawase, S.,** On the structure of the cytoplasmic-polyhedrosis virus of the silkworm, *Bombyx mori, J. Invertebr. Pathol.,* 19, 279, 1972.

48. **Kawamoto, F., Kawase, S., and Asai, J.,** Fine structure of the cytoplasmic-polyhedrosis virus of the silkworm, *Bombyx mori, Jpn. J. Appl. Entomol. Zool.,* 19, 133, 1975.

49. **Hatta, T. and Francki, R. I. B.,** Similarity in the structure of cytoplasmic polyhedrosis virus, leafhopper A virus and Fuji disease virus particles, *Intervirology,* 18, 203, 1982.

50. **Miura, K., Fujii-Kawata, I., Iwata, H., and Kawase, S.,** Electron-microscopic observation of a cytoplasmic-polyhedrosis virus from the silkworm, *J. Invertebr. Pathol.,* 14, 262, 1969.

51. **Lewandowski, L. J. and Traynor, B. L.,** Comparison of the structure and polypeptide composition of three double-stranded ribonucleic acid-containing viruses (Diplornaviruses): cytoplasmic polyhedrosis virus, wound tumor virus, and reovirus, *J. Virol.,* 10, 1053, 1972.

52. **Hayashi, Y., Kawarabata, T., and Bird, F. T.,** Isolation of a cytoplasmic-polyhedrosis virus of the silkworm, *Bombyx mori, J. Invertebr. Pathol.,* 16, 378, 1970.

53. **Ikegami, S., Sasaki, S., Higaki, T., Itoh, N., and Shinagawa, N.,** Coupled transcription-translation of silkworm cytoplasmic polyhedrosis virus injected into oocytes of the frog, *Xenopus laevis, J. Biochem.,* 103, 19, 1988.

54. **Galinski, M. S., Kingan, T., Rohrmann, G. F., Martignoni, M. E., and Beaudreau, G. S.,** Characterization of a cytoplasmic polyhedrosis virus infecting *Manduca sexta, Intervirology,* 21, 167, 1984.

55. **Longworth, J. F. and Spilling, C. R.,** A cytoplasmic polyhedrosis of the larch sawfly, *Anoplonyx destructor, J. Invertebr. Pathol.,* 15, 276, 1970.

56. **Francki, R. I. B. and Boccardo, G.,** The plant Reoviridae, in *The Reoviridae,* Joklik, W. K., Ed., Plenum Press, New York, 1983, chap. 10.

57. **Plus, N.,** The reoviruses of Trypetidae, Drosophilidae and Muscidae. A review, in *Fruit Flies of Economic Importance,* Cavalloro, R., Ed., Commission European Communities, Roma, 1987.

58. **Galinski, M. S., Chow, K. C., Rohrman, G. F., Pearson, G. D., and Beaudreau, G. S.,** Size determination of *Orgyia pseudotsugata* cytoplasmic polhedrosis virus genomic RNA, *Virology,* 123, 328, 1982.

59. **Fujii-Kawata, I., Miura, K., and Fuke, M.,** Segments of genome of viruses containing double-stranded ribonucleic acid, *J. Mol. Biol.,* 51, 249, 1970.

60. **Miura, K., Fujii, I., Sakaki, T., Fuke, M., and Kawase, S.,** Double-stranded ribonucleic acid from cytoplasmic polyhedrosis virus of the silkworm, *J. Virol.,* 2, 1211, 1968.

61. **Furuichi, Y. and Miura, K.,** Identity of the 3'-terminal sequences in ten genome segments of silkworm cytoplasmic polyhedrosis virus, *Virology,* 55, 418, 1973.

62. **Furuichi, Y. and Miura, K.,** A blocked structure at the 5′-terminus of mRNA of cytoplasmic polyhedrosis virus, *Nature (London)*, 253, 374, 1975.

63. **Shimotohno, K. and Miura, K.,** Transcription of double-stranded RNA in cytoplasmic polyhedrosis virus *in vitro*, *Virology*, 53, 283, 1973.

64. **Smith, R. E. and Furuich, Y.,** Gene mapping of cytoplasmic polyhedrosis virus of silkworm by the full-length mRNA prepared under optimized conditions of transcription *in vitro*, *Virology*, 103, 279, 1980.

65. **Furuich, Y.,** Allosteric stimulatory effect of S-adenosylmethionine on the RNA polymerase in cytoplasmic polyhedrosis virus, *J. Biol. Chem.*, 256, 483, 1981.

66. **Mertens, P. P. C. and Payne, C. C.,** Comparison of the virus-associated polymerases of cytoplasmic polyhedrosis virus type 1 and 2, *Intervirology*, 22, 104, 1984.

67. **Wertheimer, A. M., Chen, S.-Y., Borchardt, R. T., and Furuich, Y.,** S-adenosyl methionine and its analogs: structural features correlated with synthesis and methylation of mRNAs of cytoplasmic polyhedrosis virus, *J. Biol. Chem.*, 255, 5924, 1980.

68. **Nishimura, A. and Hosaka, Y.,** Electron microscopic study of RNA of cytoplasmic polyhedrosis virus of the silkworm, *Virology*, 38, 550, 1969.

69. **Kavenoff, R., Klotz, L. C., and Zimm, B. H.,** On the nature of chromosome-sized DNA molecule, *Cold Spring Harbor Symp. Quant. Biol.*, 38, 1973, 1.

70. **Yazaki, K. and Miura, K.,** Relation of the structure of cytoplasmic polyhedrosis virus and the synthesis of its messenger RNA, *Virology*, 105, 467, 1980.

71. **Yazaki, K., Mizuno, A., Sano, T., Fujii, H., and Miura, K.,** A new method for extracting circular and supercoiled genome segments for cytoplasmic polyhedrosis virus, *J. Virol. Methods*, 14, 275, 1986.

72. **Dai, R., Wu, A., Shen, X., Qian, L., and Sun, Y.,** Isolation of genome-enzymic complex from cytoplasmic polyhedrosis virus of silkworm *Bombyx mori*, *Scientia Sinica Ser. B*, 25, 29, 1982.

73. **Dai, R., Aizhen, W., and Yukun, S.,** The protein subunits of the double-stranded RNA dependent RNA polymerase and methyltransferase of the cytoplasmic polyhedrosis virus of silkworm, *Bombyx mori*, *Scientia Sinica Ser. B*, 29, 1268, 1986.

74. **Wu, A. and Sun, Y.,** Isolation and reconstitution of the RNA replicase of the cytoplasmic polyhedrosis virus of silkworm, *Bombyx mori*, *Theor. Appl. Genet.*, 72, 662, 1986.

75. **Anagnou-Veroniki, M., Bergoin, M., and Veyrunes, J. C.,** personal communication.

76. **Arnott, H. J., Smith, K. M., and Fullilove, S. L.,** Ultrastructure of a cytoplasmic polyhedrosis virus affecting the monarch butterfly. *Danaus plexippus*. I. Development of virus and normal polyhedra in the larva, *J. Ultrast. Res.*, 24, 479, 1968.

77. **Anthony, D. W., Hazard, E. I., and Crosby, S. W.,** A virus disease in *Anopheles quadrimaculatus*. *J. Invertebr. Pathol.*, 22, 1, 1973.

78. **Hukuhara, T. and Yamaguchi, K.,** Ultrastructural investigation on a strain of a cytoplasmic polyhedrosis with nuclear inclusions, *J. Invertebr. Pathol.*, 22, 6, 1973.

79. **Mori, H., Sasaki, T., Minobe, Y., Miyajima, S., and Kawase, S.,** Difference of proteins from inclusion bodies formed in the nucleus and cytoplasm of the cytoplasmic polyhedrosis virus-infected midgut in the silkworm, *Bombyx mori, J. Invertebr. Pathol.*, 50, 26, 1987.

80. **Arella, M., Belloncik, S., and Devauchelle, G.,** Protein synthesis in a *Lymantria* cell line infected by cytoplasmic polyhedrosis virus, *J. Virol.*, 52, 1024, 1984.

81. **Belloncik, S. and Bellemare, N.,** Polyèdres du CPV d'*Euxoa scandens* (Lep.:Noctuidae) produits *in vivo* et sur cellules cultivées *in vitro*. Études comparatives, *Entomophaga*, 25, 199, 1980.

82. **Granados, R. R., McCarthy, J. W., and Naughton, M.,** Replication of a cytoplasmic polyhedrosis virus in an established cell line of *Trichoplusia ni*, *Virology*, 59, 584, 1974.

83. **Quiot, J. M. and Belloncik, S.,** Caractérisation d'une polyédrose cytoplasmique chez le lépidoptère *Euxoa scandens*, Riley (Noctuidae, Agrotinae), Etudes *in vivo* et *in vitro*, *Arch. Virol.*, 55, 145, 1977.

84. **Longworth, J. P.,** The replication of a cytoplasmic polyhedrosis virus from *Chrysodeixis eriosoma* (Lepidoptera:Noctuidae) in *Spodoptera frugiperda* cells, *J. Invertebr. Pathol.*, 37, 54, 1980.

85. **Belloncik, S.,** Cytoplasmic polyhedrosis viruses-Reoviridae, *Adv. Virus Res.*, 37, 173, 1989.

86. **Federici, B. A. and Hazard, E. I.,** Iridovirus and cytoplasmic polyhedrosis virus in the freshwater daphnid *Simocephalus expinosus*, *Nature (London)*, 254, 327, 1975.

87. **Morel, G.,** Un virus cytoplasmique chez le scorpion *Buthus occitanus* Amoreux, *C. R. Acad. Sci. Paris*, 280, 2893, 1975.

88. **Meyers, T. R.,** A reo-like virus isolated from juvenile american oysters *(Crassostrea virginica)*, *J. Gen. Virol.*, 43, 203, 1979.

89. **Hill, B. J.,** Properties of a virus isolated from the bivalve mollusc *Tellina tenuis* (Da Costa), in *Wildlife Diseases*, Page, L. A., Ed., Plenum Press, New York, 1976, 445.

90. **Bovo, G., Ceshia, G., Giorgetti, G., and Vanelli, M.,** Isolation of an IPN-like virus from adult Kuruma shrimp *Penaeus japonicus)*, *Bull. Eur. Assoc. Fish. Pathol.*, 4, 121, 1984.

91. **Meyers, T. R.,** Experimental pathogenicity of reovirus 13p2 for juvenile american oysters *Crassostrea virginica* (Gmelin) and bluegill fingerlings *Lepomis macrochirus* (Rafinesque), *J. Fish Dis.*, 3, 187, 1980.

92. **Mokhtar-Maamouri, F., Lambert, A., Maillard, C., and Vago, C.,** Infection virale chez un Plathelminthe parasite, *C. R. Acad. Sci. Paris,* 283, 1249, 1976.
93. **Vago, C.,** A virus disease in crustacea, *Nature (London),* 209, 1290, 1966.
94. **Bonami, J. R.,** Recherches sur la paralysie virale du crustacé décapode *Macropipus depurator* L., *Rev. Trav. Inst. Pêches Marit.,* 37, 387, 1973.
95. **Johnson, P. T. and Bodammer, J. E.,** A disease of the blue crab, *Callinectes sapidus,* of possible viral etiology, *J. Invertebr. Pathol.,* 26, 141, 1975.
96. **Johnson, P. T.,** A viral disease of the blue crab, *Callinectes sapidus:* histopathology and differential diagnosis, *J. Invertebr. Pathol.,* 29, 201, 1977.
97. **Bonami, J. R., Comps, M., and Veyrunes, J. C.,** Etude histopathologique et ultrastructurale de la paralysie virale du crabe *Macropipus depurator* L., *Rev. Trav. Inst. Pêches Marit.,* 40, 139, 1976.
98. **Bonami, J. R.,** Recherches sur les infections virales des crustacés marins: étude des maladies à étiologie simple et complexe chez les décapodes des côtes françaises, Thèse Doct. Etat, Univ. Sci. Tech. Languedoc, Montpellier, France, 1980.
99. **Bonami, J. R. and Hill, B. J.,** unpublished data.
100. **Mari, J. and Bonami, J. R.,** Les infections virales du crabe *Carcinus mediterraneus* Czerniavski, 1884, in *Pathology in Marine Aquaculture,* Vivares, C. P., Bonami, J. R., and Jaspers, E., Eds., European Aquaculture Society, Special Publication No. 9, Bredene, Belgium, 1986, 283.
101. **Mari, J.,** Recherches sur les maladies virales du crustacé décapode marin *Carcinus mediterraneus* Czerniavski 1844, Thèse Doct. Sci. Biol., Univ. Sci. Tech. Languedoc, Montpellier, France, 1987.
102. **Bonami, J. R.,** Les maladies virales des crustacés et des mollusques, *Océnis,* 3, 154, 1977.
103. **Mari, J. and Bonami, J. R.,** A reolike virus of the mediterranean shore crab *Carcinus mediterraneus, Dis. Aquat. Org.,* 3, 107, 1987.
104. **Tsing, A. and Bonami, J. R.,** A new virus disease in the shrimp *Penaeus japonicus,* in *Pathology in Marine Aquaculture,* Vivares, C. P., Bonami, J. R., and Jaspers, E., Eds., European Aquaculture Society, Special publication No. 9, Bredene, Belgium, 1986, 295.
105. **Tsing, A. and Bonami, J. R.,** A new viral disease of the tiger shrimp, *Penaeus japonicus* Bate, *J. Fish Dis.,* 10, 139, 1987.
106. **Lightner, D. V., Redman, R. M., Bell, T. A., and Brock, J. A.,** An idiopathic proliferative disease syndrome of the midgut and ventral nerve in the Kuruma prawn *Penaeus japonicus* Bate, cultured in Hawaii, *J. Fish Dis.,* 7, 183, 1984.
107. **Tsing, A., Lightner, D. V., Bonami, J. R., and Redman, R. M.,** Is "gut and nervous syndrome" (GNS) of viral in origin, in the tiger shrimp *Penaeus japonicus* Bate? *2nd Int. Conf. EAFP,* Montpellier, France, 1985, 91.
108. **Devauchelle, G. and Vago, C.,** Particules d'allure virale dans les cellules de l'estomac de la seiche, *Sepia officinalis* L., (Mollusques, Céphalopodes), *C. R. Acad. Sci. Paris,* 272, 894, 1971.
109. **Meyers, T. R. and Hirai, K.,** Morphology of a reo-like virus isolated from juvenile american oysters *(Crassostrea virginica), J. Gen. Virol.,* 46, 249, 1980.
110. **Mari, J. and Bonami, J. R.,** The W2 virus infection of the crustacea *Carcinus mediterraneus:* a reovirus disease, *J. Gen. Virol.,* 69, 561, 1988.
111. **Winton, J. R., Lannan, C. N., Fryer, J. L., Hedrick, R. P., Meyers, T. R., Plumb, J. A., and Yamamoto, T.,** Morphological and biochemical properties of four members of a novel group of reoviruses isolated from aquatic animals, *J. Gen. Virol.,* 68, 353, 1987.

Chapter 16

BIRNAVIRIDAE

Jean-Robert Bonami and Jean R. Adams

TABLE OF CONTENTS

I. BIRNAVIRUSES OF INSECTS

A. INTRODUCTION

Birnaviridae is a new family recently established by the International Committee on Taxonomy of Viruses (ICTV).[1] It includes bisegmented dsRNA nonenveloped icosahedral viruses of approximatively 60 nm in diameter. The type species is Infectious pancreatic necrosis virus (IPNV), strain VR299.[1]

1. Members of the Group

The members of the group infecting insects includes *Drosophila* X virus (DXV) and probably a *Culicoides* virus. Recent pertinent reviews on viruses infecting *Drosophila* have been presented by Brun and Plus[2] and Plus.[3] DXV has been isolated from a natural population of *Culicoides* sp. but not from natural populations of *Drosophila*.[4,5]

2. Gross Pathology

DXV was first observed in studies involving serial passages of σ-virus.[5] Control flies also showed sensitivity to CO_2. Negative stains of ground tissues revealed the presence of icosahedral particles rather than the bullet-shaped particles characteristic of σ-virus.[5] DXV was also found contaminating *Drosophila* and *Ceratitis capitata* established cell lines.[6,7] Plus[6] proposed tests to check for viruses in supernates and sera by injecting them into flies followed by checking for symptoms of disease. If the flies appeared healthy, three successive passages were performed with a check of the final extract by electron microscopy. Plus found that 32 of the 42 *Drosophila* cell lines examined in the period of 1974 to 1978 were contaminated with four endogenous viruses, i.e., *Drosophila* A, P, and C picornaviruses and DXV. The DXV was found only in *Drosophila* cell cultures and was suspected to have come from embryonic bovine sera.[4] Further studies on the origin of the endogenous viruses supported earlier conclusions that the possible sources were bovine sera and embryos or larvae.[6,8] The contamination was eliminated by the use of virus-free *Drosophila* stocks to start the cell lines and serum screened on *Drosophila* flies for cell line maintenance.[3]

B. VIRUS STRUCTURE AND COMPOSITION
1. Particle Structure

Drosophila X virus was first isolated and described by Teninges et al.[5] and Teninges.[9] Negatively stained icosahedral virus particles (see Figure 1a) measure 72 ± 1 nm (point to point) or 62 nm (edge to edge). DXV bands at 1.345 g/ml in CsCl and has an average molecular weight of 81.2×10^6 Da. Its hydrodynamic diameter is 70.4 nm.[5,10] Other biochemical properties are described by Dobos et al.[10] Particles observed in thin sections of infected tissues contain a dense core of 32 nm surrounded by a single outer shell.[5] No biochemical information is available on *Culicoides* virus.

2. Genome Structure

The *Drosophila* X virus genome contains two segments of dsRNA which are composed of different nucleotide sequences. Segment A has a molecular weight of 2.3×10^6 Da while segment B has a molecular weight of 2.2×10^6 Da.[11] The virus capsid is composed of 4 to 6 polypeptides.[5,9-11] Genome segment A encodes polypeptides with molecular weights of 67 kDa, 34 kDa, and 27 kDa while segment B encodes the 10-kDa polypeptide.[12] Proteolytic processing of 67 kDa generates a 49-kDa polypeptide which is further processed to produce a major capsid polypeptide (VP45).[12] Absolute electron microscope length measurements and molecular weight determinations indicated that the dsRNA was 0.97 μm giving a molecular weight of 2.2×10^6 Da.[13] An RNA polymerase was extracted from DXV-infected flies and characterized by Bernard.[14]

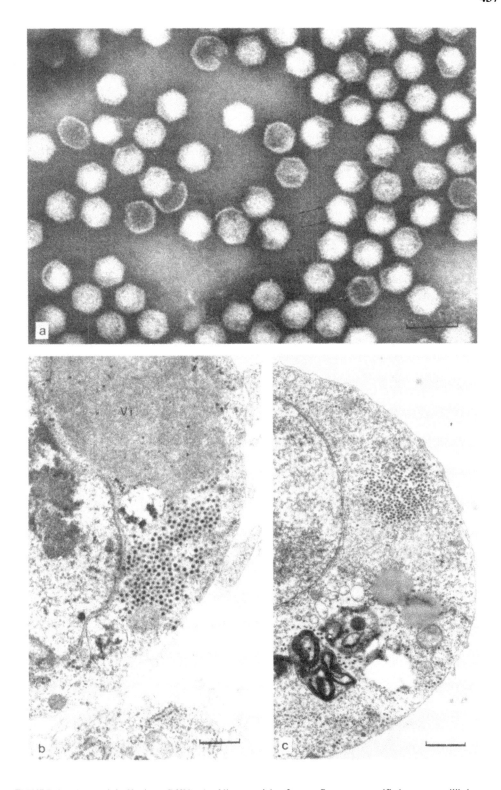

FIGURE 1. *Drosophila* X virus (DXV). (a) Virus particles from a fly extract purified on an equilibrium sucrose gradient. Arrows indicate the four morphological units on an edge of the particle. (PTA; bar = 100 nm.) (b) Ovary follicle cell, viroplasm (Vi) appears as masses of granular material. (Bar = 0.5 μm). (c) Cytoplasmic virus particles of an *in vitro* cultured cell, 8 hpi. (Bar = 1 μm.) (Courtesy of D. Teninges.)

C. CYTOPATHOLOGY

Drosophila X virus has been isolated from natural populations of *Culicoides* sp. but no reports of histopathologic examinations have been presented.[4] DXV has not been observed in laboratory or natural populations of *Drosophila* and therefore may not be a *Drosophila* virus. However, it may be transmitted horizontally.[2,8] *Drosophila* flies inoculated by injection show CO_2 sensitivity, and anoxia, which is a sensitivity to oxygen starvation by exposure to nitrogen.[5] Viral replication occurs in the cytoplasm of gut cells, tracheal matrix, muscle, and ovarian follicles where crystalline arrays of virus particles are formed (see Figure 1b).[5] Virus particles have also been observed in the brain. Infected cells contain diffuse masses of material which is probably viroplasm.[5]

Cell cultures of *Drosophila* and *Ceratitis capitata* are susceptible to DXV (see Figure 1c).[6,7] DXV was isolated from two or three cell lines of *C. capitata* by injection into virus-free *Drosophila* of culture medium from *C. capitata* cell lines. DXV from *C. capitata* was pathogenic in its own species but did not multiply in *Drosophila* sp.[7] A symptomless nonpathogenic mutant of DXV (DXV$_2$) has been isolated from haploid cell lines.[8] It does not induce the typical anoxia symptoms in *Drosophila* but is serologically related to DXV and is biochemically identical as revealed by comparisons of viral RNAs in electrophoresis studies.[8]

II. BIRNAVIRUSES OF INVERTEBRATES OTHER THAN INSECTS

A. INTRODUCTION

The major difference between birnaviruses of insects and birnaviruses of invertebrates other than insects, consists in the method of virus detection. None of them has been isolated directly from the host or observed in ultrathin section; all, without exception, have been evidenced using vertebrate cell lines (fishes)[15-17] as enhancing means, rather than invertebrate cell lines as for birnaviruses of insects.

1. Members of the Group

All members of the group have been evidenced in marine invertebrates (mollusks and crustaceans) and the model was first isolated from a wild population of the bivalve mollusk *Tellina tenuis* with a disease of the digestive gland[15] and called *Tellina* virus (TV). Later on, various strains of this virus have been evidenced from natural populations of oysters (*Crassostrea gigas* and *Ostrea edulis*),[18] hard clams (*Mercenaria mercenaria*),[18] limpets (*Patella vulgata*),[18] winkles (*Littorina littorea*),[18] crabs (*Carcinus maenas*,[18] and *Macropipus depurator*[16,16a]) and shrimps (*Penaeus japonicus*).[17]

2. Gross Pathology

As emphasized before, Birnaviridae of mollusks and crustaceans, have been evidenced only by passage on fish cell lines, particularly using the BF2 cell line which is the most susceptible.[18] Although some of them have been isolated from diseased animals (*T. tenuis* and *P. japonicus*),[15,17] no relationship between the virus and the disease has been demonstrated. As the other Birnaviridae have been found in apparently healthy animals, it seems that they are not pathogenic for their host, or the hosts can be interpreted as vectors and not as permanent hosts.

B. VIRUS STRUCTURE AND COMPOSITION

TV was the first to be purified and studied. At the level of particle size and structure, compared with oyster virus (OV) and crab virus (CV), no difference was evidenced.[10,16] The virus is an isometric hexagonal particle (see Figure 2a), and no distinct outer capsid

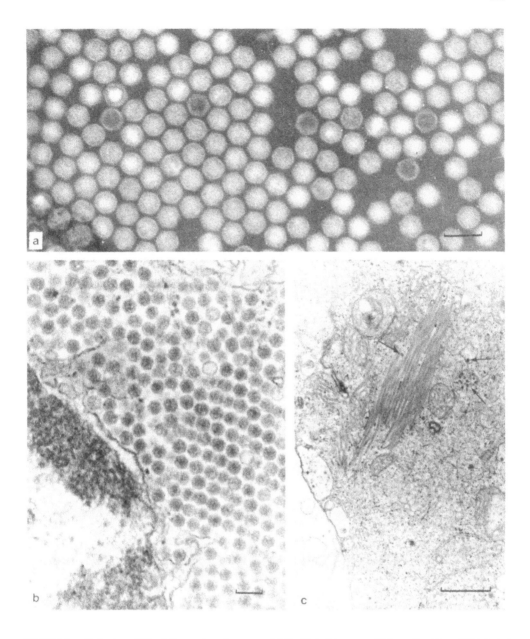

FIGURE 2.   Crab virus (CV) of *Macropipus depurator*. (a) Purified virions on a ClCs gradient from CV-infected BF2 cells. Morphological subunits of the capsid are clearly evidenced both in empty or full particles. (PTA; bar = 100 nm.) (b) Cytoplasmic paracrystalline arrangement of CV in BF2-infected cell. (Bar = 100 nm.) (c) Tubular structures associated with CV development in BF2-infected cells (virions are arrowed). (Bar = 1 μm.)

layer was observed. Its average diameter is 58 to 60 nm (side to side). Hydrodynamic diameters of OV and TV have been determined as 66.0 and 65.0 nm, respectively.[10] The size of CV is 72 nm (point to point) in average, and it bands at 1.34 g/ml in CsCl,[16] which is a little higher density value than that reported for TV (1.32 g/ml).[19]

At the level of polypeptide composition of virions, some differences are noted between the different viruses and also the values given by different authors for the same virus, as seen below:

Virus	OV[10]	TV[10]	TV[19]	CV[16]
Polypeptides	(MV $\times$ 10³)			
VP1	90	90	110	82
VP2	60	57	67	72
VP3	29		40	40
VP4	27	27		

In the case of CV isolated from *M. depurator*, some smaller particles (about 20 to 30 nm in diameter) have been observed in negatively stained preparations, associated with tubular empty structures, variable in diameter (40 to 70 nm), formed with the association of subunits in a helicoidal arrangement.[16] The presence of such tubules was also reported in the case of other Birnaviruses such as infectious bursal disease virus (IBDV) of chickens,[20,21] DXV[5] and (IPNV) of salmonid fishes.[22] Small particles have been reported also in IBDV[20,21] and in TV preparations.[19]

According to these authors, such formations (i.e., small particles and tubules) could correspond to degradation of virions and/or an abnormal self-assembling of viral proteins in excess.

The TV, OV, and CV genomes contain two segments of dsRNA.[10,16,23] The molecular weight of RNA segments of the sole CV has never been determined. In the case of TV and OV, the molecular weights of RNA segments are 2.4 to 2.2 and 2.6 to 2.2 $\times$ 10⁶ Da, respectively.[10]

C. CYTOPATHOLOGY

Cytopathological studies to determine morphogenesis and viral structure have been conducted in fish cell lines since these viruses have never been observed in naturally infected host tissues.

In the case of TV, the optimum growth temperature in BF cells is about 15°C; the virus produces plaques of 1.5 mm in diameter in the 3 days postinfection (dpi) under 0.5% agarose overlay.[19] Cytopathic effect (CPE) includes formation of small necrotic plaques in the cell monolayer, followed by an increasing destruction of cells. The culture is finally destroyed after 4 to 5 dpi.[16]

Virions develop in the cytoplasm, sometimes forming paracrystalline arrays (see Figure 2b); some particles, scattered in cytoplasm are found associated with tubule formations (see Figure 2c), similar in size to those observed in negatively stained virus extracts.[16]

Experimental infections of the hosts have been done;[15,16,24,25] no significant mortality was observed. Only in Japanese oyster (*C. gigas*) inoculated with TV, some paracrystalline arrays of virions (identical in size and morphology with TV) have been found in the cytoplasm of cells of the digestive gland,[15] although no death has occurred. In the other cases, only disorganization of tissues have been reported, without evidence of virions.[16,24,25]

At the level of serological properties, the different isolates of TV, OV, and CV have been previously classified into three serogroups on the basis of studies of their cross-reaction in neutralization tests with one of the reference serotypes of IPN virus (Sp, Ab, VR299).[18] Recent studies now indicate that they fall only into two serogroups, and no relationship exists between TV and IPNV, according to neutralization tests.[26]

ADDENDUM

Since the preparation of this manuscript, a proposed new member of Birnaviridae, pathogenic to an invertebrate other than insect, was recently found associated with reared-rotifer (*Brachionus plicatilis*) mortalities,[27,28] and was subsequently characterized.[29]

REFERENCES

1. **Francki, R.,** personal communication.
2. **Brun, G. and Plus, N.,** The viruses of *Drosophila,* in *The Biology and Genetics of Drosophila,* Ashburner, M. and Wright, T. R. F., Eds., Academic Press, New York, 1980, 625.
3. **Plus, N.,** Further studies on the origin of the endogenous viruses of *Drosophila melanogaster* cell lines, in *Invertebrate Systems in vitro,* Kurstak, E., Maramorosch, K., and Dubendorfer, A., Eds., Elsevier/North-Holland, Amsterdam, 1980, 435.
4. **Mialhe, E., Croizier, G., Veyrunes, J. C., Quiot, J. M., and Rieb, J. P.,** Etude d'un virus isolé d'une population naturelle de *Culicoides sp.* (Diptera:Ceratopogonidae), *Annu. Virol., (Inst. Pasteur),* 134, 73, 1983.
5. **Teninges, D., Ohanessian, A., Richard-Molard, C., and Contamine, D.,** Isolation and biological properties of *Drosophila* X virus, *J. Gen. Virol.,* 42, 241, 1979.
6. **Plus, N.,** Endogenous viruses of *Drosophila melanogaster* cell lines: their frequency, identification and origin, *In Vitro,* 14, 1015, 1978.
7. **Plus, N., Veyrunes, J. C., and Cavalloro, R.,** Endogenous viruses of *Ceratitis capitata* Wied. "J.R.C. Ispra strain" and of *C. capitata* permanent cell lines, *Annu. Virol., (Inst. Pasteur),* 132, 91, 1981.
8. **Plus, N., Veyrunes, J. C., Croizier, L., and Debec, A.,** A symptomless *"Drosophila* X" virus from haploid *Drosophila* cell lines and from foetal calf serum: a further indication of the exogenous origin of this virus, *Annu. Virol., (Inst. Pasteur),* 134, 293, 1983.
9. **Teninges, D.,** Protein and RNA composition of the structural components of *Drosophila* X virus, *J. Gen. Virol.,* 45, 641, 1979.
10. **Dobos, P., Hill, B. J., Hallet, R., Kells, D. T. C., Becht, H., and Teninges, D.,** Biophysical and biochemical characterization of five animal viruses with bisegmented double-stranded RNA genomes, *J. Virol.,* 32, 593, 1979.
11. **Nagy, E. and Dobos, P.,** Synthesis of *Drosophila* X virus proteins in cultured *Drosophila* cells, *Virology,* 134, 358, 1984.
12. **Nagy, E. and Dobos, P.,** Coding assignments of *Drosophila* X virus genome segments: *In vitro* translation of native and denatured virion dsRNA, *Virology,* 137, 58, 1984.
13. **Revet, B. and Delain, E.,** The *Drosophila* X virus contains a 1-μm double-stranded RNA circularized by a 67-kd terminal protein; high-resolution denaturation mapping of its genome, *Virology,* 123, 29, 1982.
14. **Bernard, J.,** *Drosophila* X virus RNA polymerase: tentative model for *in vitro* replication of the double-stranded virion RNA, *J. Virol.,* 33, 717, 1980.
15. **Hill, B. J.,** Properties of a virus isolated from the bivalve mollusc *Tellina tenuis* (Da Costa), in *Wildlife Diseases,* Page, L. A., Ed., Plenum Press, New York, 1976, 445.
16. **Clotilde, F. L.,** Recherches sur un virus de *Macropipus depurator* Linné, isolé à l'aide de lignées cellulaires de poissons, Thèse Doct. 3éme cycle, Univ. Sci. Tech. Languedoc, Montpellier, France, 1984.
16a. **Hill, B. J. and Bonami, J. R.,** unpublished data.
17. **Bovo, G., Ceshia, G., Giorgetti, G., and Vanelli, M.,** Isolation of an IPN-like virus from adult Kuruma shrimp *(Penaeus japonicus), Bull. Eur. Assoc. Fish Pathol.,* 4, 121, 1984.
18. **Hill, B. J.,** Molluscan viruses: their occurrence, culture and relationships, *Proc. 1st Int. Coll. Invertebr. Pathol.,* Kingston, Ontario, 1976, 25.
19. **Underwood, B. O., Smale, C. J., Brown, F., and Hill, B. J.,** Relationship of a virus from *Tellina tenuis* to infectious pancreatic necrosis virus, *J. Gen. Virol.,* 36, 93, 1977.
20. **Almeida, J. D. and Morris, R.,** Antigenically-related viruses associated with infectious bursal disease, *J. Gen. Virol.,* 20, 369, 1973.
21. **Harkness, J. W., Alexander, D. J., Pattison, M., and Scotta, A. C.,** Infectious bursal disease agent: morphology by negative stain electron microscopy, *Arch. Virol.,* 48, 63, 1975.
22. **Moss, L. H. and Gravell, M.,** Ultrastructure and sequential development of infectious pancreatic necrosis virus, *J. Virol.,* 3, 52, 1969.
23. **Bonami, J. R. and Hill, B. J.,** Etude comparée d'un virus isolé chez le mollusque *Tellina tenuis,* du virus P du crustacé *Macropipus depurator* et du virus de la nécrose pancréatique infectieuse des poissons, *Haliotis,* 10, 25, 1980.
24. **Hill, B. J. and Alderman, D. J.,** Observations on the experimental infection of *Ostrea edulis* with two molluscan viruses, *Haliotis,* 8, 297, 1977.
25. **Hill, B. J., Way, K., and Alderman, D. J.,** Further investigations into the pathogenicity of IPN-like viruses from oysters, *Proc. 3rd Int. Col. Invertebr. Pathol.,* Brighton, U.K., 273, 1982.
26. **Hill, B. J.,** personal communication.
27. **Comps, M., Menu, B., Breuil, G., and Bonami, J. R.,** Virus isolation from mass cultivated rotifers, *Brachionus plicatilis, Bull. Eur. Assoc. Fish Pathol.,* 10, 37, 1990.

28. **Comps, M., Menu, B., Breuil, G., and Bonami, J. R.,** Viral infection associated with rotifers mortalities in mass culture, *Aquaculture,* 93, 1, 1991.
29. **Comps, M., Mari, J., Poisson, F., and Bonami, J. R.,** Biophysical and biochemical properties of the RBV, an unusual Birnavirus pathogenic for Rotifers, *J. Gen. Virol.,* 72, 1229, 1991.

Chapter 17

RHABDOVIRIDAE

G. Brun

TABLE OF CONTENTS

The Rhabdoviridae are a family of viruses whose classification is based upon virion morphology. All the members of this family which have been studied in detail, whether vertebrate or plant viruses, were shown to have several biological characteristics in common which are now used to define this family, e.g., their genome is a single-stranded RNA with negative polarity and the virion contains five proteins for which functional homology has been established. However, there are many other cases for which only the virion or virus-like particles have been described; so they are classified as possible or probable members of the family.[1] Two reviews have recently been published by Bishop[2] and Wagner.[3] These reviews contain basic information and a complete bibliography.

I. A GLOBAL VISION OF RHABDOVIRUSES AND THEIR DISTRIBUTION AMONG ARTHROPODS

Rhabdoviruses are either vertebrate or plant pathogenic agents, but most of the former and almost all of the latter are transmitted by vector insects in which they multiply. Their vertical (transovarian) transmission in insects was shown in several cases, and they were isolated from different arthropods. We may, therefore, consider that most of them are also invertebrate viruses.

Their detection in arthropods trapped in nature depends on an economic or, eventually, medical interest. Thus, almost all the Rhabdoviruses discovered in mosquitoes or other biting arthropods occurred during epidemiological investigations of arboviruses. The detection technique consists in inoculating some crushed insect extract into the brain of newborn mice, or in spreading the extract on a vertebrate cell culture. Of course, these methods can only reveal viruses capable of multiplying at around 37°C and destroying vertebrate cells. Viruses that multiply only in invertebrate hosts are not detected with this method. The discovery of Rhabdoviruses with only an invertebrate host, especially if they are not or only slightly pathogenic for the latter, has only been coincidental. This is the case for a possible rhabdovirus found in the crab *Callinectes sapidus*[4] (see Chapter 24, "Unclassified Viruses of Crustacea"), or for the first detection of the *Drosophila* sigma virus, agent of the hereditary CO_2 sensitivity.[5] Therefore, the current vision of the distribution of the members of this viral family in the world of invertebrates may be quite partial. It is very likely, for instance, that with the growth of aquacultural techniques for sea decapods, hatcheries will be infected by some new rhabdovirus in the future; this seems to be possible after the Jahromi discovery.[4]

A full catalog of the vertebrate rhabdoviruses found in arthropods was reported in the article by Shope and Tesh.[6] All the members of the Vesiculovirus genus, whose prototype is the Indiana vesicular stomatitis virus (VSV), were found in biting arthropods (mosquitoes, midges, sandflies, mites, ticks) as well as in houseflies (New Jersey VSV). The only exception is the Piry virus. Nevertheless, it was experimentally shown that it multiplied best in *Drosophila*.[7,8] Thus, this exception is probably due to chance. In experimental conditions, the Indiana and New Jersey VSVs multiply in *Aedes, Culex,* and *Toxorhinchites* mosquitoes and in *Drosophila melanogaster* without developing any pathological symptoms. They only develop a sensitivity to CO_2 similar to the symptom due to σ-virus.[9-12] Mosquitoes experimentally inoculated with these viruses or with Chandipura, transmit the viruses to laboratory animals.[6] Transovarian transmission of Indiana and Chandipura was observed in Phlebotomine sandflies.[13,14] These facts, and observations made on the epizootics of vesicular stomatitis from which cattle suffer in the U.S. from time to time, seem to confirm the role of vector insects in the evolution of these epizootics. However, the responsible vector insects could not be clearly designated.[6]

On the other hand, the importance of the role played by infected domestic or wild mammals in the natural cycle of these viruses is challenged by some authors. Shope and Tesh[6] point out that direct transmission from one mammal to another is limited and that the viremia often remains too slight to reinfect other hematophagous insects. If this is so, then mammals would not play an amplifying role and could be no more than superficial indicators of the infection level of unknown reservoir hosts. Nevertheless, our experience leads us to believe that an alternating passage through vertebrate and invertebrate hosts is important for the conservation of the multiplying capabilities in both types of hosts.

On the other hand, a molecular study of the genetic diversity of New Jersey VSV isolates collected during the 1982 through 1985 epizootics in the U.S. (Southwest and foothills of the Rockies) and during the same periods in Mexico and Central America, areas where the infection is endemic, gives some insight to the evolution of epizootics and the conservation of the viruses in a natural environment.[15,16] Samples taken from cattle, horses, pigs, or insects, during a given epidemic are quite homogeneous. This suggests that although insects probably play a role in disease transmission, infection by direct contact between animals may be more frequent than previously thought. On the contrary, viral types with dozens of genetic changes can coexist and remain together in enzootic areas. These viral types may be conserved for some time in reservoir insects, maybe in a state of hereditary persistent infection (e.g., the σ-virus). However, contrary to the author's suggestion,[15] we believe that the maintenance of a persistent infection in a usual insect host does not provide the conditions for a rapid evolution of the virus (the persistent infections in vertebrate cells to which he refers, that are an exception, create very peculiar selection pressures). For us, the cause of the rapid diversification of vesiculoviruses in nature is the occasional passage through unusual vertebrate or invertebrate hosts and the return to the normal reservoir host or to the normal invertebrate-vertebrate host cycle, owing to selection pressures in different directions. Emphasis should be put on the wide spectrum of potential VSV arthropod hosts, as observed in the experimental multiplication of Indiana in the leafhopper, *Peregrinus maidis*, vector of a plant rhabdovirus (maize mosaic virus).[17]

The second vertebrate rhabdovirus genus defined is the Lyssavirus whose prototype is rabies virus. It differs from the Vesiculovirus genus by its physicochemical properties (in particular, the electrophoretic migration characteristics) and the number of molecules in the virion of the protein called NS for VSVs (called M1 for rabies). Rabies and other closely related viruses like Lagos bat virus do not multiply in insects or in cultured insect cells. However, Mokola, an African virus, serologically more distant but which produces a similar pathology, experimentally multiplies in the *Aedes aegypti* mosquito. A serological affinity has been observed between Mokola and two African viruses, Obodhiang and Kotonkan, which were found in insects (*Mansonia* mosquitoes and *Culicoides* midges) and were classified as lyssaviruses.[18,19]

Bovine ephemeral fever virus, which has some economical importance in Africa, southern Asia, and Australia, and similar viruses have been found in mosquitoes and *Culicoides* midges. These viruses experimentally multiply in mosquitoes and mosquito cells. The role of insect vectors in the disease transmission seems very probable.[6,20]

Fish rhabdoviruses[21] can be divided into two groups. Some of them, such as viral hemorrhagic septicemia (VHSV) of trouts and infectious hematopoietic necrosis virus (IHNV) of salmonids have a protein electrophoretic migration pattern similar to that of rabies. The others, such as spring viremia of carp virus (SVCV), and pike fry disease virus (PFDV), have patterns similar to those of vesiculoviruses. The first two, like rabies, do not multiply in *Drosophila*. The other two multiply causing the CO_2 sensitivity symptom like the vesiculoviruses and σ-viruses.[22,23] The concordance between the electrophoretic pattern of proteins and the multiplying capability in *Drosophila* has been pointed out. Spring viremia of carp virus may be transmitted by *Argulus foliaceus*, an aquatic hematophagous arthropod.[6]

Lizard viruses experimentally multiply in mosquitoes: one of them was also found in phlebotomine sandflies.[6] Shope and Tesh[6] also mention a long list of rhabdoviruses of vertebrates, many of which were found in mosquitoes, sandflies, midges, or ticks. (See also Reference 24.)

Almost all plant rhabdoviruses[19,25,27] are transmitted by vector insects, aphids, or leafhoppers. Contrary to vesiculoviruses, which have a large spectrum of natural arthropod hosts, plant rhabdoviruses have very specific vector insects. For instance, there are two different serotypes of potato yellow dwarf virus (PYDV). Each one is transmitted by a different leafhopper: *Aceratagallia sanguinolenta* and *Agallia constricta*.[25,28,29] Vector specificity was also found in aphid-transmitted viruses.[27] Some other plant rhabdoviruses are lettuce necrotic yellow virus (LNYV) whose vectors are the *Hyperomyzus lactucae* and *H. caruellinus* aphids, Sonchus yellow net virus (SYNV), transmitted by *H. lactucae* and *Macrosiphum euphorbia*, and maize mosaic virus (MMV) whose vector is the leafhopper *Peregrinus maedis*. Two types may be distinguished according to the electrophoretic pattern of the proteins. In the first type, the pattern is similar to vesiculovirus, while the second type is similar to the rabies pattern. There is no relation between this characteristic and aphid or leafhopper transmission.[30]

Plant rhabdoviruses multiply in vector insects after contamination by absorption of a viral suspension through a membrane or by injection. After an incubation period, the vector becomes capable of infecting a healthy plant. It remains a lifelong transmitter. Transovarian transmission in the vector has been shown for several of these viruses.[25,27]

Drosophila σ-virus[5,31] is nonpathogenic and noncontagious. It is maintained exclusively by hereditary transmission through a persistent infection in the female germinal cells (resulting in stabilized maternal lines). Transmission by males issued from these lines, with a possible contamination of the ovocytes, can reestablish a persistent infection in the germinal cells. This mechanism can compensate for accidental losses or losses due to other circumstances (rise of temperature[5] or selection in favor of noninfected flies, especially during winter[32]) and can maintain a balance in natural populations. σ-Virus is often found in natural *Drosophila melanogaster* populations. It also has been found in populations of different American species, particularly in *D. affinis* and *D. athabasca*. Although CO_2 sensitivity symptoms were observed, it is not known whether the viruses which were found in different species (probably isolated for a long period) are serologically different or not.[5]

Two possible rhabdoviruses, specifically from the mite *Panonychus ulmi* and the insect *Oryctes rhinoceros*, are also mentioned in the 1973 Knudson review,[33] but they are not mentioned in the more recent reviews.

II. RHABDOVIRUS MORPHOLOGY AND SITES OF BUDDING[27,33,34,35]

Rhabdovirus virions are enveloped in a membrane derived from the cellular membrane through which they bud (see Figures 1 and 2). The virions are either bullet-shaped, with one rounded tip and one flat tip (often penetrated with contrasting agents) or bacilliform with both tips rounded. It is generally considered that the bullet shape is the normal animal rhabdovirus shape, whereas native plant viruses are bacilliform. Indeed, the following is always observed for plant viruses in sections. When stain penetrates the particle, the nucleocapsid is seen as a helix. The bacilliform virions of plant rhabdoviruses seem not perfectly symmetrical; one tip of the envelope seems empty and is likely the fragile one.[36] Possibly some bullet shape observations could be due to deterioration during extraction or staining.

Detergents remove this envelope, leaving a skeleton which keeps its bullet shape. If enough salt is added, the nucleocapsid is released and is no longer coiled (see Figure 1d). These patterns, always obtained when these treatments are applied, may sometimes be observed in fragile viruses or mutants after negative staining (see Figure 1 concerning σ-virus which is morphologically similar to VSV).

The morphogenesis of animal viruses takes place entirely in the cytoplasm. Maturation occurs either through the plasma membrane toward the outer surface of the cell or through

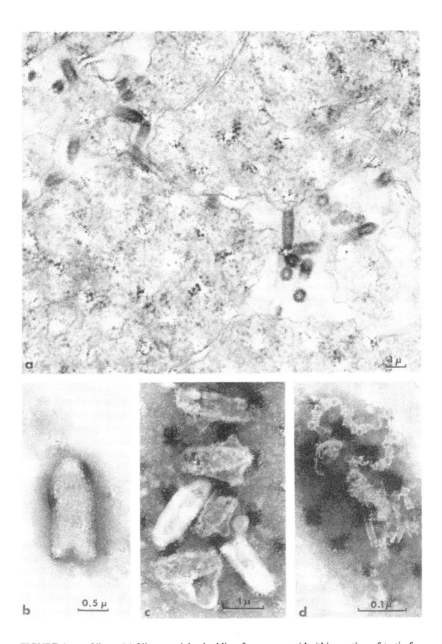

FIGURE 1. σ-Virus. (a) Virus particles budding from spermatids (thin section of testis from stabilized male). (b) A virion from infected cell culture supernatant fluid seen in negative contrast. (c) Virions from the thermolabile ts9 mutant, virions are seen losing their envelope and revealing a coiled nucleocapsid which has 32 turns. (d) An unwound nucleocapsid. (From Teninges, D., Contamine, D., and Brun, G., in *Rhabdoviruses,* Vol. 3, Bishop, D. H. L., Ed., CRC Press, Boca Raton, FL, 1980, 113. With permission.)

the membranes of the endoplasmic reticulum or intracytoplasmic vacuoles. The maturation site depends on the virus and on the cell. For slow maturing viruses, the accumulation of nucleocapsids in the cytoplasm may form minor or major inclusions. Although vesiculo-viruses are usually found budding frequently from the plasma membrane, some internal maturations[37] and viroplasms[38] have been observed. As for rabies, intracytoplasmic maturation is observed in the nervous centers on the periphery of large inclusions (Negri bodies).

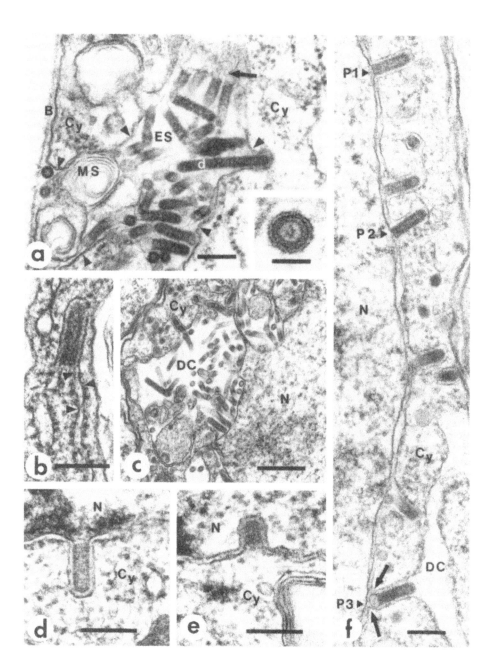

FIGURE 2.   Some accumulation and assembly sites of MMV particles in the insect vector *Peregrinus maidis*. (a) Particles accumulated in an extracellular space (ES) between basal lamina (B) and basal infoldings of the plasma membrane (arrowheads) in a salivary gland secretory cell, inset shows cross section of a particle. (Bar = 50 nm.) (b) A virus particle attached to membranes of the rough endoplasmic reticulum. (c) Particles accumulated within perinuclear dilated cisternae (DC) in cytoplasm of a brain cell. (Bar = 500 nm.) (d) and (e) Particles with intermediate length attached to and probably budding from the inner (d) or the outer (e) nuclear membranes in salivary gland cells. (f) Particles apparently budding through the inner nuclear membrane of a brain cell; P3 is completely detached and released into dilated cisternae (DC) that are connected to the perinuclear space. Cytoplasm (Cy); Nucleus (N). (Bar = 200 nm, except for inset of [a] and for [c].) (From Ammar, E. D. and Nault, L. R., *Intervirology,* 24, 33, 1985. With permission.)

In fox salivary glands, massive maturation occurs on the plasma membrane with accumulation of virions in the salivary ducts, which is obviously favorable for the survival of a virus which is essentially transmitted by biting.[34]

The morphogenesis of plant rhabdoviruses differs from that of animal rhabdoviruses. Moreover, there seem to be several types. In one type (SYNV, PYDV), nucleocapsid formation occurs in the nucleus. Virion budding occurs on the inner membrane, and virions accumulate in the perinuclear space, after which because of the continuity between the outer nuclear membrane and the endoplasmic reticulum, they reach cytoplasmic vesicles. In the second type, (LNYV), maturation seems to occur on the outer nuclear membrane and on the nearest endoplasmic reticulum. The third type, barley yellow striate mosaic virus (BYSMV), is more like the animal rhabdoviruses since its maturation occurs on the periphery of intra-cytoplasmic viroplasms.[27] Although results of observations of the same viruses in vector insect tissues are similar, there are some differences according to the insect tissue. For instance, budding of the maize mosaic virus (MMV), which matures only on plant nuclear membranes,[39] occurs on the inner nuclear membrane in leafhopper brain, on the cytoplasmic membranes in secreting vesicles, or even on the plasma membrane in the salivary glands (see Figure 2).[40]

The size of the virions is clearly an important parameter for rhabdovirus classification, since it is directly related to the length of the nucleocapsid, hence to the length of the genome RNA. Estimations, however, depend on both the integrity of the virions measured (presence of envelope and spikes) and on technical conditions. Numerous animal rhabdoviruses (vesiculoviruses, lyssaviruses, fish viruses, bovine ephemeral fever, σ-virus) are about 180 nm in length. Some are shorter (Kern Canyon, 132 nm) and one group, including Flanders and Hart Park which were detected in both birds and mosquitoes, are longer (between 220 and 230 nm). The presumed crab rhabdovirus is much shorter (115 nm in section) and it is also bacilliform. Plant rhabdoviruses seem to be longer in general (ranging from 200 to 400 nm) and to vary in size more depending on the virus.[27,34]

III. MOLECULAR BIOLOGY OF RHABDOVIRUSES

The biology of Indiana VSV in vertebrate cells has been thoroughly studied at the molecular level. Its essential modalities must of course be the same in all the host cells; the peculiarities of VSV biology in insect cells, studied in *Drosophila* cells, will be described in Section V. It now seems possible to apply the Indiana VSV model to other rhabdoviruses in regard to basic processes except for some details. These could be important for the biology of a peculiar virus, its virus-host relationships, and its induced symptoms, but only represent minor variations. A brief description of Indiana VSV molecular biology will be given in this chapter of the reviews[41-44] where the reader can find original publications.

Once the virus has penetrated the cell, either by fusion of the viral envelope with the plasma membrane or by endocytosis, the nucleocapsid is released into the cytoplasm. The nucleocapsid includes the viral genome, a single-stranded RNA molecule (3.6×10^6 Da, 11,161 nucleotides) with negative polarity, protected by N protein (49 kDa, approximately 1,250 molecules per virion) which is strongly attached to it. Two other proteins, NS (29 kDa, approximately 450 molecules) and L (241 kDa, 50 molecules) are also bound to the nucleocapsid. Viral genome is then sequentially transcribed in six RNAs of positive polarity, according to the order: 3′-leader-N-NS-M-G-L-5′. The last 59 nucleotides of the 5′ end are not transcribed. The five protein messengers are capped and polyadenylated by the viral polymerase itself (an association of L and NS proteins) which is capable of carrying out the transcription and all subsequent operations *in vitro*. At each intergene the polymerase stops, and there is a probability for the RNA synthesis not to be carried on further. Owing to this attenuation phenomenon, the viral messenger concentration gradient drops between N and L.

The viral messengers are then translated by the cellular machinery. Viral protein synthesis is necessary, for the following stage, i.e., replication. It is believed that the concentration of N protein modifies the way the viral polymerase functions. A nucleation site on the RNA leader is responsible for the encapsidation by N, which starts during RNA synthesis. The presence of N then blinds the polymerase so that it cannot read the signals for transcription stop, capping, and polyadenylation. This leads to the synthesis of an encapsidated complete positive RNA strand. This nucleocapsid with an RNA complementary to the genome (antigenome) becomes the template for the second step of replication, the synthesis of new viral genomes. Its 3' end, complementary to the 5' end of the genome, also has a site of polymerase fixation and RNA synthesis initiation as well as the complement of another encapsidation site. This site, at the 5' end of the neosynthesized negative strand, is apparently more efficient than its counterparts at the 5' end of the positive strand since a secondary transcription persists in balance with replication. There are also much fewer synthesized nucleocapsids containing an antigenome than there are nucleocapsids containing a genome. Although there is a significant amount of nucleocapsids containing an antigenome in the cytoplasm, they are not used to form virions.

Two other proteins play a role during maturation: M (26 kDa, 1800 molecules per virion) and G (75 kDa, 1200 molecules). M protein (matrix protein) is thought to be responsible for the shape of the virion because of its binding with the nucleocapsid. Skeletons obtained from virion treatment with detergents contain M protein. The complete virion is formed by budding on cellular membranes transformed by the insertion of G protein (glycoprotein) which forms the spikes. The possible roles of G and M proteins in the budding process are discussed in Reference 35. In contrast to other viral proteins which are synthesized on free ribosomes, the G protein is synthesized by the membrane bound ribosomes. It undergoes several maturation processes, including glycosylation, during its transfer to membranes until it reaches the virion budding site. Protein G is the major antigenic factor; it determines type specificity and stimulates the production of neutralizing antibodies.

Viral proteins are multifunctional. Protein L is clearly the main polymerase element for carrying out RNA-synthesis operations and capping (insert of a G[5']ppp[5'] group and its methylation) and for messenger polyadenylation (probably by slippage on the 7 U sequence at the end of each viral gene). The phosphorylated protein NS plays a role in the elongation of the synthesized viral RNA but it also forms complexes with N to ensure its storage by avoiding self-assembling. N, an essential element of the nucleocapsid, is important for transcription. Protein M has not only a structural role but also a regulating role in the transcription. Spike protein G, whose presence on the membrane determines the maturation site, is important for recognizing the cell adsorption sites, and may well contribute to membrane fusion.

The replication of rhabdoviruses, like that of many other RNA viruses, accidentally produces defective genomes which mature in truncated particles (T particles). Long-term use of massive inocula leads to the selection of highly interfering forms called defective interfering (DI) particles. A description of VSV DIs and their emergence mechanisms is given by Holland.[45] Hosts are considered to play a role in the DI spectrum produced by a virus. Till now DI particles had never been observed in rhabdovirus-infected plants, but very recently some DI have been found in plants chronically infected by SYNV.[46]

IV. σ- AND VESICULOVIRUS BIOLOGY IN *DROSOPHILA* AND *DROSOPHILA* CELLS AS A MODEL FOR RHABDOVIRUS BIOLOGY IN INSECT HOSTS

The *Drosophila* σ-virus, agent of hereditary CO_2 sensitivity, is well-known and has been extensively studied.[5,31] At first, it was considered as a biological curiosity. Later,

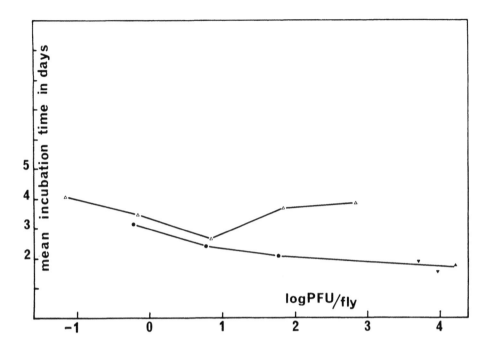

FIGURE 3. Mean incubation time of Piry virus in *Drosophila* ("Paris" strain) at 25°C plotted as a function of the inoculum titer (1 PFU = about 20 IU in D). Different symbols refer to different assays. Virus was multiplied in chick embryo cells (CEC). (△) experiment in which viral suspension was from massive CEC inoculum and contained DI particles which interfere with normal virus (unpublished data).

similarities were noticed between its relationships with *Drosophila,* the relationships of arboviruses with their invertebrate host and with those of plant viruses with their insect vectors. We believe that the σ-relationships with *Drosophila* represent a whole inventory of the possible relationships of a rhabdovirus with its vector or an insect host.

As in the case of insect vectors, no contamination of healthy insects by mere contact with an infected insect has been observed. *Drosophila* adults are experimentally infected by injecting a suspension of σ-virus or a vesiculovirus in the abdomen. After a given incubation period, the CO_2 sensitivity symptom appears in the flies. This symptom is revealed by exposing the flies to pure CO_2, usually at 12°C: the noninfected flies wake up from the induced narcosis, whereas the infected ones remain paralyzed (with Indiana VSV, this symptom may take special forms). Bussereau[47,48] showed that this paralysis is due to the infection of the thoracic nervous center and more precisely to the viral maturation in it. Observations of σ-mutants or variants lead us to believe that this symptom is caused by the modification of some nervous cell membranes, when a viral protein, G presumably, is inserted into them. When not exposed to a pure CO_2 atmosphere, infected *Drosophila* survive normally although the infection is lifelong. σ- or Piry-infected *Drosophila* remain CO_2 sensitive, but the symptom disappears for most of the other vesiculoviruses.[11]

The mean incubation time increases with the dilution of the inoculum until it reaches the limit where a fraction of flies are not infected. The law of the decrease of the infected fraction as a function of dilution confirms the hypothesis that infectious particles are independent, meaning that a virus infecting unit may be defined in the used *Drosophila* strain. The variations of the mean incubation time as a function of the inoculum titer (see Figure 3) give us some information on the viral invasion in the fly. This information is obtained only through an intermediate relay: the invasion of the thoracic nervous center, where the site of the symptom is located. Experiments show that this organ is never directly contaminated after an abdominal injection. The evolution of the fly viral yield (obtained by titration

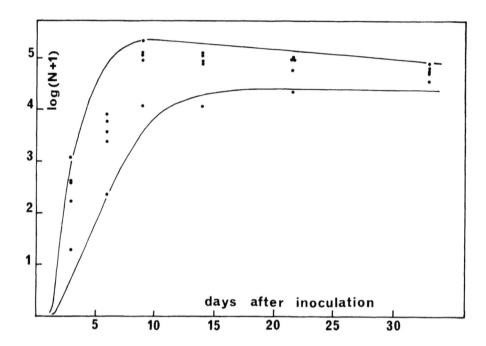

FIGURE 4.   Development of virus produced in individual flies (''Paris'' strain) inoculated with 5 IU of Piry virus at 20°C.

of crushed insect extract) as a function of time, also supplies a more global information about fly invasion (see Figure 4).

When cultured *Drosophila* cells are infected, there is a brief period of high production followed by a decreasing production and fixing of a state of persistent infection with low virus production. This state does not affect the life of the cell, which normally divides. In inoculated insects, all the cells may be affected by this state of persistent infection, provided that they could be reached. Germinal cells cannot be reached by σ-virus in adult insects. They may, however, be infected by σ-virus in a way which will be explained later.

The first demonstration of such a state of persistent infection by σ-virus and a large part of the knowledge concerning it were obtained by the study of flies which perpetuate the persistent infection state through their germinal cells. The males transmit the infection to a part of their offspring. The females transmit it to all their offspring (except for a few accidental cases). Normally, there are enough viral genomes present in the egg for all the cells, and, in particular, the germinal cells, to be infected upon segmentation. This perpetuates the persistent infection of germ cells from generation to generation in a maternal line, which is considered stabilized for σ-virus. The existence of stabilized maternal lines with viruses defective for a maturation function proves that viral information is transmitted by cellular continuity, without any reinfection being necessary. Curing experiments (by temperature shift or EMS mutagen) and genetic drift of virus in stabilized maternal lines have proven that the persistent infection state is perpetuated in the germinal cell line by a population of viral genomes independent of the chromosomes. Its probable cytoplasmic location has not been formally demonstrated. Therefore, viral genomes defective for a replication function (similar to VSV DI genomes) can probably persist through complementation by wild type genomes in this population; that would explain the lower yield of stabilized flies.

It is clear that the efficiency of the generation transmitting mechanism in a stabilized maternal line has been promoted by natural selection, since σ-virus cannot be transmitted otherwise. We believe that this efficiency, and maybe even more so the limitation of the viral multiplication capability are imposed by the virus survival conditions in its natural environment (it must not be harmful for its carrier). These imposed conditions have fashioned σ-virus as peculiar (virus slower than vesiculoviruses,[49] special refractory genes system). Indeed, whenever σ-virus is experimentally multiplied, either in cultured cells or by successive inoculations of flies, there is a selection of variants which are more or less pathogenic for the embryo or which disturb ovogenesis. Ohanessian[50] injected vesiculoviruses (Indiana and Piry) into eggs; here too, a pathogenic effect on the embryo was observed.[50a]

One of the characteristics of σ-virus is its capability to infect the cysts during ovogenesis in inoculated females. The cysts are protected by a barrier through which only g + viruses can get; g − mutants, normally perpetuated in stabilized maternal lines, cannot infect the cysts. However, there may be complementation of a g − virus by a g + virus to break through the barrier. Ohanessian[50] shows that the σ-virus can allow Piry to do so. She concludes that this complementation might be due to phenotypic mixing, an exchange of G proteins, because it is probably the only possibility for complementation between far-related rhabdoviruses.[50] Bregliano's[51] experiments have shown this barrier to be on the ovariole sheath, and suggested that g − virus maturation takes place within the ovariole sheath cells, while g + virus buds through the external membrane and is able to infect the cysts.

Spermatozoid-infected eggs and eggs infected with still cysts are different from those laid by stabilized females. The flies hatched from them do not have the same characteristics as those hatched from stabilized flies. They also differ from inoculated flies. It was pointed out earlier that there are barriers in adult insects which totally isolate male germinal cells and which partially isolate ovary cysts and cephalic and thoracic nervous centers. These barriers, due to cellular differentiation, appear either during embryogenesis or during adult morphogenesis. For both categories of infected eggs, there is a race between the viral invasion of the organism and the building up of the barriers. The outcome depends on the stage when eggs are infected and on the genetic characteristics of the virus. With σ-virus, stabilized females can generally be found again, at least when the cysts have been infected, and they will be at the origin of new stabilized maternal lines.

For vector-conveyed plant or vertebrate rhabdoviruses, insects contaminated by eating an infected plant or animal are homologous to *Drosophila* inoculated with σ-virus. There are many examples of viral infection transmitted by a female to her progeny (transovarian transmission). However, although possible, there are no clear examples of a situation equivalent to that of stabilized maternal lines. When a virus has a very wide range of invertebrate hosts, such as VSVs, its relationships with these hosts may not be the same according to the species. So the repertoire of possible situations can range from the nonbiting insect which is a dead end for the virus (possibly the case of the housefly for New Jersey VSV) to a reservoir species where the virus would be conserved for some time through stabilized maternal lines before being transmitted back to a vertebrate host. It is believed that for a virus having many possible hosts and biological cycles which can vary at random, this is very favorable for a rapid genetic diversification. This has been observed in Indiana and New Jersey VSV serotypes.

An important fact revealed during the study of σ-virus is the existence of *Drosophila* refractory (ref) genes, whose products have an effect on the biology of the virus.[52] These genes were revealed when restrictive alleles were discovered in natural populations or in some laboratory strains. Ref genes which affect the multiplication of Piry virus, and apparently all vesiculoviruses, have also been found. The pattern of ref genes known to be active on Piry and vesiculoviruses and the pattern known to be active on σ-virus are not the same. Similar genes which have an effect on multiplication in the vector of certain plant rhabdoviruses have also been found.[53] Contamine et al.[54,55] conclude (by applying temperature shift techniques to viral mutants adapted to restrictive alleles and which, meanwhile, have become ts) that the products of the five ref genes known to be active on σ-virus, all

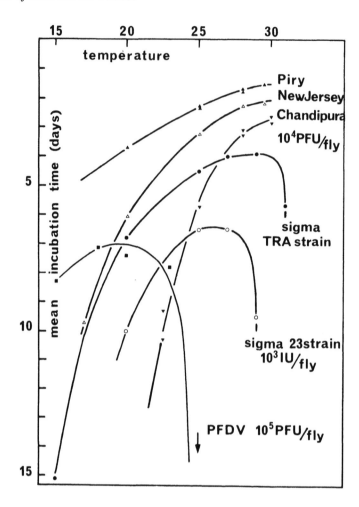

FIGURE 5.   Mean incubation time in *Drosophila* at different temperatures in degrees C° for different rhabdoviruses. Piry, New Jersey, Chandipura in hybrid (vestigial × Champetières) flies (personal data) show different degrees of cold sensitivity. Pike fry disease virus (PFDV) (data from [22]) and σ virus (data from Contamine cited in Reference 22) in hybrid (Oregon × ebony or sepia × ebony) flies show different degrees of temperature sensitivity ("23" strain is derived from nature, "TRA" strain has been selected for temperature resistance through successive passages).

act during the first steps after penetration, probably during primary transcription or replication. However, ref genes can most likely play other roles as well. For example, ovariole sheath cell differentiation seems to modify maturation conditions for g − σ-mutants. We know that cellular differentiation means that some genes are set in action or else stop; those affecting σ-virus maturation can thus be considered as ref genes. Genes which modify the viral cycle in vertebrate and insect cells can also be considered as ref genes. This explains why we find the discovery of *Drosophila* ref genes very important and the research done on their role in σ-virus and vesiculovirus cycles, not only for the comprehension of rhabdoviruses but also for providing models for the whole science of virology as well.

V. VSV (INDIANA SEROTYPE) MOLECULAR BIOLOGY IN *DROSOPHILA* CELLS

The first difference between the reaction of *Drosophila* cells to VSV infection and the

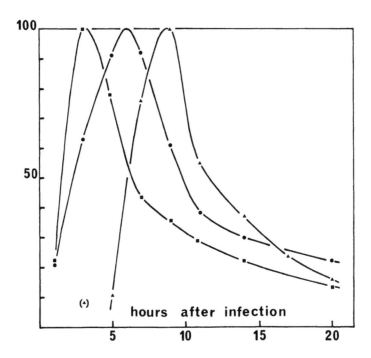

FIGURE 6.   VSV molecular synthesis in *Drosophila* cells at 25°C. ■ Kinetics of mRNA synthesis, ● of protein synthesis, and ▲ of infectious particle production expressed in percent of their maximal values. (The production point at 3 h is considered from our experience as due to loosening.) (Redrawn from Blondel, D., Petitjean, A. M., Dezelee, S., and Wyers, F., *J. Virol.*, 62, 277, 1988.)

reaction of vertebrate cells resides in cell survival. No cytopathogenic effect is observed in infected *Drosophila* cells whereas there is complete inhibition of cellular synthesis 2 h after infection in vertebrate cells. This inhibition appears to be principally due to the RNA leader. Synthesized in the cytoplasm, it is carried to the vertebrate cell nucleus where its concentration increases during the first 2 h. The RNA leader inhibits the *in vitro* transcription by II and III polymerases; it is believed to have the same effect *in vivo*.[41] Neosynthesized N protein also seems to play a part in the inhibition of cellular synthesis.[56] In *Drosophila* cells, the synthesis of the RNA leader seems less abundant than in vertebrate cells. On the other hand, there are very few molecules in the nucleus.[57]

However, although the absence of a cell-destroying mechanism is necessary, it is probably not enough to allow normal survival. At the beginning, the viral production of infected *Drosophila* cells is almost the same as in vertebrate cells, then it slows down quickly, and decreases until it reaches a stable level which characterizes the state of persistent infection. The sequence of molecular events during the 20 h following the infection of *Drosophila* cells with VSV was described by Blondel et al.[58] The synthesis rate of viral RNA messengers is the fastest to evolve: it increases during the first 3 hpi at 25°C, then reaches a maximum rate before declining (see Figure 6). The variation of the protein synthesis rate is very similar but not at the same moment (maximum rate after 6 h). This may be explained by the limited life span of the messengers. Moreover, the evolution of the production rate of infected particles (maximum reached after 9 h) may be due to the preceding rate. Thus, the control seems to bear on transcription.

What is the control mechanism? It is possibly similar to the transcription control mechanism observed in vertebrate cells which is due to M protein. It might be more efficient in

Drosophila cells since M protein matures differently in *Drosophila* cells than in vertebrate cells.[59] Replication has also been studied in *Drosophila* cells.[57,58] These authors observed incomplete encapsidated viral RNAs with positive polarity, apparently still bound to their template. Thus, it seems to be a control affecting replication. Is its mechanism different or not from the transcription control mechanism? The last hypothesis is possible because the second control mechanism might affect the first stage of replication which has the same viral genome template as transcription. On the contrary, the second step of replication — genome synthesis from antigenomes — does not seem to be affected. The number of viral genomes per cell 20 h after infection is high (estimated at 950).[58] This estimation, however, is made before an equilibrium is reached, since these cells divide about every 24 h.

A deficit of viral messenger translation was observed compared to the translation in vertebrate cells. This deficit is higher for G protein, which is synthesized in specific conditions, on the endoplasmic reticulum. It is not known whether this deficit is merely due to competition with cellular messengers whose synthesis is not stopped or is the result of a special mechanism. Moreover, G integration in the virion is lower compared to its synthesis (*Drosophila* virions are G-deficient). These facts seem unrelated with the state of persistent infection. They may be due to a characteristic specific to *Drosophila* cells.[60,61]

VI. CONCLUSION: EVOLUTIONARY FAMILY RELATIONSHIPS BETWEEN RHABDOVIRUSES ACCORDING TO CURRENT KNOWLEDGE OF NUCLEOTIDE SEQUENCES

Thanks to current molecular biology data, we are beginning to have some precise information about the family structure of Rhabdoviridae. A surprise was the discovery of a sixth RNA messenger in the infectious hematopoietic necrosis virus (IHNV), which encodes for a small protein (12 kDa), not included in the virion.[62] The corresponding gene is located between the G and L protein genes on the IHNV genome whose gene sequence is otherwise identical to those of VSV and rabies.[63] The rabies nucleotide sequence reveals a long sequence of 423 nucleotides (which is apparently a ''pseudogene'')[64] between G and L protein genes. An intergenic sequence of 21 nucleotides between G and L genes should also be mentioned for the New Jersey virus, with a pseudosignal of transcription recovery[65] whereas the corresponding Indiana sequence is limited to the two nucleotides of all other intergenic sequences.

The VSV Indiana nucleotide sequence has been completely determined.[41] The gene sequences and the amino acid sequences of the New Jersey corresponding proteins have also been determined. Amino acid homology between proteins of the two viruses is 68% for N,[66] 62% for M,[67] 51% for G,[68] and only 32% for NS[69] proteins. This difference between the evolution of viral proteins probably reflects a difference in the functional constraints of these proteins. A very high homology rate exists in the Vesiculovirus genus for RNA leader sequences.[70] Chandipura N protein is 50 and 51% homologous, respectively, to that of VSV (New Jersey) and VSV (Indiana) N proteins; Chandipura NS protein is 23 and 21% homologous, respectively, to VSV (New Jersey) and VSV (Indiana) NS proteins.[71] For less-related viruses such as the spring viremia of carp virus (SVCV), the homology with Indiana and New Jersey VSVs is very high for the leader but drops to 23 and 24% respectively, for M protein.[72,73] This homology is approximately 20% between G proteins of VSV, rabies,[68] and IHNV.[74] This is about the same rate between G proteins of σ-virus and Indiana or new Jersey VSVs (21%). This homology goes down to 13% with rabies, suggesting that σ-virus is closely related to the vesiculoviruses.[75] The authors also noticed a codon deficit containing CG doublet, characteristic of vertebrates, but not present in the messengers known in *Drosophila*. These data are not compatible with the idea that σ-virus derives directly from a rhabdovirus ancestor, which could be an insect virus. It may mean, however, that σ-virus takes its origin from a vesiculovirus if viral contamination is possible between dipterans of different species through a nonspecific biting parasite.[5]

The sequence of the Sonchus yellow net virus (SYNV), a plant rhabdovirus, is currently being determined.[76-78] An appropriate probe has evidenced a RNA leader with 139 to 144 nucleotides, almost three times as long as the known vertebrate virus RNA leaders (about 50 nucleotides). The sequences of N and M_2 proteins (the last one homologous of the VSVs NS phosphoprotein) have practically no homology with the sequences of the corresponding vertebrate virus proteins. Only a small region of M_2 protein presents a certain homology with NS. On the other hand, a clear similarity between the polyadenylation signals and the resumption of transcription signals probably reflects a functional analogy of the viral polymerases.

A special mention should be made for the comparison of the sequences of Indiana and New Jersey L protein[65] due to the size and functional importance of this protein. The average homology is 65%. However, the authors mention six regions where the homology reaches 85 to 95%, and they suggest that these are the functionally important regions. Moreover, two of these show a certain homology with two regions of the polymerases of the paramyxoviruses, Sendai and Newcastle disease. Paramyxoviruses, like rhabdoviruses are viruses with a single-molecule antimessenger RNA genome, and their replication strategy is the same. A certain similarity of the gene order coding for homologous functional proteins of the paramyxovirus and rhabdovirus was also observed.[79] There are more proteins coded by the paramyxoviruses genome and more proteins in the virion than for the rhabdoviruses. However, some rhabdoviruses may be more complex than those studied up to now. For instance, Flanders virus, which possesses a longer virion than the VSV, contains eight proteins; two of them may be issued from cuts.[80] It will be easier to measure the difference between this virus and the other animal rhabdoviruses once the messengers will be isolated and the proteins located in the virion.

REFERENCES

1. **Matthews, R. E. F.**, Classification and nomenclature of viruses, *Intervirology*, 17, 1, 1982.
2. **Bishop, D. H. L., Ed.**, *Rhabdoviruses*, Vol. 3, CRC Press, Boca Raton, FL, 1979.
3. **Wagner, R. R., Ed.**, *The Rhabdoviruses*, Plenum Press, New York, 1987.
4. **Jahromi, S. S.**, Occurrence of rhabdovirus-like particles in the blue crab, *Callinectes sapidus*, *J. Gen. Virol.*, 36, 485, 1977.
5. **Brun, G. and Plus, N.**, The viruses of *Drosophila*, in *Genetics and Biology of Drosophila*, Vol. 2, Ashburner, M. and Wright, T. R. F., Eds., Academic Press, London, 1980, 625.
6. **Shope, R. E. and Tesh, R. B.**, The ecology of rhabdoviruses that infect vertebrates, in *The Rhabdoviruses*, Wagner, R. R., Ed., Plenum Press, New York, 1987, 509.
7. **Bussereau, F.**, The CO_2 sensitivity induced by two rhabdoviruses Piry and Chandipura in *Drosophila melanogaster*, *Ann. Microbiol. (Inst. Pasteur)*, 126B, 389, 1975.
8. **Brun, G.**, Etude génétique des relations du virus Piry avec la Drosophile: I. Sélection de mutants du virus restreints dans leur capacité d'envahir la drosophile (mutants agD), *Ann. Inst. Pasteur*, 138, 195, 1987.
9. **Bussereau, F.**, Etude du symptôme de la sensibilité au CO_2 produit par le virus de la Stomatite vésiculaire chez *Drosophila melanogaster*. I. VSV de sérotype New Jersey et le virus Cocal, *Ann. Inst. Pasteur*, 121, 223, 1971.
10. **Bussereau, F.**, Etude du symptôme de la sensibilité au CO_2 produit par le virus de la Stomatite vésiculaire chez *Drosophila melanogaster*. II. VSV de sérotype Indiana, *Ann. Inst. Pasteur*, 122, 1029, 1972.
11. **Bussereau, F.**, Etude du symptôme de la sensibilité au CO_2 produit par le virus de la Stomatite vésiculaire chez *Drosophila melanogaster*. III. Souches de différents sérotypes, *Ann. Microbiol. (Inst. Pasteur)*, 124A, 535, 1973.
12. **Rosen, L.**, Carbon dioxide sensitivity in mosquitoes infected with sigma, vesicular stomatitis and other rhabdoviruses, *Science*, 207, 989, 1980.

13. **Tesh, R. B., Chianotis, B. N., and Johnson, K. M.,** Vesicular stomatitis virus (Indiana serotype): transovarial transmission by phlebotomine sandflies, *Science,* 175, 1477, 1972.

14. **Tesh, R. B. and Modi, G. B.,** Growth and transovarial transmission of Chandipura virus (Rhabdoviridae, Vesiculovirus) in *Phlebotomus papatasi, Am. J. Trop. Med. Hyg.,* 32, 621, 1983.

15. **Nichol, S. T.,** Molecular epizootiology and evolution of vesicular stomatitis virus New Jersey, *J. Virol.,* 61, 1029, 1987.

16. **Nichol, S. T.,** Genetic diversity of enzootic isolates of vesicular stomatitis virus New Jersey, *J. Virol.,* 62, 572, 1988.

17. **Lastra, J. R. and Esparza, J.,** Multiplication of vesicular stomatitis virus in the leafhopper *Peregrinus maidis,* a vector of a plant rhabdovirus, *J. Gen. Virol.,* 32, 139, 1976.

18. **Brown, F. and Crick, J.,** Natural history of the rhabdoviruses of vertebrates and invertebrates, in *Rhabdoviruses,* Vol. 1, Bishop, D. H. L., Ed., CRC Press, Boca Raton, FL, 1979, 1.

19. **Frazier, C. L. and Shope, R. E.,** Serologic relationships of animal rhabdoviruses, in *Rhabdoviruses,* Vol. 1, Bishop, D. H. L., Ed., CRC Press, Boca Raton, 1979, 43.

20. **Della-Porta, A. J. and Snowdon, W. A.,** Bovine ephemeral fever, in *Rhabdoviruses,* Vol. 3, Bishop, D. H. L., Ed., CRC Press, Boca Raton, FL, 1980, 167.

21. **Roy, P.,** Fish rhabdoviruses, in *Rhabdoviruses,* Vol. 3, Bishop, D. H. L., Ed., CRC Press, Boca Raton, FL, 1980, 193.

22. **Bussereau, F., de Kinkelin, P., and Le Berre, M.,** Infectivity of fish rhabdoviruses for *Drosophila melanogaster, Ann. Microbiol. (Inst. Pasteur),* 126A, 389, 1975.

23. **Bussereau, F. and Contamine, D.,** Infectivity of two rhabdoviruses in *Drosophila melanogaster:* rabies and Isfahan viruses, *Ann. Inst. Pasteur (Virologie),* 131E, 3, 1980.

24. **Karabatsos, N.,** *International catalogue of arboviruses including certain other viruses of vertebrates,* 3rd ed., American Society of Tropical Medicine and Hygiene, San Antonio, Texas, 1985.

25. **Francki, R. I. B.,** Plant Rhabdoviruses, *Adv. Virus Res.,* 18, 257, 1973.

26. **Francki, R. I. B. and Randles, J. W.,** Rhabdoviruses infecting plants, in *Rhabdoviruses,* Vol. 3, Bishop, D. H. L., Ed., CRC Press, Boca Raton, FL, 1980, 135.

27. **Jackson, A. O., Francki, R. I. B., and Zuidema, D.,** Biology, structure and replication of plant rhabdoviruses, in *The Rhabdoviruses,* Wagner, R. R., Ed., Plenum Press, New York, 1987, 427.

28. **Black, L. M.,** Specific transmission of varieties of potato yellow dwarf virus by related insects, *Am. Potato J.,* 18, 231, 1941.

29. **Black, L. M.,** Biological cycles of plant viruses in insect vectors, in *The Viruses,* Vol. 2, Burnet, F. M. and Stanley, W. M., Eds., Academic Press, New York, 1959, 157.

30. **Dale, J. L. and Peters, D.,** Protein composition of the virions of five plant rhabdoviruses, *Intervirology,* 16, 86, 1981.

31. **Teninges, D., Contamine, D., and Brun, G.,** Drosophila sigma virus, in *Rhabdoviruses,* Vol. 3, Bishop, D. H. L., Ed., CRC Press, Boca Raton, FL, 1980, 113.

32. **Fleuriet, A.,** Effect of overwintering on the frequency of flies infected by the rhabdovirus sigma in experimental populations of *Drosophila melanogaster, Arch. Virol.,* 69, 253, 1981.

33. **Knudson, D. L.,** Rhabdoviruses, in *Comparative Virology Symposium,* Brown, F. and Twisley, T. M., Eds., *J. Gen. Virol.,* 20 suppl., 1973, 105.

34. **Murphy, F. A. and Harrison, A. K.,** Electron microscopy of the rhabdoviruses of animals, in *Rhabdoviruses,* Vol. 1, Bishop, D. H. L., Ed., CRC Press, Boca Raton, FL, 1979, 65.

35. **Dubois-Daley, M., Holmes, K. V., and Rentier, B.,** Assembly of enveloped RNA viruses, Springer-Verlag, New York, 1984, chap 2.

36. **Peters, D. and Kitajima, E. W.,** Purification and electron microscopy of sowthistle yellow vein virus, *Virology,* 41, 135, 1970.

37. **Zee, Y. C., Hackett, A. J., and Taleus, L.,** Vesicular stomatitis virus maturation sites in six different host cells, *J. Gen. Virol.,* 7, 95, 1970.

38. **Zajac, B. A. and Hummeler, K.,** Morphogenesis of the nucleoprotein of vesicular stomatitis virus, *J. Virol.,* 6, 243, 1970.

39. **McDaniel, L. L., Ammar, E. D., and Gordon, D. T.,** Sites of assembly and ultrastructure of a hawaiian isolate of maize mosaic virus in maize plants, *Phytopathology,* 75, 1167, 1985.

40. **Ammar, E. D. and Nault, L. R.,** Assembly and accumulation sites of maize mosaic virus in its planthopper vector, *Intervirol.,* 24, 33, 1985.

41. **Rose, J. and Schubert, M.,** Rhabdovirus genome and their products, in *The Rhabdoviruses,* Wagner, R. R., Ed., Plenum Press, New York, 1987, 129.

42. **Emerson, S. U.,** Transcription of vesicular stomatitis virus, in *The Rhabdoviruses,* Wagner, R. R., Ed., Plenum Press, New York, 1987, 245.

43. **Wertz, G. W., Davis, N. L., and Patton, J.,** The role of proteins in vesicular stomatitis virus RNA replication, in *The Rhabdoviruses,* Wagner, R. R., Ed., Plenum Press, New York, 1987, 271.

44. **Pal, R. and Wagner, R. R.,** Rhabdovirus membrane and maturation, in *The Rhabdoviruses,* Wagner, R. R., Ed., Plenum Press, New York, 1987, 75.

45. **Holland, J. J.,** Defective interfering rhabdoviruses, in *The Rhabdoviruses,* Wagner, R. R., Ed., Plenum Press, New York, 1987, 297.

46. **Ismael, I. D. and Milner, J. J.,** Isolation of defective interfering particles of sonchus yellow net virus from chronically infected plants, *J. Gen. Virol.,* 69, 999, 1988.

47. **Bussereau, F.,** Etude du symptôme de la sensibilité au CO_2 produit par le virus sigma chez la drosophile. I. Influence du lieu d'inoculation sur le delai d'apparition du symptôme. *Ann. Inst. Pasteur,* 118, 367, 1970.

48. **Bussereau, F.,** Etude du symptôme de la sensibilité au CO_2 produit par le virus sigma chez la drosophile. II. Evolution comparée du rendement des centres nerveux et de divers organes après inoculation dans l'abdomen et dans le thorax, *Ann. Inst. Pasteur,* 118, 626, 1970.

49. **Richard-Molard, C., Blondel, D., Wyers, F., and Dezelee, S.,** Sigma virus: growth in *Drosophila melanogaster* cell culture: purification, protein composition and localization, *J. Gen. Virol.,* 65, 91, 1984.

50. **Ohanessian, A.,** Vertical transmission of the Piry rhabdovirus by sigma-infected Drosophila melanogaster females, *J. Gen. Virol.,* 70, 209, 1989.

50a. **Ohanessian, A.,** personal communication.

51. **Bregliano, J. C.,** Contribution à l'étude des relations entre le virus sigma et son hôte. Etude du caractère g−, *Ann. Inst. Pasteur,* 117, 325, 1969.

52. **Gay, P.,** Les gènes de la drosophile qui interviennent dans la multiplication du virus sigma, *Mol. Gen. Genet.,* 159, 269, 1968.

53. **Nagaraj, A. N. and Black, L. M.,** Hereditary variation in the ability of a leafhopper to transmit two unrelated plant viruses, *Virology,* 16, 152, 1962.

54. **Contamine, D.,** Role of the *Drosophila* genome in sigma virus multiplication. I. Role of the ref(2)pP gene: selection of host adapted mutants at the non permissive allele P^P, *Virology,* 114, 474, 1981.

55. **Coulon, P. and Contamine, D.,** Role of the *Drosophila* genome in sigma virus multiplication. II. Host spectrum variants among the haP mutants, *Virology,* 123, 381, 1982.

56. **Dunigan, D. D., Baird, S., and Lucas-Lenard, J.,** Lack of correlation between the accumulation of plus-stand leader RNA and the inhibition of protein and RNA synthesis in vesicular stomatitis virus infected mouse L cells, *Virology,* 150, 231, 1986.

57. **Dezelee, S., Blondel, D., Wyers, F., and Petitjean, A. M.,** Vesicular stomatitis virus in *Drosophila melanogaster* cells: lack of leader RNA transport into the nuclei and frequent abortion of the replication step, *J. Virol.,* 61, 1391, 1987.

58. **Blondel, D., Petitjean, A. M., Dezelee, S., and Wyers, F.,** Vesicular stomatitis virus in *Drosophila melanogaster* cells: regulation of viral transcription and replication, *J. Virol.,* 62, 277, 1988.

59. **Blondel, D., Dezelee, S., and Wyers, F.,** Vesicular stomatitis virus growth in *Drosophila melanogaster* cells. II. Modifications of viral protein phosphorylation, *J. Gen. Virol.,* 64, 1793, 1983.

60. **Wyers, F., Richard-Molard, C., Blondel, D., and Dezelee, S.,** Vesicular stomatitis virus growth in *Drosophila melanogaster* cells: G protein deficiency, *J. Virol.,* 33, 411, 1980.

61. **Wyers, F., Blondel, D., Petitjean, A. M., and Dezelee, S.,** Restricted expression of viral glycoprotein in vesicular stomatitis virus-infected *Drosophila melanogaster* cells, *J. Gen. Virol.,* 70, 213, 1989.

62. **Kurath, G. and Leong, J.,** Characterization of infectious hematopoietic necrosis virus mRNA species reveals a nonvirion rhabdovirus protein, *J. Virol.,* 53, 462, 1985.

63. **Kurath, G., Ahern, K. G., Pearson, G. D., and Leong, J. C.,** Molecular cloning of the six mRNA species of infectious hematopoietic necrosis virus, a fish rhabdovirus, and gene order determination by R-loop mapping, *J. Virol.,* 53, 469, 1985.

64. **Tordo, N., Poch, O., Ermine, A., Keith, G., and Rougeon, F.,** Walking along the rabies genome: is the large G-L intergenic region a remnant gene?, *Proc. Natl. Acad. Sci. U.S.A.,* 83, 3914, 1986.

65. **Feldhaus, A. L. and Lesnaw, J. A.,** Nucleotide sequence of the L gene of vesicular stomatitis virus (New Jersey): identification of conserved domains in the New Jersey and Indiana L protein, *Virology,* 163, 359, 1988.

66. **Banerjee, A. K., Rhodes, D. P., and Gill, D. S.,** Complete nucleotide sequence of the mRNA coding for the N protein of vesicular stomatitis virus (New Jersey serotype), *Virology,* 137, 432, 1984.

67. **Gill, D. S. and Banerjee, A. K.,** Complete nucleotide sequence of the matrix protein mRNA of vesicular stomatitis virus (New Jersey serotype), *Virology,* 150, 308, 1986.

68. **Gallione, C. J. and Rose, J. K.,** Nucleotide sequence of a cDNA clone encoding the entire glycoprotein from the New Jersey serotype of vesicular stomatitis virus, *J. Virol.,* 46, 162, 1983.

69. **Gill, D. S. and Banerjee, A. K.,** Vesicular stomatitis virus NS proteins: structural similarity without extensive sequence homology, *J. Virol.,* 55, 60, 1985.

70. **Giogi, C., Blumberg, B., and Kolakofsky, D.,** Sequence determination of the + leader RNA regions of the vesicular stomatitis virus Chandipura, Cocal and Piry serotype genomes, *J. Virol.,* 46, 125, 1983.

71. **Masters, P. S. and Banerjee, A. K.,** Sequences of Chandipura virus N and NS genes: evidence for high mutability of the NS gene within vesiculoviruses, *Virology,* 157, 298, 1987.

72. **Roy, P., Gupta, K. C., and Kiuchi, A.,** Characterization of spring viremia of carp virus mRNA species and the 3′ sequence of the viral RNA, *Virus Res.,* 1, 189, 1984.

73. **Kiuchi, A. and Roy, P.,** Comparison of the primary sequence of spring viremia of carp virus M protein with that of vesicular stomatitis virus, *Virology,* 134, 238, 1984.

74. **Koener, J. F., Passavant, C. W., Kurath, G., and Leong, J.,** Nucleotide sequence of a cDNA clone carrying the glycoprotein gene of infectious hematopoietic necrosis virus, a fish rhabdovirus, *J. Virol.,* 61, 1342, 1987.

75. **Teninges, D. and Bras-Herreng, F.,** Rhabdovirus sigma, the hereditary CO_2 sensitivity agent of *Drosophila:* nucleotide sequence of a cDNA clone encoding the glycoprotein, *J. Gen. Virol.,* 68, 2625, 1987.

76. **Zuidema, D., Heaton, L. A., Hanau, R., and Jackson, A. O.,** Detection and sequence of plus-strand leader RNA of sonchus yellow net virus, a plant rhabdovirus, *Proc. Nat. Acad. Sci. U.S.A.,* 83, 5019, 1986.

77. **Zuidema, D., Heaton, L. A., and Jackson, A. O.,** Structure of the nucleocapsid protein gene of sonchus yellow net virus, *Virology,* 159, 373, 1987.

78. **Heaton, L. A., Zuidema, D., and Jackson, A. O.,** Structure of the M2 protein gene of sonchus yellow net virus, *Virology,* 161, 234, 1987.

79. **Pringle, C. R.,** Rhabdovirus genetics, in *The Rhabdoviruses,* Wagner, R. R., Ed., Plenum Press, New York, 1987, 167.

80. **Boyd, K. R. and Witaker-Dowling, P.,** Flanders virus replication and protein synthesis, *Virology,* 163, 349, 1988.

Chapter 18

TOGAVIRIDAE AND FLAVIVIRIDAE*

Philip M. Grimley

TABLE OF CONTENTS

*   The opinions or assertions contained herein are the private views of the author and should not be construed as official or necessarily reflecting the views of the Uniformed Services University of the Health Sciences or the Department of Defense.

Dedication

The author dedicates this chapter to the memory of Dr. Nat Young, a devoted arbovirologist and pathologist, whose wit and gentle humor inspired many friends and colleagues.

I. INTRODUCTION

The Togaviridae and Flaviviridae are closely related families of enveloped RNA viruses. Togavirions measure 60 to 65 nm in diameter; flavivirions measure 35 to 45 nm in diameter. They possess icoshoderal nucleocapsids with 32 capsomers.[1,2] Historically, the currently defined families of Togaviridae, Flaviviridae, and Bunyaviridae were grouped ecologically as arthropod-borne "arboviruses", with evidence for transmission by blood-sucking insects.[3-5] The clinical grouping of arboviruses retains a useful epidemiological concept; however, it is now well recognized that a number of RNA viruses with the structural, biochemical, and molecular biological attributes of togaviruses or flaviviruses are transmitted directly among vertebrate hosts.

Electron microscopy demonstrated that bunyaviruses comprise a structurally distinct family of arboviruses: symmetry of their nucleocapsids is helical rather than icosohedral.[2] While the togaviruses and flaviviruses share closer physical similarities, they differ sufficiently both in molecular replication strategies and in specifics of morphogenesis to warrant taxonomic separation as two large families.[5-8] Comprehensive reviews of Togavirus and/or Flavivirus molecular biology, molecular biochemistry, and structural development have been published within the last 2 years,[2,6-9] and there is an important earlier review.[10] In this atlas, current knowledge is briefly summarized as a context for comparing the key ultrastructural features of Togavirus and Flavivirus morphogenesis.

A. TOGAVIRIDAE

The family Togaviridae embraces four major genera: *Alphavirus, Rubivirus, Pestivirus,* and *Arterivirus.* Within the genus *Alphavirus,* all 25 known members are mosquito transmitted. Alphaviruses infect a broad range of vertebrate hosts, including rodents, wild birds, reptiles, and amphibians.[11] The infections in nature are generally subclinical, yet some wild hosts sustain sufficiently high viremic titers to provide reservoirs for mosquito infections and for incidental transmission to humans.[4,11] In humans, the majority of alphavirus infections also are inapparent and asymptomatic.[3] Asymptomatic alphavirus infections may be due to multifactorial effects of immune responses,[12,13] efficient induction of interferon,[14] production of defective interfering virus,[13,15,16] or local constraints on virus dissemination.[17,18] Symptomatic disease manifestations typically include febrile illness, exanthem, flu-like arthritic aches, and mild encephalitis,[3] but some alphavirus infections can produce myocarditis, hepatitis, or other visceral damage.[4,12] Fulminant disease with severe morbidity or fatal complications is always a risk[12,17] and all alphavirus strains must be considered potentially virulent in field or laboratory environments.[3]

Other than the genus *Alphavirus,* togaviruses are not arthropod-borne. The genus *Rubivirus* includes only rubella virus, which is potentially teratogenic during early gestation.[19] The genus *Pestivirus* includes several agents of diarrheal diseases in farm animals. The genus *Arterivirus* is represented only by equine arteritis virus. Lactic dehydrogenase virus of mice is classified as a togavirus, but its relationship to defined genera is not yet settled.

B. FLAVIVIRIDAE

The family Flaviviridae currently includes 30 mosquito-borne viruses, 12 tick-borne viruses, and 22 viruses with no known insect vector. Some of the latter agents, such as

Dakar bat Modoc, or Rio Bravo viruses spread directly among small rodents or bats[5] and fail to replicate in arthropods or arthropod cell cultures.

In general, flavivirus infections present a more severe health threat to humans, subhuman primates, and domestic livestock than do the Alphavirus infections.[3,4] Flavivirus infections are more often complicated by severe encephalitis, secondary hemorrhagic manifestations, or tissue necrosis with damage to vital functions of the liver or kidney.[3,4,12,17] Indeed, the family name derives from yellow fever which causes jaundice secondary to hepatic damage; and yellow fever was the first virus disease of humans subjected to rigorous etiologic and epidemiological investigation[5].

Traditionally, serological surveys were the primary source of epidemiological data concerning arbovirus infections, and seroconversion remains practically useful in clinical diagnosis of togavirus or flavivirus infections.[3] The serological tests include hemagglutination inhibition, complement fixation, or plaque-reduction neutralization. Serology has become increasingly powerful in flavivirus classification since the development of monoclonal antibodies which recognize specific glycoprotein determinants on virus envelopes.[20] At present flaviviruses are divided into seven antigenic subgroups and are ecologically subclassified by vector of transmission.[6]

II. INSECT INFECTIONS

Hematophagous arthropods typically become infected with alpha togaviruses or flaviviruses following ingestion of a blood meal from viremic animal or human hosts.[21] Some recent experiments have also suggested the possibility of arthropod-to-arthropod infection during cofeeding on a nonviremic animal.[22] Immediate transmission from arthropod to vertebrate can be mechanical, due to bloody contamination of arthropod mouthparts; however, biological transmission after multiplication of virus in an arthropod vector is the epidemiologically significant route.[3,21] Under conditions of experimental oral exposure, propagation of virus infections in mosquito or tick vectors requires a period of extrinsic incubation and this "eclipse phase" is temperature dependent.[23,24]

Virus replication and amplification begins in epithelial cells lining the posterior mesenteron[23,25-27] and possibly the abdominal fat body.[27] Infection can be retarded at the virus penetration or attachment stage by genetically controlled host factors;[23] nevertheless, biochemically specific host cell viroceptors[28] appear unlikely in view of the broad host and vector ranges.[4,23,29] Virus dissemination in mosquitos occurs via the hemolymph, so that multiple tissues, including ganglia, oviducts, thoracic muscle, and salivary glands become heavily infected within several days.[30] After 3 d to 2 weeks, alphavirus replicates in, and is secreted from, the salivary glands.[30,31a] This exocrine emergence of virions ensures infection of vertebrate hosts during subsequent feeding cycles: arthropod saliva acts as an anticoagulant when introduced subdermally during the exploratory probing prior to bloodsucking.[32]

Cytopathic effects of alphaviruses in insect vectors are evidently minimal;[25,31a] although central nervous system infection can occur[27] and lesions occurred in the midgut of *Culiseta melanura* 2 to 5 days after viral infection with Eastern equine encephalitis virus.[31b] *In vitro*, cytopathic effects of alphavirus or flavivirus infections of mosquito cells also occur relatively late[33] and the infections tend to be persistent.[23,29,34] Defective-interfering particles may modulate production of functional virus.[35]

III. MOLECULAR BIOLOGY AND REPLICATION CYCLE

A. PENETRATION OF VIRUS
The enveloped RNA genomes of Togaviridae and Flaviviridae are single "plus" strands

of approximately 12 kilobases and can function directly as mRNA. In some cases, the naked RNA has been proven infectious under experimental conditions.

Togaviruses infect cells by adsorption to the host cell surface[18,26,36] uptake into clathrin coated pits, and fusion with lysosomes.[37] A possible role for immunochemically specific receptors remains controversial.[39] Associated changes in surface membrane fluidity occur.[38] Viral nucleic acid is released into the cytoplasmic compartment after fusion of the virion envelope with the lysosomal limiting membrane.[37] Acidic pH within the lysosome evidently triggers final penetration.[37,40]

Flaviviruses or flavivirus-antibody complexes also enter cells by adsorbtion and coated pits.[41] Endocytosis delivers the virions to lysosomes where membrane fusion probably occurs at acidic pH.[42,43] Antibody (IgG)-dependent enhancement of flavivirus infections can occur in cells which bear Fc receptors,[41] so that increased virus yield from macrophages or monocytes may be a factor in clinical disease expression.[6,16,41]

B. GENOME REPLICATION AND TRANSCRIPTION
1. Alphaviruses

After penetration, the parental plus strand acts as a messenger to direct the synthesis of a 200- to 230-kDa polyprotein precursor. The large polyprotein is proteolytically cleaved to yield several nonstructural proteins required for an RNA-dependent RNA polymerase complex.[2] This strategy facilitates the copying of full length complementary minus strands from the virion RNA. Under optimized experimental conditions, copying of minus strands can be completed within about 4 h. It must precede replication of new plus strand 42 S genomic RNA and of a new subgenomic, polycistronic 26 S mRNA. The latter is made in great excess compared to genomic 42 S RNA[44,45] and directs synthesis of a second polyprotein which subsequently cleaves to produce all of the virion structural proteins (see Reference 2 for a detailed analysis of alphavirus proteins).

Steps in both transcription and replication of several enveloped RNA viruses appear to be membrane associated[2,28] and membrane-associated RNA replication complexes have been identified in alphavirus infections.[46,47] Morphogenetically this is reflected by development of vesicles, attached to the interior of cytoplasmic vacuoles (CPV-1) or at the cell surface (see Section V, "Ultrastructural Morphogenesis").

2. Flaviviruses

The parental virion RNA serves as template for synthesis of complementary minus strand 42 to 44 S RNA which then replicates new plus strands. The completed new plus strands serve either as messengers for the translation of virion structural proteins or as replicative molecules for encapsidation into progeny virions. Unlike the togaviruses and similar only to the picornaviruses, flaviviruses produce no subgenomic forms of mRNA.[6,7] The ratio of plus- to minus-strand RNA equilibrates at about 10:1.[48]

The mechanisms for determining amplification of plus-strand flavivirus RNA and the sources of polymerase enzyme complexes which are essential both for minus- and plus-strand replication are not fully resolved.[6,7] RNA polymerase activity has been associated with internal cytomembranes, particularly the perinuclear membranes where a large proportion of viral RNA synthesis occurs;[7,49] and there is some evidence for recycling of an RNA replicative intermediate with a minus-strand template.[50] Early morphogenesis of membrane vesicles may be related to a membranous replication complex,[51] but there is yet no direct evidence (see Section V, "Ultrastructural Morphogenesis").

C. NUCLEOCAPSID ASSEMBLY AND VIRION ENVELOPMENT
1. Alphaviruses

A nucleocapsid C protein is translated on free polysomes in the cytoplasm, then rapidly unites with 42 S genomic RNA. This is a relatively specific process, and in contrast to cells infected with icosohedral DNA viruses, no empty capsids form during togavirus assembly. The C protein is evidently an efficient nucleic acid binder and it can self assemble *in vitro* with isolated genomic RNA.[52] Completion of nucleocapsid assembly occurs in close proximity to cytomembranes, and with progression of infection, the nucleocapsids accumulate on the cytoplasmic faces of vacuoles derived from the rough endoplasmic reticulum (RER).

Virion envelopment may proceed either by nucleocapsid budding at the cell surface or within cisternae of subcellular compartments (endoplasmic reticulum or Golgi apparatus).[2,10,28] Two transmembranous envelope glycoproteins and a peripheral glycoprotein interact to form the virion peplomers and envelope spikes.[2,45] This involves a complex sequence of biosynthetic and cleavage steps in which a polyprotein precursor is translated from 26 S mRNA on the RER and translocated across the cisternal membrane (see Reference 2 for details). The precursor is cleaved to yield the envelop hemagglutinin (E1) and a p62 protein. Core oligosaccharides are added cotranslationally in the RER and nascent glycoproteins are transported through the Golgi apparatus where high-mannose oligosaccharides are trimmed and terminal sugar groups are added.

The final alphavirion maturation step involves cleavage of p62 at the plasma membrane or along the Golgi membranes to yield another envelope protein (E2) and spike glycoprotein (E3). Glycoprotein E3 does not appear to be essential for infectivity of some alphaviruses.[53] Host glycosyl transferases determine the precise molecular composition of the envelope protein oligosaccharides and the host membrane lipid content is reflected in the envelope lipids.[10,35,54] Thus, details of alphavirus biochemistry may differ depending upon replication in arthropod or vertebrate hosts.[35]

2. Flaviviruses

A full-sized genomic RNA is the only flavivirus RNA thus identified on polysomes of infected cells; however, production of a stable polyprotein precursor has not been established. It seems likely that coordinate cleavage of a transient polyprotein occurs during translation;[6,7,55] however, biosynthesis of individual polypeptides by a nonuniform translation process has not been excluded.

The assembly of flavivirus nucleocapsids is another major puzzle: unlike alphavirus morphogenesis, there has been no ultrastructural evidence for intermediate accumulation of virion structural precursors. Only completed enveloped virions are detected on the surfaces of the cells or within the cisternae of the endoplasmic reticulum or Golgi apparatus. If conventional biophysical "budding" of nucleocapsids does occur through the membrane bilayers, it is evidently too rapid for capture by current methods of cell fixation or freeze fracture.[28,56] An alternative mechanism might involve cotranslational assembly to the "outside" of cell compartments. These questions probably will not be answered until the train of molecular events in flavivirus encapsidation and envelopment has been more precisely resolved, perhaps by means of genetically engineered and cloned viral RNA.[57]

IV. MACROMOLECULAR ORGANIZATION OF VIRIONS

A. ALPHAVIRUSES

Virions are spherical particles of 60 to 65 nm in diameter. In ultrathin sections, the virus core is relatively electron dense, but shows a filamentous internal structure at high magnifications.[58] RNA cyclization is observed *in vitro*.[8] In highly purified preparations, the cubic (icosohedral) symmetry of nucleocapsids may be resolved and the peplomers also show cubic symmetry.[9,58] An indistinct fuzzy coat probably represents low molecular weight E3 glycoprotein.[2] In freeze-fracture preparations, the periphery of the virus envelope is covered with 4-nm subunits with a center-to-center spacing of 6 nm.[59]

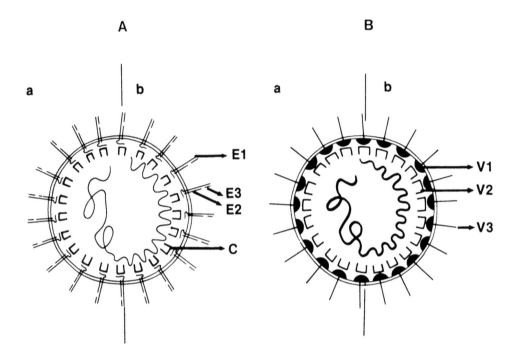

FIGURE 1.   (A) A model of alphavirion organization[2] which indicates the locations of three envelope glycoproteins; E1 and E2 are transmembrane glycoproteins, whereas E3 is peripheral. The cytoplasmic domain
of E2 interacts with the C protein subunits in the icosohedral nucleocapsid and this results in a cubic symmetrical
arrangement of peplomers of the virion envelope. The arrangement of the genomic RNA is not known:
hemisphere (a) diagrams a relatively free RNA strand; hemisphere (b) diagrams a regular interaction of the
virion RNA with capsid C protein subunits. (B) A model of flavivirion,[2] based upon the presumptive locations
of two envelope proteins; the glycoprotein V3 forms the peplomers, the nonglycosylated protein V1 lies interior
to the lipid bilayer, and V2 is the nucleocapsid protein. Again, the arrangement of the genomic RNA is not
known: hemisphere (a) diagrams a relatively free RNA strand; hemisphere (b) diagrams a regular interaction
of the virion RNA with V2 capsid subunits. (From Dubois-Dalcq, M., Holmes, K. V., and Rentier, B.,
Assembly of Enveloped RNA Viruses, Springer-Verlag, Vienna, 1984, 235. Reprinted by permission of author
and Springer-Verlag.)

The virion is composed of single-stranded genomic RNA, and 250 copies each of the
capsid C protein and three envelope glycoproteins (E1, E2, E3).[9] The formation of an
icosohedral nucleocapsid structure from genomic RNA and the C protein probably depends
upon electrostatic interactions between the RNA and basic amino acids.[2] The C protein is
evidently oriented so that the carboxy-terminal end interacts with envelope glycoproteins
E1 and E2 as illustrated in Figure 1A. The smaller E3 glycoprotein is associated with
peplomers and may form spikes. Strong transmembrane interactions between peplomers and
nucleocapsids evidently account for the icosohedral exterior symmetry (T = 4) of togaviruses.[43]

B. FLAVIVIRUSES

The sperhical virions measure 36 to 44 nm in diameter. The virion core is electron dense
and measures 27 nm in diameter. Peplomers may be clearly resolved in purified virions and
from a thin capsule (see Figure 1B). At high magnification, some flaviviruses reveal circular
subunits (7 nm in diameter) on the outer envelope.[60] During flavivirus infections, donut
shaped 14-nm structures may be released from the envelope (see Figure 2); these represent
a slowly sedimenting hemagglutinin.[2]

The virion contains single-stranded RNA and three structural proteins (V1, V2, V3).

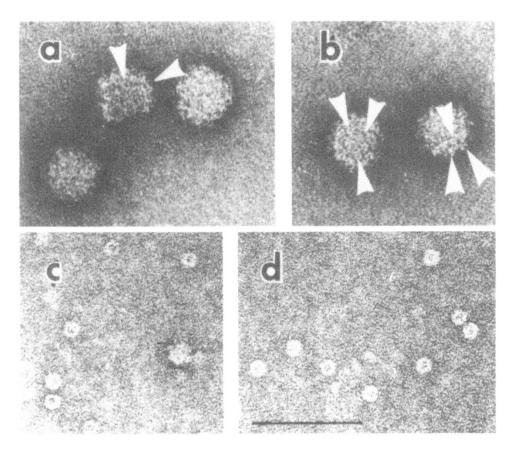

FIGURE 2. Negatively stained preparations of flavivirons (Dengue virus) and subviral structures. (a) and (b) Show circular 7-nm diameter subunits (arrowheads) on the surface of the viral envelopes. (c) and (d) Show slowly sedimenting particles which are released during virus infections and are associated with hemagglutination; they contain V3, V1, and V2 proteins in a 14-nm diameter ring configuration (Magnification × 140,000.) (Courtesy of Dr. Monica Dubois-Dalcq.[2]) (From Smith, T. J., Brandt, W. E., Swanson, J. L., McGown, J. M., and Buescher, E. L., *J. Virol.*, 5, 524, 1970; reproduced with permission of the American Society for Microbiology).

Protein V2 is the nucleocapsid protein and binds to virus RNA. Protein V1 lies beneath the envelope membrane in intact virions and is not glycosylated. Protein V3 is glycosylated and comprises the peplomers. It is the virus hemagglutinin and may influence binding to host cells.[2,61]

V. ULTRASTRUCTURAL MORPHOGENESIS

The molecular biology and morphogenesis of alphaviruses and flaviviruses have been studied most extensively in mammalian systems, both in tissue cultures and intact animals. In recent years, techniques for investigating arbovirus infections in mosquito cell cultures have become widely accessible;[29,33,62] as yet, however, there have been relatively few ultrastructural morphogenetic studies *in vitro*.[63-66] Ultrastructural investigation of virus development in intact insects is technically challenging, but there have been several comprehensive descriptions of alphavirus infections.[25,27,30] A major problem in interpreting the results of arthropod investigations is that the *in vivo* virus developmental kinetics depend upon extrinsic incubation conditions and results are difficult to compare with warm-blooded animals.[23-25,65]

In vitro, the ultrastructural features of Togavirus and/or Flavivirus infections thus far described appear sufficiently similar in mammalian and invertebrate cells as to indicate morphogentically identical, chronologically parallel events.[65] In this chapter, therefore, the author has relied upon mammalian cell studies for the bulk of illustrative materials.

A. ALPHAVIRUSES
1. Early Membrane Vesicles

In many insect and vertebrate cells, the earliest evidence of alphavirus replication is formation of 50-nm diameter membranous vesicles with irregular electron-dense internal strands (see Figures 3 through 6). These characteristic vesicles line internal vacuolar membranes or develop on patches of the surface membrane.[65,67] The membranes delimiting vesicles exhibit a bilaminar electron density in thin sections;[65,67] however, freeze-cleavage has indicated an absence of protein particles on the interior fracture face.[68] Overall phospholipid biosynthesis is inhibited during alphavirus infection,[69] so that both the individual vesicles and the vesicle-lined vacuoles presumably arise from preexisting cytomembranes or catabolic products.

The potential biological significance of membranous vesicles in togavirus infections was deduced in cells infected by Semliki Forest virus:[46,70] the numerous vesicle-lined vacuoles, measuring 1 to 2 μm in diameter, were designated cytopathic vacuoles, Type 1 (CPV-1), in order to emphasize their relatively early appearance, and formation prior to accumulation of viral nucleocapsids.[70] Membranous vacuoles surrounded by nucleocapsids were designated (CPV-2). The association of CPV-1 with alphavirus infections has been repeatedly observed, both *in vivo* and *in vitro,* and in many types of infected vertebrate or mosquito cells under various incubation conditions[1,27,28,31,64,65,67,68,70-73] (see Table 1). As with any other ultrastructural sampling, detection of CPV-1 is dependent upon size of the structures, thickness of the sections, numbers or sections examined, and binominal statistics.[1] Thus, some reports of apparently negative findings[18,64] could reflect reduced diameter of the CPV-1, such as noted in mosquito cells,[65] relatively late times of sampling, or a statistically insufficient sampling.

Formation of CPV-1 in tissue cultures did not depend upon multiplicity of infection or conditions of virus passage and was not related to production of defective interfering virus.[15,67,71] CPV-1 have been observed in infected tissues of intact mammals and mosquitos[1,28,31,72,73a] (see Figures 6B and 7; Table 1). The asynchrony of *in vivo* infections makes chronological correlation of morphogenetic and biochemical molecular events difficult. Infection of mouse striated muscle proved more useful in this regard, since the alphavirus inoculum could be locally contained and titers of progeny virus were high after a single cycle to replication.[17] Under the latter conditions, CPV-1 were found as early as 7 h (see Figure 6B).[28]

Coordinate cell fractionation and high resolution autoradiographic experiments have indicated that CPV-1 are sites of viral RNA synthesis[46,70] and very recent evidence[73b] shows an association with alphavirus replicase. The chronology of their appearance in tissue cultures corresponds to the interval when complete minus-strand replication of the alphavirus genome occurs. CPV-1 do not develop during guanidine inhibition of virus RNA replication[64] or when temperature sensitive alphavirus mutants are incubated at the nonpermissive temperature.[71] The fine electron-dense strands within vesicles lining CPV-1 (see Figures 3 to 5) could represent a form of nucleic acid. A fascinating parallel has been reported in plant cells infected by plus-strand RNA viruses: early vesicles correspond in size to those of CPV-1[74-76] and internal dense material reacts cytochemically as a double-stranded RNA.[74]

In thin sections, CPV-1 occasionally are enwrapped by profiles of RER (see Figure 6B). This may reflect a mechanism for efficient transfer of nascent 26 S mRNA to polysome translation sites or of progeny 42 S virion RNA to sites of C protein capsid assembly. The fate of CPV-1 and membrane vesicles is uncertain; however, some ultimately fuse with primary lysosomes and undergo degradation.[67] Expulsion from cells of mosquito species with low susceptibility to alphavirus infection has been suggested.[65]

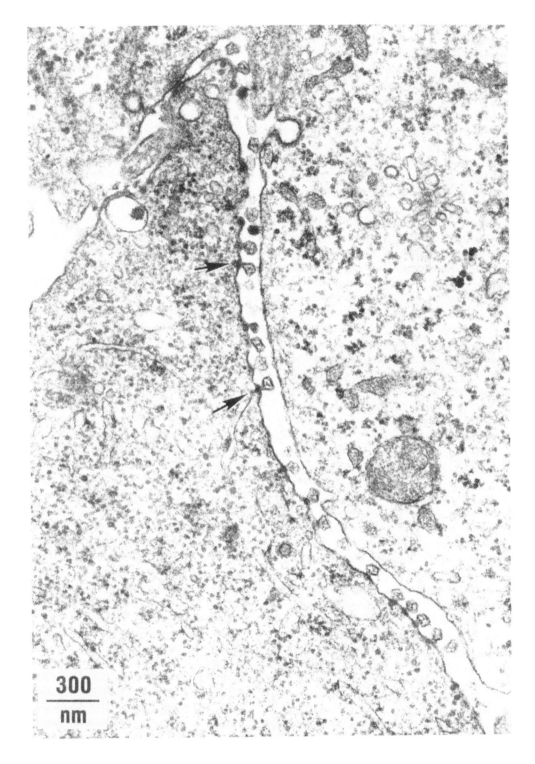

FIGURE 3.   Section of Vero monkey kidney cells infected with the alphavirus WEE for 6 h. Numerous membrane vesicles hang by delicate pedicles (arrows) from the surface of the cell to left. These vesicles measure 50 nm in diameter and exhibit an irregular interior density. (Magnification × 47,500.)

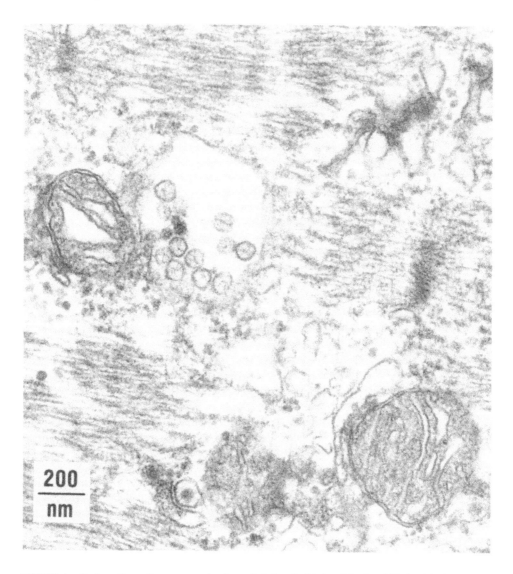

FIGURE 4.   Section of weanling mouse striated muscle infected with the alphavirus SFV for 8 h. The cluster of spherical membrane vesicles appears to be associated with the sarcotubular reticulum. A threadlike density is evident within each vesicle. (Magnification × 67,500.)

2. Assembly of Nucleocapsids

The formation of nucleocapsids occurs very efficiently and rapidly after the translation of C protein (14×10^3 mol wt). Basic amino acids of the C protein chain have a strong charge affinity for nucleic acids and protein-nucleic acid subunits assembly rapidly in the cytoplasm[2,8] — possibly in close association with ribosomes.[77] Binding of nucleocapsids to the cytoplasmic face of the plasma membrane (see Figure 8) or internal cell membranes (see Figures 9 and 10A) appears to be governed by the membrane content of a p62 glycoprotein which is the precursor of envelope proteins E2 and E3.[2] Since vacuolar membranes of Golgi or other origins typically become surrounded by nucleocapsids relatively late in the infectious cycle they have been designated "cytopathic vacuoles, Type 2" (CPV-2).[70] Accumulation occurred in gut epithelial cells of the vector *Culiseta melanura* infected with Eastern equine encephalitis virus.[31b]

In cells treated with monensin, cleavage of p62 is prevented, and nucleocapsids pref-

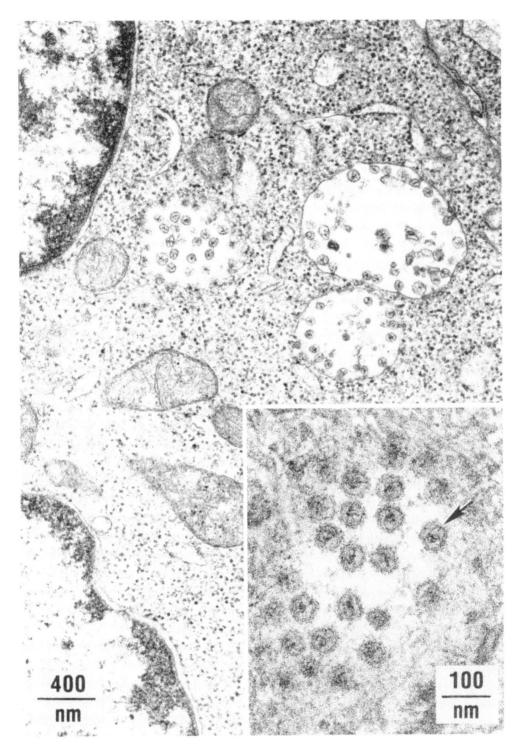

FIGURE 5.   Section of mouse LMTK-cells infected with the alphavirus SFV for 12 h. Spherical membrane vesicles line the interior of cytopathic vacuoles, Type 1 (CPV-1). (Magnification × 40,000.) (Inset) A higher magnification of CPV-1 vesicles in SFV-infected chicken embryo cells (4 h) shows the membrane bilayer and internal dense structure. (Magnification × 130,000.) (Inset from Grimley, P. M., Levin, J. G., Berezesky, I. K., and Friedman, R. M., *J. Virol.*, 10, 492, 1972; reproduced with permission of the American Society for Microbiology).

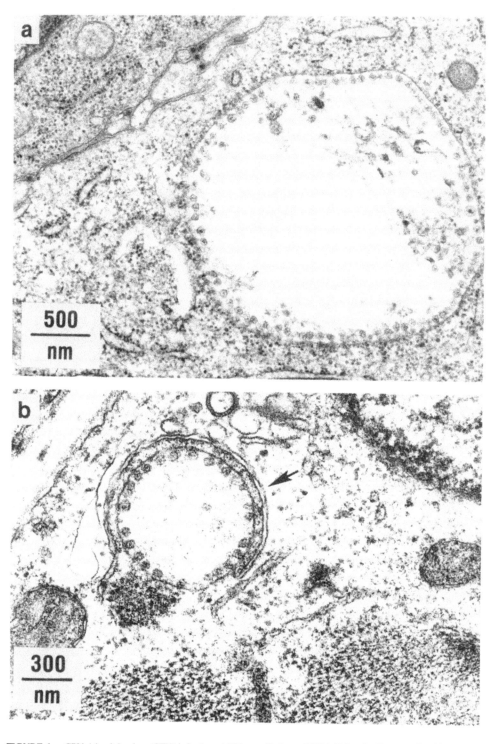

FIGURE 6.   CPV-1 in alphavirus (SFV) infection. (a) Unusually large CPV-1 (2.5 m in diameter) lined by vesicles in mouse L929 cell, 12 h after virus adsorbtion. (Magnification × 32,500.) (b) CPV-1 in weanling mouse skeletal muscle, 7 h after virus inoculation. It is partially surrounded by a profile of endoplasmic reticulum (arrowhead). (Magnification × 48,000.) ([b] from Grimley, P. M. and Demsey, A., in *Pathobiology of Cell Membranes II*, Academic Press, New York, 1980, 94. With permission.)

TABLE 1
CPV-1 in Alphavirus Infections

Species	Tissue	Cell line	Temp (°C)	Virus	1st noted (h)	Ref.
Aedes aegypti		AA-20A	28	SFV[a]	4	65
Aedes pseudocutellaris		AP-61	28	SFV	4	65
Aedes triseriatus	Salivary gland		27	EEE[b]	Day 13	31
Anopheles stephensi		AS-43	28	SFV	4	65
Chicken		Embryo	30	SFV	14	71
Chicken		Embryo	37	SFV	2—3	67
Chicken		Embryo	37	Sindbis	3—4	67
Chicken		Embryo	37	WEE[c]	6	67
Hamster		Baby kidney	37	SFV	3—6	47, 67
Hamster		Baby kidney	37	Sindbis	3—6	67
Human[d]		HeLa	37	SFV	8	—
Monkey		Vero	37	Ross River	8	64
Monkey		Vero	37	SFV	4	65
Monkey		Vero	37	WEE	6	—
Mouse	Brain (neurons)		37	EEE	48	72
Mouse	Brain (neurons)		37	SFV	36—40	73
Mouse	Striated muscle		37	SFV	7	1, 17
Mouse		L	37	SFV	3—4	67
Mouse		L	37	Sindbis	3—6	67

[a] Semliki Forest.
[b] Eastern equine encephalomyelitis.
[c] Western equine encephalomyelitis.
[d] Grimley, P. M. and Young, N. A., unpublished data.

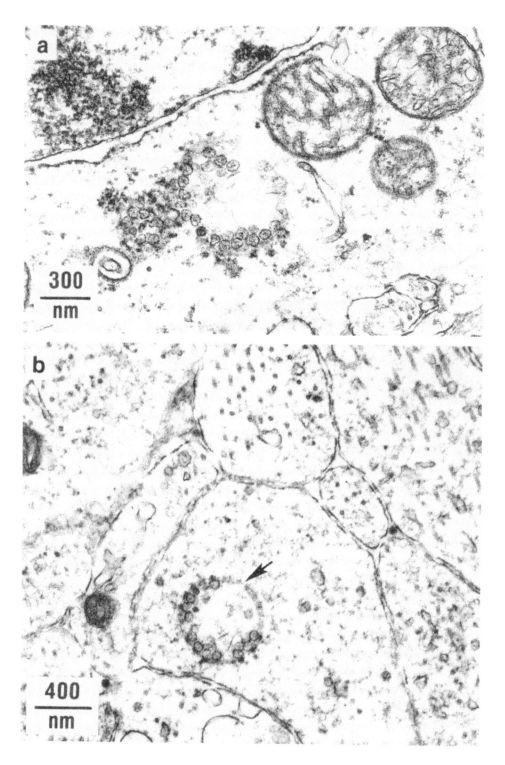

FIGURE 7.   Alphavirus (SFV) infection of newborn mouse brain at 36 h after inoculation. (a) Paranuclear cluster of spherical membrane vesicles in neuron. (Magnification × 47,500.) (b) CPV-1 in neuraxon of neurophil (Magnification × 40,000.) ([a] From Grimley, P. M. and Friedman, R. M., *Exp. Mol. Pathol.*, 12, 1, 1970. With permission.)

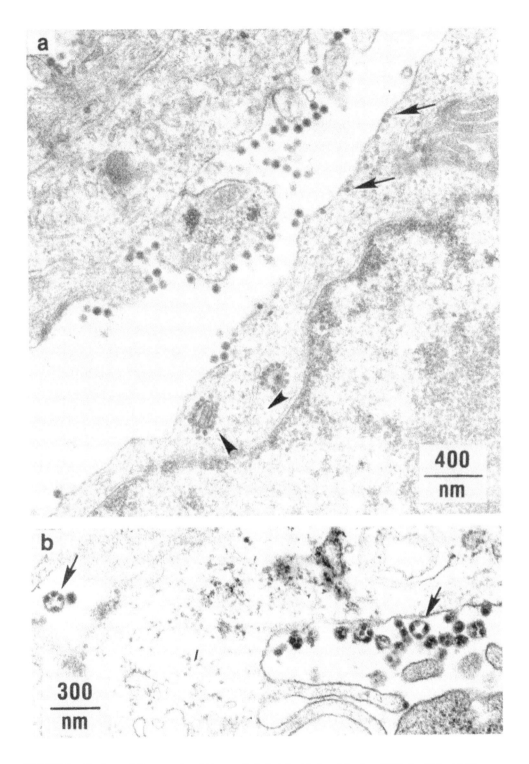

FIGURE 8. Sections of Vero monkey kidney cells infected with the alphavirus WEE for 12 h. (a) Virus nucleocapsids are attached just interior to plasma membrane budding sites (arrows lower left); other nucleocapsids surround small vacuoles (arrowheads) forming cytopathic vacuoles, Type 2 (CPV-2). (Magnification × 40,000.) (b) Higher magnification illustrating multiple nucleocapsids (arrow) within a shared envelope (multiploid virions). (Magnification × 47,500.)

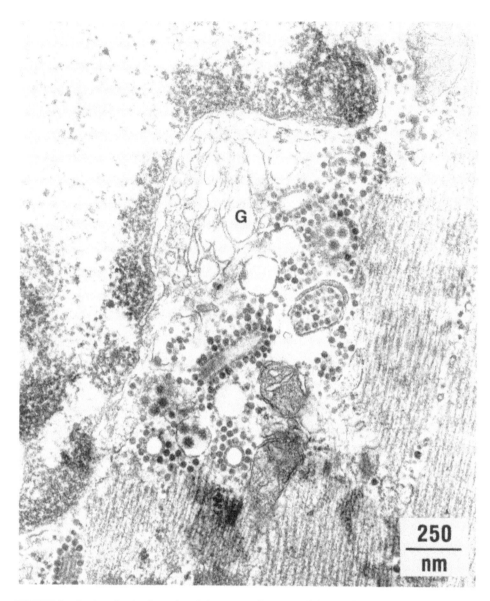

FIGURE 9.   Section of striated muscle cell from a weanling mouse infected with alphavirus (SFV) for 12 h. Nucleocapsids have accumulated in the perinuclear region and surround vacuoles originating from endoplasmic reticulum or sarcoplasmic reticulum (CPV-2). Saccules in region of Golgi complex (G) appear relatively less involved. (Magnification × 60,000.) (From Grimley, P. M. and Demsey, A., *Pathobiology of Cell Membranes II*, Academic Press, New York, 1980, 94. With permission.)

erentially accumulate around Golgi membranes and bud internally rather than at cell surfaces.[78] This effect suggests that an apparently preferential accumulation of nucleocapsids on internal cytomembranes of some cell types, such as neurons (Figure 10B),[73] could be related to subtle deficiencies in envelope protein biosynthesis or maturation.[2] Conversely, massive stockpiles of envelope glycoproteins in late infection may cause a dramatic accumulation of excessive nucleocapsids around every available membrane interface (Figure 9).

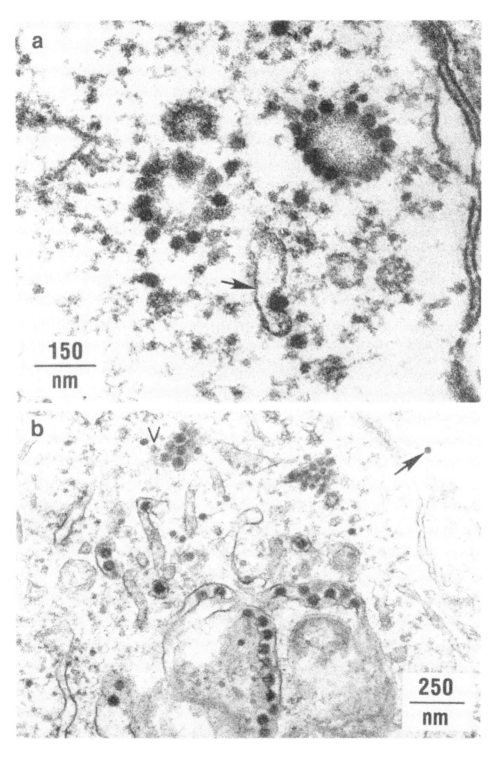

FIGURE 10.   Sections of newborn mouse brain infected with alphavirus (SFV). (a) High magnification of CPV-2 and initial budding of nucleocapsid (arrow) though internal cytomembrane of neuron. (Magnification × 112,000.) (b) Multiple budding virions in region of Golgi cytomembranes and accumulation of complete virions in region of Golgi cytomembranes and accumulation of complete virions in distended vacuole (V) of neuron; arrow points to a free nucleocapsid in the cytosol. (Magnification × 60,000.) (From Grimley, P. M. and Friedman, R. M., *Exp. Mol. Pathol.*, 12, 1, 1970. With permission.)

3. Envelopment and Release of Alphavirions

Enveloped RNA viruses all reutilize portions of the host cell plasma membrane or internal cytomembranes in construction of their lipoprotein exterior.[2,79] Thus, lipid content of the virus envelope consistently reflects the proportions of host cell membrane lipids,[10,35,46,79] while the envelope proteins are predominantly or entirely of viral gene origin.[2,10,45]

In cells both of vertebrates and mosquitos, alphavirus nucleocapsids emerge into the exterior space by budding through host membranes.[18,28,31,59] This process may be observed ultrastructurally and occurs at the cell surface (Figure 8A) or into cisternae of the endoplasmic reticulum or of the Golgi complex (Figures 10B and 11). The cisternae are potential extracellular[28] compartments. At the cell surface, emergence of alpha virions is patchy and probably reflects the topographic distribution of envelope glycoproteins.[59] Cytoplasmic flow dynamics evidently play some role in migration of nucleocapsids toward the cytomembranes, so that virus preferentially buds from microvillus projections.[28,80,81] Effects of flow dynamics are most evident in the central nervous system where nucleocapsids can be identified in neuropil axoplasm (see Figure 12), very peripheral to sites of viral protein biosynthesis.[73,81]

Cleavage of the p62 glycoprotein is required for successful alphavirus budding and apparently depends upon a membrane associated host protease.[2,10,45] Other factors, including adequate quantities of E1 must be a factor, since release is inhibited in low ionic strength media,[82] despite p62 cleavage.[2] The essential factors can now be dissected by molecular genetic techniques with temperature sensitive mutants.[83,84] Ultrastructural freeze-cleavage studies have shown that the 7 to 10-nm particle clusters of the internal fracture face of the plasma membrane disappear during virus budding. Fusion of the inner and outer leaflets of the cell membrane, as the virus emerges, is a two step process.[59]

B. FLAVIVIRUSES
1. Early Membrane Changes

The morphogenetic events in flavivirus infected cells have been relatively difficult to analyze. Even under conditions of exerimental synchronicity with high multiplicities of infection, very few ultrastructural changes occur before 12 h. In sequential electron microscopic studies of carefully sampled, flavivirus-infected, mammalian cell cultures provided by Dr. Nathaniel A. Young, the author observed that membranous, spherical vesicles, 100 to 120 nm in diameter, consistently appeared just before or coincidentally with the first appearance of progeny virions (see Figures 13 and 14).[1,28] They were always found intracellularly, often within cisternae of the endoplasmic reticulum (see Figure 14). Similarly the relatively early formation of spherical vesicles has been noted in a few other studies of insect or mammalian cells[63,85,86] infected by flaviviruses, and they have been variously described as globoid bodies,[90] or smooth membrane structures[86] with a size range from 70 to 150 nm in diameter. Circumstantially, the many observations of such vesicles in flavivirus infections,[63,66,89-99] both in vertebrate and arthropod cells (see Table 2), suggest a significant biological role.

A membrane associated replication complex has been detected in flavivirus infected cells[51] and nuclear envelope cytomembranes are involved in the viral replicative processes.[49,87,88] It is, therefore, tempting to speculate that the spherical membrane vesicles may be ultrastructurally analogous to the 50-nm diameter spherical vesicles lining CPV-1 in alphavirus infection. Viewed in the broader perspective of plus-strand RNA viruses, smooth membranous vesicles appear to play a critical role in the replication of several plant viruses,[74-76] and may occur during coronavirus[2] infections. A direct correlation of ultrastructural and molecular biochemical events nevertheless remains to be established in flavivirus infections.

2. Emergence of Flavivirions

In all cases reported, flavivirions are found completely enveloped either on the surface

FIGURE 11. Section of weanling mouse striated muscle cell infected with alphavirus (SFV). Mature enveloped virions distend profiles of sarcoplasmic reticulum. (Magnification × 70,000.)

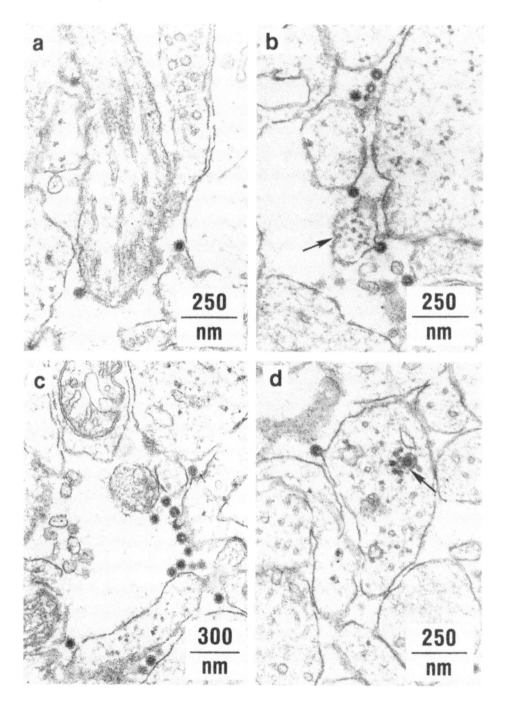

FIGURE 12.   Alphavirus (SFV) infection of newborn mouse brain. Multiple sections of neuropil show budding of virions from (a) neuraxon, magnification × 60,000, and (b) nerve endings, magnification × 60,000. Note: neurotubules (arrow) in nerve terminal. (c) CPV-2 with central virion innerve terminal (arrow), magnification × 60,000 and (d) spherical membranous vesicles, magnification × 47,500 (compare to Figure 3). (From Grimley, P. M. and Friedman, R. M., *Exp. Mol. Pathol.*, 12, 1, 1970. Reproduced with permission of Academic Press, New York).

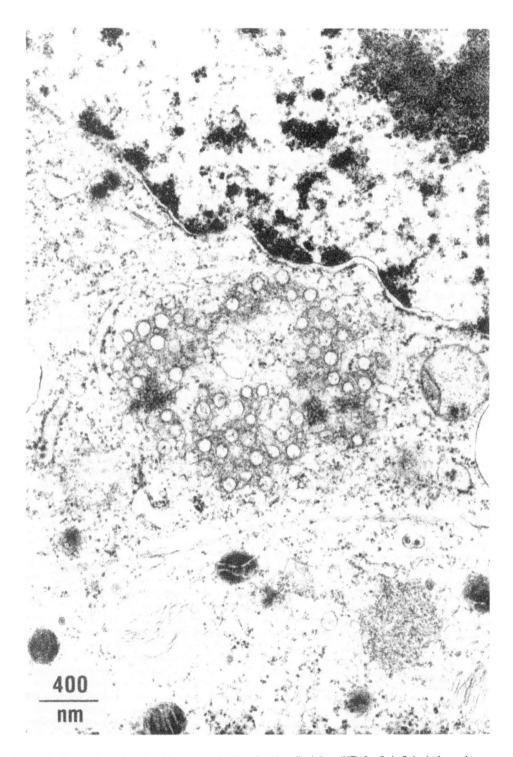

FIGURE 13. Section of *Aedes albopictus* cell infected with a flavivirus (YF) for 3 d. Spherical membranous vesicles massed in the paranuclear region appear similar to those observed in alphavirus infections (see Figures 3, 4), but measure up to 120 nm diameter. Note threadlike interior densities. Vesicles are surrounded by dense granular membranous material (see Figures 14, 16, 17). (Magnification × 40,000.)

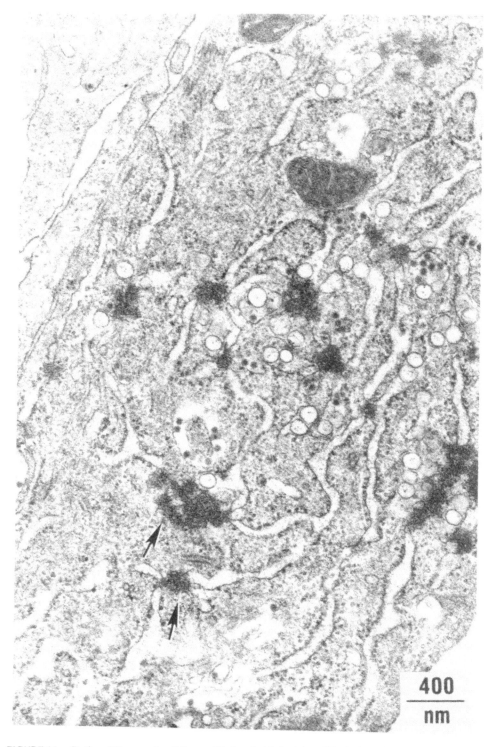

FIGURE 14.   Section of Vero monkey kidney cell infected with flavivirus (YF) for 24 h. Spherical vesicles similar to those in Figure 13 are located within cisternae of the endoplasmic reticulum. Intracisternal virions are also evident. Note irregular zones of electron dense granular membranes forming ''nodes'' (arrowheads) between interconnecting profiles of endoplasmic reticulum (also see Figure 22). (Magnification × 40,000.)

TABLE 2
Smooth Membrane Vesicles in Flavivirus Infections

Species	Tissue	Cell line	Temp (°C)	Virus	1st noted (h)	Ref.
Aedes aegypti		Grace	30	JEV[a]	48	63
Aedes aegypti		Grace	30	MVE[b]	Day 9	63
Aedes albopictus		Singh	27	Kunjin	16	66
Aedes albopictus[c]		Singh	31	YF[d]	Day 3	—
Chicken		Embryo	37	West Nile		93
Human		Hep2(epi)	37	Ilheus	24	98
Human		HeLa	37	Spring-Summer	24—48	89
Monkey		Vero	37	Dengue	18	85
Monkey		Vero	37	JEV	15	86
Monkey[c]		Vero	37	SLE[e]		—
Monkey[c]		Vero	37	YF	11	—
Mouse	Brain (neurons)		37	CEE[f]	96	95
Mouse	Brain (neurons)		37	JEV	48	90
Mouse	Brain (neurons)		37	Langat	72	94
Mouse	Brain (neurons)		37	OHF[g]	96	96
Mouse	Brain (neurons)		37	SLE[e]	96	91
Mouse	Brain (neurons)		37	West Nile	Day 5	92
Mouse	Brain (neurons)		37	YF	Day 8—10	95
Mouse	Brain (neurons)		37	Zimmern	96	99
Rabbit[c]		MA-111	37	YF	36	—

[a] Japanese encephalitis.
[b] Murray Valley encephalomyelitis.
[c] Grimley, P. M. and Young, N. A., unpublished data.
[d] Yellow fever.
[e] St. Louis encephalitis.
[f] Central European encephalitis.
[g] Omsk hemorrhagic fever.

of infected cells (see Figure 15), or within the cisternae of the endoplasmic reticulum or Golgi apparatus (see Figures 16 and 17). Functionally, therefore, the flavivirions are always found exterior to the cytoplasmic compartment. Naked flavivirus nucleocapsids have not been definitively identified by electron microscopy, and no experimental attempts to interrupt morphogenesis or release of flavivirions have succeeded thus far in revealing an intermediate assembly stage similar to that of togaviruses. It is possible that the small size of flavivirus nucleocapsids fails to distinguish them from cytosolic polysomes or membrane-associated ribosomes (see Figure 16); however, it now seems most likely that the replication strategy of flaviviruses involves a uniquely coordinated translation and assembly in close association with cytomembranes at sites of release.

VI. SUBCELLULAR CYTOPATHOLOGIC EFFECTS

A. CYTOPLASMIC INCLUSIONS

Apparently unique types of cytoplasmic inclusions have been reported in mammalian cells infected by alphaviruses or flaviviruses. These include compact masses of 30-nm diameter coated tubules (see Figure 18) in cells infected by alphaviruses[73] and bundles or sheets of banded filaments or tubular structures (see Figures 19 and 20) in flavivirus infected cells.[1,2] These formations occur at relatively late times during infection and do not appear directly linked to morphogenetic events. More likely, they represent accumulations of excessively produced viral proteins or possibly RNA-protein complexes. In alphavirus infections, straight or curved rod structures commonly accumulate within CPV-2 (see Figure 21) as infection progresses.[1,73,100] These smooth rods measure 50 nm in diameter, have a dense double layered coat and a less dense 22-nm core. They possibly represent incomplete viral coat materials. In flavivirus infections, 28-nm diameter microhelices may form within the endoplasmic reticulum. These are straight or curved, measure up to 0.8 μm in length and exhibit an 11-nm periodicity.[99]

B. CYTOMEMBRANE REORGANIZATIONS

Cytomembrane changes are observed ultrastructurally during both alphavirus and flavivirus infections.[28] In view of an early inhibition of cellular lipid biosynthesis,[28,69] all such changes must be presumed to reflect reorganizations of preexisting cytomembranous elements or a differential utilization of cytomembrane precursor pools. The early spherical vesicles which are a common feature of both alphavirus and flavivirus infections in mammalian and insect cells have been fully discussed in the preceding text. In addition, both late alphavirus and late flavivirus infections are characterized by intracisternal invaginations of tubular membranes and redundant membranous arrays associated with the endoplasmic reticulum[1,28,66,73,91,92,98,101] (see Figures 14, 16, 21, 22). Recently, it has been shown that similar intracisternal cytomembrane changes known generically as "tubuloreticular inclusions" are induced in human and nonhuman primate cells by Type 1 interferons:[102,103] a loose tubuloreticular form develops in human cells,[101] while more geometric paracrystalline tubular membrane assays develop in subhuman primate cells (see Figure 23).[103,104] It now seems probable that the intracisternal membrane reorganizations observed in alphavirus and/or flavivirus infections of many different cell types and many species reflect a secondary autocrine or paracrine effect of virus-induced interferon.

VII. SUMMARY

Togaviridae and Flaviviridae are closely related families of enveloped RNA viruses. The virions are spherical with icosodedral nucleocapsids and appear to be internally organized in a similar manner. Many members of these virus families are transmissible by arthropods

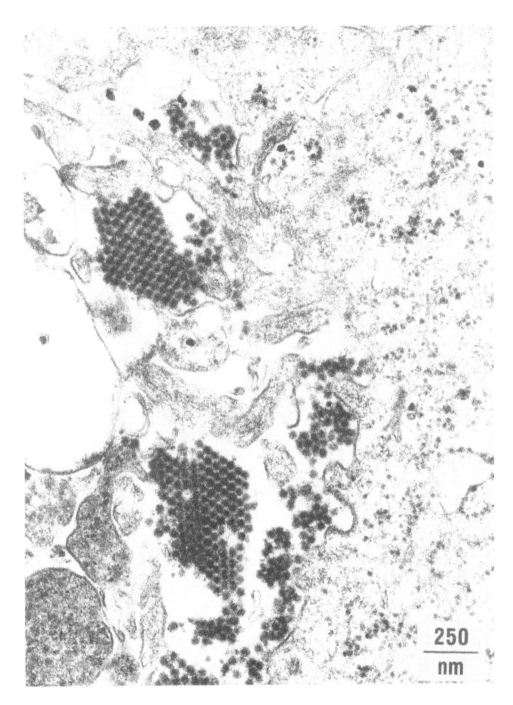

FIGURE 15.   Flavivirions (YF) accumulated on exterior of Vero monkey kidney cell at 85 h after infection. The paracrystalline arrangement of virions reflects an underlying cubic symmetry of peplomers and nucleocapsids. (Magnification × 60,000.)

and have a wide host and temperature range. These virus families nevertheless differ significantly in replication strategies: togavirus RNA transcribes a subgenomic messenger RNA for structural proteins, while flavivirus RNA transcribes only complete self copies. In togavirus infections, polyprotein precursors are enzymatically cleaved into discrete nonstructural or structural proteins, but polyprotein precursors have not been identified in flavivirus

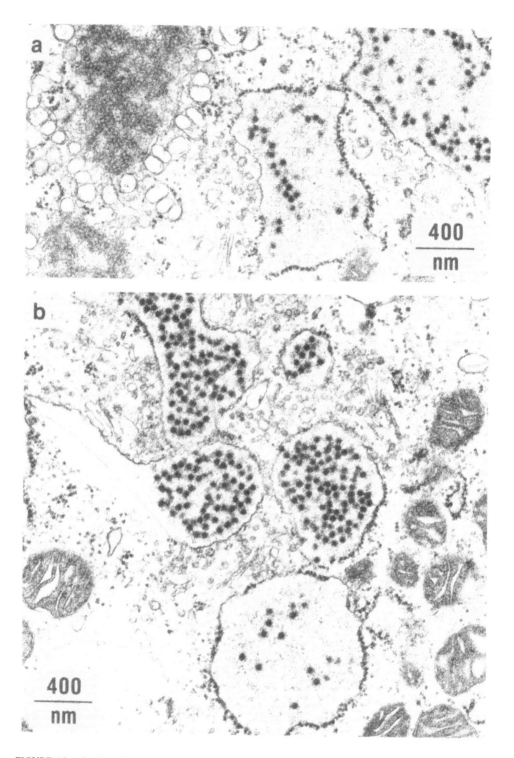

FIGURE 16.   Section of *Aedes albopictus* cell from a culture chronically infected with a flavivirus (YF) for 30 d. (a) Spherical vesicles surround a mass of dense granular membranes and virions have accumulated within distended cisternae of endoplasmic reticulum. (Magnification × 40,000.) (b) Masses of virions within distended endoplasmic reticulum. Dense particles surrounding the endoplasmic reticulum appear to represent ribosomes rather than nucleocapsids, and thus differ from CPV-2 in alphavirus infections. (Magnification × 40,000.) ([a] From Grimley, P. M. and Henson, D. E., in *Diagnostic Electron Microscopy*, © Copyright 1983 by John Wiley & Sons, New York, 1983, 1. Reprinted by permission of John Wiley & Sons, Inc.)

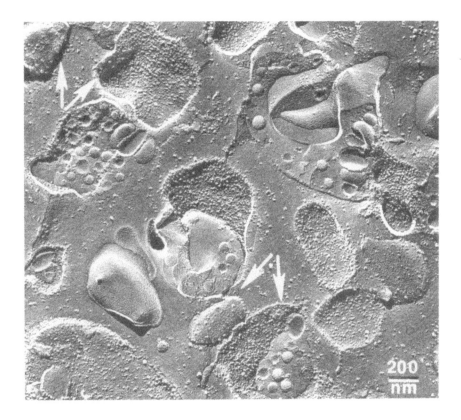

FIGURE 17.   Flavivirus (Dengue) infection of LLC-MK2 monkey kidney cell. Freeze fracture preparation shows mature virions within cytoplasmic vacuoles derived from the endoplasmic reticulum. Arrows indicate reverse polarity of vacuolar membranes. No budding stages can be identified. (Magnification × 40,000.) (From Grimley, P. M. and Demsey, A., *Pathobiology of Cell Membranes II*, Academic Press, New York, 1980, 94. With permission.)

infected cells. Morphogenesis of both togaviruses and flaviviruses is cytomembrane dependent, and vesicles observed during early stages of development may be related to RNA replicative or transcriptional events. Nucleocapsid precursors of virion cores can be detected ultrastructurally in thin sections of togavirus infected cells, but are not identifiable in flavivirus infected cells. The final steps of togavirus emergence and release can be observed and interrupted experimentally, but flavivirus emergence has not been physically observed. In both cases, mature virions eventually may be found within cisternae of intracellular membrane systems or at the cell surface. The sites of emergence may be governed both by dynamics of cytoplasmic flow and conditions of envelope glycoprotein localization in cytomembranes.

ACKNOWLEDGMENTS

The author expresses special appreciation to Dr. Robert M. Friedman who was a "plus strand" for many of the studies herein illustrated as well as to Dr. Kathryn V. Holmes who critically appraised the emergent text. Successful completion depended upon contributions by Dr. Monique Dubois-Dalcq, Dr. Anthony Demsey, and Dr. Dorothy Feldman. Transcription was by Ms. Elinore Dunphy.

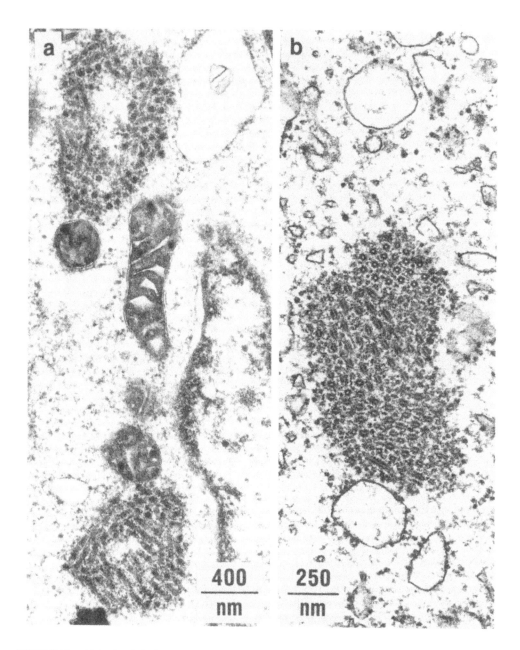

FIGURE 18.   Alphavirus infection. (a) Vero monkey kidney cell infected with alphavirus WEE for 12 h shows compact aggregates of 30 nm in diameter coated tubules in cytoplasm. (Magnification × 40,000.) (b) Newborn mouse brain infected with alphavirus SFV shows a similar tubular aggregate and lucent tubule cores of 80 to 90 Å are visible at this magnification. (Magnification × 57,000.) ([b] From Grimley, P. M. and Friedman, R. M., *Exp. Mol. Pathol.*, 12, 1, 1970. With permission.)

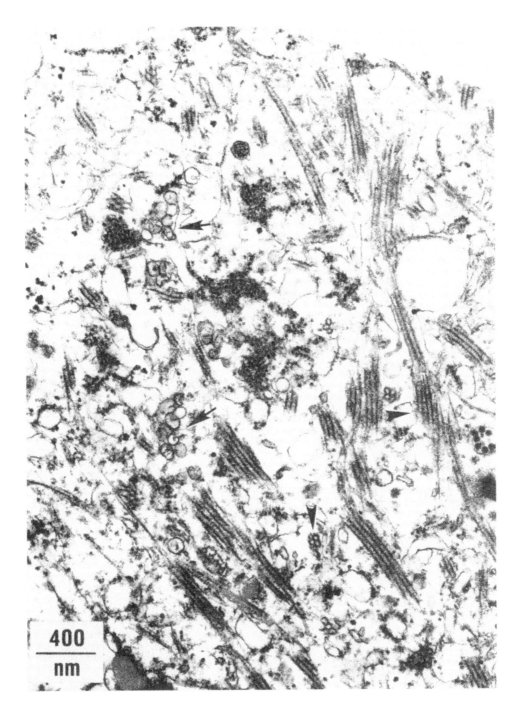

FIGURE 19. Flavivirus (YF) infection of monkey kidney cell (Vero) at 59 h. Bundles of straight tubules (33 nm in diameter) accumulate as infection progresses. Note a few cross sections (arrowheads). Clusters of spherical vesicles persist (arrows). (Magnification × 40,000.)

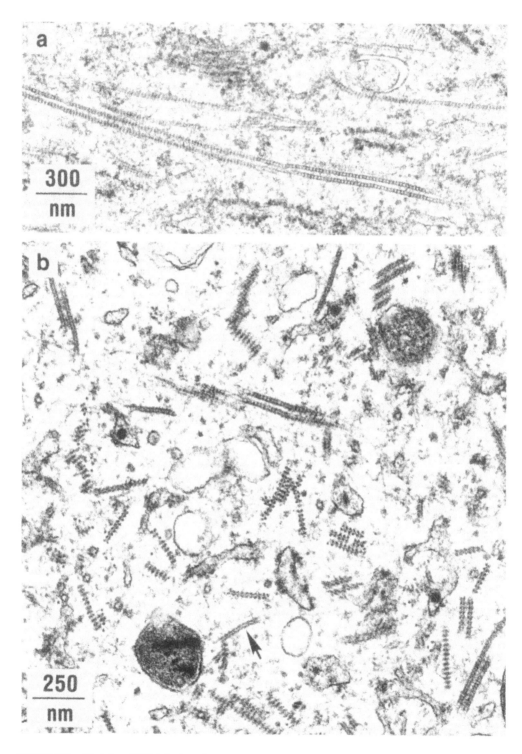

FIGURE 20.   Flavivirus (SLE) infection of Vero monkey kidney cells at 84 h. Bundles of banded filaments (16 to 18-nm periodicity) and 16-nm diameter tubular structures (arrow) accumulate in large aggregates. ([a] Magnification × 48,000; [b] Magnification × 60,000.)

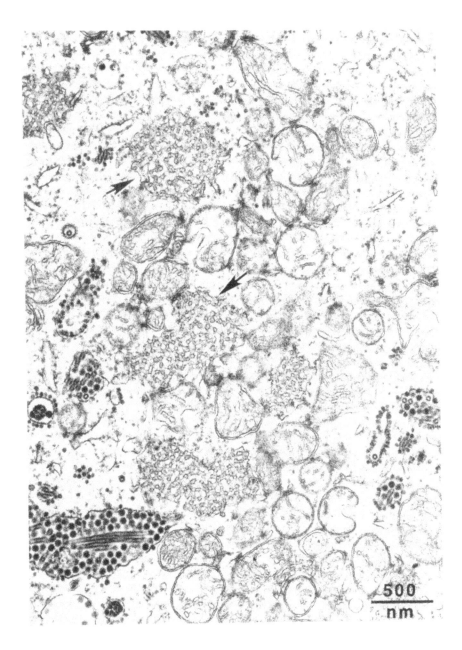

FIGURE 21.   Section of neuron from weanling mouse brain infected with the alphavirus SFV. Arrays of 25-nm membranous tubules (arrow) distend cisternae of the endoplasmic reticulum. These apppear similar to tubuloreticular inclusions induced by interferon (see Figure 23 and References 102 to 104). Rod structures (50 nm in diameter) are noted in the large vacuole with mature virions at lower left, and there are several CPV-2. (Magnification × 34,500.) (From Grimley, P. M. and Demsey, A., in *Pathobiology of Cell Membranes II*, Academic Press, New York, 1980, 94. With permission.)

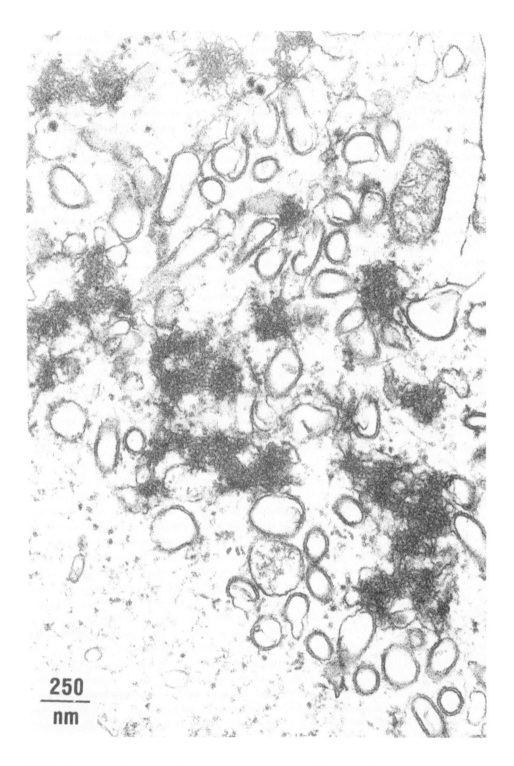

FIGURE 22.   Section of MA-111 rabbit cell infected with flavivirus (YF) for 45 h. Spherical membranous vesicles appear to have distended (up to 140 nm) and lost their interior density. Note the large zones of dense granular membrane-associated material connecting cisternal profiles. Even at this magnification, the latter material fails to exhibit a discrete substructure, but the distribution suggests disorganized membrane products related to the interferon-induced tubuloreticular inclusions (see Figures 21, 23 and References 101 to 104). (Magnification × 60,000.)

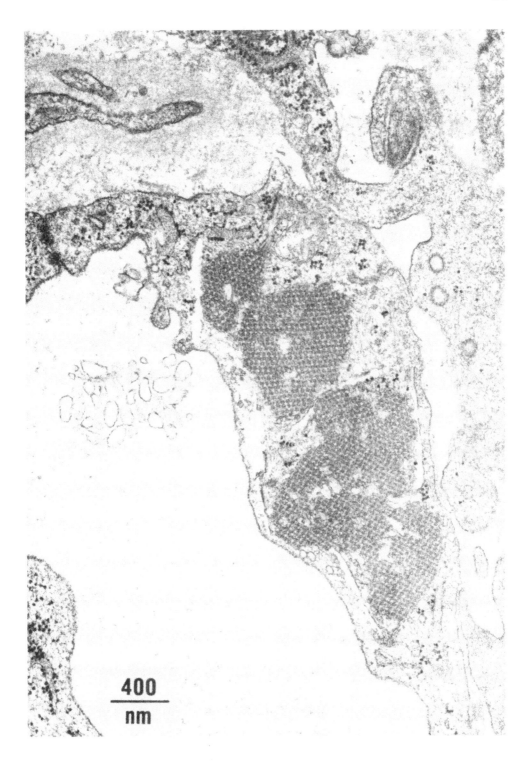

FIGURE 23.   Placental capillaries of rhesus monkey treated for 7 d with high systemic doses of alpha interferon.[104] Paracrystalline membranous inclusions develop within cisternae of the endoplasmic reticulum. Similar intracisternal "tubuloreticular inclusions" are induced by alpha-interferon in human cells, but the tubules are loosely arranged.[102,103] (Magnification × 40,400.) (From Feldman, D., Hoar, R. M., Niemann, W. H., Valentine, T., Cukierski, M., and Hendrickx, A. G., *Am. J. Obstet. Gynecol.*, 155, 413, 1986. With permission.)

REFERENCES

1. **Grimley, P. M. and Henson, D. E.**, Electron microscopy in virus infections, in *Diagnostic Electron Microscopy*, Trump, B. F. and Jones, R. T., Eds., John Wiley & Sons, New York, 1983, 1.
2. **Dubois-Dalcq, M., Holmes, K. V., and Rentier, B.**, *Assembly of Enveloped RNA Viruses*, Springer-Verlag, Vienna, 1984, 235.
3. **Downs, W. G.**, Arboviruses, in *Viral Infections of Humans*, Evans, A. S., Ed., Plenum Press, New York, 1976, 71.
4. **Shope, R. E.**, Medical significance of togaviruses: an overview of diseases caused by togaviruses in man and in domestic and wild vertebrate animals, in *The Togaviruses: Biology, Structure, Replication*, Schlesinger, R. W., Ed., Academic Press, New York, 1980, 47.
5. **Porterfield, J. S.**, Comparative and historical aspects of the Togaviridae and Flaviviridae, in *The Togaviridae and Flaviviridae*, Schlesinger, S. and Schlesinger, M. J., Eds., Plenum Press, New York, 1986, 1.
6. **Brinton, M. A.**, Replication of flaviviruses, in *The Togaviridae and Flaviviridae*, Schlesinger, S. and Schlesinger, M. J., Eds., Plenum Press, New York, 1986, 327.
7. **Westaway, E. G.**, Flavivirus replication strategy, *Adv. Virus Res.*, 33, 45, 1987.
8. **Strauss, E. G. and Strauss, J. H.**, Structure and replication of the alphavirus genome, in *The Togaviridae and Flaviviridae*, Schlesinger, S. and Schlesinger, M. J., Plenum Press, New York, 1986, 35.
9. **Harrison, S. C.**, Alphavirus structure, in *The Togaviridae and Flaviviridae*, Schlesinger, S. and Schlesinger, M. J., Eds., Plenum Press, New York, 1986, 21.
10. **Pfefferkorn, E. and Shapiro, D.**, Reproduction of togaviruses, in *Comprehensive Virology Reproduction: Small and Intermediate RNA Viruses*, Vol. 2, Fraenkel-Conrat, H. and Wagner, R. E., Eds., Plenum Press, New York, 1974, 171.
11. **Chamberlain, R. W.**, Epidemiology of arthropod-borne togaviruses: the role of arthropods as hosts and vectors and of vertebrate hosts in natural transmission cycles, in *The Togaviruses: Biology, Structure, Replication*, Schlesinger, R. W., Ed., Academic Press, New York, 1980, 175.
12. **Peters, C. J. and Johnson, K. M.**, Hemorrhagic fever viruses, in *Concepts in Viral Pathogenesis*, Notkins, A. L. and Oldstone, M. B. A., Eds., Springer-Verlag, New York, 1984, 325.
13. **Brinton, M. A., Blank, K. J., and Nathanson, N.**, Host genes that influence susceptibility to viral diseases, in *Concepts in Viral Pathogenesis*, Notkins, A. L. and Oldstone, M. B. A., Eds., Springer-Verlag, New York, 1984, 71.
14. **Stanton, G. J. and Baron, S.**, Interferon and viral pathogenesis, in *Concepts in Viral Pathogenesis*, Notkins, A. L. and Oldstone, M. B. A., Eds., Springer-Verlag, New York, 1984, 3.
15. **Levin, J. G., Ramseur, J. M., and Grimley, P. M.**, Host effect on arbovirus replication: appearance of defective interfering particles in murine cells, *J. Virol.*, 12, 1401, 1973.
16. **Grimley, P. M.**, Arbovirus encephalitis: which road traveled by makes all the difference?, *Lab. Invest.*, 48, 369, 1983.
17. **Grimley, P. M. and Friedman, R. M.**, Arboviral infection of voluntary striated muscles, *J. Infect. Dis.*, 122, 45, 1970.
18. **Brown, D. T. and Conderay, L. D.**, Replication of alphaviruses in mosquito cells, in *The Togaviridae and Flaviviridae*, Schlesinger, S. and Schlesinger, M. J., Eds., Plenum Press, New York, 1986, 171.
19. **Henson, D. A. and Grimley, P. M.**, Virus infections of the fetus, in *Principles of Obstetrics*, Caplan, R. M., Ed., Williams and Wilkins, Baltimore, 1982, 218.
20. **Roehrig, J. T.**, The use of monoclonal antibodies in studies of the structural proteins of togaviruses and flaviviruses, in *The Togaviridae and Flaviviridae*, Schlesinger, S. and Schlesinger, M. J., Eds., Plenum Press, New York, 1986, 251.
21. **Chamberlain, R. W. and Sudia, W. D.**, Mechanism of transmission of viruses by mosquitoes, *Annu. Rev. Entomol.*, 6, 371, 1961.
22. **Jones, L. D., Davies, C. R., Steele, G. M., and Nuttall, P. A.**, A novel mode of arbovirus transmission involving a non-viremic host, *Science*, 237, 775, 1987.
23. **Hardy, J. L., Houck, E. J., Kramer, L. D., and Reeves, W. C.**, Intrinsic factors affecting vector competence of mosquitoes for arboviruses, *Ann. Rev. Entomol.*, 28, 229, 1983.
24. **Gushchina, E. A., Aristova, V. A., Gromashevskii, V. L., Voltsit, O. V., and Gushchin, B. V.**, Electron microscopy study of the midgut of ticks after experimental infection with Karshi virus, *Vopr. Virusol.*, 29, 235, 1984.
25. **Scott, T. W., Hildreth, S. W., and Beaty, B. J.**, The distribution and development of Eastern Equine encephalitis virus in its enzootic mosquito vector, *Culiseta melanura*, *Am. J. Trop. Med. Hyg.*, 33, 300, 1984.
26. **Houck, E. J., Kramer, L. D., Hardy, J. L., and Chiles, R. E.**, Western equine encephalitis virus. *In vivo* and morphogenesis in mosquito mesenteronal epithelial cells, *Virus Res.*, 2, 123, 1985.

27. **Weaver, S. C.,** Electron microscopic analysis of infection patterns for Venezuelan equine encephalomyelitis virus in the vector mosquito, *Culex (Melanoconion) taeniopus, Am. J. Trop. Med. Hyg.,* 35, 624, 1986.

28. **Grimley, P. M. and Demsey, A.,** Functions and alterations of cell membranes during active virus infection, in *Pathobiology of Cell Membranes II,* Trump, B. F. and Arstila, A. U., Eds., Academic Press, New York, 1980, 94.

29. **Stollar, V.,** Approaches to the study of vector specificity for arboviruses-model systems using cultured mosquito cells, *Adv. Virus Res.,* 33, 327, 1987.

30. **Scott, T. W. and Burrage, T. G.,** Rapid infection of salivary glands in *Culiseta melanura* with Eastern Equine encephalitis virus: an electron microscopic study, *Am. J. Trop. Med. Hyg.,* 33, 961, 1984.

31a. **Whitfield, S. G., Murphy, F. A., and Sudia, W. D.,** Eastern Equine encephalomyelitis virus: an electron microscopic study of aedes triseriatus (Say) salivary gland infection, *Virology,* 43, 110, 1971.

31b. **Weaver, S. C., Scott, T. W., Lorenz, L. H., Lerdthusnee, K., and Romoser, W. S.,** Togavirus-associated pathologic changes in the midgut of a natural mosquito vector. *J. Virol.,* 62, 2083, 1988.

32. **Ribera, J. M. C.,** Role of saliva in blood-feeding by arthropods, *Annu. Rev. Entomol.,* 32, 463, 1987.

33. **El Mekki, A. A. and van der Groen, G.,** A comparison of indirect immunofluorescence and electron microscopy for the diagnosis of some haemorrhagic viruses in cell cultures, *J. Virol. Meth.,* 3, 61, 1981.

34. **Simizu, B.,** Inhibition of host cell macromolecular synthesis following Togavirus infection, in *Comprehensive Virology,* Vol. 2, Fraenkel-Conrat, H. and Wagner, R. E., Eds., Plenum Press, New York, 1984, 465.

35. **Stollar, V.,** Togaviruses in cultured arthropod cells, in *The Togaviruses: Biology, Structure, Replication,* Schlesinger, R. W., Ed., Academic Press, New York, 1980, 584.

36. **Pierce, J. S., Strauss, E. G., and Strauss, J. H.,** Effect of ionic strength on the binding of Sindbis virus to chick cells, *J. Virol.,* 13, 1030, 1974.

37. **Kielian, M. and Helenius, A.,** Entry of alphaviruses, in *The Togaviridae and Flaviviridae,* Schlesinger, S. and Schlesinger, M. J., Eds., Plenum Press, New York, 1986, 91.

38. **Levanon, A., Kohn, A., and Inbar, M.,** Increase in lipid fluidity of cellular membrane induced by adsorbtion of RNA and DNA viruses, *J. Virol.,* 22, 353, 1977.

39. **Oldstone, M. B. A., Tishon, A., Dutko, F. J., Kennedy, S. I. T., Holland, J. J., and Lampert, P. W.,** Does the major histocompatibility complex serve as a specific receptor for Semliki Forest virus?, *J. Virol.,* 34, 256, 1980.

40. **Edwards, J., Mann, E., and Brown, D. T.,** Conformational changes in Sindbis virus envelope proteins accompanying exposure to low pH, *J. Virol.,* 45, 1090, 1983.

41. **Gollins, S. W. and Porterfield, J. S.,** Flavivirus infection enhancement in macrophages: an electron microscopic study of viral cellular entry, *J. Gen. Virol.,* 66, 1969, 1985.

42. **Gollins, S. W. and Porterfield, J. S.,** pH-dependent fusion between the flavivirus West Nile and liposomal model membranes, *J. Gen. Virol.,* 67, 157, 1986.

43. **Kimura, T., Gollins, S. W., and Porterfield, J. S.,** The effect of pH on the early interaction of West Nile virus with P388D1 cells, *J. Gen. Virol.,* 67, 2423, 1986.

44. **Sawicki, S. G. and Sawicki, D. L.,** The effect of overproduction of nonstructural proteins on alphavirus plus-strand and minus-strand RNA synthesis, *Virology,* 152, 507, 1986.

45. **Wirth, D. F., Katz, F., Small, B., and Lodish, H. F.,** How a single Sindbis virus mRNA directs the synthesis of one soluble protein and two integral membrane glycoproteins, *Cell,* 10, 253, 1977.

46. **Friedman, R. M., Levin, J. G., Grimley, P. M., and Berezesky, I. K.,** Membrane associated replication complex in arbovirus infection, *J. Virol.,* 10, 504, 1972.

47. **Peränen, J. and Kääränen, L.,** Biogenesis of type I cytopathic vacuoles in Semliki forest virus-infected BHK cells, *J. Virol.,* 65, 1623, 1991.

48. **Cleaves, G. R., Ryan, T. E., and Schlesinger, R. W.,** Identification and characterization of type 2 dengue virus replicative intermediate and replicative form RNAs, *Virology,* 111, 73, 1981.

49. **Grun, J. B. and Brinton, M. A.,** Characterization of West Nile virus RNA-dependent RNA polymerase and cellular terminal adenylyl and uridylyl transferases in cell-free extracts, *J. Virol.,* 60, 1113, 1986.

50. **Chu, P. W. G. and Westaway, E. G.,** Replication strategy of Kunjin virus: evidence for recycling role of replicative form RNA as template in semiconservative and asymmetric replication, *Virology,* 140, 68, 1985.

51. **Qureshi, A. A. and Trent, D. W.,** Saint Louis encephalitis viral ribonucleic acid replication complex, *J. Virol.,* 9, 565, 1972.

52. **Wengler, G.,** The mode of assembly of alphavirus cores implies a mechanism for the disassembly of the cores in the early stages of infection, *Arch. Virol.,* 94, 1, 1987.

53. **Simizu, B., Hashimoto, K., and Ishida, I.,** A variant of Western equine encephalitis with nonglycosylated E3 protein, *Virology,* 125, 99, 1983.

54. **Lenard, J.,** Lipids of alphaviruses, in *The Togaviruses: Biology, Structure, Replication,* Schlesinger, R. W., Ed., Academic Press, New York, 1980, 335.

55. **Rice, C. M., Strauss, E. G., and Strauss, J. H.,** Structure of the flavivirus genome, in *The Togaviridae and Flaviviridae,* Schlesinger, S. and Schlesinger, M. J., Eds., Plenum Press, New York, 1986, 251.

56. **Demsey, A., Steers, R. L., Brandt, W. E., and Veltri, B. J.,** Morphology and development of Dengue-2 virus employing the freeze fracture and thin-section techniques, *J. Ultrastr. Res.,* 46, 103, 1974.

57. **Rice, C. M., Levis, R., Strauss, J. H., and Huang, H. V.,** Production of infectious RNA transcripts from Sindbis virus cDNA clones: mapping of lethal mutations, rescue of a temperature-sensitive marker, and *in vitro* mutagenesis to generate defined mutants, *J. Virol.,* 61, 3809, 1987.

58. **Simpson, R. W. and Hauser, R. E.,** Structural differentiation of group A arboviruses based on nucleoid morphology in ultrathin sections, *Virology,* 34, 568, 1968.

59. **Brown, D. T., Waite, M. R. F., and Pfefferkorn, E. R.,** Morphology and morphogenesis of Sindbis virus as seen with freeze-etching techniques, *J. Virol.,* 10, 524, 1972.

60. **Wengler, G., Wengler, G., Nowak, T., and Wahn, K.,** Analysis of the influence of proteolytic cleavage on the structural organization of the surface of the West Nile flavivirus leads to the isolation of a protease-resistant E protein oligomer, *Virology,* 160, 210, 1987.

61. **Winkler, G., Heinz, F. X., and Kunz, C.,** Studies on the glycosylation of flavivirus E proteins and the role of carbohydrate in antigenic structure, *Virology,* 159, 237, 1987.

62. **Lazar, A., Silberstein, L., Reuveny, S., and Mizrahi, A.,** Microcarriers as a culturing system of insect cells and insect viruses, *Dev. Biol. Stand.,* 66, 315, 1987.

63. **Filshie, B. K. and Rehacek, J.,** Studies of the morphology of Murray Valley encephalitis and Japanese encephalitis viruses growing in cultured mosquito cells, *Virology,* 34, 435, 1968.

64. **Raghow, R. S., Grace, T. D. C., Filshie, B. K., Bartley, W., and Dalgarno, L.,** Ross River virus replication in cultured mosquito and mammalian cells, *J. Gen. Virol.,* 21, 109, 1973.

65. **Lehane, N. J. and Leake, C. J.,** A kinetic and ultrastructural comparison of alphavirus infection of cultured mosquito and vertebrate cells, *J. Trop. Med. Hyg.,* 85, 229, 1982.

66. **Ng, M. L.,** Ultrastructural studies of Kunjin virus-infected *Aedes albopictus* cells, *J. Gen. Virol.,* 68, 577, 1987.

67. **Grimley, P. M., Levin, J. G., Berezesky, I. K., and Friedman, R. M.,** Specific membranous structures associated with the replication of group A arboviruses, *J. Virol.,* 10, 492, 1972.

68. **Virtanen, I. and Wartiovaara, J.,** Virus-induced cytoplasmic membrane structures associated with Semliki Forest virus infection studied by the freeze-etching method, *J. Virol.,* 13, 222, 1974.

69. **Vance, D. and Lam, J.,** Sindbis virus inhibits phosphatidyl choline biosynthesis in BHK-21 cells, *J. Virol.,* 16, 1075, 1975.

70. **Grimley, P. M., Berezesky, I. K., and Friedman, R. M.,** Cytoplasmic structures associated with an arbovirus infection: loci of viral ribonucleic acid synthesis, *J. Virol.,* 2, 1326, 1968.

71. **Tan, K. B.,** Electron microscopy of cells infected with Semliki Forest virus temperature sensitive mutants: correlation of ultrastructural and physiological observations, *J. Virol.,* 5, 632, 1970.

72. **Murphy, F. A. and Whitfield, S. W.,** Eastern Equine encephalitis virus infection: electron microscopic studies of mouse central nervous system, *Exp. Mol. Pathol.,* 13, 131, 1970.

73a. **Grimley, P. M. and Friedman, R. M.,** Development of Semliki Forest virus in mouse brain, *Exp. Mol. Pathol.,* 12, 1, 1970.

73b. **Froshauer, S., Kartenbeck, J., Helenius, A.,** Alphavirus RNA replicase is located on the cytoplasmic surface of endosomes and lysosomes, *J. Cell Biol.,* 107, 2075, 1988.

74. **Hatta, T. and Francki, R. I. B.,** Cytopathic structures associated with tonoplasts of plant cells infected with cucumber mosaic and tomato aspermy viruses, *J. Gen. Virol.,* 53, 343, 1981.

75. **Francki, R. I. B.,** Responses of plant cells to virus infection with special reference to the sites of RNA replication, in *Positive Strand RNA Viruses,* Brinton, M. A. and Rueckert, R. R., Eds., Alan R. Liss, New York, 1987, 423.

76. **Van Kammen, A. and Mellema, J. E.,** Comoviruses, in *The Atlas of Insect and Plant Viruses: Including Mycoplasmaviruses and Viriods,* Maramorosch, K., Ed., Academic Press, New York, 1977, 167.

77. **Soderlund, H. and Ulmanen, I.,** Transient association of Semliki Forest virus capsid protein with ribosomes, *J. Virol.,* 24, 907, 1977.

78. **Griffiths, G., Quinn, P., and Warren, G.,** Dissection of the Golgi complex. I. Monensin inhibits the transport of viral membrane proteins from medial to trans Golgi cisternae in baby hamster kidney cells infected with Semliki forest virus, *J. Cell. Biol.,* 96, 835, 1983.

79. **Quigley, J. P., Rifkin, D. B., and Reich, E.,** Phospholipid composition of Rous Sarcoma virus, host cell membranes and other enveloped RNA viruses, *Virology,* 46, 106, 1971.

80. **Higashi, N., Matsumoto, A., Tabata, K., and Nagatomo, Y.,** Electron microscopic study of development of Chikungunya virus in Green Monkey kidney stable (Vero) cells, *Virology,* 33, 55, 1967.

81. **Grimley, P. M., Michelitch, J. H., and Friedman, R. M.,** Influence of subcellular organization on the development and dissemination of an arbovirus, *Am. J. Pathol.,* 59, 48a, 1970.

82. **Waite, M., Brown, D., and Pfefferkorn, E.,** Inhibition of Sindbis virus release by media of low ionic strength: an electron microscope study, *J. Virol.,* 10, 537, 1972.

83. **Lindquist, B. H., DiSakvim, H. M., Rice, C. M., Strauss, J. H., and Strauss, E. G.,** Sindbis virus mutant ys20 of complementation group E contains a lesion in glycoprotein E2, *Virology,* 151, 10, 1986.

84. **Vrati, S., Faragher, S. G., Weir, R. C., and Dalgarno, L.,** Ross virus mutant with a deletion in the E2 gene: properties of the virion, virus-specific macromolecule synthesis and attenuation of virulence for mice, *Virology,* 151, 222, 1986.

85. **Matsumara, T., Stollar, V., and Schlesinger, R. W.,** Study on the nature of Dengue virus. V. Structure and development of Dengue virus in Vero cells, *Virology,* 46, 344, 1971.

86. **Leary, K. and Blair, C. D.,** Sequential events in the morphogenesis of Japanese B encephalitis virus, *J. Ultrastruct. Res.,* 72, 123, 1980.

87. **Stohlman, S. A., Wisseman, C. L., Eylar, D. R., and Silverman, D. J.,** Dengue-virus associated modifications of host cell membranes, *J. Virol.,* 16, 1017, 1975.

88. **Zebovitz, E., Leong, J. K. L., and Doughty, S. E.,** Involvement of the host cell nuclear envelope membranes in the replication of Japanese encephalitis virus, *Infect. Immun.,* 10, 204, 1974.

89. **Kovac, W., Kunz, C. H., and Stockinger, L.,** Die elektroenenmikroskopische Darstellung des Virus der Fruhsommer-Meingoencephalits (FSME) in HeLa Zellen, *Arch. Gesante Virusforsch.,* 11, 544, 1962.

90. **Yasuzumi, G. and Tsubo, I.,** Analysis of the development of Japanese B encephalitis, *J. Ultrastruct. Res.,* 12, 304, 1965.

91. **Murphy, F. A., Harrison, A. K., Gray, G. W., and Whitfield, S. W.,** St. Louis encephalitis virus infection of mice. Electron microscopic studies of central nervous system, *Lab. Invest.,* 19, 652, 1968.

92. **Tamalet, J., Toga, M., Chippaux-Hyppolite, C., Choux, R., and Cesarini, J-P.,** Encephalite experimental a virus West-Nile de la souris: aspects ultrastructuraux du systeme nerveux central, *C. R. Acad. Sci. Paris,* 269d, 668, 1969.

93. **Chippaux-Hippolite, C., Choux, R., Olmer, H., and Tamalet, J.,** Multiplication du virus West-Nile dans les cultures cellulaires d'embryon de poulet, observee en microscopie electronique, *C. R. Acad. Sci. Paris,* 270d, 3162, 1970.

94. **Boulton, P. S. and Webb, H. E.,** An electron microscopic study of Langat virus encephalitis in mice, *Brain,* 94, 411, 1971.

95. **Blinzinger, K.,** Comparative electron microscopic studies of several experimental group B arbovirus infections of the murine CNS (CEE virus, Zimmern virus, Yellow fever virus), *Ann. Inst. Pasteur,* 123, 497, 1972.

96. **Shestapalova, N. M., Reingold, V. N., Gagarina, A. V., Kornilova, E. A., Popov, G. V., and Chumakov, M. P.,** Electron microscopic study of the central nervous system in mice infected by Omsk Hemorrhagic Fever (OHF) virus, *J. Ultrastruct. Res.,* 40, 458, 1972.

97. **Tam, S. D-W., Labzoffsky, N. A., and Hamvas, J. J.,** Morphogenesis of Yellow Fever virus in mouse brain, *Arch. Gesante Virusforsch.,* 36, 372, 1972.

98. **Tandler, B., Erlandson, R. A., and Southam, C. M.,** Unusual membrane formations in HEp-2 cells infected with Ilheus virus, *Lab. Invest.,* 28, 217, 1973.

99. **Blinzinger, K., Muller, W., and Anzil, A. P.,** Microhelices in the endoplasmic reticulum of murine neurons infected with a Group B arbovirus, *Arch. Gesamte Virusforsch.,* 35, 194, 1971.

100. **Erlandson, R. A., Babcock, V. I., Southam, C. M., Bailey, R. B., and Shipkey, F. H.,** Semliki Forest virus in HEp-2 cell cultures, *J. Virol.,* 1, 996, 1967.

101. **Grimley, P. M. and Schaff, Z.,** Significance of tubuloreticular inclusions in the pathobiology of human diseases, in *Pathobiology Annual, Vol. 6,* Ioachim, H. L., Ed., Appleton-Century-Crofts, New York, 1976, 221.

102. **Grimley, P. M., Rutherford, M. N., Kang, Y-H., Williams, T., Woody, J. N., and Silverman, R. M.,** Formation of tubuloreticular inclusions in human lymphoma cells compared to the induction of 2'-5'-oligoadenylate synthetase by leucocyte interferon in dose-effect and kinetic studies, *Cancer Res.,* 144, 3480, 1984.

103. **Schaff, Z., Gerety, R. J., Grimley, P. M., Iwarson, S. A., Jackson, D. R., and Tabor, E.,** Ultrastructural and cytochemical study of hepatocytes and lymphocytes during experimental non-A, non-B infections in chimpanzees, *J. Exp. Pathol.,* 2, 225, 1985.

104. **Feldman, D., Hoar, R. M., Niemann, W. H., Valentine, T., Cukierski, M., and Hendrickx, A. G.,** Tubuloreticular inclusions in placental chorinic villi of rhesus monkeys after maternal treatment with interferon, *Am. J. Obstet. Gynecol.,* 155, 413, 1986.

Chapter 19

BUNYAVIRIDAE

Michèle Bouloy

TABLE OF CONTENTS

I. THE FAMILY BUNYAVIRIDAE

A. INTRODUCTION

The Bunyaviridae family comprises more than 200 arboviruses (arthropod-borne viruses) formerly known as viruses of the Bunyamwera supergroup.[1-3] Orginally, the classification into 11 serogroups (13 at the present time) was based on the relationships between these viruses which could be distinguished by neutralization, hemagglutination inhibition or complement fixation tests. Serological characteristics of the members of the Bunyaviridae family have been reviewed in detail.[4-7] These viruses are ubiquitous in the world and have been isolated from various arthropods, mosquitoes, ticks, and sandflies as well as from vertebrates including man. A number of bunyaviruses cause severe disease such as encephalitis or hemorrhagic fevers in man and animals. In many infections classical symptoms are headache, fever, and muscle pain.

B. GENERAL CHARACTERISTICS

All members of the Bunyaviridae family share the same morphological features: the virus particles are spherical (80 to 110 nm in diameter) and possess a membrane envelope from which protrude polypeptide spikes 5 to 10 nm long. These data were obtained by the classical negative staining method. More recently, by ultrarapid freezing in layers of vitreous ice, purified preparations of bunyaviruses have been reexamined under the electron microscope.[8] Precise measurements of LaCrosse virions showed heterogeneity in diameter (75 to 115 nm) with 10 nm long spikes protruding from a 4 nm thick membrane envelope (see Figure 1). In bunyaviruses, no periodicity has been seen at the surface of the virion. This observation is in contrast to results obtained with Uukuniemi[9] and Punta Toro[10] viruses which exhibit a distinct surface lattice arrangement. The Bunyaviridae replicate in the cytoplasm[5] of infected cells and no nuclear step takes place during the viral cycle.[11] Virions are formed by budding into vesicles of the Golgi complex (see Figure 2). No budding has been observed at the plasma membrane. The particles are composed of four proteins: two glycoproteins, G1 and G2, which form the protruding spikes on the viral envelope, a nucleocapsid protein N, and a minor large polypeptide L. The genome consists of three single-stranded RNA segments which, in association with numerous copies of the N protein and a few copies of the L protein, form three circular nucleocapsids (see Figure 3).[12-14]

In 1973 when the family was recognized, it included only the *Bunyavirus* genus (approximately 90 viruses) and other possible members (50 bunyavirus-like viruses) which were morphologically similar to bunyaviruses but not serologically related to them. Later, as molecular analyses had been developed, several genera were proposed. At the present time the family comprises five genera: *Bunyavirus, Nairovirus, Phlebovirus, Uukuvirus,* and *Hantavirus,*[15,16] the properties of which will be reported in the following sections.

Almost all bunyaviruses have been isolated from mosquitoes; only a few have been obtained from ticks or phlebotomines. The converse situation is observed for phleboviruses, uukuviruses, and nairoviruses.[3-7] Unlike the other members of the family, hantaviruses have not been isolated from arthropods and appear to be directly transmitted via urine, feces, and saliva of infected rodents, rats *(Rattus),* mice *(Apodemus),* and voles *(Clethrionomys* and *Microtus).*[17] The life cycles of bunyaviruses involve alternate replication in vertebrates and arthropods. The cycles are different depending on the host. When the virus infects vertebrate cell lines, it usually causes a cytopathic effect leading to cell death. In mosquito cells, however, it establishes a persistent infection. In *vivo,* the infection causes little or no histopathologic changes in the arthropods, and the virus is passed onto the next generation by transovarial and venereal transmission.

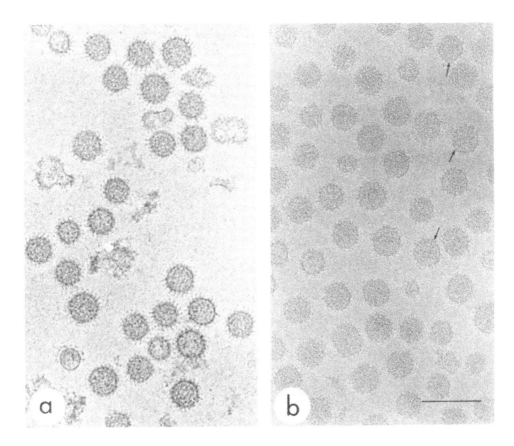

FIGURE 1. Transmission electron micrographs of vitrified-hydrated La Crosse virions taken (a) large defocus value at which spikes are visible, and (b) small defocus value, at which some of the membrane bilayers are visible (arrows). Bar, 200 nm. (From Talmon, Y., Venkataram Prasad, B. V., Clerx, J. P. M., Wah Chiu, G. J. W., and Hewlett, M. J., *J. Virol.*, 61, 2319, 1979. With permission.)

II. THE *BUNYAVIRUS* GENUS

Bunyaviruses are abundantly represented (more than 150 viruses) in the Bunyaviridae family, classified into 13 serogroups. The prototype Bunyamwera virus is transmitted by Aedes, Mansonia, and Culex mosquitoes.

A. VIRION
1. Proteins

The virion is composed of four proteins,[6,7,18,19] the large polypeptide, L; two envelope glycoproteins, G1 and G2; and the nucleoprotein, N. In addition, the viral particle contains lipids and carbohydrates (23 and 7%, respectively).

Obijeski et al.[6,18,19] have estimated the stoichiometric composition of LaCrosse virus: G1 and G2 glycoproteins are present in about equal amounts, approximately 650 molecules per virus particle, the major nucleocapsid protein N is present in about 2100 molecules per virus particle and the large protein L in about 25 molecules per virus particle. The two glycoproteins are present at the surface and function as type-specific immunizing antigen.[20] In LaCrosse virus, G1 appears to be the hemagglutinin and antibodies directed against G1 possess neutralizing activity as well as hemagglutination inhibition.[21] No G2 specific monoclonal antibody has been isolated, and so far no viral function has been attributed to it.[22]

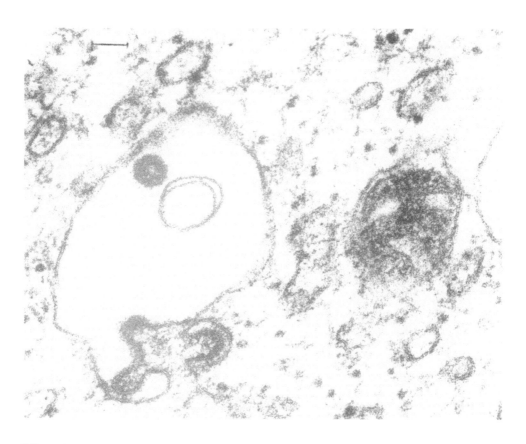

FIGURE 2.   Ultrathin section of BHK21 cells infected with Lumbovirus. Cells were harvested at 10 h after infection. (Bar = 100 nm.) (From Samso, A. and Bouloy, M., unpublished results. With permission.)

The G1 and/or G2 protein(s) play(s) a role in infectivity, since specific removal by protease digestion reduced the specific infectivity of the particle by about 10^5 compared to the untreated control.[18] By genetic analyses Bishop and Shope[7], Gentsch et al.[23-26], and Janssen et al.[27] have shown that the M RNA segment was the major determinant of virulence. Cell-to-cell fusion activity is carried by G1 protein and might influence the tropism of the virus.

The N protein represents the principal complement fixation antigen[28] and is the major protein component of the viral nucleocapsid. A minor protein is also present in the nucleocapsid, the L protein. It is encoded by the L RNA segment and is presumed to be the RNA dependent RNA polymerase associated with the virion.

2. Genome

The genome consists of three single-stranded RNA molecules of negative polarity named large (L), medium (M), and small (S) according to their size.[19,25,29] The genomic RNAs are present in circular forms under moderately denaturing conditions.[19,30] The circularization is due to the inverted complementary sequences (at least 10 to 15 nucleotides) at the 3' and 5' ends.[31-33] The terminal sequences are conserved in the three segments and also within members of each genus (see Table 1). The L RNA segment (2.7 to 3.1 $\times$ 10^6 Da) codes for the L protein, the M segment (1.8 to 2.3 $\times$ 10^6 Da) codes for the two glycoproteins G1 and G2 and for the nonstructural polypeptide NSM. The S segment (0.28 to 0.50 $\times$ 10^6 Da) codes for the nucleoprotein N and a nonstructural protein named NSS, the function of which is not yet known.

The nucleotide sequence of the complete S segment has been determined for at least

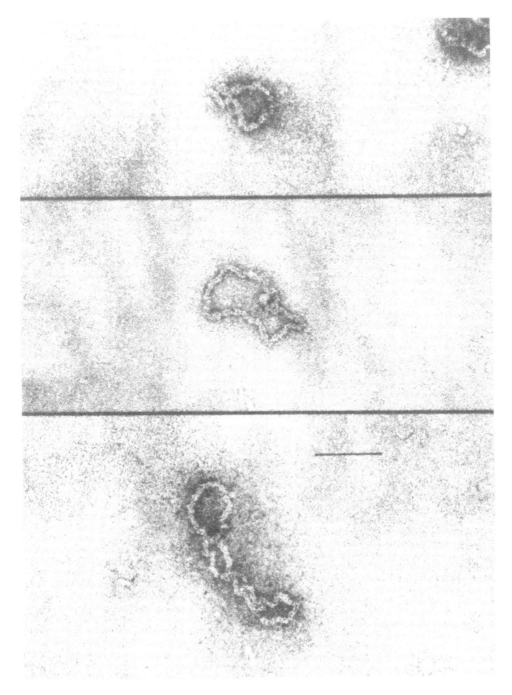

FIGURE 3. Nucleocapsids isolated from Lumbo virus. (Bar = 100 nm.) (From Samso, A., Bouloy, M., and Hannoun, C., *C. R. Acad. Sci. (Paris)*, D280, 779, 1975. With permission.)

four bunyaviruses: LaCrosse and snowshoe hare viruses[34-36] (California serogroup), Aino virus[37] (Simbu serogroup), and Germiston virus[38] (Bunyamwera serogroup). Open reading frames (ORF) are found only in the sequences complementary to the viral RNA segment (cRNA). The N and NSS proteins are read from two overlapping ORFs. Germiston virus S RNA segment contains a third ORF which could code for a 75 amino acid long polypeptide. Experiments are in progress to determine whether this ORF is expressed in infected cells.

TABLE 1
Conserved 3′ and 5′ Terminal Sequences within Each Genus

Genus	3′ terminus	5′ terminus
Bunyavirus	UCAUCACAUGA	UCGUGUGAUGA
Phlebovirus	UGUGUUUC	CCAGAAACACA
Uukuvirus	UGUGUUUCU	ACAGAAACACA
Nairovirus	AGAGAUUCUUU	
Hantavirus	AUCAUCAUCUG	ACAGAUGAUGAU

Note: Data from References 34—41, 70—72, 80, 92, 96, 97.

TABLE 2
Percentage Homology Between N and NSS Proteins of Several Bunyaviruses

	N protein			NSS protein		
	AINO	**SSH**	**LAC**	**AINO**	**SSH**	**LAC**
GER	44.2	47.6	46.8	26.6	37.0	45.3
LAC	45	90.6		35.0	87.0	
SSH	44.6			38.5		

Note: GER, Germiston virus; LAC, LaCrosse virus; SSH, snowshoe hare virus; AINO, Aino viruses.

From Gerbaud, S., Vialat, P., Pardigon, N., Wychowski, C., Girard, M., and Bouloy, M., *Virus Res.*, 8, 1, 1987. With permission.

Computer analysis indicates that the N proteins and, to a lesser extent, the NSS proteins are conserved within viruses of different serogroups (see Table 2). Tightly associated with RNA molecules in the nucleocapsids, the N proteins must share conserved regions involved in RNA-protein interactions.

The sequence of the M RNA has been determined for Bunyamwera,[39] snowshoe hare,[40] LaCrosse,[41] and Germiston viruses.[41a] There is a long ORF in the mRNA molecule that contains sufficient information to code for a large polypeptide (160 to 162 kDa) which could be the precursor to the three proteins encoded by the M segment, the two glycoproteins and the NSM protein. The predicted amino acid sequences of the polyproteins of these viruses as well as their hydropathy plots have been compared. A high level of homology is observed between viruses belonging to the same serogroup. Several hydrophobic regions are conserved at the internal regions and at the C and N termini. The hydrophobic region at the N terminus probably functions as a signal peptide for the translocation. The C terminus hydrophobic domain is likely to be anchored into membranes. However, the critical question of how the polyprotein is processed to generate G1, G2, and NSM protein is still unsolved.

The complete sequence of L RNA segment has not yet been reported. Approximately 80% of Bunyamwera virus L RNA has been sequenced and reveals the presence of a presumably single large ORF[41b] in the cRNA.

B. REPLICATION

In order to obtain insight into the biology of these arboviruses of the Bunyaviridae family, it is important to study the replication on a molecular basis in vertebrate and invertebrate cells which are both natural hosts. Extensive studies in vertebrate cells have been reported but information related to virus multiplication in mosquito cells is much more limited. Only a small number of bunyaviruses has been studied to provide information on the replication. Among these, one should mention LaCrosse, snowshoe hare, and Lumbo viruses belonging to the California complex, Germiston virus of the Bunyamwera serogroup and Akabane virus of the Akabane serogroup.

1. In Vertebrate Cells

Bunyaviruses multiply to relatively high titers (10^7 to 10^9 pfu/ml) in vertebrate cell lines of various origins: Vero BHK21, MDCK, MDBK, primary chick embryo fibroblasts, XTC-2 *(Xenopus laevis)*, etc. After adsorption of the virus to the cell surface, the genomic RNA must be transcribed into cRNA molecules which serve as messengers to synthesize proteins involved in early functions. This step called primary transcription is achieved by the RNA-dependent RNA polymerase as is the case in negative strand RNA viruses.[42-45] This appears to be obligatory, since viral RNA extracted from virions is not infectious, and only cRNA but not virion RNA has been found in the polysomes.[46] Several laboratories have shown that primary (as well as secondary) transcription is greatly inhibited in the presence of protein synthesis inhibitors such as cycloheximide, puromycin, and anisomycin.[45,47-50] Therefore, primary transcription appears to require ongoing protein synthesis. Indeed, Bellocq and Kolakofsky[44,51] and our laboratory[51a] have shown that authentic mRNAs are synthesized in *in vitro* transcription assays containing purified virus (LaCrosse or Germiston virus) only when actively translating reticulocyte lysate is added. A model has been proposed for LaCrosse virus transcription. Ribosomes loaded on the mRNA immediately after initiation of its synthesis would prevent interactions between the nascent chain and the template. This eliminates the synthesis of small transcripts which is due to pausing of the transcriptase and/or premature termination.

The three L, M, and S messenger RNAs are unique and have been well characterized in cells infected by LaCrosse,[52] snowshoe hare,[53,54] and Germiston[45,55] viruses. They are subgenomic complementary RNAs, where the sequence corresponding to some 60 to 115 bases of the 5' end of the template is lacking at the 3' end of the molecules. When compared with the complete transcripts, also called antigenomes or replicative intermediates, the mRNAs, but not the antigenomes, contain 5' extensions of approximately 15 nucleotides in length which are heterogeneous in sequence and not coded by the virus. They probably represent host primers used to initiate the transcription by a mechanism similar to that of influenza virus.[56,57]

An interesting problem is posed by the dicistronic S mRNA (possibly tricistronic in Germiston virus) in which two (or three) different AUGs must be recognized to initiate the synthesis of their respective proteins. In most cases, the leaky scanning model proposed by Kozak explains the results.[58]

In several cell free systems, both N and NSS proteins are efficiently synthesized when S mRNA is added but very little, if any, protein is made by M mRNA. From the sequence data, one would expect to detect a protein of high molecular weight which must be the precursor to the three proteins G1, G2, and NSM. Nevertheless, such a product has never been found either *in vitro* or in infected cells.[59]

2. In Insect Cells

Attempts to study replication in mosquito cells or in intact mosquitoes have been few. Whereas vertebrate cells die during infection, mosquito cells are susceptible to virus growth in a persistent state which does not impair cell life. No change in host cell DNA, RNA, and protein synthesis is observed.

The presence of defective interfering particles has been demonstrated and could explain the establishment of the persistent infection.[60-62] Persistence is also accompanied by the generation of viruses with variable genetic and phenotypic characteristics as reported by Elliot.[62] Bunyamwera virus released from persistently infected cells produced an altered plaque morphology and temperature sensitivity. Molecular analysis also showed that in one case, the persistent infection produces a virus with a nucleocapsid protein altered in its electrophoretic migration. Moreover, the RNA profile is modified compared to that of vertebrate cell infection; the S RNA segment is oversynthesized, while the M and L RNAs are underreplicated.

Viral antigens have been classified by serological tests: neutralization and hemagglutination-inhibition, which both define the serotype, and complement fixation, which may coincide with viruses of the same or a totally different serotype. Neutralization and hemagglutination epitopes are carried by the G1 protein, coded by the M segment and the major complement-fixing antigen is the nucleocapsid protein N, coded by the S segment. It is, therefore, tempting to postulate that recombination has occurred in nature generating natural reassortants.[7] Bishop and his group[23-26,63] have carried out extensive genetic studies. Using temperature sensitive mutants of LaCrosse, snowshoe hare, and Tahyna viruses obtained by *in vitro* mutagenesis, they have demonstrated that the genome of these viruses can recombine by segment reassortment. These studies established the coding relationships of each segment which are now confirmed by sequence data of cloned cDNA. Similar conclusions were also obtained by other groups.[66,67] The genetic studies originally developed with LaCrosse and snowshoe hare viruses in mammalian cells have been extended to live *Aedes triseriatus* mosquitoes. Beaty et al.[66] have shown that the M segment determines the efficiency of dissemination from the midgut to infect secondary target organs after oral transmission in mosquitoes. Simultaneous dual infections result in the generation and transmission of reassortant viruses.[67] Dual bunyavirus infection in mosquitoes must be simultaneous since secondary infection with viruses of the same group is specifically inhibited, while the mosquitoes remain permissive to heterologous viruses.[68] The induction of virus-specific interference is not yet understood, but suggests that natural bunyavirus reassortment in mosquitoes must be limited.

III. THE *PHLEBOVIRUS* GENUS

Phlebotomus fever is caused by the Sicilian sandfly fever phlebovirus which is the prototype of the genus.[3] Rift Valley fever virus infects man, sheep, and other ruminants. Infection is sometimes fatal in man and induces abortions in ruminants. Many phleboviruses have been isolated from phlebotomines and occasionally from mosquitoes. There are at least 30 viruses in the genus, all serologically related and, unlike bunyaviruses, forming a single serological group. Several of them including Rift Valley fever and Punta Toro viruses, have been analyzed on a molecular basis.

A. VIRION

Structurally, phleboviruses are comparable to bunyaviruses.[10] The major components are the nucleocapsid protein (20 to 22 kDa) and two glycoproteins with similar apparent molecular weight (62 kDa). The genome is the tripartite RNA with segments L, M, and S. In early studies it has been observed that the S RNA segments of Karimabad and Punta Toro viruses (0.75 to 0.8×10^6 kDa) are significantly larger than those of LaCrosse or snowshoe hare viruses (0.45 to 0.5×10^6 Da).[69] At present this observation is substantiated by the coding capacity and the novel coding strategy of this segment.[70] Indeed, sequencing of the S RNA segment of Punta Toro virus has revealed that, although the nucleoprotein N (27 kDa) is coded in the sense complementary to the virus RNA, a nonstructural protein NSS (29 kDa) is coded in the viral sense. Unlike bunyaviruses, phleboviruses have only one ORF in the S cRNA. Based on the definition of positive and negative viral RNA genomes, the term of ambisense strategy has been proposed for phleboviruses.

The M segments of Rift Valley fever[71] and Punta Toro[72] viruses have been cloned and sequenced. As in other members of the Bunyaviridae, there is a single ORF in the cRNA. The predicted gene product is a polyprotein precursor to the viral glycoproteins G1 and G2. Limited amino terminal sequence analyses of the glycoproteins have indicated the gene order, the potential cleavage sites in the precursor, and the existence of a putative nonstructural protein NSM (30 or 17 kDa). Moreover, the N (but not the C) termini of G1 and G2 proteins are positioned in the polypeptide precursor, so that another small polypeptide (5.8 kDa) representing the intergenic region between G1 and G2 proteins might be generated. These two nonstructural polypeptides have still to be identified in infected cells. In addition, the polyprotein precursor has never been detected during infection; this suggests that it must be processed very rapidly during its synthesis. At least one of the glycoproteins contains epitopes involved in neutralization.

The L RNA segment is approximately 2.3×10^6 Da in length. Although no direct evidence is available, it is presumed to code for the component of the viral transcriptase.

B. REPLICATION

Unexpectedly, the strategy of replication of phleboviruses differs considerably from that of other members of the Bunyaviridae studied so far, due to the existence of the ambisense S genomic RNA. This novel strategy is utilized also by the members of the Arenaviridae family to express the nucleocapsid protein NP and the glycoprotein precursor. Analysis of intracellular RNAs has shown the existence of two unique mRNAs, each one expressing the N or the NSS protein.[73,74] The NSS mRNA, however, occurs as a secondary transcription, since it requires the synthesis of the antigenomic RNA molecule that it uses as a template. Proteins N and NSS are translated from subgenomic mRNAs corresponding to the 3′ or 5′ half of the S RNA segment, respectively. Like bunyaviruses, phleboviruses contain a transcriptase which is active *in vitro* and probably initiates cRNA messenger (N mRNA) synthesis with host RNA primers. Indeed, messengers of viral complementary sense and coding for the N protein contain 5′ heterogeneous extensions representing nonviral primer sequences. Probably, viral sense mRNA coding for the NSS protein also contains these 5′ sequences. The 3′ ends of the N and NSS mRNAs have been precisely analyzed. Transcription stops in a hairpin region localized in the center of the molecule.

In mosquito cells (AP-61) derived from *Aedes pseudocutellaus* persistent infection has been established.[75]

IV. THE *UUKUVIRUS* GENUS

Members of the Uukuvirus genus (at least eight viruses) belong to a single serotype. Uukuniemi virus, isolated from ticks like most of the other uukuviruses, is the prototype and almost exclusively, has served to determine the characteristic features of uukuviruses. None of the uukuviruses have been associated with disease in man.[3,7]

A. VIRION

The particle is composed of the four characteristic proteins, L (180 to 200 kDa), N (25 kDa), G1, and G2, the two latter being of roughly equal sizes (75 and 65 kDa, respectively) like those of phleboviruses.[76] The nucleocapsids are circular[77] and contain three unique RNA segments L (2.4×10^6 Da), M (1.1×10^6 Da), and S (0.5×10^6 Da).[78] The RNA molecules are also present in circular form[79] due to inverted complementary sequences at the 3′ and 5′ ends of the molecules.[80]

Electron microscopy studies of the virion indicate the presence of glycoprotein spikes at the surface which are regularly clustered forming an icosahedral surface lattice of penton-hexons (T = 12, P = 3).[9] This type of architecture has also been described for phlebovirus but seems to differ in bunyaviruses.

Recently, Röhnholm and Pettersson[80] have reported the sequence of the M segment of Uukuniemi virus. A single ORF comprising 1008 amino acids was found in the cRNA sense strand which is the polyprotein precursor to the glycoproteins G1 and G2. Surprisingly, the 3' and 5' terminal sequences are identical to those of the M segment of the phleboviruses Punta Toro and Rift Valley fever, while the amino acid sequences of the polyproteins share only a weak homology, confirming that there exists no serological cross-reactivity between uuku- and phleboviruses.

B. REPLICATION

Like bunyaviruses, Uukuniemi virus contains an RNA dependent RNA polymerase,[81] indicating that the genome is transcribed into cRNAs which serve as mRNA. The S mRNA is smaller in size than the full length cRNA, but it has not yet been characterized further. It is capable of inducing *in vitro* the synthesis of the nucleocapsid protein N and a nonstructural protein NSS (30 kDa). The M mRNAs isolated from uukuvirus infected cells are efficiently translated *in vitro* into a polyprotein precursor (p110) which can be further processed if dog pancreas microsomes are added.[82] This distinguishes uukuviruses from other members of the Bunyaviridae family (see earlier sections). The processing of the polyprotein precursor generates the two glycoproteins G1 and G2.[85] From the amino acid sequence determination of N terminal of these two glycoproteins, it appears that G1 represents the N terminus (after cleavage of a probable signal peptide). As the C terminus has not yet been localized, it is not known whether the processing generates a small 81 amino acid long intergenic peptide. The maturation of uukuniemi virus in the Golgi complex has been well documented.[84-87] Proteins G1 and G2 appear to be good models for studying protein localization in the Golgi apparatus.

Although a tick cell line is available, no data have been reported on the replication in insect cells.

V. THE *NAIROVIRUS* GENUS

Crimean-Congo hemorrhagic fever virus is the prototype.[88-90] The genus comprises at least 19 viruses categorized into six serogroups. Many nairoviruses are pathogenic for humans and animals. Crimean hemorrhagic fever has occasional fatal outcomes. Dhori virus causes human disease, due to laboratory infection with fever and with symptoms of meningo encephalitis. Nairobi sheep disease virus is associated with high mortality in ruminants. These viruses are predominantly transmitted by ticks.

Present data are restricted to a few members: Qalyub,[91,92] Dugbe,[93] and CloMor[94] viruses. By comparison with the L RNA of other members of the Bunyaviridae family, the L RNA of Qalyub is apparently 50% larger. The nucleocapsid protein N is approximately twice the size of that of bunyaviruses. Several large glycosylated polypeptides are absent from the viral particles but present in infected cells. Their origin and functions are not known.

VI. THE *HANTAVIRUS* GENUS

Hantaan virus has recently been classified as the prototype of the *Hantavirus* genus. It is the etiologic agent of Korean hemorrhagic fever. Viruses antigenically related to Hantaan virus and responsible for hemorrhagic fever associated with renal syndrome have a broad geographic distribution.[17] Hantaviruses have never been isolated from arthropods but only from wild or laboratory rats.

A. VIRION

Hantaan virus possesses four proteins, two glycoproteins G1 (68 kDa) and G2 (55 kDa), a nucleocapsid protein N of molecular weight (50 kDa), larger than that of bunyaviruses, and the large polypeptide L (200 kDa).[15,95] Two of the segments of the tripartite genome, namely, S (0.6×10^6 Da) and M (1.2×10^6 Da) have been cloned and sequenced.[96,97]

The S segment is of negative polarity as one ORF is present in the cRNA strand. However, compared to other Bunyaviridae, the S segment is much larger. This reflects the fact that it codes for an unusually large nucleoprotein N (50 kDa). In addition, hantaviruses differ from other Bunyaviridae by the fact that the S segment codes only for the N protein and no nonstructural protein.

The M segment contains a single long ORF in the cRNA strand. The predicted gene product (126 kDa) is the precursor to the glycoproteins G1 and G2, the partial amino acid sequences of which have been determined. The position of the proteins in the polyprotein was determined using antisera raised against synthetic peptides, allowing to map the C terminus of each glycoprotein. The results indicate that the M segment does not code for a nonstructural protein similar to the NS_M of bunyaviruses.

B. REPLICATION

Messenger RNAs are probably synthesized by a transcriptase similar to that of bunyavirus as a polymerase activity has been observed *in vitro*.[15] No glycoprotein precursor has been demonstrated in infected cells suggesting that the processing must be cotranslational.

ACKNOWLEDGMENTS

The author thanks Dr. M. Hewlett for electron micrographs used in the text. Work performed in this laboratory was supported in part by Grant INSERM No. 861003.

ADDENDUM

Since this review was written, several papers contributing to a better understanding of the molecular biology of Bunyaviridae have been published. The L segment of Bunyamwera virus (bunyavirus) has been cloned and sequenced; in the viral complementary sense, it codes for a single large ORF corresponding to the L protein.[98] No significant homology was found with any viral polymerase except to a low level with Influenza virus P_{B1} protein. The two glycoproteins and the nonstructural protein NS_M of snowshoe hare Bunyavirus have been mapped within the ORF encoded by the M segment.[99] The precursor to the glycoproteins of the Rift Valley fever phlebovirus has been synthesized in an *in vitro* system.[100] The sequence of the S segment of the Uukuniemi uukuvirus has been reported.[101] Surprisingly, it contains two ORFs, one in the virion sense and the other in the viral complementary sense, indicating that Uukuviruses utilize the ambisense strategy. The S segment of the Nairovirus Dugbe has been shown to contain one ORF in the viral complementary sense, a situation similar to that of Hantaviruses.[102]

REFERENCES

1. **Porterfield, J. S., Casals, J., Chumakov, M. P., Gaidamovich, S. Ya, Hannoun, C., Holmes, I. H., Horzinek, M. C., Mussgay, M., and Russel, P. K.,** Bunyaviruses and Bunyaviridae, *Intervirology,* 2, 270, 1973.
2. **Porterfield, J. S., Casals, J., Chumakov, M. P., Gaidamovich, S. Ya., Hannoun, C., Holmes, I. H., Horzinek, M. C., Mussgay, M., Oker-Blom, N., and Russell, P. K.,** Bunyaviruses and Bunyaviridae, *Intervirology,* 6, 13, 1975.
3. **Bishop, D. H. L., Calisher, C. H., Casals, J., Chumakov, M. P., Gaidamovich, S. Ya., Hannoun, C., Lvov, D. K., Marshal, I. D., Oker-Blom, N., Petterssen, R. F., Porterfield, J. S., Russell, P. K., Shope, R. E., and Westaway, E. G.,** Bunyaviridae, *Intervirology,* 14, 125, 1980.
4. **Casals, J.,** Arboviruses: incorporation in a general system of virus classification, in *Comparative Virology,* Maramorosch, K. and Kurstak, E., Eds., Academic Press, New York, 1971, 307.
5. **Murphy, F. A., Harrison, A. K., and Whitfield, S. G.,** Bunyaviridae: morphologic and morphogenetic similarities of Bunyamwera serologic supergroup viruses and several other arthropod-borne viruses, *Intervirology,* 1, 297, 1973.
6. **Obijeski, J. F. and Murphy, F. A.,** Bunyaviridae: recent biochemical developments, *J. Gen. Virol.,* 37, 1, 1977.
7. **Bishop, D. H. L. and Shope, R. E.,** Bunyaviridae, in *Comprehensive Virology,* Vol. 14, 1979.
8. **Talmon, Y., Venkataram Prasad, B. V., Clerx, J. P. M., Wah Chiu, G. J. W., and Hewlett, M. J.,** Electron microscopy of vitrified-hydrated LaCrosse virus, *J. Virol.,* 61, 2319, 1987.
9. **von Bonsdorff, C. H. and Pettersson, R.,** Surface structure of Uukuniemi virus, *J. Virol.,* 16, 1296, 1975.
10. **Smith, J. and Pifat, D.,** Morphogenesis of sandfly fever viruses, (Bunyaviridae family), *Virology,* 121, 61, 1982.
11. **Rossier, C., Patterson, J., and Kolakofsky, D.,** La Crosse virus small genome mRNA is made in the cytoplasm, *J. Virol.,* 58, 647, 1986.
12. **Pettersson, R. F. and von Bonsdorff, C. H.,** Ribonucleoproteins of Uukuniemi virus are circular, *J. Virol.,* 15, 386, 1975.
13. **Obijeski, J. F., Bishop, D. H. L., Murphy, F. A., and Palmer, F. L.,** Segmented genome and nucleocapsid of LaCrosse virus, *J. Virol.,* 20, 664, 1976.
14. **Samso, A., Bouloy, M., and Hannoun, C.,** Présence de ribonucléoprotéines cirulaires dans le virus Lumbo (Bunyavirus), *C. R. Acad. Sci., Paris, Ser. D,* 280, 779, 1975.
15. **Schmaljohn, C. S. and Dalrymple, J. M.,** Analysis of Hantaan virus RNA. Evidence for a new genus of Bunyaviridae, *Virology,* 131, 482, 1983.
16. **Mahy, B. and Kolakovsky, D.,** *The Biology of Negative Strand Viruses,* Elsevier, 1987.
17. **Schmaljohn, C. S., Hasty, S. E., Dalrymple, J. M., LeDuc, J. W., Lee, H. W., von Bonsdorff, C. H., Brummer-Korvenkontio, M., Vaheri, A., Tsai, T. F., Regnery, H. L., Goldgaber, D., and Lee, P. W.,** Antigenic and genetic properties of viruses linked to hemorrhagic fever with renal syndrome, *Science,* 227, 1041, 1985.
18. **Obijeski, J. F., Bioshop, D. H. L., Murphy, F. A., and Palmer, F. L.,** Structural proteins of LaCrosse virus, *J. Virol.,* 19, 985, 1976.
19. **Gentsch, J. R., Bishop, D. H. L., and Obijeski, J. F.,** The virus particle nucleic acids and proteins of four bunyaviruses, *J. Gen. Virol.,* 34, 257, 1977.
20. **Kingsford, L. and Hill, D. W.,** The effect of proteolytic cleavage of LaCrosse virus G1 glycoprotein on antibody neutralization, *J. Gen. Virol.,* 64, 2147, 1983.
21. **Gonzalez-Scarano, F., Shope, R. E., Calisher, C. H., and Nathanson, N.,** Characterization of monoclonal antibodies against the G1 and N proteins of LaCrosse and Tahyna, two California serogroup bunyaviruses, *Virology,* 120, 42, 1982.
22. **Gonzalez-Scarano, F.,** LaCrosse virus G1 glycoprotein undergoes a conformational change at the pH of fusion, *Virology,* 140, 209, 1985.
23. **Gentsch, J. and Bishop, D. H. L.,** Recombination and complementation between temperature sensitive mutants of the bunyavirus, snowshoe hare virus, *J. Virol.,* 20, 351, 1976.
24. **Gentsch, J. R., Rozhon, E. J., Klimas, R. A., El Said, L. H., Shope, R. E., and Bishop, D. H. L.,** Evidence from recombinant bunyavirus studies that the M RNA gene products elicit neutralizing antibodies, *Virology,* 102, 190, 1980.
25. **Gentsch, J. R. and Bishop, D. H. L.,** Small virus RNA segment of bunyavirus codes for viral nucleocapsid protein, *J. Virol.,* 30, 417, 1978.
26. **Gentsch, J. R. and Bishop, D. H. L.,** M viral RNA segment of bunyaviruses codes for two glycoproteins, G1 and G2, *J. Virol.,* 30, 767, 1979.
27. **Janssen, R. S., Nathanson, N., Endres, M. J., and Gonzalez-Scarano, F.,** Virulence of LaCrosse virus is under polygenic control, *J. Virol.,* 59, 1, 1986.

28. **Lindsey, H. S., Klimas, R. A., and Obijeski, J. F.,** LaCrosse virus soluble cell culture antigen, *J. Clin. Microbiol.,* 6, 618, 1977.

29. **Bouloy, M., Krams-Ozden, S., Horodniceanu, F., and Hannoun, C.,** Three segment RNA genome of Lumbo virus (Bunyavirus), *Intervirology,* 2, 173, 1974.

30. **Samso, A., Bouloy, M., and Hannoun, C.,** Mise en évidence de molécules d'acide ribonucléique circulaire dans le virus Lumbo (Bunyavirus), *C. R. Acad. Sci.,* (Paris), D282, 1653, 1976.

31. **Obijeski, J. F., McCauley, J., and Skehel, J.,** Nucleotide sequences at the termini of LaCrosse virus RNAs, *Nucl. Acids Res.,* 8, 2431, 1980.

32. **Pardigon, N., Vialat, P., Girard, M., and Bouloy, M.,** Panhandles and hairpin structures at the termini of Germiston virus RNAs (Bunyavirus), *Virology,* 122, 191, 1982.

33. **Patterson, J. L., Kolakofsky, D., Holloway, B. P., and Obijesky, J. F.,** Isolation of the ends of LaCrosse virus RNA as a double-stranded structure, *J. Virol.,* 45, 882, 1983.

34. **Bishop, D. H. L., Gould, K. G., Akashi, H., and Clerx-van Haaster, C. M.,** The complete sequence and coding content of snowshoe hare bunyavirus small (S) viral RNA species, *Nucl. Acids Res.,* 10, 3707, 1982.

35. **Cabradilla, C. D., Holloway, B. P., and Obijesky, J. F.,** Molecular cloning and sequencing of the LaCrosse virus S RNA, *Virology,* 128, 463, 1983.

36. **Akashi, H. and Bishop, D. H. L.,** Comparison of the sequences and coding of LaCrosse and snowshoe hare bunyaviruses S RNA species, *J. Virol.,* 45, 1155, 1983.

37. **Akashi, H., Gay, M., Ihara, T., and Bishop, D. H. L.,** Localized conserved regions of the S RNA gene products of bunyaviruses are revealed by sequence analyses of the Simbu serogroup Aino virus, *Virus Res.,* 1, 51, 1984.

38. **Gerbaud, S., Vialat, P., Pardigon, N., Wychowski, C., Girard, M., and Bouloy, M.,** The S segment of the germiston virus RNA genome can code for three proteins, *Virus Res.,* 8, 1, 1987.

39. **Lees, J. F., Pringle, C. R., and Elliot, R. M.,** Nucleotide sequence of the Bunyamwera virus M RNA segment: conservation of structural features in the Bunyavirus glycoprotein gene product, *Virology,* 148, 1, 1986.

40. **Eshita, Y. and Bishop, D. H. L.,** The complete sequence of the M RNA of snowshoe hare bunyavirus reveals the presence of internal hydrophobic domains in the viral glycoprotein, *Virology,* 137, 227, 1984.

41. **Grady, L. J., Sanders, M. L., and Campbell, W. P.,** The sequence of the M RNA of an isolate of LaCrosse virus, *J. Gen. Virol.,* 68, 3057, 1987.

41a. **Pardigon, N.,** unpublished data.

41b. **Elliot, R.,** personal communication.

42. **Bouloy, M. and Hannoun, C.,** Studies on Lumbo virus replication. I. RNA-dependent RNA polymerase associated with virions, *Virology,* 62, 258, 1976.

43. **Patterson, J. L., Holloway, B., and Kolakofsky, D.,** LaCrosse virions contain a primer-stimulated RNA polymerase and a methylated cap-dependent endonuclease, *J. Virol.,* 52, 215, 1984.

44. **Bellocq, C., Raju, R., Patterson, J., and Kolakofsky, D.,** Translational requirement of LaCrosse virus S mRNA synthesis: *in vitro* studies, *J. Virol.,* 61, 87, 1987.

45. **Gerbaud, S., Pardigon, N., Vialat, P., and Bouloy, M.,** The S segment of the Germiston bunyavirus genome: coding strategy and transcription, in *The Biology of Negative Strand Viruses,* Mahy, B. and Kolakofsky, D., Eds., Elsevier, Amsterdam, 1987, 191.

46. **Bouloy, M. and Hannoun, C.,** Studies on Lumbo virus replication 2. Properties of viral ribonucleoproteins and characterization of messenger RNAs, *Virology,* 71, 363, 1976.

47. **Raju, R. and Kolakofsky, D.,** Inhibitors of protein synthesis inhibit both LaCrosse virus S-mRNA and S genome synthesis, *in vivo, Virus Res.,* 5, 1, 1986.

48. **Raju, R. and Kolakofsky, D.,** Translational requirement of LaCrosse virus S-mRNA synthesis: *in vivo* studies, *J. Virol.,* 61, 96, 1987.

49. **Pattnaik, A. K. and Abraham, G.,** Identification of four complementary RNA species in Akabane virus-infected cells, *J. Virol.,* 47, 452, 1983.

50. **Abraham, B. and Pattnaik, A. K.,** Early RNA synthesis in Bunyamwera virus-infected cells, *J. Gen. Virol.,* 64, 1277, 1983.

51. **Bellocq, C. and Kolakofsky, D.,** The translational requirement for LaCrosse virus S-mRNA synthesis: a possible mechanism, *J. Virol.,* 61, 3960, 1987.

51a. **Bouloy, M.,** unpublished results.

52. **Patterson, J. L. and Kolakovsky, D.,** Characterization of LaCrosse virus small-genome transcripts, *J. Virol.,* 49, 680, 1984.

53. **Bishop, D. H. L., Gay, M. E., and Matsuoko, Y.,** Nonviral heterogeneous sequences are present at the 5' end of one species of snowshoe hare bunyavirus S complementary RNA, *Nucl. Acids Res.,* 10, 3703, 1983.

54. **Eshita, Y., Ericson, B., Romanowski, V., and Bishop, D. H. L.,** Analyses of the mRNA transcription processes of snowshoe hare Bunyavirus S and M RNA species, *J. Virol.,* 55, 681, 1985.

55. **Bouloy, M., Vialat, P., Girard, M., and Pardigon, N.,** A transcript from the S segment of the Germiston bunyavirus is uncapped and codes for the nucleoprotein and a nonstructural protein, *J. Virol.,* 49, 717, 1984.

56. **Bouloy, M., Plotch, S. J., and Krug, R. M.,** Globin mRNAs are primer for the transcription of influenza virus RNA *in vitro, Proc. Natl. Acad. Sci. U.S.A.,* 75, 4886, 1978.

57. **Bouloy, M., Plotch, S. J., and Krug, R. M.,** Both the 7-methyl and the 2'-O-methyl groups in the cap of mRNA strongly influence its ability to act as primer for influenza virus RNA transcription, *Proc. Natl. Acad. Sci. U.S.A.,* 77, 3952, 1980.

58. **Kozak, M.,** Bifunctional messenger RNAs in eukaryotes, *Cell,* 47, 481, 1986.

59. **Elliot, R. M.,** Identification of nonstructural proteins encoded by viruses of the Bunyamwera serogroup (family Bunyaviridae), *Virology,* 143, 119, 1985.

60. **Kascsak, R. J. and Lyons, M. J.,** Bunyamwera virus. II. The generation and nature of defective interfering particles, *Virology,* 89, 539, 1978.

61. **Newton, S. E., Short, N. J., and Dalgarno, L.,** Bunyamwera virus replication in cultured *Aedes albopictus* (mosquitoes) cells: establishment of a persistent viral infection, *J. Virol.,* 39, 1015, 1981.

62. **Elliot, R. M. and Wilkie, M.,** Persistent infection of *Aedes albopictus* C6/36 cells by Bunyamwera virus, *Virology,* 150, 21, 1986.

63. **Fuller, F. and Bishop, D. H. L.,** Identification of virus coded non-structural proteins in bunyavirus infected cells, *J. Virol.* 41, 643, 1982.

64. **Pringle, C. R., Lees, J. F., Clark, W., and Elliott, R. M.,** Genome subunit reassortment among bunyaviruses analysed by dot hybridization using molecularly cloned complementary DNA probes, *Virology,* 135, 244, 1984.

65. **Ozden, S. and Hannoun, C.,** Isolation and preliminary characterization of temperature-sensitive mutants of Lumbo virus, *Virology,* 84, 210, 1978.

66. **Beaty, B. J., Miller, B. R., Shope, R. E., Rozhon, E. J., and Bishop, D. H. L.,** Molecular basis of bunyavirus per os infection of mosquitoes: role of the middle-sized RNA segment, *Proc. Natl. Acad. Sci. U.S.A.,* 79, 1295, 1982.

67. **Beaty, B. J., Rozhon, E. J., Gensemer, P., and Bishop, D. H. L.,** Formation of reassortant Bunyaviruses in dually infected mosquitoes, *Virology,* 111, 662, 1981.

68. **Beaty, B. J., Bishop, D. H. L., Gay, M., and Fuller, F.,** Interference between Bunyaviruses in *Aedes triseratus* mosquitoes, *Virology,* 127, 83, 1983.

69. **Robeson, G., El Said, L., Brandt, W., Dalrymple, J., and Bishop, D. H. L.,** Biochemical studies on the phlebotomus fever group viruses (Bunyaviridae family), *J. Virol.,* 30, 339, 1979.

70. **Ihara, T., Akashi, H., and Bishop, D. H. L.,** Novel coding strategy (ambisense genomic RNA) revealed by sequence analysis of Punta Toro phlebovirus S RNA, *Virology,* 136, 2393, 1984.

71. **Collett, M. S., Purchio, A. F., Keegan, K., Frazier, S., Hays, W., Anderson, D. K., Parker, M. D., Schmaljohn, C., Schmidt, J., and Darlrymple, J. M.,** Complete nucleotide sequence of the M RNA segment of Rift valley fever virus, *Virology,* 144, 228, 1985.

72. **Ihara, S., Smith, J., Dalrymple, J. M., and Bishop, D. H. L.,** Complete sequence of the glycoproteins and M RNA of Punta Toro phlebovirus compared to those of Rift valley fever virus, *Virology,* 144, 246, 1985.

73. **Emery, V. C. and Bishop, D. H. L.,** Characterization of Punto Toro S mRNA species and identification of an inverted complementary sequence in the intergenic region of Punta Toro phlebovirus ambisense S RNA that is involved in mRNA transcription termination, *Virology,* 156, 1, 1987.

74. **Overton, H. A., Ihara, T., and Bishop, D. H. L.,** Identification of the N and NSS proteins coded by ambisense S RNA of Punta Toro phlebovirus using monospecific antisera raised to baculovirus expressed N and NSS proteins, *Virology,* 157, 338, 1987.

75. **Nicoletti, L. and Verami, P.,** Growth of the phlebovirus Toscana in mosquito *(Aedes pseudo scutellaris)* cell line (AP-61): establishment of a persistent infection, *Arch. Virol.,* 85, 35, 1985.

76. **Pettersson, R. and Kääriäinen, L.,** The ribonucleic acids of Uukuniemi virus, a non-cubical tick-borne arbovirus, *Virology,* 56, 608, 1973.

77. **Pettersson, R. F. and von Bonsdorff, C. H.,** Ribonucleoproteins of Uukuniemi virus are circular, *J. Virol.,* 15, 386, 1975.

78. **Pettersson, R. F., Hewlett, M. J., Baltimore, D., and Coffin, J. M.,** The genome of Uukuniemi virus consists of three unique RNA segments, *Cell,* 11, 51, 1977.

79. **Hewlett, M. J., Pettersson, R. F., and Baltimore, D.,** Circular forms of Uukuniemi virion RNA. An electron microscopic study, *J. Virol.,* 21, 1085, 1977.

80. **Röhnholm, R. and Pettersson, R. F.,** Complete nucleotide sequence of the M RNA segment of Uukuniemi virus encoding the membrane glycoproteins G1 and G2, *Virology,* 160, 1987.

81. **Ranki, M. and Pettersson, R.,** Uukuniemi virus contains an RNA polymerase, *J. Virol.,* 16, 1420, 1975.

82. **Ulmanen, I., Seppala, P., and Pettersson, R. F.,** *In vitro* translation of Uukuniemi virus-specific RNAs. Identification of a non-structural protein and a precursor to the membrane glycoproteins, *J. Virol.,* 37, 72, 1981.

83. **Kuismanen, E.,** Posttranslational processing of Uukuniemi virus glycoproteins G1 and G2, *J. Virol.,* 51, 806, 1984.

84. **Kuismanen, E., Hedman, K., Saraste, J., and Pettersson, R. F.,** Uukuniemi virus maturation: accumulation of virus particles and viral antigens in the Golgi complex, *Mol. Cell. Biol.,* 2, 1441, 1982.

85. **Pesonen, M., Kuismanen, E., and Pettersson, R. F.,** Monosaccharide sequence of protein-bound glycans of Uukuniemi virus, *J. Virol.,* 41, 390, 1982.

86. **Kuismanen, E., Bang, B., Hurme, M., and Pettersson, R. F.,** Uukuniemi virus maturation: immunofluorescence microscopy with monoclonal glycoprotein-specific antibodies, *J. Virol.,* 51, 137, 1984.

87. **Gahmberg, N., Kuismanen, E., Keränen, S., and Pettersson, R. F.,** Uukuniemi virus glycoproteins accumulate in and cause morphological changes of the Golgi complex in the absence of virus maturation, *J. Virol.,* 57, 899, 1986.

88. **Donets, M. A. and Chumakov, M. P.,** Several characteristics of the virions of the viruses Crimean hemorrhagic fever and Congo, in *Voprosi Medical Virology,* Transactions of the Institute of Poliomyelitis and Viral Encephalitis, Moscow, 1975, 287.

89. **Chumakov, M. P. and Donets, M. A.,** Virion and subvirion constituents of Crimean hemorrhagic fever virus, *Int. Virol.,* 3, 193, 1975.

90. **Korolev, M. B., Donets, M. A., Rubin, S. G., and Chumakov, M. P.,** Morphology and morphogenesis of Crimean hemorrhagic fever virus, *Arch. Virol.,* 50, 169, 1976.

91. **Clerx, J. P. M. and Bishop, D. H. L.,** Qalyub virus, a member of the newly proposed *Nairovirus* genus (Bunyaviridae), *Virology,* 108, 361, 1981.

92. **Clerx, J. P. M., Casals, J., and Bishop, D. H. L.,** Structural characteristics of nairoviruses (genus Nairovirus Bunyaviridae), *J. Gen. Virol.,* 55, 165, 1981.

93. **Cash, P.,** Polypeptide synthesis of Dugbe virus, a member of the Nairovirus genus of the Bunyaviridae, *J. Gen. Virol.,* 66, 141, 1985.

94. **Watret, G. E. and Elliot, R. M.,** The proteins and RNA specified by CloMor virus, a Scottish Nairovirus, *J. Gen. Virol.,* 66, 2513, 1985.

95. **Elliot, L. H., Kiley, M. P., and McCormick, J. B.,** Hantaan virus: identification of virion proteins, *J. Gen. Virol.,* 65, 1285, 1984.

96. **Schmaljohn, C. S., Jennings, G. B., Hay, J., and Dalrymple, J. M.,** Coding strategy of the S genome segment of Hantaan virus, *Virology,* 155, 633, 1986.

97. **Schmaljohn, C. S., Schmaljohn, A. L., and Dalrymple, J.,** Hantaan virus M RNA: coding strategy, nucleotide sequence and gene order, *Virology,* 157, 31, 1987.

98. **Elliot, R. M.,** *Virology,* 173, 426, 1989.

99. **Fazakerley, J. K., Gonzalez-Scarano, F., Strickler, J., Dietzschold, B., Karush, F., and Nathanson, N.,** *Virology,* 167, 422, 1988.

100. **Suzich, J. A. and Collett, M. S.,** *Virology,* 164, 476, 1988.

101. **Simons, J. F., Hellman, U., and Pettersson, R. F.,** *J. Virol.,* 64, 247, 1990.

102. **Ward, V. K., Marriott, A. C., El-Ghorr, A. A., and Nuttall, P. A.,** *Virology,* 175, 518, 1990.

Chapter 20

Infectious Flacherie Virus

H. Watanabe

TABLE OF CONTENTS

I. INTRODUCTION

A. MEMBERS OF THE GROUP (HOST RANGE)

Flacherie diseases of the silkworm, *Bombyx mori,* are major factors causing serious loss of cocoon production in sericultural countries. Flacherie is a general term for silkworm diseases with flacidity symptoms. These diseases were generally considered to be caused by the abnormal multiplication of bacteria in the larval gut lumen, but in 1960, Yamazaki and colleagues[1] described an infectious disease exhibiting flacherie symptoms which they suspected was caused by a virus. Later, Aizawa and colleagues[2] isolated a nonoccluded virus from silkworm with flacherie symptoms and called it the infectious flacherie virus (IFV). Kawase and colleagues[3] characterized the IFV and suggested that the virus was a member of the Picornaviridae.

IFV isolated from the silkworm is also infectious to the mulberry wild silkworm, *Bombyx mandarina,* an insect pest of field mulberry.[4,5] Watanabe and colleagues[6] found, however, that many of the mulberry pyralids, *Glyphodes pyloalis,* infesting mulberry plantations, are infected with a nonoccluded virus in a chronic state in their midgut epithelia, which is serologically identical to the *Bombyx* IFV. When silkworm larvae were inoculated with the *Glyphodes* nonoccluded virus, they developed histopathologically a typical infection of IFV, suggesting that *Bombyx* IFV originated from the *Glyphodes* nonoccluded virus. Epizootiologically, the mulberry pyralid is a common habitual host of IFV.

B. GROSS PATHOLOGY

When silkworm larvae are infected perorally with IFV, they gradually lose their appetites and usually die after 10 to 12 d with body flacidity as a major symptom. The target of IFV is the goblet cell of the midgut epithelium, and the virus multiplies in the cytoplasm, eventually disintegrating the cell. On dissection, the diseased larva shows an alimentary canal pale yellow in color and devoid of most of its content. This sign is similar to those of densonucleosis virus infection and of common flacherie diseases caused by the abnormal multiplication of intestinal bacteria. Accordingly, for the accurate diagnosis of IFV infection, a histochemical[7] and serological techniques such as gel immunodiffusion test,[8] single radial immunodiffusion,[9] electrosyneresis,[10] fluorescent antibody test,[11] enzyme-linked immunosorbent assay (ELISA),[12] and latex agglutination test[13] have been developed.

Younger larvae are more susceptible to IFV than older larvae. When the older larva infected with IFV undergoes metamorphosis, the virus scarcely multiplies in the newly formed pupal midgut and the infected pupa, which is immune to a lethal infection, develops into an adult.[14] The LD_{50}-time curve of virus multiplication in larvae fed the virus varies with the lapse of time showing a decreasing, an eclipse, an increasing, and a stationary phase.[15] However, the IFV infection is significantly inhibited when larvae are fed mulberry leaves sprayed with aqueous solution of 5-fluorouracil[16] (5×10^{-2} or $1 \times 10^{-2}\%$) or an aqueous solution of 0.01% guanidine hydrochloride (GH)[17] prior to and after virus inoculation. Choi and colleagues[18] suggested GH inhibits accumulation of IFV-specific translatable mRNA and structural polypeptides in the silkworm midgut.

In an IFV-infected larva exposed to a high temperature (37°C), virus multiplication and synthesis of virus antigen is greatly inhibited.[19] This thermal inhibition has resulted in an effective method of thermal therapy in fifth instar larvae for the control of the infectious flacherie disease.[20]

II. VIRUS STRUCTURE AND COMPOSITION

A. PARTICLE STRUCTURE

The virus particles are isometric and about 27 ± 2 nm in diameter[3] (see Figure 1). They

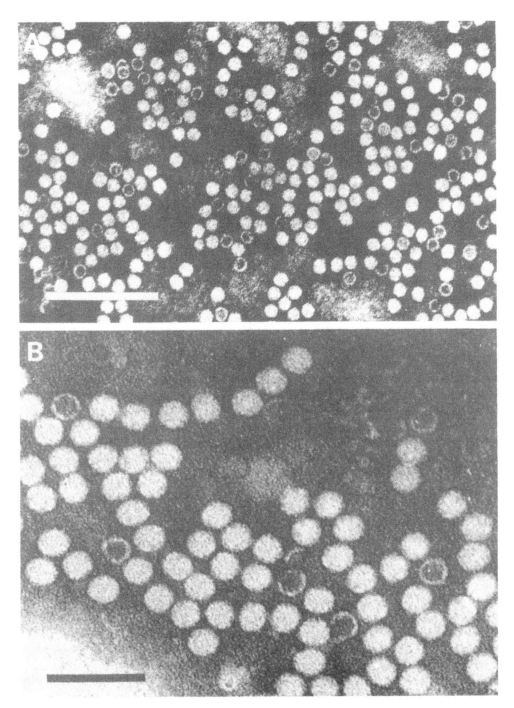

FIGURE 1. Purified infectious flacherie virus (IFV) stained with phosphotungstic acid. (A) (Bar = 200 nm; courtesy of Dr. S. Kawase, Nagoya University, Japan.) (B) (Bar = 100 nm; courtesy of Dr. M. Matsui, National Research Institute of Vegetables, Ornamental Plants and Tea, Japan.)

have a sedimentation of 183 S, a buoyant density in CsCl of 1.375 g/ml,[21] and contain 28.5% RNA with no pairing base ratio, GACU = 23.3:30.6:18.8:27.5.[22]

The IFV has four major structural proteins and a few minor ones. Molecular weights of the major proteins are 35.2 K (VP1), 33.0 K (VP2), 31.2 K (VP3), and 11.6 K (VP4). Amino acid compositions of VP1, VP2, and VP3 are similar but that of VP4 is somewhat different. The isoelectric points of the four major proteins are 7.7 (VP1), 6.7 (VP2), 4.8 (VP3), and 5.5 (VP4), respectively.[23]

IFV agglutinates mouse erythrocytes but not chicken, rat, goat, sheep, and human erythrocytes.[24]

The infectivity of purified IFV is reduced at pH 2, but unaffected at pH 3 to 12. The purified IFV shows resistance to pronase, trypsin, and pepsin at 30°C for 24 h. The virus is inactivated at 70°C for 20 min and by the irradiation of ultraviolet light (2537 Å) for 120 min at 40 cm from the light source.[25]

B. GENOME STRUCTURE

The IFV genome consists of a single molecule of single-stranded RNA, which has a molecular weight of 2.4×10^6 Da.[3] The genomic RNA is of positive polarity which contains a small viral protein covalently linked to the 5′ terminus, and a poly (A) tract at the 3′ terminus.[26,27]

Hashimoto and colleagues[28] found that the 38 S IFV-RNA acted as an efficient mRNA in a wheat germ extract and a rabbit reticulocyte lysate translation system. In either cell-free translation system, the sum of molecular weights of translation products far exceeded the coding capacity of the virus genome, suggesting the occurrence of proteolytic cleavage of large primary products to smaller polypeptides and/or premature termination of translation. The highest molecular weight product of 200 K (i.e., polyprotein-like product), which corresponded to the coding capacity of the full length of RNA, was translated in both systems. In time course experiments, a polypeptide with a molecular weight of 130 K (immunoprecipitable with antiserum to IFV) was the most abundant of the translation products after incubation for 30 min and 1 h. The molecular weight of 130 K corresponded to the molecular weights of the four major viral proteins. An additional 2 h incubation yielded products comigrating with each structural protein of IFV, indicating that the processing occurred with good fidelity. Peptide mapping of these polypeptides showed that the two *in vitro* systems translated the same cistron in the viral RNA and that the smaller polypeptide was a part of the 130 K product.

III. CYTOPATHOLOGY

Inoue and Ayuzawa[15] investigated the distribution of virus antigen in the silkworm larva infected with an IFV with the fluorescent antibody technique. Specific fluorescence was observed in the midgut, but not in the silk gland, Malpighian tubules, trachea, brain, ganglion, ovary, testis, blood cell, integument, and fat body. In the midgut epithelium-specific fluorescence was localized in the cytoplasm and not in the nuclei of goblet cells. In a later period of infection, small fluorescent bodies from 3 to 10 μm in diameter occurred in the cytoplasm of columnar cells, and exhibited more intense fluorescence along the periphery than in the center. These bodies had shapes and sizes similar to the basophilic bodies that stained with pyronine and frequently appeared in the cytoplasm of columnar cells in the midgut infected with IFV.[7]

Autoradiographic studies with [³H]-uridine as a RNA precursor indicated that the goblet cells infected with IFV actively incorporated the label material into their nuclei; subsequently, most of the label seemed to diffuse and accumulate in the cytoplasm around the goblet cavity.[29]

Electron microscope observations[30] on the midgut infected with IFV revealed that, in an early stage of infection, membrane-bound vesicles which were specifically induced by the virus infection were formed in the cytoplasm of goblet cell (see Figure 2). In the portion adjacent to the specific vesicles, virus particles appeared. The specific vesicles were almost round, about 100 to 300 nm in diameter, and contained filamentous structures. The nature

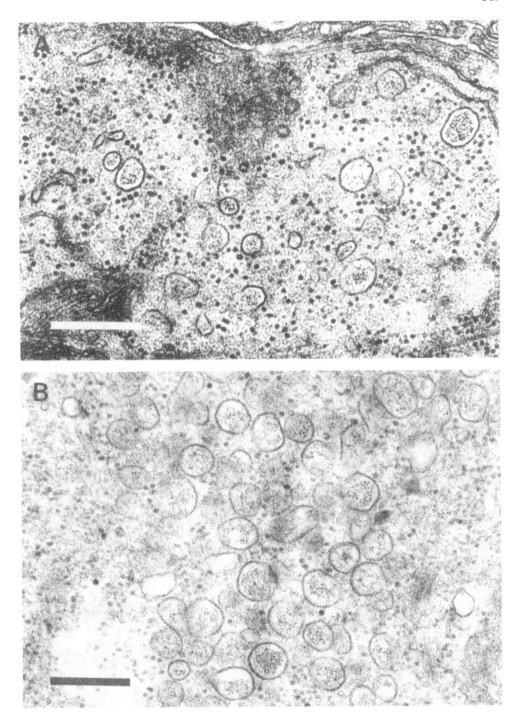

FIGURE 2.   An early stage of IFV infection in the goblet cell of *Bombyx mori*. (A) Virus particles appear in the cytoplasm and accompany specific vesicles. (Bar = 500 nm.) (B) Accumulation of specific vesicles. (Bar = 500 nm.) (Courtesy of Dr. M. Matsui, National Research Institute of Vegetables, Ornamental Plants and Tea, Japan.)

of these structures was uncertain but it resembled nucleic acid. Similar vesicles are known to be formed in the cytoplasm of vertebrate cells infected with picornaviruses such as poliovirus,[31] mengovirus,[32,33] and echovirus.[34]

At a late stage of infection, the cytoplasm of the goblet cell was filled with virus particles and specific vesicles (see Figure 3A). Ribosomes, endoplasmic reticula, and glycogen gran-

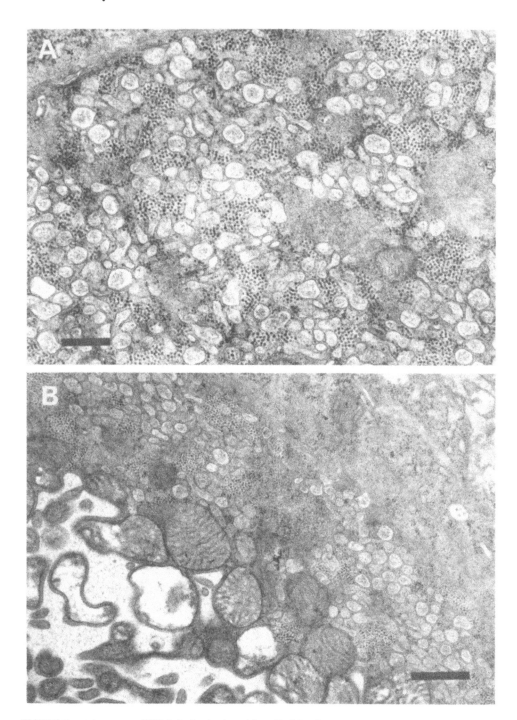

FIGURE 3. A late stage of IFV infection in the goblet cell of *Bombyx mori*. (A) The cytoplasm is filled with virus particles as well as specific vesicles. (Bar = 500 nm.) (B) Normal cell organelles disappear and microvilli and mitochondria are morphologically altered. (Bar = 500 nm.) (Courtesy of Dr. M. Matsui, National Research Institute of Vegetables, Ornamental Plants and Tea, Japan.)

ules in the cytoplasm disappeared, and the mitochondria swelled and became round (see Figure 3B). Later, these mitochondria as well as the microvilli disintegrated into membranous structures. The goblet cells shrunk in size, became round, and were discharged into the midgut lumen (see Figure 4A). Some of the degenerated goblet cells, however, were enclosed

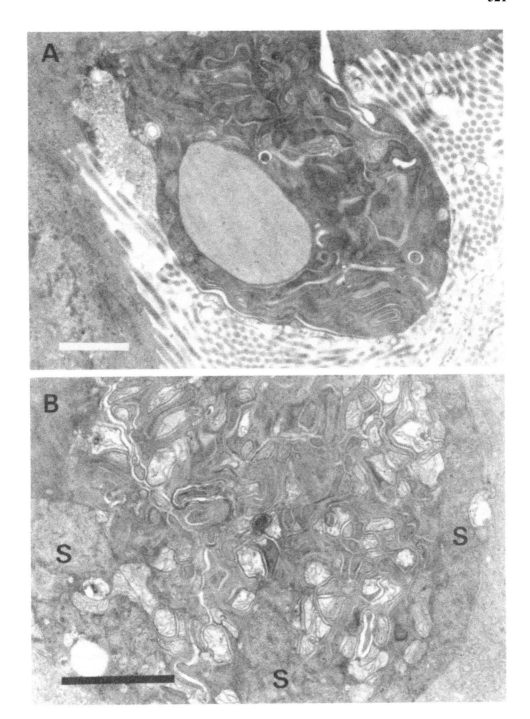

FIGURE 4. Terminal stage of IFV infection in the midgut epithelium of *Bombyx mori*. (A) A virus-infected, small round goblet cell being discharged into the midgut lumen. (Bar = 2 μm.) (B) A disintegrated goblet cell, the so-called basophilic body, enclosed within a columnar cell. Masses of virus particles are packed in the periphery (S) of the body and membranous structures occupy the centrosphere region. (Bar = 2 μm.) (Courtesy of Dr. M. Matsui, National Research Institute of Vegetables, Ornamental Plants and Tea, Japan.)

within the adjacent columnar cells and were presumably phagocytized. The small round, degenerated goblet cell engulfed in a columnar cell was identical to the small fluorescent body or the basophilic body previously described (see Figure 4B). The newly formed goblet cells produced at molt were rapidly infected with the virus and the larvae eventually died.

REFERENCES

1. **Yamazaki, H., Sakai, E., Shimodaira, M., and Yamada, T.,** Studies on the infectious flacherie (F-type) silkworm, *Bombyx mori* L., *Bull. Nagano-ken Sericult. Exp. Stn.,* 61, 1, 1960.
2. **Aizawa, K., Furuta, Y., Kurata, K., and Sato, F.,** On the etiologic agent of the infectious flacherie of the silkworm, *Bombyx mori* (Linnaeus), *Bull. Sericult. Exp. Stn.,* 19, 223, 1964.
3. **Kawase, S., Hashimoto, Y., and Nakagaki, M.,** Characterization of flacherie virus of the silkworm, *Bombyx mori, J. Sericult. Sci. Jpn.,* 49, 477, 1980.
4. **Matsui, M. and Watanabe, H.,** Electron microscope observations on the midgut of *Theophila mandarina* infected with a flacherie virus of the silkworm, *Bombyx mori, J. Sericult. Sci. Jpn.,* 43, 467, 1974.
5. **Inoue, H., Sato, F., and Ayuzawa, C.,** Infection of an infectious-flacherie virus of *Bombyx mori* to *Bombyx mandarina, Sanshi-Kenkyu,* 92, 120, 1974.
6. **Watanabe, H., Kurihara, Y., Wang, Y. X., and Shimizu, T.,** Mulberry pyralid, *Glyphodes pyloalis:* habitual host of nonoccluded viruses pathogenic to the silkworm, *Bombyx mori, J. Invertebr. Pathol.,* 52, 401, 1988.
7. **Iwashita, Y. and Kanke, E.,** Histopathological diagnosis of diseased larva infected with flacherie virus of the silkworm, *Bombyx mori* Linnaeus, *J. Sericult. Sci. Jpn.,* 38, 64, 1969.
8. **Sekijima, Y.,** Studies on serological diagnosis of the infectious flacherie in the silkworm, *Bombyx mori* L. I. Demonstration of specific antigen in extract from the silkworm larvae infected with flacherie virus by precipitation reaction, *J. Sericult. Sci. Jpn.,* 40, 49, 1971.
9. **Seki, H. and Sekijima, Y.,** Detection of the specific antigen of the infectious flacherie virus in the silkworm, *Bombyx mori* L., by the single radial immunodiffusion method, *J. Sericult. Sci. Jpn.,* 45, 13, 1976.
10. **Sekijima, Y. and Ohtsu, A.,** Detection of specific antigen of the infectious flacherie virus of the silkworm, *Bombyx mori* L. by electrosyneresis, *Appl. Entomol. Zool.,* 11, 356, 1976.
11. **Inoue, H. and Ayuzawa, C.,** Studies on diagnosis of the infectious flacherie in the silkworm, *Bombyx mori,* by fluorescent antibody technique, *Bull. Sericult. Exp. Sta.,* 25, 21, 1971.
12. **Shimizu, S.,** Enzyme-linked immunosorbent assay for the detection of the flacherie virus of the silkworm, *Bombyx mori, J. Sericult. Sci. Jpn.,* 51, 370, 1982.
13. **Shimizu, S., Ohba, M., Kanda, K., and Aizawa, K.,** Latex agglutination test for the detection of the flacherie virus of the silkworm, *Bombyx mori, J. Invertebr. Pathol.,* 42, 151, 1983.
14. **Inoue, H.,** Effect of the metamorphosis of the silkworm, *Bombyx mori,* on the multiplication of a flacherie virus, *J. Sericult. Sci. Jpn.,* 46, 20, 1977.
15. **Inoue, H. and Ayuzawa, C.,** Fluorescent antibody study of an infectious flacherie of the silkworm, *Bombyx mori, J. Invertebr. Pathol.,* 19, 227, 1977.
16. **Miyajima, S. and Kawase, S.,** The inhibitory effect of 5-fluorouracil on the outbreak of infectious flacherie in the silkworm, *Bombyx mori, J. Sericult. Sci. Jpn.,* 34, 359, 1965.
17. **Kawase, S. and Miyajima, S.,** Inhibitory effect of guanidine on the incidence of infectious flacherie of the silkworm, *Bombyx mori, J. Sericult. Sci. Jpn.,* 51, 341, 1982.
18. **Choi, H. K., Kobayashi, M., and Kawase, S.,** Inhibition of the accumulation of virus-specific translatable mRNA and structural polypeptides by guanidine hydrochloride in the midgut of the silkworm, *Bombyx mori,* infected with infectious flacherie virus, *J. Invertebr. Pathol.,* 53, 392, 1989.
19. **Inoue, H., Ayuzawa, C., and Kawamura, Jr., A.,** Effect of high temperature on the multiplication of infectious flacherie virus in the silkworm, *Bombyx mori, Appl. Entomol. Zool.,* 7, 155, 1972.
20. **Inoue, H. and Tanada, Y.,** Thermal therapy of the flacherie virus disease in the silkworm, *Bombyx mori, J. Invertebr. Pathol.,* 29, 63, 1977.
21. **Ayuzawa, C.,** Studies on the infectious flacherie of the silkworm, *Bombyx mori* L. I. Purification of the virus and some of its properties, *J. Sericult. Sci. Jpn.,* 41, 338, 1972.
22. **Kawase, S., Suto, C., Ayuzawa, C., and Inoue, H.,** Chemical properties of infectious flacherie virus of the silkworm, *Bombyx mori* L. (Lepidoptera: Bombycidae), *Appl. Entomol. Zool.,* 9, 100, 1974.
23. **Hashimoto, Y. and Kawase, S.,** Characteristics of structural proteins of infectious flacherie virus from the silkworm, *Bombyx mori, J. Invertebr. Pathol.,* 41, 68, 1983.

24. **Suto, C. and Kawase, S.,** Hemagglutination with flacherie virus of the silkworm, *Bombyx mori, J. Invertebr. Pathol.,* 19, 290, 1971.
25. **Inoue, H.,** Effects of physical and chemical treatments on the properties of infectious flacherie virus of the silkworm, *Bombyx mori, J. Sericult. Sci. Jpn.,* 41, 279, 1972.
26. **Hashimoto, Y., Watanabe, A., and Kawase, S.,** Preliminary studies of terminal structures of infectious flacherie virus RNA, *Microbiologica,* 7, 91, 1984.
27. **Hashimoto, Y., Watanabe, A., and Kawase, S.,** Evidence for the presence of a genome-linked protein in infectious flacherie virus, *Arch. Virol.,* 90, 301, 1986.
28. **Hashimoto, Y., Watanabe, A., and Kawase, S.,** *In vitro* translation of infectious flacherie virus RNA in a wheat germ and a rabbit reticulocyte system, *Biochim. Biophys. Acta,* 781, 76, 1984.
29. **Watanabe, H.,** Autoradiographic studies on the nucleic acid synthesis in the midgut of the silkworm, *Bombyx mori* L., under normal and virus-infected conditions, *J. Sericult. Sci. Jpn.,* 36, 371, 1967.
30. **Matsui, M.,** Electron microscope study on the flacherie virus infection in the goblet cell of midgut of the silkworm, *Bombyx mori* L., *J. Sericult. Sci. Jpn.,* 42, 11, 1973.
31. **Dales, S., Eggers, H. J., Tamm, I., and Palade, G. E.,** Electron microscope study of the formation of poliovirus, *Virology,* 26, 379, 1965.
32. **Amako, K. and Dales, S.,** Cytopathology of mengovirus infection. I. Relationship between cellular disintegration and virulence, *Virology,* 32, 184, 1967.
33. **Amako, K. and Dales, S.,** Cytopathology of mengovirus infection. II. Proliferation of membranous cisternae, *Virology,* 32, 201, 1967.
34. **Skinner, M. S., Halperen, S., and Harkin, J. C.,** Cytoplasmic membrane-bound vesicles in echovirus 12-infected cells, *Virology,* 36, 241, 1968.

Chapter 21

VIRUSES OF HONEY BEES

Brenda V. Ball and Leslie Bailey

TABLE OF CONTENTS

I. INTRODUCTION

Until about the middle of this century only one virus disease of honey bees, sacbrood of larvae, had been recognized. Knowledge of honey bee viruses has now greatly increased and a further 19 have been identified, including some serologically related strains. Only two of these, the filamentous virus and *Apis* iridescent virus, have DNA genomes. The rest are RNA viruses and most have particles 30 nm in diameter, thus resembling many of the viruses familiar to animal and plant pathologists. These small isometric viruses of bees, however, are a very heterogeneous group with characteristics that differentiate them from other morphologically similar virus particles. Attempts have been made to classify them, especially with some other mammalian and insect viruses, but apart from the iridescent virus of *Apis cerana* from the Himalayan regions, none fit convincingly into any such scheme.

A few morphologically novel virus types have been identified, including the cause of the longest recognized disease of bees — paralysis. This virus with its associated 17-nm particle has also provided the first well-documented evidence of satellitism in insects.

The signs of some virus diseases, especially paralysis, have frequently been attributed to common and relatively easily seen nonviral parasites which cause no overt disease. Several of the viruses, however, are closely associated with and augment the pathogenicity of one or more of these parasites, and a few depend on the relationship for their infectivity. Similar associations may perhaps be anticipated between the viruses and parasites of other insect species.

II. FILAMENTOUS VIRUS

A. GENERAL CHARACTERISTICS
1. Members of the Group

The filamentous virus of honey bees has been detected in *Apis mellifera* from almost every continent. It has not been assigned to any of the existing taxonomic groups of viruses but most closely resembles the nonoccluded virions of the Baculoviridae[1] in having a rod-shaped nucleocapsid that is enveloped and contains up to 15% by weight of double-stranded DNA. Particles are ellipsoidal, measuring 450×150 nm with a buoyant density in CsCl of 1.28 g/cm^3.[2] The virus differs markedly from the Baculoviridae, however, in the great length and flexuous nature of its nucleocapsid (3000×40 nm), the low density in CsCl of the nucleocapsid (1.36 g/cm^3) and the apparent M_r of its DNA (12×10^6) compared with 1.47 g/cm^3 and 58 to 110×10^6, respectively, for the Baculoviridae.

The 12 proteins of the bee virus and the range of their relative molecular weight (13 to 70×10^3) resemble those of the virus from *Oryctes rhinoceros* which also comprises 12 proteins (M_r 10 to 76×10^3). They may be contrasted with the larger number and size range of the 18 and 20 proteins of M_r 16 to 105×10^3 and 11 to 200×10^3 of the viruses from braconid[4] and ichneumonid[5] wasps, respectively. Virus-like particles seen in ultrathin sections of *Tenebrio molitor*[6] and *Scolytus scolytus*[7] appear similar to those of the filamentous virus of bees although their apparent nucleocapsids are only 500×25 nm, and none of their physicochemical characteristics have been determined.

2. Gross Pathology

Filamentous virus multiplies in the fat body and ovarian tissue of adult bees.[8] The hemolymph of severely infected individuals becomes milky white with particles, but there are no other symptoms and there is no evidence that the lives of bees are shortened by infection. The virus can multiply when injected into bees[8,9] but it infects bees most readily when fed to adults together with spores of the microsporidian *Nosema apis* with which it is almost invariably associated in nature. In Britain it is probably the most common but least pathogenic of all honey bee viruses.

B. VIRUS STRUCTURE AND COMPOSITION

1. Particle Structure

When examined in the electron microscope after negative staining with ammonium molybdate, particles that have been exposed to solutions of only low ionic strength appear ellipsoidal and contain a single-coiled filament (see Figure 1) which emerges partly or entirely when stained with neutral phosphotungstate (see Figure 2a). There is no evidence of the "tail-like projections" seen in preparations of the virus from *Oryctes rhinoceros*[3] or the calyx fluid particles of braconid and ichneumonid wasps.[10] The enveloped nucleocapsids appear to have a helical substructure (see Figure 2b).

Intact virions have 12 structural proteins, four of them major, of M_r 13 to 70 $\times$ 10^3. Proteins P37 and P23 are major components of the envelope, P40 is the major filament protein, and P13 is approximately equally divided between the two. Of the eight minor components, P69, -33 and -19 are associated with the envelope and P59, -31, and -16 with the filaments. P50 and P38 have not been assigned because they have not been detected in the fractionated virus.

The buoyant densities of whole particles are 1.28, 1.29, and 1.26 g/cm^3 in solutions of CsCl, KBr, and sucrose, respectively.[2]

2. Genome Structure

The filaments released by NP-40 treatment and suspended in 0.5 *M* KCl have an absorption spectrum typical of nucleoprotein with an E_{260}:E_{280} ratio of 1.42 compared to a ratio of 1.26 for the intact particles. Assuming typical extinctions for protein and nucleic acid, this suggests the filaments contain about 11% DNA with an apparent M_r of 12 $\times$ 10^6.

Determination of the buoyant density in CsCl of filaments fixed in 1% HCHO gives a tridentate peak with values of 1.360, 1.365, and 1.368 g/cm^3. All of the DNA from Sarkosyl-treated virions appears to be covalently closed according to its buoyant density of 1.61 g/cm^3 in CsCl + ethidium bromide.[2] Although this value suggests that the DNA is in a superhelical form, electron microscopy of CsCl-purified material spread on basic protein monolayers shows that many molecules are linear and uniform in length. It is not inconceivable that covalently closed circular DNA molecules open during preparation for microscopy, but the uniformity of their length and the absence of many apparently circular molecules may mean that the intact particles contain linear DNA and that abnormal buoyant densities were obtained. A more detailed study is necessary to understand these conflicting results.

C. CYTOPATHOLOGY

Virions with folded nucleocapsids have been seen in ultrathin sections of all tissues examined but virogenesis has been observed only in fat body and ovarian cells.[8] The nuclear membrane of many infected cells apparently breaks down and particles are visible among spongelike masses of virogenic stroma. Infected cells eventually rupture to release virions into the hemolymph (see Figure 3).

III. *APIS* IRIDESCENT VIRUS

A. GENERAL CHARACTERISTICS

1. Members of the Group

Apis iridescent virus (AIV) was originally isolated from dead or moribund adult individuals of *Apis cerana*, the Eastern hive bee, collected in Kashmir[11] and, subsequently, in Northern India. The physical properties of AIV more closely resemble those of the smaller particles of the iridovirus genus[1] than of those mainly represented by the larger mosquito iridescent viruses. However, AIV fails to multiply when injected into larvae of *Galleria*

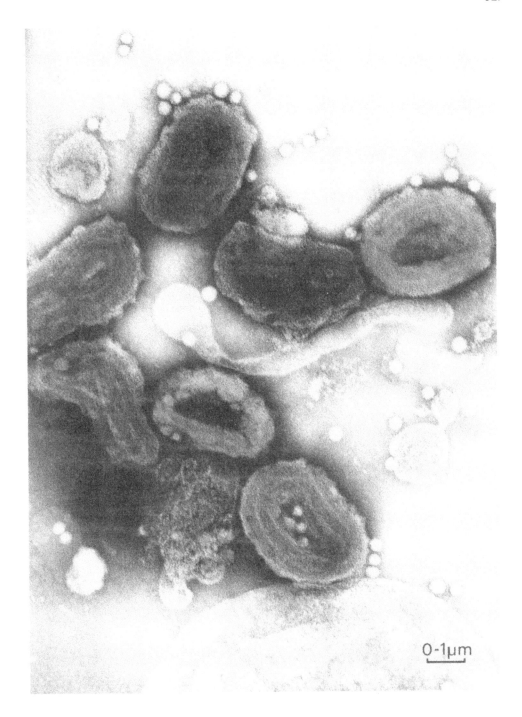

FIGURE 1.   Crude preparation of filamentous virus particles. (Negatively stained with ammonium molybdate.)

mellonella, unlike the majority of iridoviruses. Moreover, AIV is only distantly related serologically to *Tipula, Sericesthis, Chilo, Wiseana,* and *Opogonia* iridescent viruses.[11]

2. Gross Pathology

The most striking and consistent sign of infection is the unusual inactivity of the colonies, especially in summer. Such colonies frequently form several small detached clusters of

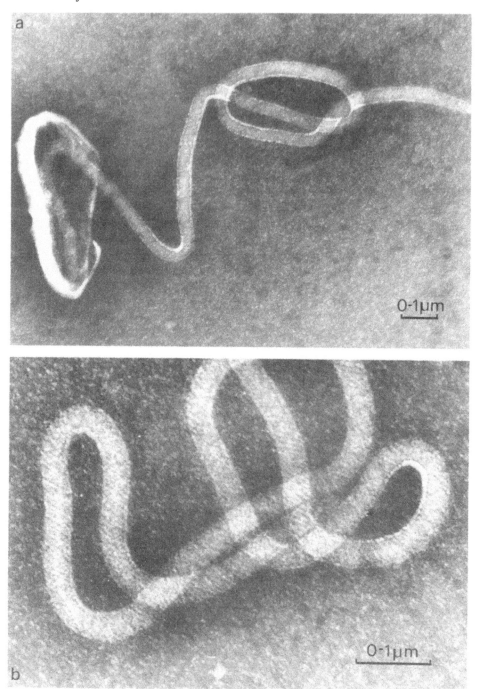

FIGURE 2.   (a) Particle and (b) nucleocapsid of filamentous virus. (Negatively stained with sodium phosphotungstate.)

flightless individuals and often lose many bees crawling on the ground. AIV will multiply in various tissues of the European honey bee, *Apis mellifera*, forming crystals which appear bright blue-violet when illuminated with white light.

B. VIRUS STRUCTURE AND COMPOSITION

AIV particles are 150 nm in size. The outer icosahedral shell of the capsid is composed

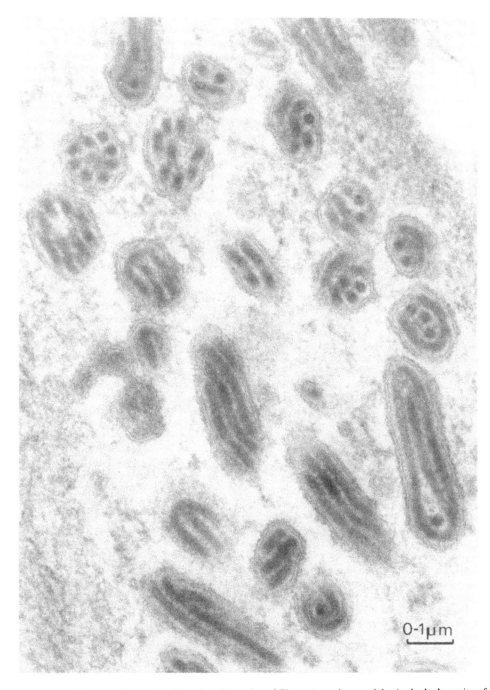

FIGURE 3. Electron micrograph of an ultrathin section of filamentous virus particles in the body cavity of an adult bee. (R. Milne.)

of isometric subunits, 7 nm in diameter, that become detached from disintegrating capsids after prolonged storage in suspension. These are frequently arranged in trisymmetrons (see Figure 4b) each composed of 55 subunits. Particles contain DNA and have a sedimentation coefficient of 2216 S and a buoyant density in CsCl of about 1.32 g/cm^3.[11]

C. CYTOPATHOLOGY

Ultrathin sections of infected tissues show the accumulation of mature virus particles in

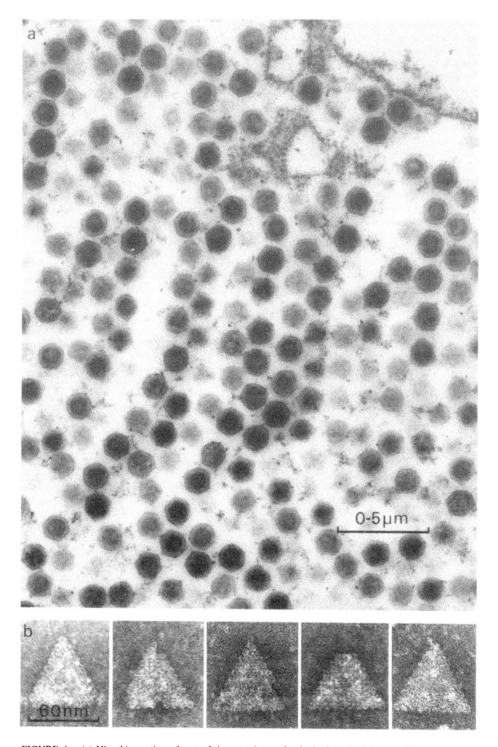

FIGURE 4.   (a) Ultrathin section of part of the cytoplasm of a fat body cell of *Apis mellifera* infected with *Apis* iridescent virus (AIV); (b) trisymmetrons from AIV.

the cytoplasm of cells; none have been seen in the nucleus (see Figure 4a). Early in infection of hypopharyngeal cells, particles are often found in areas of rough endoplasmic reticulum (see Figure 5). In later stages in fat body cells (see Figure 6) maturing virus particles are associated with large areas of "virogenic stroma".[12] Eventually the greatly enlarged cells become packed with crystalline arrays of AIV particles and no cell organelles remain.

IV. CLOUDY WING VIRUS

A. GENERAL CHARACTERISTICS
1. Members of the Group

Cloudy wing virus, with particles only 17 nm in diameter, is included in a group of "mini" viruses[13] with some characteristics of the Picornaviridae. The members of this group, however, are very heterogeneous. Whereas chronic paralysis virus associate (CPVA)[14] and a particle associated with a virus of *Antherea eucalypti*[15] appear to be satellite viruses, the crystalline array virus (CAV) of the grasshopper *Melanoplus bivittatus*,[16] and cloudy wing virus (CWV)[14] seem to multiply independently of any other virus. None are serologically related to any other.

2. Gross Pathology

Although the wings of severely infected honey bees often lose their transparency, this is not a reliable symptom. Diagnosis can be done with certainty only by serology; sufficient virus multiplies in the heads and thoraxes of infected individuals to be readily detected in this way. CAV is infective by injection and *per os* but CWV did not multiply when fed to caged bees or injected into the hemolymph.[14] In nature CWV infection is probably airborne, and the virus is transmitted directly between individuals, especially when they are confined closely together. The lives of infected individuals are shortened, and colonies in which most bees are infected soon become inactive and die.[17]

B. VIRUS STRUCTURE AND COMPOSITION

Particles are isometric and indistinguishable from CPVA when negatively stained with phosphotungstate and examined in the electron microscope (see Figure 7a). The capsid of CWV is composed of a single polypeptide of M_r 19×10^3 as determined by polyacrylamide-gel electrophoresis. Particles have a buoyant density of 1.38 g/cm^3 in CsCl and a sedimentation coefficient of 49 S. The single-stranded RNA has an estimated M_r of 0.45×10^6.[14]

C. CYTOPATHOLOGY

The small size of CWV particles makes their identification uncertain in ultrathin tissue sections except when they form crystalline arrays. Such crystals often occur between the sarcolemmae of fibers of infected thoracic muscle in the region of the tracheoles (see Figure 8). Initial infection and replication may well be in the tracheal matrix, as reported for the CAV of grasshoppers,[18] and especially among the muscle fibrils of the thorax where the main inhalatory spiracles occur. Occasionally, particles are evident within the cisternae of membranous vesicles (see Figure 7b) enclosing a densely stained amorphous material which also occurs around the tracheoles. No other gross histopathologic differences between infected and uninfected muscle have been observed.

V. CHRONIC PARALYSIS VIRUS AND CHRONIC PARALYSIS VIRUS ASSOCIATE

A. GENERAL CHARACTERISTICS
1. Members of the Group

Chronic paralysis virus (CPV) has been detected serologically in extracts of *Apis mellifera*

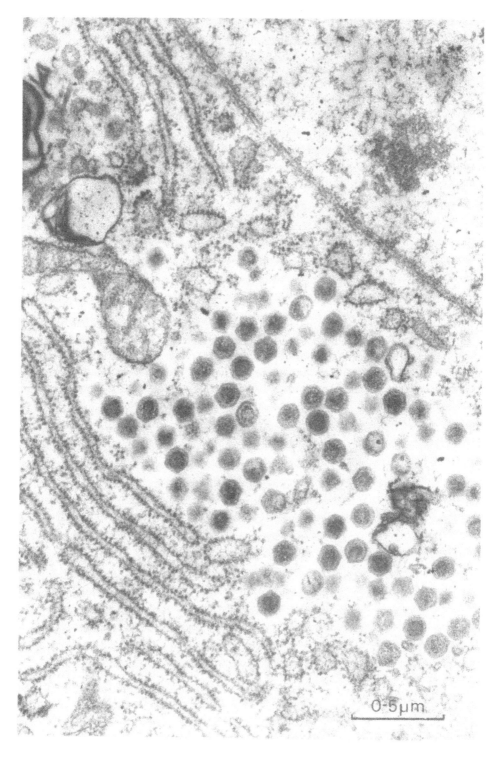

FIGURE 5.   Maturation of AIV particles in areas of rough endoplasmic reticulum in the cytoplasm of a hypopharyngeal gland cell.

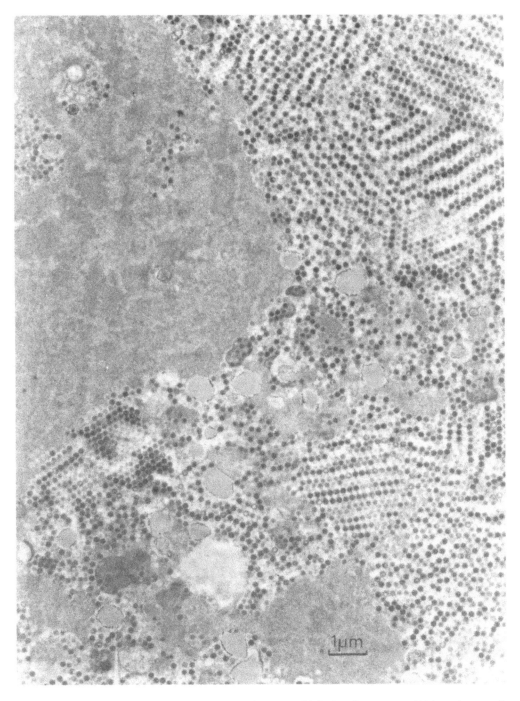

FIGURE 6. Ultrathin section of a fat body cell at a late stage of infection. Dense areas of "virogenic stroma" and crystalline arrays of AIV particles fill the cytoplasm.

found with "bee paralysis" symptoms on almost every continent.[19] Particles are anisometric, mostly ellipsoidal in outline and very variable in length. An apparently morphologically similar virus has been isolated from *Drosophila*[20] although the particles of *Drosophila* RS virus are regular in size, measuring 52×32 nm and have a buoyant density in CsCl of 1.26 g/cm^3 compared to 1.33 g/cm^3 for CPV. The two particles are unrelated serologically and will not multiply in the host of the other virus.[20]

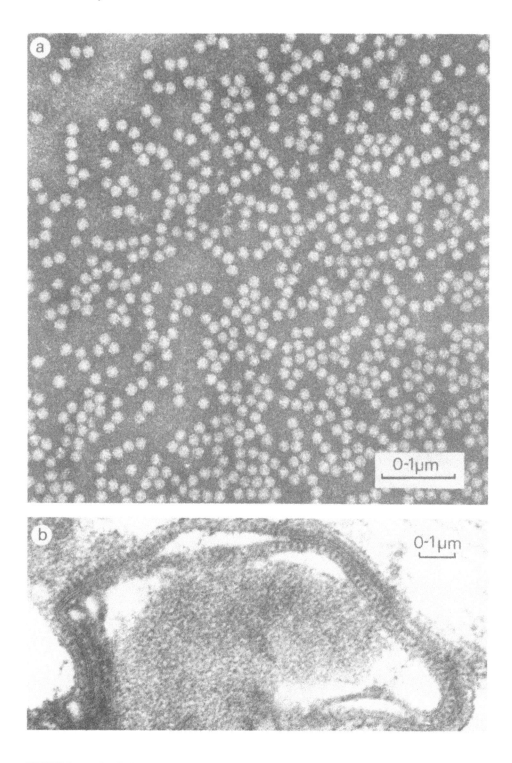

FIGURE 7.   (a) Purified preparation of cloudy wing virus (CWV) particles (negatively stained with sodium phosphotungstate). (b) Membranous vesicle between the fibrils of CWV-infected thoracic muscle. Virus-like particles are evident within the cisternae.

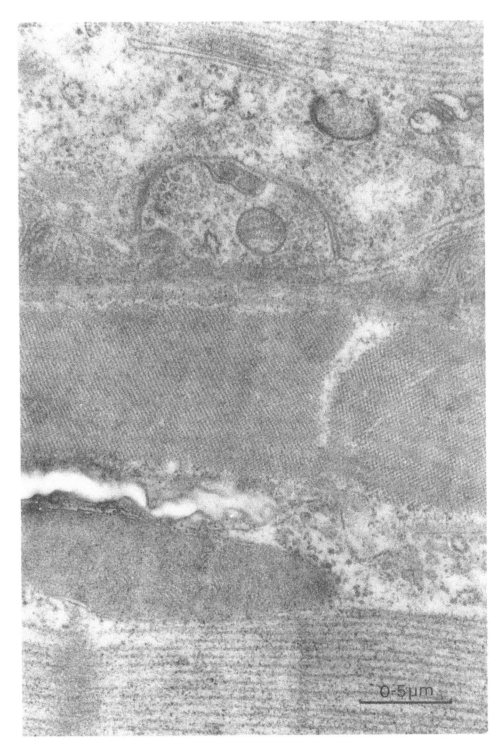

FIGURE 8.   Ultrathin section of thoracic muscle of an adult bee infected with CWV showing (top to bottom), muscle fibril, sarcoplasm, neuromuscular junction, crystal of particles, tracheole, sarcosome, and muscle fibril.

Chronic paralysis virus associate (CPVA) has 17 nm isometric particles which are serologically unrelated to CPV but are frequently found associated with it in natural infections.[14] This seems analogous to the relationship between the virus from *Antherea eucalypti* and its associated 13 nm particle.[15] In this instance, however, no clear evidence of interdependence for replication has been demonstrated, whereas CPVA cannot multiply when injected into bees unless CPV is included in the inoculum. The amount of CPVA that replicates in individuals varies considerably but it is negatively associated with the amount, lengths of particles, and maximum sedimentation coefficient of the CPV. CPVA is the first well-documented example of a satellite virus in insects.

2. Gross Pathology

Adult honey bees infected with CPV may exhibit one of two distinct sets of symptoms or syndromes, which are probably determined by inherited factors.[21] Adults showing one of these syndromes often have a bloated abdomen, caused by distension of the honey sac with liquid. Affected individuals do not fly but crawl on the ground beneath their colonies and exhibit an abnormal trembling motion of the wings and body. Bees showing the second syndrome are initially able to fly but they become almost hairless, appearing dark or almost black, and shiny. They suffer nibbling attacks by other bees and are hindered from returning to their colony by guard bees. In a few days they begin trembling, remain flightless, and soon die.

CPV can readily be transmitted by injection, but in nature it is probably transmitted contagiously between crowded live adult bees via the cytoplasm of broken cuticular hairs. It multiplies most abundantly in honey bees maintained at 30°C; however, infected bees die earlier, within a few days, when kept at 35°C.[22]

Infection is very common. Most, or perhaps all, colonies in Britain are permanently infected but usually only a few individual bees in them become paralytic at any one time. The severity of infection is positively correlated with the number of colonies per unit area;[23] increased colony density decreases foraging activity and thus increases close bodily contact between bees within their colonies.

B. VIRUS STRUCTURE AND COMPOSITION
1. Particle Structure

CPV particles stained with neutral sodium phosphotungstate or uranyl acetate and examined in the electron microscope appear mostly ellipsoidal with a small irregular protuberance at one end. A wide range of shapes and sizes of particles is possible which may include rings, figure eights, branching rods, and lengths up to 640 nm (see Figure 9a). Three or four components are resolved in sucrose gradients or by analytical untracentrifugation with sedimentation coefficients of 82, 97 to 106, 110 to 124, and 125 to 136 S.[21] Samples of these components contain particles that vary considerably in size but have modal lengths of about 30, 40, 55, and 65 nm, respectively. All have a modal width of about 22 nm. The great deviation from the mean in the lengths of particles from each component, measured from electron micrographs, suggests that the lengths of particles change when they are prepared for electron microscopy. Particles may also appear flattened in shadow preparations, causing central concavities which may account for the ring-shaped particles that are sometimes seen. Despite the diversity of particle sizes all have the same buoyant density of 1.33 g/cm^3 in CsCl. The capsids are composed of a single coat polypeptide of M$_r$ 23.5 × 10^3.

The isometric 17 nm particles of CPVA tend to form crystalline arrays when purified and examined in the electron microscope by negative staining with neutral sodium phosphotungstate (see Figure 9b). CPVA has a buoyant density of 1.38 g/cm^3 in CsCl and a sedimentation coefficient of 41 S. Capsids have a single polypeptide of M$_r$ 15 × 10^3.[14]

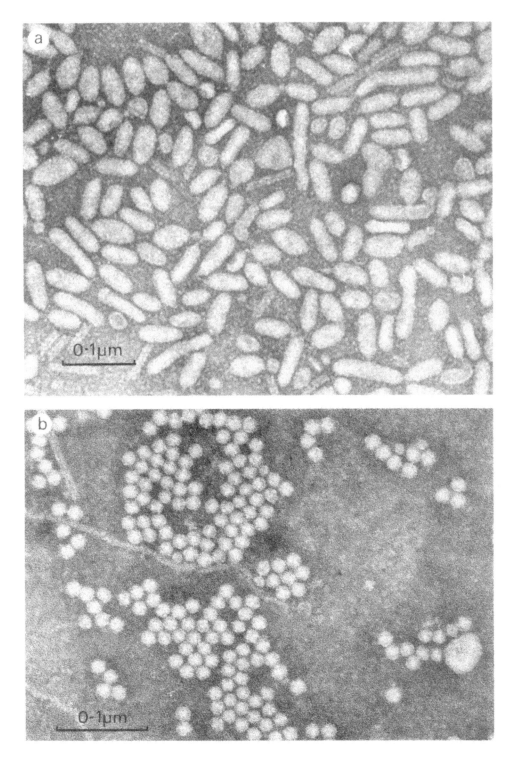

FIGURE 9. Purified preparations of (a) chronic paralysis virus and (b) chronic paralysis virus associate (negatively stained with sodium phosphotungstate).

2. Genome Structure

CPV has five single-stranded RNA components; two larger RNAs designated 1 (M_r 1.35 $\times$ 10^6; 4200 nucleotides) and 2 (M_r 0.9 $\times$ $10.^6$; 2800 nucleotides) and three smaller RNAs designated 3a, 3b, and 3c, each with M_r 0.35 $\times$ 10^6 (1100 nucleotides).[24] CPV adopts a flexible packaging arrangement in which individual RNA species may be packaged separately or various combinations incorporated together.

CPVA has three single-stranded RNA components designated A, B, and C each with M_r of 0.35 $\times$ 10^6 (1100 nucleotides).[24] Gel electrophoretic and T_1 fingerprint analyses have shown that RNAs 3a, 3b, and 3c of CPV are very similar to and probably identical to RNAs A, B, and C, respectively (i.e., these three RNAs appear to be encapsidated in both CPV and CPVA particles). The efficient encapsidation of CPVA RNA by CPV may be due to the presence of common sequences in CPV RNA 2.

CPVA interferes with the replication of CPV; the total amount of CPV RNAs 1 and 2 is greatly reduced, and the proportion of CPVA RNA encapsidated in CPV particles greatly increases as CPVA replication increases.

C. CYTOPATHOLOGY

Numerous particles, morphologically similar to those of CPV can be seen in honey bees suffering from paralysis. These particles appear in densely but randomly packed groups either free or within vesicles in the cytoplasm of thoracic and abdominal ganglia, of the gut, and of the mandibular and hypopharyngeal glands.[25,26] However, they do not appear in the cytoplasm of fat or muscle tissue.[27]

VI. BEE VIRUS X AND BEE VIRUS Y

A. GENERAL CHARACTERISTICS
1. Members of the Group

Bee viruses X (BVX) and Y (BVY) have featureless isometric particles 35 nm in diameter containing single-stranded RNA. Both have similar physicochemical characteristics, and they are distantly related serologically.[28] These two particles superficially resemble those of the Nudaurelia-β group[1] but their physical properties do not correspond sufficiently to suggest any relationship.

2. Gross Pathology

BVX and BVY have similar biological properties; they occur only in the alimentary tract of adult bees and multiply when ingested by caged individuals kept at 30°C but not at 35°C. There are no known signs of infection. Neither virus multiplies when injected into adult bees or pupae. The most striking difference between the two viruses lies in their epizootiology. Whereas BVX is usually detectable only during the winter, BVY is found most frequently about May or June.[29] Moreover, in nature, BVY almost invariably occurs in individual bees in association with the microsporidian *Nosema apis,* and in laboratory tests BVY infects bees much more readily when they ingest the virus with spores of *N. apis* than without.[9] By contrast BVX often multiplies alone, although it frequently occurs in association with *Malpighamoeba mellificae* in bees found dead in the field.[9] However, this association may be due to a considerably greater virulence of the pathogens when they occur together than when alone.

B. VIRUS STRUCTURE AND COMPOSITION

Particles of bee viruses X and Y are indistinguishable by electron microscopy (see Figure 10), and they have the same sedimentation rate of 187 S. However, a mixture of the two viruses separates into two bands when centrifuged to equilibrium in CsCl. BVX has the

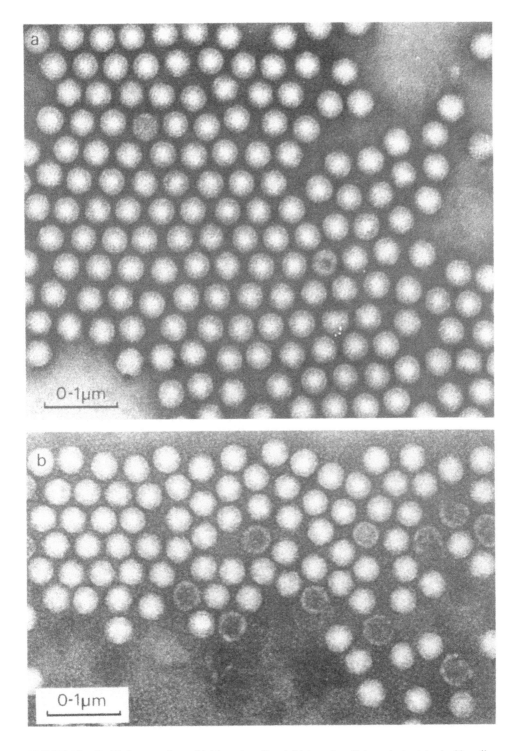

FIGURE 10. Purified preparations of (a) bee virus X and (b) bee virus Y (negatively stained with sodium phosphotungstate).

slightly higher density of 1.355 g/cm³ compared to that of 1.347 g/cm³ for BVY, and its capsid protein has a relative molecular weight larger by more than 2000 (52300 ± 350).[28]

Differences between the genomes of these two viruses have not been determined. No cytopathological effects have been reported.

VII. SACBROOD VIRUS

A. GENERAL CHARACTERISTICS
1. Members of the Group

The type of strain of sacbrood virus (SBV)[32] isolated from the European honey bee, *Apis mellifera,* is physicochemically distinguishable from a strain identified in the Eastern hive bee, *Apis cerana,* from Thailand (TSBV), but is closely related to it serologically.[33] SBV is serologically unrelated to any other honey bee virus, and there are no other known hosts. Longworth[13] proposed sacbrood virus as a possible member of the enterovirus group of the Picornaviridae, but its properties seem to distinguish it from the other members and it remains unassigned to any taxonomic group.

2. Gross Pathology

Whereas healthy bee larvae pupate 4 d after they have been sealed in their cells, larvae with sacbrood fail to pupate and remain stretched on their backs with their heads toward the cell capping. Fluid then accumulates between the body of a diseased larva and its unshed skin and the body color of the larva changes from pearly white to a pale yellow. Finally, after it has died a few days later, the larva becomes dark brown and dries down to a flattened gondola-shaped scale. Although sacbrood is a well-known disease of the larval honey bee, laboratory experiments have established that the virus multiplies in adult bees causing certain behavioral changes but no obvious pathological signs.[34,35] The virus accumulates especially in the hypopharyngeal glands of workers and in the brains of drone (male) bees.

B. VIRUS STRUCTURE AND COMPOSITION

SBV and TSBV have naked, featureless isometric particles 30 nm in diameter (see Figure 11) containing single-stranded RNA of M_r 2.8 × 10⁶, and having identical buoyant densities in CsCl of 135 g/cm³. Unlike SBV particles, however, those of TSBV aggregate when exposed to buffers of low molarity and then sediment quickly during low-speed centrifugation.[33] TSBV extracted in the absence of EDTA also becomes susceptible to penetration by neutral phosphotungstate stain and particles appear empty when viewed by electron microscopy (see Figure 11b). Emptying occurs more quickly when Mg^{2+} is added. SBV also produces empty particles in the presence of Mg^{2+} but is much less sensitive than TSBV.

TSBV sediments at 160 S in 0.1 M KCl but at only 150 S in 0.01 M phosphate and produces three close but well-defined bands of protein on 5% SDS acrylamide gels, with apparent relative molecular weight of approximately 30, 34, and 39 × 10³.[33] By contrast, SBV sediments at 160 S in 0.01 M phosphate and presents one broad protein band of M_r range 29 to 34 × 10³ which only separates into three bands on gels of higher concentration.

C. CYTOPATHOLOGY

Larvae killed by sacbrood virus each contain about 10¹³ particles, many of which occur in fluid between the final larval skin and the body of the prepupa. Many particles, sometimes in crystalline array, have been seen in the cytoplasm of fat, muscle, and tracheal-end cells of larvae that had ingested sacbrood virus under laboratory conditions.[36] Similar particles have also been seen in the cytoplasm of the fat bodies of apparently healthy adult bees that had been injected with the virus.[37]

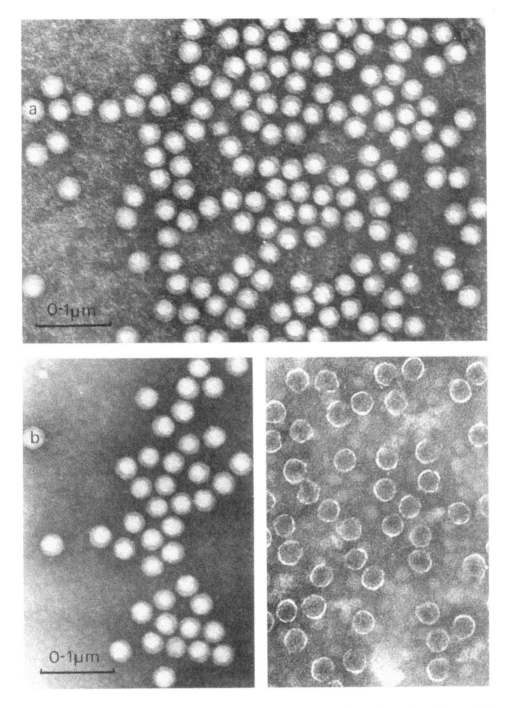

FIGURE 11.   Sodium phosphotungstate stained preparations of (a) sacbrood virus from *Apis mellifera* and (b) full and empty particles of Thai sacbrood virus from *Apis cerana*.

VIII. BLACK QUEEN CELL VIRUS

A. GENERAL CHARACTERISTICS
1. Members of the Group

Black queen cell virus (BQCV) has isometric 30 nm particles with many of the physical characteristics of the enterovirus group of the Picornaviridae.[1] It is the only bee virus of this type in which four capsid proteins have been detected; three with a relative molecular weight of approximately 30×10^3 and a low relative molecular protein of 6×10^3.[30] The virus superficially resembles *Drosophila* C virus and Cricket paralysis virus,[13] having the same buoyant density in CsCl and an RNA with approximately the same relative molecular weight. BQCV, however, is serologically unrelated to either of these particles or to any other virus isolated from honey bees.

2. Gross Pathology

BQCV was first identified as a cause of death of queen larvae or prepupae after they had been sealed in their cells.[31] In the early stages of infection prepupae have a pale yellow appearance and a tough saclike skin, resembling larvae killed by sacbrood virus; but unlike sacbrood virus, BQCV does not readily multiply when ingested by young worker larvae and does not usually multiply alone in adult bees. However, it multiplies abundantly in adults when these are also infected with the microsporidian *Nosema apis* and it then shortens their life, more than when they are infected with *N. apis* alone, although it causes no outward sign of disease.[9]

B. VIRUS STRUCTURE AND COMPOSITION

The isometric featureless particles, 30 nm in diameter, are morphologically stable in buffer solutions of pH 4 or above (see Figure 12a). BQCV sediments at 151 S, has a buoyant density in CsCl of 1.34 g/cm³ and contains a single genomic RNA of M_r 2.8×10^6.[30] The viral coat protein comprises four polypeptides of M_r 34, 32, 29, and 6×10^3.

Nothing is known of its cytopathological effects.

IX. ACUTE PARALYSIS VIRUS

A. GENERAL CHARACTERISTICS
1. Members of the Group

Acute paralysis virus (APV) is a common inapparent infection of honey bees in Britain. The 30 nm particles have physicochemical characteristics similar to those of sacbrood virus (SBV) but the two viruses are serologically unrelated. APV is stable below pH 4, but unlike those mammalian picornaviruses which are acid stable, the buoyant density of APV increases with increasing pH.[31] Two isolates of APV from honey bees in Europe are closely related to the British type strain but are serologically distinguishable.[38] APV has also been detected in *Bombus* species[39] and is the only honey bee virus that appears to have an alternative host in nature.

2. Gross Pathology

In Britain APV has never been associated with disease or mortality in nature, but the virus occurs commonly in seemingly healthy bees, particularly during the summer.[29] In continental Europe APV has been detected as a cause of mortality of both adult bees and brood in honey bee colonies infested with the parasitic mite *Varroa jacobsoni*. The mite damages tissues during feeding and may release APV from cells in which it is normally contained. Once APV enters the hemolymph of adult bees it rapidly becomes systemic and lethal, and *V. jacobsoni* can then also act as a virus vector, transferring APV from severely infected to healthy individuals.[38]

Bees injected with approximately 100 particles of APV develop trembling and semi-paralytic symptoms after 2 to 3 d and die 1 or 2 d later. Injected bees are killed quickly at 30°C but they live longer, and yet more virus accumulates within them, when they are kept at 35°C after inoculation. This is in direct contrast with chronic paralysis virus under the same conditions.[22]

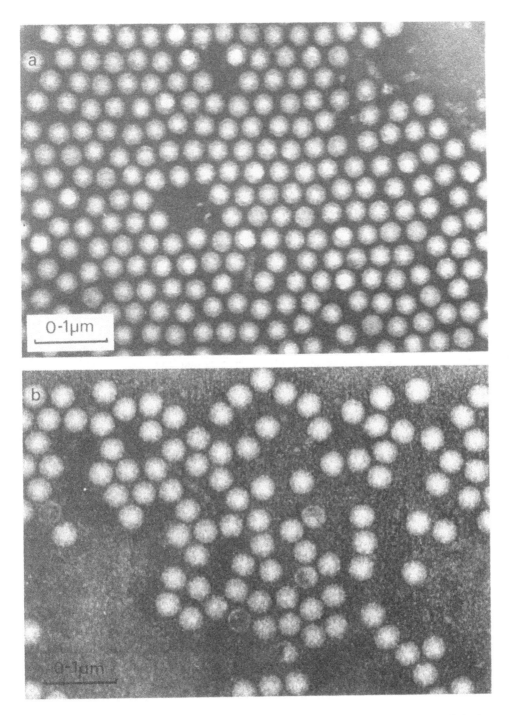

FIGURE 12. Purified preparations of (a) black queen cell virus in buffer of pH 5 and (b) acute paralysis virus negatively stained with sodium phosphotungstate.

B. VIRUS STRUCTURE AND COMPOSITION

APV has naked featureless particles 30 nm in diameter, some appearing empty when negatively stained with neutral sodium phosphotungstate (see Figure 12b). The virus sediments in two fractions corresponding to full and empty particles of 160 and 80 S. High pH values increase the buoyant density of APV from 1.37 g/cm^3 in CsCl at pH 7.0 to 1.42 to

1.43 g/cm³ at pH 9.0.[31] Capsids comprise three polypeptides of M_r[38] 35, 33, and 24 × 10³ and the RNA genome has a base ratio (A:C:G:U) of 30.3:20.5:18.8:30.4.[40]

C. CYTOPATHOLOGY

No detailed cytopathological studies have been undertaken but crystalline arrays of particles have been seen in the cytoplasm of fat body cells,[41] the brain, and hypopharyngeal glands[22] of acutely paralyzed bees.

X. KASHMIR BEE VIRUS

A. GENERAL CHARACTERISTICS

1. Members of the Group

Kashmir bee virus (KBV) was first identified in adults of the Eastern hive bee, *Apis cerana,* from the northern and western regions of India,[31] but several serologically related strains were later isolated from the European bee, *Apis mellifera,* from widely separated areas of Australia.[42] The virus has also been detected in *A. mellifera* in New Zealand[43] but is not known to occur elsewhere in the world. All isolates failed to react with an antiserum to a physically similar virus from *Gonometa podocarpi.*[13]

2. Gross Pathology

KBV commonly persists as an inapparent infection in both adult bees and pupae, but occasionally is associated with severe mortality of bees and brood in the field.[42] When injected into adults of *Apis mellifera* or rubbed onto their bodies, it multiplies quickly and profusely, killing them within 3 d. The virus causes no symptoms when fed to adults or larvae.[43]

B. VIRUS STRUCTURE AND COMPOSITION

KBV from *A. cerana* (see Figure 13a) and the four Australian isolates from *A. mellifera* are indistinguishable by electron microscopy, sedimentation coefficient (167 to 170 S), or buoyant density in CsCl (1.37 to 1.38 g/cm³).[42] All Australian isolates are less closely related serologically to the type strain from Kashmir than they are to each other. This is reflected in the analysis of the capsid proteins of each isolate on 5% SDS polyacrylamide gels (see Table 1). Whereas the Australian isolates were not distinguishable from each other by the patterns of their proteins, as each contains parts of the same five-band pattern, the proteins of Kashmir virus differ measurably from them. The RNA genome of KBV has not been characterized.

C. CYTOPATHOLOGY

A detailed ultrastructural study of KBV replication in honey bee pupae by Dall[44] showed that the virus multiplied in most tissues, apparently within cytoplasmic membrane-bound vesicles. No evidence of KBV multiplication was found in tissues of the nervous system. Infected cells showed a marked cytopathological response which included condensation of chromatin, disruption of cytoplasmic organelles, and formation of lobules of cytoplasm at the cell margin.

XI. ARKANSAS BEE VIRUS

A. GENERAL CHARACTERISTICS

1. Members of the Group

Arkansas bee virus (ABV) was originally detected in honey bees in the U.S. by injecting apparently healthy adult bees with extracts of locally gathered pollen loads.[45] In view of the

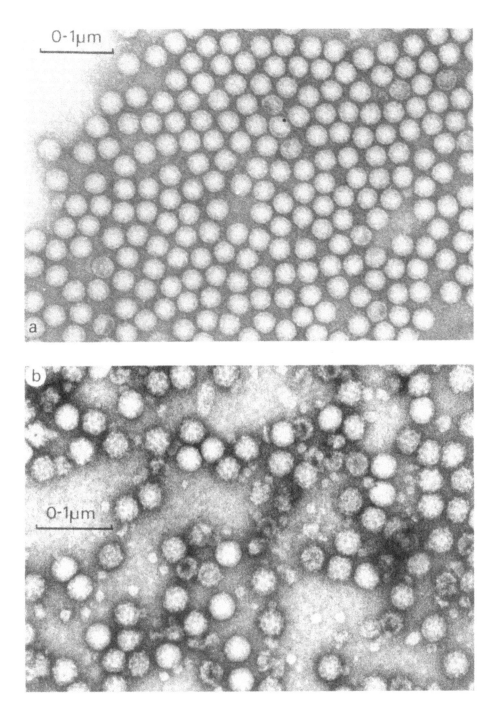

FIGURE 13. (a) Purified preparation of Kashmir bee virus from *Apis cerana* (negatively stained with sodium phosphotungstate) and (b) semipurified preparation of Arkansas bee virus (negatively stained with sodium phosphotungstate) at pH 4.

origin of ABV it was tested as a possible plant virus and conflicting reports were published on its ability to multiply in *Chenopodium* species.[21,46] ABV was proposed as a possible member of the Nodaviridae[1] which has some physicochemical similarities to the plant-infecting Dianthovirus group; however, recent evidence[47] suggests that ABV should not be included in either taxonomic group and its status remains unresolved.

TABLE 1
**Molecular Weights of Kashmir Bee Virus Proteins Estimated on 5%
SDS-Acrylamide Gels**

Virus	VP1	VP2	VP3	VP4	VP5
Kashmir isolate	41,100	37,300	24,500	—	—
South Australia isolate (fresh)	44,300	39,300	35,600	—	25,000
South Australia isolate (aged)	—	39,950	—	33,950	25,000
New South Wales isolate (fresh)	44,500	—	35,400	—	24,950
New South Wales isolate (aged)	—	40,250	—	32,800	25,050
Queensland 1 isolate	—	39,800	—	32,950	24,800
Queensland 2 isolate	—	40,300	36,400	—	25,250

2. Gross Pathology

Bees injected with ABV die after about 14 d. There are no clearly recognizable symptoms. Although ABV commonly persists as an inapparent infection, the virus has been detected directly by serology in honey bees from colonies in California. In nature ABV always seems associated with a newly identified picorna-like virus named Berkley bee picorna-virus (BBPV).[47]

B. VIRUS STRUCTURE AND COMPOSITION

Purified preparations of ABV particles stained with phosphotungstate at pH 4 and examined in the electron microscope appear mostly intact with some evidence of surface structure (see Figure 13b). When centrifuged on sucrose gradients the major virus peak of approximately 128 S was typically preceded by a minor component of 170 S. This faster sedimenting component was composed of dimers. Particles have a density of 1.37 g/cm^3 in CsCl.

ABV was originally thought to contain two genomic RNAs of different molecular weight but recent studies[47] suggest that the virus is often present in mixed infections with a morphologically similar particle (BBPV) which sediments at the same rate as ABV dimers. ABV and the recently isolated BBPV are serologically unrelated, but were not distinguished originally because of the production of a mixed antiserum. Lommel et al.[47] have determined that ABV has a single genomic RNA of M_r 1.8 $\times$ 10^6 which excludes it from both the Nodaviridae and the plant Dianthovirus group.

XII. EGYPT BEE VIRUS

A. GENERAL CHARACTERISTICS
1. Members of the Group

Egypt bee virus (EBV), originally isolated from *Apis mellifera* from Egypt has isometric 30 nm particles containing single-stranded RNA.[42] Serologically distantly related isolates have recently been identified in honey bees from several countries but these have not yet been fully characterized and their true identity and interrelationships await further studies.[48]

2. Gross Pathology

Honey bee pupae injected with purified preparations of EBV die at a late stage, usually just before they are due to metamorphose, about 7 or 8 d after inoculation. Attempts to propagate the virus by injection into adult bees have been unsuccessful. Nothing is known of its natural history.

B. VIRUS STRUCTURE AND COMPOSITION

Semipurified preparations of EBV are unstable in buffers of low molarity or below pH

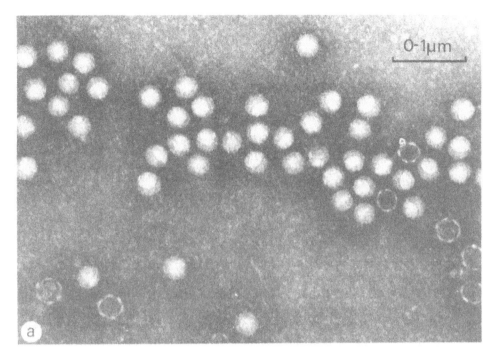

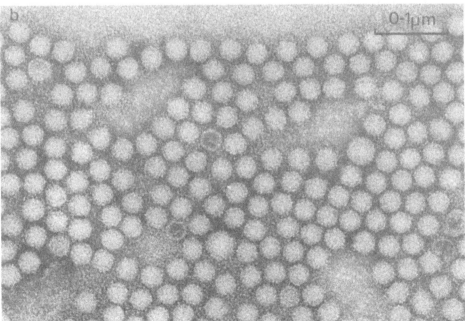

FIGURE 14. (a) Purified preparation of Egypt bee virus. Empty particles penetrated by sodium phosphotungstate stain show the six symmetrically arranged thickenings of the capsid. (b) Purified preparation of slow paralysis virus (negatively stained with sodium phosphotungstate).

5. Under these conditions many capsids, typically with six clearly defined thickenings can be seen penetrated by stain (see Figure 14a). Virus capsids are composed of three polypeptides of M_r 41, 30, and 25 × 10^3. Purified particles sediment at about 165 S in 0.01 M phosphate or 0.1 M KCl, and their buoyant density in CsCl is 1.37 g/cm^3. The absorbance ratio (A_{260}:A_{280}) of 1.74 and high buoyant density indicate an RNA content of 20 to 25%.[42]

XIII. SLOW PARALYSIS VIRUS

A. GENERAL CHARACTERISTICS

1. Members of the Group

Serological relationships between slow paralysis virus (SPV) and physicochemically similar particles from *Gonometa podocarpi* and *Drosophila* (Virus A) have been reported[13] but are unconfirmed. SPV is unrelated to any other virus isolated from honey bees.

2. Gross Pathology

SPV was originally found as an inapparent infection of adult honey bees in Britain and no disease has been associated with it in nature. Bees injected with SPV die after about 12 d, typically suffering paralysis of the anterior two pairs of legs for a day or two before death.[45]

B. VIRUS STRUCTURE AND COMPOSITION

SPV has featureless isometric particles 30 nm in diameter (see Figure 14b) which sediment at 176 S and have buoyant density in CsCl of 1.37 g/cm^3.[45] The virus has three capsid proteins of M$_r$ 27, 29, and 46 $\times$ 10^3 and contains RNA. Purified particles aggregate spontaneously in the agar of immunodiffusion plates when these contain 0.85% sodium or potassium chloride at pH 5.8 to 8.8, sometimes producing "Liesgang" rings. The virus particles aggregate less readily in 0.01% KCl and are prevented from such aggregation when mixed with an equal volume of normal rabbit serum.

REFERENCES

1. **Matthews, R. E. F.,** Classification and nomenclature of viruses. Fourth report of the International Committee on Taxonomy of Viruses, *Intervirology*, 17, 1, 1982.
2. **Bailey, L., Carpenter, J. M., and Woods, R. D.,** Properties of a filamentous virus of the honey bee, *Virology*, 114, 1, 1981.
3. **Payne, C. C., Compson, D., and De Looze, S. M.,** Properties of the nucleocapsids of a virus isolated from *Oryctes rhinoceros, Virology*, 77, 269, 1977.
4. **Krell, P. J. and Stoltz, D. B.,** Unusual baculovirus of the parasitoid wasp *Apanteles melanoscelus:* isolation and preliminary characterization, *J. Virol.*, 29, 269, 1977.
5. **Krell, P. J. and Stoltz, D. B.,** Virus-like particles in the ovary of an ichneumonid wasp: purification and preliminary characterization, *Virology*, 101, 408, 1980.
6. **Thomas, D. and Gouranton, J.,** Development of virus-like particles in the crystal-containing nuclei of the midgut cells of *Tenebrio molitor, J. Invertebr. Pathol.*, 25, 159, 1975.
7. **Arnold, M. K. and Barson, G.,** Occurrence of virus-like particles in midgut epithelial cells of the large elm bark beetle, *Scolytus scolytus, J. Invertebr. Pathol.*, 29, 373, 1977.
8. **Clark, T. B.,** A filamentous virus of the honey bee, *J. Invertebr. Pathol.*, 32, 332, 1978.
9. **Bailey, L., Ball, B. V., and Perry, J. N.,** Association of viruses with two protozoal pathogens of the honey bee, *Ann. Appl. Biol.*, 103, 13, 1983.
10. **Stoltz, D. B. and Vinson, S. B.,** Viruses and parasitism in insects, *Adv. Virus Res.*, 24, 125, 1979.
11. **Bailey, L., Ball, B. V., and Woods, R. D.,** An iridovirus from bees, *J. Gen. Virol.*, 31, 459, 1976.
12. **Xeros, N.,** Development of the *Tipula* iridescent virus, *J. Insect. Pathol.*, 6, 261, 1964.
13. **Longworth, J. F.,** Small isometric viruses of invertebrates, *Adv. Virus Res.*, 23, 103, 1978.
14. **Bailey, L., Ball, B. V., Carpenter, J. M., and Woods, R. D.,** Small virus-like particles in honey bees associated with chronic paralysis virus and with a previously undescribed disease, *J. Gen. Virol.*, 46, 149, 1980.
15. **Bellet, A. J. D., Fenner, F., and Gibbs, A. J.,** The viruses, in *Viruses and Invertebrates*, Gibbs, A. J., Ed., Elsevier/North-Holland, Amsterdam, 1973, 41.
16. **Jutila, J. W., Henry, J. E., Anacker, R. L., and Brown, W. R.,** Some properties of a crystalline-array virus (CAV) isolated from the grasshopper *Melanoplus bivittatus* (Say) (Orthoptera; Acrididae), *J. Invertebr. Pathol.*, 15, 225, 1970.

17. **Bailey, L.,** Viruses of honey bees, *Bee World,* 63, 165, 1980.
18. **Henry, J. E. and Oma, E. A.,** Ultrastructure of the replication of the grasshopper crystalline array virus in *Schistocerca americana* compared with other picornaviruses, *J. Invertebr. Pathol.,* 21, 273, 1973.
19. **Bailey, L.,** Recent research on honey bee viruses, *Bee World,* 56, 55, 1975.
20. **Plus, N. and Veyrunes, J. C.,** RS virus of *Drosophila,* in *Progress in Invertebrate Pathology,* Proc. Int. Coll. on Invertebr. Pathol. Society for Invertebr. Pathol., Prague, 153, 1979.
21. **Bailey, L.,** Viruses attacking the honey bee, *Adv. Virus Res.,* 20, 271, 1976.
22. **Bailey, L. and Milne, R. G.,** The multiplication regions and interaction of acute and chronic bee-paralysis viruses in adult honey bees, *J. Gen. Virol.,* 4, 9, 1969.
23. **Bailey, L., Ball, B. V., and Perry, J. N.,** Honey bee paralysis: its natural spread and its diminished incidence in England and Wales, *J. Apic. Res.,* 22, 191, 1983.
24. **Overton, H. A., Buck, K. W., Bailey, L., and Ball, B. V.,** Relationships between the RNA components of chronic bee-paralysis virus and those of chronic bee-paralysis virus associate, *J. Gen. Virol.,* 63, 171, 1982.
25. **Giauffret, A., Duthoit, J. L., and Caucat, M. J.,** Histological study of nervous tissue from honey bees infected with "black disease", *Bull. Apic. Doc. Sci. Tech. Inf.,* 9, 221, 1966.
26. **Giauffret, A., Duthoit, J. L., and Tostain-Caucat, M. J.,** Ultrastructure of cells of bees infected with the virus of black disease paralysis. Study of the cellular inclusions, *Bull. Apic. Doc. Sci. Tech. Inf.,* 13, 115, 1970.
27. **Lee, P. E. and Furgala, B.,** Chronic bee paralysis virus in the nerve ganglia of the adult honey bee, *J. Invertebr. Pathol.,* 7, 170, 1965.
28. **Bailey, L., Carpenter, J. M., Govier, D. A., and Woods, R. D.,** Bee virus Y, *J. Gen. Virol.,* 51, 405, 1980.
29. **Bailey, L., Ball, B. V., and Perry, J. N.,** The prevalence of viruses of honey bees in Britain, *Ann. Appl. Biol.,* 97, 109, 1981.
30. **Bailey, L. and Ball, B. V.,** *Honey Bee Pathology,* Academic Press, London, 1991, chap. 3.
31. **Bailey, L. and Woods, R. D.,** Two more small RNA viruses from honey bees and further observations on sacbrood and acute bee-paralysis viruses, *J. Gen. Virol.,* 37, 175, 1977.
32. **Bailey, L., Gibbs, A. J., and Woods, R. D.,** Sacbrood virus of the larval honey bee (*Apis mellifera* Linnaeus), *Virology,* 23, 425, 1964.
33. **Bailey, L., Carpenter, J. M., and Woods, R. D.,** A strain of sacbrood virus from *Apis cerana, J. Invertebr. Pathol.,* 39, 264, 1982.
34. **Bailey, L. and Fernando, E. F. W.,** Effects of sacbrood virus on adult honey bees, *Ann. Appl. Biol.,* 72, 27, 1972.
35. **Bailey, L.,** The multiplication and spread of sacbrood virus of bees, *Ann. Appl. Biol.,* 63, 483, 1969.
36. **Lee, P. E. and Furgala, B.,** Electron microscopic observations on the localization and development of sacbrood virus, *J. Invertebr. Pathol.,* 9, 178, 1967.
37. **Lee, P. E. and Furgala, B.,** Viruslike particles in adult honey bees (*Apis mellifera* Linnaeus) following injection with sacbrood virus, *Virology,* 32, 11, 1967.
38. **Ball, B. V.,** Acute paralysis virus isolates from honey bee colonies infested with *Varroa jacobsoni, J. Apic. Res.,* 24, 115, 1985.
39. **Bailey, L. and Gibbs, A. J.,** Acute infection of bees with paralysis virus, *J. Insect. Pathol.,* 6, 395, 1964.
40. **Newman, J. F. E., Brown, F., Bailey, L., and Gibbs, A. J.,** Some physico-chemical properties of two honey bee picornaviruses, *J. Gen. Virol.,* 19, 405, 1973.
41. **Furgala, B. and Lee, P. E.,** Acute bee paralysis virus, a cytoplasmic insect virus, *Virology,* 29, 346, 1966.
42. **Bailey, L., Carpenter, J. M., and Woods, R. D.,** Egypt bee virus and Australian isolates of Kashmir bee virus, *J. Gen. Virol.,* 43, 641, 1979.
43. **Anderson, D.,** Viruses of New Zealand honey bees, *N.Z. Beekeeper,* 188, 8, 1985.
44. **Dall, D. J.,** Multiplication of Kashmir bee virus in pupae of the honey bee, *Apis mellifera, J. Invertebr. Pathol.,* 49, 279, 1987.
45. **Bailey, L. and Woods, R. D.,** Three previously undescribed viruses from the honey bee, *J. Gen. Virol.,* 25, 175, 1974.
46. **Cole, A. and Morris, T. J.,** Evidence for the replication of an invertebrate virus in plants, *Phytopathology,* 69, 914, 1979.
47. **Lommel, S. A., Morris, T. J., and Pinnock, D. E.,** Characterization of nucleic acids associated with Arkansas bee virus, *Intervirology,* 23, 199, 1985.
48. **Ball, B. V.,** unpublished observations, 1987.

Chapter 22

UNCLASSIFIED VIRUSES OF INSECTS

Jean R. Adams and David Guzo

TABLE OF CONTENTS

I. INTRODUCTION

There are many reports of unclassified viruses and virus-like particles (VLPs) in insects. Excellent reviews have appeared on RNA viruses of insects[1-3] and viruses of *Drosophila*[4] in which sections on VLPs have been included. Many of these VLPs have been isolated from insects in mass rearing facilities or observed as microorganisms causing cytopathology in ultrastructural studies of insect tissues. A partial list of reports in the literature is summarized in Tables 1 and 2. Several illustrations of various VLP morphologies will serve to illustrate this large group. However, only a few have the requisite biochemical data for classification (see Table 1).[5-12]

II. SPHERICAL VIRUSES

Some spherical VLPs possess similarities to viruses in recognized groups but differ in one or more biochemical or morphological characteristics. Kawino virus, for example, might be included in the enterovirus subgroup of the Picornaviridae given its physicochemical similarities to polio virus; however, the viral RNAs lack the poly (A) tracts which seem to be necessary for infectivity of some vertebrate picornaviruses.[5] The kelp fly virus[8] is a small virus (29 nm in diameter) with distinctive surface projections and physicochemical properties which distinguish it from other small insect viruses. The HPS-1 virus, approximately 36 nm in diameter, possesses an RNA genome, and contains a major coat protein of 120 kDa but no lipid.[6] Partial denaturation mapping studies of the HPS-1 genome revealed that the dsRNA molecules are homogeneous in size and sequence organization. Electron microscopic examinations of aqueous (nondenaturing) spreads of HPS-1 genomic RNA confirmed the conclusion from agarose gel electrophoresis that the single segments are 6 kb in size.[6] Chronic stunt virus (CSV), a ssRNA virus, isolated from the navel orangeworm, *Amyelois transitella*, is a possible member of the Caliciviridae.[9] The *Culicoides* virus[10] is similar in some physicochemical characteristics to the bisegmented dsRNA animal virus group described in Matthews' "Classification and Nomenclature of Viruses".[12] This virus group also includes *Drosophila* X and has been described in detail in the chapter on the Birnaviridae.

The *Rhopalosiphum padi* virus (RhPV) was first identified from laboratory reared clones of *R. padi* and *Schizaphis graminum* which were maintained for plant virus transmission studies.[13] It utilizes plants as reservoirs allowing the virus to survive outside the host.[14] Plants act as passive reservoirs since RhPV does not replicate in plants. RhPV has been observed to replicate only in the cytoplasm of posterior midgut, hindgut, and hemocoel of *R. padi* and *S. graminum*. Cytopathological effects included a loss of ribosomes of the endoplasmic reticulum and development of 200-nm membrane vesicles.[11]

Tables 2 to 4 summarize further information about VLPs revealed in electron microscopic examinations of infected insect tissues.[15-107] Groups are presented according to virus particle morphology. Examples have been selected from each group to illustrate the different morphologies.

III. SPHERICAL AND ICOSAHEDRAL VLPs

The smallest viruses in this group are the cricket crystalline array virus (CAV) and the *Antheraea* satellite virus.[15-18] The CAV has been responsible for high mortalities in grasshopper nymphs which were hatched from eggs collected in Montana. Symptoms of this disease were loss of appetite and vigor and a general lack of coordination. CAV has been shown to be infectious p.o. and by intrahemocoelic injection. Susceptible tissues include trachea, muscle, pericardial cells, and dorsal vessel (see Figure 1). The aggregation of virions into rod-shaped crystals 2 × 10 μm greatly aids in the diagnosis of this disease in

TABLE 1
Unclassified Spherical Viruses Isolated from Insects

Virus or VLP	Size (nm)	N.A.	Mol. wt.	Host (order)[a]	Ref.
Kawino	28	ssRNA		*Mansonia uniformis* (D)	5
HPS-1	36	ssRNA	6×10^6	*Drosophila* cell line (D)	6
	40	ssRNA	1.9×10^6	*Pseudaplusia includens* (L)	7
Kelp fly	29	ssRNA	1.42×10^6	*Chaetocoelopa sydneyensis* (D)	8
CSV	38	ssRNA	2.5×10^6	Navel orangeworm, *Amyelois transitella* (L)	9
	54	dsRNA	2.5×10^6 2.6×10^6	*Culicoides* (D)	10
RhPV	27	dsRNA	7.6×10^6 1.7×10^6	*Rhopalosiphum padi* (H)	11, 12

[a] D = Diptera; H = Homoptera; L = Lepidoptera.

TABLE 2
Unclassified Spherical Viruses and Virus-Like Particles Observed in Insect Tissues

Size (nm)	Site of infection N	Site of infection C	Tissues[a] infected	Additional information	Host (order)[b]	Ref.
13		+	TR, MU, PC, DV	CAV, RNA	*Melanoplus bivattatus* (Or)	15
15			PC, DV	Crystals 2—10 μm long	*Schistocerca americana* (Or)	16,17
14	+		MG	*Antheraea* satellite, S-44	*Antheraea eucalypti* (L)	18,19
20		+	MT, ES	RNA, crystalline array	*Triatoma infestans* (He) *Panstrongylus megistus* (He)	20,21
24—26		+	S	Crystalline array	*Diabrotica virgifera* (C)	22
27			GH	ssRNA, chronic stunt	Navel orangeworm, *Amyelois transitella* (L)	23—26
23.5					*Phormia, Calliphora* larvae (D)	27
24				DVV, buoyant density = 1.35	*Ceratitis* (D)	28
28	+		MG		Large elm bark beetle, *Scolytus scolytus* (C)	29
30			NS		Termite, *Nasulitermes exitiosus* (I)	30
30			MG		*Triatoma infestans* (HE)	31
30	+	+	HY, MG, TR	Crystals	*Culicoides cavaticus* (D)	32
31—33			CC		*Drosophila* (D)	33
32	+		MG		*Antheraea helena* (L)	18
32—35	+		MG	Crystalline array	*Glossina palpalis gambiensis* (D)	34
33				CTV-2	*Culex tarsalis* (D)	35
35—39	+		MG	Crystalline array	*Glossina fuscipes-fuscipes* (D)	36
38				RNA	*Gonometa podocarpi* (L)	37

TABLE 2 (continued)
Unclassified Spherical Viruses and Virus-Like Particles Observed in Insect Tissues

Size (nm)	Site of infection		Tissues[a] infected	Additional information	Host (order)[b]	Ref.
	N	C				
38	+		MG		*Simulium vittatum* (D)	38
38—40		+	SG	Intranuclear fibrils in vesicles	*Triatoma infestans* (He)	39,40
39			MG		*Hyalophora cecropia* (L)	41
40					*Hyalophora cecropia* (L)	42
38 × 45		+	MG	Oval	*Drosophila melanogaster* (D)	43
38 × 42	+		FB, H, GC	Oval	*Drosophila melanogaster* (D)	44
40—50	+		GC		*Drosophila melanogaster* (D)	45
38	+		MG, NC, PA, TR, CT	Crystalline arrays	*Drosophila melanogaster* (D)	46
42					*Actius selene* (L)	42
40	+		MG	Quasi-crystalline array, rare	*Lucilia cuprina* (D)	47
43	+		MG		*Drosophila* (D)	48
42—46		+	CC, MG	Mostly hollow particles; crystalline inclusion	*Antheraea eucalypti* (L), *Aedes aegypti* (D), *Drosophila melanogaster* (D)	49
46—50		+	MY		*Acheta domesticus* (Or)	50
48—52		+	SG, MG	In vesicles	Tsetse fly, *Glossina morsitans* (D)	51,52
45—50		+	S	Crystalline array	*Heliothis virescens* (L)	53
70	+				*Heliothis virescens* (L)	53
40	+		TR		*Drosophila melanogaster* (D)	54
50	+		FB		*Drosophila melanogaster* (D)	54
60		+	MRO		*Drosophila melanogaster* (D)	54
50	+		Cell cultures		*Drosophila melanogaster* (D)	55 56
55	+		ID		*Drosophila* (D)	57
50		+	MRO		*Typhlocyba douglasi* (Ho)	58
50	+		MG	RNA, crystal	*Antheraea eucalypti* (L)	59
62		+	MG, MU, FB, HY, H, NS	Crystalline array	*Goeldichironomous holoprasinus* (D)	60
67	+		FG	Crystal	Yellow mealworm, *Tenebrio molitor* (C)	61
60—75			FB, P	Wassersucht, RNA	*Melolontha* sp. (C)	62
70—75	+	+	H	Enveloped	Bean leaf beetle, *Cerotoma trifurcata* (C)	63
70	+	+	H	Enveloped	Spotted cucumber beetle, *Diabrotica undecimpunctata* (C)	64

TABLE 2 (continued)
Unclassified Spherical Viruses and Virus-Like Particles Observed in Insect Tissues

Size (nm)	Site of infection N	Site of infection C	Tissues[a] infected	Additional information	Host (order)[b]	Ref.
70	+	+	AG, E, FB, H, HY, MG, MXG, MU, MT, NS, O, T		*Epilachna varivestis* (C)	65
80			*Aedes* CC	RNA, buoyant density 1.15	*Culex tritaeniorhynchus* (D)	66
37		+	MRO	Rect. aggregates, empty particles	*Drosophila virilis* (D)	67
37—40		+	H	Polyhedral (nonenveloped), crystalline array	Bean leaf beetle, *Cerotoma trifurcata* (C)	63
56	+		MG	Hexagonal	*Drosophila melanogaster* (D)	68
74		+	MG		*Drosophila melanogaster* (D)	68
40—65		+	MG	Hexagonal	Fruitfly, *Dacus tryoni* (D)	69
				Spherical inclusions 1.21—2.87 μm, diam		
120—140	+		C	No nucleic acids?	*Nemeritis canescens* (Hy)	70,71
60		+	H, O, MRO, TR	RNA virus, S virus	*Drosophila simulans* (D)	72

[a] AG = accessory gland; C = calyx; CC = cell culture; CT = connective tissue; DV = dorsal vessel; E = eggs; ES = extracellular spaces; FB = fat body; FG = foregut; GC = glia cells of brain; GH = granular hemocytes; H = hemocytes; HY = hypodermis; ID = imaginal discs; MG = midgut; MXG = maxillary gland; MRO = male reproductive organ; MT = Malpighian tubule; MU = muscle; MY = myocardium; NC = nurse cells; NS = nervous system; O = ovarioles; P = plasmatocytes; PC = pericardial cells; S = sperm; SG = salivary glands; T = testes; TR = tracheal matrix.
[b] C = Coleoptera; D = Diptera; He = Hemiptera; Hy = Hymenoptera; I = Isoptera; L = Lepidoptera.

the light microscope. The *Antheraea* satellite virus (14 nm in diameter) and another RNA virus of 32 nm in diameter were isolated from *Antheraea helena* larvae. Sections of midgut cells showed the viral particles together within large vesicles in the cytoplasm of infected cells.[18]

The chronic stunt virus of navel orangeworms is readily transmitted perorally. Chronic infections produced stunted larvae, pupae, and adults.[23-26] The cytoplasm of infected granular hemocytes is filled with crystalline arrays of virus particles (see Figure 2A). Ovaries of infected moths are severely stunted, and egg production is greatly reduced but no evidence of transovum transmission has been detected.[24] Biochemical studies have shown that the virus is ssRNA with a genome of 2.5×10^6 Da.[9] Electron microscopic observations of purified fractions have revealed that there are two sizes of viral particles, a 38-nm cupped particle, which sometimes accumulates in crystalline arrays in host granular hemocytes, and a 28-nm particle which occurs predominantly in the frass and apparently results from proteolytic digestion of the 38-nm particle[9] (see Figure 2B). The two different sized particles are both almost equally infectious perorally in navel orangeworm larvae.[26] Hoffman and Hillman[26] observed that the alteration of the capsid protein was not sufficient to cause a significant change in pathogenicity as measured by the *in vivo* assay method and caution

TABLE 3
Unclassified Bacilliform Viruses and Virus-like Particles Observed in Insect Tissues

Size (nm)	Site of infection N	Site of infection C	Tissues[a] infected	Additional information	Host (order)[b]	Ref.
35 × 150	+		MG	Protein crystals	Whirlagig beetle, *Gyrinus natator* L. (C)	73
16—20 × 50—55		+	HY, CN, NL		*Acheta domesticus* (Or)	74
16—24 × 25—124		+	MT, MU, TR, MG		*Acheta domesticus* (Or)	75—77
16—18 × 25—40			MG, FB			
18 × 53		+	MY		*Acheta domesticus* (Or)	46
65 × 650—700			SG	DNA	*Merodon equestris* (D)	78
16—25 × 55—110		+	MG	RNA	Mexican bean beetle, *Epilachna varivestis* (C)	79
25 × 66—166	+		DO	Longest = 1.15 μm	*Epilachna varivestis* (C)	79
87 × 177	+		FB	Hypertrophied, nuclei DNA	*Gryllus campestris* (Or), *Gryllus bimaculatus, Teleogryllus oceanicus* (Or), *Teleogryllus commodus*	80
31 × 160—170		+	MG		Leafhopper, *Euscelis plebejus* (Ho)	81
27—33		+	SG	Males sterile, tranovarial transmission	Tsetse fly, *Glossina pallidipes* (D)	82—87
57 × 700—1300						88
75—106 × 260—300	+		MG	Paracystalline arrays	*Chaoborus crystallinus* (D)	89
52 × 230	+		MG, H		*Diabrotica undecimpunctata* (C)	90
60—70 × 250—300			AGF	Crystalline arrays	*Biosteres longicaudatus*	91

[a] CN = cercal nerve; DO = dorsal oocytes; FB = fat body; H = hemocytes; HY = hypodermis; MG = midgut; MT = Malpighian tubules; MU = muscle; N = nervous system; NL = neural lamella; MY = myocardium; S = spermatocyst; SG = salivary glands; TR = tracheal matrix; AGF = accessory gland filaments.

[b] C = Coleoptera; D = Diptera; Ho = Homoptera; Or = Orthoptera.

other investigators to consider this possibility when isolating nonoccluded viruses from extracts of infected tissues or frass and attempting to identify them on the basis of preliminary morphological or sedimentation characteristics.

Some larvae of *Culicoides cavaticus* are infected with virus-like particles approximately 30 nm in diameter. Particles are observed in hypodermal cells (see Figure 3A), between the cuticle and hypodermis, and on the surface of muscle bundles (see Figure 3A). Gut and tracheal matrix cell nuclei and cytoplasm often also contain tightly packed crystals of VLPs[32] (see Figure 3B). This is one of the four spherical viruses listed in Table 2 that may replicate in both nucleus and cytoplasm of susceptible cells.

Many viruses and VLPs have been reported in *Drosophila* spp.[4] Some unclassified VLPs are listed in Table 2.[15-72] A spiked virus and *Drosophila* RS virus are shown in Figure 4A and B. Small nonoccluded viruses forming crystalline arrays in midguts of lepidopteran larvae have been reported by Longworth and Harrap.[42] These viruses were later found to be similar to viruses of the Nudaurelia-β virus group.[108]

Three types of virus-like particles have been reported in the sheep blowfly, *Lucilia cuprina*.[47] The most abundant particles, termed VLP1 (28 nm × 38 nm) were present as filamentous and oval forms within vesicles in the corpus allatum, midgut, muscle, tracheal matrix, and oenocytes near fat body, and in the glial cells of corpus cardiacum and thoracic ganglion. In the corpus allatum oval forms occurred in basal lamina and extracellular spaces as well as in tubules of the smooth endoplasmic reticulum (SER). Binnington et al.,[47] also noted that accumulation of virions in infected cells often resulted in distention of the SER and transformation of mitochondria in the corpus allatum. Further studies with the VLPs from *L. cuprina* which transformed the mitochondria into vesicle containing bodies, also produced vesiculated mitochondria when inoculated into a *Drosophila* cell line.[109] Similarities to the transformations of mitochondria which occur in virus-infected plant cells were discussed.[109] The oval form of VLP1 which measured approximately 20 × 28 nm in negative stained preparations, appeared to be similar in morphology to chronic bee paralysis virus or RS virus of *Drosophila*. The VLP2s occurred mainly in the guts of older flies, were larger and not enclosed within vesicles, and did not appear to produce cytopathic effects as were observed in tissues infected with VLP1. One form (pleomorphic filaments about 60 nm in diameter) generally occurred in the apex of gut cells; the spheres (60 nm) were scattered in gut cells and occasionally in tracheal matrix. The VLP3, about 40 nm in diameter, occurred infrequently.[47]

Recently a reolike virus has been shown to be the probable cause of a heritable abnormality in bristle formation (the S character) in *Drosophila simulans*.[72] The virus particle, designated the *Drosophila* S virus, is approximately 60 nm in diameter with an electron-dense core surrounded by a double shell. Particles were found in the cytoplasm of hypodermal cells, in the tormogen and trichogen cells associated with abnormal bristles, and in the cells of male and female gonads. Both male and female flies are able to transmit the virus to progeny, but the rates of transmission are apparently not very high.

The 50-nm virus-like particles, isolated from *Drosophila melanogaster* cell cultures by Saigo et al.[55] and Shiba et al.,[56] have been identified as retrovirus-like particles which contain five kilobases of RNA (VLPH-RNA) homologous, at the sequence level,[110-112] to the transposable element *copia*. The VLPs and *copia* are, therefore, probably the viral particle and proviral forms, respectively, of the *copia* element. Interestingly, the *copia* element has a priming mechanism for replication which is unlike that of vertebrate retroviral replication, i.e., a cellular transfer RNA primer synthesis of the first DNA strand. In the case of *copia* retrovirus-like particles, the first DNA extension starts from an internal site (two nucleotides after the anticodon loop), rather than the 3' end of a tRNA.[112] The VLP used half of the tRNA$_i^{Met}$ molecule (nucleotides 1 to 39) as a primer for the RNA-directed DNA synthesis. The primer binding site (PBS) of VLPH-RNA was located next to the 5'LTR (left-hand long terminal repeat) with no nucleotide inserts between the 5'LTR and PBS.[112]

The few reports of spherical viruses infecting Hemiptera include those of Dolder and Mello[20,21,39,40] in which crystalline arrays containing 20 nm viral particles were observed in the cytoplasm of the Malpighian tubules (see Figure 5) and in extracellular spaces between the plasma membrane and the multilayered basement lamina. Infection in the salivary glands of affected individuals, however, involved 38- to 40-nm viral particles in vesicles in the cytoplasm (see Figure 4C), viral particles between the plasma membranes of adjacent cells, and the presence of abnormal intranuclear fibrils. Virus particles also appeared to be discharged into the lumen of the salivary glands.[40] A small isometric virus 30 nm in diameter has also recently been isolated from midgut tissues of *Triatoma infestans*.[31] Further studies[113]

TABLE 4
Other Unclassified Viruses and Virus-like Particles Observed in Insect Tissues

Virus or VLP	Size (nm)	Site of infection N	Site of infection C	Tissues[a] infected	Additional information	Host (order)[b]	Ref.
Flexuous rods	25	+		MG	DNA, no external sign of disease	Yellow mealworm, *Tenebrio molitor* (C)	92—97
Ovoid bodies	200 × 400						
Flexuous rod + Envelope	25, 60						
Filamentous Nucleocapsid	20	+		C	Apical region	*Didegma terebrans* (Hy), *Microplitis croceipe* (Hy), *Cotesia hyphantriae* (Hy), *Cotesia congregata* (Hy)	98,99
+ Envelope	30						
Whole VLP lengths up to 2 mm	65 ×						
Flexuous rods lengths up to 485 nm	22	+	+	MG	Trilaminar membrane	Large elm bark beetle, *Scolytus scolytus* (C)	29
Ovoid bodies	173 × 314						
Nucleocapsid diam.	30—39	+		HY, TR	Parasite *Cotesia marginiventris*	*Spodoptera frugiperda*, (L), (HY) *S. exigua*, (L) *S. ornithogalli*, (L) *Heliothis zea*, (L) *Trichoplusia ni* (L)	100
+ Envelope lengths up to 865—890 nm	50—70						
Nucleocapsid diam +	35—39	+		O	Repl. in epithelium of common and lateral oviducts	*Cotesia marginiventris* (H)	101
Single envelope	60—70						
Double envelope	90						
Oval form with lengths up to 1 mm	20 × 28		+	CA, GC-CC, GC-TG, HY, MU, OE, TR	VLP1, spherical filamentous forms in vesicles, no CPE (cytopathic effect)	*Lucilia cuprina* (D)	43

Capsid	Size	+/-	Tissue	Description	Host	Ref
Paramyxovirus-like (PLP), quasi-spherical	60	+	MG-apical part of cells	VLP2, spherical and filamentous forms, not in vesicles, CPE	*Lucilia cuprina* (D)	43
	112—160	+	SG, NG, O, LM		Planthopper, *Peregrinus maidis* (Ho)	102
Quasi-cylindrical nucleocapsid	100 × 800	+	C, MU, MG, FG	dsDNA, 125 kbp	*Cotesia melanoscela* (Hy)	103,104
Small cylindrical nucleocapsids?	25 × 50	+	C	Small nucleocapsids surrounded by unit membranes	*Bathyplectes anurus*, (Hy) *B. curculionis*	105—107

[a] C = calyx; CA = corpus allatum; GC-CC = glial cells-corpus cardiacum; GC-TG = glial cells-thoracic ganglion; HY = hypodermis; LM = leg muscles; MG = midgut; MU = muscles; O = ovarioles; OE = oenocytes; NG = nerve ganglia; SG = salivary gland; TR = tracheal matrix.

[b] C = Coleoptera; Ho = Homoptera; Hy = Hymenoptera; L = Lepidoptera.

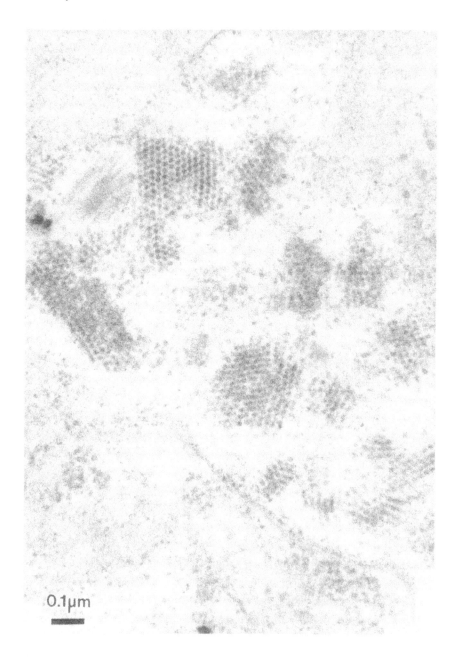

FIGURE 1.   Section of a pericardial cell of *Schistocerca americana* showing crystals containing crystalline array (CAV) within a vesicle in the cytoplasm. (Courtesy of Dr. John E. Henry.)

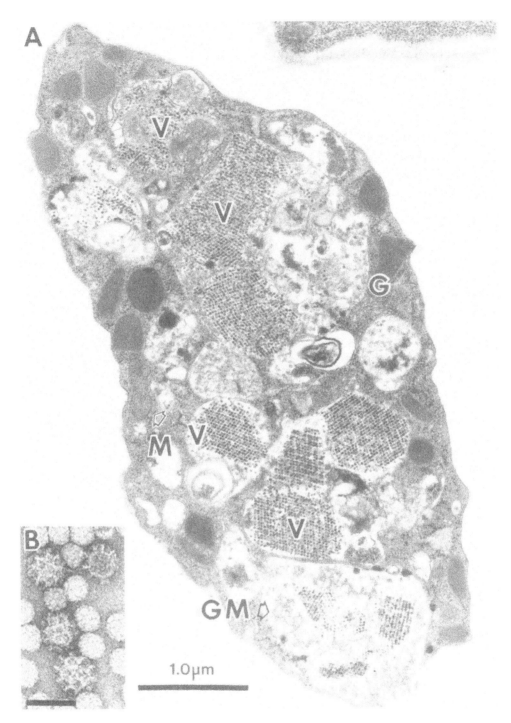

FIGURE 2. (A) Granular hemocyte from the navel orangeworm, *Amyelois transitella*, filled with aggregations and arrays of chronic stunt virus (CSV) within vesicles. Note the viroplasmic bodies at the top and the granular material associated with arrays of CSV at the bottom of the figure. Many structured and unstructured granules (G) are present: mitochondrion (M); virus (V). (B) Unfractionated CSV particles recovered from frass containing cupped particles (38 nm in diameter) and smooth particles (28 nm in diameter). (Bar = 50 nm.) (Figure 2A from Kellen, W. R. and Hoffman, D. F., *J. Invertebr. Pathol.*, 38, 52, 1981. With permission.) (Figure 2B, courtesy of Dr. Bradley Hillman.)

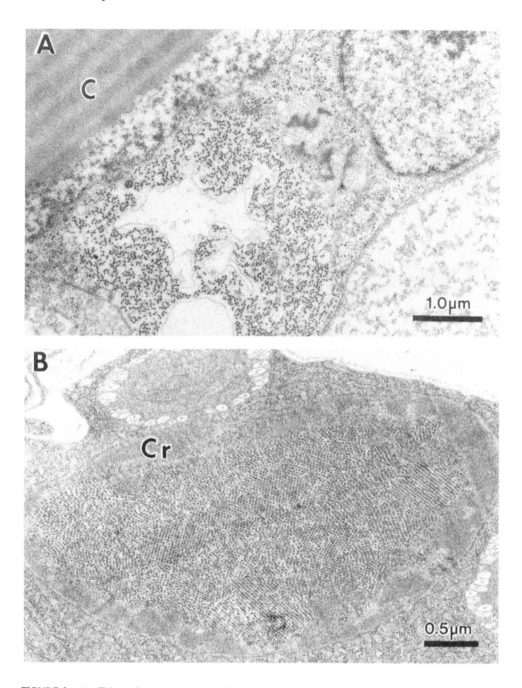

FIGURE 3. (A) Thin section through cuticle (C) and infected hypodermal cell of larva of *Culicoides cavaticus* showing numerous free virus-like particles. (B) Section of a nucleus of hypodermis showing typical crystalline array of virus-like particles. Note the marginated chromatin (Cr). (From Clark, T. B. and O'Grady, J. J., *J. Invertebr. Pathol.*, 26, 415, 1975. With permission.)

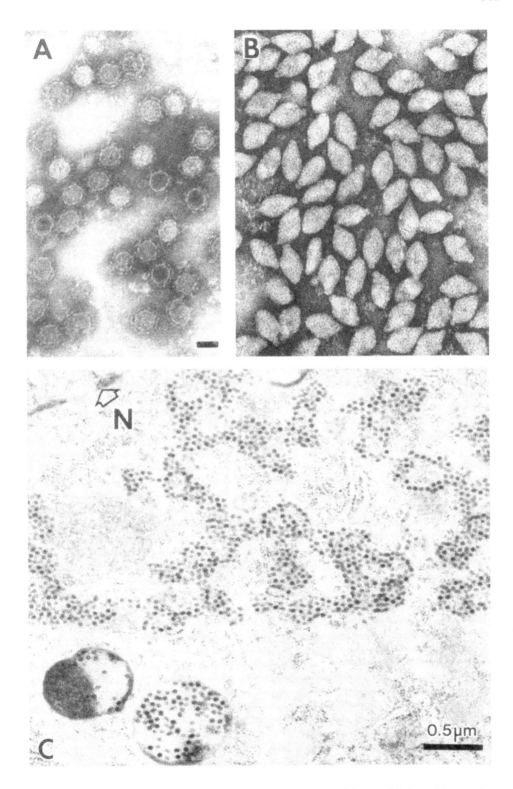

FIGURE 4. (A) G virus of *Drosophila* showing complete and empty particles. (B) RS virus of *Drosophila*. (Bar = 25 nm.) ([A] and [B] courtesy of Dr. Nadine Plus.) (C) Arrays of virus-like particles observed free in the cytoplasm of a salivary gland cell of *Triatoma infestans*. They also appear free in the salivary gland secretion of this insect. The particles are 38 to 40 nm in diameter; nucleus (N). (Courtesy of Dr. Maria L. S. Mello.)

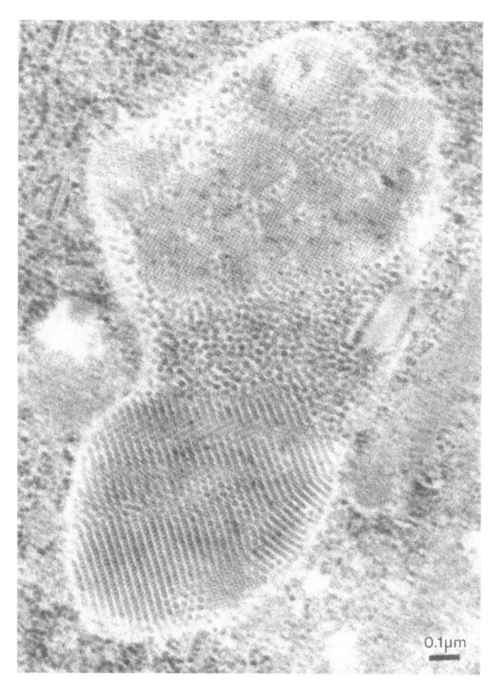

FIGURE 5.   Paracrystalline array of virus-like particles in the cytoplasm of a Malpighian tubule of *Panstrongylus megistus* subjected to Bernhard's EDTA technique. The viral particles are nearly 20 nm in diameter. (From Dolder, H. and Mello, M. L. S., *Cell. Mol. Biol.*, 23, 199, 1978. With permission.)

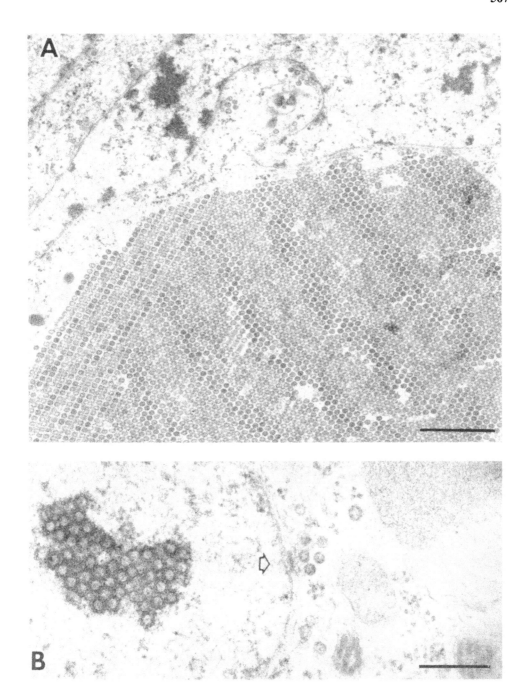

FIGURE 6. (A) Paracrystalline arrangement of virus-like particles (VLPs) in spermatocyst cell of adult *Heliothis virescens*. Small groups of these 45- to 50-nm particles are clustered around the outer nuclear pore in top-center of micrograph. (Bar = 0.4 μm.) (B) Grapelike cluster of 70-nm "coated" VLPs in nucleus of *H. virescens* spermatocyst cell. The spherical, central portion of these nuclear VLPs is similar in size and appearance to the 45- to 50-nm particles clustered outside of the nuclear pore (↑). (Bar = 0.3 μm.) (Degrugillier, M. E., *J. Invertebr. Pathol.*, 54, 281, 1989. With permission.)

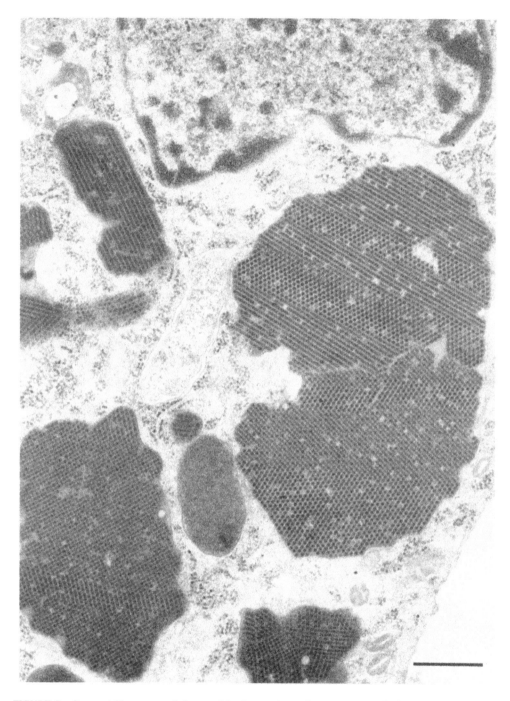

FIGURE 7.   Paracrystalline arrays of virus particles in cytoplasm of spermatocys cell of *Diabrotica virgifera*. Size of particles is 24 to 26 nm. (Bar = 0.5 μm.) (Degrugillier, M. E., Degrugillier, S. S., and Jackson, J. J., *J. Invertebr. Pathol.*, 7, 50, 1991. With permission.)

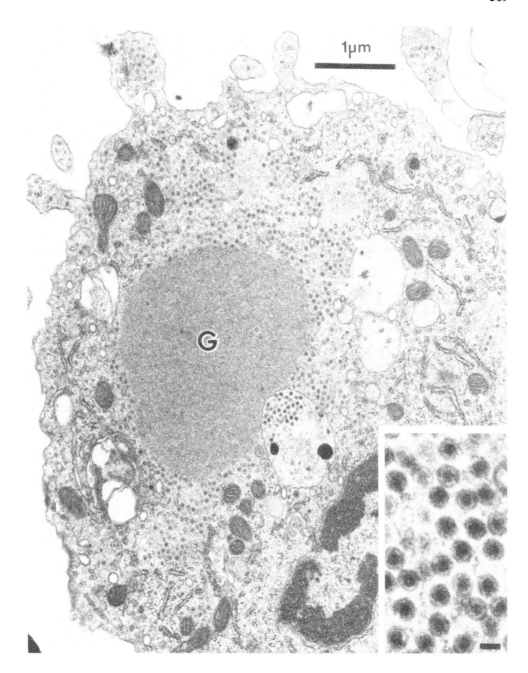

FIGURE 8. Section of hemocyte of spotted cucumber beetle, *Diabrotica undecimpunctata*, containing virus particles (about 70 nm diameter) surrounding the granular body (G) and dispersed throughout the cytoplasm. (Inset) Virus particles. (Bar = 50 nm.) (From Kim, K. S., *J. Invertebr. Pathol.*, 36, 292, 1980. With permission.)

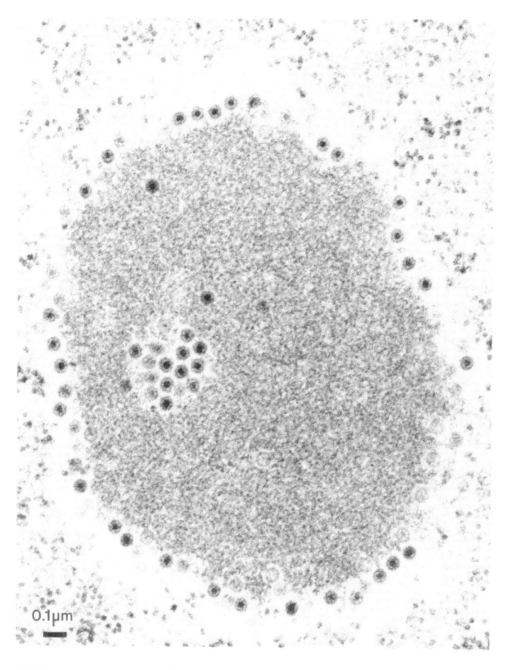

FIGURE 9. Section of viroplasm-like inclusion in nurse cell in the germarium of the Mexican bean beetle, *Epilachna varivestis* showing virus particles (about 70 nm) around the periphery and within the dense matrix. (From Kitajima, E. W., Kim, K. S., Scott, H. A., and Gergerich, R. C., *J. Invertebr. Pathol.*, 46, 83, 1985. With permission.)

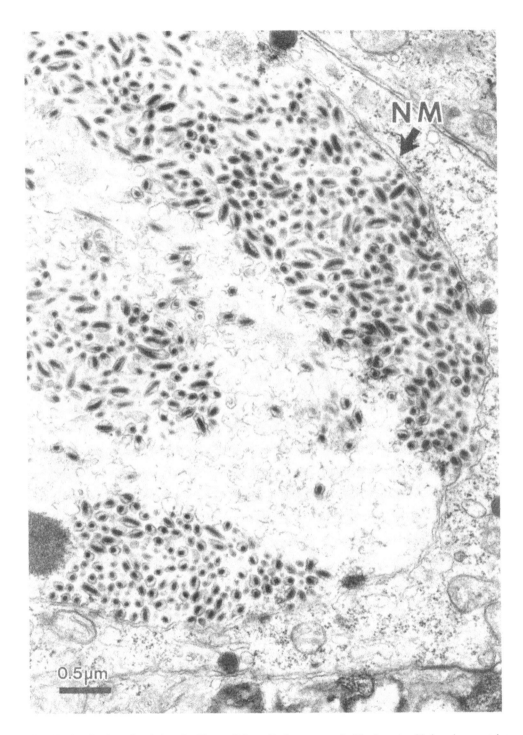

FIGURE 10. Section of an infected midgut cell from *Gyrinus natator* L. The hypertrophied nucleus contains numerous virus particles; nuclear membrane (NM). (Courtesy of Dr. Jean Gouranton.)

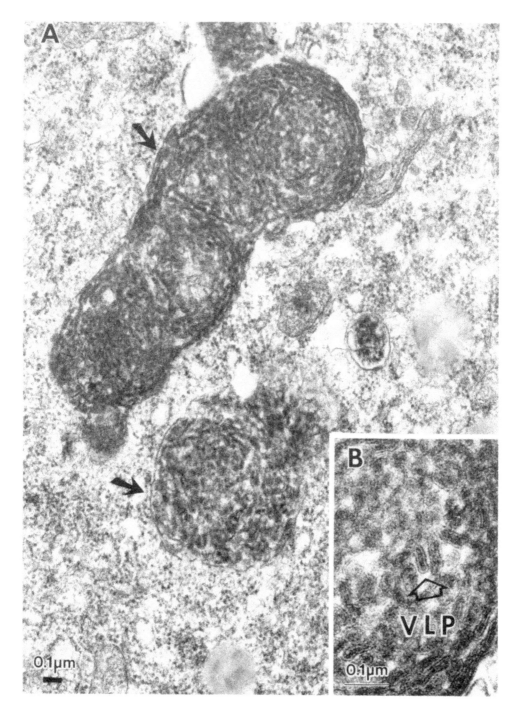

FIGURE 11.   Section of infected midgut cell from *Epilachna varivestis*. Numerous enveloped virus-like bacilliform particles (VLP) are contained within vesicles (arrows). (Insert, bar = 100 nm.)

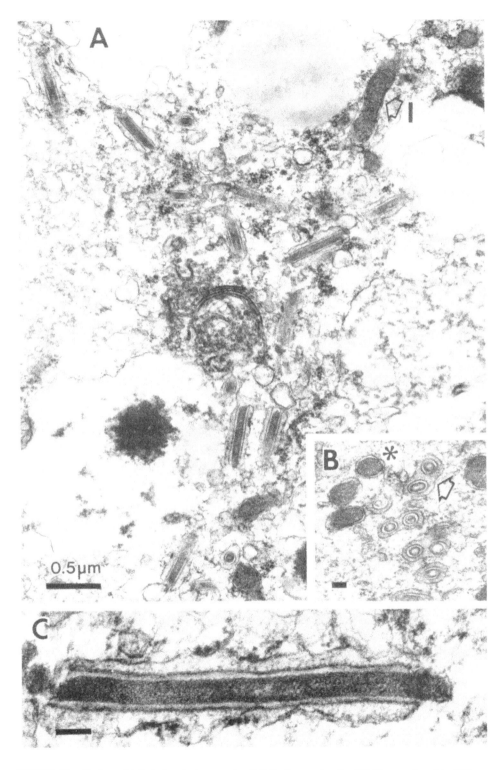

FIGURE 12. Section of infected maturing oocyte of *Epilachna varivestis*. (A) The enveloped bacilliform virus-like particles (VLP) contain what appears to be a hollow core. Note immature virus forms (I). (B) Cross sections of VLP. The three dense bodies (*) are immature forms while the hollow enveloped forms (arrow) are the mature forms. (Bar = 100 nm.) (C) Section of mature form of VLP. The nucleocapsid often protrudes from the outer envelope. The longest VLP measured 1.15 μm. ([C] from Adams, J. R., Tompkins, G. J., Wilcox, T. A., and Schroder, R. F., *J. Invertebr. Pathol.*, 33, 111, 1979. With permission.)

on this picorna-like virus isolated from *Triatoma infestans* indicated that the viral genome contains single-stranded RNA molecule with a relative molecular weight of 3×10^6 Da and polypeptides of 39 K, 33 K, and 45 K, supporting its proposed classification in the Picornaviridae.

Paracrystalline arrays of VLPs (about 45 to 50 nm in diameter) have been found in the cytoplasm of spermatocyst cells of *Heliothis virescens* while 70-nm particles occurred in the nuclei (see Figure 6A and B).[53] A number of unclassified viruses and VLPs have been reported from coleopteran insects.[22,29,61-65,73,79,90,92-97] Paracrystalline arrays of virus particles (about 24 to 26 nm in diameter) were observed in the cytoplasm of spermatocyst cells of *Diabrotica virgifera* (see Figure 7).[22] Observations on the Wassersucht disease of *Melolontha* sp. have revealed the presence of particles (60 to 75 nm in diameter) in the fat body and plasmatocytes.[62] Recently, reovirus-like particles approximately 70 nm in diameter have been found in the cytoplasm of hemocytes in the spotted cucumber beetle *Diabrotica undecimpunctata*[64] (see Figure 8) and the bean leaf beetle *Ceratoma trifurcata*[63] as well as in the cytoplasm and nuclei of many tissues in the Mexican bean beetle, *Epilachna varivestis* (see Figure 9).[65] The VLPs associated with the Mexican bean beetle appear to be a latent virus which also infects the host reproductive tissues and is transmitted vertically in infected beetle populations. However, no pathologies associated with this virus have been observed.

Morphogenesis of an unusual spherical VLP has been observed in the nuclei of calyx cells in the ichneumonid parasitoid *Nemeritis canescens*.[70] The particles, which are approximately 120 to 140 nm in diameter and consist of an electron-dense core surrounded by a unit membrane derived by *de novo* synthesis, appear to bud from one or more "dense bodies" in the nuclei of infected cells. Particles subsequently enter the lumen of the oviduct in large numbers, coat the parasitoid eggs and are injected into host *(Ephestia kuehniella)* insects during parasitization. The particulate "coat" surrounding the parasitoid eggs has been implicated in protecting the eggs from immune recognition in host larvae. Interestingly, preliminary histochemical tests have failed to detect the presence of nucleic acids in the particles.[71]

IV. BACILLIFORM VIRUSES AND VLPs

Most of the nonoccluded baculovirus-like viruses have been presented in Chapter 9. Other bacilliform VLPs are shown in Table 4. The earliest report of bacilliform VLPs in Coleoptera was in the whirligig beetle, *Gyrinus natator* L., in which midgut nuclei were filled with enveloped bacilliform particles, containing DNA, 35 nm by up to 150 nm.[73] The viral envelope also contained a distinctive spiny coat. The first morphological signs of infection were networks of endoplasmic reticulum from which the ribosomes were released; subsequently mature virions were produced from viral DNA fibrils and membranes which appeared in the nuclei of infected cells (see Figure 10).[73]

Small bacilliform VLPs were first observed in Mexican bean beetle adults during tests for possible transmission of *Autographa californica* nuclear polyhedrosis virus (NPV). The bacilliform VLPs (16 to 25 × 55 to 110 nm), containing what appeared to be a hollow core, occurred in vesicles in the cytoplasm of the anterior and posterior midgut columnar cells. These vesicles stained negatively with Feulgen and positively for RNA with the methyl green pyronin technique (see Figure 11).[79] Larger VLPs, (25 × 60 to 75 nm) also containing what appeared to be a hollow 25 nm core, have been observed in stunted developing oocytes beneath the endo- and exochorion (see Figure 12).[79]

Nonoccluded baculovirus-like particles have recently been described from larvae of the phantom midge, *Chaoborus crystallinus*.[89] The viral rods measured 75 to 106 × 260 to 300 nm, were surrounded by a unit membrane derived by *de novo* synthesis, and appeared to be arranged in paracrystalline arrays in the nuclei of infected cells. Particle morphogenesis

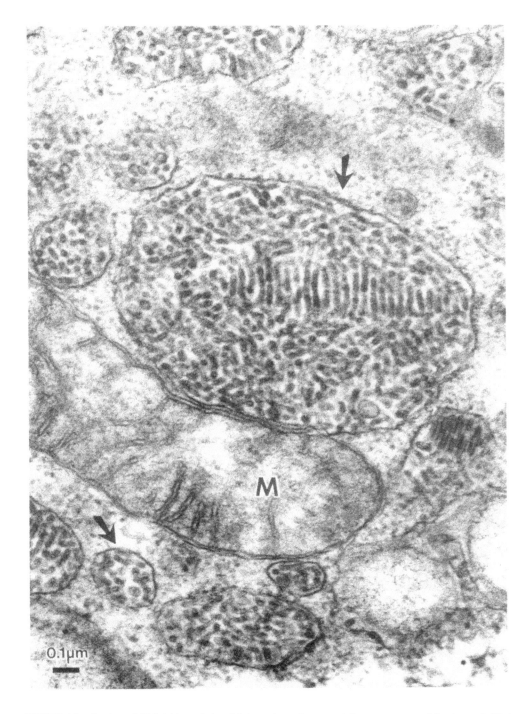

FIGURE 13. Section of Malpighian tubule of *Acheta domesticus* containing numerous vesicles (arrows) filled with virus-like particles. Note mitochondria (M).

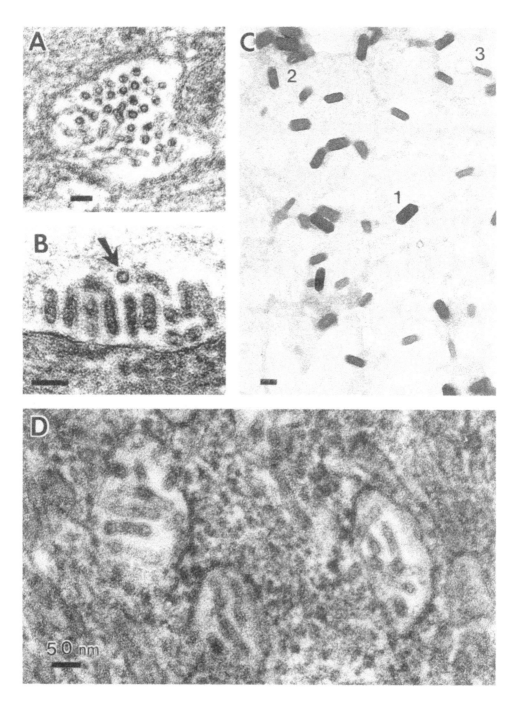

FIGURE 14.   (A) and (B) Sections of muscle tissue from *Acheta domesticus* containing a vesicle filled with virus-like particles (VLP). (A) Cross section of VLP; note what appears to be a hollow core (arrow). (B) Longitudinal sections of VLP and one VLP in cross section (arrow). (C) Purified isolated VLP from diseased crickets. There are three size ranges of VLP based on differences in diameters (1—3). The smallest VLPs (3) are comparable in size to the VLP shown in (A), (B), and (D). Note that the VLPs appear to be pointed on both ends. (D) Section of infected fat body tissue of *Acheta domesticus*. (Bar [A]—[C] = 50 nm.)

appeared to be limited to nuclei of the midgut epithelium with heavily infected midguts showing a yellow discoloration.

In addition to the cricket paralysis virus, which is discussed in Chapter 13, there are several reports of bacilliform VLPs which occasionally cause high mortalities in hatching and early instar cricket larvae in mass rearing facilities. Infection with these VLPs also appears to significantly reduce the life span of adult crickets.[75,76] Since these reports our laboratory has isolated similar viral particles from diseased crickets which were collected from Georgia, Louisiana, Arkansas, and North Carolina. This virus has also been found to cause disease in crickets in South Carolina and Louisiana by Carner[77] who found infection in the cytoplasm of midgut, trachea, and muscle cells. Diseased crickets exhibited symptoms of twitching and shaking of metathoracic legs. The smallest bacilliform cricket virus particles in infected tissues measured 16 to 24 × 25 to 124 nm and 16 to 18 × 25 to 40 nm and contained a hollow core. The small VLPs have been reported in the epithelial sheath of the cercal nerve, neural lamella,[74] and in vesicles in Malpighian tubules, tracheal matrix, fat body,[75,76] and myocardium.[50] Three sizes of viral particles were observed by electron microscopy following isolation and purification procedures modified from those reported previously (see Figures 13 and 14).[75,76] The size ranges of the smallest isolated particles (17 to 26 × 48 to 72 nm) were comparable to those observed most frequently in sectioned tissues. The largest isolated viral particles (53 to 60 × 132 to 139 nm) were not observed in the sectioned tissues. The VLPs found in Mexican bean beetle midguts (see Figure 11) were of similar morphology and dimensions as the smallest VLPs observed in diseased crickets (see Figures 13 and 14).

Another bacilliform virus with particle sizes of 80 to 100 × 145 to 240 nm has been isolated from crickets.[80] The single rod-shaped nucleocapsid is approximately 66 × 162 nm, contains DNA, and resembles baculoviruses (Group C).[12,80]

Virus-like rods (VLRs) (31 × 160 to 170 nm) associated with mycoplasma have been identified in mycoplasma-infected phloem cells of plants and in midgut cells of leafhoppers.[81] In sectioned tissues some VLRs were attached to the single trilaminar membrane of mycoplasmas while other tissues revealed aggregates of parallel rows of VLRs.[81]

A nonoccluded filamentous virus (CmFV) was found by Hammetal[100] which replicated in the epithelial tissues of the common and lateral oviducts and rarely in the small nests of cells in the calyx of *Cotesia marginiventris*. The enveloped nucleocapsids measured 60 to 70 nm in diameter for single-enveloped particles and 90 nm for double-enveloped particles while the nucleocapsids measured 35 to 39 nm in diameter. CmFV was transmitted to larvae of *Spodoptera frugiperda* during parasitization in addition to a polydna virus (PV) containing multiple nucleocapsids per envelope which replicated in the apical region of the calyx adjacent to the ovarioles.[100]

V. FLEXUOUS RODS

Flexuous rods and ovoid bodies have been found in nuclei of midgut epithelial cells of *Tenebrio molitor*.[92] The viral rods measure about 25 nm in diameter with lengths up to 0.5 mm; ovoid bodies measure approximately 200 × 450 nm. Further studies by other investigators of similar flexuous rods in *Tenebrio molitor* have revealed that the flexuous rods developed in crystal containing nuclei of the midgut epithelial cells and that the protein in the crystals is apparently not used in the synthesis of the virus-like particles (see Figure 15).[93-97] Autoradiographic studies indicate that the flexuous rods, measuring 25 nm in diameter, are enclosed in an envelope and contain DNA. Ovoid bodies also occurred in infected nuclei (see Figure 17).[93] Formations of flexuous rods and ovoid bodies have also been reported in *Scolytus scolytus* (see Figures 16 and 17). These resulted after inoculations of *Oryctes rhinoceros* baculovirus.[29] The role, if any, of the baculovirus inoculation in inducing formation of the flexuous VLPs was not determined. This finding is interesting

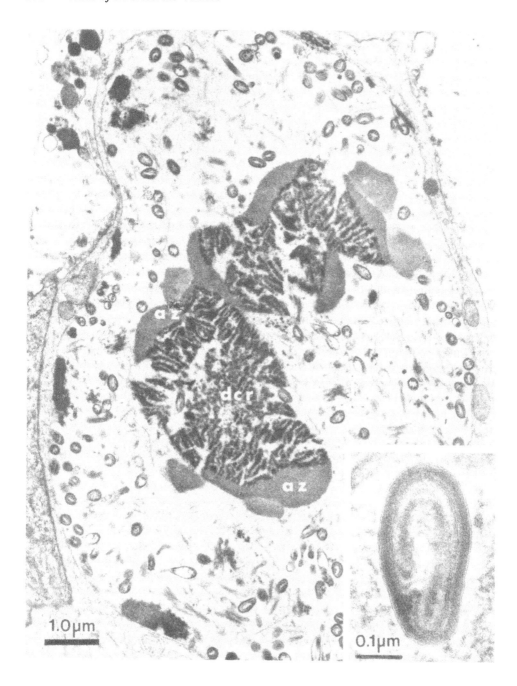

FIGURE 15.   Virus-like particles multiplying in the nucleus of a crystal containing midgut cell of *Tenebrio molitor*. The disintegrated crystal (dcr) is surrounded by a characteristic amorphous zone (az). (Inset) High magnification of VLP. (From Thomas, D. and Gouranton, J., *J. Invertebr. Pathol.*, 25, 159, 1975. With permission.)

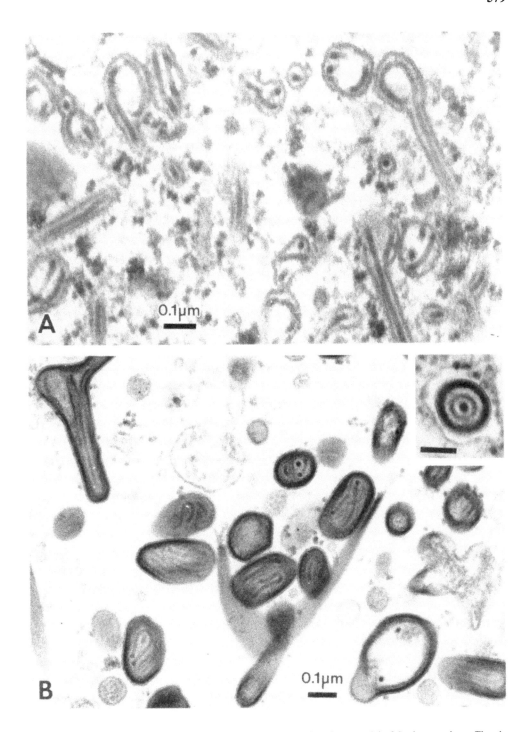

FIGURE 16. Virus-like particles (flexuous rods) multiplying in the midgut nuclei of *Scolytus scolytus*. The elm bark beetles were inoculated with virus extracted from infected *Oryctes rhinoceros*. (A) Flexuous rods looped within loosely fitting unit-membrane envelopes. (B) Ovoid bodies in longitudinal section. (Inset) Cross section of flexuous rod within its envelope. (Bar = 100 nm.) (From Arnold, M. K. and Barson, G., *J. Invertebr. Pathol.*, 29, 373, 1977. With permission.)

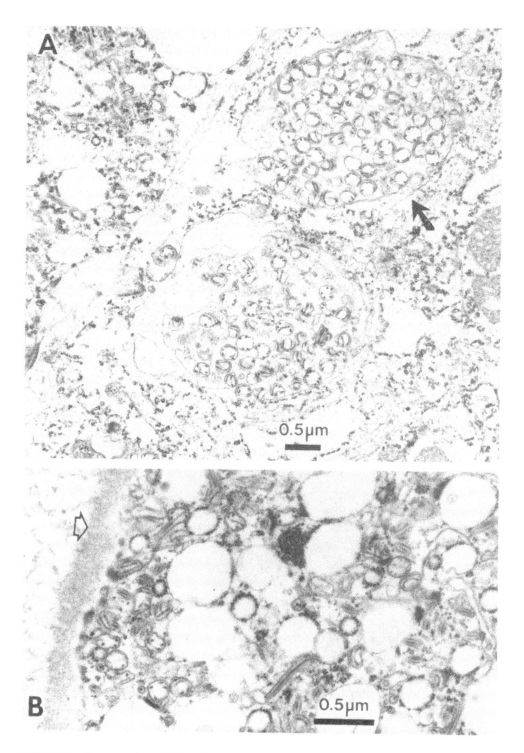

FIGURE 17.   Flexuous rods multiplying in the midgut cells of *Scolytus scolytus*. Beetles were inoculated with virus extracted from infected *Oryctes rhinoceros*. (A) Note vesicles containing enveloped rods in the cytoplasm. (B) Vacuolated region of cytoplasm containing enveloped rods near the basement membrane (arrow) of a columnar epithelial cell. (From Arnold, M. K. and Barson, G., *J. Invertebr. Pathol.*, 29, 373, 1977. With permission.)

in that it illustrates the problems of accurately identifying the causal agents of disease in insects which may have succumbed to the effects of multiple pathogens.

VI. FILAMENTOUS VIRUSES

There are many reports of viruses isolated from insect parasites that are injected in the act of stinging the insect host.[98,114-120] Most of these viruses, which are not filamentous in structure, have been grouped in the Polydnaviridae and are presented in detail in Chapter 10. In addition to polydnaviruses, however, a long filamentous virus has recently been reported in the ichneumonid wasp, *Diadegma terebrans*. This virus has been observed to replicate in large hypertrophied nuclei in the apical region of the female calyx.[9] The virions contained an electron-dense core 20 nm in diameter with an electron-dense outer layer making the virion diameter 30 nm with lengths up to 1 μm. Some of the virions in the cytoplasm were naked and others were surrounded by a typical trilaminar membrane giving the virions a total diameter of 65 nm. The mature virions occurred in areas of the endoplasmic reticulum that have been transformed into many vesicles. Similar long filamentous viral particles associated with polydnaviruses have been described in the braconid parasitoids *Cotesia congregata* and *Microplitis croceipes*.[99]

Filamentous viruses were discovered recently in five species of noctuid larvae parasitized by the braconid, *Cotesia marginiventris*, in tests with ascovirus.[100] No polydnavirus capsids or ascoviruses were observed in the filamentous virus-infected nuclei. The enveloped nucleocapsids measure 50 to 70 nm in diameter; virions with double envelopes measure 85 to 98 nm in diameter. The nucleocapsids are 30 to 39 nm in diameter with a core 21 to 24 nm in diameter by 865 to 890 nm long. These viruses replicate in hypertrophied nuclei of hypodermis and tracheal matrix cells and are also found in the cytoplasm of infected cells (see Figure 18).[100] The senior author has also observed a filamentous virus similar to that described by Styer et al.[100] associated with ascovirus in stunted *Heliothis virescens* larvae collected by E. A. Stadelbacher in Mississippi. A nonoccluded filamentous virus (CmFV) was found by Hamm et al.[101] which replicated in the epithelial tissues of the common and lateral oviducts and rarely in the small nests of cells in the calyx of *Cotesia marginiventris*. The enveloped nucleocapsids measured 60 to 70 nm in diameter for single-enveloped particles while the nucleocapsids measured 35 to 39 nm in diameter. CmFV was transmitted to larvae of *Spodoptera frugiperda* during parasitization in addition to a polydnavirus (PV) containing multiple nucleocapsids per envelope which replicated in the apical region of the calyx adjacent to the ovarioles.[101]

Unusual filamentous virus-like particles have been found associated with hyperplastic salivary glands and reproductive tract abnormalities in *Glossina pallidipes* Austen in Kenya (see Figure 19). Salivary glands of infected individuals were enlarged up to four times the normal diameter while viral infection in host gonadal tissues resulted in degeneration of the testes in males and ovarian pathologies in females.[82-87] Virions purified from these tissues measured 57 × 700 to 1300 nm and contained dsDNA which appeared to be linear from analysis by electrophoresis in agarose gels, ethidium bromide-cesium chloride gradient centrifugation and electron microscopy (see Figure 20). Restriction enzyme analyses have failed to provide a clear profile of DNA fragments indicating that the DNA may be heterogeneous in size.[88] The *G. pallidipes* VLPs could be transmitted p.o. or by intrahemocoelic injection.[86] Transovarial transmission seemed likely since crosses between infected males and females produced infected offspring with a shortened life span. Viral infection also produced a sex ratio distortion with crosses between infected individuals producing sterile, predominantly male offspring.[87] Additionally, studies on infected *G. pallidipes* and *G. m. morsitans*, which included electron microscopic examinations, revealed that 83% of the enlarged glands were also infected with rickettsia-like organisms.[121] Recent ultrastructural evidence has indicated

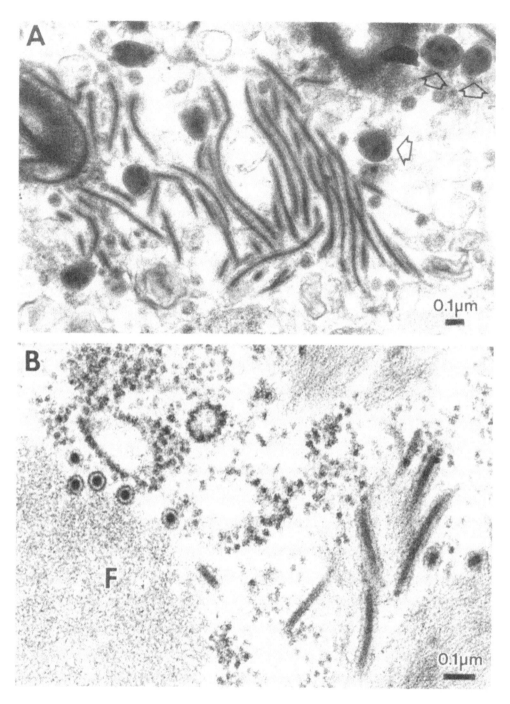

FIGURE 18.   (A) Section of common oviduct between epithelial cell and cuticle of *Cotesia marginiventris* showing filamentous virus and polydnavirions (arrows). (Courtesy of Dr. Eloise L. Styer.) (B) Section of hypodermal cell of *Heliothis zea* 3 d p.t. showing cross and longitudinal sections of *Cotesia marginiventris* filamentous virus; fibrous inclusions (F). (From Styer, E. L., Hamm, J. J., and Nordlund, D. A., *J. Invertebr. Pathol.*, 50, 302, 1987. With permission.)

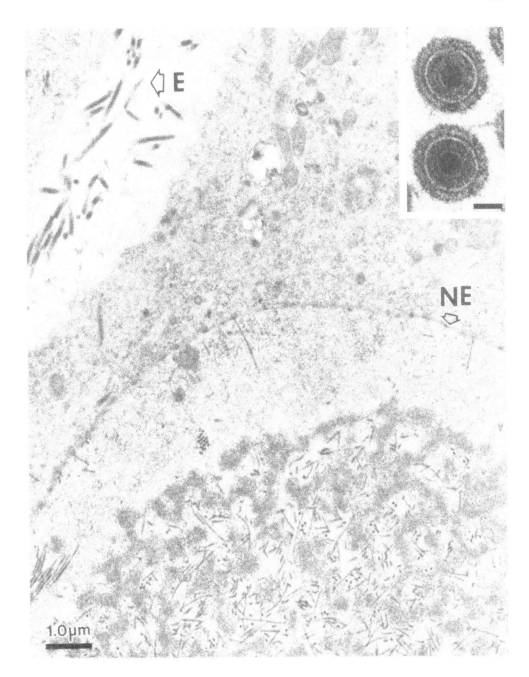

FIGURE 19. Section of hyperplastic salivary gland cells of *Glossina pallidipes* containing virus-like rods (VLR) in the nucleus within the virogenic stroma and near the nuclear envelope (E). Enveloped VLP (E) occur in the cytoplasm and intercellular spaces. (Inset) Cross section of VLP showing an enveloped nucleocapsid within an outer envelope. (Bar = 50 nm.) (From Jaenson, T. G. T., *Trans. R. Soc. Trop. Med. Hyg.*, 72, 234, 1978. With permission.)

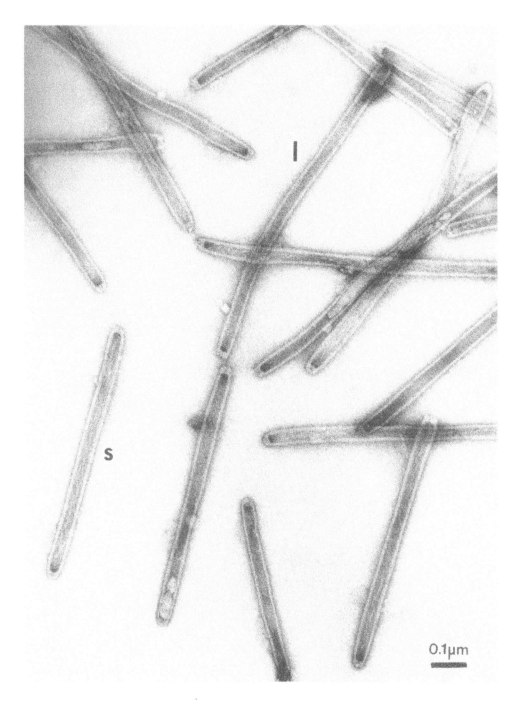

FIGURE 20.   Morphology of purified virus particles extracted from hypertrophied salivary glands of the tsetse fly, *Glossina pallidipes*. Measurement of virus particle size revealed that particle lengths fell broadly into two size classes. "Short" particles (S) averaged 869 ± 41 nm and "long" (l) particles averaged 1175 ± 48 nm. The mean particle width was 57 ± 7 nm for both particle types. The current knowledge of the morphology and biochemical properties of the virus suggest that it cannot be placed in any of the existing taxonomic groupings of DNA viruses. (From Odindo, M. O., Payne, C. C., Crook, N. E., and Jarrett, P., *J. Gen. Virol.*, 67, 527, 1986. With permission.)

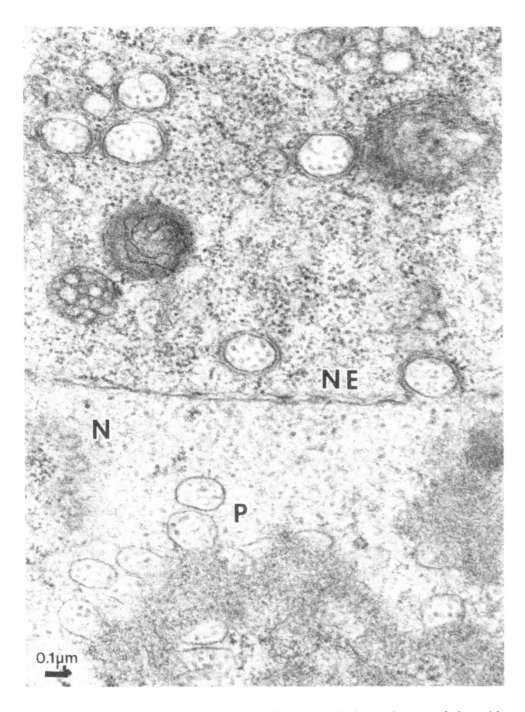

FIGURE 21.   Section of a calyx cell from an adult female *B. curculionis* showing morphogeneous of calyx particles (P) in the cell nucleus (N) and budding of particles through the nuclear envelope (NE). Particles lose any membranes acquired by budding through the nuclear envelope prior to entry into the calyx lumen.

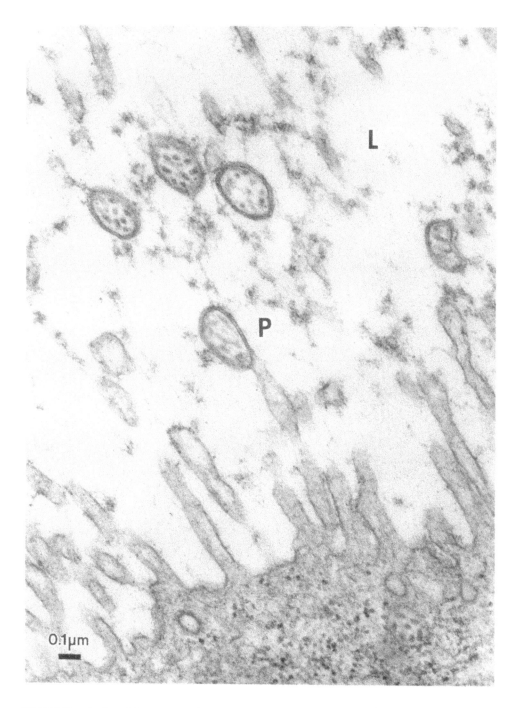

FIGURE 22.   Section of the periphery of a calyx epithelial cell from *B. curculionis* showing budding of a calyx particle (P) into the lumen (L) of the lateral oviduct. Particles acquire an envelope derived from the cell cytoplasmic membrane during this budding process.

that transorum transmission of this virus occurs since it was found in the germ cell cystocyte clusters, follicles, nurse cells, and oocyte cytoplasm of *G. pallidipes* females.[122]

VII. OTHER VIRUS-LIKE PARTICLES

An unusual VLP has been found associated with nuclei of the calyx epithelium in the alfalfa weevil parasitoids *Bathyplectes curculionis* and *B. anurus* (see Figures 21 and 22).[105,106] Each VLP appears to consist of a number of small cylindrical nucleocapsids (approximately 25 × 50 nm) enveloped in a single unit membrane. Several aspects of the morphogenesis of these particles were similar to that of ichneumonid polydnaviruses; however, unlike the polydnaviruses, the *B. curculionis* VLPs were not produced in large numbers and did not accumulate to form a "calyx fluid" in the lumen of the wasp oviduct.[107]

Various tissues of certain strains of the braconid endoparasitoid, *Cotesia melanoscela,* in addition to being infected with a polydnavirus, also support the replication of a second, as yet unclassified virus.[103,104] Nucleocapsids of the second virus, designated CmV2, are quasi-cylindrical in shape, approximately 100 × 800 nm in size, and are surrounded by two unit membranes. The viral genome appears to consist of a single dsDNA molecule of approximately 125 kilobase pairs (kbp). Both male and female *C. melanoscela* wasps are infected with CmV2; however, the viral DNA appears to be vertically transmitted through the maternal germ line. No pathologies associated with the CmV2 in adult wasps have been demonstrated.

Replication of CmV2, along with the *C. melanoscela* polydnavirus, also occurs in the nuclei of the female wasp calyx epithelial cells and the CmV2 is likewise injected into host insects during parasitization. The CmV2 has been shown to replicate primarily in nuclei of fat body and hemocytes from several hosts parasitized by *C. melanoscela,* although it is not pathogenic for these species. Successful replication of the CmV2 *in vitro* has been achieved using cell lines derived from *Lymantria dispar* and *Estigmene acrea.*[104]

Virus particles which have been described as "paramyxovirus-like" (PLPs) have been found in the delphacid planthopper, *Peregrinus maidus,* a vector of maize stripe virus.[102] If this indeed is a paramyxovirus, it is the first report of a virus in this family infecting insect tissues. The PLPs are quasi-spherical, bound by a single membrane and covered with surface projections; PLP diameters range from 112 to 160 nm but larger particles are as large as 363 nm in diameter. PLPs were found in the salivary glands, nerve ganglia, ovarioles, and leg muscles. In addition to the PLPs, the *P. maidus* were also infected with *Staphylococcus* and rickettsia-like organisms.[102]

Many diseased insects collected in the field are victims of multiple pathogens. Often unclassified viruses or VLPs, in combination with identified pathogens, result in increased mortality, stunting, or other abnormalities which effectively prevent the host insects from proceeding through their normal life cycles. Hopefully, some future efforts will be directed into more thoroughly characterizing these VLPs with the goal of utilizing them in effective insect control programs.

ADDENDUM

Further studies by Kokwaro et al.[123] on laboratory-reared male *Glossina morsitans morsitans* with hypertrophied salivary glands revealed an infection of rickettsia-like organisms in addition to the rod-shaped virus particles shown in Figures 19 to 20. A 60-nm diameter nonenveloped icosahedral virus was reported by Poinar et al.[124] in the grass grub, *Costelytra zealandica.* The virus occurred predominantly in the midgut region close to the midgut-foregut junction. Cytopathic effects were observed in the endoplasmic reticulum which became vesiculated.

Wechsler et al.[125] recently isolated a 30-nm diameter nonenveloped icosahedral dsRNA virus from cells (CuVa) cultured from one laboratory colony of the biting midge, *Culicoides variipennis*. Virions occurred singly and in crystalline arrays of the cytoplasm of infected cells.

ACKNOWLEDGMENTS

The authors gratefully acknowledge the contributions of electron micrographs from each contributor as noted in the figure legends, also including Darlene F. Hoffmann who made a special effort to assist us in providing the electron micrograph in Figure 2.

REFERENCES

1. **Longworth, J. F.,** Small isometric viruses of invertebrates, *Adv. Virus Res.,* 23, 103, 1978.
2. **Reavy, B. and Moore, N. F.,** The replication of small RNA-containing viruses of insects, *Microbiologica,* 5, 63, 1982.
3. **Moore, N. F. and Tinsley, T. W.,** The small RNA-viruses of insects, *Arch. Virol.,* 72, 229, 1982.
4. **Brun, G. and Plus, N.,** The viruses of *Drosophila,* in *The Biology and Genetics of Drosophila,* Ashburner, M. and Wright, T. R. F., Eds., Academic Press, New York, 625, 1980.
5. **Pudney, M., Newman, J. F. E., and Brown, F.,** Characterization of Kawino virus, an enterolike virus isolated from the mosquito *Mansonia uniformis* (Diptera:Culicidae), *J. Gen. Virol.,* 40, 433, 1978.
6. **Scott, M. P., Fostel, J. M., and Pardue, M. L.,** A new type of virus from cultured *Drosophila* cells: characterization and use in studies of the heat-shock response, *Cell,* 22, 929, 1980.
7. **Chao, Y. C., Scott, H. A., and Young, III, S. Y.,** An icosahedral RNA virus of the soybean looper *(Pseudoplusia includens), J. Gen. Virol.,* 64, 1835, 1983.
8. **Scotti, P. D., Gibbs, A. J., and Wrigley, N. G.,** Kelp fly virus, *J. Gen. Virol.,* 30, 1, 1976.
9. **Hillman, B., Morris, T. J., Kellen, W. R., Hoffman, D., and Schlegel, D. E.,** An invertebrate calicilike virus: evidence of partial virion disintegration in host excreta, *J. Gen. Virol.,* 60, 115, 1982.
10. **Mialhe, E., Croizier, G., Veyrunes, J. C., Quiot, J. M., and Rieb, J. P.,** Study of a virus isolated from a natural population of *Culicoides sp.* (Diptera:Ceratopogonidae), *Ann. Virol.,* 134, 73, 1983.
11. **Gildow, F. E. and D'Arcy, C. J.,** Cytopathology and experimental host range of *Rhoplosiphum padi* virus, a small insometric RNA virus infecting cereal grain aphids, *J. Invertebr. Pathol.,* 55, 245, 1990.
12. **Matthews, R. E. F.,** Classification and nomenclature of viruses. Fourth Report of the International Committee on Taxonomy of Viruses, *Intervirology,* 17, 1, 1982.
13. **D'Arcy, C. J., Burnett, P. A., Hewings, A. D., and Goodman, R. M.,** Purification and characterization of a virus from aphid *Rhopalosiphum padi, Virology,* 112, 346, 1981.
14. **Gildow, F. E. and D'Arcy, C. J.,** Barley and oats as reservoirs of an aphid virus and the influence on barley yellow dwarf transmission, *Phytopathology,* 78, 811, 1988.
15. **Jutila, J. W., Henry, J. E., Anacker, R. I., and Brown, W. R.,** Some properties of a crystalline-array virus (CAV) isolated from the grasshopper *Melanoplus bivittatus* (Say) (Orthoptera:Acrididae), *J. Invertebr. Pathol.,* 15, 225, 1970.
16. **Henry, J. E.,** Development of the crystalline-array virus (CAV) in cultures of dorsal vessels from *Schistocerca americana* (Orthopera:Acrididae), *J. Invertebr. Pathol.,* 19, 325, 1972.
17. **Henry, J. E. and Oma, E. A.,** Ultrastructure of the replication of the grasshopper crystalline array virus in *Schistocerca americana* compared with other picornaviruses, *J. Invertebr. Pathol.,* 21, 273, 1973.
18. **Brzostowski, H. W. and Grace, T. D. C.,** Observations of the infectivity of RNA from *Antheraea* virus (AV), *J. Invertebr. Pathol.,* 16, 277, 1970.
19. **Bellett, A. J. D., Fenner, F., and Gibbs, A. J.,** The viruses, in *Viruses and Invertebrates,* Gibbs, A. J., Ed., North-Holland, Amsterdam, 41, 1973.
20. **Dolder, H. and Mello, M. L. S.,** Virus-like particles in the Malpighian tubes of blood-sucking hemipterans, *Cell. Mol. Biol.,* 23, 299, 1979.
21. **Mello, M. L. S., Rodrigues, V. L. C. C., and Dolder, H.,** Viral infection in bloodsucking hemipterans, *J. Submicroscopic. Cytol.,* 12, 85, 1980.
22. **Degrugillier, M. E., Degrugillier, S. S., and Jackson, J. J.,** Nonoccluded cytoplasmic virus particles and rickettsia-like organisms in testes and spermathecae of *Diabrotica virgifera, J. Invertebr. Pathol.,* 57, 50, 1991.

23. **Kellen, W. R. and Hoffman, D. F.**, A pathogenic nonoccluded virus in hemocytes of the navel orangeworm, *Amyelois transitella* (Pyralidae:Lepidoptera), *J. Invertebr. Pathol.*, 38, 52, 1981.

24. **Kellen, W. R. and Hoffman, D. F.**, Dose-mortality and stunted growth responses of larvae of the navel orangeworm *Amyelois transitella* infected by chronic stunt virus, *Environ. Entomol.*, 11, 214, 1982.

25. **Kellen, W. R. and Hoffman, D. F.**, Longevity and fecundity of adult *Amyelois transitella* (Lepidoptera:Pyralidae) infected by two small RNA viruses, *Environ. Entomol.*, 12, 1542, 1983.

26. **Hoffman, D. F. and Hillman, B.**, Observations on the comparative pathogenicity of intact and degraded forms of a calicivirus of *Amyelois transitella*, *J. Invertebr. Pathol.*, 43, 422, 1984.

27. **David, W. A. L.**, A noninclusion virus of blowfly larvae, *J. Invertebr. Pathol.*, 13, 310, 1969.

28. **Plus, N., Veyrunes, J. C., Cavallow, K.**, Endogenous viruses of *Ceratitus capitata* Wied "J. R. C. Ispar" strain, and of *C. capitata* permanent cell lines, *Ann. Virol.*, 132, 91, 1981.

29. **Arnold, M. K. and Barson, G.**, Occurrence of virus-like particles in midgut epithelial cells of the large elm bark beetle, *Scolytus scolytus*, *J. Invertebr. Pathol.*, 29, 373, 1977.

30. **Gibbs, A. J., Gay, F. J., and Wetherly, A. H.**, A possible paralysis virus of termites, *Virology*, 40, 1063, 1970.

31. **Muscio, O. A., Latorre, J. L., and Schodeller, E. A.**, Small nonoccluded viruses from triatomine bug *Triatoma infestans* (Hemiptera:Reduviidae), *J. Invertebr. Pathol.*, 49, 218, 1987.

32. **Clark, T. B. and O'Grady, J. J.**, Nonoccluded viruslike particles in larvae of *Culicoides cavaticus* (Diptera:Ceratopogonidae), *J. Invertebr. Pathol.*, 26, 415, 1975.

33. **Gross, R. H., Hormann, R. E., and Saxe, J. C.**, Purification and partial characterization of virus-like particles from Schneider line 2 *Drosophila* cells, *Arch. Biochem. Biophys.*, 207, 455, 1981.

34. **Stiles, J. K., Molyneux, D. H., and Wallbanks, K. R.**, Viruslike particles in *Glossina palpalis gambiensis* (Diptera:Glossinidae), *Ann. Soc. Belge Méd. Trop.*, 68, 161, 1988.

35. **Richardson, J., Sylvester, E. S., Reeves, W. C., and Hardy, J. L.**, Evidence of two inapparent nonoccluded viral infections of *Culex tarsalis*, *J. Invertebr. Pathol.*, 23, 213, 1974.

36. **Jenni, L. and Steiger, R.**, Virus-like particles of *Glossina fuscipes fuscipes*, *Acta Trop.*, 31, 177, 1974.

37. **Harrap, K. A., Longworth, J. F., Tinsley, T. W., and Brown, K. W.**, A noninclusion virus of *Gonometa podocarpi* (Lepidoptera:Lasiocampidae), *J. Invertebr. Pathol.*, 8, 270, 1966.

38. **Charpentier, G., Back, C., Garzon, S., and Strykowski, H.**, Observations on a new intranuclear virus-like particle infecting larvae of a black fly *Simulium vittatum* (Diptera:Simuliidae), *Dis. Aquatic Organisms*, 1, 147, 1986.

39. **Mello, M. L. S.**, Viral particles in the salivary glands of *Triatoma infestans.*, in *Proc. Seventh Eur. Congr. on Electron Microsc.*, Vol. 2, Brederoo, P. and dePriester, W., Eds., North-Holland, Amsterdam, 1980, 504.

40. **Mello, M. L. S. and Dias, C. A.**, Detection of viral particles in the salivary glands on the hemipteran *Triatoma infestans* Klug, *Cell. Mol. Biol.*, 27, 459, 1982.

41. **Peters, D. and Staal, G. B.**, Viruslike particles from *Hyalophora cecropia* larvae succumbing from an intestinal disease, *J. Invertebr. Pathol.*, 11, 330, 1968.

42. **Longworth, J. F. and Harrap, K. A.**, A nonoccluded virus isolated from four saturniid species, *J. Invertebr. Pathol.*, 10, 139, 1968.

43. **Akai, H., Gateff, E., Davis, L. E., and Schneiderman, H. A.**, Virus-like particles in normal and tumorous tissues of *Drosophila*, *Science*, 157, 810, 1967.

44. **Philpott, D. E., Weibel, H., Atlan, H., and Miquel, J.**, Viruslike particles in the fat body, oenocytes, and central nervous tissue of *Drosophila melanogaster* imagoes, *J. Invertebr. Pathol.*, 14, 31, 1969.

45. **Herman, M. M., Johnson, M., and Miquel, J.**, Viruslike particles and related filaments in neurons and glia of *Drosophila melanogaster* brain, *J. Invertebr. Pathol.*, 17, 442, 1971.

46. **Rae, P. M. M. and Green, M. M.**, Viruslike particles in adult *Drosophila melanogaster*, *Virology*, 34, 187, 1968.

47. **Binnington, K. C., Lockie, E., Hines, E., and Van Gerwen, A. C. M.**, Fine structure and distribution of three types of virus-like particles in the sheep blowfly, *Lucilia cuprina* and associated cytopathic effects, *J. Invertebr. Pathol.*, 49, 175, 1987.

48. **Miquel, J., Bensch, K. G., and Philpott, D.**, Viruslike particles in the tissues of normal and γ irradiated *Drosophila melanogaster*, *J. Invertebr. Pathol.*, 19, 156, 1972.

49. **Filshie, B. K., Grace, T. D. C., Poulson, D. F., and Rehacek, J.**, Viruslike particles in insect cells of three types, *J. Invertebr. Pathol.*, 9, 271, 1967.

50. **Tadkowski, T. M. and McCann, F. V.**, Ultrastructure of two viruslike particles in the myocardium of *Acheta domesticus*, *J. Invertebr. Pathol.*, 36, 119, 1980.

51. **Jenni, L.**, Viruslike particles in a strain of *Glossina morsitans centralis* Machado 1970, *Trans. R. Soc. Trop. Med. Hyg.*, 67, 295, 1973.

52. **Jenni, L. and Steiger, R. F.**, Viruslike particles in the tsetse fly *Glossina morsitans* spp., *Rev. Suisse Zool.*, 81, 663, 1986.

53. **Degrugillier, M. E.,** Virus-like particles in testes of *Heliothis virescens, H. subflexa,* and backcross males derived from hybridization of *H. virescens* males × *H. subleexa* females, *J. Invertebr. Pathol.,* 54, 281, 1989.

54. **Felluga, G., Jonsson, V., and Liljeros, M. R.,** Ultrastructure of new viruslike particles in *Drosophila, J. Invertebr. Pathol.,* 17, 339, 1971.

55. **Saigo, K., Shiba, T., and Miyake, T.,** Virus-like particles of *Drosophila melanogaster* containing t-RNA and 5 S ribosomal RNA. I. Isolation and purification from cultured cells and detection of low molecular weight RNAs in the particles, in *Invertebrate Systems in Vitro,* Kurstak, E., Maramorosch, K., and Dubendorfer, A., Eds., Elsevier/North-Holland, Amsterdam, 1980, 411.

56. **Shiba, T., Saigo, K., and Miyake, T.,** Virus-like particles of *Drosophila melanogaster* containing t-RNA and 5 S ribosomal RNA. II. Isolation and characterization of low molecular weight RNAs, in *Invertebrate Systems in Vitro,* Kurstak, E., Maramorosch, K., and Dubendorfer, A., Eds., Elsevier/North-Holland, Amsterdam, 1980, 425.

57. **Wehman, H. J. and Brager, M.,** Viruslike particles in *Drosophila:* constant appearance in imaginal discs in vitro, *J. Invertebr. Pathol.,* 18, 127, 1971.

58. **Maillet, P. L. and Folliot, R.,** Sur la presence d'um virus dans le testicule chez un insecte homoptère, *C. R. Acad. Sci., Paris, Ser. D.,* 264, 2828, 1967.

59. **Grace, T. D. C. and Mercer, E. H.,** A new virus of the saturniid *Antheraea eucalypti,* Scott, *J. Invertebr. Pathol.,* 7, 241, 1965.

60. **Frederici, B. A., Granados, R. R., Anthony, D. W., and Hazard, E. I.,** An entomopox virus and nonoccluded virus-like particles in larvae of the chirmonomid *Goeldrichironomus holoprasinus, J. Invertebr. Pathol.,* 23, 117, 1974.

61. **Zeikus, R. D. and Steinhaus, E. A.,** Teratology of the beetle *Tenebrio molitor.* V. Ultrastructural changes and viruslike particles in the foregut epithelium of pupal-winged adults, *J. Invertebr. Pathol.,* 14, 115, 1969.

62. **Krieg, A. and Huger, A.,** A virus disease of the coleopterous insects, *J. Insect Pathol.,* 2, 274, 1960.

63. **Kim, K. S. and Scott, H. P.,** Two types of virus-like particles in the bean leaf beetle, *Cerotoma trifurcata, J. Invertebr. Pathol.,* 31, 77, 1978.

64. **Kim, K. S.,** Cytopathology of spotted cucumber beetle hemocytes containing virus-like particles, *J. Invertebr. Pathol.,* 36, 292, 1980.

65. **Kitajima, E. W., Kim, K. S., Scott, H. A., and Gergerich, R. C.,** Reovirus-like particles and their vertical transmission in the mexican bean beetle, *Epilachna varivestis* (Coleoptera:Coccinellidae), *J. Invertebr. Pathol.,* 46, 83, 1985.

66. **Igarashi, A., Sasao, F., Fukai, K., and Wada, T.,** An enveloped virus isolated from field-caught *Culex tritaeniorhynchus* by *Aedes albopictus* clone c6/36 cell cultures causing extensive cytopathic effect, *Trop. Med.,* 28, 209, 1986.

67. **Tandler, B.,** Viruslike particles in testis of *Drosophila virilis, J. Invertebr. Pathol.,* 20, 214, 1972.

68. **Gartner, L. P.,** Viruslike particles in the adult *Drosophila* midgut, *J. Invertebr. Pathol.,* 20, 364, 1972.

69. **Moussa, A. Y.,** A new cytoplasmic inclusion virus from Diptera in the Queensland fruitfly, *Dacus tryoni* (Frogg), *J. Invertebr. Pathol.,* 32, 77, 1978.

70. **Rotheram, S.,** The surface of the egg of a parasitic insect. II. The ultrastructure of the particulate coat on the egg of *Nemeritis, Proc. R. Soc. London, Ser. B,* 183, 195, 1973.

71. **Bedwin, O.,** An insect glycoprotein: a study of the particles responsible for the resistance of a parasitoid's egg to the defense reactions of its insect host, *Proc. R. Soc. London, Ser. B,* 205, 271, 1979.

72. **Louis, C., Fopez-Ferber, M., Comendados, M., Plus, N., Kuhl, G., and Baker, S.,** *Drosophila* S virus, a hereditary reolike virus, probable agent of the morphological S character in *Drosophila simulans, J. Virol.,* 62, 1266, 1988.

73. **Gouranton, J.,** Development of an intranuclear nonoccluded rod-shaped virus in some midgut cells of an adult insect, *Gyrinus natator* L. (Coleoptera), *J. Ultrastruct. Res.,* 39, 281, 1972.

74. **Morrissey, R. E. and Edwards, J. S.,** Viruslike particles in peripheral nerve of the house cricket, *J. Invertebr. Pathol.,* 29, 247, 1977.

75. **Adams, J. R., Dougherty, E. M., and Heimpel, A. M.,** Rhabdovirus-like particles isolated from crickets: isolation, pathogenesis and characterization, *4th Int. Congr. Virol.,* The Hague, 573, 1978.

76. **Adams, J. R., Dougherty, E. M., and Heimpel, A. M.,** Pathogenesis of rhabdovirus like particles in the house cricket, *Acheta domesticus, J. Invertebr. Pathol.,* 35, 314, 1980.

77. **Carner, G. R.,** personal communication, 1978, 1986.

78. **Amargier, A., Lyon, J. P., Vago, C., Meynadier, G., and Veyrunes, J. C.,** Discovery and purification of a virus in gland hyperplasia of insects: a study on *Merodon equestris* F. (Dipera, Syrphidae), *C. R. Acad. Sci. Paris,* 289, 481, 1980.

79. **Adams, J. R., Tompkins, G. J., Wilcox, T. A., and Schroder, R. F.,** Two virus-like particles in the mexican bean beetle, *Epilachna varivestis, J. Invertebr. Pathol.,* 33, 111, 1979.

80. **Huger, A. M.**, A new virus disease of crickets (Orthoptera:Gryllidae) causing macronucleosis of fat body, *J. Invertebr. Pathol.*, 45, 108, 1985.
81. **Giannotti, J., Devauchelle, G., Vago, C., and Marchoux, G.**, Rod-shaped virus-like particles associated with degenerating mycoplasma in plant and insect vector, *Ann. Phytopathol.*, 5, 461, 1973.
82. **Jaenson, T. G. T.**, Virus-like rods associated with salivary gland hyperplasia in tsetse *Glossina pallidipes*, *Trans. R. Soc. Trop. Med. Hyg.*, 72, 234, 1978.
83. **Otieno, L. H., Kokwaro, E. D., Chimtawi, M., and Onyango, P.**, Prevalence of enlarged salivary glands in wild populations of *Glossina pallidipes* in Kenya with a note on the ultrastructure of the affected organ, *J. Invertebr. Pathol.*, 36, 113, 1980.
84. **Odindo, M. O.**, Incidence of salivary gland hypertrophy in field populations of the tsetse *Glossina pallidipes* on the south Kenyan coast, *Insect Sci. Appl.*, 3, 59, 1982.
85. **Odindo, M. O. and Amutalla, P. A.**, Distribution pattern of the virus of *Glossina pallididpes* Austen in a forest ecosystem, *Insect Sci. Appl.*, 7, 79, 1980.
86. **Odindo, M. O., Sabwa, D. M., and Amutalla, P. A.**, Preliminary test on the transmission of virus-like particles to the tsetse *Glossina pallidipes*, *Insect Sci. Appl.*, 2, 219, 1981.
87. **Jaenson, T. G. T.**, Sex ratio distortion and reduced lifespan of *Glossina pallidipes* infected with the virus causing salivary gland hyperplasia, *Entomol. Exp. Appl.*, 41, 265, 1986.
88. **Odindo, M. O. and Payne, C. C.**, Properties of a novel DNA virus from the tsetse fly *Glossina pallidipes*, *J. Gen. Virol.*, 67, 527, 1986.
89. **Larsson, R.**, Baculovirus-like particles in the midgut epithelium of the phantom midge, *Chaoborus crystallinus* (Diptera, Chaoborideae), *J. Invertebr. Pathol.*, 44, 178, 1984.
90. **Kim, K. S. and Kitajima, E. W.**, Nonoccluded baculovirus and filamentous virus-like particles in the spotted cucumber beetle, *Diabrotica undecimpunctata* (Coleoptera:Chrysomelid), *J. Invertebr. Pathol.*, 43, 234, 1984.
91. **Lawrence, P. O. and Akin, D.**, Virus-like particles from the poison glands of the parasitic wasp *Menoptera braconidae*, *Can. J. Zool.*, 68, 539, 1990.
92. **Devauchelle, G. and Vago, C.**, Presence de particules d'allure virale dans les noyaux des cellules de l'intestin moyendu coléoptère *Tenebrio molitor* (Linné), *C. R. Acad. Sci. Paris*, 269, 1142, 1969.
93. **Thomas, D. and Gouranton, J.**, Development of viruslike particles in the crystal-containing nuclei of the midgut cells of *Tenebrio molitor*, *J. Invertebr. Pathol.*, 25, 159, 1975.
94. **Devauchelle, G.**, Inclusions cristallines et particules d'allure viral dans les noyaux des cellules de l'intestin moyen du coléoptère *Tenebrio molitor* (L.), *J. Ultrastruct. Res.*, 33, 263, 1970.
95. **Thomas, D. and Gouranton, J.**, Isolement des cristaux intranucléaires de l'intestin moyen du ver de farine *Tenebrio molitor* L. et observations au microscope à balayage, *J. Microsc. (Paris)*, 14, 125, 1972.
96. **Thomas, D. and Gouranton, J.**, Durée de formation des cristaux protéiques intranucléaires de l'intestin moyen de *Tenebrio molitor*, *J. Insect Physiol.*, 19, 515, 1973.
97. **Thomas, D. and Gouranton, J.**, Etude au moyen de précurseurs marqués, de la synthèse de cristaux proteiques intranucléaires chez *Tenebrio molitor* L., *J. Microsc. (Paris)*, 16, 287, 1973.
98. **Stoltz, D. B. and Vinson, S. B.**, Viruses and parasitism in insects, *Adv. Virus Res.*, 24, 125, 1979.
99. **Krell, P. J.**, Replication of long virus-like particles in the reproductive tract of the ichneumonid wasp *Diadegma terebrans*, *J. Gen. Virol.*, 68, 1477, 1987.
100. **Styer, E. L., Hamm, J. J., and Nordlund, D. A.**, A new virus associated with the parasitoid *Cotesia marginiventris* (Hymenoptera:Braconidae): replication in noctuid host larvae, *J. Invertebr. Pathol.*, 50, 302, 1987.
101. **Hamm, J. J., Styer, E. L., and Lewis, W. J.**, Comparative virogenesis of filamentous virus and polydnavirus in the female reproductive tract of *Cotesia marginiventris* (Hymenoptera:Braconidae), *J. Invertebr. Pathol.*, 55, 357, 1990.
102. **Ammar, E. D., Nault, L. R., Styer, W. E., and Saif, Y. M.**, Staphylococcus, paramyxovirus-like, rickettsia-like and other structures in *Peregrinus maidis* (Homoptera, Delphacidae), *J. Invertebr. Pathol.*, 49, 209, 1987.
103. **Stoltz, D. B. and Faulkner, G.**, Apparent replication of an unusual virus-like particle in both a parsitoid wasp and its host, *Can. J. Microbiol.*, 24, 1509, 1978.
104. **Stoltz, D. B., Krell, P. J., Cook, D., MacKinnon, E. A., and Lucarotti, D. J.**, An unusual virus from the parasitic wasp *Cotesia melanoscela*, *Virology*, 162, 311, 1988.
105. **Hess, R. T., Poinar, Jr., G. O., Etzel, L., and Merritt, C. C.**, Calyx particle morphology of *Bathyplectes anurus* and *Bathyplectes curculionis* (Hymenoptera:Ichneumonidae), *Acta Zool. (Stockholm)*, 61, 111, 1980.
106. **Stoltz, D. B., Krell, P. J., and Vinson, S. B.**, Polydisperse viral DNA's in ichneumonid ovaries: a survey, *Can. J. Microbiol.*, 27, 123, 1981.
107. **Guzo, D.**, unpublished observations, 1982.
108. **Longworth, J. F.**, personal communication, 1988.
109. **Binnington, K. C.**, Structural transformation of blowfly mitochondria by a putative virus: similarities with virus-induced changes in plant mitochondria, *J. Gen. Virol.*, 68, 201, 1987.

110. **Shiba, T. and Saigo, K.,** Retrovirus-like particles containing RNA homologous to the transposable element *copia* in *Drosophila melananogaster, Nature (London),* 302, 119, 1983.

111. **Emori, Y., Shiba, T., Kanaya, S., Inouye, S., Yaki, S., and Saigo, K.,** The nucleotide sequences of *copia* and *copia*-related RNA in *Drosophilla* virus-like particles, *Nature (London),* 315, 773, 1985.

112. **Kikuchi, Y., Yumiko, A., and Shiba, T.,** Unusual printing mechanism of RNA directed DNA synthesis in *copia* retrovirus-like particles of *Drosophila, Nature (London),* 323, 824, 1986.

113. **Muscio, O. A., LaTorre, J. L., and Scodeller, E. A.,** Characterization of triatoma virus, a picorna-like virus isolated from the triatomine bug *Triatoma infestans, J. Gen. Virol.,* 69, 2929, 1988.

114. **Stoltz, D. B., Vinson, S. B., and MacKinnon, E. A.,** Baculovirus-like particles in the reproductive tracts of female parasitoid wasps, *Canadian J. Microbiol.,* 22, 1013, 1976.

115. **Stoltz, D. B. and Vinson, B. S.,** Baculovirus-like particles in the reproductive tracts of female parasitoid wasps. II. The genus *Apanteles, Can. J. Microbiol.,* 23, 23, 1977.

116. **Krell, P. J. and Stoltz, D. B.,** Unusual baculovirus of the parasitoid wasp *Apanteles melanocelus:* isolation and preliminary characterization, *J. Virol.,* 29, 1118, 1979.

117. **Krell, P. J. and Stoltz, D. B.,** Virus-like particles in the ovary of an ichneumonid wasp: purification and preliminary characterization, *Virology,* 101, 408, 1980.

118. **Stoltz, B. D.,** A putative baculovirus in the ichneumonid parasitoid *Mesoleius tenthredinis, Can. J. Microbiol.,* 27, 116, 1981.

119. **Krell, P. J. and Summers, M. D.,** Virus with a multipartite superhelical DNA genome from the ichneumonid parasitoid *Campoletis sonorensis, J. Virol.,* 43, 859, 1982.

120. **Stoltz, D. B., Guzo, D., and Cook, D.,** Studies on polydnavirus transmission, *Virology,* 15, 120, 1986.

121. **Ellis, D. S. and Maudlin, I.,** Salivary gland hyperplasia in wild caught tsetse from Zimbabwe, *Entomol. Exp. Appl.,* 45, 167, 1987.

122. **Jura, W. G. Z. O., Otieno, L. H., and Chimtawi, M. M. B.,** Ultrastructural evidence for trans-ovum transmission of the DNA virus of tsetse, *Glossina pallidipes* (Diptera:Glossinidae), *Curr. Microbiol.,* 18, 1, 1989.

123. **Kokwaro, E. D., Nyindo, M., and Chimtawi, M.,** Ultrastructural changes in salivary glands of tsetse, *Glossina morsitans morsitans,* infected with virus and rickettsia-like organisms, *J. Invertebr. Pathol.,* 56, 337, 1990.

124. **Poinar, R. H., Poinar, G. O., and Jackson, T.,** Electron microscope evidence of a new virus in the grass grub, *Costelytra zealandica* (Coleoptera: Scarabaeicae), *J. Invertebr. Pathol.,* 57, 137, 1991.

125. **Wechsler, S. J., McHolland, L. E., and Wilson, W. C.,** A RNA virus in cells from *Culicoides variipennis, J. Invertebr. Pathol.,* 57, 200, 1991.

Chapter 23

UNCLASSIFIED VIRUSES OF MOLLUSCA

Michel Comps and Jean-Robert Bonami

TABLE OF CONTENTS

In addition to the viruses causing epizootics, such as the iridoviruses, agents of the gill disease, and mass mortality of the Portugese oyster *(Crassostrea angulata),*[1] some viruses, or lesions suspected to be viral, have been observed, often fortuitously, on several marine molluscans. Their taxonomic position is undefined or uncertain.

I. REOVIRUS TYPE

An icosahedral enveloped virus, 78 nm in size, is described in *Sepia officinalis.* Virions develop at the periphery of electron-dense virogenic area and form paracrystalline arrays; tubular structures may be associated with the viral lesion. It has some of the characteristics of a reovirus.[2]

In this group, it is fitting to cite the reovirus 13p2 isolated from the oyster, *C. virginica.*[3] It was demonstrated, however, that this virus does not induce real pathogenic effects in the oyster.[4]

The properties of these viruses are reported in Chapter 15.

II. HERPESVIRUS TYPE

The first virus known in a marine bivalve is a herpes-type virus, found in oysters *(Crassostrea virginica),* subjected to experimental conditions of high temperature[5] (see Figure 1 C and D). The virus replicates in intranuclear lesions which are Feulgen positive. The nonenveloped particles are icosahedral and 90 nm in size, and possess an eccentrically located nucleoid; some virions exhibit filaments at their periphery. Enveloped particles measure 200 × 250 nm. Tubules, 45 to 55 nm in diameter, are also found in the intranuclear inclusions.

An epizootiological study has shown a correlation between high mortality and high prevalence of the viral infection. According to Farley,[6] *Ostrea edulis* and *Mercenaria merceneria* may be also affected by herpes-type viruses.

III. RETROVIRUS TYPE

By its morphological and physical characteristics, the virus, causative agent of hemic neoplasia in the soft clam *(Mya arenaria),* shows similarities to the B-type retrovirus[7] (see Figure 1 E and F). The isolated viral particles are pleomorphic and 120 nm in size; enveloped, they contain an excentric, 60- to 70-nm nucleoid. The purified virions exhibit a density ranging from 1.17 to 1.18 g/cm³. The absorbance ratio 260/280 nm is 1.23.

Experimental induction of neoplasia was obtained by inoculation with purified virions, isolated from neoplastic tissue.

Similar hemic neoplasia occurring in the bay mussel, *Mytilus edulis,* have been reported by Elston and Kent,[8] but the causes of the presumed viral disease are not yet established.

Finally, virus-like particles, resembling C-type particles from baboon and human placenta, have been described by Farley[6] in digestive cells of *Crassostrea virginica.* The virions, 100 to 110 nm, are enveloped with an electron-lucent center; peplomer-like structures are also mentioned by the author.[9]

IV. OTHER VIRUSES

From histological examinations (light and electron microscopy), several cases of viral infections have been reported in various species of marine bivalves.

An enveloped DNA virus (53 nm in size), considered as a possible member of the Papovaviridae family, was described by Farley[10,11] in gametogenic epithelium of *C. virginica*

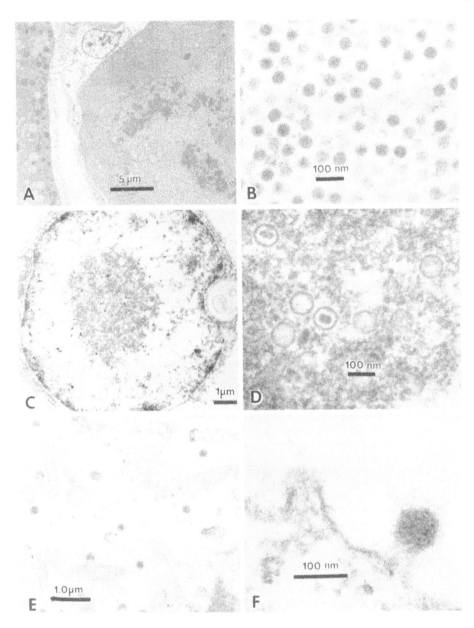

FIGURE 1. (A) Viral gametocyte hypertrophy in *Crassostrea virginica*. Low magnification electron micrograph of virus infected gametocyte from female oyster. Virions are confined to the intranuclear inclusion which is strongly Feulgen positive — indicative of a DNA virus (the cell to the left is a normal ovum). (B) Higher magnification of virus from (A). Virions average 50 to 55 nm in diameter and are isometric with five- and six-sided configurations suggestive of icosahedral symmetry. Morphology, size, developmental strategy, and presence of DNA within the viroplasm suggests that this virus belongs in the family Papovaviridae and the genus *Papillomavirus*. (C) Herpes-type virus in *Crassostrea virginica*. Cowdry type A intranuclear inclusion in oyster hemocyte. Immature, nonenveloped nucleocapsids are seen within the inclusion. Inclusions are Feulgen variable indicating the presence of DNA during the development of viroplasm. (D) Higher magnification of the herpes-type virus. Full and empty isometric five- and six-sided nucleocapsids are seen — some showing toroid structure indicative of herpes virus. Nucleocapsids average 90 nm in diameter. The combination of characteristics suggests that this virus is probably a distant relative of the herpes viruses. (E) Retrovirus-like particles in *Crassostrea virginica*. Virus-like particles (90 to 115 nm) budding from microvillar plasma membranes of digestive tubule epithelial cells in an oyster from Chincoteague Bay, VA. (F) Higher magnification of retrovirus-like particle. Peplomer-like structures, budding from the plasma membrane, size and electron-lucid centers are morphological features suggestive of retrovirus; however, no other characterization has been accomplished. This is a first and only occurrence with no overt indication of disease. (Courtesy of C. Austin Farley.)

(see Figure 1 A and B). Another nonenveloped virus (40 to 45 nm in diameter) associated with cell hypertrophy in *Mya arenaria* is also mentioned in this group.[12]

According to Farley,[6] the Togaviridae could be represented in the marine molluscans, by a virus of the oyster, *Ostrea lurida,* (enveloped particles, 50 nm in size, presumably RNA). This same author furthermore in his scheme of classification,[13] proposes as Paramyxoviridae a presumed agent developing virus-like inclusions in *Mya arenaria* (teratomatous glandular tissue).

Lastly, the family of Pedoviridae is noted to classify the phages found in a chlamydial parasite of *Mercenaria mercenaria*[14] and in a mycoplasma of *Tellina tenuis.*[15]

REFERENCES

1. **Comps, M.,** Epizootic diseases of oysters associated with viral infections, *Am. Fish. Soc. Spec. Publ.,* 18, 23, 1988.
2. **Devauchelle, G. and Vago, C.,** Particules d'allure virale dans les cellules de l'estomac de la seiche, *Sepia officinalis* L. (Mollusques, Cephalopodes), *C. R. Acad. Sci. Paris,* 272, 894, 1971.
3. **Meyers, T. R.,** A reo-like virus isolated from juvenile American oysters *(Crassostrea virginica), J. Gen. Virol.,* 43, 203, 1979.
4. **Meyers, T. R.,** Experimental pathogenicity of reovirus 13p2 for juvenile American oysters *Crassostrea virginica* (Gmelin) and bluegill fingerlings *Lepomis macrochirus* (Rafinesque), *J. Fish Dis.,* 3, 187, 1980.
5. **Farley, C. A., Banfield, W. G., Kasnic, Jr., B., and Foster, W. S.,** Oyster herpes-type virus, *Science,* 759, 1972.
6. **Farley, C. A.,** Viruses and viruslike lesions in marine mollusks, *Mar. Fish. Rev.,* 40, 18, 1978.
7. **Oprandy, J. J., Chang, P. W., Pronovost, A. D., Cooper, K. R., Brown, R. S., and Yates, V. J.,** Isolation of a viral agent causing hematopoietic neoplasia in the soft-shell clam, *Mya arenaria, J. Invertebr. Pathol.,* 38, 45, 1981.
8. **Elston, R. A. and Kents, M. L.,** Biology of the hemic neoplasia in the bay mussel, *Mytilus edulis, 2nd Int. Colloq. Pathol. Mar. Aquacult.,* Gloucester Point, VA, U.S., 85, 1988.
9. **Farley, C. A.,** Electron microscopy of virus infections in oysters *(Crassostrea virginica), IX Annu. Meet. Soc. Invertebr. Pathol.,* Corvallis, Oregon, 1975, 28.
10. **Farley, C. A.,** Ultrastructural observations on epizootic neoplasia and lytic virus infection in bivalve mollusks, *Prog. Exp. Tumor Res.,* 20, 283, 1976.
11. **Farley, C. A.,** Viral gametocyte hypertrophy in oysters. Identification Leaflet No. 25, *Identification Leaflets for Diseases and Parasites of Fish and Shellfish,* Sinderman, C. J., Ed., Int. Counc. for Exploration of the Sea, DK-1261, Copenhagen, Denmark, 1, 1985.
12. **Farley, C. A.,** Proliferative disorders in bivalve mollusks, *Mar. Fish. Rev.,* 38, 30, 1976.
13. **Farley, C. A.,** Phylogenetic relationships between viruses, marine invertebrates and neoplasia, in *Phyletic Approaches to Cancer,* Dawe, C. J., Harshbarger, J. C., Kondo, S., Sugimura, T., and Takayama, S., Eds., Japan Science Societies, Tokyo, 1981, 75.
14. **Harshbarger, J. C., Chang, S. C., and Otto, S. W.,** Chlamydiae (with phages), mycoplasmas, and rickettsiae in Chesapeake Bay bivalves, *Science,* 196, 666, 1977.
15. **Buchanan, J. S.,** Electron microscopic observations of virus-like particles in digestive gland of a marine bivalve mollusc, *Tellina tenuis* (da Costa), *16th Annu. Meet. Soc. Invertebr. Pathol.,* 1973, 28.

Chapter 24

UNCLASSIFIED VIRUSES OF CRUSTACEA

Jean-Robert Bonami and Donald V. Lightner

TABLE OF CONTENTS

I. INTRODUCTION

Within general virology, the virology of crustacea constitutes a comparatively recent and not very advanced field. To date, numerous viruses or virus-like particles, associated or not with diseases, have been described in crustacea.[1-5] A few of them have been characterized and proposed as members of Baculoviridae[6] and Reoviridae.[7] They are mentioned in the previous chapters (see Chapters 6, 9, and 15). But most of them, insufficiently characterized, particularly at the biochemical level and/or genome analysis, can be considered as unclassified viruses. However, authors have provisionally related them for convenience to known virus families.[8]

Thus, virus particles have been essentially reported from crabs and prawns and related — for some of them only on the basis of morphological observations — to Bunyaviridae, Rhabdoviridae, Paramyxoviridae, Herpesviridae, Picornaviridae and Parvoviridae (see Table 1).

II. CRUSTACEAN VIRUSES RELATED TO BUNYAVIRIDAE

A. CRAB HEMOCYTOPENIC VIRUS (CHV)

CHV is the name given to a virus responsible for a transmissible abnormal blood clotting of wild crabs.[9] It has been found only in captive wild crabs, *Carcinus maenas*,[10] from the Brittany coasts (Roscoff, France). This infection is not necessarily lethal and recovery has been frequently observed without disappearance of virions in blood. An autointerference mechanism has been hypothetized to explain this phenomenon and the low observed mortality rates.[11-13]

Signs of infection are those of abnormal clotting of the hemolymph, characterized by the *in vitro* lack of clot formation and a reduced number of circulating hemocytes.

Viral particles are enveloped, 55 to 125 nm in diameter (see Figure 1). They develop in cytoplasmic vesicles associated with the Golgi apparatus of hemocytes, particularly the granulocytes.[9,13]

B. S VIRUS

S virus is the name of a virus causing crab diseases in the genus Portunidae and Carcinidae.

1. Host Range and Symptoms

S virus has been found in wild populations of the crab, *Macropipus depurator*[14,15] and *Carcinus mediterraneus*,[16,17] caught in the Mediterranean area from France, as well as in *Carcinus maenas*[16] in Brittany (Roscoff). It has mostly been studied in *M. depurator* where it produces experimentally high mortality rates. Similar sized and shaped particles have been reported in the Y organ of *C. mediterraneus* by other authors.[17]

Clinical symptoms of the disease are consistent with anorexia, reduced activity, and progressive weakness leading to the death of infected animals.

2. Virus Structure and Composition

Negatively stained virus particles (see Figure 2a) exhibit an important pleomorphism; some of them are spherical or ovoid in shape and others are bacilliform, with a size range from 80 to 320 nm. Virions possess a 20 nm thick envelope covered with spikes sometimes arranged in a well-organized superficial structure (see Figure 2b and c).[19]

The genome is formed with a ssRNA, and a preliminary study indicates the presence of at least three segments. Virion proteins are made of two major polypeptides (72 and 48 $\times 10^3$) and five minor polypeptides (114, 81, 59, 35, and 30.5 $\times 10^3$ Da).[1]

In sucrose gradient, the density of S virions is of 1.13 to 1.17 g/cm^3.

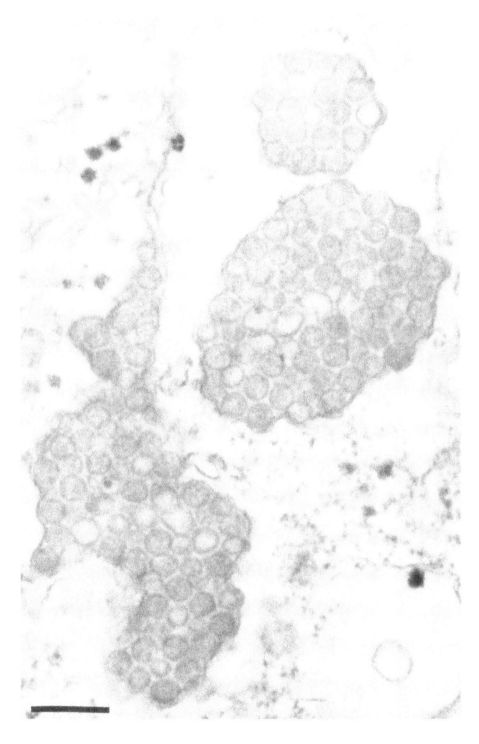

FIGURE 1.   Crab hemocytopenic virus (CHV) of *C. maenas*. Enveloped viral particles in cytoplasmic vesicles of an infected hemocyte. (TEM; bar = 0.3 μm.) (Courtesy of F. K. Bang.)

TABLE 1
Unclassified Viruses of Crustacea

Virus or Virus-like Particles								Hosts		
Name	Related to	Size (nm)	Cell location[a]	Tissue tropism	Envelope[b]	Nucleic acid[c]	Infectivity tests[d]	Family	Name	Ref.
CHV	Bunyaviridae	55/125	C	Hemocytes	+	?	+	Carcinidae	*Carcinus maenas*	9—13
S	Bunyaviridae	80/320	C	Hemocytes, connective cells	+	ssRNA	+	Portunidae Carcinidae	*Macropipus depurator Carcinus maenas Carcinus mediterraneus*	1,14—17,19
RhVA (EGV2)	Rhabdoviridae	20—30 × 110—170	N	Gastric mill Swann cells, hemocytes, mandibular organ	?	?	−	Portunidae Carcinidae	*Callinectes sapidus Carcinus mediterraneus*	16,20—22
RhVB (EGV1) YOV	Rhabdoviridae	50—70 × 100—170	C	Mandibular gland, hemocytes, connective cells, Y organ	+	?	−	Portunidae Carcinidae	*Callinectes sapidus Carcinus maenas Carcinus mediterraneus*	4,16,21,24,25
EHV	Paramyxo- and Orthomyxoviridae	105—194 × 300	C	Hematopoietic organ	+	?	−	Portunidae	*Callinectes sapidus*	26
—	Herpesviridae	100—110	N-C	Germinal, testicular area	+	?	−	Xanthidae	*Rhithropanopeus harrisii*	27
HLV	Herpesviridae	185—241	N	Hemocytes, hematopoietic cells	+	?	−	Portunidae	*Callinectes sapidus*	4,5,28
—	Herpesviridae	140—165	N	Bladder, antennal gland, podocytes	−	?	−	Lithodidae	*Paralithodes platypus*	29

Virus	Classification	Size (nm)	Cell location[a]	Tissue/cells	Envelope[b]	Nucleic acid[c]	Infectivity[d]	Family	Species	Ref.
IHHN	Picornaviridae or Parvoviridae	20—22	C	Gills, cuticular epidermis, connective tissue gonads, hematopoietic tissue	—	?	—	Penaeidae	Penaeus stylirostris, P. vannamei, P. monodon, P. semisulcatus	30—35
CBV	Picornaviridae	30	C	Mesodermic and ectodermic cells	—	RNA[e]	—	Portunidae	Callinectes sapidus	4,5,23
—	Picornaviridae	25—29	C	NR[f]	—	?	—	Grapsidae	Hemigrapsus oregonensis	36
—	Reoviridae	58—61	C	Epidermal layer hepatopancreas	—	?	—	Entoniscidae, Grapsidae, Entoniscidae	Portunion conformis, P. conformis	
F N	Picornaviridae	31 24	C	Connective cells, hemocytes	—	?	—	Portunidae	Macropipus depurator	1,37
—	Picorna- or Parvoviridae or phages	23	?	Cardiac cells	—	?	—	Penaeidae	Penaeus aztecus	38
HPV	Parvoviridae	22—24	N	Hepatopancreas	—	DNA	—	Penaeidae	Penaeus merguiensis, P. semisulcatus, P. orientalis, P. esculentus, P. monodon, P. penicillatus	34,39—41
PC84	Parvoviridae	29—31	N-C	Connective tissue, fibroblasts and myoepithelial cells	—	DNA	+	Carcinidae	Carcinus mediterraneus	16,17,42

a   Cell location in the nucleus (N) or/and in the cytoplasm (C).
b   Presence of envelope (+); absence (−); nonestablished (?).
c   Nature of nucleic acid not established (?).
d   Infectivity established (+); not established or not done (−).
e   Feulgen negative reaction.
f   Not reported.

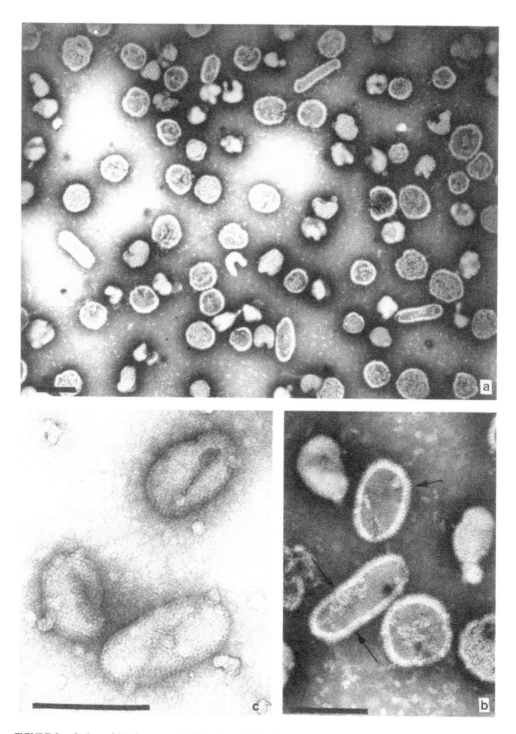

FIGURE 2.   S virus of *M. depurator*. (a) Negative staining (PTA) of purified virions. Note the pleomorphism of the particles, more or less PTA-penetrated; (b) higher magnification showing the spikes on the envelope (arrows); (c) uranyl-acetate-stained virions exhibiting the implantation of spikes on the envelope. (Bars = 0.2 μm.)

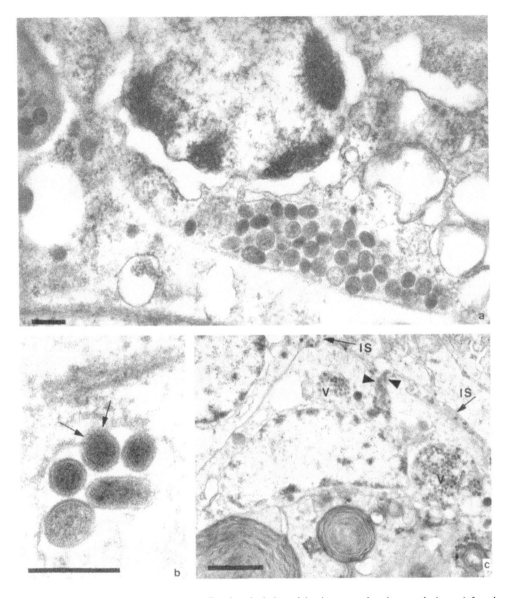

FIGURE 3. S virus of *M. depurator*. (a) Enveloped viral particles in a cytoplasmic vacuole in an infected connective cell; (b) morphogenesis by budding (arrows) in a cytoplasmic vesicle. (TEM; bars = 0.3 μm.) (c) Release of virions from cytoplasmic vacuoles (V) to the intercellular spaces where they accumulate (IS) by fusion of the vacuole with the plasma membrane (arrowheads). (TEM; bar = 2 μm.)

3. Cytopathology

S virus develops in cytoplasmic vesicles of connective tissue and blood cells. Virus particles in sections are spherical (100 to 120 nm) or ovoid (120 to 200 nm length), depending on the section (see Figure 3a).

Morphogenesis occurs by budding in such vesicles, and virions are released in intracellular spaces where they accumulate, by opening virus-containing vesicles on the cell surface (see Figure 3 b and c).[1,19]

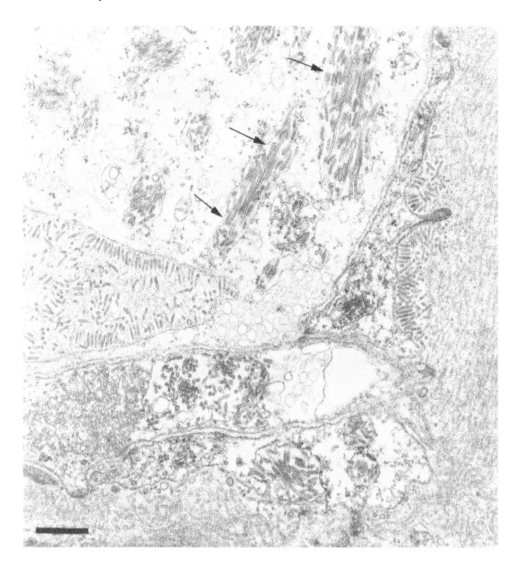

FIGURE 4.   Extra- and intracellular rhabdo-like virus A (RhVA) in epicardium of *C. sapidus*. Note segmentation taking place in the endoplasmic reticulum (arrows). (TEM; bar = 0.5 μm.) (Courtesy of P. T. Johnson.)

III. CRUSTACEAN VIRUSES RELATED TO RHABDOVIRIDAE-PARAMYXOVIRIDAE

Based on their location in host tissues and morphology, numerous virus-like particles, accidentally observed in crabs, have been related to Rhaboviridae and/or Paramyxoviridae. Some confusion in the literature exists because several authors have described the same agent in the same crab species, but in different organs, different names have been attributed to each agent. Prevalence, pathogenicity, and effect are unknown in all cases.

A. RHABDO-LIKE VIRUS A (RhVA)

Viral particles are bacilliform, 20 to 30 nm in diameter and 110 to 170 nm in length. They were first observed in muscles of the gastric mill of the blue crab, *Callinectes sapidus*, and encountered as well in Swan cells, hemocytes, blood vessel cells, and in extracellular spaces (see Figure 4).[20] Other authors[21,22] have described the same type of particles in cells of the mandibular organ, which they had errantly identified as the ecdysal gland, and then incorrectly called ecdysal gland virus 2 (EGV2). Johnson[23] described similar particles in different organs of the blue crab and called them RhVA. Viral particles showing structural similarities have been noted in *Carcinus mediterraneus* from the French Mediterranean coasts.[16]

The particles develop in the interlamellar space of the nuclear envelope and form kinds of clusters in the cytoplasm. Some filamentous particles, reaching 600 nm in length have been reported.

B. RHABDO-LIKE VIRUS B (RhVB)

Extracellular particles, ovoid or bacilliform, have been found associated with the basal membrane of the mandibular gland of *Callinectes sapidus*.[21] Called initially EGV1 (ecdysal gland virus 1), they have been renamed RhVB.[4] They are 50 to 70 nm in diameter and 100 to 170 nm in length, and possess characteristic projections on the envelope.

Similar virus-like particles (YOV) have been reported in the Y-organ of *Carcinus maenas* from Brittany coasts (France).[24] Particles are located between the plasma membrane of glandular cells and the basal membrane which covered the organ (see Figure 5a and b). Virions are enveloped and morphogenesis occurs by budding of the plasma membrane of gland cells. Development is closely associated with the presence of dense cytoplasmic area, containing a high density of ribosomes and dense filaments interpreted as nucleocapsids.

Such virus-like formations have also been reported on the surface of hemocytes, connective cells and secretory epithelium in *C. maenas* from French Channel coasts[25] and in *C. mediterraneus* from the French Mediterranean coast.[16]

C. ENVELOPED HELICAL VIRUS (EHV)

Observed in the hematopoietic organ of the blue crab, *Callinectes sapidus*, of the East Coast of the U.S.[26] these particles are ovoid to rod-shaped, 105 to 194 nm in diameter, with a length of 300 nm for elongated forms (see Figure 6a). Virions develop by budding from the plasma membrane (see Figure 6b), explaining the extracellular location of mature virions observed between the plasma membrane of cells and the basal lamina surrounding lobes of hemopoietic tissue.

Cytoplasm of infected cells contains nucleocapsids, 24 nm in diameter and granular inclusions. Authors have indicated morphological and morphogenetical similarities with the Rhabdo-like virus of the Y organ of *C. maenas*, but relate EHV with the Paramyxoviridae or Orthomyxoviridae rather than with the Rhabdoviridae.

IV. CRUSTACEAN VIRUSES RELATED TO HERPESVIRIDAE

Three types of virus-like particles, related by various authors to the Herpesviridae family have been reported in crabs.

In the course of an ultrastructural study of the germinal testicular area of the crab, *Rhithropanopeus harrisii*, virus-like particles were noted.[27] Located in nuclei of mesodermic cells, these particles (75 to 80 nm in diameter) appeared most often in the vicinity of the nuclear envelope which exhibited large evaginated areas. In the cytoplasm, these particles were present, but enveloped, with a diameter of 100 to 110 nm. They seem to be released from the nucleus into the cytoplasm through evaginations of the nuclear envelope.

The second type, named herpes-like virus (HLV) infects the blue crab, *Callinectes sapidus*, from the Chesapeake Bay.[28] It develops in wild populations of crabs with an estimated prevalence of about 13%. It seems to be epizootic for juvenile crab populations and to cause high mortality rates.[4,5]

Signs of infection are consistent with a milky appearance of hemolymph which does

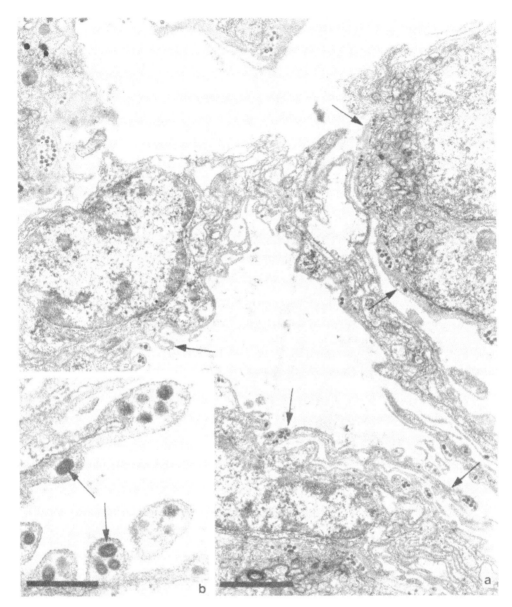

FIGURE 5.   Y-organ virus of *C. maenas*. (a) The viral particles are located between the plasma membrane of the glandular cells and the basal lamina (arrows) which covered the organ. (TEM; bar = 2 μm.) (b) Higher magnification; envelope of the particles is clearly visible (arrows). (TEM; bar = 0.5 μm.) (Courtesy of C. Chassard-Bouchaud and M. Hubert.)

not clot and exhibit a reduced number of hemocytes. Infected cells (hemocytes and hematopoietic cells) show a hypertrophied nucleus and large refringent Feulgen positive cytoplasmic inclusions. Enveloped virions, hexagonal in shape and 185 to 241 nm in diameter, develop in hypertrophied nucleic (see Figure 7 a). The envelope is double layered. The capsid, 127 to 136 nm in diameter, contains a cylindrical nucleoid, 78 nm in diameter and 124 to 136 nm in length (see Figure 7 b).[4]

The third type of herpes-like virus has been described as pathogen of the Alaskan blue king crab, *Paralithodes platypus*.[29] Since the late 1970s, important declines of the natural populations of this species of the Bering Sea had been recorded and investigations have led to evidence of a viral infection.

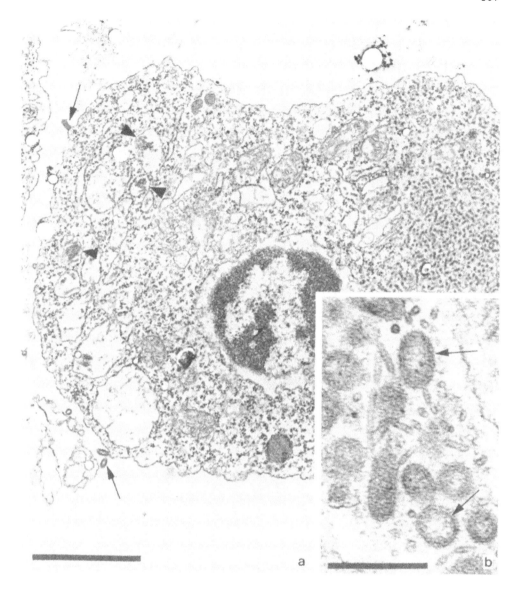

FIGURE 6. Enveloped helical virus (EHV) of *C. sapidus*. (a) Slightly infected hemocyte showing the budding of particles (arrows) on the plasma membrane surface; note the presence of RhVA in cytoplasm (arrowheads). (TEM; bar = 2 μm.) (b) Higher magnification of released EHV particles associated with RhVA; the trilaminar structure of the envelope of EHV particles is clearly evidenced (arrows). (TEM; bar = 0.2 μm.) (Courtesy of P. T. Johnson.)

Histological symptoms are consistent with a nuclear hypertrophy associated with chromatin margination and presence of eosinophilic ground substance (Cowdry type A inclusion) with or without eosinophilic inclusion bodies in the bladder and the antennal gland labyrinthal epithelium; podocytes were also infrequently infected, but without inclusion bodies.

Virions, hexagonal in shape, develop in the ground substance (viroplasm) of hypertrophied nuclei. Although this agent was morphologically close to the HLV of *C. sapidus*,[28] the authors interpret its structure differently; they consider the external trilaminar layer of the particle to be a capsid made of two distinct electron-dense layers, resulting in an unenveloped nucleocapsid of 140 to 165 nm in diameter; this hexagonal nucleocapsid contains a nucleoid consisting of an electron-dense cylinder 55 to 60 nm × 90 to 105 nm, surrounded by a toroidal structure of 90 to 105 nm.

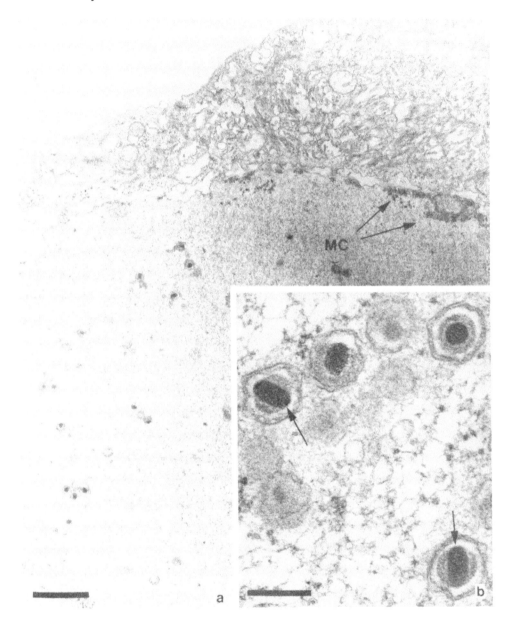

FIGURE 7.   Herpes-like virus (HLV) of *C. sapidus:* infected hemocyte. (a) The hypertrophied nucleus contains some virions and exhibit a marginated chromatin (MC). (TEM; bar = 1 μm.) (b) Higher magnification of the nucleoplasm containing enveloped virions. Note the cylindrical nucleoid (arrows). (Bar = 0.2 μm.) (Courtesy of P. T. Johnson.)

The presence of inclusion bodies in the virogenic stroma — often with frayed ends, particularly in heavily infected nuclei — seems to be a characteristic of this herpes-like virus infection; but their significance, origin, and function are unclear.[29]

V. CRUSTACEAN VIRUSES RELATED TO PICORNAVIRIDAE

A. INFECTIOUS HYPODERMAL AND HEMATOPOIETIC NECROSIS VIRUS (IHHNV)

Infectious hypodermal and hematopoietic necrosis (IHHN) is a virus disease commonly observed in farm-raised marine shrimps and prawns of the genus Penaeus.[30-35]

1. Host and Geographic Range

IHHN has been found only in penaeid shrimps and prawns reared in the Americas, Southeast Asia, and the Middle East. Natural infections have been observed in *Penaeus stylirostris* and *P. vannamei*, which are American Pacific species, and in *P. monodon* and *P. semisulcatus*, which are IndoPacific species. IHHNV has been experimentally shown to infect other penaeid species, and it probably can infect all penaeids.[34] However, only in *P. stylirostris*, and to a lesser extent in *P. monodon*, has IHHNV been found to cause serious disease losses.[32,34] The intercontinental spread of this disease with live shrimp importations for aquaculture purposes has been well documented.[32-34]

The original source and distribution of IHHNV in wild penaeid stocks (or other crustaceans) is not known, nor has IHHN disease been demonstrated in wild populations.[34] However, only in Malaysia and the Philippines has IHHN been observed in captive wild shrimps, suggesting that the virus may be enzootic in wild shrimp populations in that region.

2. Gross Pathology

Gross signs are not IHHN specific, but juvenile *P. stylirostris* with acute IHHN show a marked reduction in food consumption; moderate to severe fouling of the cuticle, appendages, and gills by epicommensal bacteria, algae, and protozoa; and a "mottling" (due to white to buff-colored petechial spots in the subcutis) of the cuticle, especially of the edges of the carapace and tergal plates of the abdomen. This mottling fades rapidly in acutely ill shrimp, and moribund *P. stylirostris* may be more bluish than normal; moribund *P. monodon* may range from bluish to almost black. Tank-cultured juvenile *P. stylirostris* have been observed to swim lethargically near the water surface, and then suddenly become motionless, roll over, and sink slowly to the culture tank bottom. Only severely stressed juvenile *P. vannamei* display clinical IHHN disease, and under most aquaculture conditions this species seems to carry the virus without any apparent ill effect.[30-32]

3. Virus Structure

IHHN disease of penaeid shrimp is caused by a nonenveloped, glycerol tolerant, very small (20 to 22 nm in diameter) icosahedral virus which is replicated in the cytoplasm of host cells (see Figures 9 through 11).[32,35] The density of IHHNV, as determined by density gradient centrifugation is cesium chloride gradients, is in the range of 1.34 to 1.40 g/ml. Preliminary results from electrophoresis of the proteins extracted from purified virus preparations showed one major and two minor capsid proteins of approximately 36, 49, and 35 kDa, respectively. Analysis of the nucleic acid extracted from purified IHHN virions with ribonucleases and deoxyribonucleases have given inconclusive results. However, Feulgen staining of tissue sections containing IHHNV nuclear and cytoplasmic inclusions suggest IHHNV to not contain DNA. Although not completely characterized, all available data indicate that IHHN virus is related to the Picornaviridae,[32,35] but the possibility that it belongs to the Parvoviridae cannot be ruled out.

4. Cytopathology

Current diagnostic procedures for IHHN disease are dependent upon the histological demonstration of prominent Cowdry type A eosinophilic proteinic intranuclear inclusion bodies in target tissues of infected shrimp that have been preserved in fixatives such as Davidson's acetic acid-formalin-alcohol and Bouin's solutions, or in Giemsa-stained hemocytes from shrimps with acute infections (see Figure 8a to c). Cytoplasmic inclusion bodies, presumed to be viral areas, are also observed occasionally, but these are not as prominent a histological feature of the disease as are the proteinic intranuclear bodies (see Figure 8d). Generalized to multifocal necrosis and inflammation of gill, cuticular epidermis, connective tissue, gonad, and hematopoietic tissues are typical in shrimps with clinical IHHN.[30-32]

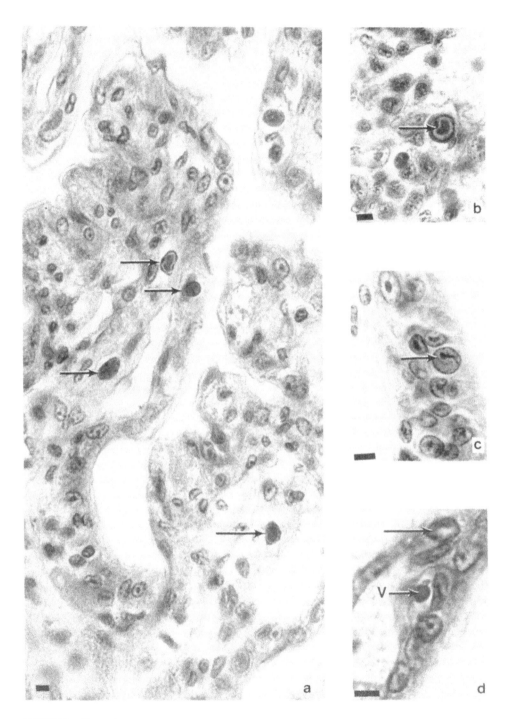

FIGURE 8.   Photomicrographs of a typical IHHNV-caused eosinophilic intranuclear inclusion bodies (arrows) in histological sections of: (a) gill, (b) hematopoietic tissue, (c) foregut cuticular hypodermis, and (d) gill, also showing a basophilic cytoplasmic inclusion body (V), which is an IHHN virus paracrystalline array. (H.E.; bars = 5 μm.)

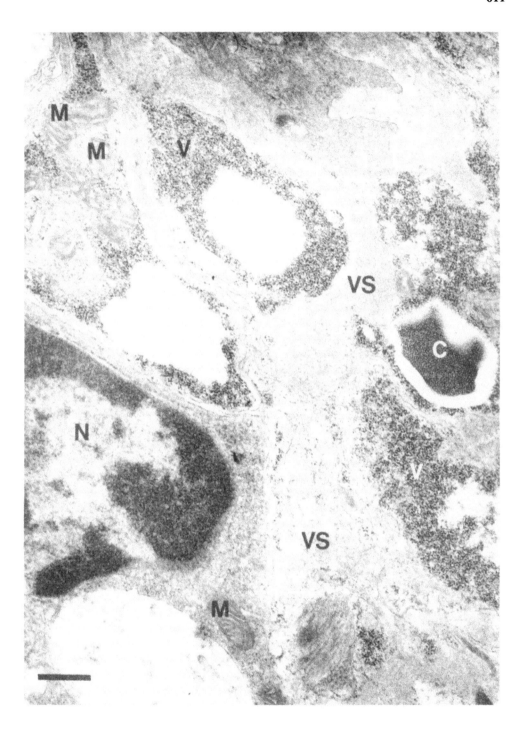

FIGURE 9. Transmission electron micrograph of a section of gills from an IHHN-infected juvenile *Penaeus stylirostris*. The nucleus (N) of the gill epidermal cell shown surrounded by the cell cytoplasm that except for mitochondria (M) is virtually filled with masses of virogenic stroma (VS) and masses of IHHN virus particles (V), some in a paracrystalline array (C). (Bar = 250 nm.)

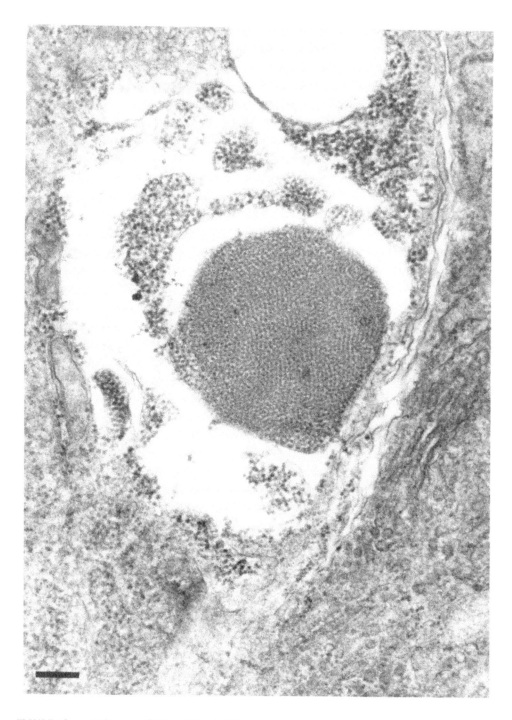

FIGURE 10.   A higher magnification TEM of IHHN virions in disorganized cytoplasmic masses and in a para-crystalline array. Note the apparent smaller size of the virus in the arrays compared to the larger "free" virions. (Bar = 125 nm.)

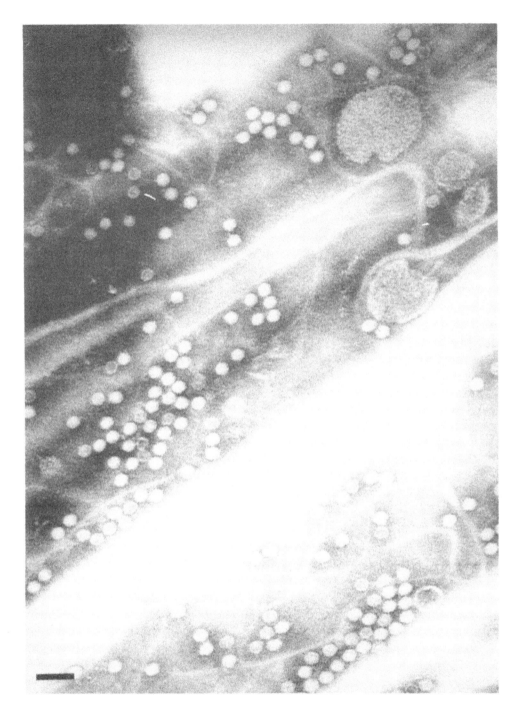

FIGURE 11. Transmission electron micrograph of negative stained (2% PTA) purified preparation of IHHN virions from cesium chloride density gradient centrifugation. Individual virus particles measure 20 to 22 nm in diameter. (Bar = 40 nm.)

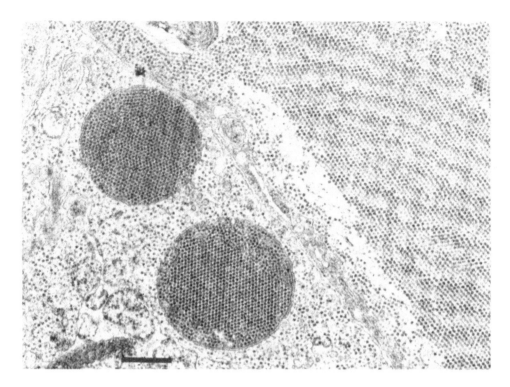

FIGURE 12.   Chesapeake Bay virus (CBV) of *C. sapidus*. Paracrystalline arrays of virions in cytoplasm of infected cells; some are enclosed in vesicle-like structures. (TEM; bar = 0.5 μm.) (Courtesy of P. T. Johnson.)

B. CHESAPEAKE BAY VIRUS (CBV)

CBV disease has been described in the blue crab, *Callinectes sapidus,* from the Chesapeake Bay area.[23] Infected animals exhibit abnormal behavior described as "head downwards" and uncoordinated swimming and movements. Virus particles develop in the cytoplasm of mesodermic and ectodermic cells, particularly in epidermis, nervous system, retina, anterior and posterior gut epithelium, gill epithelium, and sometimes in hemocytes and in the hemopoietic organ.[4,5] Infected cells were hypertrophied with a Feulgen negative cytoplasm. Virions, icosahedral in shape and 30 nm in diameter, form cytoplasmic paracrystalline arrays often associated with membranous structures (see Figure 12).[23]

C. OTHERS

In the course of a study on parasitosis of the crab, *Hemygrapsus oregonensis,* viruslike particles (25 to 29 nm in diameter) arranged in paracrystalline arrays have been observed in the cytoplasm of infected cells.[36] The capsid is formed with three major polypeptides, and the sedimentation coefficient has been estimated at 160 to 165 S. These particles also are present in the parasite, *Portunion conformis,* associated here with larger particles (58 to 61 nm) and both developing in the cytoplasm, especially of the epidermal layer and hepatopancreas.

Two other particles, F and N virus (31 and 24 nm in diameter, respectively) have been found associated together in the same animal; they are pathogenic to the crab, *Macropipus depurator,* from the French Mediterranean area.[37] Icosahedral in shape, these particles develop in the cytoplasm of connective cells and hemocytes, but they have never been observed simultaneously in cytoplasm of the same cell. In infected cells, F particles form large cytoplasmic masses, without a peculiar arrangement; however, they are often associated with membranous systems, whereas N particles display cytoplasmic paracrystalline areas in an electron-dense matrix.[1,37]

Finally, aggregates of virus-like particles have been reported in cardiac cells of *Penaeus aztecus*.[38] These particles, 23 nm in diameter, have been found associated with cellular debris in cytoplasmic vacuoles of phagocytes attached to the myocardium. The authors relate these particles to Picornaviridae or Parvoviridae, but do not exclude the possibility of phages derived from phagocytized and partially lysed bacteria, rickettsia, or chlamydia.

VI. CRUSTACEAN VIRUSES RELATED TO PARVOVIRIDAE

A. HEPATOPANCREATIC PARVO-LIKE VIRUS (HPV)
Hepatopancreatic parvo-like virus (HPV) is the name given to a presumed virus-caused disease of cultured and wild shrimps and prawns of the genus *Penaeus*.[34,39-41]

1. Host and Geographic Range
HPV has been found in cultured and captive wild penaeid shrimp from the IndoPacific coasts of Africa and Australia, and in Asia from the Middle East to the Yellow Sea.[34,39-41] In Brazil, it has recently been observed in shrimp imported from Asia. All penaeids are probably susceptible to HPV, but disease due to infection has only been observed in *Penaeus merguiensis, P. semisulcatus, P. orientalis, P. penicillatus, P. esculentus,* and *P. monodon*.[34,40,41] HPV seems to cause disease and mortalities in the midjuvenile stages in the penaeid species where it has been most studied. HPV infections have been diagnosed in populations of cage and tank-cultured *P. merguiensis,* and in raceway-cultured *P. semisulcatus,* where its presence has been associated with high accumulative mortalities.[39]

2. Gross Pathology
Gross signs of HPV infection include atrophy of the hepatopancreas, poor growth rate, anorexia, reduced preening activity and consequent increased tendency for surface and gill fouling by epicommensal organisms, occasional opacity of abdominal muscles, and secondary infections by opportunistic pathogens like *Vibrio* spp. and *Fusarium solani*. Mortalities accompanied by these signs most often occur in the juvenile stages.[39]

3. Virus Structure and Cytopathology
HPV is a small (22 to 24 nm in diameter) DNA-containing parvo-like virus. While HPV has not been isolated from infected shrimp nor has it been experimentally transmitted, the cytopathology of affected host cells suggest that the lesions are due to virus infection by a parvo-like virus.[34,39]

HPV lesions in the hepatopancreas are present in the E (mitotic) cells and adjacent R and F cells of the distal portions of the hepatopancreatic tubules. Affected cells possess a single prominent basophilic (with H. & E. stains), Feulgen-positive intranuclear inclusion body in a hypertrophied nucleus (see Figures 13 and 14). Consequent lateral displacement and compression of the host cell nucleolus and chromatin margination are also prominent characteristics of HPV-infected cells. Transmission electron microscopy of HPV intranuclear inclusions shows them to contain 22 to 24 nm in diameter virus-like particles (see Figure 15). Paracrystalline arrays of these particles have not been observed.[34,39]

B. PC84 VIRUS
This parvo-like virus, called PC84, has been found in captive *Carcinus mediterraneus* collected from the Languedoc lagoons (France).[17] To date, the host range is limited only to this crab species. Experimentally, it induces a lethal disease. Clinical symptoms include weakness, anorexia, and lethargy. Experimental crabs died between 10 and 25 d after being inoculated.

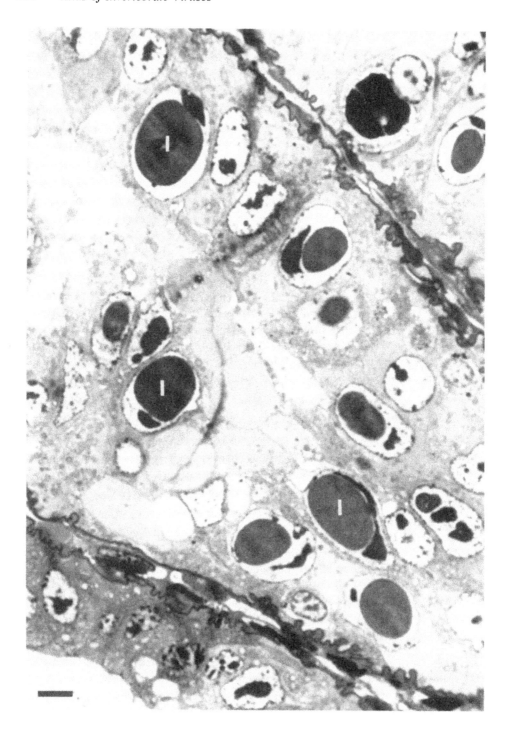

FIGURE 13.   Histological section of hepatopancreas tubules of *P. merguiensis* that are heavily infected with HPV. Dense basophilic HPV inclusion bodies (I) are present within markedly hypertrophied host cell nuclei. (Toluidine blue-O; bar = 5 μm.)

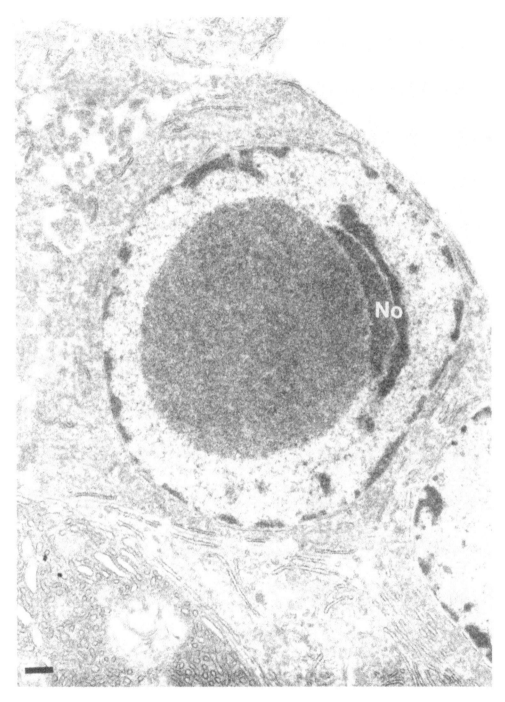

FIGURE 14.   TEM of an HPV-infected hepatopancreatocyte of *P. orientalis*, that contains an HPV intranuclear inclusion body. The host cell nucleolus (No) has been displaced laterally by the developing inclusion body. (Bar = 0.5 μm.)

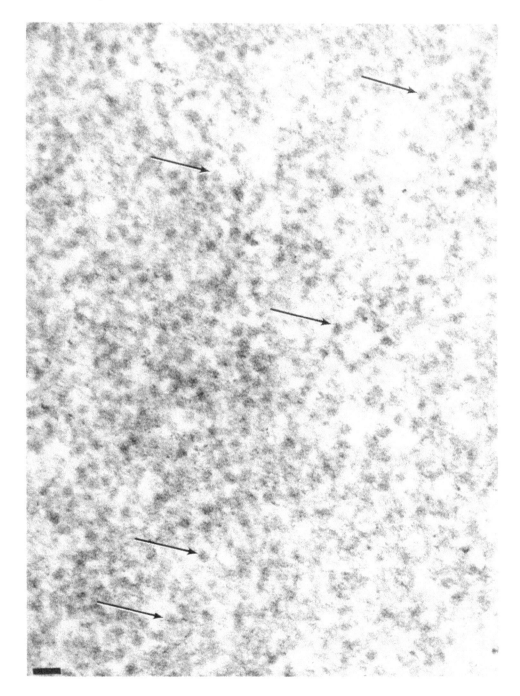

FIGURE 15. High magnification TEM of virogenic stroma within an HPV inclusion body. Profiles of some 22 to 24 nm diameter virus-like particles are clearly angular (arrows). (Bar = 50 nm.)

1. Virus Structure and Composition

In section, viral particles were nearly spherical, nonenveloped and 23 to 27 nm in diameter (see Figure 16a). In isopycnic centrifugation in cesium chloride gradients, three bands were separated. The first one corresponds to empty particles. The two others, named L (light) and H (heavy) particles, have been purified. L and H particles each contain nucleic acid, and their density was determined to be 1.255 and 1.339 g/cm³, respectively.[16,42]

a. Virus Structure

In negative staining, all types of particles (empty, L, and H) exhibit a constant diameter of 29 to 31 nm with an icosahedral symmetry. The capsid (4 to 5 nm thick) displays subunits suggesting capsomers and surrounds, in empty particles, an electron-dense center, 20 to 21 nm in diameter (see Figure 16b and c).

L particles appear to be complete virions compared with H particles which are paradoxically electron dense, showing different degrees of damage and seeming to be more fragile than H particles. In fact, H particles seem to be dense precursors or immature virions.[16,42]

b. Biochemical Composition

The genome of the two types of PC84 particles (H and L) consists of DNA. L particles exhibit four polypeptides with a molecular weight of 84, 80, 42, and 26 × 10³, whereas H particles possess two polypeptides with a molecular weight of 80 and 42 × 10³.

The DNA has been determined as at least partially double stranded *in situ,* and it has been hypothesized that this could be an arrangement of a ssDNA molecule possessing self-complementary sequences of nucleotides to give a hair-loop-like structure in a single virus particle. The genome is made of only one segment with an estimated molecular weight of 1.4 × 10⁶.[16,42]

2. Cytopathology

PC84 virus develops in the connective tissue where it induces cell necrosis. Feulgen positive nuclei without noticeable hypertrophy are scattered in the connective tissues of diseased animals. Lesions are more specifically located in fibroblastic and myoepithelial cells. In infected cells, virions 23 to 27 nm in diameter are present in the nucleus and in the cytoplasm (see Figure 16a). In moderately infected cells, the nucleoplasm contains virions and a few virions are assembled in the cytoplasm around the nuclear membrane. In more heavily infected cells, nuclei contain numerous virions located centrally in clear areas. In parallel, cytoplasmic particles are more numerous. This seems to indicate a release from the nucleus to the cytoplasm; however, this process through nuclear pores has never been observed.[16,42] L particles, therefore, seem to correspond to cytoplasmic particles and H virions to nuclei particles. The release of virions from host cell nuclei to the cytoplasm could be interpreted as a maturation phenomenon and correspond to the acquisition of supplementary polypeptides.[16,42]

The last stage of infection is cellular lysis which releases virions into intercellular spaces. In this case, virions are often arranged in paracrystalline clusters, which have never been observed in infected nuclei or cytoplasm.

ADDENDUM

Since the preparation of this chapter, recent studies on the IHHN-virus characterization[43] indicate clearly that this agent must be now considered as a Parvovirus rather than a Picornavirus.

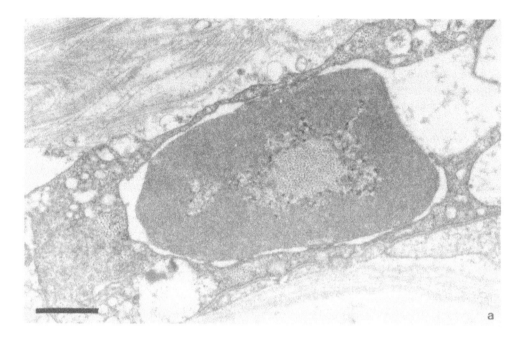

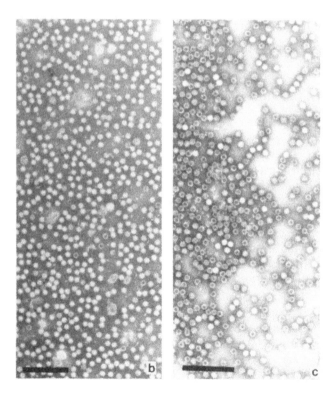

FIGURE 16.   PC84 virus of *C. mediterraneus*. (a) Section of an infected connective cell; virions are located both in the nucleus (centrally in clear area) and in the cytoplasm. (TEM; bar = 1 μm.) (b) and (c) Negatively stained (PTA) light (L) and heavy (H) particles, respectively, after purification on a cesium chloride density gradient. (Bars = 0.3 μm.) (From Mari, J. and Bonami, J. R., *J. Invertebr. Pathol.*, 51, 145, 1988. With permission.)

ACKNOWLEDGMENTS

We thank Mr. J. Luciani for his technical assistance.

REFERENCES

1. **Bonami, J. R.,** Recherches sur les infections virales des crustacés marins: étude des maladies à étiologie simple et complexe chez les décapodes des côtes françaises, thèse Doct. Etat, Univ. Sci. Tech. Languedoc, Montpellier, France, 1980.
2. **Bonami, J. R.,** Les affections virales des crevettes Penaeides, *Océanis,* 13, 233, 1987.
3. **Couch, J. A.,** Viral diseases of invertebrates other than insects, in *Pathogenesis of Invertebrate Microbial Diseases,* Davidson, E. W., Ed., Allanheld, Osmun, Totowa, NJ, 1981, 127.
4. **Johnson, P. T.,** Diseases caused by viruses, rickettsiae, bacteria and fungi, in *The Biology of Crustacea,* Provenzano, A. J., Ed., Academic Press, New York, 1983, 1.
5. **Johnson, P. T.,** Viral diseases of marine invertebrates, *Helgol. Meeresunters,* 37, 65, 1984.
6. **Summers, M. D.,** Characterization of shrimp Baculovirus, *E.P.A. Ecol. Res. Ser.,* 600, 77, 1977.
7. **Mari, J. and Bonami, J. R.,** The W2 virus infection of the crustacea *Carcinus mediterraneus:* a reovirus disease, *J. Gen. Virol.,* in press, 1988.
8. **Matthews, R. E. F.,** Classification and nomenclature of viruses, *Intervirology,* 17, 1, 1982.
9. **Hoover, K. L.,** The effect of a virus infection on the hemocyte population in *Carcinus maenas,* Sc. D. thesis, John Hopkins University, Baltimore, 1977.
10. **Bang, F. B.,** Transmissible disease, probably viral in origin, affecting the amoebocytes of the European shore crab, *Carcinus maenas, Infect. Immun.,* 3, 617, 1971.
11. **Bang, F. K.,** Pathogenesis and autointerference in a virus disease of crabs, *Infect. Immun.,* 9, 1057, 1974.
12. **Bang, F. K.,** Crustacean diseases responses, in *The Biology of Crustacea,* Provenzano, A. J., Ed., Academic Press, New York, 1983, 114.
13. **Hoover, K. L. and Bang, F. B.,** Immune mechanisms and disease response in a virus disease of *Carcinus, Viruses Environ.,* 26, 515, 1978.
14. **Bonami, J. R. and Vago, C.,** A virus of a new type pathogenic to crustacea, *Experientia,* 27, 1363, 1971.
15. **Bonami, J. R., Vago, C., and Duthoit, J. L.,** Une maladie virale chez les crustacés décapodes due à un virus d'un type nouveau, *C. R. Acad. Sci., Paris, Ser. D,* 274, 3087, 1971.
16. **Mari, J.,** Recherches sur les maladies virales du crustacé décapode marin *Carcinus mediterraneus* Czerniavski 1884, thèse Doct. Sci. Biol., Univ. Sci. Tech. Languedoc, Montpellier, France, 1987.
17. **Mari, J. and Bonami, J. R.,** Les infections virales du crabe *Carcinus mediterraneus* Czerniavski, 1884, in *Pathology in Marine Aquaculture,* Vivares, C. P., Bonami, J. R., and Jaspers, E., Eds., European Aquaculture Society, Special Publication No. 9, Bredene, Belgium, 1986, 283.
18. **Zerbib, C., Andrieux, N., and Berreur-Bonnenfant, J.,** Données préliminaires sur l'ultrastructure de la glande de mue (organe Y) chez le crabe *Carcinus mediterraneus* sain et parasité par *Sacculina carcini, C. R. Acad. Sci., Paris,* 281, 1167, 1975.
19. **Bonami, J. R., Veyrunes, J. C., Cousserans, F., and Vago, C.,** Ultrastructure, développement et acide nucléique du virus S du crustacé décapode *Macropipus depurator* L., *C. R. Acad. Sci., Paris,* 280, 359, 1975.
20. **Jahromi, S. S.,** Occurrence of Rhabdovirus-like particles in the blue crab, *Callinectes sapidus, J. Gen. Virol.,* 36, 485, 1977.
21. **Yudin, A. I. and Clark, Jr., W. H.,** Two virus-like particles found in the ecdysial gland of the blue crab, *Callinectes sapidus, J. Invertebr. Pathol.,* 32, 219, 1978.
22. **Yudin, A. I. and Clark, Jr., W. H.,** A description of rhabdovirus-like particles in the mandibular gland of the blue crab, *Callinectes sapidus, J. Invertebr. Pathol.,* 33, 133, 1979.
23. **Johnson, P. T.,** Viral diseases of the blue crab, *Callinectes sapidus, Mar. Fish. Rev.,* 40, 13, 1978.
24. **Chassard-Bouchaud, C., Hubert, M., and Bonami, J. R.,** Particules d'allure virale associées à l'organe Y du crabe *Carcinus maenas* (Crustacé, Décapode), *C. R. Acad. Sci., Paris,* 282, 1565, 1976.
25. **Bazin, F.,** La régénération de la patte chez le crabe *Carcinus maenas* L., thèse Doct. Etat, Univ. Caen, 1984.
26. **Johnson, P. T. and Farley, E. A.,** A new enveloped helical virus from the blue crab *Callinectes sapidus, J. Invertebr. Pathol.,* 35, 90, 1980.

27. **Payen, G. and Bonami, J. R.,** Mise en évidence de particules d'allure virale associées aux noyaux des cellules mésodermiques de la zone germinative testiculaire au crabe *Rhithropanopeus harrisii* Gould (Brachyoure, Xanthide), *Rev. Trav. Inst. Pêches Marit.,* 43, 361, 1979.

28. **Johnson, P. T.,** A herpeslike virus from the blue crab, *Callinectes sapidus, J. Invertebr. Pathol.,* 27, 419, 1976.

29. **Sparks, A. K. and Morado, J. F.,** A herpes-like virus disease in the blue king crab *Paralithodes platypus, Dis. Aquat. Org.,* 1, 115, 1986.

30. **Bell, T. A. and Lightner, D. V.,** IHHN virus: infectivity and pathogenicity studies in *Penaeus stylirostris* and *Penaeus vannamei, Aquaculture,* 38, 185, 1984.

31. **Bell, T. A. and Lightner, D. V.,** IHHN disease of *Penaeus stylirostris:* effects of shrimp size on disease expression, *J. Fish Dis.,* 10, 165, 1987.

32. **Lightner, D. V., Redman, R. M., and Bell, T. A.,** Infectious hypodermal and hematopoietic necrosis, a newly recognized virus disease of penaeid shrimp, *J. Invertebr. Pathol.,* 42, 62, 1983.

33. **Lightner, D. V., Redman, R. M., and Bell, T. A.,** Detection of IHHN virus in *Penaeus stylirostris* and *P. vannamei* imported into Hawaii, *J. World Maricult. Soc.,* 14, 212, 1983.

34. **Lightner, D. V., Redman, R. M., Williams, R. R., Mohney, L. L., Clerx, J. P. M., and Brock, J. A.,** Recent advances in penaeid virus disease investigations, *J. World Maricult. Soc.,* 16, 267, 1985.

35. **Lightner, D. V., Mohney, L. L., Williams, R. R., and Redman, R. M.,** Glycerol tolerance of IHHN virus of penaeid shrimp, *J. World Aquaculture Soc.,* 18, 194, 1987.

36. **Kuris, A. M., Poinar, G. O., Hess, R., and Morris, T. J.,** Virus particles in an internal parasite, *Portunion conformis* (Crustacea:Isopoda:Entoniscidae), and its marine crab host, *Hemigrapsus oregonensis, J. Invertebr. Pathol.,* 34, 26, 1979.

37. **Bonami, J. R.,** Les maladies virales des crustacés et des mollusques, *Océanis,* 3, 154, 1977.

38. **Foster, C. A., Farley, E. A., and Johnson, P. T.,** Virus-like particles in cardiac cells of the brown shrimp, *Penaeus aztecus* Ives, *J. Submicrosc. Cytol.,* 13, 723, 1981.

39. **Lightner, D. V. and Redman, R. M.,** A parvo-like virus disease of penaeid shrimp, *J. Invertebr. Pathol.,* 45, 47, 1985.

40. **Paynter, J. L., Lightner, D. V., and Lester, R. J. G.,** Prawn virus from juvenile *Penaeus esculentus,* in *2nd Aust. Natl. Prawn Sem.,* NPS2, Rothlisberg, P. C., Hill, B. J., and Staples, D. J., Eds., Cleveland, Australia, 1985, 61.

41. **Colorni, A., Samocha, T., and Colorni, B.,** Pathogenic viruses introduced into Israel mariculture systems by imported penaeid shrimp, *Bamidgeh,* 39, 21, 1987.

42. **Mari, J. and Bonami, J. R.,** PC84, a parvo-like virus from the crab *Carcinus mediterraneus:* pathological aspects, ultrastructure of the agent, and first biochemical characterization, *J. Invertebr. Pathol.,* 51, 145, 1988.

43. **Bonami, J. R., Trumper, B., Mari, J., Brehelin, M., and Lightner, D. V.,** Purification and characterization of the infectious hypodermal and haematopoietic necrosis virus of penaeid shrimps, *J. Gen. Virol.,* 71, 2657, 1990.

Chapter 25

UNCLASSIFIED VIRUSES OF ARACHNIDA

Gilles Morel

TABLE OF CONTENTS

I. INTRODUCTION

Many viruses have been discovered in Arachnida but most of them, present in blood-sucking Acarina (ticks), are transmitted to a vertebrate host and belong to Arbovirus groups. Only six nonoccluded viruses are strictly associated with other Arachnida. They have been encountered in *Scorpionidae* (scorpions),[1] *Aranea* (spiders),[2,3] and *Acarina* (spider mites).[4-6] Due to the lack of genome analysis, they are still unclassified but their morphological and developmental characteristics show probable relationships with three virus families: Baculoviridae, Adenoviridae, and Reoviridae.

II. SCORPION VIRUS

Only one cytoplasmic virus has been observed in *Scorpionidae*. It develops in *Buthus occitanus* Am. (Buthidae).[1]

A. PATHOLOGY

The virus causes lethal disease; death occurs from 2 to 5 months after *per os* contamination. Symptoms are not specific of a viral disease; they affect behavior. Diseased scorpions stop eating and show a decreasing activity a few weeks before death. Virions are regularly released through feces from the twentieth day of the disease.

Histological studies do not allow recognition of the infected cells in the hepatopancreas. There is neither nuclear change nor particular cytoplasm alteration until lysis of the host cells. The whole cells of the hepatopancreas are progressively destroyed and eliminated through the digestive tract.

B. VIRUS STRUCTURE

Negatively stained particles have a diameter of 60 to 70 nm. They are icosahedral with a spherical appearance (see Figure 1a) or a hexagonal outline. Cores 40 nm wide are observed in altered or in penetrated particles. Capsids 15 nm thick appear composed of ring-shaped structures 11 nm wide and in crude extracts, fibrillary projections 20 nm long are observed at the vertices of the particles (see Figure 1b). Together with virions, flattened tubelike structures, 100 to 110 nm wide, are present before purification; generally these structures contain virus-like particles (Figure 1c and d). They are constituted by hexagonally packed subunits 10 to 11 nm center-to-center spacing and look like tubular structures associated with Rotavirus (Reoviridae).[7]

C. CYTOPATHOLOGY

The viruses are always observed in the cytoplasm (see Figure 2) where they have circular outlines 60 nm in diameter. Regressive staining points out the presence of RNA in the virus.

Morphogenesis of virions occurs in large virogenic stromata from 5 to 10 μm in diameter. Early viroplasms are moderately electron dense and contain a homogeneous, finely granular material. Numerous polyribosomes are present in these viroplasms (see Figure 3a). Development of the virogenic centers results in rearrangement of electron-dense material in networks (see Figure 3b). Progressively, viral assembly occurs in this network (see Figure 3c). Late virogenic stromata exhibit many virions together with residual electron-dense material (see Figures 2 and 3d).

Tubular structures containing virions are frequently associated with the virus in the periphery of late viroplasms (see Figure 4). They are comparable to intracytoplasmic tubelike structures described in plant reovirus like maize rough dwarf virus.[8,9]

D. RELATIONSHIPS WITH CLASSIFIED VIRUSES

General morphology and morphogenesis of this RNA-containing virus suggest it is related

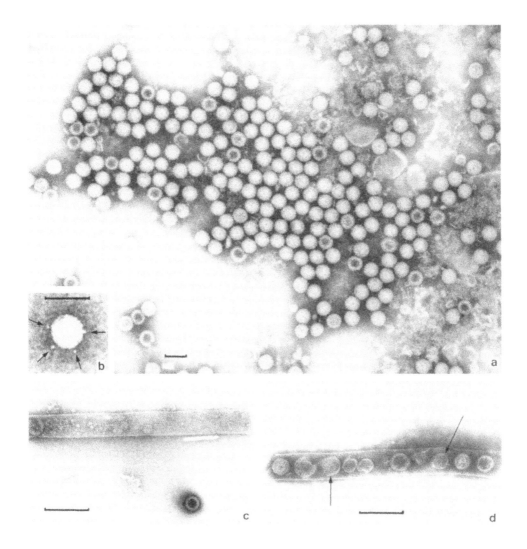

FIGURE 1. Cytoplasmic virus of the scorpion *Buthus occitanus* (negatively stained with neutral phosphotungstic acid [PTA]). (a) Partly purified preparation with full and empty particles. (Bar = 100 nm.) (b) Projections (arrows) are observed at the vertices of the particle in crude extract. (Bar = 100 nm.) (c) and (d) Tubelike structures associated with virus (arrows). (Bar = 200 nm.)

to Reoviridae. Nonoccluded and without the characteristic spikes observed in cytoplasmic polyhedrosis virus, it shows more affinity with other Reoviridae such as Phytoreovirus or Rotavirus.

III. SPIDER VIRUSES

Two nuclear viruses are known from *Araneae*. They have been isolated from *Pisaura mirabilis* Cl. (Pisauridae).[2,3] One is a naked icosahedral virus, and the other has an enveloped rod-shaped nucleocapsid and looks like Baculovirus.

A. PATHOLOGY

These two viruses cause low level infections which cannot be recognized through particular symptoms. Contaminated spiders are discernible only because they release viruses through feces.

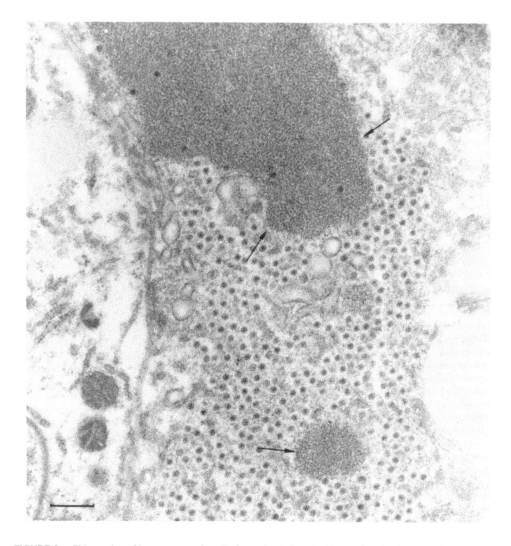

FIGURE 2.   Thin section of hepatopancreatic cell of scorpion infected with cytoplasmic virus. Residual material (arrows) is present beside mature virions. (Bar = 200 nm.)

Histological examination shows alteration of hepatopancreatic cells. In both diseases, nuclei are hypertrophied. Development of baculovirus-like particles takes place in large ovoid nuclei while morphogenesis of icosahedral virus occurs in regularly spherical nuclei. Lysed infected cells are released into the lumen of hepatopancreatic diverticulum and hypertrophied nuclei are present in the posterior digestive tract.

B. VIRUS STRUCTURE
1. Baculovirus-like Particles

Negatively stained particles are ovoid, 175 × 120 nm, and show a characteristic coffee bean appearance, a result of their shape and of the phosphotungstic acid deposit along their long axis (see Figure 5a). Penetrated virions have a U-shaped nucleocapsid, 50 nm wide, enclosed in a thin envelope.

Rod-shaped nucleocapsids 350 × 50 nm are liberated when the envelope is disrupted (see Figure 5b). They exhibit, like U-shaped nucleocapsids in penetrated particles, a regular helicoidal arrangement of subunits.

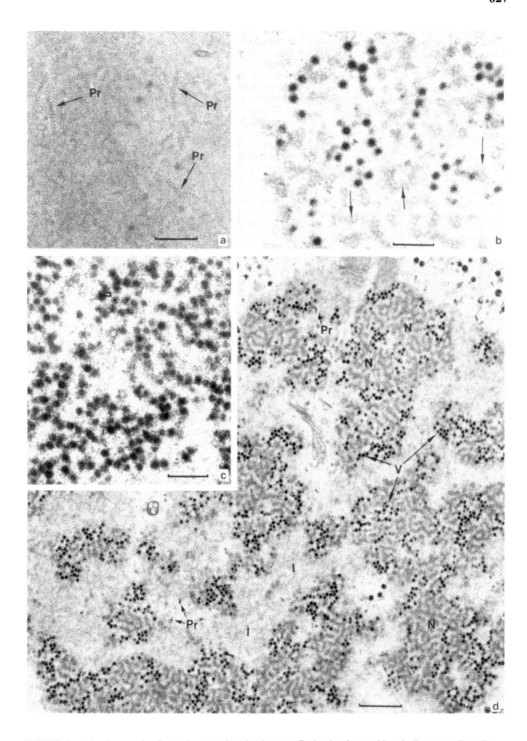

FIGURE 3. Morphogenesis of scorpion cytoplasmic virus. (a) Early viroplasm with polyribosomes (Pr). (Bar = 200 nm.) (b) Formation of a network of moderately electron-dense material (arrows) and of the early particles. (Bar = 200 nm.) (c) Virions have taken the place of network. (Bar = 200 nm.) (d) Large intracytoplasmic viroplasm with residual disorganized area (I) containing polyribosomes (Pr), network of electron-dense material (N) and assembled viral particles (V). (Bar = 0.5 μm.)

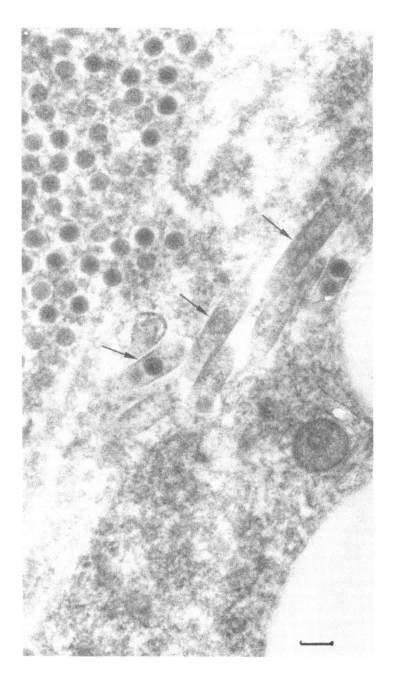

FIGURE 4.   Tubelike structures (arrows) containing scorpion cytoplasmic virus associated with late viroplasm. (Bar = 100 nm.)

2. Icosahedral Virus

Negatively stained virions are icosahedral, 75 to 80 nm in diameter. They have regular angular outlines (see Figure 8). Capsomers cannot be distinguished on the surface of the particles.

C. CYTOPATHOLOGY
1. Baculovirus-like Particles

The whole morphogenesis of virions occurs in the nucleus (see Figure 6). Early stages

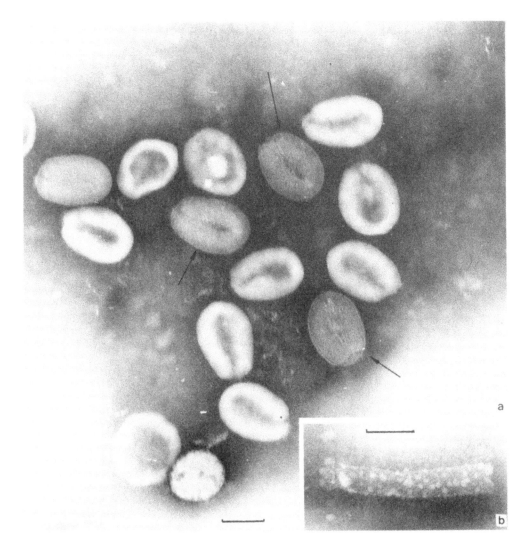

FIGURE 5. Spider baculovirus-like particles (negatively stained preparation [PTA]) showing: (a) coffee bean appearance of virions with U-shaped nucleocapsid (arrows) in penetrated particles; (bar = 100 nm); (b) rod-shaped nucleocapsid outside envelope (bar = 100 nm). ([a] From Morel, G., Bergoin, M., and Vago, C., *C. R. Acad. Sci. Paris, Ser. D*, 285, 933, 1977. With permission.)

of morphogenesis give straight rod-shaped nucleocapsids 300 nm by 40 nm. They appear within granular electron dense material (see Figure 7a). Envelopment takes place directly in the virogenic stromata by *de novo* synthesis of the envelopes. First, one end of the nucleocapsid is capped by diffuse material (see Figure 7b); then, this less electron-dense material spreads progressively along the nucleocapsid. Bending of the nucleocapsids apparently occurs at the end of the envelopment process, and only after this bending the envelope presents the characteristic trilaminar aspect of a membrane-like structure. Intranuclear membrane-like formations sometimes appear without associated nucleocapsids.

2. ICOSAHEDRAL VIRUS

Finely granular virogenic stromata form in the center of the nucleus and grow until they constitute the major part of the nucleus (see Figure 9a). Developed viroplasms are paraspherical and host chromatin is pushed back along the nuclear envelope.

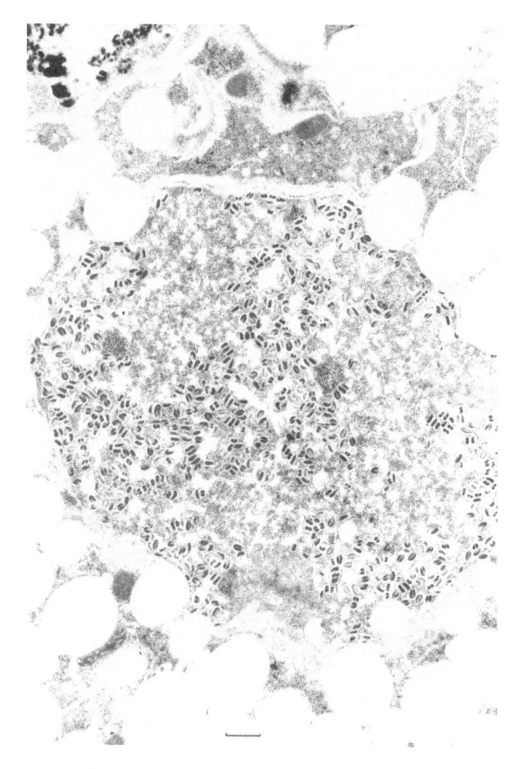

FIGURE 6.   Thin section of hepatopancreatic cell of spider *Pisaura mirabilis* Cl. Nucleus of infected cell contains numerous enveloped particles with U-shaped nucleocapsids. (Bar = 0.5 μm.) (From Morel, G., Bergoin, M., and Vago, C., *C. R. Acad. Sci. Paris, Ser. D*, 285, 933, 1977. With permission.)

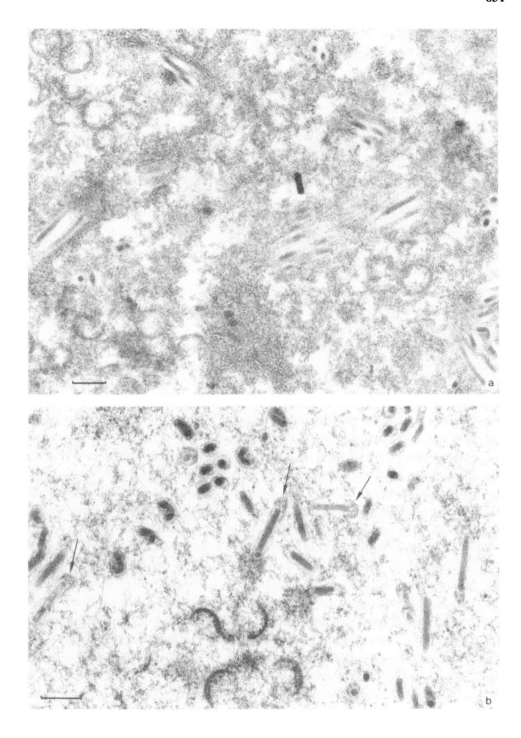

FIGURE 7. Morphogenesis of spider baculovirus-like particles. (a) In early stages of morphogenesis, straight, nonenveloped nucleocapsids appear. (Bar = 200 nm.) (b) Envelopment begins by capping (arrows) at one end of the rod; completely enveloped particles have U-shaped nucleocapsid. (Bar = 200 nm.)

FIGURE 8. Spider ico-
sahedral virus. (Nega-
tively stained particles
[PTA]). (Bar = 100 nm.)

Morphogenesis of virions takes place principally inside viroplasms but some viruses appear on the surface of the virogenic stromata. These external viruses regroup to form an accumulation of regularly packed viruses (see Figure 9a). In the same way, viruses regroup in crystalline arrays inside viroplasms (see Figure 9b). They constitute many groups which later form an irregular network inside residual virogenic stromata.

During cell lysis, nuclear envelopes disappear and cytoplasms become disorganized but the paraspherical shape of viroplasms is preserved (see Figure 10). Virions disperse in the meshes of the network; they leave viroplasms and spread to the cytoplasm.

D. RELATIONSHIPS WITH CLASSIFIED VIRUSES

The enveloped rod-shaped virus, in spite of the bending of its nucleocapsid, appears to be related with Baculoviridae like the nonoccluded Baculovirus of the subgroup C.[10] The icosahedral virus is probably related to the Adenoviridae based on its morphology, size, and cytopathology in infected nuclei of susceptible tissues.

IV. SPIDER MITE VIRUSES

A few icosahedral or rod-shaped virus-like particles have been observed in the European red mite *(Panonychus ulmi)*[4,6] and in the citrus red mite *(P. citri)*.[5,11] For the moment, only three rod-shaped particles are retained as viruses and cause diseases in the mites.[11] Two of them have been found in *P. ulmi* and one in *P. citri*.

A. PATHOLOGY

The viruses cause lethal diseases in the mites. *P. citri* virus develops in nuclei of midgut

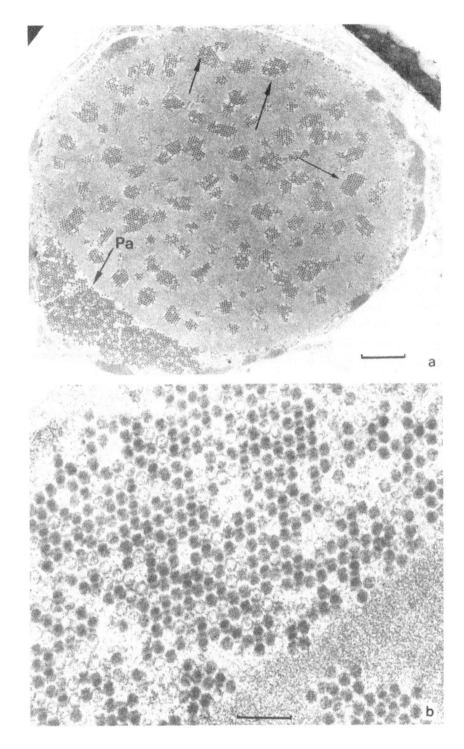

FIGURE 9. Thin section of hepatopancreatic cell of spider *Pisaura mirabilis* Cl. infected with icosahedral virus. Virus morphogenesis has occurred in the nucleus. (a) Paracrystalline arrays inside (arrows) or outside (Pa) paraspherical viroplasm. (Bar = 1 μm.) (b) Details of paracrystalline array of virions showing numerous empty particles. (Bar = 200 nm.)

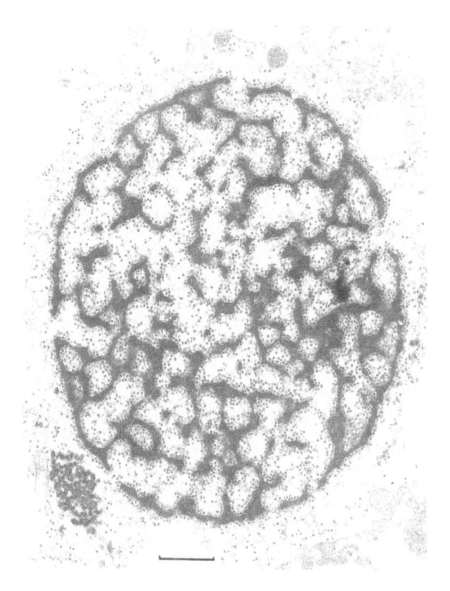

FIGURE 10.   Viroplasm in a disorganized cell. Icosahedral virions leave the viroplasm through meshes in the network. (Bar = 1 μm.)

cells, and the presence of birefringent crystals in cells is associated with the disease.[12] Diarrhea is characteristic in this intestinal disease. *P. ulmi* viruses develop in nuclei of fat body cells, but no particular symptoms are associated with the disease.

B. VIRUS STRUCTURE AND CYTOPATHOLOGY

Only *P. citri* virus has been observed after purification.[13] Negatively stained, it is enveloped with a rod-shaped nucleocapsid 206 × 81 nm. Both *P. citri* and *P. ulmi* viruses develop in the nucleus. Intranuclear morphogenesis leads to the formation of enveloped nucleocapsids. The size of viral particles is 266 × 111 nm (nucleocapsid 194 × 58 nm) for *P. citri* virus[5] and 200 × 90 nm (nucleocapsid 150 × 38 nm)[4] or 205 × 65 nm (170 × 155 nm)[6] for *P. ulmi* viruses. Reed et al.[5] have observed that *P. citri* virus takes on a second envelope by budding through the nuclear membrane. During this process, rods are singly or multiply enclosed.

C. RELATIONSHIPS WITH CLASSIFIED VIRUSES

These three viruses of mites are probably related to Baculovirus of the subgroup C as suggested by Reed et al.[5]

ACKNOWLEDGMENTS

I thank Dr. J. R. Bonami for help in the preparation of this chapter and Mr. J. Luciani for his technical assistance.

REFERENCES

1. **Morel, G.,** Un virus cytoplasmique chez le scorpion *Buthus occitanus* Amoreux, *C. R. Acad. Sci. Paris,* 280, 2893, 1975.
2. **Morel, G., Bergoin, M., and Vago, C.,** Mise en évidence chez l'araignée *Pisaura mirabilis* Cl. d'un nouveau type de virus apparenté aux Baculovirus, *C. R. Acad. Sci. Paris,* 285, 933, 1977.
3. **Bergoin, M. and Morel, G.,** Studies on three viruses of arachnids, in: *Progress in Invertebrate Pathology,* Weiser, J., Ed., Agricultural College, Prague, Czechoslovakia, 21, 1978.
4. **Bird, F. T.,** A virus disease of the European red mite *Panonychus ulmi* (Koch), *Can. J. Microbiol.,* 13, 1131, 1967.
5. **Reed, D. K. and Hall, I. M.,** Electron microscopy of a rod-shaped non-inclusion virus infecting the citrus red mite, *J. Invertebr. Pathol.,* 20, 272, 1972.
6. **Sidor, C. and Grbic, V.,** Diseases of red mite (*Panonychus ulmi* Koch) provoked by microorganisms in region of Vojvodina (Yugoslavia), *Plant Prot.,* Beograd, 151, 69, 1980.
7. **Holmes, I. H., Ruck, B. J., Bishop, R. F., and Davidson, G. P.,** Infantile enteritis viruses: morphogenesis and morphology, *J. Virol.,* 16, 937, 1975.
8. **Vidano, C.,** Phases of Maize Rough Dwarf virus multiplication with vector *Laodelphax striatellus* (Fallen), *Virology,* 41, 218, 1970.
9. **Shikata, E.,** Plant reovirus group, in *The Atlas of Insect and Plant Viruses,* Maramorosch K., Ed., Academic Press, New York, 1977, 377.
10. **Matthews, R. E. F.,** Classification and nomenclature of viruses, *Intervirology,* 17, 1, 1982.
11. **Reed, D. K. and Desjardins, P. R.,** Isometric virus-like particles from citrus red mites, *Panonychus citri, J. Invertebr. Pathol.,* 31, 188, 1978.
12. **Reed, D. K., Hall, I. M., Rich, J. E., and Shaw, J. G.,** Birefringent crystal formation in citrus red mites associated with noninclusion virus disease, *J. Invertebr. Pathol.,* 20, 170, 1972.
13. **Reed, D. K. and Desjardins, P. R.,** Morphology of a nonoccluded virus isolated from citrus red mite, *Panonychus citri, Experientia,* 38, 468, 1982.

Chapter 26

VIRUSES FROM BEDBUGS

N. F. Moore, S. M. Eley, R. A. Gardner, and D. H. Molyneux

TABLE OF CONTENTS

I. INTRODUCTION

Bedbugs are obligate hematophagus insects associated with socially deprived human habitation. They can transmit *Trypanosoma cruzi,* the causative agent of Chagas' disease, under experimental conditions, but have not been conclusively demonstrated to be the natural vector of a specific mammalian disease. The normal vectors of *T. cruzi* are Triatomine bugs. Repeated feeding during the bedbug nymphal and adult stages in socially deprived domestic environments would suggest that this insect could transmit the bunyaviruses and reoviruses normally transmitted by other biting insects — this has been extensively examined. In the knowledge that small amounts of blood passed from one individual to another can transmit the causative virus of AIDS the possibility arises that bedbugs could act as mechanical vectors.

II. EVIDENCE FOR VIRUSES ASSOCIATED WITH BEDBUGS

Surprisingly, there are only two reports to suggest that bedbugs may transmit human pathogenic viruses. Williams et al.[1] implicated bedbugs *Strictimex parvus* and *Cimex insuetus* as vectors of Kaeng Khoi bunyavirus in Central Thailand and Wills et al.[2] detected the surface antigen of hepatitis B virus in a number of *C. hemipterus* from bedding in Senegalese villages.

While examining the transmission of bat trypanosomes by the bedbug *(Cimex lectularius)* and the bat bug *(C. pipistrelli),* large numbers of particles, which appeared to have the morphology of viruses, were seen in the epithelial cells of the ventriculus of the bedbug.[3] The virus particles were identified in numerous cytoplasmic arrays. Figure 1 shows a composite figure of a number of electron micrographs and demonstrates the large numbers of arrays present. An enlargement of a typical array is shown in Figure 2. Attempts were made to introduce the virus into a tissue culture system to facilitate its further study. The virus did not replicate in cell cultures of BHK-21, Vero, *Xenopus laevis* (XTC-2), *Drosophila melanogaster, Aedes albopictus,* primary chick embryo fibroblasts, or mouse embryo fibroblasts. No mortality, pathology, or histopathology occurred after intracerebral inoculation of suckling mice.[3]

III. VIRUS STRUCTURE AND COMPOSITION

Virus particles were purified by differential centrifugation and sucrose gradient separation. By electron microscopy both electron-lucent and electron-opaque particles were visible (see Figure 3). The average diameter of the particles was 56.4 nm, and the double-shelled appearance of the virions was similar to that found with reoviruses, though intact reovirus particles are slightly larger.[4] To further characterize the virus particles the nucleic acids were extracted and shown to consist of 11 segments of dsRNA with molecular weights ranging from 0.15×10^6 to 2.75×10^6 (see Figure 4). Comparison with other members of the family *Reoviridae* indicates that only the rotaviruses contain 11 segments of RNA.[4,5]

The morphology of the purified virions, however, is not similar to that of rotaviruses, and there are no known arthropod vectors of rotaviruses. The virus RNA was electrophoresed in parallel with the RNA of reovirus Type 1, an orbivirus and a cytoplasmic polyhedrosis virus[3,6,7] and was distinctly different. In addition the genomic size distribution pattern is dissimilar to the published pattern of the main genera within the *Reoviridae*. Obviously the 11 segments could be the result of a mixed infection, and in the absence of a method of plaque purification in a susceptible culture system this possibility cannot be eliminated. Without further studies and the identification of susceptible hosts it cannot be determined whether this virus is being vectored or is a genuine virus of the bedbug.

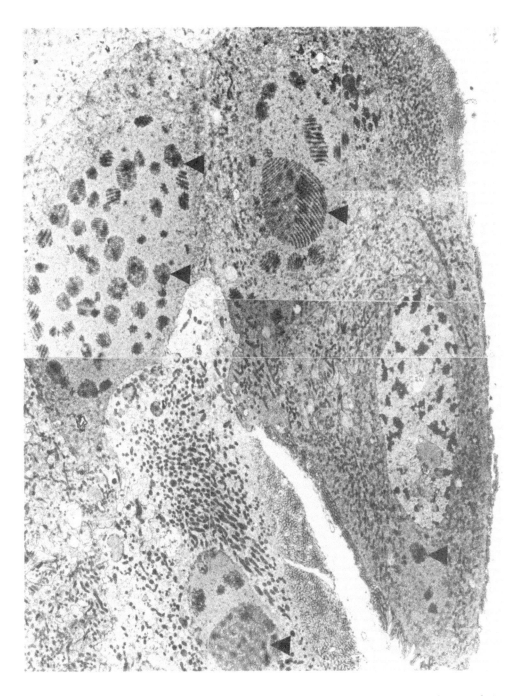

FIGURE 1. Composite picture of several electron micrographs showing the large number of arrays of virus particles present in the cells lining the gut. Some of the arrays are indicated by arrowheads.

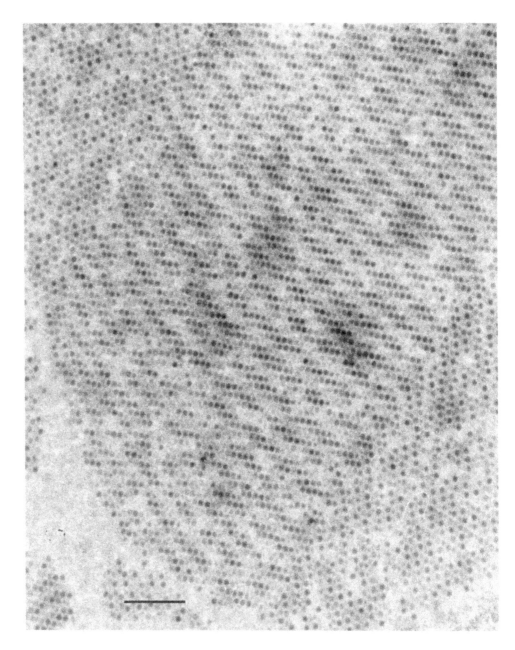

FIGURE 2.   Enlargement of a typical array of virus particles. (Bar = 1000 nm.)

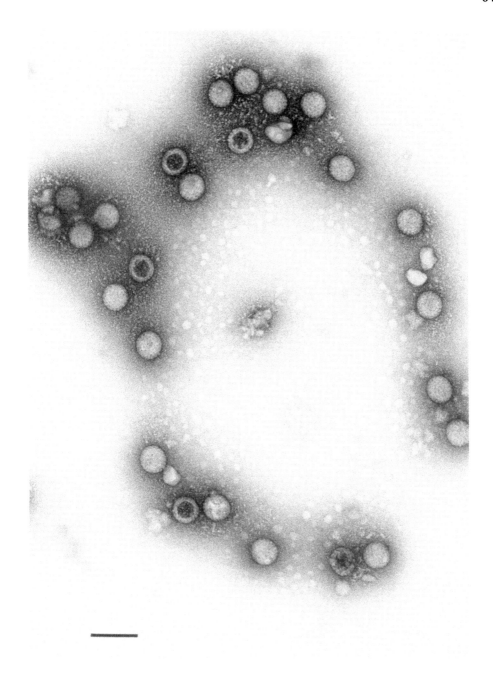

FIGURE 3. Electron micrograph of purified virus particles. The preparation was negatively stained with 2% (w/v) aqueous uranyl acetate for 1 min and examined on a JEOL 100CX electron microscope with an accelerating voltage of 100 kV. (Bar = 100 nm.) Empty particles and intact virions are apparent.

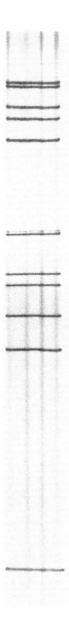

FIGURE 4.   Double-stranded RNA genome of the *Cimex* virus electrophoresed on a 10% (SDS-free) polyacrylamide gel (silver stained by standard methods).

REFERENCES

1. **Williams, J. E., Imlarp, S., Top, Jr., F. H., Cavenaugh, D. C., and Russell, P. K.,** Kaeng-Khoi virus from naturally infected bedbugs Cimicidae and immature free-tailed bats. *Bull. Organ. Mondiale Sante (Switzerland),* 53, 365, 1976.

2. **Wills, W., Larouze, B., London, W. T., Millman, I., Werner, B. G., Ogson, W., Pourtaghva, M., Diallo, S., and Blumberg, B. S.,** Hepatitis B virus in bedbugs *Cimex hemipterus* from Senegal, *Lancet,* (North American ed.), 2, 217, 1977.

3. **Eley, S. M., Gardner, R., Molyneux, D. H., and Moore, N. F.,** A reovirus from the bedbug, *Cimex lectularius, J. Gen. Virol.,* 68, 195, 1987.

4. **Joklik, W. K.,** Reproduction of reoviridae, in *Comprehensive Virology,* Fraenkel-Conrat, H. and Wagner, R. R., Eds., Plenum Press, New York, 1974, 231.

5. **Matthews, R. E. F.,** Classification and nomenclature of viruses, *Intervirology,* 17, 1, 1982.

6. **Pullin, J. S. K. and Moore, N. F.,** Gene assignments of a cytoplasmic polyhedrosis virus (type 2) from *Nymphalis io, Microbiologica,* 8, 131, 1985.

7. **Spence, R. P., Eley, S. M., Nuttall, P. A., Pullin, J. S. K., and Moore, N. F.,** Replication and polypeptide synthesis of Mill Door/79, an orbivirus isolated from ticks from a seabird colony in Scotland, *J. Virol.,* 53, 705, 1985.

APPENDIX

SECTION 1
KEY REFERENCES ON TECHNIQUES FOR LIGHT AND ELECTRON
MICROSCOPY OF PATHOGENS AND TISSUES

I. TEXTS

The references listed below supplement Chapter 2. Some references, however, are repeated for quick access to the subject category.

1. **Agar, A. W., Alderson, R. H., and Chescoe, D.,** Principles and Practice of Electron Microscope Operation, *Practical Methods in Electron Microscopy,* Vol. 2, Glauert, A. M., Ed., North-Holland, Amsterdam, 1974.

2. **Aldrich, H. C. and Todd, W. J., Eds.,** *Ultrastructure Techniques for Microorganisms,* Plenum Press, New York, 1986.

3. **Bradbury, S.,** *Introduction to the Optical Microscope,* Royal Microscopy Society Microscopy Handbook, Oxford University Press, Oxford, 1989.

4. **Chapman, S. K.,** *Maintaining and Monitoring the Transmission Electron Microscope,* Royal Microscopy Society Microscopy Handbook, Oxford University Press, Oxford, 1986.

5. **Dawes, C. J.,** *Biological Techniques in Electron Microscopy,* Barnes and Noble, New York, 1971.

6. **Dawes, C. J.,** *Introduction to Biological Microscopy: Theory and Techniques,* Ladd Research Industries, Burlington, VT, 1988.

7. **Gabriel, B. L.,** *Biological Electron Microscopy,* Van Nostrand Reinhold, New York, 1983.

8. **Glauert, A. M., Ed.,** *Practical Methods in Electron Microscopy,* Vol. 2, North-Holland, Amsterdam, 1974.

9. **Griffith, J. D., Ed.,** *Electron Microscopy in Biology,* Vol. 2, Wiley-Interscience Publ., John Wiley & Sons, New York, 1982.

10. **Hayat, M. A.,** *Principles and Techniques of Electron Microscopy Biological Applications,* Vol. 1, Van Nostrand Reinhold, New York, 1970.

11. **Hayat, M. A.,** *Basic Electron Microscopy Techniques,* Van Nostrand Reinhold, New York, 1972.

12. **Hayat, M. A.,** *Principles and Techniques of Electron Microscopy. Biological Applications,* Vol. 2, 1972; Vol. 3, 1973; Vol. 4, 1974; Vol. 5, 1975; Vol. 6, 1976; Vol. 7, 1977; Vol. 8, 1978; Vol. 9, 1978; Van Nostrand Reinhold, New York.

13. **Hawkes, P. W.,** *The Beginnings of Electron Microscopy,* Academic Press, Orlando, FL, 1985.

14. **Johnson, Jr., J. E.,** Transmission and scanning electron microscopy, in *Current Trends in Morphological Techniques,* Vol. 1, Johnson, J., Ed., CRC Press, Boca Raton, FL, 213, 1981.

15. **Kay, D. H.,** *Techniques for Electron Microscopy,* F. A. Davis, Philadelphia, 1965.

16. **Koehler, J. K., Ed.,** *Advanced Techniques in Biological Electron Microscopy,* Vol. 1, Springer-Verlag, New York, 1973.

17. **Koehler, J. K., Ed.,** *Advanced Techniques in Biological Electron Microscopy, Specific Ultrastructural Probes,* Vol. 2, Springer-Verlag, New York, 1978.

18. **Koehler, J. K., Ed.,** *Advanced Techniques in Biological Electron Microscopy,* Vol. 3, Springer-Verlag, New York, 1986.

19. **Klomparens, K. L., Flegler, S. L., and Hooper, G. R.,** *Procedures for Transmission and Scanning Electron Microscopy for Biological and Medical Science,* Ladd Research Industries, Burlington, VT, 1986.

20. **Ladd, M. W.,** *The Electron Microscope Handbook. Specimen Preparation and Related Laboratory Procedures,* Ladd Research Industries, Burlington, VT, 1973.

21. **Pease, D. C.,** *Histological Techniques for Electron Microscopy,* Academic Press, New York, 1964.

22. **Ploem, J. S. and Tanka, H. J.,** *Introduction to Fluorescence Microscopy,* Royal Microscopy Society Microbiology Handbook, Oxford University Press, Oxford, 1987.

23. **Racker, D. K.,** *Transmission Electron Microscopy-Methods of Application,* Charles C Thomas, Springfield, 1983.
24. **Ruska, E.,** The development of the electron microscope and of electron microscopy, Nobel lecture, *EMSA Bull.,* 18(2), 53, 1988.
25. **Siegel, B. M., Ed.,** *Modern Developments in Electron Microscopy,* Academic Press, New York, 1964.
26. **Siegel, B. M.,** The physics of the electron microscope, in *Modern Developments in Electron Microscopy,* Siegel, B. M., Ed., Academic Press, New York, 1964, 1.
27. **Sjostrand, F. S.,** *Electron Microscopy of Cells and Tissues. Instrumentation and Techniques,* Vol. 1, Academic Press, New York, 1967.
28. **Sommerville, J. and Scheer, U., Eds.,** *Electron Microscopy in Molecular Biology–A Practical Approach,* IRL Press, Oxford, 1987.
29. **Wischnitzer, S.,** The electron microscope, in *Principles and Techniques of Electron Microscopy,* Vol. 3, Hayat, M. A., Ed., Van Nostrand Reinhold, New York, 1973, 1.
30. **Wischnitzer, S.,** *Introduction to Electron Microscopy,* 3rd ed., Pergamon Press, Elmsford, NY, 1981.

The Electron Microscopy Society of America business office has an annual updated reference list on publications in electron microscopy. The address is: EMSA Business Office, P. O. Box EMSA, Woods Hole, MA 02543, U.S.

II. PREPARATIVE TECHNIQUES FOR LIGHT MICROSCOPY

A. GENERAL

1. **Barbosa, P.,** *Manual of Basic Techniques in Insect Histology,* Autumn Publishers, Amherst, 1974.
2. **Bennett, H. S., Wyrick, A. D., Lew, S. W., and McNeil, J. H.,** Science and art in preparing tissues embedded in plastic for light microscopy, with special reference to glycol metharcylate, glass knives and simple stains, *Stain Technol.,* 51, 71, 1976.
3. **Bissing, D. R.,** Haupt's gelatin adhesive with formalin for affixing paraffin sections to slides, *Stain Technol.,* 49, 116, 1974.
4. **Bourgenois, N., Braspenninckx, E., Wijnenn, M., and Buyssens, N.,** The agar-paraffin embedding technique applied to small diagnostic biopsies, *Stain Technol.,* 57, 251, 1982.
5. **Bracegirdle, B.,** The development of biological preparative techniques for light microscopy, 1839—1989, *J. Microsc. (Oxford),* 155, 307, 1989.
6. **Butler, J. K.,** Methods for improved light microscope microtomy, *Stain Technol.,* 54, 53, 1979.
7. **Cole, Jr., M. B. and Sykes, S. M.,** Glycol methacrylate in light microscopy: a routine method for embedding and sectioning animal tissues, *Stain Technol.,* 49, 387, 1974.
8. **Dichiara, J. and Koofmans, H. A.,** New method for reducing static electricity during parafin sectioning, *Stain Technol.,* 52, 353, 1977.
9. **Fink, S.,** Some new methods for affixing sections to glass slides. I. Aqueous adhesives, *Stain Technol.,* 62, 27, 1987.
10. **Gerrits, P. O.,** A simple method to handle tacky glycol methacrylate blocks, *Stain Technol.,* 58, 120, 1983.
11. **Hammond, G. F. and Beckman, B.,** Piccolyte investments for better section cutting, *Stain Technol.,* 53, 113, 1978.
12. **Humason, G. L.,** *Animal Tissue Techniques,* W. H. Freeman, San Francisco, 1962.
13. **King, R. A.,** *Plastic (GMA) Microtomy: A Practical Approach,* Olio-2, Marietta, GA, 1983.
14. **Luna, L. G., Ed.,** *Manual of Histologic Staining Methods of the Armed Forces Institute of Pathology,* 3rd ed., McGraw-Hill, New York, 1960.
15. **Richards, P. R.,** A technique for obtaining ribbons of epoxy and other plastic sections for light microscope histology, *J. Microsc. (Oxford),* 122, 213, 1980.
16. **Scadding, S. R.,** Elimination of static electricity during paraffin sectioning, *Stain Technol.,* 351, 1978.

B. DIAGNOSIS-INVERTEBRATE DISEASES

1. **Kalmakoff, J. and Longworth, J. F., Eds.,** *Microbial Control of Insect Pests,* N.Z. Dep. of Sci. and Ind. Res. Bull., 228, 1980.
2. **Poinar, Jr., G. O. and Thomas, G. M.,** *Laboratory Guide to Insect Pathogens and Parasites,* Plenum Press, New York, 1984.
3. **Thomas, G. M.,** Diagnostic techniques, in *Insect Diseases,* Vol. 1, Cantwell, G. E., Ed., Marcel Dekker, New York, 1974, 1.
4. **Weiser, J.,** *An Atlas of Insect Diseases,* W. Junk B. V. Publ., The Hague, Netherlands, 1977.
5. **Weiser, J. and Briggs, J. D.,** Identification of pathogens in *Microbial Control of Insects and Mites,* Burges, H. D. and Hussey, N. W., Eds., Academic Press, London, 1971, 66.

C. MICROWAVE-AIDED TECHNIQUE

1. **Boon, M. E. and Kok, L. P.**, *Microwave Cookbook of Pathology*, Coulomb Press, Leyden, 1988.
2. **Conder, G. A. and Williams, J. F.**, The microwave oven: a novel means of decontaminating parasitological specimens and glassware, *J. Parasitol.*, 69, 181, 1983.
3. **Kok, L. P., Visser, P. E., and Boon, M. E.**, Histoprocessing with the microwave oven: an update, *Histochemistry*, 20, 323, 1988.
4. **Leong, A. S-Y. and Gove, D. W.**, Microwave techniques for tissue fixation, processing and staining, *EMSA Bull.*, 20, 61, 1990.
5. **Leong, A. S-Y., Milios, J., and Duncis, C. G.**, Antigen preservation in microwave-irradiated tissues: a comparison with formaldehyde fixation, *J. Pathol.*, 156, 157, 1988.
6. **Moran, R. A., Nelson, F., Jagirdar, J., and Paronetto, F.**, Application of microwave irradiation to immunohistochemistry: preservation of antigens of the extracellular matrix, *Stain Technol.*, 63, 263, 1988.
7. **Sanborn, M. R., Wan, S. K., and Bulard, R.**, Microwave sterilization of plastic tissue culture vessels for reuse, *Appl. Environ. Microbiol.*, 44, 960, 1982.
8. **Van de Kant, H. J. G., Boon, M. E., and de Rooij, D. G.**, Microwave-aided technique to detect bromodeoxyuridine in s-phase cells using immunogold-silver staining plastic-embedded sections, *Histochemistry*, 20, 335, 1988.
9. **Wild, P., Krahenbuhl, M., and Schraner, E. M.**, Potency of microwave irradiation during fixation for electron microscopy, *Histochemistry*, 91, 213, 1989.

D. AUTORADIOGRAPHY

1. **Baker, J. R. J.**, *Autoradiography: A Comprehensive Overview*, Royal Microscopy Society Microscopy Handbook 18, Oxford University Press, Oxford, 1989.
2. **Korr, H. and Schmidt, H.**, An improved procedure for background correction in autoradiography, *Histochemistry*, 88, 407, 1988.
3. **Tacha, D. E. and Richards, T. C.**, Comparison of glycol metacrylate and paraffin-embedded tissues for identification of tritium-labeled cells after emulsion radioautography, *Histochemistry*, 4, 59, 1981.
4. **Anon.**, Technical Information, *Nuclear Research Materials — A Range of Materials for Autoradiography*, Ilford Limited Scientific Products Business, Mobberley, England, 1984.
5. **Anon.**, *Photographic Procedures for Light and Electron Microscope Autoradiography*, Kodak Public M6-110, Eastman Kodak, Rochester, NY, 1986.
6. **Anon.**, *Autoradiography at the Light Microscope Level*, Kodak Tech. Bits, Public. P-3-88-1, Eastman Kodak, Rochester, NY, 1988.

E. *IN SITU* HYBRIDIZATION

1. **Brahic, M., Haase, A. T., and Cash, E.**, Simultaneous *in situ* detection of viral RNA and antigens, *Proc. Natl. Acad. Sci. U.S.A.*, 81, 5445, 1984.
2. **Brigati, D. J., Myerson, D., Leang, J. J., Spalholz, B., Travis, S. Z., Fong, C. K. Y., Hsiung, G. D., and Ward, D. C.**, Detection of viral genomes in cultured cells and paraffin-embedded tissue sections using biotin-labeled hybridization probes, *Virology*, 126, 32, 1983.
3. **Calisher, C. H., Auvinen, P., Mitchell, C. J., Rice, C. M., and Hukkanen, V.**, Use of enzyme immunoassay and nucleic acid hybridization for detecting sindbis virus in infected mosquitoes (J V M 00611), *J. Virol. Methods*, 17, 229, 1987.
4. **Chesselet, M-F., Ed.**, *In Situ Hybridization Histochemistry*, CRC Press, Boca Raton, FL, 1990.
5. **Cullen, B. R.**, A rapid and simple procedure for the preparation of homogenously labeled DNA probes, *Gene*, 43, 305, 1986.
6. **Danscher, G. and Rytter Norgaard, J. O.**, Ultrastructural autometallography: a method for silver amplification of catalytic metals, *J. Histochem. Cytochem.*, 33, 706, 1985.
7. **Gall, J. G. and Pardue, M. L.**, Formation and detection of RNA-DNA hybrid molecules in cytological preparations, *Proc. Acad. Natl. Sci., U.S.A.*, 63, 378, 1969.
8. **Gendelman, H. E., Moench, T. R., Nanayan, O., Griffin, D. E., and Clements, J. E.**, A double labeling technique for performing immunocytochemistry and *in situ* hybridization in virus infected cell cultures and tissues, *J. Virol. Methods*, 2, 93, 1985.
9. **Guelin, M. and Lepesant, K.**, *In situ* hybridization: a routine method for parallel localization of DNA sequences and of their transcripts in consecutive paraffin sections with the use of [^{3}H]-labelled nick-translated cloned DNA probes, *Biol. Cell*, 53, 1, 1985.
10. **Haase, A. T., Blum, H., Stowring, L., Geballe, A., Brahic, M., and Jensen, R.**, Hybridization analysis of viral infection at the single cell level in *Clinical Laboratory Molecular Analysis*, Nakamura, T., Ed., Grune & Stratton, 1985, 247.
11. **Janssen, H. P., Van Loon, A. M., Meddens, M. J., Herbrink, P., Lindeman, J., and Quint, W. G. V.**, Comparison of *in situ* DNA hybridization and immunological staining with conventional virus isolation for the detection of human cytomegalovirus infection in cell cultures, *J. Virol. Methods*, 17, 311, 1987.

12. **John, H. A., Birnstiel, M. L., and Jones, K. W.,** RNA-DNA hybrids at the cytological level, *Nature (London)*, 233, 582, 1969.

13. **Lo, C. W.,** Localization of low abundance DNA sequences in tissue sections by *in situ* hybridization, *J. Cell Sci.*, 81, 143, 1986.

14. **McAllister, H. A. and Rock, D. L.,** Comparative usefulness of tissue fixatives for *in situ* viral nucleic acid hybridization, *J. Histochem. Cytochem.*, 33, 1026, 1985.

15. **McDougall, J. K., Myerson, D., and Beckmann, A. M.,** Detection of viral DNA and RNA by *in situ* hybridization, *J. Histochem. Cytochem.*, 34, 33, 1986.

16. **Mullink, H., Walboomers, J. M. M., Raap, A. K., and Meyer, C. J.,** Two colour DNA *in situ* hybridization for the detection of two viral genomes using non-radioactive probes, *Histochemistry*, 91, 195, 1989.

17. **Raap, A. K., Marijner, J. G. J., Vrolijk, J., and Van der Ploeg, M.,** Denaturation, renaturation and loss of DNA during *in situ* hybridization procedures, *Cytometry*, 7, 235, 1986.

18. **Van de Kant, H. J. G., Boon, M. E., and DeRooij, D. G.,** Microwave-aided technique to detect bromodeoxyuridine in S-phase cells using immunogold-silver staining and plastic-embedded sections, *Histochem. J.*, 20, 335, 1988.

F. IMMUNOCYTOCHEMISTRY

1. **Blissard, G. W. and Rohrmann, G. F.,** Location, sequence, transcriptional mapping and temporal expression of the gp64 envelope glycoprotein gene of the *Orgyia pseudotsugata* multicapsid nuclear polyhedrosis virus, *Virology*, 170, 537, 1989.

2. **Bourne, J. A.,** *Handbook of Immunoperoxidase Staining Methods*, Dako, Santa Barbara, 1983.

3. **Cole, Jr., M. B. and Ellinger, J.,** Glycol methacrylate in light microscopy: nucleic acid cytochemistry, *J. Microsc. (Oxford)*, 123, 75, 1981.

4. **DeWaele, M.,** Colloidal gold as marker for the detection of cell surface antigens, *Acta Histochem.*, (Suppl. 35), 53, 1988.

5. **DeWaele, M., DeMey, J., Renmans, W., Labeur, C., Reynaert, R. H., and Van Camp, B.,** An immunogold-silver staining method for detection of cell-surface antigens in light microscopy, *J. Histochem. Cytochem.*, 34, 935, 1986.

6. **Kuhlmann, W. D.,** *Immuno Enzyme Techniques in Cytochemistry*, Verlagchemie, Weinheim, 1984.

7. **Polak, J. M. and Van Noorden, S.,** *An Introduction to Immunocytochemistry: Current Techniques and Problems*, Royal Microscopy Society Microscopy Handbook II, rev. ed., Oxford University Press, Oxford, 1988.

8. **Seno, S., Akita, M., and Hsueh, C. I.,** A new method of the immunohistochemical detection of cellular antigens for light and electron microscopy, *Histochemistry*, 91, 449, 1989.

9. **Sternberger, L.,** *Immunocytochemistry*, Prentice-Hall, Englewood Cliffs, NJ, 1974.

III. PREPARATIVE TECHNIQUES FOR TRANSMISSION ELECTRON MICROSCOPY (TEM)

A. PREPARATION OF SUPPORT FILMS ON SPECIMEN GRIDS

1. **Aebi, U. and Pollard, T. D.,** A glow discharge unit to render electron microscope grids and other surfaces hydrophilic, *J. Electron Microsc. Tech.*, 7, 29, 1987.

2. **Baumeister, W. and Hahn, M.,** Specimen supports, in *Principles and Techniques of Electron Microscopy*, Vol. 8, Hayat, M. A., Ed., Van Nostrand Reinhold, New York, 1978, 1.

3. **Edwards, C. A., Walker, G. K., and Avery, J. K.,** A technique for achieving consistent release of formvar film from clean glass slides, *J. Electron. Microsc. Tech.*, 1, 204, 1984.

4. **Edwards, R. and Price, R.,** Butvar B-98 resin as a section adhesive, *Stain Technol.*, 57, 50, 1982.

5. **Handley, D. A. and Olsen, B. R.,** Butvar B-98 as a thin support film, *Ultramicroscopy*, 4, 479, 1979.

6. **Howard, K. S.,** The preparation of formvar grids for transmission electron microscopy, *EMSA Bull.*, 13(1), 96, 1983.

B. DIAGNOSTIC ELECTRON MICROSCOPY

1. **Almeida, J. D.,** Practical aspects of diagnostic electron microscopy, *Yale J. Biol. Med.*, 53, 5, 1980.

2. **Dickerson, G. R.,** *Diagnostic Electron Microscopy: A Text/Atlas*, Igaku: Shoin, New York, 1988.

3. **Doane, F. W.,** Virus morphology as an aid for rapid diagnosis, *Yale J. Biol. Med.*, 53, 19, 1980.

4. **Doane, F. W.,** Immunoelectron microscopy in diagnostic virology, *Ultrastruct. Pathol.*, 11, 681, 1987.

5. **Dvorak, A. M. and Monahan-Early, R. A.,** *Diagnostic Ultrastructural Pathology EM Pathology Monograph #1*, Philips Electronic Instruments, New Jersey, 1986.

6. **Dvorak, A. M.,** Procedural guide to specimen handling for the ultrastructural pathology service laboratory, *J. Electron Microsc. Tech.,* 6, 255, 1987.
7. **Hsiung, G. D.,** *Diagnostic Virology,* 3rd ed., Yale University Press, New Haven, 1982.
8. **Huang, S. N. and Chan, J.,** Application of electron immunocytochemistry in diagnostic pathology, *EMSA Bull.,* 15(2), 65, 1985.
9. **Johannessen, J. V.,** *Diagnostic Electron Microscopy,* McGraw-Hill, New York, 1987.
10. **Miller, S. E.,** Detection and identification of viruses by electron microscopy, *J. Electron Microsc. Tech.,* 4, 265, 1986.
11. **Miller, S. E.,** Diagnostic virology by electron microscopy, *ASM News,* 54, 475, 1988.
12. **Ramaekers, F. C. S., Voo, J. S. G. P., Huijsmans, A. C. L. M., Pol, M. R. J., and Salet, V. D.,** Immunohistochemistry as an aid in diagnostic cytopathology, *Advances in the Immunohistochemistry,* De Lellis, R. A., Ed., Raven Press, New York, 1988, 133.
13. **Trump, B. F. and Jones, R. T.,** *Diagnostic Electron Microscopy,* John Wiley & Sons, New York, Vol. 1, 1978; Vol. 2, 1979; Vol. 3, 1980; Vol. 4, 1983.

C. NEGATIVE STAINING

1. **Alain, R., Nadon, F., Sequin, C., Payment, P., and Trudel, M.,** Rapid virus subunit visualization by direct sedimentation of samples on electron microscope grids, JVM 00594, *Arch. Anat. Microsc. Morphol. Exp.,* 75, 209, 1986.
2. **Aldrich, H. C. and Coleman, S. E.,** Localization of nucleic acids in sections, in *Ultrastructure Techniques for Microorganisms,* Aldrich, H. C. and Todd, W. J., Eds., Plenum Press, New York, 1986, 447.
3. **Arcidiacono, A., Stasiak, A., and Koller, T.,** Protein-free specimen preparation of nucleic acids and nucleic acid-protein complexes, *Proc. 7th Eur. Congr. Electron Microsc.,* Vol. 2, Brederoo, P. and de Priester, W., Eds., North-Holland, Amsterdam, 1980, 516.
4. **Barth, O. M.,** The use of polylysine during negative staining of viral suspensions, *J. Virol. Methods,* 11, 22, 1985.
5. **Bradley, D. E.,** A study of the negative staining process, *J. Gen. Microbiol.,* 29, 503, 1962.
6. **Breese, Jr., S. S.,** Visualization of viruses by electron microscopy, in *Electron Microscopy in Biology,* Vol. 2, Griffith, J. D., Ed., John Wiley & Sons, New York, 1982, 273.
7. **Hall, C. E.,** In studies on biological macromolecules, in *Modern Developments in Electron Microscopy,* Siegel, B. M., Ed., Academic Press, New York, 1964, 395.
8. **Harris, J. R. and Horne, R. W.,** *Electron Microscopy of Proteins. Viral Structure,* Vol. 5, Academic Press, London, 1986.
9. **Horne, R. W. and Wildy, P.,** An historical account of the development and applications of the negative staining technique to the electron microscopy of viruses, *J. Microsc. (Oxford),* 117, 103, 1979.
10. **Hayat, M. A.,** *Principles and Techniques of Microscopy, Biological Applications,* 3rd ed., CRC Press, Boca Raton, FL, 1989.

D. NUCLEIC ACID EXAMINATIONS

1. **Acetarin, J., Catlemalm, E., and Villiger, W.,** Developments of new Lowicryl® resins for embedding biological specimens at even lower temperatures, *J. Microsc. (Oxford),* 143, 81, 1986.
2. **Arcidiacono, A., Stasiak, A., and Koller, T.,** Protein-free specimen preparation of nucleic acids and nucleic acid-protein complexes, *Proc. 7th Eur., Congr. Electron Microsc.,* Vol. 2, Brederoo, P. and Priester, W., Eds., North-Holland, Amsterdam, 1980, 516.
3. **Aldrich, H. C. and Coleman, S. E.,** Localization of nucleic acids in sections in *Ultrastructure Techniques for Microorganisms,* Aldrich, H. C. and Todd, W. J., Eds., Plenum Press, New York, 1986, 447.
4. **Benada, O.,** A modified BAC method for the preparation of large number of DNA samples for electron microscopy, *J. Electron Microsc. Tech.,* 5, 381, 1987.
5. **Brack, C.,** DNA electron microscopy, *Crit. Rev. Biochem.,* 10, 113, 1981.
6. **Inman, R. B.,** Partial denaturation mapping of DNA determined by electron microscopy, in *Electron Microscopy in Biology,* Vol. 2, Griffith, J. D., Ed., John Wiley & Sons, New York, 1982, 237.
7. **Koller, T. Meer, M., Muller, M., and Muhlethaler, K.,** Electron microscopy of selectively stained molecules, in *Principles and Techniques of Electron Microscopy,* Vol. 3, Hayat, M. A., Ed., Van Nostrand Reinhold, New York, 1975, 55.
8. **Miller, M. F.,** Particle counting of viruses in *Principles and Techniques of Electron Microscopy. Biological Applications,* Vol. 4, Hayat, M. A., Ed., Van Nostrand Reinhold, New York, 1974, 89.
9. **Smith, K. O. and Melnick, J. L.,** A method for staining virus particles and identifying their nucleic acid type in the electron microscope, *Virology,* 17, 480, 1962.
10. **Vollenweider, H. Q., Sogo, J. M., and Koller, T.,** A routine method for protein-free spreading of double and single stranded nucleic acid molecules, *Proc. Natl. Acad. Sci. U.S.A.,* 72, 83, 1975.
11. **Whitaker, H. K. and Alderson, C.,** The use of negative contrast electron microscopy (NCEM) for diagnosis of viral infections in animals, *Am. Assoc. Vet. Lab. Diagnosticians, 23rd Annu. Proc.,* 1980, 321.

12. **Williams, R. C.,** Use of polylysine for adsorption of nucleic acids and enzymes to electron microscope specimen films, *Proc. Natl. Acad. Sci. U.S.A.,* 74, 3211, 1977.

E. QUANTITATION OF VIRUS PARTICLES

1. **Gregory, B. G., Ignoffo, C. M., and Shapiro, M.,** Nucleopolyhedrosis of *Heliothis:* morphological description of inclusion bodies and virions, *J. Invertebr. Pathol.,* 14, 186, 1969.
2. **Martignoni, M. E.,** Separation of two types of viral inclusion bodies by isopycnic centrifugation, *J. Virol.,* 1, 646, 1967.
3. **Miller, II, M. F.,** Virus particle counting by electron microscopy, in *Electron Microscopy in Biology,* Vol. 1, Griffith, J. D., Ed., John Wiley & Sons, New York, 1982, 305.
4. **Miller, II, M. F.,** Particle counting of viruses, in *Principles and Techniques of Electron Microscopy, Biological Applications,* Vol. 4, Hayat, M. A., Ed., Van Nostrand Reinhold, New York, 1974, 89.

F. QUANTITATIVE TECHNIQUES

1. **Bahr, G. F., Engler, W. F., and Mazzone, H. M.,** Determination of the mass of viruses by quantitative electron microscopy, *Q. Rev. Biophys.,* 9, 459, 1976.
2. **Bahr, G. F. and Zeiter, E. H., Eds.,** *Quantitative Electron Microscopy Laboratory Investigation,* Vol. 14, Williams & Wilkins, Baltimore, 1965.
3. **Taylor, D. L. and Wang, Y. L.,** Fluorescence microscopy of living cells in culture. Quantitative fluorescence microscopy-imaging and spectroscopy, in *Methods in Cell Biology,* Vol. 30 (Part B), Academic Press, San Diego, 1989, 503.
4. **Williams, M. A.,** *Quantitative Methods in Biology. Practical Methods in Electron Microscopy.* Vol. 6 (Part 2), Glauert, A. M., Ed., Elsevier, Amsterdam, 1977.

G. PREPARATION OF CELLS, TISSUES, AND VIRUSES FOR TEM
1. Preparation of Small Samples

1. **Arnold, A. R. and Boor, P. J.,** Improved transmission electron microscopy (TEM) of cultured cells through a floating sheet method, *J. Ultrastruct. Mol. Struct.,* 94, 30, 1986.
2. **Asafo-Adjei, E., Bernard, E. F., and Hase, T.,** Growing and embedding monolayer cells *in situ* in Beem capsule caps for light and electron microscopy, *J. Electron Microsc. Tech.,* 5, 367, 1987.
3. **Caceci, T. and Herron, M. A.,** Centrifuge adapters for processing small volume, low concentration cell suspensions, *J. Electron Microsc. Tech.,* 2, 263, 1985.
4. **Carson, J. and Lee, R. M.,** A simple method for the handling and orientation of small specimens for electron microscopy, *J. Microsc. (Oxford),* 126, 201, 1982.
5. **Dalen, H. and Neuralainen, T. J.,** Direct epoxy embedding for vertical sectioning of cells grown as a monolayer on a millipore filter, *Stain Technol.,* 43, 217, 1968.
6. **Day, J. W.,** A beem capsule chamber-pipette for handling small specimens for electron microscopy, *Stain Technol.,* 49, 406, 1974.
7. **Delcarpio, J. B., Underwood, E. G., and Moses, R. L.,** Light, scanning, and transmission electron microscopy of a single population of detergent-extracted cardiac myocytes *in vitro, J. Electron Microsc. Tech.,* 12, 24, 1989.
8. **Deter, R. L.,** Electron microscopic evaluation of subcellular fractions obtained by ultracentrifugation, in *Principles and Techniques of Electron Microscopy,* Vol. 3, Hayat, M. A., Ed., Van Nostrand Reinhold, New York, 1973, 199.
9. **Eakin, R. M. and Westfall, J. A.,** Dissection and oriented embedding of small specimens for ultramicrotomy, *Stain Technol.,* 40, 13, 1965.
10. **Flickinger, C. J.,** Methods for handling small numbers of cells for electron microscopy, in *Methods in Cell Physiology,* Prescott, D. M., Ed., Academic Press, New York, 1966, 311.
11. **Gwynn, I. and Meredith, R. W. J.,** A simple preparation for microscopical investigations of monolayered cells at specific cell-cycle stages, *J. Microsc.,* 118, 187, 1980.
12. **Hayunga, E.,** A specimen holder for dehydrating and processing very small tissue, *Trans. Am. Microsc. Soc.,* 96, 156, 1977.
13. **Heusser, R. C. and Van de Veide, R. L.,** Using a micro-product vial to process minute quantities of free floating cells for electron microscopy, *EMSA Bull., 11(2)* 66, 1981.
14. **Hong, H. and Barta, J. R.,** Processing small biological specimens for transmission electron microscopy, *J. Electron Microsc. Tech.,* 3, 377, 1986.
15. **Hogan, M. E., Hassouna, H. I., and Klomparen, K. L.,** Sectionable microcarrier beads: an improved method for the preparation of cell monolayers for electron microscopy, *J. Electron Microsc. Tech.,* 5, 159, 1987.
16. **Janisch, R.,** Oriented embedding of single cell organisms, *Stain Technol.,* 49, 151, 1974.
17. **Kawamoto, K., Hirano, A., and Herz, F.,** Simplified *in situ* preparation of cultured cell monolayers for electron microscopy, *J. Histochem. Cytochem.,* 28, 178, 1980.

18. **Kellenberger, E. and Ryter, A.**, In Bacteriology, in *Modern Developments in Electron Microscopy*, Siegel, B. M., Ed., Academic Press, New York, 1964, 335.

19. **Kingsley, R. E. and Cole, N. L.**, Preparation of cultured mammalian cells for transmission and scanning electron microscopy using aclar film, *J. Electron Microsc. Tech.*, 10, 77, 1988.

20. **Koukal, M., Trebichavsky, I., Nyklieck, O., and Smolova, M.**, Polyester film as a advantageous tool for light and ultrastructural examination of single cells infected with viruses, *Acta Virol.*, 31, 83, 1987.

21. **Kushida, H.**, New device for handling small blocks of tissue during fixation, dehydration and embedding, *J. Electron Microsc.*, 22, 279, 1973.

22. **Langbein, L.**, Simple devices for the preparation of specimens for scanning and transmission electron microscopy, *Acta Histochem.*, 81, 239, 1987.

23. **Larramendi, P. C. H.**, Method for obtaining a cross section layer of tissue culture for EM without pelleting or re-embedding, *J. Electron Microsc. Tech.*, 10, 119, 1988.

24. **LaVail, III, M. H.**, A method of embedding selected areas of tissue cultures for electron microscopy, *Texas Rep. Biol. Med.*, 26, 215, 1968.

25. **Louw, J., Williams, K., Harper, I. S., and Walfe-Coote**, Electron dense artefactual: the role of ethanol, uranyl acetate and phosphate buffer, *Stain Technol.*, 243, 1990.

26. **Malamed, S.**, Micro-centrifugal processing of isolated mitochondria for electron microscopy, *Proc. 5th Int. Congr. Electron Microsc.* Vol. 2, Philadelphia, 1962, 00-4.

27. **Malamed, S.**, Use of a microcentrifuge for preparation of isolated mitochondria and cell suspensions for electron microscopy, *J. Cell Biol.*, 18, 696, 1963.

28. **Mereau, M., Dive, D., and Vivier, E.**, A new rapid method for collection and preparation of cell suspensions for electron microscopy, *Experientia*, 38, 282, 1982.

29. **Miller, G. J. and Jones, A. S.**, A simple method for the preparation of selected tissue culture cells for transmission electron microscopy, *J. Electron Microsc. Tech.*, 5, 385, 1987.

30. **Pentz, S., Vergani, G., Amthor, H., Hörler, H., and Rich, I.**, A method for electron microscopic preparation of cultured cells monolayer in a new test chamber (TCSC-1), *Microsc. Acta*, 84, 117, 1981.

31. **Plantholt, B. A.**, Embedding cells suspensions in agar, *EMSA Bull.*, 15(2), 1985.

32. **Postek, M. T., Kirk, W. L., and Cox, E. R.**, A container for the processing of delicate organisms for scanning or transmission electron microscopy, *Trans. Am. Microsc. Soc.*, 93, 265, 1974.

33. **Price, Z. H.**, Correlation light and electron microscopy of single cultured cells, in *Principles and Techniques of Electron Microscopy*, Vol. 4, Hayat, M. A., Ed., Van Nostrand Reinhold, New York, 1974, 45.

34. **Purdy-Ramos, S. I., Hawkey, L. A., and Forbes, M. S.**, A simple method for mass embedment of microvessels for cross sectioning, *Stain Technol.*, 59, 243, 1984.

35. **Purdy-Ramos, S. I., Hawkey, L. A., and Forbes, M. S.**, Embedment of specifically oriented, multiple small samples, *EMSA Bull.*, 15(1), 118, 1985.

36. **Redmond, B. L. and Bob, C. F.**, A turbulence-free device for processing small, delicate samples for electron microscopy, *J. Electron Microsc. Tech.*, 5, 379, 1987.

37. **Ridgway, R. L. and Chestnut, M. H.**, Processing small tissue specimens in acrylic resins for ultramicrotomy: improved handling and orientation, *J. Electron Microsc. Tech.*, 1, 205, 1984.

38. **Robbins, E. and Jentzsch, G.**, Rapid embedding of cell culture monolayers and suspensions for electron microscopy, *J. Histochem. Cytochem.*, 15, 181, 1967.

39. **Sabbath, M., Anderson, B., and Ioachim, H. L.**, Cultures grown and embedded *in situ* on spurr discs, a low viscosity epoxy resin for electron microscopy, *Exp. Cell Res.*, 80, 486, 1973.

40. **Samish, M., Adams, J. R., and Vaughn, J. L.**, A method for serial sampling of cell cultures for examination by light or electron microscopy, *Stain Technol.*, 60, 112, 1985.

41. **Shea, T. B. and Berry, S. E.**, A simple method for *in situ* embedding of monolayer cultures for electron microscopy, *J. Tissue Cult. Methods*, 8, 41, 1983.

42. **Siklos, L., Rozsa, M., and Zambori, J.**, A simple method for correlative light, scanning electron microscopic and X-ray microanalytical examination of the same section, *J. Microsc. (Oxford)*, 142, 107, 1986.

43. **Taylor, D. G.**, A technique for embedding small organisms for electron microscopy, *R. Soc. Trop. Med. Hyg.*, 63, 543, 1969.

44. **Zagury, D., Pappas, G. D., and Marcus, P. I.**, Preparation of cell monolayers for combined light and electron microscopy: staining in blocks, *J. Microsc.*, 7, 287, 1968.

2. Orientation of Microorganisms

1. **Griffin, M., Bilello, L., Mapp, P., and Witkus, R.**, A procedure for handling microorganisms during preparation for electron microscopy, *J. Electron Microsc. Tech.*, 7, 61, 1987.

2. **Nordbring-Hertz, B., Veenhuis, M., and Harder, W.**, Dialysis membrane technique for ultrastructural studies of microbial interactions, *Appl. Environ. Microbiol.*, 47, 195, 1984.

3. **Pavlova, I. B. and Katz, L. N.**, A new method of preparing the specimens of microorganisms for electron microscopy, *Microbiology*, 33, 537, 1964.

4. **Reymond, O. L. and Pickett-Heaps, J. D.,** A routine flat embedding method for electron microscopy of microorganisms allowing selection and precisely oriented sectioning of single cells by light microscopy, *J. Microsc. (Oxford),* 130, 79, 1983.

5. **Smith, K. A. and Gehle, W. D.,** Pelleting viruses and virus-infected cells for thin section electron microscopy, *Proc. Soc. Exp. Biol. Med.,* 130, 1117, 1969.

6. **Takagi, I., Sato, T., and Yamada, K.** A rapid method for embedding fractionated samples for electron microscopy, *J. Electron Microsc.,* 28, 316, 1979.

7. **Whitehouse, R. L. S., Bénichou, J. C., and Ryter, A.,** Procedure for the longitudinal orientation of rodshaped bacteria and the production of a high cell density of procaryotic and eucaryotic cells in thin sections for electron microscopy, *Biol. Cell.,* 30, 155, 1977.

3. Fixation

1. **Baur, P. S. and Stacey, T. R.,** The use of PIPES buffer in the fixation of mammalian and marine tissues for electron microscopy, *J. Microsc. (Oxford),* 109, 315, 1977.

2. **Bone, Q. and Denton, E. J.,** The osmotic effects of electron microscope fixatives, *J. Cell Biol.,* 49, 571, 1971.

3. **Bone, O. and Ryan, K. P.,** Osmolarity of osmium tetroxide and glutaraldehyde fixatives, *J. Histochem.,* 4, 331, 1972.

4. **Brown, J. N.,** The avian erythrocyte: a study of fixation for electron microscopy, *J. Microsc. (Oxford),* 104, 293, 1975.

5. **Bullock, G. R.,** The current status of fixation for electron microscopy: a review, *J. Microsc. (Oxford),* 133, 1, 1984.

6. **Coetzee, J. and Van der Merve, C. F.,** Some characteristics of the buffer vehicle in glutaraldehyde based fixatives, *J. Microsc. (Oxford),* 146, 143, 1987.

7. **Fawcett, D. W.,** In Histology and Cytology, in *Modern Developments in Electron Microscopy,* Siegel, B. M., Ed., Academic Press, New York, 1964, 257.

8. **Federman, Q. and Steven, V.,** A new device and method for rapid processing of cytologic smears, histologic sections and cell cultures for electron microscopy, *Acta Cytol.,* 28, 509, 1984.

9. **Gipson, I. and Scott, H. A.,** An electron microscope study of effects of various fixatives and thin-section enzyme treatments on a nuclear polyhedrosis virus, *J. Invertebr. Pathol.,* 26, 171, 1975.

10. **Glauert, A. M.,** The fixation and embedding of biological specimens, *Techniques for Electron Microscopy,* Kay, D. H., Ed., F. A. Davis, Philadelphia, 1965, 166.

11. **Glauert, A. M.,** Fixation, dehydration and embedding of biological specimens, in *Practical Methods in Electron Microscopy,* Vol. 3 (Part 1), Glauert, A. M., Ed., North-Holland, Amsterdam, 1974, 1.

12. **Hayat, M. A.,** *Fixation for Electron Microscopy,* Academic Press, New York, 1981.

13. **Hayat, M. A.,** Glutaraldehyde: Role in electron microscopy, *Micron Microsc. Acta,* 17, 115, 1986.

14. **Hayat, M. A.,** *Basic Techniques for Electron Microscopy,* Academic Press, Orlando, FL, 1986.

15. **Hayat, M. A.,** *Principles and Techniques of Electron Microscopy. Biological Applications,* 3rd ed., CRC Press, Boca Raton, FL, 1989.

16. **Hundgen, M., Weissenfels, N., and Schafer, D.,** The effect of osmolality on the fixation quality of 8 different aldehydes, *Cytobiology,* 4, 62, 1971.

17. **Johnson, T. J. A.,** Aldehyde fixatives: quantification of acid producing reactions, *J. Electron Microsc. Tech.,* 2, 129, 1985.

18. **Kellenberger, E. and Ryter, A.,** In bacteriology, in *Modern Developments in Electron Microscopy,* Siegel, B. M., Ed., Academic Press, New York, 1964, 335.

19. **Kingsley, R. E. and Cole, N. L.,** Osmometry: a comparison of methods, *EMSA Bull.,* 17(2), 74, 1987.

20. **Klomparens, K. L.,** Alternatives to standard fixatives, *EMSA Bull.,* 17(1), 83, 1987.

21. **LaFountain, Jr., J. R., Zobel, C. R., Thomas, H. R., and Galbreath, C.,** Fixation and staining of f-actin and microfilaments using tannic acid, *J. Ultrastruct. Res.,* 58, 78, 1977.

22. **Litman, R. B. and Barnett, R. J.,** The mechanism of the fixation of tissue components by osmium tetroxide via hydrogen bonding, *J. Ultrastruct. Res.,* 38, 63, 1972.

23. **Mangum, C. P. and Johansen, K.,** The colloid osmotic pressures of invertebrate body fluids, *J. Exp. Biol.,* 63, 661, 1975.

24. **Maser, M. P., Powell, III, T. E., and Philpott, C. W.,** Relationships among pH osmolality, and concentration of fixative solutions, *Stain Technol.,* 42, 175, 1967.

25. **Minassian, H. and Huang, S-N,** Effect of sodium azide on the ultrastructural preservation of tissues, *J. Microsc. (Oxford),* 177, 242, 1979.

26. **Mizuhira, V. and Futaesaku, Y.,** On the new approach of tannic acid and digitonin to the biological fixatives, *Proc. Annu. Meet. Electron Microsc. Soc. Am.,* 29, 494, 1971.

27. **Pinto, R. M., Jofre, J., and Bosch, A.,** A simple method for the cultivation of cell monolayers for electron microscopy studies, *Stain Technol.,* 65, 51, 1990.

28. **Porter, K. R.,** Electron microscopy of cultured cells, in *The American Association of Anatomists 1888—1987,* Williams & Wilkins, Baltimore, 1988, 59.
29. **Sabatini, D. D., Bensch, K., and Barnett, R. J.,** Cytochemistry and electron microscopy. The preservation of cellular ultrastructure and enzymatic activity by aldehyde fixation, *J. Cell Biol.,* 17, 19, 1963.
30. **Short, J. A. and Walker, P. D.,** Techniques for transmission electron microscopy, in *Microbial Ultrastructure,* Fuller, R. and Lovelook, D. W., Eds., Academic Press, London, 1976, 1.
31. **Takahashi, M., Suzuki, H., and Yamamoto, I.,** A formaldehyde-osmium tetroxide mixture and a formaldehyde-gultaraldehyde-osmium tetroxide mixture as fixative for electron microscopy, *J. Electron Microsc.,* 27, 149, 1978.
32. **Todd, M. E. and Tokito, M. K.,** Improved ultrastructural detail in tissues fixed with potassium permanganate, *Stain Technol.,* 56, 335, 1981.
33. **Visscher, G. E. and Argentieri, G. J.,** Fundamentals of fixation: an introductory overview, *EMSA Bull.,* 17(2), 83, 1987.
34. **Wagner, R. C.,** The effect of tannic acid on electron images of capillary endothelial cell membranes, *J. Ultrastruct. Res.,* 57, 139, 1976.
35. **Winborn, W. B.,** Paraformaldehyde and s-collidine–a fixative for preserving large blocks for electron microscopy, *Texas Rep. Biol. Med.,* 28, 347, 1970.

a. Microwave-Aided Fixation

1. **Leong, A. S-Y. and Gove, D. W.,** Microwave techniques for tissue fixation, processing and staining, *EMSA Bull.,* 20, 61, 1990.
2. **Login, G. R. and Dvorak, A. N.,** Microwave fixation provides excellent preservation of tissue, cells and antigens for light and electron microscopy, *Histochemistry,* 20, 373, 1988.
3. **Login, G. R., Dwyer, B. K. and Dvorak, A. M.,** Rapid primary microwave- osmium fixation. I. Preservation of structure for electron microscopy in seconds, *J. Histochem. Cytochem.,* 38, 755, 1990.
4. **Van Dort, J. B., De Bruijn, W. C., Schneijdenberg, Ch. T. W. M., Boon, M. E., and L. P. Kok.,** Preservation of structure and cytochemical reactivity at the ultrastructural level, using microwave irradiation, *Histochem. J.,* 20, 365, 1988.

4. Dehydration and Embedding
a. General

1. **Chang, J. P.,** A new technique for separation of coverglass substrate from epoxy-embedded specimens for electron microscopy, *J. Ultrastruct. Res.,* 37, 370, 1971.
2. **Davis, D. B.,** Handling epoxy resins, *EMSA Bull.,* 16(1), 99, 1986.
3. **Doane, F. W., Anderson, N., Chao, J., and Noonan, A.,** Two-hour embedding procedure for intracellular detection of viruses by electron microscopy, *Appl. Microbiol.,* 27, 407, 1974.
4. **Freeman, J. A. and Spurlock, B. O.,** A new epoxy embedment for electron microscopy, *J. Cell Biol.,* 13, 437, 1962.
5. **Frösch, D. and Westphal, C.,** Melamine resins and their application in electron microscopy, *Electron Microsc. Rev.,* 2, 231, 1989.
6. **Hooper, G. R. and Bath, J. E.,** Pressure resin infiltration of aphids for electron microscopy, *Stain Technol.,* 48, 167, 1973.
7. **Johnson, Jr., J. E.,** Problems associated with dehydration and embedding of biological samples for electron microscopy, *Chesapeake Microsc. J.,* 1, 12, 1988.
8. **Ingram, F. D. and Ingram, M. J.,** Freeze-dried, plastic embedded tissue preparation: a review, *Scanning Electron Microsc. Proc. IV,* 147, 1980.
9. **Kondo, H. and Ushiki, T.,** Polyethylene glycol (PEG) embedding and subsequent de-embedding as a method for the correlation of light microscopy, scanning microscopy, and transmission electron microscopy, *J. Electron Microsc. Tech.,* 2, 457, 1985.
10. **Kurtz, S. M.,** A new method for embedding tissues in Vestopal W., *J. Ultrastruct. Res.,* 5, 468, 1961.
11. **Kushida, H.,** Embedding method for electron microscopy in biology, *Tokai J. Exp. Clin. Med.,* 10, 557, 1985.
12. **Kushida, H., Kushida, T., and Aita, S.,** An improved method for embedding with Quetol 651 and ERL 4206 for stereoscopic observation of thick sections under a 400KV(JEM-4000EX) transmission electron microscope, *JEOL News,* 25E, 7, 1987.
13. **Langenberg, W. G., Schroder, H. F., Welch, A. B., and Cook, G. E.,** Epoxy film separable from glass surfaces for selective light and electron microscopy of tissue and *in situ* grown cells, *Stain Technol.,* 47, 303, 1972.
14. **Leduc, E. H. and Bernhard, W.,** Recent modifications of the glycol methacrylate embedding procedure, *J. Ultrastruct. Res.,* 19, 196, 1967.
15. **Leduc, E. H. and Holt, S. J.,** Hydroxypropyl methacrylate, a new water miscible embedding medium for electron microscopy, *J. Cell Biol.,* 26, 137, 1965.

16. **Mollenhauer, H. H.,** Plastic embedding mixtures for use in electron microscopy, *Stain Technol.,* 39, 111, 1964.

17. **Mollenhauer, H. H. and Droleskey, R. E.,** Some characteristics of epoxy embedding resins and how they affect contrast, cell organelle size, and block shrinkage, *J. Electron Microsc. Tech.,* 2, 557, 1985.

18. **Pease, D. C.,** Substitution techniques, in *Advanced Techniques in Biological Electron Microscopy,* Koehler, J. K., Ed., Springer-Verlag, New York, 1973.

19. **Ringo, D. L., Read, D. B., and Cota-Robles, E. H.,** Glove materials for handling epoxy resins, *J. Electron Microsc. Tech.,* 1, 417, 1984.

20. **Spaur, R. C. and Moriarty, G. C.,** Improvements of glycol methacrylate. I. Its use as an embedding medium for electron microscopic studies, *J. Histochem. Cytochem.,* 25, 163, 1977.

21. **Spurlock, B. O., Kattine, V. C., and Freeman, J. A.,** Technical modifications in maraglas embedding, *J. Cell Biol.,* 17, 203, 1963.

22. **Spurr, A. R.,** A low-viscosity epoxy resin embedding medium for electron microscopy, *J. Ultrastruct. Res.,* 26, 31, 1969.

23. **Sutton, J. S.,** *In situ* embedding in epoxy resin of tissues cultured in Leighton tubes, selection of single cells for electron microscopy, *Stain Technol.,* 40, 151, 1965.

24. **Trune, D. R.,** Silicone coating of plastic block holders facilitates their reuse in glycol methacrylate histology, *Stain Technol.,* 63, 121, 1988.

b. Low Temperature Embedding

1. **Acetarin, J-D., Catlemalm, E., and Villiger, W.,** Developments of new Lowicryl® resins for embedding biological specimens at even lower temperatures, *J. Microsc. (Oxford),* 143, 81, 1986.

2. **Acetarin, J. D., Werner, V., and Catlemalm, E.,** A new-heavy metal-containing resin for low-temperature embedding and imaging of unstained sections of biological material, *J. Electron Microsc. Tech.,* 4, 257, 1986.

3. **Altman, L. G., Schnieder, B. G., and Papermaster, D. S.,** Rapid embedding of tissues in Lowicryl® K4M for immunoelectron microscopy, *J. Histochem. Cytochem.,* 32, 1217, 1984.

4. **Armbruster, B. L., Earlemalm, E., Chivotti, R., Garavito, L. R. M., Hobot, J. A., Kellenberger, E., and Villinger, W.,** Specimen preparation for electron microscopy using low temperature embedding resins, *J. Microsc. (Oxford),* 126, 77, 1982.

5. **Armbruster, B. L. and Kellenberger, E.,** Low-temperature embedding, in *Ultrastructure Techniques for Microorganisms,* Aldrich, H. C. and Todd, W. J., Eds., Plenum Press, New York, 1986, 267.

6. **Ashford, A. E., Allaway, W. G., Gubler, F., Lennon, A., and Sleegers, J.,** Temperature control in Lowicryl® K4M and glycol methacrylate during polymerization: is there a low temperature embedding method?, *J. Microsc. (Oxford),* 144, 107, 1986.

7. **Bou-Gharios, G., Adams, G., Moss, J., Shore, I., and Olsen, I.,** A simple technique for *in situ* embedding of monolayer cultures in Lowicryl® K4M, *J. Microsc. (Oxford),* 150, 161, 1987.

8. **Catlemalm, E., Garavito, R. M., and Villiger, W.,** Resin development for electron microscopy and an analysis of embedding at low temperature, *J. Microsc., (Oxford),* 126, 123, 1982.

9. **Chappard, D.,** Uniform polymerization of large blocks in glycol methacrylate at low temperature with special reference to enzyme histochemistry, *Mikroskopie,* 42, 148, 1985.

10. **Hayat, M. A.,** *Principles and Techniques of Electron Microscopy. Biological Applications,* 3rd ed., CRC Press, Boca Raton, FL, 1989.

11. **Robards, A. W. and Sleytr, U. B.,** *Low Temperature Methods in Biological Electron Microscopy. Practical Methods in Electron Microscopy,* Vol. 10, Glauert, A. M., Ed., Elsevier, Amsterdam, 1985.

12. **Simpson-Gomes, A. and Simon, G. T.,** Flat mold embedding with LR White and Lowicryl® K4M, *J. Electron Microsc. Tech.,* 13, 266, 1989.

13. **Simon, G. T., Thomas, J. A., Chorneyko, K. A., and Carlemalm, E.,** Rapid embedding in Lowicryl® K4M for immunoelectron microscopic studies, *J. Electron Microsc. Tech.,* 6, 317, 1987.

14. **Steinbrecht, R. A. and Zierold, K., Eds.,** *Cryotechniques in Biological Electron Microscopy,* Springer-Verlag, Berlin, 1987.

15. **Volker, W. and Frick, B.,** A simple device for low temperature polymerization of Lowicryl® K4M resin, *J. Microsc. (Oxford),* 138, 91, 1985.

5. Ultramicrotomy

1. **Alonso, J. R. and Aijón, J.,** A modified watchmakers forceps for optimal transfer of thin and semithin sections, *Stain Technol.,* 63, 376, 1988.

2. **Barnard, T.,** Thin frozen-dried cryosections and biological X-ray microanalysis, *J. Microsc. (Oxford),* 126, 317, 1982.

3. **Basgen, J. M.,** A technique for evaluating ultramicrotomes, *EMSA Bull.,* 15(2), 48, 1985.

4. **Butcher, W. I., Schmidt, R. E., Elias, F. A., and Hammond, M. J.,** A rapid method for resectioning of semithin large epoxy sections for electron microscopy, *Micron,* 10, 141, 1979.

5. **Chandler, W. B. and Schoenwolf, G. C.,** Wrinkle-free plastic sections for light microscopy, *Stain Technol.,* 58, 238, 1983.
6. **Chien, K.,** New approaches to microtomy, *EMSA Bull.,* 9(2), 11, 1979.
7. **Christian, A.,** Thick sectioning with a dry knife, *EMSA Bull.,* 15(1), 113, 1985.
8. **Couve, E.,** Controlled mounting of serial sections for electron microscopy, *J. Electron Microsc. Tech.,* 3, 453, 1986.
9. **Cramer, C. T.,** An ultrathin sectioning alignment tool with application to cell monolayers, *J. Electron Microsc. Tech.,* 11, 172, 1989.
10. **Dalen, H.,** Vertical sectioning. A. Cells on millipore filters, in *Tissue Culture Methods and Applications,* Kruse, Jr., P. F., and Patterson, Jr., M. K., Eds., Academic Press, New York, 1973, 443.
11. **Davison, E. A. and Rieder, C. L.,** The use of glass fibers as an alternative to hair tools, *J. Electron Microsc. Tech.,* 2, 259, 1985.
12. **Fahrenback, W. H.,** Continuous serial thin sectioning for electron microscopy, *J. Electron Microsc. Tech.,* 1, 387, 1984.
13. **Forbes, M. S.,** Dog hairs as section manipulators, *EMSA Bull.,* 16(2), 67, 1986.
14. **Fraser, T. W.,** The anti-static pistol as an aid to ultrathin sectioning, *J. Microsc. (Oxford),* 106, 97, 1976.
15. **Friedman, M. M.,** A simple plexiglas holder that secures epoxy specimen blocks for precise trimming in a dissecting microscope, *J. Electron Microsc. Tech.,* 2, 281, 1985.
16. **Grimley, P. M.,** Selection for electron microscopy of specific areas in large epoxy tissue sections, *Stain Technol.,* 40, 259, 1965.
17. **Johnson, P. C.,** A rapid setting glue for resectioning and remounting epoxy embedding tissue, *Stain Technol.,* 51, 275, 1976.
18. **Keen, L. N., Reynolds, R., Whittle, W. L., and Kruse, Jr., P. F.,** Vertical sectioning. B. Cells in plastic flasks, in *Tissue Culture-Methods and Applications,* Kruse, Jr., P. F. and Patterson, Jr., M. K., Eds., Academic Press, New York, 1973, 448.
19. **Larramendi, P. C. H.,** Method for obtaining thin sections for EM from different areas of the same block, *J. Electron Microsc. Tech.,* 12, 417, 1989.
20. **Mollenhauer, H. H.,** A simple charge neutralization device for ultrathin sectioning, *J. Electron Microsc. Tech.,* 4, 173, 1986.
21. **Mollenhauer, H. H. and Bradfute, O. E.,** Comparisons of surface roughness of sections cut by diamond, sapphire and glass knives, *J. Electron Microsc. Tech.,* 6, 81, 1987.
22. **Nicholson, P. W.,** A device for static elimination in ultramicrotomy, *Stain Technol.,* 53, 237, 1978.
23. **Nyhlen, L.,** Modified method for preparing teflon-tipped probes for manipulation of ultrathin sections, *Stain Technol.,* 50, 365, 1975.
24. **Porter, K. R.,** Ultramicrotomy, in *Modern Developments in Electron Microscopy,* Siegel, B. M., Ed., Academic Press, New York, 1964, 119.
25. **Reid, N.,** Ultramicrotomy, in *Practical Methods in Electron Microscopy,* Vol. 3, (Part 2), Glauert, A. M., Ed., North-Holland, Amsterdam, 1974, 217.
26. **Reymond, O. L.,** The diamond knife "semi": a substitute for glass or conventional diamond knives in the ultramicrotomy of thin and semi-thin sections, *Basic Appl. Histochem.,* 30, 494, 1986.
27. **Rostgaard, J. and Kvortrup, K.,** Ultrathin sectioning for electron microscopy: the distilled water in the knife through many extract phosphatase reaction products from the sections, *J. Microsc. (Oxford),* 156(2) 253, 1989.
28. **Sakai, T.,** Relation between thickness and interference colors of biological ultrathin sections, *J. Electron Microsc.,* 29, 369, 1980.
29. **Slabe, T. J., Rasmussen, S. T., and Tandler, B.,** A simple method for improving glass knives, *J. Electron Microsc. Tech.,* 15, 316, 1990.
30. **Troyer, D. and Wollert, B.,** Trimming selected areas of embedments for electron microscopy-dead simple, *Stain Technol.,* 57, 289, 1982.
31. **Ward, R. T.,** Some observations on glass-knife making, *Stain Technol.,* 52, 305, 1977.
32. **Zelechowska, M. G. and Potworowski, E. F.,** Improved adhesion of ultrathin sections to filmless grids, *J. Electron Microsc. Tech.,* 2, 389, 1985.

6. Cryofixation and Cryoultramicrotomy
1. **Doubochet, J., Adrian, M., Chang, J. J., Homo, J. C., Lepault, J., and McDowall, A. W.,** Cryo-electron microscopy of vitrified specimens, *Q. Rev. Biophys.,* 21, 129, 1988.
2. **Frederick, P. M., Bomans, P. H. H., Busing, W. M., Odeslius, R., and Hax, W. M. A.,** Vapor fixation for immunocytochemistry and X-ray microanalysis on cryoultramicrotome sections, *J. Histochem. Cytochem.,* 322, 636, 1984.
3. **Frederick, P. M., Busing, W. M., and Persson, A.,** Concerning the nature of the cryosectioning process, *J. Microsc. (Oxford),* 125, 167, 1982.

4. **Hayat, M. A.**, *Basic Techniques in Transmission Electron Microscopy*, Academic Press, Orlando, FL, 1986.

5. **Hayat, M. A.**, *Principles and Techniques of Electron Microscopy. Biological Applications*, 3rd ed., CRC Press, Boca Raton, FL, 1989.

6. **Humbel, B. and Muller, M.**, Freeze substitution and low temperature embedding, in *The Science of Biological Specimen Preparation*, Muller, M., Becker, R. P., and Wolosewick, J. J., Eds., SEM, AMF O'Hare, Chicago, 1986, 175.

7. **Ichikawa, M., Sasaki, K., and Ichikawa, A.**, Optimal preparatory procedures of cryofixation for immunocytochemistry, *J. Electron Microsc. Tech.*, 12, 88, 1989.

8. **Karp, R. D. and Silcox, J. C.**, Cryoultramicrotomy: evidence against melting and the use of a low temperature cement for specimen orientation, *J. Microsc. (Oxford)*, 125, 157, 1982.

9. **Parsons, D., Belotto, D. J., Schultz, W. W., Buja, M., and Hagler, H. K.**, Toward routine cryoultramicrotomy, *EMSA Bull.*, 14(2), 49, 1984.

10. **Roomans, G. M., Wei, X., and Seveus, L.**, Cryoultramicrotomy as a preparative method for X-ray microanalysis in pathology, *Ultrastruct. Pathol.*, 3, 65, 1982.

11. **Roos, N. and Morgan, A. J.**, *Cryopreparation of Thin Biological Specimens for Electron Microscopy: Methods and Applications*, Oxford University Press/Royal Microscopical Society, Oxford, 1990.

12. **Simard, R.**, Cryoultramicrotomy, in *Principles and Techniques of Electron Microscopy*, Vol. 6, Hayat, M. A., Ed., Van Nostrand Reinhold, New York, 1976, 290.

13. **Sitte, H., Edelmann, L., and Neumann, K.**, Cryofixation without pretreatment at ambient pressure, in *Cryotechniques in Biological Electron Microscopy*, Steinbrecht, R. A. and Zierold, K., Eds., Springer-Verlag, Berlin, 1987, 87.

14. **Sitte, H., Neumann, K., and Edelmann, L.**, Cryosectioning according to Tokuyasu vs. rapid-freezing, freeze-substitution and resin embedding, in *Immuno-Gold Labeling in Cell Biology*, Verkleij, A. J. and Leunissen, J. L. M., Eds., CRC Press, Boca Raton, FL, 1989, 63.

15. **Sitte, H., Neumann, K., and Edelmann, L.**, Cryofixation and cryosubstitution for routine work in transmission electron microscopy, in *The Science of Biological Specimen Preparation*, Muller, M., Becker, R. A., Boyde, A., and Wolosewick, J. J., Eds., SEM, AMF O'Hare, Chicago, 1986, 103.

16. **Steinbrecht, R. A. and Muller, M.**, Freeze-substitution and freeze-drying, in *Cryotechniques in Biological Electron Microscopy*, Steinbrecht, R. A. and Zierold, K., Eds., Springer-Verlag, Berlin, 1987, 149.

17. **Steinbrecht, R. A. and Zierold, K., Eds.**, *Cryotechnology in Biological Electron Microscopy*, Springer-Verlag, Berlin, 1987.

18. **Tokuyasu, K. T.**, A technique for ultracryotomy of cell suspensions and tissues, *J. Cell Biol.*, 57, 551, 1973.

19. **Tokuyasu, K. T.**, Application of cryoultramicrotomy to immunocytochemistry, *J. Microsc. (Oxford)*, 143, 139, 1986.

20. **Verkleij, A. J. and Leunissen, J. L. M., Eds.**, *Immuno-Gold Labeling in Cell Biology*, CRC Press, Boca Raton, FL, 1989.

21. **Williamson, F. A.**, A device for the reproducible spreading of cryoultramicrotome sections, *J. Microsc. (Oxford)*, 116, 265, 1979.

7. Staining
a. Thick Sections

1. **Anon.**, *Kodak Products for Electron Microscopy*, Kodak Pub. JJ-284, Eastman Kodak, Rochester, NY.

2. **Bennett, H. S., Wyrick, A. D., Lee, S. W., and McNeil, J. H.**, Science and art in preparing tissues embedded in plastic for light microscopy, with special reference to glycol methacrylate, glass knives and simple stains, *Stain Technol.*, 51, 71, 1976.

3. **Braak, F.**, A method of firmly attaching 4—10 μm thick araldite serial sections to glass slides for light microscopic staining procedures, *Stain Technol.*, 52, 54, 1977.

4. **Chang, S. C.**, Hematoxylin-eosin staining of plastic-embedded tissue sections, *Arch. Pathol.*, 93, 351, 1972.

5. **Crowley, H. H. and Leichtling, B. H.**, Elimination or reduction of wrinkles in semithin epoxy sections by vacuum drying, *Stain Technol.*, 64, 221, 1989.

6. **Giammara, B. L. and Hanker, J. S.**, Epoxy-slide embedment of cytochemically stained tissue and cultured cells for light and electron microscopy, *Stain Technol.*, 61, 51, 1986.

7. **Hogan, D. L. and Smith, G.**, Gentle, consistent and rapid solvation of araldite or epon from 2μ thick sections, *EMSA Bull.*, 13(2), 102, 1983.

8. **Humphrey, C. D. and Pittman, F. E.**, A simple methylene blue-azure II basic fuchsin stain for epoxy-embedded tissue sections, *Stain Technol.*, 40, 9, 1974.

9. **Kushida, H., Kushida, T., and Iijima, H.**, An improved method for both light and electron microscopy of identical sites in semi-thin tissue sections under 200 kv transmission electron microscope, *J. Electron Microsc.*, 34, 438, 1985.

10. **McGee, R. S. M. and Smale, N. B.,** On colouring open-embedded tissue sections with sudan black B or nile blue A for light microscopy, *Q. J. Microsc. Sci.,* 104, 109, 1963.

11. **Pizzolato, T. D.,** Preparation and flattening of thick epoxy sections for light microscopy, *Can. J. Bot.,* 54, 2405, 1976.

12. **Richardson, R. L., Hinton, D. M., and Campion, D. R.,** An improved method for storing and using stains in electron microscopy, *J. Electron Microsc. Tech.,* 32, 216, 1986.

13. **Ridgway, R. L.,** Flat adherent well contrasted semithin plastic sections for light microscopy, *Stain Technol.,* 61, 253, 1986.

14. **Sato, T. and Shamoto, M.,** A simple rapid polychrome stain for epoxy-embedded tissue, *Stain Technol.,* 48, 223, 1973.

15. **Slater, M.,** Adherence of LR white sections to glass slides for silver enhancement immunogold labeling, *Stain Technol.,* 65, 297, 1989.

16. **Smith, N. A., Kilpatrick, J., and Bain, R.,** Staining large epoxy resin embedded histologic sections by a simplified hematoxylin and eosin method using heat, *Stain Technol.,* 60, 59, 1988.

17. **Spurlock, B. O., Skinner, M. S., and Kattine, A. A.,** A simple rapid method for staining epoxy-embedded specimens for light microscopy with the polychromatic stain Paragon-1301, *Am. J. Clin. Pathol.,* 46, 252, 1966.

18. **Tandler, B.** Improved slides of semithin sections, *J. Electron Microsc. Tech.,* 14, 285, 1990.

19. **Troyer, H. and Babich, E.,** A hematoxylin and eosin-like stain for glycol methacrylate embedded tissue sections, *Stain Technol.,* 56, 39, 1981.

20. **von Hees, H. and Rothbacher, I.,** On cutting and staining of thick sections, *Mikroskopie,* 42, 267, 1986.

21. **Weyda, F.,** Adaptation of Mallory's trichrome stain to insect tissue epoxy sections, *Z. Mikrosk.-Anat. Forsch.,* 96, 79, 1982.

b. Thin Sections

1. **Anon.,** *Kodak Products for Electron Microscopy,* Kodak Publ. JJ-284, Eastman Kodak, Rochester, NY, 1981.

2. **Avery, S. W. and Ellis, E. A.,** Methods for removing uranyl acetate precipitate from ultra-thin sections, *Stain Technol.,* 53, 137, 1978.

3. **Boyles, J.,** The use of mordants and osmium to stain lipids, *EMSA Bull.,* 14(2), 73, 1984.

4. **Cardamone, Jr., J. J.,** A simple and inexpensive multiple grid staining device, *EMSA Bull.,* 12(1), 78, 1982.

5. **Craig, S.,** Useful tips for staining grids, *EMSA Bull.,* 14(2), 106, 1984.

6. **Ellis, E. A. and Anthony, D. W.,** A method for removing precipitate from ultrathin sections resulting from glutaraldehyde-osmium tetroxide fixation, *Stain Technol.,* 54, 282, 1979.

7. **Erickson, P. A., Anderson, D. H., and Fisher, S. K.,** Use of uranyl acetate *en bloc* to improve tissue preservation and labeling for post-embedding immunoelectron microscopy, *J. Electron Microsc. Tech.,* 5, 314, 1987.

8. **Feldman, D. G.,** A method of staining thin sections with lead hydroxide for precipitate-free sections, *J. Cell Biol.,* 15, 592, 1962.

9. **Frasca, J. M. and Parks, V. R.,** A routine technique for double staining ultrathin sections using uranyl and lead salts, *J. Cell Biol.,* 25, 157, 1965.

10. **Galey, F. and Nilsson, S. E. G.,** A new method for transferring sections from the liquid surface of the trough through staining solutions to the supporting film of a grid, *J. Ultrastruct. Res.,* 14, 405, 1966.

11. **Godkin, S. E.,** Improved staining boxes with fast, uniform staining of ultrathin sections on grids, *Stain Technol.,* 52, 265, 1977.

12. **Hayat, M. A.,** *Positive Staining for Electron Microscopy,* Van Nostrand Reinhold, New York, 1975.

13. **Hanaichi, T., Sato, T., Iwamoto, T., Malavasi-Yamashiro, J., Hoshino, M. H., and Mizuno, N.,** A stable lead by modification of Sato's method, *J. Electron Microsc.,* 35, 304, 1986.

14. **Heinrich, H.,** Precipitate-free lead citrate staining of thin sections, *J. Electron Microsc. Tech.,* 2, 275, 1985.

15. **Järvilehto, M. and Harjula, R.,** A staining syringe for multiple EM grids, *J. Electron Microsc. Tech.,* 4, 171, 1986.

16. **Karnovsky, M. J.,** Simple methods for "staining with lead" at high pH in electron microscopy, *J. Biophys. Biochem. Cytol.,* 11, 729, 1961.

17. **Ko, N-J., Edwardson, J. R., and Zettler, F. W.,** An efficient procedure for staining electron microscope grids, *J. Electron Microsc. Tech.,* 3, 375, 1986.

18. **Kuo, J.,** A simple method for removing stain precipitates from biological sections for transmission electron microscopy, *J. Microsc. (Oxford),* 120, 221, 1980.

19. **Kuo, J., Husca, G. L., and Lucas, L. N. D.,** Forming and removing stain precipitates on ultrathin sections, *Stain Technol.,* 56, 199, 1981.

20. **Kushida, M. and Frijita, K.,** Simultaneous double staining, *J. Electron Microsc.,* 16, 323, 1967.

21. **Lewis, P. R. and Knight, D. P.,** *Staining Methods for Sectioned Material. Practical Methods for Electron Microscopy,* Vol. 5, (Part 1), Glauert, A. M., Ed., North-Holland, Amsterdam, 1977.

22. **Mollenhauer, H. H.,** Poststaining sections for electron microscopy, *Stain Technol.,* 49, 305, 1974.

23. **Mollenhauer, H. H.,** Poststaining sections for electron microscopy: alternate procedures, *Stain Technol.,* 50, 292, 1975.

24. **Mollenhauer, H. H. and Morre, D. J.,** Contamination of thin sections, cause and elimination, in *Electron Microscopy, 1978,* Vol. 2, Sturgess, J. M., Kalins, C. K., Ottensmeyer, F. P., and Simon, G. T., Eds., Imperial Press, Ontario, 1978, 78.

25. **Mollenhauer, H. H.,** Contamination of thin sections: some observation on the cause and elimination of "embedding pepper", *J. Electron Microsc. Tech.,* 5, 59, 1987.

26. **Neiss, W. F.,** A simple method for thorough rinsing of ultrathin sections in cytochemistry, *Stain Technol.,* 62, 201, 1987.

27. **Normann, T. C.,** Staining thin sections with lead hydroxide without contamination by precipitated lead carbonate, *Stain Technol.,* 39, 50, 1964.

28. **Parsons, D. F. and Darden, E. B.,** A technique for the simultaneous dirt-free lead staining several electron microscopes grids of thin sections, *J. Biophys. Biochem. Cytol.,* 8, 835, 1960.

29. **Pylypas, S. P. and Karim, A. C.,** A method for staining ultrathin sections on slotted grids, *J. Electron Microsc. Tech.,* 37, 163, 1988.

30. **Reynolds, E. S.,** The use of lead citrate at high pH as an electron opaque stain in electron microscopy, *J. Cell Biol.,* 17, 208, 1963.

31. **Rich, S. A., Anzola, M. C., and Gibbons, W. E.,** Uranyl acetate staining of thick sections for electron microscope autoradiography, *J. Electron Microsc. Tech.,* 6, 309, 1987.

32. **Riva, A.,** A simple and rapid staining method for enhancing the contrast of tissues previously treated with uranyl acetate, *J. Microscopie,* 19, 105, 1974.

33. **Rowden, G.,** A method of obtaining stained electron-microscope specimens free from contamination, *J. Microsc.,* 89, 229, 1969.

34. **Sato, T.,** A modified method for lead staining of thin sections, *J. Electron Microsc.,* 17, 62, 1968.

35. **Stetson, D. L. and Morrone, M.,** A flow-through staining system for use with multiple EM grids, *J. Electron Microsc. Tech.,* 13, 154, 1989.

36. **Tandler, B.,** Improved uranyl acetate staining for electron microscopy, *J. Electron Microsc. Tech.,* 16, 81, 1990.

37. **Uphoff, C., Raber, B. T., and Cole, Jr., T. B.,** Tannic acid in routine staining of thin sections, *J. Electron Microsc. Tech.,* 1, 419, 1984.

38. **Venable, J. H. and Coggleshall, R.,** A simplified lead citrate stain for use in electron microscopy, *J. Cell Biol.,* 25, 407, 1965.

39. **Walton, J.,** Lead aspartate, an *en bloc* contrast stain particularly useful for ultrastructural enzymology, *J. Histochem. Cytochem.,* 27, 1337, 1979.

H. HIGH VOLTAGE ELECTRON MICROSCOPY

1. **Hama, K.,** High voltage electron microscopy, in *Advanced Techniques in Biological Electron Microscopy,* Koehler, J. K., Ed., Springer-Verlag, New York, 1973, 275.

2. **Johnson, Jr., J. E., Hirsch, P., Fujita, H., Schimizu, R., and Thomas, G.,** Eds., *High Resolution and High Voltage Electron Microscopy,* Alan R. Liss, New York, 1986.

3. **Nierzwicki-Baner, S. A.,** High-voltage electron microscopy, in *Ultrastructure Techniques for Microorganisms,* Aldrich, H. C. and Todd, W. J., Eds., Plenum Press, New York, 1986, 297.

4. **Porter, K. R.,** Electron microscopy in medical research and diagnosis present and future directions. 7. High voltage electron microscopy, *J. Electron Microsc. Tech.,* 4, 142, 1986.

I. EM AUTORADIOGRAPHY

1. **Baker, J. R. J.,** *Autoradiography: A Comprehensive Overview,* Royal Microscopy Society Microscopy Handbook 18, Oxford University Press, Oxford, 1989.

2. **Blackett, N. M. and Parry, D. M.,** A simplified method of "hypothetical grain" analysis of electron microscope autoradiographs, *J. Histochem. Cytochem.,* 25, 206, 1977.

3. **Burry, R. W. and Lasher, R. S.,** Freeze-drying of unfixed monolayer cultures for electron microscope autoradiography, *Histochemistry,* 58, 259, 1978.

4. **Katsumoto, T., Takagi, A., Hirano, A., and Kurimura, T. A.,** New device for electron microscope autoradiography of whole cultured cells, *Experientia,* 36, 689, 1980.

5. **Langager, J. M., Howard, G. A., and Baylink, D. J.,** An improved technique for rapid autoradiography of cells and tissue sections, *Histochemistry,* 75, 523, 1982.

6. **Williams, M. A.,** Autoradiography: its methodology at the present time, *J. Microsc. (Oxford),* 128, 79, 1982.

J. IMMUNOCYTOCHEMISTRY

1. General

1. **Almeida, J. D., Skelly, J., Howard, C. R., and Zuckerman, A. J.**, The use of markers in immune electron microscopy, *J. Virol. Methods*, 2, 169, 1981.
2. **Baigent and Müller, G.**, Carbon-based immunocytochemistry. A new approach to the immunostaining of epoxy-resin-embedded material, *J. Microsc. (Oxford)*, 158, 73, 1989.
3. **Christensen, A. K. and Komorowski, T. E.**, The preparation of ultra thin frozen sections for immunocytochemistry at the electron microscope level, *J. Electron Microsc. Tech.*, 2, 497, 1985.
4. **Clancy, M. J.**, The formation of buffers and media for enzyme histochemistry, *Histochem. J.*, 19, 27, 1987.
5. **Gerrits, P. O., Horobin, R. W., and Hardonk, M. J.**, Use of tissue-free glycol methacrylate sections as semi-permeable membranes: a simple way to shorten incubation times and to improve localization in enzyme histochemistry, *J. Histochem. Cytochem.*, 37, 173, 1989.
6. **Hogan, D. L. and Smith, G. H.**, Unconventional application of standard light and electron immunocytochemical analysis to aldehyde-fixed araldite-embedded tissues, *Histochem. Cytochem.*, 30, 1301, 1982.
7. **Kohn, A. and Katz, D.**, Immunochemistry of viruses, in *Immunocytochemistry of Viruses. The Basis for Serodiagnosis and Vaccines*, Van Regenmortel, M. H. V. and Neurath, A. R., Eds., Elsevier, Amsterdam, 1985, 71.
8. **Kondo, H.**, Polyethylene glycol (PEG) embedding and subsequent de-embedding as a method for the structural and immunocytochemical examination of biological specimens by electron microscopy, *J. Electron Microsc. Tech.*, 1, 227, 1984.
9. **Kuhlmann, W. D.**, Ultrastructural immunoperoxidase cytochemistry, *Prog. Histochem. Cytochem.*, 10, 1, 1977.
10. **Kurstak, E., Tijssen, P., van den Hurk, J., Kurstak, C., and Morisset, R.**, Detection by immunoperoxidase and ELISA of arbovirus antigens and antibodies *in vivo* and *in vitro* systems, in *Invertebrate Systems in Vitro*, Kurstak, E., Maramorosch, K., and Dubendorfer, A., Eds., Elsevier/North-Holland, Amsterdam, 1980, 365.
11. **Larramendi, P. C. H.**, Rapid communication: a good support for the thick frozen sections for immunolabeling when post embedding in plastic for EM is required, *J. Electron Microsc. Tech.*, 10, 123, 1988.
12. **Perkins, W. D. and Koehler, J. K.**, Antibody-labeling techniques, in *Advanced Techniques in Biological Electron Microscopy II*, Koehler, J. K., Ed., Springer-Verlag, Berlin, 1978, 39.
13. **Petrali, J. P.**, PAP postembedding on-the-grid staining, *EMSA Bull.*, 13(2), 104, 1983.
14. **Petrali, J. P., Hinton, D. M., Moriarty, G. C., and Sternberger, L. A.**, The unlabeled antibody enzyme method of immunocytochemistry. Quantitative Comparison of sensitivities with and without peroxidase-antiperoxidase complex, *J. Histochem. Cytochem.*, 22, 782, 1974.
15. **Petrusz, P., Sar, M., Ordronneau, P., and DiMilo, P.**, Specificity in immunocytochemical staining, *J. Histochem. Cytochem.*, 24, 1110, 1976.
16. **Polak, J. M. and Van Noorden, S.**, *An Introduction to Immunocytochemistry, Current Techniques and Problems. Royal Microscopy Society Microscopy Handbook II*, rev. ed., Oxford University Press, Oxford, 1988.
17. **Romano, E. L. and Romano, M.**, Historical aspects, in *Immunolabeling for Electron Microscopy*, Polak, J. M. and Varndell, I. M., Eds., Elsevier, Amsterdam, 1984, 3.
18. **Sternberger, L. A., Donati, E. J., Cuculis, J. J., and Petrali, J. P.**, Indirect immunouranium technique for staining of embedded antigens in electron microscopy, *Exp. Mol. Pathol.*, 4, 112, 1965.
19. **Takes, P. A., Krug, R., and Kewley, S.**, Microwave technology in immunohistochemistry application to avidin-biotin staining of diverse antigens, *J. Histotechnol.*, 12, 95, 1989.
20. **Williams, M. A.**, *Autoradiography and Immunocytochemistry. Practical Methods in Electron Microscopy*, Vol. 6 (Part 1), Glauert, A. M., Ed., Elsevier/North-Holland, Amsterdam, 1977.

2. Colloidal Gold

1. **Balslev, Y. and Hansen, G. H.**, Preparation and use of recombinant protein G-gold complexes as markers in double labeling immunocytochemistry, *Histochem. J.*, 21, 449, 1989.
2. **Barta, J. R. and Corbin, J.**, Use of immunogold-silver staining to visualize antibody-antigen complexes on LR White embedded tissues prior to electron microscopy, *J. Electron Microsc. Tech.*, 16, 83, 1990.
3. **Beesley, J. E.**, *Colloidal Gold: A New Perspective for Cytochemical Marking, Royal Microscopy Society Microscopy Handbook 17*, Oxford University Press, Oxford, 1989.
4. **Bendayan, M.**, Electron microscopial localization of nucleic acids by means of nuclease gold complexes, *Histochem. J.*, 13, 699, 1981.
5. **Bendayan, M.**, Ultrastructural localization of nucleic acid by the use of enzyme-gold complexes, *J. Histochem. Cytochem.*, 29, 531, 1981.
6. **Bendayan, M.**, Ultrastructural localization of nucleic acids by the use of enzyme-gold complexes: influence of fixation and embedding, *Biol. Cell.*, 43, 153, 1982.

7. **Bendayan, M.,** Double immunocytochemical labeling, applying the protein A gold technique, *J. Histochem. Cytochem.,* 30, 81, 1982.

8. **Bendayan, M.,** Enzyme-gold electron microscopic cytochemistry: a new affinity approach for the ultra-structural localization of macromolecules, *J. Electron Microsc. Tech.,* 1, 349, 1984.

9. **Bendayan, M. and Duhr, M-A.,** Modification of the protein A-gold immunocytochemical technique for the enhancement of its efficiency, *J. Histochem. Cytochem.,* 34, 569, 1986.

10. **Bendayan, M., Nanci, A., and Kan, F. W. K.,** Effect of tissue processing on colloidal gold cytochemistry, *J. Histochem. Cytochem.,* 35, 983, 1987.

11. **Bendayan, M. and Puvion, E.,** Ultrastructural detection of RNA: complementarity of high-resolution autoradiography and of RNAase-gold method, *J. Ultrastruct. Res.,* 83, 274, 1983.

12. **Bendayan, M. and Puvion, E.,** Ultrastructural localization of nucleic acids through several cytochemical techniques on osmium-fixed tissues: comparative evaluation of the different labelings, *J. Histochem. Cytochem.,* 32, 1185, 1984.

13. **Bendayan, M. and Zollinger, M.,** Ultrastructure localization of antigenic sites on osmium-fixed tissues applying the protein A-gold technique, *J. Histochem. Cytochem.,* 31, 101, 1983.

14. **Benhamou, N.,** Ultrastructural localization of DNA on ultrathin sections of resin-embedded tissues by the lactoferrin-gold complex, *J. Electron Microsc. Tech.,* 12, 1, 1989.

15. **Birrell, B. G., Hedberg, K. K., and Griffith, O. H.,** Pitfalls of immunogold labeling: analysis by light microscopy, transmission electron microscopy and photoelectron microscopy, *J. Histochem. Cytochem.,* 35, 843, 1987.

16. **Brooks, E. M. and Binnington, K. C.,** Gold labeling with wheat germ agglutinin and RNAse on osmicated tissue embedded in epoxy resin, *J. Histochem. Cytochem.,* 37, 1557, 1989.

17. **Cox, D. P. and Schroff, P. D.,** Immunogold staining reagents: a sensitive technique for biomedical research, *Am. Biotechnol. Lab.,* 4, 18, 1986.

18. **DeWaele, M.,** Colloidal gold as marker for the detection of cell surface antigens, *Acta Histochemica,* Suppl. 35, 53, 1988.

19. **Dirrenberger, M. B.,** Removal of background label in immunocytochemistry with the apolar lowicryls by using washed protein A-gold precoupled antibodies in a one step procedure, *J. Electron Microsc. Tech.,* 11, 109, 1989.

20. **Escolar, G., Sauk, J. J., Bravo, M. L., Krumwiede, M., and White, J. G.,** Development of a simple embedding procedure allowing immunocytochemical localization at the ultrastructural level, *J. Histochem. Cytochem.,* 36, 1579, 1988.

21. **Gough, K. H. and Shukla, D. D.,** Further studies on the use of protein-A in immune electron microscopy for detecting virus particles, *J. Gen. Virol.,* 51, 415, 1980.

22. **Groppoli, T. J.,** The indirect colloidal gold conjugated, antibody labeling procedure; practical aspects, *EMSA Bull.,* 19(2), 103, 1989.

23. **Hainfield, J. F.,** A small gold-conjugated antibody label: improved resolution for electron microscopy, *Science,* 236, 450, 1987.

24. **Hayat, M. A.,** *Colloidal Gold. Principles, Methods and Applications,* Vols. 1 and 2, Academic Press, San Diego, CA, 1989.

25. **Hayat, M. A.,** *Colloidal Gold. Principles, Methods and Applications,* Vol. 3, Academic Press, San Diego, CA, 1990.

26. **Horisberger, M.,** The gold method as applied to lectin cytochemistry in transmission and scanning electron microscopy, in *Techniques in Immunocytochemistry,* Vol. 3, Bullock, G. R. and Petrusz, P., Eds., Academic Press, Orlando, FL, 1985, 155.

27. **Horisberger, M. and Rosset, J.,** Colloidal gold, a useful marker for transmission and scanning electron microscopy, *J. Histochem. Cytochem.,* 25, 295, 1977.

28. **Kjeldsberg, E.,** Demonstration of calicivirus in human faeces by immunosorbent and immunogold-labeling electron microscopy methods, *J. Virol. Methods,* 14, 321, 1986.

29. **Lucoq, J. M. and Roth, J.,** Colloidal gold and colloidal silver-metallic markers for light microscope histochemistry, in *Techniques in Immunocytochemistry,* Vol. 3, Bullock, G. R. and Petrusz, P., Eds., Academic Press, Orlando, FL, 1985, 203.

30. **Mar, H., Tsukada, T., Gown, A. M., Wight, T. N., and Baskin, D. G.,** Correlative light and electron microscopic immunocytochemistry on the same section with colloidal gold, *J. Histochem. Cytochem.,* 35, 419, 1987.

31. **Mar, H. and Wight, T. N.,** Colloidal gold immunostaining on deplasticized ultra-thin sections, *J. Histochem. Cytochem.,* 36, 1387, 1988.

32. **Marchetti, A., Bistocchi, M., and Tognetti, A. R.,** Silver enhancement of protein A-gold probes on resin-embedded ultrathin sections. An electron microscope localization of mouse mammary tumor virus (MMTV) antigens, *Histochemistry,* 86, 371, 1987.

33. **Murti, K. G., Portner, A., Troughton, K., and Deshpande, K.,** Localization of proteins on viral nucleocapsids using immunoelectron microscopy, *J. Electron Microsc. Tech.,* 2, 139, 1985.

34. **Namork, E. and Heier, H. E.,** Silver enhancement of gold probes (S-40 nm): single and double labeling of antigenic sites on cell surfaces imaged with backscattered electrons, *J. Electron Microsc. Tech.,* 11, 102, 1989.
35. **Namork, E., Heier, H. E., and Falleth, E.,** Double labeling of cell surface antigens imaged with backscattered electrons, *J. Electron Microsc. Tech.,* 6, 87, 1987.
36. **Pathak, R.K. and Anderson, R. G. W.,** Use of dinitrophenol-IgG conjugates to detect sparse antigens by immunogold labeling, *J. Histochem. Cytochem.,* 37, 69, 1989.
37. **Seno, S., Akita, M., and Hsueh, C. L.,** A new method of the immunohistochemical detection of cellular antigens for light and electron microscopy, *Histochemistry,* 91, 449, 1989.
38. **Silver, M. M. and Hearn, A. S.,** Postembedding immunoelectron microscopy using protein A-gold, *Ultrastruct. Pathol.,* 11, 693, 1987.
39. **Van Lent, J. W. M. and Verduin, B. J. M.,** Detection of viral protein and particles in thin sections of infected plant tissue using immunogold labeling, in *Developments in Applied Biology. 1. Developments and Applications in Virus Testing,* Jones, R. A. C. and Torrance, L., Eds., AAB/Levenham Press, London, 1987, 193.
40. **Weiland, F.,** Ultrastructural visualization of virus-antibody binding by means of colloidal gold-protein-A, *Ann. Virol.,* 132, 549, 1981.

3. *In situ* Hybridization

1. **Binder, M.,** *In situ* hybridization at the electron microscope level, *Scanning Microsc.,* 1, 331, 1987.
2. **Puvion-Dutilleul, F. and Puvion, E.,** Ultrastructural localization of viral DNA in thin sections of herpes simplex virus type 1 infected cells by *in situ* hybridization, *Eur. J. Cell Biol.,* 49, 99, 1989.

K. FREEZE FRACTURE, FREEZE-ETCH

1. **Bullivant, S.,** Freeze-etching and freeze-fracturing, in *Advanced Techniques in Biological Electron Microscopy,* Koehler, J. K., Ed., Springer-Verlag, New York, 1973, 67.
2. **Collins, T. R., Bartholomew, J. C., and Calvin, M.,** A simple method for freeze-fracture of monolayer cultures, *J. Cell Biol.,* 67, 904, 1975.
3. **Falcieri, E., Mariani, A. R., Del Coco, R., Facchini, A., and Maraldi, N. M.,** A high yield technique for freeze-fracturing of small fractions of isolated cells, *J. Submicrosc. Cytol. Pathol.,* 20, 623, 1988.
4. **Hui, S. W.,** *Freeze-Fracture Studies of Membranes,* CRC Press, Boca Raton, FL, 1989.
5. **Martin-Willison, J. H. and Rowe, A. J.,** *Replica, Shadowing and Freeze-Etching Techniques. Practical Methods in Electron Microscopy,* Vol. 8, Glauert, A. M., Ed., Elsevier, Amsterdam, 1980.
6. **Menco, B. P. M.,** A survey of ultra-rapid cryofixation methods with particular emphasis on applications to freeze-fracturing, freeze-etching and freeze substitution, *J. Electron Microsc. Tech.,* 4, 177, 1986.
7. **Pauli, B. U., Weinstine, R. S., Soble, L. W., and Alroy, J.,** Freeze-fracture of monolayer cultures, *J. Cell Biol.,* 72, 763, 1977.
8. **Pinto da Silva, P. and Kan, F. W. K.,** Label-fracture: a method for high resolution labeling of cell surfaces, *J. Cell Biol.,* 99, 1156, 1984.
9. **Prescott, L. and Brightman, M. W.,** A technique for the freeze-fracture of tissue culture, *J. Cell Sci.,* 30, 37, 1978.
10. **Pumplin, D. W., Luther, P. W., Samuelsson, S. J., Ursitti, J. A., and Strong, J.,** Quick-freeze, deep-etch replication of cells in monolayers, *J. Electron Microsc. Tech.,* 14, 342, 1990.
11. **Rash, J. E. and Hudson, C. S., Eds.,** *Freeze-Fracture: Methods, Artifacts and Interpretations,* Raven Press, New York, 1979.
12. **Sjöstrand, F. S.,** The interpretation of pictures of freeze-fractured biological material, *J. Ultrastruct. Res.,* 69, 378, 1979.
13. **Sjöstrand, F. S.,** Low temperature techniques applied for CTEM and STEM analysis of cellular components at a molecular level, *J. Microsc. (Oxford),* 128, 279, 1982.

L. X-RAY MICROANALYSIS AND ENERGY LOSS SPECTROSCOPY

1. **Budd, P. M. and Goodhew, P. J.,** *Light-Element Analysis in the Transmission Electron Microscope, Royal Microscopy Society Microscopy Handbook, 16,* Oxford University Press, Oxford, 1988.
2. **Chandler, J. A.,** *X-Ray Microanalysis in the Electron Microscope. Practical Methods in Electron Microscopy,* Vol. 5 (Part 2), Glauert, A. M., Ed., Elsevier, New York, 1977.
3. **Goldstein, J. I., Newbury, D. E., Echlin, P., Joy, D. C., Fiori, C., and Lifshin, E.,** *Scanning Electron Microscopy and X-Ray Microanalysis: A Text for Biologists, Materials Scientists and Geologists,* Plenum Press, New York, 1984.
4. **Hayat, M. A., Ed.,** *X-Ray Microanalysis in Biology,* University Park Press, Baltimore, 1980.
5. **Howitt, D. G., Medlin, D. L., and Walker, T. M.,** Radiation induced signal losses in analytical electron microscopy, *EMSA Bull.,* 18(2), 69, 1988.
6. **Hren, J. J.,** Analytical electron microscopy: prospects and pitfalls, *EMSA Bull.,* 9(1), 16, 1979.

7. **Hutchinson, T. E. and Somlyo, A. P., Eds.,** *Microprobe Analysis of Biological Systems,* Academic Press, New York, 1981.
8. **Joy, D. C.,** An introduction to analytical electron microscopy and its application to biology, *EMSA Bull.,* 9(1), 9, 1979.
9. **Joy, D. C., Ronnig, Jr., A. D., and Goldstein, J. I., Eds.,** *Principles of Analytical Electron Microscopy,* Plenum Press, New York, 1986.
10. **Morgan, A. J.,** *X-Ray Microanalysis in Electron Microscopy for Biologists,* Royal Microscopy Society Microscopy Handbook, 5, Oxford University Press, Oxford, 1985.
11. **Newbury, D. E., Joy, D. C., Echlin, P., Fiori, C., and Goldstein, J. I.,** *Advanced Scanning Electron Microscopy and X-Ray Microanalysis,* Plenum Press, New York, 1986.
12. **Roman, R. J. and Hatchell, M.,** Energy-dispersive X-ray microanalysis of aqeuous biologic samples on bulk supports, *J. Electron Microsc. Tech.,* 1, 141, 1984.
13. **Roomans, G. M. and Shelburne, J. D.,** *Basic Methods in Biological X-Ray Microanalysis,* Scanning Electron Microscopy, AMF, O'Hare, Chicago, 1983.
14. **Roos, N. and Morgan, A. J.,** *Cryopreparation of Thin Biological Specimens for Electron Microscopy: Methods and Applications,* Oxford University Press/Royal Microscopical Society, Oxford, 1990.
15. **Somlyo, A. P., Ed.,** Recent advances in electron and light optical imaging in biology and medicine, *Ann. N.Y. Acad. Sci.,* Vol. 483, 1986.
16. **Woldseth, R.,** *All You Ever Wanted To Know About X-Ray Energy Spectrometry,* 1st ed., Kevex, California, 1973.
17. **Zaluzec, N. J.,** A beginner's guide to X-ray analysis in an analytical electron microscope. I. Quantification K-factor and si (Li) detector calculations, *EMSA Bull.,* 14(2), 67, 1984.
18. **Zaluzec, N. J.,** A beginner's guide to X-ray analysis in an analytical electron microscope. II. Quantification using absorption and fluorescence corrections, *EMSA Bull.,* 14(2), 61, 1984.
19. **Zaluzec, N. J.,** A beginner's guide to X-ray analysis in analytical electron microscope. III. Additional topics in AEM-based X-ray microanalysis, *EMSA Bull.,* 15(2), 67, 1985.
20. **Zaluzec, N. J.,** A beginner's guide to energy loss spectroscopy in an analytical electron microscope. I. Basic principles, *EMSA Bull.,* 15(2), 72, 1985.
21. **Zaluzec, N. J.,** A beginner's guide to energy loss spectroscopy. II. Electron spectrometers, *EMSA Bull.,* 16(2), 58, 1986.
22. **Zierold, K. and Hagler, H. K., Eds.,** *Electron Probe Microanalysis. Applications in Biology and Medicine,* Springer-Verlag, Berlin, 1989.
23. **Zierold, K. and Schafer, D.,** Preparation of cultured and isolated cells for X-ray microanalysis, *Scanning Microsc.,* 2, 1775, 1988.

M. QUANTITATION TECHNIQUES

1. *Cold Spring Harbor Symposia on Quantitative Biology. XLVI. Organization of the Cytoplasm,* Vol. 1, Cold Spring Harbor Laboratory, Cold Spring Harbor, NY, 1982.
2. *Cold Spring Harbor Symposia on Quantitative Biology. XLVI. Organization of the Cytoplasm,* Vol. 2, Cold Spring Harbor Laboratory, Cold Spring Harbor, NY, 1982.
3. **Peachy, L. D.,** Stereoscopic electron microscopy: principles and methods, *EMSA Bull.,* 8(1), 15, 1978.
4. **Sterling, P.,** Quantitative mapping with the electron microscope, in *Principles and Techniques of Electron Microscopy. Biological Applications,* Vol. 5, Hayat, M. A., Ed., Van Nostrand Reinhold, New York, 1975, 1.
5. **Weibel, E. B.,** Stereological methods for morphometric cytology, *Int. R.W. Rev. Cytol.,* 26, 235, 1969.
6. **Weibel, E. B.,** Stereological Techniques for electron microscopic morphometry, in *Principles and Techniques of Electron Microscopy: Biological Applications,* Vol. 3, Hayat, M. A., Ed., Van Nostrand Reinhold, New York, 1973, 239.

IV. PREPARATIVE TECHNIQUES FOR SCANNING ELECTRON MICROSCOPY

A. FIXATION AND PREPARATION OF SPECIMENS

1. **Abandowitz, H. M. and Geissinger, H. D.,** Preparation of cells from suspensions for correlative scanning electron and interference microscopy, *Histochemistry,* 45, 89, 1975.
2. **Anderson, T. R.,** Techniques for the preservation of the fine structure in preparing specimens for the electron microscope, *Trans. N.Y. Acad. Sci.,* 13, 130, 1951.
3. **Boyde, A.,** Histological and cytological methods for the SEM in biology and medicine, in *Scanning Electron Microscopy,* Wells, O. C., Ed., McGraw-Hill, New York, 1974, 308.

4. **Boyde, A. and Vesely, P.**, Comparison of fixation and drying procedure for preparation of some cultured cell-lines for examination in the SEM, in *Scanning Electron Microscopy, Part II*, Johari, O. and Corvin, I., Eds., IIT Research Institute, Chicago, 1972, 265.

5. **Boyde, A. and Wood, C.**, Preparation of animal tissues for surface scanning electron microscopy, *J. Microsc. (Oxford)*, 90, 221, 1969.

6. **Bram, R. A. and Anthony, D. W.**, Scanning electron microscopy of cultured cells of *Aedes aegypti*, *J. Invertebr. Pathol.*, 21, 107, 1973.

7. **Brunk, U., Collins, V. P., and Arro, E.**, The fixation, dehydration, drying and counting of cultured cells for SEM, *J. Microsc. (Oxford)*, 123, 121, 1981.

8. **Cole, Jr., T. B. and John, D. T.**, Preparing amoebae for scanning electron microscopy, *J. Elec. Microsc. Tech.*, 2, 277, 1985.

9. **Corvellec, M. M., Lalague, E. D., Pelletier, S., Haviernick, S., Derghazarian, C., and Cousineau, G. H.**, Tannic acid and thiocarbohydrazide-mediated osmium tetroxide binding in preparation of human leucocytes for SEM observation, *J. Microsc. (Oxford)*, 132, 229, 1983.

10. **Gabriel, B. L.**, Handling free-living cells, in *Biological Scanning Electron Microscopy*, Van Nostrand Reinhold, New York, 1982, 123.

11. **Garland, C. D., Lee, A., and Dickson, M. R.**, The preservation of surface-associated micro-organisms prepared for scanning electron microscopy, *J. Microsc. (Oxford)*, 116, 227, 1979.

12. **Goldstein, J. I., Newbury, D. E., Echlin, P., Joy, D. C., Fiori, C., and Lifshin, E.**, *Scanning Electron Microscopy and X-Ray Microanalysis: A Text for Biologists, Materials Scientists and Geologists*, Plenum Press, New York, 1984.

13. **Geissinger, H. D. and Abandowitz, H. M.**, Correlated transmission electron microscopy (TEM) and scanning electron microscopy (SEM) of single cells, *Mikroskopie*, 32, 17, 1976.

14. **Hayat, M. A.**, *Principles and Techniques of Scanning Electron Microscopy, Biological Applications*, Vol. 1, 1974; Vol. 2, 1974; Vol. 3, 1975; Vol. 4, 1975; Vol. 5, 1976; Van Nostrand Reinhold, New York.

15. **Hawkes, J. W. and Stehr, C. M.**, Ultrastructural studies of marine organisms: a manual of techniques and applications, *Norelco Rep.*, 27, 27, 1980.

16. **Kurtzman, C. P., Baker, F. L., and Smiley, M. J.**, Specimen holder to critical-point dry microorganisms for scanning electron microscopy, *Appl. Microbiol.*, 24, 708, 1974.

17. **Lewis, J. C. and Roberts, M. K.**, Preparation of particulate samples for scanning electron microscopy, *EMSA Bull.*, 6(2), 17, 1976.

18. **Meller, S. M., Coppe, M. R., Ito, S., and Waterman, R. E.**, Transmission electron microscopy of critical point dried tissue after observation in the scanning electron microscope, *Anat. Rec.*, 176, 245, 1973.

19. **Müller, M., Becker, R. P., Boyde, A., and Wolosewick, J. J., Eds.**, *The Science of Biological Specimen Preparation*, SEM, AMF, O'Hare, Chicago, 1986.

20. **Murphy, J. A. and Roomans, G. M., Eds.**, *Preparation of Biological Specimens for Scanning Electron Microscopy*, Scanning Electron Microscopy, AMF, O'Hare, Chicago, 1984.

21. **Porter, K. R.**, The cytoplasm and its matrix, in *Cells and Tissues: A Three-Dimensional Approach by Modern Techniques in Microscopy, Progress in Clinical and Biological Research*, Vol. 295, Motta, P. M., Ed., Alan R. Liss, New York, 1989, 15.

22. **Revel, J. P., Barnard, T., and Haggis, G. H. Eds.**, *Science of Biological Specimen Preparation*, SEM, AMF, O'Hare, Chicago, 1984.

23. **Roli, J. and Flood, P. R.**, A simple method for the determination of thickness and grain size of deposited films as used on non-conductive specimens for scanning electron microscopy, *J. Microsc. (Oxford)*, 112, 359, 1978.

24. **Safa, A. R. and Tseng, M. T.**, A simple method for scanning electron microscope preparation of cells grown in multiwell culture plates, *Stain Technol.*, 57, 107, 1982.

25. **Sanders, K. R.**, A technique for mounting specimens for SEM, *EMSA Bull.*, 18(2), 91, 1988.

26. **Schroeter, D., Spiess, E., Paweletz, N., and Benke, R.**, A procedure for rupture-free preparation of confluently grown monolayer cells for scanning electron microscopy, *J. Electron Microsc. Tech.*, 1, 219, 1984.

27. **Seviour, R. J., Pethica, L. M., and McClure, S.**, A simple modified procedure for preparing microbial cells for scanning electron microscopy, *J. Microbiol. Methods*, 3, 1, 1984.

28. **Sugimura, M.**, Scanning and transmission electron microscopy of individual cells, *J. Electron Microsc. Tech.*, 22, 367, 1973.

29. **Tanaka, K., Mitsushima, A., Kashima, Y., Nakadera, T., and Fukudome, H.**, Ultra-high resolution scanning electron microscopy of biological materials, in *Cells and Tissues: A Three Dimensional Approach by Modern Techniques in Electron Microscopy, Progress in Clinical and Biological Research*, Vol. 295, Motta, P. M., Ed., 1989, 21.

30. **Taylor, P. G.**, A container handling small specimens during preparation and examination in the scanning electron microscopy (SEM), *J. Microsc. (Oxford)*, 105, 335, 1976.

31. **Varnum, E. and Weiss, R. L.,** A method for mounting wet specimens for scanning electron microscopy, *J. Electron Microsc. Tech.,* 2, 269, 1985.

32. **Watson, L. P., McKee, A. E., and Merrell, B. R.,** Preparation of microbiological specimens for scanning electron microscopy, *Scanning Electron Microsc.,* 2, 45, 1980.

33. **Weiss, R. L.,** A polylysine binding method for scanning electron microscopy, *J. Electron Microsc. Tech.,* 1, 95, 1984.

34. **Witcomb, M. J.,** The suitability of various adhesives as mounting media for scanning electron microscopy, *J. Microsc. (Oxford),* 12, 289, 1981.

35. **Witcomb, M. J.,** The suitability of various adhesives as mounting media for scanning electron microscopy. II. General purpose glues, *J. Microsc. (Oxford),* 139, 75, 1985.

36. **Wolff, A. M.,** The use of carbon-coated formvar films as bacterial adhesion substrates for scanning electron microscopy, *J. Electron Microsc. Tech.,* 10, 315, 1988.

37. **Wells, O. C.,** *Scanning Electron Microscopy,* McGraw-Hill, New York, 1974.

B. CRITICAL POINT DRYING

1. **Bork, T.,** An inexpensive holder for critical point drying numerous flat samples, *EMSA Bull.,* 18(2), 91, 1988.

2. **Bassett, L. A. and Pendergrass, R. E.,** A method for handling free cells through critical point drying, *J. Microsc. (Oxford),* 109, 311, 1975.

3. **Cohen, W. D.,** Simple magnetic holders for critical point drying of microspecimen suspensions, *J. Microsc. (Oxford),* 108, 221, 1976.

4. **Collin, S. M. and Wray, W. A.,** Multisample chamber for dehydration and critical point drying, *J. Electron Microsc. Tech.,* 1, 199, 1984.

5. **Gabriel, B. L.,** Critical point drying, in *Beginning Scanning Electron Microscopy,* Van Nostrand Reinhold, New York, 1982, 96.

6. **Newell, D. G. and Roath, S.,** A container for processing small volumes of cell suspensions for critical point drying, *J. Microsc. (Oxford),* 104, 321, 1975.

7. **Strout, G. W. and Russell, S. D.,** A micro-sample critical point drying device for small SEM and TEM specimens, *J. Electron Microsc. Tech.,* 14, 175, 1990.

C. ALTERNATIVE DRYING OF TISSUES

1. **Kennedy, J. R., Williams, R. W., and Gray, J. P.,** Use of Peldri II, a fluorocarbon solid at room temperature, as an alternative to critical point drying for biological tissues, *J. Electron Microsc. Tech.,* 11, 117, 1989.

2. **Lamoreaux, W.,** Prevention of outgassing when coating tissues dried with hexamethyldisilazane, (HMDS), *EMSA Bull.,* 18(1), 91, 1988.

3. **Nation, J. L.,** A new method using hexamethyldisilazane for preparation of soft insect tissues for scanning electron microscopy, *Stain Technol.,* 58, 347, 1983.

D. COLLOIDAL GOLD LABELING

1. **Anon.,** *Biotechnology and Bioapplications of Colloidal Gold,* Scanning Microscopy International, 1988.

2. **Brown, S. S. and Revel, J. P.,** Cell surface labeling for the scanning electron microscopy, in *Advanced Techniques in Biological Electron Microscopy, Vol. 2,* Koehler, J. K., Ed., Springer Verlag, Berlin, 1978, 65.

3. **De Harven, E. and Leung, R.,** A novel approach for scanning electron microscopy of colloidal gold-labeled cell surfaces, *J. Cell Biol.,* 99, 53, 1984.

4. **De Harven, E., Soligo, D., and Christensen, H.,** Double labeling of cell surface antigens with colloidal gold markers, *Histochem. J.,* 22, 18, 1990.

5. **Hodges, G. M., Southgate, J., and Toulson, E. C.,** Colloidal gold — a powerful tool in scanning electron microscope immunocytochemistry: an overview of bioapplications, *Scanning Microsc.,* 1, 301, 1987.

6. **Horisburger, M.,** The gold method as applied to lectin cytochemistry in transmission and scanning electron microscopy, in *Techniques in Immunocytochemistry, Vol. 3,* Bullock, G. R. and Petrusz, P., Eds., Academic Press, Orlando, FL, 1985, 155.

7. **Horisberger, M. and Rosset, J.,** Colloidal gold, a useful marker for scanning electron microscopy, *J. Histochem. Cytochem.,* 25, 295, 1977.

8. **Kjeldsberg, E.,** Demonstration of calicivirus in human faeces by immunosorbent and immunogold-labeling electron microscopy methods, *J. Virol. Methods,* 14, 321, 1986.

9. **Molday, R. S. and Maher, P.,** A review of cell surface markers and labeling techniques for scanning electron microscopy, *Histochem. J.,* 12, 273, 1980.

10. **Park, K.,** Factors affecting efficiency of colloidal gold staining: pH-dependent stability of protein-gold conjugates, *Scanning Microsc., Suppl.,* 3, 15, 1989.

11. **Yasuda, K., Aiso, S., Nagatani, T., and Yamada, M.,** Direct observations of immunoreactive sites and antibody molecules by ultra-high resolution scanning electron microscope, *J. Electron Microsc. Tech.*, 12, 155, 1989.

V. MISCELLANEOUS

A. CALIBRATION

1. **Agar, A. W.,** Calibration of microscopes for magnifications and resolution, *Microsc. Anal.*, July, 1988.

B. SAFETY

1. **Barber, V. C.,** General safety features of an electron microscope laboratory, *EMSA Bull.*, 14(2), 79, 1984.
2. **Barber, V. C. and Clayton, D. L., Eds.,** *Electron Microscopy Safety Handbook*, 1st ed., San Francisco Press, San Francisco, 1985.
3. **Gambling, T. M.,** A primer on biological safety for the electron microscopist, *EMSA Bull.*, 18(2), 82, 86.
4. **Smithwick, E. B.,** Cautions common sense, and rational for the electron microscopy laboratory, *J. Electron Microsc. Tech.*, 2, 193, 1985.

C. PHOTOGRAPHY

1. **Anon.,** *Electron Microscopy and Photography*, Kodak Public. P-236, Eastman Kodak, Rochester, NY, 1973.
2. **Anon.,** *Scientific Imaging with Kodak Films and Plates*, Kodak Public. P-315, Eastman Kodak, Rochester, NY, 132, 1987.
3. **Anon.,** *Photography Through the Microscope*, Kodak Public. P-2, Eastman Kodak, Rochester, NY, 96, 1985.
4. **Anon.,** *Kodak Products for Electron Micrography*, Kodak Public. N-923, Eastman Kodak, Rochester, NY, 1986.
5. **Anon.,** *Kodak Products for Imaging in the Biological Sciences*, Kodak Public. M6-1, Eastman Kodak, Rochester, NY, 1986.
6. **Anon.,** *Quality Enlarging with Kodak B/W Papers*, Kodak Public. G-1, Eastman Kodak, Rochester, NY, 1982.
7. **Anon.,** *Kodak Scientific Imaging Products*, Kodak Public. L-10, Eastman Kodak, Rochester, NY, 78, 1989.
8. **Farnell, G. C. and Flint, R. B.,** Photographic aspects of electron microscopy, in *Principles and Techniques of Electron Microscopy*, Vol. 5, Hayat, M. A., Ed., Van Nostrand Reinhold, New York, 1975, 19.
9. **Goddard, R. H. and Nelson, R. K.,** The use of kodak technical pan film for electron micrograph projection slides, *EMSA Bull.*, 13(1), 96, 1983.
10. **Lamvik, M. K. and Davilla, S.,** Calibration methods for quantitative image processing in electron microscopy, *J. Electron Microsc. Tech.*, 11, 101, 1989.
11. **Mollenhauer, H. H. and Droleskey, R. E.,** Degradation of negatives during storage — a case report, *J. Electron Microsc. Tech.*, 1, 313, 1975.
12. **Orndoff, M. and Anderson, R.,** Extended capabilities for the electron microscope laboratory darkroom, *EMSA Bull.*, 10(2), 15, 1980.
13. **Premsela, H. F., Nieuwenhuizen, J. M., and Schotanus, B.,** *Photography Techniques for Electron Microscopy*, 2nd ed., Electron Optics Guide, No. 1, Philips Industries, Eindhoven, Netherlands, 1982.
14. **Puchtler, H., McGowan, A. L., and Meloan, S. N.,** Why do some stains fade? II. Effect of dye structure, molecular orbital systems, metals, substrates, moisture, and chemical reactions, *J. Histotechnol.*, 12, 129, 1989.
15. **Weiss, R. L.,** Mounting and removal of electron micrographs for publication, *J. Electron Microsc. Tech.*, 1, 415, 1984.
16. **Wergin, W. P. and Pooley, C. D.,** Photographic and interpretive artifacts, in *Artifacts in Biological Electron Microscopy*, Crang, R. F. E., and Klomparens, K. L., Eds., Plenum Press, New York, 1988, 175.

D. BIOLOGICAL STAINS

1. **Anon.,** *Eastman Biological Stains and Related Products*, JJ-281, Eastman Kodak, Rochester, NY, 1977.
2. **Baker, J. R.,** *Cytological Technique*, 4th ed., John Wiley & Sons, New York, 1963.
3. **Malinin, G. I.,** Permanent fluorescent staining of nucleic acids in isolated cells, *J. Histochem. Cytochem.*, 26, 1018, 1978.
4. **Puchtler, H., McGowan, A. L., and Meloan, S. N.,** Why do some stains fade?. I. History, fastness ratings, light absorption and fluorochromes, *J. Histotechnol.*, 12, 57, 1989.

5. **Puchtler, H., McGowan, A. L., and Meloan, S. N.,** Why do some stains fade?. II. Effect of dye structure, molecular orbital systems, metals, substrates, moisture and chemical reactions, *J. Histotechnol.,* 12, 129, 1989.

E. DETECTION OF VIRUSES IN SINGLE LARVAE

1. **Hamelin, C., Lavallée, C., and Belloncik, S.,** A simplified method for the characterization of nuclear polyhedrosis virus genomes, *FEMS Microbiol. Lett.,* 60, 233, 1989.
2. **Keating, S. T., Burand, J. P., and Elkington, J. S.,** DNA hybridization assay for detection of gypsy moth nuclear polyhedrosis virus in infected gypsy moth (*Lymanthria dispar* L.) larvae, *Appl. Environ. Microbiol.,* 55, 2749, 1989.
3. **Kelly, D. C., Edward, M. L., Evans, H. F., and Robertson, J. S.,** The use of the enzyme linked immunosorbent assay to detect a nuclear polyhedrosis virus in *Heliothis armigera* larvae, *J. Gen. Virol.,* 40, 465, 1978.
4. **Ma, M., Burkholder, J. K., Webb, R. E., and Hsu, H. T.,** Plastic-bead Elisa: an inexpensive epidemiological tool for detecting gypsy moth *(Lepidoptera: Lymantriidae),* nuclear polyhedrosis virus, *J. Econ. Entomol.,* 77, 537, 1984.
5. **Morris, J. J., Vail, P. V., and Collier, S. S.,** An RNA virus in *Autographa californica* nuclear polyhedrosis preparations: detection and identification, *J. Invertebr. Pathol.,* 38, 201, 1981.
6. **Street, D. A. and McGuire, M. R.,** Microbial control of rangeland grasshoppers: new techniques for the detection of entomopathogens, *Mont. Agresearch,* Montana, 1988.
7. **Van Beek, N. A. M., Flore, P. H., Wood, H. A., and Hughes, P. R.,** Rate of increase of *Autographa californica* nuclear polyhedrosis virus in *Trichopulsia ni* larvae determined by DNA: DNA hybridization, *J. Invertebr. Pathol.,* 55, 85, 1990.
8. **Volkman, L. E. and Falon, L. A.,** Use of monoclonal antibody in an enzyme-linked immunosorbent assay to detect the presence of *Trichoplusia ni* (Lepidoptera: Noctuidae) S nuclear polyhedrosis in *T. ni* larvae, *J. Econ. Entomol.,* 75, 868, 1986.
9. **Young, S. Y., Yearian, W. C., and Scott, H. A.,** Detection of nuclear polyhedrosis virus infection in *Heliothis* spp. by agar gel double diffusion, *J. Invertebr. Pathol.,* 26, 309, 1975.

F. CENTRIFUGATION

1. **Birnie, G. D. and Rickwood, D., Eds.,** *Centrifugal Separations in Molecular and Cell Biology,* Butterworths, London, 1978.
2. **Hinton, R. and Dobrota, M.,** *Density Gradient Centrifugation, Laboratory Techniques in Biochemistry and Molecular Biology,* Work, T. S. and Work, E., Eds., North-Holland, Amsterdam, 1976.
3. **Price, C. A.,** *Centrifugation in Density Gradients,* Academic Press, New York, 1982.
4. **Rickwood, D., Ed.,** *Centrifugation. A Practical Approach,* 2nd ed., IRL Press, Oxford, 1984.
5. **Scotti, P. D. and Longworth, J. F.,** Purification of polyhedra from insect extracts contaminated with a small isometric virus, *J. Invertebr. Pathol.,* 32, 216, 1978.
6. **Sheeler, P.,** *Centrifugation in Biology and Medical Science,* John Wiley & Sons, New York, 1981.
7. **Vanden Berghe, D. A.,** A comparison of various density-gradient media for the isolation and characterisation of animal diseases, in *Iodinated Density Gradient Media. A Practical Approach,* Rickwood, D., Ed., IRL Press, Oxford, 1983, 175.

G. ATLASES

1. Invertebrate Viruses

1. **Liang, D. G., Cai, Y. N., Lin, B. Y., Zhang, Q. L., Hu, Y. Y., He, H. J., and Zhao, K. B.,** *The Atlas of Insect Viruses in China,* Human Science and Technology Press, Hunan, China, 1986.
2. **Maramorosch, K., Ed.,** *The Atlas of Insect and Plant Viruses,* Academic Press, New York, 1977.
3. **Tchuchrity, M. G.,** *Ul'trastructura Virusov Cheshuckrylykh-Vrediteli Rastenii* (Atlas), Shtiintsa Publ., Kishinev, 1982.
4. **Weiser, J.,** *An Atlas of Insect Diseases,* Dr. W. Junk B. V. Publ., The Hague, Netherlands, 1977.

2. Ultrastructural Cytology and Cytopathology

1. **Fawcett, D. W.,** In histology and cytology, in *Modern Developments in Electron Microscopy,* Siegel, B. M., Ed., Academic Press, New York, 1964, 257.
2. **Fawcett, D. W.,** *An Atlas of Fine Structure. The cell, its Organelles and Inclusions,* 2nd ed., W. B. Saunders, Philadelphia, 1981.
3. **Fuller, R. and Lovelock, D. W., Eds.,** *Microbial Ultrastructure,* Academic Press, London, 1976.
4. **Ghadially, F. N.,** *Ultrastructural Pathology of the Cell,* Butterworths, London, 1977.
5. **Ghadially, F. N.,** *Ultrastructural Pathology of the Cell and Matrix,* Vols. 1 and 2, 3rd ed., Butterworths, London, 1988.
6. **Porter, K. R. and Bonneville, M. A.,** *An Introduction to Fine Structure of Cells and Tissues,* Lea & Febiger, Philadelphia, 1964.

7. **Porter, K. R.,** The cytomatrix, a short history of its study, *J. Cell Biol.,* 99, 3, 1984.

3. Invertebrate Tissues

1. **Davidson, E. W., Ed.,** *Pathogenesis of Invertebrate Microbial Diseases,* Allanheld, Osmun Publ., Towata, NJ, 1981.
2. **Erickson, E. M.,** *A Scanning Electron Microscopy Atlas of the Honey Bee,* Iowa State University Press, Ames, IA, 1986.
3. **Gibbs, A. J., Ed.,** *Viruses and Invertebrates. North Holland Research Monographs, Frontiers of Biology,* Vol. 31, North-Holland, Amsterdam, 1973.
4. **King, R. C. and Akai, H., Eds.,** *Insect Ultrastructure,* Vol. 1, Plenum Press, New York, 1982.
5. **King, R. C. and Akai, H., Eds.,** *Insect Ultrastructure,* Vol. 2, Plenum Press, New York, 1984.
6. **Smith, D. S.,** *Insect Cells: Their Structure and Function,* Oliver & Boyd, Edinburgh, 1968.
7. **Weiser, J.,** *An Atlas of Insect Diseases,* 2nd ed., Academia Publ. House, Czechoslovakia Acad. of Sciences, Prague, 1977.

SECTION 2
STAINING TECHNIQUES FOR LIGHT MICROSCOPY

A Negative Stain for Light Microscopy
(Used by A. M. Heimpel)

Nigrosin — 10 g distilled water — 100 ml

Place in boiling water bath for 30 min; then add as a preservative 0.5 ml formaldehyde. Filter twice through double filter paper and store in serological test tubes, placing about 5 ml per tube. Store in refrigerator until used.

Place about 50 μl drop (or two drops from plastic Pasteur pipette) of sample on a clean slide. Sterilize small loop. Take one loop or part of a loop of nigrosine and mix thoroughly with sample. Sterilize loop. Allow to air dry or heat dry and examine under oil immersion.

Modified Azan Staining Technique for Occlusion Body Viruses (NPV, CPV, GV, and Entomopox virus)
(From Hamm, J. J., *J. Invertebr. Pathol.,* 8, 125, 1966.)

Solution I — Dissolve 0.1 g of azocarmine G in 100 ml of distilled water and boil the solution for 5 min. Allow to cool and add 2 ml of glacial acetic acid. Filter before use.

Solution II — Dissolve in 100 ml of distilled water: 1.0 g phosphotungstic acid, 0.1 g aniline blue (water soluble), 0.5 g orange G, and 0.2 g fast green FCF.

Staining Procedure

1. Toluene via alcohols to water
2. 50% Acetic acid, 5 min
3. Distilled water rinse, 2 min
4. Azocarmine (Solution I), 15 min
5. Dist water rinse, 5 s
6. Aniline,* 1% in 95% alcohol, 30 s
7. Dist water rinse, 5 s (should be changed often)
8. Counterstain (Solution II), 15 min
9. 50% Alcohol, 10 s
10. Absolute alcohol, 2 changes, 30 s each
11. Toluene, 2 changes
12. Mount in neutral, synthetic mounting medium

Results

Virus inclusion bodies — red
Epicuticle — red
Endocuticle — blue
Muscle — light blue to blue-green
Epidermal cells — yellowish-green

---

* Aniline should be distilled or highly purified and kept in freezer.

Fat body cells — yellowish-green with darker green nuclei
Nerve tissue — light blue
Silk gland — green, contents red or blue
Midgut epithelium — green and blue

References on General Stains for Insect Pathogens
1. **Poinar, Jr., G. O. and Thomas, G. M.**, *Laboratory Guide to Insect Pathogens and Parasites*, Plenum Press, New York, 1984.
2. **Weiser, J. and Briggs, J. D.**, Identification of pathogens, in *Microbial Control of Insects and Mites*, Burges, H. D. and Hussey, N. W., Eds., Academic Press, London, 1971, 13.
3. **Wigley, P. J.**, Practical: diagnosis of virus infections-staining of insect inclusion body viruses, in *Microbial Control of Insect Pests*, Kalmakoff, J. and Longworth, J. F., Eds., N.Z. Dep. Sci. and Industrial Res., DSIR Bull., 228, 1980, 35.
4. **Wigley, P. J.**, Practical: Counting micro-organisms, in *Microbial Control of Insect Pests*, Kalmakoff, J. and Longworth, J. F., Eds., N.Z. Dep. Sci. and Industrial Res., DSIR Bull., 228, 1980, 29.
5. **Smirnoff, W. A.**, Rapid staining of polyhedra from ledidopterous and hymenopterous hosts, *J. Insect Pathol.*, 3, 218, 1961.
6. **Smirnoff, W. A. and O'Connell, D. C.**, Detection of polyhedra from insect virus diseases on filter membranes, *Stain Technol.*, 37, 207, 1962.
7. **Palmieri, J. R., Kelderman, W., and Sullivan, J. T.**, A staining technique for smears containing inclusion body viruses, *J. Invertebr. Pathol.*, 31, 264, 1978.
8. **Huger, A.**, Methods for staining capsular virus inclusion bodies typical of granuloses of insects, *J. Insect Pathol.*, 3, 338, 1961.
9. **Kaupp, W. J. and Sohi, S. S.**, Quantitation of insect viruses in *Viral Insecticides for Biological Control*, Maramorosch, K. and Sherman, K. E., Eds., Academic Press, New York, 1985, 675.

SECTION 3
MORPHOLOGICAL GUIDE ON DIAMETERS OF VIRUS PARTICLES

Family	N.A.	diam (nm)[a]	diam (nm)[b]	Morphology
Parvoviridae	DNA	18—26		Icosahedral
Picornaviridae	RNA	22—30		Icosahedral
Nodaviridae	RNA	29		Icosahedral
Caliciviridae	RNA	35—39		Icosahedral
Flaviviridae	RNA	35—45	36—44	Icosahedral
Papovaviridae	DNA	45—55		Icosahedral
Birnaviridae	RNA	60		Icosahedral
Togaviridae	RNA	60—65	60—65	Spherical
Reoviridae	RNA	60—80		Icosahedral
Coronaviridae	RNA	75—150	60—160	Spherical
Retroviridae	RNA	80—100	100	Spherical
Bunyaviridae	RNA	80—110	80—100	Spherical
Orthomyxoviridae	RNA	80—120	80—100	Pleomorphic, spherical or filamentous
Herpesviridae	DNA	120—200		Spherical
Iridoviridae	DNA	125—300		Icosahedral
Paramyxoviridae	RNA	150 +	200—300	Spherical, pleomorphic
Baculoviridae	DNA	40—60 × 200—400		Rod shaped, bacilliform
Entomopoxvirinae	DNA	170—250 × 300—400		Brick shaped, spherical
Rhabdoviridae	RNA	50—95 × 130—380	70 × 160—170	Bullet shaped

[a]   Matthews, R. E. F., *Intervirology*, 17, 1, 1982.
[b]   Dubois-Dalcq, M., Holmes, K. V., and Rentier, B., *Assembly of Enveloped RNA Viruses*, Springer-Verlag, Wien, 1984.

(See also Doane, F. W., Virus morphology as an aid for rapid diagnosis, *Yale J. Biol. Med.*, 53, 19, 1980.)

SECTION 4
SEVERAL FIXATION AND EMBEDDING SCHEDULES FOR TRANSMISSION ELECTRON MICROSCOPY

A. SAMPLE BASELINE FIXATION PROTOCOL*

1. Immerse 1-mm cubes of tissues in 5 to 10 vol of 3% (purified) glutaraldehyde in 0.1 M sodium phosphate buffer, pH 7.0 (40 min at room temperature or extend time at lower temperatures).
2. Rinse briefly, then again for 5 min in 0.1 M sodium phosphate buffer, (pH 6.8).
3. Immerse tissues in 5 to 10 vol of 1% osmium (pH 6.8), prechilled in ice bath, for 30 min.
4. Transfer directly to 30% ethanol, rinse quickly, and proceed through a graded series of cold ethanols (50, 70, 80, and 95%) for 15 min each.
5. Bring sample in 95% ethanol to room temperature and make four changes in 100% ethanol.
6. Further processing will depend on the embedding medium chosen.

*Bowers, B. and Maser, M., in *Artifacts in Biological Electron Microscopy*, Crang, R. F. E. and Klomparens, K. L., Eds., Plenum Press, New York, 1988, 13.

B. FIXATION AND EMBEDDING SCHEDULE FOR INSECT PATHOGENS AND INSECT TISSUES, INSECT BIOCONTROL LAB, U.S. DEPARTMENT OF AGRICULTURE, BELTSVILLE, MD

1. **Fixation*** — 2.5% Glutaraldehyde in 0.05 M sodium cacodylate buffer (pH 7.2) + 0.17 M sucrose — 4 h — overnight (4°C).
2. Make three changes in wash buffer (0.05 M sodium cacodylate buffer + 0.34 M sucrose) — over 24 h at 4°C.
3. Postfix in 1% osmium tetroxide in 0.05 M sodium cacodylate buffer (4°C) — 2 h.
4. Rinse in 0.05 M sodium cacodylate buffer (4°C) — 15 min.
5. **En Bloc Stain** — 2% Uranyl acetate (aq) (4°C) — 45 min.
6. Rinse in 0.05 M sodium cacodylate buffer — 2 changes. Bring to room temperature. All succeeding changes are carried out at room temperature.
7. Dehydrate in 50, 80, 95% ethyl alcohol — 15 min each. Absolute alcohol — 2 changes — 15 min.
8. **Embedding** — Infiltrate tissues in propylene oxide: Araldite 506 mixture (1:1) — 1 h. Infiltrate tissues in propylene oxide: Araldite 506 mixture (1:3) — 1 h. Place tissues in capsules in Araldite 506 mixture.

Araldite 506 Resin Mixture

NMA (nadic methyl anhydride)	15.64 g
DDSA (dodecenyl succinic anhydride)	23.37 g
Araldite 506 resin	60.99 g
DMP-30	1.00 g

Continue infiltration in vacuum oven. Pump to about 25 in. of mercury for 15 min. Leave under vacuum overnight. Bring oven to atmosphere. Check orientation of tissues in capsules. Pump again to 25 in. of mercury for about 15 min. Turn on oven preset to 60°C. Allow polymerization to begin for 4 h with tissues under vacuum; then slowly introduce air to bring to atmosphere within 5 min. Continue polymerization at 60°C for 20 h.

*A small carrier was developed for holding tissues that do not sink readily in the fixative solutions such as insect hypodermis and insect eggs (J. R. Adams, unpublished data, 1989). It consists of a BEEM embedding capsule (size 00) with the bottom removed. The length may vary from 10 to 15 mm. Two 9 mm diam 100 mesh steel screens are inserted into top and bottom BEEM capsule lids which have had large circular or rectangular holes cut into them to allow for exchange of fluids. The specimens are submerged in the container which is placed in a larger vial of fixative solution; the top lid is then snapped on and air bubbles trapped within are removed with a Pasteur pipette, allowing the container to sink to the bottom of the vial. Because the tissues will stay submerged by Step 2, the BEEM capsule carrier can be removed just prior to postfixation (Step 3), cleaned and reused.

C. FIXATION PROTOCOLS AND EMBEDDING SCHEDULE FOR PREPARATION OF MARINE SPECIMENS FOR TRANSMISSION ELECTRON MICROSCOPY
(Technique used in the Laboratoire de Pathologie Comparée, USTL, Montpellier, France.) (By J. R. Bonami)

Stock Solutions

Sodium cacodylate buffer 0.4 M, pH 7.4
NaCl—10% in dist water
Glutaraldehyde—25%
Osmium tetroxide—4% in dist water

1. **Fixation:** 2 h to overnight at 4°C in 2.5% glutaraldehyde mixture

Glutaraldehyde—25%	2 vol
Cacodylate buffer	5 vol
NaCl—10%	3.6 vol
Dist water	9.4 vol

Cut off quickly large pieces of organs from living animals, immerse in 2.5% glutaraldehyde mixture, cut into small pieces of tissue (1 mm³), and place in vials of glutaraldehyde mixture for fixation.

2. **Washing buffer:** at least 3 changes, 1 h each at 4°C in

Cacodylate buffer	5 vol
NaCl—10%	10 vol
Dist water	9.4 vol

3. **Postfixation:** 1 to 2 h at 4°C in 1% OsO_4 mixture

OsO_4—4%	1 vol
Cacodylate buffer	1 vol
NaCl—10%	1 vol
Dist water	1 vol

4. Rinse with washing buffer at room temperature.
5. Dehydrate in 50, 75, and 95% ethyl alcohol, 15 min each. Absolute alcohol, three changes, 15 min each.
6. Infiltrate tissues with propylene oxide, two changes, 30 min each.
7. **Embedding:** Infiltrate tissues in propylene oxide-Epon mixture (1:1) 2 h. Place tissues in capsules in Epon resin 4 h to overnight at room temperature, then in oven at 60°C, 12—24 h for polymerization.

Epon 812 Resin Mixture

Epon 812	24.6 g
DDSA	12.4 g
NMA	12.4 g
DMP-30	0.6 g

Method Used for Shrimp Fixation
(From Lightner, D. et al., *J. Invertebr. Pathol.*, 42, 62, 1983.)

Shrimps are fixed live by injection into the hepatopancreas and hemocoel (using 1-ml syringe) of cold (4°C) 6% glutaraldehyde in 0.15 *M* Millonig's phosphate buffer* supplemented with 1% NaCl and 0.5% sucrose. Then, animals are immersed whole in the same fixative after the cuticle is opened for the length of the animal along the lateral midline using dissecting scissors. Whole shrimps are stored in fixative at 4°C until processed further. Tissues selected for EM are removed by dissection and postfixed in 1% osmium tetroxide in the same buffer as is used for the glutaraldehyde fixative. After dehydration through graded alcohols, the samples are embedded in Spurr's resin (Ladd Research, Burlington, VT.)

* Pease, D. C., *Histological Techniques for Electron Microscopy*, Academic Press, New York, 1964.

DISCLAIMER

ACKNOWLEDGMENTS

The technical assistance of Mchezagi Axum, Tangita Adams, Mary Fenton, LaVonne Thaden, Karen Riedl, Janice Bradsher, and Dwight Lynn is greatly appreciated.

INDEX

D

X

Y